Table of Atomic Masses*

Element	Symbol	Atomic Number	Atomic Mass
Actinium	Ac	89	(227)†
Aluminum	Al	13	26.98
Americium	Am	95	(243)
Antimony	Sb	51	121.8
Argon	Ar	18	39.95
Arsenic	As	33	74.92
Astatine	At	85	(210)
Barium	Ba	56	137.3
Berkelium	Bk	97	(247)
Beryllium	Be	4	9.012
Bismuth	Bi	83	209.0
Boron	B	5	10.81
Bromine	Br	35	79.90
Cadmium	Cd	48	112.4
Calcium	Ca	20	40.08
Californium	Cf	98	(251)
Carbon	C	6	12.01
Cerium	Ce	58	140.1
Cesium	Cs	55	132.9
Chlorine	Cl	17	35.45
Chromium	Cr	24	52.00
Cobalt	Co	27	58.93
Copper	Cu	29	63.55
Curium	Cm	96	(247)
Dysprosium	Dy	66	162.5
Einsteinium	Es	99	(252)
Erbium	Er	68	167.3
Europium	Eu	63	152.0
Fermium	Fm	100	(257)
Fluorine	F	9	19.00
Francium	Fr	87	(223)
Gadolinium	Gd	64	157.3
Gallium	Ga	31	69.72
Germanium	Ge	32	72.59
Gold	Au	79	197.0
Hafnium	Hf	72	178.5
Helium	He	2	4.003
Holmium	Ho	67	164.9
Hydrogen	H	1	1.008
Indium	In	49	114.8
Iodine	I	53	126.9
Iridium	Ir	77	192.2
Iron	Fe	26	55.85
Krypton	Kr	36	83.80
Lanthanum	La	57	138.9
Lawrencium	Lr	103	(260)
Lead	Pb	82	207.2
Lithium	Li	3	6.941
Lutetium	Lu	71	175.0
Magnesium	Mg	12	24.31
Manganese	Mn	25	54.94
Mendelevium	Md	101	(258)
Mercury	Hg	80	200.6
Molybdenum	Mo	42	95.94
Neodymium	Nd	60	144.2
Neon	Ne	10	20.18
Neptunium	Np	93	(237)
Nickel	Ni	28	58.69
Niobium	Nb	41	92.91
Nitrogen	N	7	14.01
Nobelium	No	102	(259)
Osmium	Os	76	190.2
Oxygen	O	8	16.00
Palladium	Pd	46	106.4
Phosphorus	P	15	30.97
Platinum	Pt	78	195.1
Plutonium	Pu	94	(244)
Polonium	Po	84	(209)
Potassium	K	19	39.10
Praseodymium	Pr	59	140.9
Promethium	Pm	61	(145)
Protactinium	Pa	91	(231)
Radium	Ra	88	226
Radon	Rn	86	(222)
Rhenium	Re	75	186.2
Rhodium	Rh	45	102.9
Rubidium	Rb	37	85.47
Ruthenium	Ru	44	101.1
Samarium	Sm	62	150.4
Scandium	Sc	21	44.96
Selenium	Se	34	78.96
Silicon	Si	14	28.09
Silver	Ag	47	107.9
Sodium	Na	11	22.99
Strontium	Sr	38	87.62
Sulfur	S	16	32.07
Tantalum	Ta	73	180.9
Technetium	Tc	43	(98)
Tellurium	Te	52	127.6
Terbium	Tb	65	158.9
Thallium	Tl	81	204.4
Thorium	Th	90	232.0
Thulium	Tm	69	168.9
Tin	Sn	50	118.7
Titanium	Ti	22	47.88
Tungsten	W	74	183.9
Uranium	U	92	238.0
Vanadium	V	23	50.94
Xenon	Xe	54	131.3
Ytterbium	Yb	70	173.0
Yttrium	Y	39	88.91
Zinc	Zn	30	65.38
Zirconium	Zr	40	91.22

*The values given here are to four significant figures.
†A value given in parentheses denotes the mass of the longest-lived isotope.

Brief Contents

This **Instructor's Annotated Edition** is intended to serve as your guide to any of the three student versions you may choose to use:

Introductory Chemistry: A Foundation, Second Edition
Introductory Chemistry, Second Edition
Basic Chemistry, Second Edition

The **annotations** shown in red type throughout this *Instructor's Edition* are provided by Iris Stovall of the University of Illinois. Based on many years of teaching experience in this course, the annotations furnish instructors and teaching assistants with practical solutions to problems frequently encountered by students in this course. The annotations also offer helpful tips for teaching the material in the chapter and guidelines for setting up chemistry demonstrations in class. We hope you find the annotations useful.

Introductory Chemistry: A Foundation
Instructor's Annotated Edition
Chapters 1–21

Introductory Chemistry: A Foundation
Chapters 1–21

Introductory Chemistry
Chapters 1–19

Basic Chemistry
Chapters 1–15

Introductory Chemistry
A Foundation
Second Edition

Annotated Instructor's Edition

Steven S. Zumdahl
University of Illinois

with annotations by

Iris K. Stovall
University of Illinois

D.C. Heath and Company

Lexington, Massachusetts   Toronto

Address editorial correspondence to:

D. C. Heath and Company
125 Spring Street
Lexington, MA 02173

Editorial Director: Kent Porter Hamann
Developmental Editor: Barbara Withington Meglis
Production Editor: Cormac Joseph Morrissey
Designer: Alwyn R. Velásquez
Photo Researcher: Martha L. Shethar
Art Editor: Janet Theurer, Alwyn R. Velásquez
Production Coordinator: Lisa Merrill
Permissions Editor: Margaret Roll
Project Consultant: Richard E. Morel

Cover: Charlotte Raymond / Photo Researchers, Inc.

Published simultaneously in Canada.

Printed in the United States of America.

International Standard Book Number: 0−669−32856−1

Library of Congress Catalog Number: 92−71435

10 9 8 7 6 5 4 3 2 1

To Eunice, Whitney, Leslie, and Scott

PREFACE

Our goal for this revision is to continue to make introductory chemistry accessible to students who have little or no background in chemistry. The central question that formed the basis of the first edition of this text is still the focus, What can we do to make this material understandable and interesting to students?

We as chemical educators are discouraged by how little our students really seem to understand about our subject, and we are concerned about the dwindling number of chemistry majors. Many educators have suggested that one factor contributing to these problems is that we as chemists have made chemistry too abstract and too sterile—that is, we do not spend enough time dealing with "real chemistry" and how it applies to the lives of our students. We have addressed these issues in this text in several ways, and we hope that our approach will not only help your students understand the subject but also generate interest in science in general.

Emphasis on Reaction Chemistry

We continue to emphasize chemical reactions early in the book, leaving the more abstract material on orbitals for later chapters. In a course where students encounter chemistry for the first time and lack an appreciation for the elegance of atomic theory, it seems especially important that we present the chemical nature of matter before we discuss the theoretical intricacies of atoms and orbitals. Reactions, on the other hand, are inherently interesting to students and can help us draw the students to chemistry. In particular, reactions can form the basis for fascinating classroom demonstrations and laboratory experiments.

We have therefore chosen to emphasize reactions before going on to the details of atomic structure. Relying only on very simple ideas about the atom, Chapters 6–8 represent a thorough treatment of chemical reactions, including how to recognize a chemical change and what a chemical equation means. The properties of aqueous solutions are discussed in detail, and careful attention is given to precipitation and acid–base reactions. In addition, a simple treatment of oxidation–reduction reactions is given. The material on reactions is covered in three short chapters, allowing students to digest it in small doses with lots of end-of-chapter problems while giving instructors considerable flexibility in covering it. These three chapters should provide a solid foundation, relatively early in the course, for reaction-based laboratory experiments.

For instructors who feel it is essential to introduce orbitals early in the course, prior to chemical reactions, the chapters on atomic theory and bonding (Chapters 11–12) can be covered directly after Chapter 4. The book was written to be flexible in this way because we realize that many instructors have a strong preference about when to introduce orbitals.

Development of Problem-Solving Skills

Problem solving seems to be somewhat of a cliche in chemical education. We all want our students to develop real problem-solving skills. In fact, this was a central focus of the first edition of this text. We have maintained this feature in the second edition, and although we do not expect miracles, we feel that the approach will help students become better problem solvers.

In the first chapters we spend considerable time trying to help the student understand why learning chemistry often seems difficult. At the same time, we explain that the complexities that can make chemistry frustrating at times can also provide the opportunity to develop the problem-solving skills that are beneficial in any profession. Learning to "think like a chemist" is useful to virtually anyone. To emphasize this idea, we apply scientific thinking to some real-life problems in Chapter 1.

One reason chemistry is difficult for beginning students is that they often do not possess the mathematical skills that are required. Thus we have paid careful attention to fundamental mathematical skills, such as using scientific notation, rounding off to the correct number of significant figures, and rearranging equations to solve for a particular quantity. And we have been very careful to follow the rules we have set down, so as not to confuse the students.

Attitude plays a crucial role in achieving success in problem solving. Students must learn that a systematic, thoughtful approach to problems is better than brute force memorization. We try to establish this attitude early in the book, using temperature conversions as a vehicle in Chapter 2. Throughout the book we encourage an approach that starts with trying to represent the essence of the problem by using symbols and/or diagrams and ends with thinking about whether the answer makes sense. We approach new concepts by carefully working through the material before we give mathematical formulas or overall strategies. We try to encourage a thoughtful step-by-step approach rather than the premature use of algorithms. Once we have provided the necessary foundation, we highlight important rules and processes in skill development boxes so that students can locate them easily.

In the second edition, we have expanded the number of end-of-chapter exercises. As in the first edition, we have set up the end-of-chapter exercises in pairs. Exercises numbered in blue are in "matched pairs," meaning that the two horizontally aligned problems address similar topics. An "Additional Problems" section, marked by red numbers, includes further practice in chapter concepts as well as more challenging problems. Answers for all even-numbered exercises appear in a special section at the end of the book.

Handling the Language of Chemistry and Applications

We have gone to great lengths to make this book "student friendly" and have received enthusiastic feedback from students who've used it.

As in the first edition, we present here a very systematic and thorough treatment of chemical nomenclature. Once this framework is established, students can progress through the book comfortably.

Along with chemical reactions, applications form an important part of descriptive chemistry. Because students are interested in the impact that chemistry has on their lives, we have included many "Chemistry in Focus" boxes, which describe current applications of chemistry. These special interest boxes cover topics such as the relationship of elemental distributions to the disappearance of the dinosaurs, the application of diamond coatings to plastic, alternative fuels, and the greenhouse effect.

Visual Impact of Chemistry

Responding to instructors' requests to include graphic illustrations of chemical reactions, phenomena, and processes, our four-color design enables color to be used functionally, thoughtfully, and consistently to help students understand chemistry and to make the subject more inviting to them. In the second edition, we have modified the photo program to include only those photos that illustrate a chemical reaction or phenomenon or make a connection between chemistry and the real world.

Three Choices of Coverage

For the convenience of instructors, three versions of the second edition are available: one paperback version and two hardbound versions. *Basic Chemistry,* Second Edition, a paperback text, provides basic coverage of chemical concepts and applications through solution chemistry and has 15 chapters. *Introductory Chemistry,* Second Edition, a hardbound text, expands the coverage to 19 chapters with the addition of radioactivity and nuclear energy. Finally, *Introductory Chemistry: A Foundation,* Second Edition, a hardbound text, has 21 chapters with the final two chapters providing a brief introduction to organic and biological chemistry.

Features New to this Edition

We were pleased to receive an overwhelmingly positive response to the first edition of this text. The recommendation made most frequently by reviewers of the first edition was to leave the book alone. With that in mind, we intentionally left much of the text unchanged in the second edition. We only made changes such as the following that we hope will further enhance the text.

- We reworked any text section that seemed to bother students and needed clearer explanation.
- We updated the chemistry discussion and accompanying photos where necessary.

- We introduced a new art feature, the periodic table icons. Some of the more common elements are highlighted in periodic table icons designed to remind students about the position of selected elements in the periodic table and to allow students to become more familiar with the periodic table over the course of the semester.
- We expanded the number of end-of-chapter questions and problems. Answers to the even-numbered questions and problems are in the back of the text.
- In response to reviewer requests, we streamlined the photo program to include only photos that are pedagogically important to the discussion.

Supplements for the Text

A comprehensive teaching and learning package accompanies all three versions of the text:

Study Guide by Iris Stovall
Solutions Guide by James F. Hall
Complete Solutions Guide by James F. Hall
Introductory Chemistry in the Laboratory by James F. Hall
Instructor's Resource Guide for Introductory Chemistry in the Laboratory by
    James F. Hall
Test Item File
Computerized Testing
Heath Chemistry Lecture Demonstrations
Transparencies

Another item that may be of interest to you is our microscale lab text, *Chemical Investigations: A Laboratory Text for Introductory Chemistry* by Jerry A. Bell.

    We have worked hard to make this book and its supplements clear, interesting, and accurate. We would appreciate any comments that would make the book more useful to students and instructors.

Acknowledgments

This book represents the collaborative efforts of many talented and dedicated people to whom I owe much. I am especially grateful to my wife, Eunice, who was a true partner in this project: processing words, proofreading, offering constructive criticism, and providing support in many different ways. I also greatly appreciate the support and friendship of Kent Porter Hamann, Editorial Director, whose consummate professionalism, unreasonable optimism, and unerring judgment concerning level and content have proved invaluable in producing this book. I am also grateful to Barbara Meglis, Developmental Editor, who greatly improved the manuscript in many different ways. Dick Morel, Project Consultant, also contributed enormously to this edition through his insight and creativity in making the text better than it was before.

My thanks go also to Cory Morrissey, Production Editor, who did a really excellent job in the face of a very tight schedule. The creative design by Alwyn Velásquez maintained the airy, nonthreatening appearance of the first edition. Thanks also to Martha Shethar, Photo Researcher, for the beautiful chapter openers and other in-text photos.

Jim Hall of the University of Lowell, Project Consultant, contributed in many different ways to the success of this project. In particular, I appreciate his help with the end-of-chapter problems and his care in getting everything right.

I am grateful to Bruce Zimmerli for hiring dedicated people and for insisting on high quality. My thanks also go to Tom Flaherty, Jeff Lasser, and Jim Porter Hamann for their unswerving enthusiasm for this book and for their extraordinary knowledge of the market. I want to express special appreciation to the D. C. Heath "book reps," a bunch of bright, stimulating people who have provided much valuable feedback and a good deal of fun over the last several years.

My sincerest appreciation goes to the following users of the first edition who reviewed the text in preparation for the revision: Gerald Berkowitz, Erie Community College; John L. Burmeister, University of Delaware; Terry L. Eyrich, Merced College; John Foster, Arizona State University; Marcia L. Gillette, Indiana University-Kokomo; Catherine Hagen Howard, Texarkana College; Jeffrey A. Hurlbut, Metropolitan State College of Denver; Lawrence S. Lynn, St. Louis Community College at Meramec; William T. Nolan, Oakland Community College; Stuart A. Nowinski, Glendale Community College; Paul O'Brien, West Valley College; Virginia M. Shepler, Ferris State University; Sol Shulman, Illinois State University; William Steele, Penn State University; Bradley M. Stone, San Jose State University.

My thanks also go to the following reviewers of the first edition manuscript at various stages of development: Melvin L. Anderson, Lake Superior State University; John T. S. Andrews, Hiram College; Stanley L. Ashbaugh, Orange Coast College; Caroline L. Ayers, East Carolina University; Edith S. Bartley, Tarrant County Junior College; Paul P. Blanchette, Saint Mary's College of Maryland; Ruth J. Bowen, California State Polytechnic University; Ben B. Chastain, Samford University; James Coke, University of North Carolina at Chapel Hill; Ellene Tratras Contis, Eastern Michigan University; W. Dan Covey, L. A. Pierce College; Jerry A. Driscoll, University of Utah; Vicky D. Ellis, Gulf Coast Community College; Gordon J. Ewing, New Mexico State University; Robert D. Farina, Western Kentucky University; K. Thomas Finley, State University of New York, College at Brockport; David V. Frank, Ferris State University; Barbara B. Frohardt, Oakland Community College; Patrick M. Garvey, Des Moines Area Community College; Sue Griffin, Boston University; John Grove, South Dakota State University; Michael D. Hatlee, Briar Cliff College; H. William Hausler, Madison Area Technical College; Don B. Hilton, University of Lowell; Y. C. Jean, University of Missouri—Kansas City; Joseph M. Kanamueller, Western Michigan University; M. Christine Kerr, Montgomery College; Robert C. Kowerski, College of San Mateo; Leslie J. Lovett, West Virginia University; Douglas L. Magnus, Saint Cloud State University; David M. Manuta, Shawnee State University; Ruth P. Manuta, Shawnee State University; Valerie C. Meehan, City College of San Francisco; Kenneth E. Miller, Milwaukee Area Technical College; Patricia Milliken Wilde, Triton College; Jerry L. Mills, Texas Tech University; Steven L. Murov, Modesto Junior College; Laurie B. O'Connor, Chabot College; Lucy T. Pryde, Southwestern College; Fred Redmore, Highland Community College; Agatha Riehl, College of Saint Scholastica; Dan D.

Scott, Middle Tennessee State University; Edmund E. Sorman, Mesa College; Vernon J. Thielmann, Southwestern Missouri State University; Eugene Thomas, California State University—Chico; Frina S. Toby, Rutgers—The State University; Mary S. Vennos, Essex Community College; George H. Wahl, Jr., North Carolina State University; William J. Wasserman, Seattle Central Community College; James H. Weber, University of New Hampshire; John A. Weyh, Western Washington University; Thomas M. Willard, Florida Southern College; Linda Woodward, University of Southwestern Louisiana; Mervat E. Zewail, Cerritos Community College.

S. S. Z.

C ONTENTS

CHAPTER 4

Chemical Foundations: Elements and Atoms   91

CHAPTER 5

Elements, Ions, and Nomenclature   119

CHAPTER   12   Chemical Bonding   375

1 Chemistry: An Introduction

Chemistry deals with the natural world.

CONTENTS

Chapter One is a short introduction to the scientific method. The material can be covered during the first lecture, after introducing students to the course. The first lecture is also a good time to discuss how grades will be determined. Students will want to know whether or not their homework problems will be part of their grade, and whether answers to exams will be written out or multiple choice. You can also tell students what resources are available outside of class if they need extra help. They have all heard that chemistry is hard. Reassure them that they can succeed.

Chemistry can teach problem-solving skills. To solve chemistry problems, students must discard unnecessary pieces of data and correctly fit together relevant pieces. Solving chemistry problems will help develop problem-solving skills that will be useful long after students have finished this course.

A chemist performs quality control work.

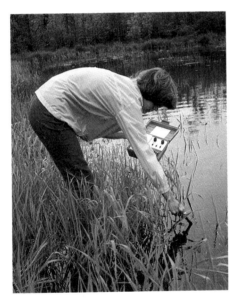

A scientist testing the acidity of a pond.

Chemical and physical changes will be discussed in Chapter 3.

Chemistry is important—there is no doubt about that. Chemistry lies at the heart of our efforts to produce new materials that make our lives safer and easier, to produce new sources of energy that are abundant and nonpolluting, and to understand and control the many diseases that threaten us and our food supplies. Even if your eventual career does not require the daily use of chemical principles, your life will be greatly influenced by chemistry. One main goal of this text, then, is to help you see how chemistry affects your life.

A major by-product of your study of chemistry is that you will become a better problem solver. One reason why chemistry has the reputation of being "tough" is that it often deals with rather complicated systems that require some effort to figure out. Although this might at first seem like a disadvantage, you can turn it to your advantage if you have the right attitude and skills. Recruiters for companies of all types maintain that one of the first things they look for in a prospective employee is the ability to solve problems. We will spend a good deal of time solving various types of problems in this book by using a systematic, logical approach that will serve you well in solving any kind of problem in any field. Keep this broader goal in mind as you learn to solve the specific problems connected with chemistry.

Although learning chemistry is often not easy, it's never impossible. In fact, anyone who is interested, patient, and willing to work can learn the fundamentals of chemistry. In this book we will try very hard to help you understand what chemistry is and how it works and to point out how chemistry applies to the things going on in your life.

Our sincere hope is that this text will motivate you to learn chemistry, make its concepts understandable to you, and demonstrate how interesting and vital the study of chemistry is.

1.1   What Is Chemistry?

AIM: To define chemistry.

Chemistry can be defined as *the science that deals with the materials of the universe and the changes that these materials undergo.* Chemists are involved in activities as diverse as examining the fundamental particles of matter, looking for molecules in space, synthesizing and formulating new materials of all types, using bacteria to produce such chemicals as insulin, and inventing new diagnostic methods for early detection of disease.

Chemistry is often called the central science—and with good reason. Most of the phenomena that occur in the world around us involve chemical changes, changes where one or more substances become different substances. Here are some examples of chemical changes:

Wood burns in air, forming water, carbon dioxide, and other substances.
A plant grows by assembling simple substances into more complex
  substances.
The steel in a car rusts.
Eggs, flour, sugar, and baking powder are mixed and baked to yield a
  cake.
The definition of the term *chemistry* is learned and stored in the brain.
Grape juice ferments to form wine.
Emissions from a power plant lead to the formation of acid rain.

As we proceed, you will see how the concepts of chemistry allow us to
understand the nature of these and other changes and thus help us manipulate
natural materials to our benefit.

1.2 Solving Problems Using a Scientific Approach

AIM: To introduce scientific thinking.

The fact that chemical changes take place as a wooden house burns is quite obvious.

One of the most important things we do in everyday life is solve problems. In fact,
most of the decisions you make every day can be described as solving problems.

It's 8:30 A.M. on Friday. Which is the best way to drive to school to
  avoid traffic congestion?
You have two tests on Monday. Should you divide your study time
  equally or allot more time to one than to the other?
Your car stalls at a busy intersection and your little brother is with you.
  What should you do next?

These are everyday problems of the type we all face. What process do we
use to solve them? You may not have thought about it before, but there are
several steps that almost everyone uses to solve problems:

1. Recognize the problem and state it clearly. Some information becomes
    known, or something happens that requires action. In science we call
    this step *making an observation.*
2. Propose *possible* solutions to the problem or *possible* explanations for
    the observation. In scientific language, suggesting such a possibility is
    called *formulating a hypothesis.*
3. Decide which of the solutions is the best or decide whether the
    explanation proposed is reasonable. To do this we search our memory
    for any pertinent information or we seek new information. In science
    we call searching for new information *performing an experiment.*

I always emphasize that solving chemistry
problems is not a completely new
experience. Students arrive in a chemistry
class having previously solved many
problems that require logical thinking. It can
be reassuring for students to hear that some
familiar ways of thinking can be useful in a
chemistry course.

A Mystifying Problem

*T*o illustrate how science helps us solve problems, consider a true story about two people, David and Susan (not their real names). Ten years ago David and Susan were healthy 40-year-olds living in California where David was serving in the Air Force. Gradually Susan became quite ill, showing flu-like symptoms including nausea and severe muscle pains. Even her personality changed: she became uncharacteristically grumpy. She seemed like a totally different person from the healthy, happy woman of a few months earlier. Following her doctor's orders, she rested and drank lots of fluids, including large quantities of coffee and orange juice from her favorite mug, part of a 200-piece set of pottery dishes recently purchased in Italy. However, she just got sicker, developing extreme abdominal cramps and severe anemia.

During this time David also became ill and exhibited symptoms much like Susan's: weight loss, excruciating pain in his back and arms, and uncharacteristic fits of temper. The disease became so debilitating that he retired early from the Air Force and the couple moved to Seat-

tle. For a short time their health improved, but after they unpacked all their belongings (including those pottery dishes), their health began to deteriorate again. Susan's body became so sensitive that she could not tolerate the weight of a blanket. She was near death. What was wrong? The doctors didn't know, but one suggested she might have porphyria, a rare blood disease.

Desperate, David began to search the medical literature himself. One day while he was reading about porphyria, a phrase jumped off the page: "Lead poisoning can sometimes be confused with porphyria." Could the problem be lead poisoning?

We have described a very serious problem with life-or-death implications. What should David do next? Overlooking for a moment the obvious response of calling the couple's doctor immediately to discuss the possibility of lead poisoning, could David solve the problem via scientific thinking? Let's use the three steps described in Section 1.2 to attack the problem one part at a time. This is important: usually we solve complex problems by breaking them down into manageable parts. We can then

assemble the solution to the overall problem from the answers we have found "piecemeal."

In this case there are many parts to the overall problem:

What is the disease?
Where is it coming from?
Can it be cured?

Let's attack "What is the disease?" first.

Observation: David and Susan are ill with the symptoms described. Is the disease lead poisoning?
Hypothesis: The disease is lead poisoning.

Italian pottery.

Experiment: If the disease is lead poisoning, the symptoms must match those known to characterize lead poisoning. Look up the symptoms of lead poisoning. David did this and found that they matched the couple's symptoms almost exactly.

This discovery points to lead poisoning as the source of their problem, but David needed more evidence.

Observation: Lead poisoning results from high levels of lead in the bloodstream.
Hypothesis: The couple have high levels of lead in their blood.
Experiment: Perform a blood analysis. Susan arranged for such an analysis, and the results showed high lead levels for both David and Susan.

This confirms that lead poisoning is probably the cause of the trouble, but the overall problem is still not solved. David and Susan are likely to die unless they find out where the lead is coming from.

Observation: There is lead in the couple's blood.
Hypothesis: The lead is in their food or drink when they buy it.

Experiment: Find out whether anyone else who shopped at the same store was getting sick (no one was). Also note that moving to a new area did not solve the problem.

Observation: The food they buy is free of lead.
Hypothesis: The dishes they use are the source of the lead poisoning.
Experiment: Find out whether their dishes contain lead. David and Susan learned that lead compounds are often used to put a shiny finish on pottery objects. And laboratory analysis of their Italian pottery dishes showed that lead was present in the glaze.

Observation: Lead is present in their dishes, so the dishes are a possible source of their lead poisoning.
Hypothesis: The lead is leaching into their food.
Experiment: Place a beverage, such as orange juice, in one of the cups and then analyze the beverage for lead. The results showed high levels of lead in drinks that had had contact with the pottery cups.

After many applications of the scientific method, the problem is solved. We can summarize the answer to the problem (David and Susan's illness) as follows: the Italian pottery they used for everyday dishes contained a lead glaze that contaminated their food and drink with lead. This lead accumulated in their bodies to the point where it interfered seriously with normal functions and produced severe symptoms. This overall explanation, which summarizes the hypotheses that agree with the experimental results, is called a *theory* in science. This explanation accounts for the results of all the experiments performed.*

We could continue to use the scientific method to study other aspects of this problem, such as

What types of food or drink leach the most lead from the dishes?
Do all pottery dishes with lead glazes produce lead poisoning?

As we answer questions using the scientific method, other questions naturally arise. By repeating the three steps over and over, we can come to understand a given phenomenon thoroughly.

---

*"David" and "Susan" recovered from their lead poisoning and are now publicizing the dangers of using lead-glazed pottery. This happy outcome is the answer to the third part of their overall problem, "Can the disease be cured?" They simply stopped eating from that pottery!

I often pose the rhetorical question "What is the difference between what you do when you solve problems and what a chemist does?" The answer I give is that, basically, there is no difference. But a chemist knows enough chemical "facts" to solve chemistry problems as well as the facts needed to solve practical, everyday problems.

As we will discover in the next section, scientists use these same procedures to study what happens in the world around us. The important point here is that scientific thinking can help you in all parts of your life. It's worthwhile to learn how to think scientifically—whether you want to be a scientist, an auto mechanic, a doctor, a politician, or a poet!

1.3   The Scientific Method

AIM: To describe the method scientists use to study nature.

In the last section we began to see how science approaches problems. In this section we will further examine this approach.

Illustrate the scientific method with some examples, not necessarily from the world of chemistry. Show students that thinking like a chemist has real-world applications. For example, Chemistry in Focus on page 4 shows that thinking like a chemist can be useful. This may help provide motivation for some students.

Science is a framework for gaining and organizing knowledge. Science is not simply a set of facts but is also a plan of action—a *procedure* for processing and understanding certain types of information. Although scientific thinking is useful in all aspects of life, in this text we will use it to understand how the natural world operates. The process that lies at the center of scientific inquiry is called the **scientific method.** As we saw in the previous section, it consists of the following steps:

Steps in the Scientific Method

Quantitative observations involve a number. Qualitative ones do not.

1. *Make observations.* Observations may be *qualitative* (the sky is blue; water is a liquid) or *quantitative* (water boils at 100 °C; a certain chemistry book weighs 4.5 pounds). A qualitative observation does not involve a number. A quantitative observation is called a **measurement** and does involve a number (and a unit, such as pounds or inches). We will discuss measurements in detail in Chapter 2.
2. *Formulate hypotheses.* An hypothesis is a *possible* explanation for the observation.
3. *Perform experiments.* An experiment is something we do to test the hypothesis. We gather new information that allows us to decide whether the hypothesis is supported by the new information we have learned from the experiment. Experiments always produce new observations, and this brings us back to the beginning of the process again.

To explain the behavior of a given part of nature, we repeat these steps many times. Gradually we accumulate the knowledge necessary to understand what is going on.

Once we have a set of hypotheses that agree with our various observations, we assemble them into a theory that is often called a *model.* A **theory** (model) is a set of tested hypotheses that gives an overall explanation of some part of nature.

It is important to distinguish between observations and theories. An observation is something that is witnessed and can be recorded. A theory is an *interpretation*—a possible explanation of *why* nature behaves in a particular way. Theories inevitably change as more information becomes available. For example, the motions of the sun and stars have remained virtually the same over the thousands of years during which humans have been observing them, but our explanations—our theories—have changed greatly since ancient times.

The point is that we don't stop asking questions just because we have devised a theory that seems to account satisfactorily for some aspect of natural behavior. We continue doing experiments to refine our theories. We generally do this by using the theory to make a prediction and then doing an experiment (making a new observation) to see whether the results bear this prediction out.

Always remember that theories (models) are human inventions. They represent our attempts to explain observed natural behavior in terms of our human experiences. We must continue to do experiments and refine our theories to be consistent with new knowledge if we hope to approach a more nearly complete understanding of nature.

As we observe nature, we often see that the same observation applies to many different systems. For example, studies of innumerable chemical changes have shown that the total weight of the materials involved is the same before and after the change. We often formulate such generally observed behavior into a statement called a **natural law.** The observation that the total weight of materials is not affected by a chemical change in those materials is called the law of conservation of mass.

You must recognize the difference between a law and a theory. A law is a summary of observed (measurable) behavior, whereas a theory is an explanation of behavior. *A law tells what happens; a theory (model) is our attempt to explain why it happens.*

In this section (which is summarized in Figure 1.1), we have described the scientific method as it might ideally be applied. However, it is important to remember that science does not always progress smoothly and efficiently. Scientists are human. They have prejudices; they misinterpret data; they can become emotionally attached to their theories and thus lose objectivity; and they play politics. Science is affected by profit motives, budgets, fads, wars, and religious beliefs. Galileo, for example, was forced to recant his astronomical observations in the face of strong religious resistance. Lavoisier, the father of modern chemistry, was beheaded because of his political affiliations. And great progress in the chemistry of nitrogen fertilizers resulted from the desire to produce explosives to fight wars. The progress of science is often affected more by the frailties of humans and their institutions than by the limitations of scientific measuring devices. The scientific method is only as effective as the humans using it. It does not automatically lead to progress.

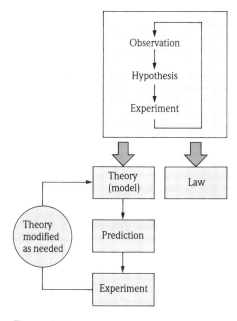

Figure 1.1
The various parts of the scientific method.

Law: A summary of observed behavior.
Theory: An explanation of behavior.

CHEMISTRY IN FOCUS

Observations, Theories, and the Planets

*H*umans have always been fascinated by the heavens—by the behavior of the sun by day and the stars by night. Although more accurately measured now thanks to precise instruments, our basic *observations* of their behavior have remained the same over the past 4000 years. However, our *interpretations* of these observations have changed dramatically. For example, around 2000 B.C. the Egyptians postulated that the sun was a boat inhabited by the god Ra, who daily sailed across the sky.

Over the years, patterns in the changes in the heavens were recognized and, through marvelous devices such as Stonehenge in England, were connected to the seasons of the year. People also noted that seven objects seemed to move against the background of "fixed stars." These objects (actually the sun, the moon, and the planets Mercury, Venus, Mars, Jupiter, and Saturn) were called the "wanderers." The planets generally seemed to move from west to east, but sometimes they seemed to slow down and even to move backwards for a few weeks.

Eudoxus, born in 400 B.C., tried to explain these observations. He imagined the earth as fixed in space and the planets as attached to a set of transparent spheres, each slightly larger than the previous one, that moved at different rates around the earth. The stars were attached to a fixed outermost sphere. This model, although clever, still did not account for the "backward" movement of some of the planets. Five hundred years later Ptolemy, a Greek scholar, worked out a plan more complex than that of Eudoxus, in which the planets were attached to the edges of spheres that "rolled around" the spheres of Eudoxus. This model accounted for the observed behavior of all the planets, including the apparent reversals in their motions.

Because of a natural human prejudice that the earth should be the center of the universe, Ptolemy's model was assumed to be correct for more than a thousand years, and its wide acceptance actually inhibited the advancement of astronomy. Finally, in 1543, the Polish cleric Nicholas Copernicus postulated that the

1.4 Learning Chemistry

AIM: To discuss the best approach for learning chemistry.

Succeeding in chemistry will require quite a bit of work on the students' part. They will need to read the textbook, learn facts, and solve problems. For their efforts they can receive more than just a grade. Succeeding in chemistry can provide benefits outside the classroom.

Chemistry courses have a universal reputation for being difficult. There are some good reasons for this. For one thing, the language of chemistry is unfamiliar in the beginning; many terms and definitions need to be memorized. As with any language, *you must know the vocabulary* before you can communicate effectively.

The Egyptian sun-god, Ra (drawn on papyrus).

earth was only one of the planets, all of which revolved around the sun. This "demotion" of the earth in status produced violent opposition to the new model, and in fact Copernicus's writings were "corrected" by religious officials before scholars were allowed to use them.

The Copernican theory persisted and was finally given a solid mathematical base by Johannes Kepler. Kepler's hypotheses were in turn further refined 36 years after his death by Isaac Newton, who recognized that the concept of gravity could account for the positions and

motions of the planets. However, even the brilliant models of Newton were discovered to be incomplete by Albert Einstein, who showed that Newton's ideas were just a part of a much more general model.

Thus the same basic observations were made for several thousand years, but the explanations—the models—changed remarkably from the Egyptians' boat of Ra to Einstein's relativity.

The lesson is that our models (theories) inevitably change and that we should expect them to do so. They can help us make scientific progress, or they can inhibit our progress if we become too attached to them. Although the fundamental observations of chemistry will remain the same, the models given in a chemistry text written in 2100 will certainly be quite different from the ones presented here.

We will try to help here by pointing out those things that it is necessary to memorize.

But memorization is only the beginning. Don't stop there or your experience with chemistry will be frustrating. Be willing to do some thinking, and learn to trust yourself to figure things out. To solve a typical chemistry problem, you must sort through the given information and decide what is really important.

One of the most important things you must realize is that chemical systems tend to be complicated—there are typically many components—and we must make approximations in describing them. Therefore trial and error play a major

For example, many stories that make news headlines are based on chemistry. Bring in some current news stories about chemistry and go over them in class. After you present the news articles, you can conclude that chemistry is important and that a thorough knowledge of general chemistry can make current events understandable.

role in solving chemical problems. In tackling a complicated system, a practicing chemist really does not expect to be right the first time he or she analyzes the problem. The usual practice is to make several simplifying assumptions and then give it a try. If the answer obtained doesn't make sense, the chemist adjusts the assumptions, using feedback from the first attempt, and tries again. The point is this: in dealing with chemical systems, do not expect to understand immediately everything that is going on. In fact, it is typical (even for an experienced chemist) *not* to understand at first. Make an attempt to solve the problem and then analyze the feedback. *It is no disaster to make a mistake as long as you learn from it.*

The only way to develop your confidence as a problem solver is to practice solving problems. To help you, this book contains examples worked out in detail. Follow these through carefully, making sure you understand each step. These examples are usually followed by a similar exercise (called a self-check exercise) that you should try on your own (detailed solutions of the self-check exercises are given at the end of each chapter). Use the self-check exercises to test whether you are understanding the material as you go along.

There are questions and problems at the end of each chapter. The questions review the basic concepts of the chapter and give you an opportunity to check whether you properly understand the vocabulary introduced. Some of the problems are really just exercises that are very similar to examples done in the chapter. If you understand the material in the chapter, you should be able to do these exercises in a straightforward way. Other problems require more creativity. These contain a knowledge gap—some unfamiliar territory that you must cross— and call for thought and patience on your part. For this course to be really useful to you, it is important to go beyond the questions and exercises. Life offers us many exercises, routine events that we deal with rather automatically, but the real challenges in life are true problems. This course can help you become a more creative problem solver.

As you do homework, be sure to use the problems correctly. If you cannot do a particular problem, do not immediately look at the solution. Review the relevant material in the text and then try the problem again. Don't be afraid to struggle with a problem. Looking at the solution as soon as you get stuck short-circuits the learning process.

Learning chemistry takes time. Use all the resources available to you and study on a regular basis. Don't expect too much of yourself too soon. You may not understand everything at first, and you may not be able to do many of the problems the first time you try them. This is normal. It doesn't mean you can't learn chemistry. Just remember to keep working and to keep learning from your mistakes, and you will make steady progress.

Finally, some demonstrations showing that chemistry is more than solving math problems would be well worth the effort. You can reinforce the practical aspect of making observations and perhaps lead into Chapter Two, Measurements and Calculations, by careful choice of demonstrations. You might consider some colorful precipitation reactions, the combustion of magnesium ribbon, or the dilution of sulfuric acid. Students can record their observations when you ask questions such as Did a solid form?, What color is the solid?, Did the magnesium burn?.

Chemistry: An Important Component of Your Education

Wﾠhat is the purpose of education? Because you are spending considerable time, energy, and money to pursue an education, this is an important question.

Some people seem to equate education with the storage of facts in the brain. These people apparently believe that education simply means memorizing the answers to all of life's present and future problems. Although this is clearly unreasonable, many students seem to behave as though this were their guiding principle. These students want to memorize lists of facts and to reproduce them on tests. They regard as unfair any exam questions that require some original thought or some processing of information. Indeed, it might be tempting to reduce education to a simple filling up with facts, because that approach can produce short-term satisfaction for both student and teacher. And of course, storing facts in the brain *is* important. You cannot function without knowing that red means stop, electricity is hazardous, ice is slippery, and so on.

However, mere recall of abstract information, without the ability to process it, makes you little better than a talking encyclopedia. Former students always seem to bring the same message when they return to campus. The characteristics that are most important to their success are a knowledge of the fundamentals of their fields, the ability to recognize and solve problems, and the ability to communicate effectively. They also emphasize the importance of a high level of motivation.

How does studying chemistry help you achieve these characteristics? The fact that chemical systems are complicated is really a blessing, though one that is well disguised. Studying chemistry will not by itself make you a good problem solver, but it can help you develop a positive, aggressive attitude toward problem solving and can help boost your confidence. Learning to "think like a chemist" can be valuable to anyone in any field. In fact, the chemical industry is heavily populated at all levels and in all areas by chemists and chemical engineers. People who were trained as chemical professionals often excel not only in chemical research and production, but also in the areas of personnel, marketing, sales, development, finance, and management. The point is that much of what you learn in this course can be applied to any field of endeavor. So be careful not to take too narrow a view of this course. Try to look beyond short-term frustration to long-term benefits. It may not be easy to learn to be a good problem solver, but it's well worth the effort.

Students pondering the structure of a molecule.

CHAPTER REVIEW

Questions and Problems

All even-numbered exercises have answers in the back of this book and solutions in the Solutions Guide.

Introduction

QUESTIONS

1. Why are you studying chemistry? Is it a required course for your major? If so, *why* do you suppose it is required? What do you hope to gain from this course? If chemistry is not a requirement, what prompted you to consider the subject as an elective?

2. Although your knowledge of chemistry may be limited at this point, you have certainly met many people who make everyday use of chemistry (physicians, pharmacists, farmers, and so on). Discuss three specific examples of such persons, and explain how they might make use of chemistry in their professions.

3. Chemistry is often associated with attempts to address such issues as the profusion of toxic wastes and nonbiodegradable plastics that pollute the countryside. How might an awareness of chemistry be used to solve these and other such problems?

1.1 What Is Chemistry?

QUESTIONS

4. This textbook provides a specific definition of chemistry: the study of the materials of which the universe is made and the transformations that these materials undergo. From your point of view at this time, how would *you* define chemistry? In your mind, what are chemicals? What do chemists do?

5. Several examples of chemical transformations are listed in this section. Give five additional examples of chemical changes.

1.2 Solving Problems Using a Scientific Approach

QUESTIONS

6. In this section three situations are described: (1) the best way to drive to school on Friday morning, (2) having two examinations on the same day, and (3) stalling your car at a busy intersection with your little brother aboard. Use the methods developed in this section to analyze these situations. What observations are made? What sorts of hypotheses and experiments might be devised?

7. For the ''Chemistry in Focus'' discussion of lead poisoning given in this section, discuss how David and Susan analyzed the situation, arriving at the theory that the lead glaze on the pottery was responsible for their symptoms.

8. Discuss several situations in which you have yourself analyzed a problem such as those presented in this section. What hypotheses did you suggest? How did you test those hypotheses?

1.3 The Scientific Method

QUESTIONS

9. What are the three operations involved in applying the scientific method? How does the scientific method help us to understand our observations of nature?

10. Which of the following are quantitative observations, and which are qualitative observations?
    a. My waist size is 31 inches.
    b. My eyes are blue.
    c. My right index finger is 1/4 inch longer than my left.
    d. The leaves of most trees are green in summer.
    e. An apple consists of over 95% water.
    f. Chemistry is an easy subject.
    g. I got 90% on my last Chem exam.

11. What is the difference between an *hypothesis* and a *theory?* How are the two similar? How do they differ?

12. What is a *natural law?* Give examples of such laws. How does a law differ from a theory?

13. Discuss several political, social, or personal considerations that might affect a scientist's evaluation of a theory. Give examples of how such external forces have influenced scientists in the past. Discuss methods by which such bias might be excluded from future scientific investigations.

1.4 Learning Chemistry

QUESTIONS

14. In some academic subjects, it may be possible to receive a good grade primarily by memorizing facts. Why is chemistry not one of these subjects?

15. Why is the ability to solve problems important in the study of chemistry? Why is it that the *method* used to attack a problem is as important as the answer to the problem itself?

16. Students approaching the study of chemistry must learn certain basic facts (such as the names and symbols of the most common elements), but it is much more important that they learn to think critically and to go beyond the specific examples discussed in class or in the textbook. Explain how learning to do this might be helpful in any career, even one far removed from chemistry.

2 Measurements and Calculations

CONTENTS

Most students, when asked, will say they are familiar with scientific notation and the metric system. Although most have been exposed to these concepts in high school, few are as comfortable with scientific notation and metric units as their responses indicate.

Two different sized erlenmeyer flasks, each containing the same volume of solution.

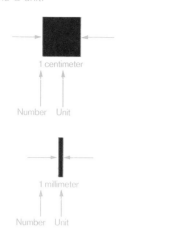

As we pointed out in Chapter 1, making observations is a key part of the scientific process. Sometimes observations are *qualitative* ("the substance is a yellow solid") and sometimes they are *quantitative* ("the substance weighs 4.3 grams"). A quantitative observation is called a **measurement.** A measurement always consists of two parts: a *number* and a *unit.* Both parts are necessary to make the measurement meaningful. For example, suppose a friend tells you that she saw a bug 5 long. This statement is meaningless as it stands. 5 what? If it's 5 millimeters, the bug is quite small. If it's 5 centimeters, the bug is quite large. If it's 5 meters, run for cover!

The point is that for a measurement to be meaningful, it must consist of both a number and a unit that tells us the scale being used.

In this chapter we will consider the characteristics of measurements and the calculations that involve measurements.

The Sun and the Earth as seen from the space shuttle.

2.1 Scientific Notation

AIM: To show how very large or very small numbers can be expressed as the product of a number between 1 and 10 and a power of 10.

The numbers associated with scientific measurements are often very large or very small. For example, the distance from the earth to the sun is approximately 93,000,000 (93 million) miles. Written out, this number is rather bulky. Scientific notation is a method for making very large or very small numbers more compact and easier to write.

To see how this is done, consider the number 125, which can be written as the product

$$125 = 1.25 \times 100$$

Because $100 = 10 \times 10 = 10^2$, we can write

$$125 = 1.25 \times 100 = 1.25 \times 10^2$$

Similarly, the number 1700 can be written

$$1700 = 1.7 \times 1000$$

and because $1000 = 10 \times 10 \times 10 = 10^3$, we can write

$$1700 = 1.7 \times 1000 = 1.7 \times 10^3$$

Scientific notation simply expresses a number as *a product of a number between 1 and 10 and the appropriate power of 10.* For example, the number 93,000,000 can be expressed as

When describing very small distances, such as the diameter of a swine flu virus (shown here magnified 16,537 times), it is convenient to use scientific notation.

$$93{,}000{,}000 = 9.3 \times 10{,}000{,}000 = \quad 9.3 \quad \times \quad 10^7$$

Number
between
1 and 10

Appropriate
power of 10
$(10{,}000{,}000 = 10^7)$

The easiest way to determine the appropriate power of 10 for scientific notation is to start with the number being represented and count the number of places the decimal point must be moved to obtain a number between 1 and 10. For example, for the number

9 3 0 0 0 0 0 0

7 6 5 4 3 2 1

we must move the decimal point seven places to the left to get 9.3 (a number between 1 and 10). To compensate for every move of the decimal point to the left, we must multiply by 10. That is, each time we move the decimal point to the left, we make the number smaller by one power of 10. So for each move of the decimal point to the left, we must multiply by 10 to restore the number to its original magnitude. Thus moving the decimal point seven places to the left means we must multiply 9.3 by 10 seven times, which equals 10^7:

$$93{,}000{,}000 = 9.3 \times 10^7 \quad \leftarrow \text{ We moved the decimal point seven}$$
places to the left, so we
need 10^7 to compensate.

Remember: whenever the decimal point is moved to the *left,* the exponent of 10 is *positive.*

We can represent numbers smaller than 1 by using the same convention, but in this case the power of 10 is negative. For example, for the number 0.010 we must move the decimal point two places to the right to obtain a number between 1 and 10:

0.0 1 0

1 2

This requires an exponent of -2, so $0.010 = 1.0 \times 10^{-2}$. Remember: whenever the decimal point is moved to the *right,* the exponent of 10 is *negative.*

Next consider the number 0.000167. In this case we must move the decimal point four places to the right to obtain 1.67 (a number between 1 and 10):

0.0 0 0 1 6 7

1 2 3 4

Moving the decimal point four places to the right requires an exponent of -4. Therefore

$$0.000167 = 1.67 \times 10^{-4} \quad \leftarrow \text{ We moved the decimal}$$
point four places
to the right.

I always begin my discussion of scientific notation with the numbers 1,000 and 5,000. It is obvious to most students that 1,000 is the same as $10 \times 10 \times 10$, or 10^3, and 5,000 is the same as $5 \times 10 \times 10 \times 10$, which can be written as 5×10^3. This is an easy way to show how the coefficient is derived.

Moving the decimal point to the left requires a positive exponent.

Some students can benefit from extra work with arithmetic using scientific notation. You can refer them to Appendix 3. In addition, I often provide students with a worksheet containing extra problems specifically written to provide drill in this area.

Moving the decimal point to the right requires a negative exponent.

Read Appendix 3 if you need a further discussion of exponents and scientific notation.

We can summarize these procedures as follows:

Using Scientific Notation

$100 = 1.0 \times 10^2$
$.010 = 1.0 \times 10^{-2}$

Left Is Positive; remember LIP.

- Any number can be represented as the product of a number between 1 and 10 and a power of 10 (either positive or negative).
- The power of 10 depends on the number of places the decimal point is moved and in which direction. The *number of places* the decimal point is moved determines the *power of 10*. The *direction* of the move determines whether the power of 10 is *positive* or *negative*. If the decimal point is moved to the left, the power of 10 is positive; if the decimal point is moved to the right, the power of 10 is negative.

| EXAMPLE 2.1 | Scientific Notation: Powers of 10 (Positive) |

Work at least one example of each kind of scientific-notation problem; convert a decimal number greater than 1 to scientific notation, a number greater than 1 in scientific notation to decimal, a decimal number less than 1 to scientific notation, and a number less than 1 in scientific notation to decimal. If you summarize in table form the rules for moving the decimal point to the right or left for each kind of problem, students can readily see how to convert from one notation to the other. Some numbers you might want to convert are 12,500; 0.000193; 2.90433×10^{-7}; and 3.954×10^3.

Represent the following numbers in scientific notation.

a.   238,000
b.   1,500,000

SOLUTION

a.   First we move the decimal point until we have a number between 1 and 10, in this case 2.38.

$$2\ 3\ 8\ 0\ 0\ 0 \qquad \text{The decimal point was moved}$$
$$5\ 4\ 3\ 2\ 1 \qquad \text{five places to the left.}$$

Because we moved the decimal point five places to the left, the power of 10 is positive 5. Thus $238,000 = 2.38 \times 10^5$.

b. $1\ 5\ 0\ 0\ 0\ 0\ 0$ The decimal point was moved
      $6\ 5\ 4\ 3\ 2\ 1$ six places to the left, so
      the power of 10 is 6.

Thus $1,500,000 = 1.5 \times 10^6$.

| EXAMPLE 2.2 | Scientific Notation: Powers of 10 (Negative) |

Represent the following numbers in scientific notation.

a.   0.00043
b.   0.089

EXAMPLE 2.2, CONTINUED

SOLUTION

a. First we move the decimal point until we have a number between 1
    and 10, in this case 4.3.

$$0.0\ 0\ 0\ 4\ 3$$

The decimal point was moved
four places to the right.

1 2 3 4

Because we moved the decimal point four places to the right, the power
of 10 is negative 4. Thus $0.00043 = 4.3 \times 10^{-4}$.

b. $0.0\ 8\ 9$ The power of 10 is negative 2 because the
decimal point was moved two places to the right

1 2

Thus $0.089 = 8.9 \times 10^{-2}$.

SELF-CHECK EXERCISE 2.1

Write the numbers 357 and 0.0055 in scientific notation. If you are having
difficulty with scientific notation at this point, reread Appendix 3.

2.2   Units

AIM: To describe the English, metric, and SI systems of
        measurement.

The **units** part of a measurement tells us what *scale* or *standard* is being used to
represent the results of the measurement. From the earliest days of civilization,
trade has required common units. For example, if a farmer from one region
wanted to trade some of his grain for the gold of a miner who lived in another
region, the two people had to have common standards (units) for measuring the
amount of the grain and the weight of the gold.

The need for common units also applies to scientists, who measure quantities such as mass, length, time, and temperature. If every scientist had her or his
own personal set of units, complete chaos would result. Unfortunately, although
standard systems of units did arise, different systems were adopted in different
parts of the world. The two most widely used systems are the **English system**
used in the United States and the **metric system** used in most of the rest of the
industrialized world.

I never ask students to memorize the exact
conversions between English and metric or
SI units. They can always look up conversions if they forget, and most books have
a list of the common ones. I provide a table
of conversion factors as they are needed
for solving problems on the exams.

The base units in Table 2.1 are not actually the most convenient to use. I tell students that we will learn more useful units of mass, length, volume and temperature later in the chapter.

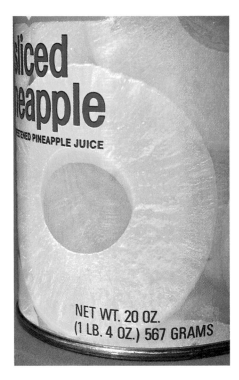

A pineapple can showing both English units (oz) and SI units (grams).

Table 2.1 Some Fundamental SI Units

Physical Quantity	Name of Unit	Abbreviation
mass	kilogram	kg
length	meter	m
time	second	s
temperature	kelvin	K

The metric system has long been preferred for most scientific work. In 1960 an international agreement set up a comprehensive system of units called the **International System** (*le Système Internationale* in French), or **SI**. The SI units are based on the metric system and units derived from the metric system. The most important fundamental SI units are listed in Table 2.1. We will discuss how to manipulate some of these units later in this chapter.

Because the fundamental units are not always a convenient size, the SI system uses prefixes to change the size of the unit. The most commonly used prefixes are listed in Table 2.2. Although the fundamental unit for length is the meter (m), we can also use the decimeter (dm), which represents one-tenth (0.1) of a meter; the centimeter (cm), which represents one one-hundredth (0.01) of a meter; the millimeter (mm), which represents one one-thousandth (0.001) of a meter; and so on. For example, it's much more convenient to specify the diameter of a certain contact lens as 1.0 cm than as 1.0×10^{-2} m.

Some metric units will already have meaning to students. They are familiar with one liter soft-drink bottles and some will know how long a 400-meter run actually is. The large and small prefixes are not commonly encountered outside of the sciences, so students should memorize the prefixes and their meanings given in Table 2.2. You might want to show an overhead transparency of the table and go over the pronunciations of the prefixes. Students are not apt to use a prefix if they feel uncertain about pronouncing the word aloud.

Table 2.2 The Commonly Used Prefixes in the Metric System

Prefix	Symbol	Meaning	Power of 10 for Scientific Notation
mega	M	1,000,000	10^6
kilo	k	1,000	10^3
deci	d	0.1	10^{-1}
centi	c	0.01	10^{-2}
milli	m	0.001	10^{-3}
micro	μ	0.000001	10^{-6}
nano	n	0.000000001	10^{-9}

2.3 Measurements of Length, Volume, and Mass

AIM: To describe the metric system for measuring length, volume, and mass.

The fundamental SI unit of length is the meter, which is a little longer than a yard (1 meter = 39.37 inches). In the metric system fractions of a meter or multiples of a meter can be expressed by powers of 10, as summarized in Table 2.3.

The meter was originally defined, in the eighteenth century, as one ten-millionth of the distance from the equator to the North Pole and then, in the late nineteenth century, as the distance between two parallel marks on a special metal bar stored in a vault in Paris. More recently, for accuracy and convenience, a definition expressed in terms of light waves has been adopted.

Table 2.3   The Metric System for Measuring Length

Unit	Symbol	Meter Equivalent
kilometer	km	1000 m or 10^3 m
meter	m	1 m or 1 m
decimeter	dm	0.1 m or 10^{-1} m
centimeter	cm	0.01 m or 10^{-2} m
millimeter	mm	0.001 m or 10^{-3} m
micrometer	μm	0.000001 m or 10^{-6} m
nanometer	nm	0.000000001 m or 10^{-9} m

The English and metric systems are compared on the ruler shown in Figure 2.1. Note that

$$1 \text{ inch } = 2.54 \text{ centimeter}$$

Other English-metric equivalences are given in Section 2.6.

A quarter is about 2.5 cm across.

Figure 2.1
Comparison of English and metric units for length on a ruler.

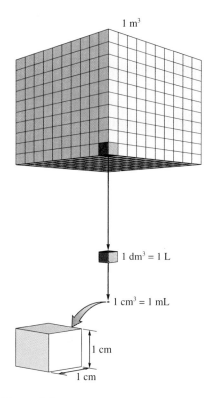

Figure 2.2
The largest drawing represents a cube which has sides 1 m in length and a volume of 1 m³. The middle-sized cube has sides 1 dm in length and a volume of 1 dm³, or 1 L. The smallest cube has sides 1 cm in length and a volume of 1 cm³, or 1 mL.

Table 2.4   The Relationship of the Liter and Milliliter

Unit	Symbol	Equivalence
liter	*L*	$1\ L\ =\ 1000\ mL$
milliliter	*mL*	$\frac{1}{1000}\ L\ =\ 10^{-3}\ L\ =\ 1\ mL$

Volume is the amount of three-dimensional space occupied by a substance. The fundamental unit of volume in the SI system is based on the volume of a cube that measures 1 meter in each of the three directions. That is, each edge of the cube is 1 meter in length. The volume of this cube is

$$1\text{ m} \times 1\text{ m} \times 1\text{ m} = (1\text{ m})^3 = 1\text{ m}^3$$

or, in words, one cubic meter.

In Figure 2.2 this cube is divided into 1000 smaller cubes. Each of these small cubes represents a volume of 1 dm³, which is commonly called the **liter** (rhymes with meter and is slightly larger than a quart) and abbreviated L.

The cube with a volume of 1 dm³ (1 liter) can in turn be broken into 1000 smaller cubes each representing a volume of 1 cm³. This means that each liter contains 1000 cm³. One cubic centimeter is called a **milliliter** (abbreviated mL), a unit of volume used very commonly in chemistry. This relationship is summarized in Table 2.4.

The *graduated cylinder* (see Figure 2.3), commonly used in chemical laboratories for measuring the volumes of liquids, is marked off in convenient units of volume (usually milliliters). The graduated cylinder is filled to the desired volume with the liquid, which then can be poured out.

Another important measurable quantity is **mass,** which can be defined as the quantity of matter present in an object. The fundamental SI unit of mass is the **kilogram.** Because the metric system, which came before the SI system, used the gram as the fundamental unit, the prefixes for the various mass units are based on the **gram,** as shown in Table 2.5.

I like to take as many different pieces of glassware for volume measurement as possible to class. Among the ones that are easiest to come by are volumetric flasks, graduated cylinders, beakers, transfer pipets, and burets. In a lecture that is otherwise filled with rules, memorization, and calculation, it can relax the atmosphere to show interesting shapes and explain how they are used. I like to compare different pieces of glassware that contain the same volumes. For example, a 1 L volumetric flask holds the same volume of liquid as a l L graduated cylinder. Yet each has a different purpose.

Table 2.5   The Most Commonly Used Metric Units for Mass

Unit	Symbol	Gram Equivalent
kilogram	kg	$1000\text{ g} = 10^3\text{ g} = 1\text{ kg}$
gram	g	1 g
milligram	mg	$0.001\text{ g} = 10^{-3}\text{ g} = 1\text{ mg}$

Table 2.6 Some Examples of Commonly Used Units	
length	A dime is 1 mm thick.
	A quarter is 2.5 cm in diameter.
	The average height of adult men is 1.8 m.
mass	A nickel has a mass of about 5 g.
	A 120-lb woman has a mass of about 55 kg.
volume	A 12-oz can of soda has a volume of about 360 mL.
	A half gallon of milk is equal to about 2 L of milk.

In the laboratory we determine the mass of an object by using a balance. A balance compares the mass of the object to a set of standard masses ("weights"). For example, the mass of an object can be determined by using a single-pan balance (Figure 2.4).

To help you get a feeling for the common units of length, volume, and mass, some familiar objects are described in Table 2.6.

100-mL graduated cylinder

Figure 2.3
A 100-mL graduated cylinder.

(a)

(b)

Figure 2.4
(a) Actual appearance of a single-pan analytical balance used in chemistry labs. (b) Schematic view showing its principle of operation. An object is placed on the balance pan, and "weights" (masses) are removed from holders above the pan to restore the balance. The mass of the object equals the mass of the removed "weights."

Although we tend to use the terms mass and weight interchangeably, they have different meanings to a chemist. I try to explain that when we "weigh" a sample in the lab, we are actually finding the mass, the quantity of matter, of the substance. Weighing refers to the pull of gravity. Using the explanation that on the moon the mass of a substance doesn't change, but the weight does, seems to help make the distinction clearer.

It is useful for students to know approximately how the common English and metric or SI units compare. I ask them to know the approximate relationship between a liter and a quart, a gram and a pound, and a kilometer and a mile.

A student working in a chemistry lab.

I like to illustrate uncertainty in measurement with some simulated measurements. I either draw on the board or a transparency some glassware for measuring volume and ask students to tell me what the correct volume measurement should be. I can illustrate that some pieces of glassware are more precise than others by comparing the number of significant figures generated by a beaker of liquid compared with a buret containing approximately the same volume of liquid.

Every measurement has some degree of uncertainty.

2.4   Uncertainty in Measurement

AIM: To show how uncertainty in measurement arises.
        To explain how to indicate a measurement's uncertainty by using significant figures.

Whenever a measurement is made, an estimate is always required. We can illustrate this by measuring the pin shown in Figure 2.5(a). We can see from the ruler that the pin is a little longer than 2.8 cm and a little shorter than 2.9 cm. Because there are no graduations on the ruler between 2.8 and 2.9, we must estimate the pin's length between 2.8 and 2.9 cm. We do this by *imagining* that the distance between 2.8 and 2.9 is broken into 10 equal divisions (Figure 2.5b) and estimating to which division the end of the pin reaches. The end of the pin appears to come about halfway between 2.8 and 2.9, which corresponds to 5 of our 10 imaginary divisions. So we estimate the pin's length as 2.85 cm. The result of our measurement is that the pin is approximately 2.85 cm in length, but we had to rely on a visual estimate, so it might actually be 2.84 or 2.86 cm.

Because the last number is based on a visual estimate, it may be different when another person makes the same measurement. For example, if five different people measured the pin, the results might be

Person	Result of Measurement
1	2.85 cm
2	2.84 cm
3	2.86 cm
4	2.85 cm
5	2.86 cm

Note that the first two digits in each measurement are the same regardless of who made the measurement; these are called the *certain* numbers of the measurement. However, the third digit is estimated and can vary; it is called an *uncertain* number. When one is making a measurement, the custom is to record all of the certain numbers plus the *first* uncertain number. It would not make any sense to try to measure the pin to the third decimal place (thousandths of a centimeter), because this ruler requires an estimate of even the second decimal place (hundredths of a centimeter).

It is very important to realize that *a measurement always has some degree of uncertainty.* The uncertainty of a measurement depends on the measuring device. For example, if the ruler in Figure 2.5 had marks indicating hundredths of a centimeter, the uncertainty in the measurement of the pin would occur in the

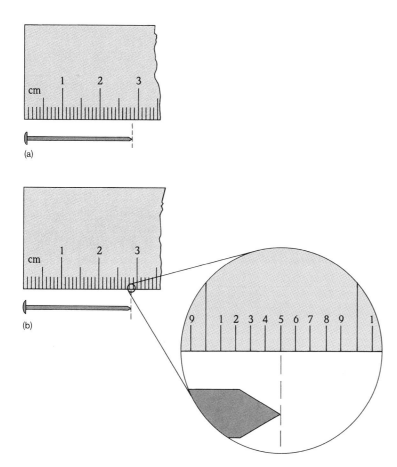

If you bring glassware to class to illustrate volume, you can also use it to help illustrate uncertainty in measurement. I usually draw the shape and markings of a representative piece on the blackboard and mark the volume between two of the etched lines. Because the volume falls between the lines, there is uncertainty involved. The exact number to be recorded for the last digit may vary from time to time and from student to student.

Figure 2.5
Measuring a pin. (a) The length is between 2.8 cm and 2.9 cm. (b) Imagine that the distance between 2.8 and 2.9 is divided into 10 equal divisions. The end of the pin occurs after about 5 of these divisions.

thousandths place rather than the hundredths place, but some uncertainty would still exist.

The numbers recorded in a measurement (all the certain numbers plus the first uncertain number) are called **significant figures.** The number of significant figures for a given measurement is determined by the inherent uncertainty of the measuring device. For example, the ruler used to measure the pin can give results only to hundredths of a centimeter. Thus when we record the significant figures for a measurement, we automatically give information about the uncertainty in a measurement. The uncertainty in the last number (the estimated number) is usually assumed to be ± 1 unless otherwise indicated. For example, the measurement 1.86 kilograms can be interpreted as 1.86 ± 0.01 kilograms, where the symbol $\pm$ means plus or minus. That is, it could be 1.86 kg $-$ 0.01 kg = 1.85 kg or 1.86 kg $+$ 0.01 kg = 1.87 kg.

Students often believe that learning about significant figures is something they must do for Chapter 2, and then can forget about for the remainder of the semester. Try to emphasize that significant figures will be part of *every* calculation for this course.

2.5 Significant Figures

AIM: To show how to determine the number of significant figures in a calculated result.

We have seen that any measurement involves an estimate and thus is uncertain to some extent. We signify the degree of certainty for a particular measurement by the number of significant figures we record.

Because doing chemistry requires many types of calculations, we must consider what happens when we do arithmetic with numbers that contain uncertainties. It is important that we know the degree of uncertainty in the final result. Although we will not discuss the process here, mathematicians have studied how uncertainty accumulates and have designed a set of rules to determine how many significant figures the result of a calculation should have. You should follow these rules whenever you carry out a calculation. The first thing we need to do is learn how to count the significant figures in a given number. To do this we use the following rules:

Rules for Counting Significant Figures

1. *Nonzero integers.* Nonzero integers *always* count as significant figures. For example, the number 1457 has four nonzero integers, all of which count as significant figures.
2. *Zeros.* There are three classes of zeros:
   a. *Leading zeros* are zeros that *precede* all of the nonzero digits. They *never* count as significant figures. For example, in the number 0.0025, the three zeros simply indicate the position of the decimal point. The number has only two significant figures, the 2 and the 5.
   b. *Captive zeros* are zeros that fall *between* nonzero digits. They *always* count as significant figures. For example, the number 1.008 has four significant figures.
   c. *Trailing zeros* are zeros at the *right end* of the number. They are significant only if the number contains a decimal point. The number one hundred written as 100 has only one significant figure, but written as 100., it has three significant figures.
3. *Exact numbers.* Often calculations involve numbers that were not obtained using measuring devices but were determined by counting: 10 experiments, 3 apples, 8 molecules. Such numbers are called *exact numbers.* They can be assumed to have an unlimited number of significant figures. Exact numbers can also arise from definitions. For

---

Students easily learn to determine the correct number of significant figures in a number as long as there are no zeros. Show examples with zeros in various locations. Students will need to work the assigned homework problems on their own before they become comfortable determining the correct number of significant figures. Some examples you might use are 1.001, 45.00, 0.0103, and 4.10×10^4.

Leading zeros are never significant figures.

Captive zeros are always significant figures.

Trailing zeros are sometimes significant figures.

Exact numbers never limit the number of significant figures in a calculation.

Students are often unsure about what kinds of numbers qualify as "exact numbers." I try to point them out as we use them so that students learn to recognize what kinds of numbers will not affect the number of significant figures in the final answer.

example, 1 inch is defined as *exactly* 2.54 centimeters. Thus in the statement 1 in. = 2.54 cm, neither 2.54 nor 1 limits the number of significant figures when it is used in a calculation.

Rules for counting significant figures also apply to numbers written in scientific notation. For example, the number 100. can also be written as 1.00×10^2, and both versions have three significant figures. Scientific notation offers two major advantages: the number of significant figures can be indicated easily, and fewer zeros are needed to write a very large or a very small number. For example, the number 0.000060 is much more conveniently represented as 6.0×10^{-5}, and the number has two significant figures no matter in which form it is written.

Significant figures are easily indicated by scientific notation.

EXAMPLE 2.3   Counting Significant Figures

Give the number of significant figures for each of the following measurements.

a. A sample of orange juice contains 0.0108 g of vitamin C.
b. A forensic chemist in a crime lab weighs a hair and records its weight as 0.0050060 g.
c. The distance between two points was found to be 5.030×10^3 ft.
d. In yesterday's bicycle race, 110 riders started but only 60 finished.

SOLUTION

a. The number contains three significant figures. The zeros to the left of the 1 are leading zeros and are not significant, but the remaining zero (a captive zero) is significant.
b. The number contains five significant figures. The leading zeros (to the left of the 5) are not significant. The captive zeros between the 5 and the 6 are significant, and the trailing zero to the right of the 6 is significant because the number contains a decimal point.
c. This number has four significant figures. Both zeros in 5.030 are significant.
d. Both numbers are exact (they were obtained by counting the riders). Thus these numbers have an unlimited number of significant figures.

Determining the number of significant figures in a number someone else has recorded is a matter of learning the rules on page 26. When students make their own measurements, such as measuring the length of a pin or recording the volume of a beaker, they should record all the numbers they can read directly from the measuring device, plus one estimated digit.

SELF-CHECK EXERCISE 2.2

Give the number of significant figures for each of the following measurements.

a. 0.00100 m
b. 2.0800×10^2 L
c. 480 Corvettes

The rules for rounding off *given* numbers are straightforward. But few students can correctly round off numbers resulting from calculations. Work an example of a multiplication or division problem, and an addition or subtraction problem to show the difference between the two. You might point out that definitions such as 1 cm = 2.54 in., and exact numbers such as 10 students in class, do not affect the number of significant figures in a calculation. Some examples you might use for determining the number of significant figures from calculations are 483.21 × 5.00, 0.002/0.041, 28.043 − 2.31, and 0.93 + 10.1.

These rules reflect the way electronic calculators round off.

Students seem to find rounding off easy to do, but are often not sure when to round off. Stress that rule 2 applies. In general, students should carry all calculator digits until all steps are completed, and then round off. When you write multiple-choice exams, it is important to follow the same steps when constructing both the correct answer and the alternative choices. Otherwise, the student's correct answer and your correct answer may differ, leading students to believe they have performed the calculations incorrectly.

Do not round off sequentially. The number 6.8347 rounded to three significant figures is 6.83, not 6.84.

Rounding Off Numbers

When you perform a calculation on your calculator, the number of digits displayed is usually greater than the number of significant figures that the result should possess. So you must "round off" the number (reduce it to fewer digits). The rules for **rounding off** follow.

Rules for Rounding Off

1. If the digit to be removed
   a. is less than 5, the preceding digit stays the same. For example, 1.33 rounds to 1.3.
   b. is equal to or greater than 5, the preceding digit is increased by 1. For example, 1.36 rounds to 1.4, and 3.15 rounds to 3.2.
2. In a series of calculations, carry the extra digits through to the final result and *then* round off.* This means that you should carry all of the digits that show on your calculator until you arrive at the final number (the answer) and then round off, using the procedures in rule 1.

We need to make one more point about rounding off to the correct number of significant figures. Suppose the number 4.348 needs to be rounded to two significant figures. In doing this, we look *only* at the *first number* to the right of the 3:

$$4.348$$

↑
Look at this
number to round off
to two significant figures.

The number is rounded to 4.3 because 4 is less than 5. It is incorrect to round sequentially. For example, do *not* round the 4 to 5 to give 4.35 and then round the 3 to 4 to give 4.4.

When rounding off, *use only the first number to the right of the last significant figure.*

Determining Significant Figures in Calculations

Next we will learn how to determine the correct number of significant figures in the result of a calculation. To do this we will use the following rules.

---

*This practice will not be followed in the worked out examples in the body of this text, because we want to show the correct number of significant figures in each step of the example.

Rules for Using Significant Figures in Calculations

1. For *multiplication or division,* the number of significant figures in the result is the same as that in the measurement with the *smallest number* of significant figures. We say this measurement is *limiting,* because it limits the number of significant figures in the result. For example, consider this calculation:

Because 1.4 has only two significant figures, it limits the result to two significant figures. Thus the product is correctly written as 6.4, which has two significant figures. Consider another example. In the division $\frac{8.315}{298}$, how many significant figures should appear in the answer?

Because 8.315 has four significant figures, the number 298 (with three significant figures) limits the result. The calculation is correctly represented as

2. For *addition or subtraction,* the limiting term is the one with the smallest number of decimal places. For example, consider the following sum:

$$
\begin{array}{ll}
12.11 & \leftarrow \text{ Two decimal places} \\
18.0 & \longleftarrow \text{ Limiting term (has one decimal place)} \\
\underline{1.013} & \\
31.123 & \boxed{\text{Round off}} \quad 31.1 \\
& \qquad\qquad\qquad \uparrow \\
& \qquad\qquad \text{One decimal place}
\end{array}
$$

If you need help in using your calculator, see Appendix 1.

The correct result is 31.1 (it is limited to one decimal place because 18.0 has only one decimal place). Consider another example:

$$
\begin{array}{r}
0.6875 \\
-0.1 \\
\hline
0.5875
\end{array}
$$

Limiting term (one decimal place)

Round off → 0.6

Note that *for multiplication and division, significant figures are counted. For addition and subtraction, the decimal places are counted.*

Now we will put together the things you have learned about significant figures by considering some mathematical operations in the following examples.

EXAMPLE 2.4   Counting Significant Figures in Calculations

Without performing the calculations, tell how many significant figures each answer should contain.

a.   5.19
     1.9
     0.842

b. $1081 - 7.25$

c. 2.3×3.14

d.   the total cost of 3 boxes of candy at $2.50 a box

SOLUTION

a.   The answer will have one digit after the decimal place. The limiting number is 1.9, which has one decimal place, so the answer has two significant figures.
b.   The answer will have no digits after the decimal point. The number 1081 has no digits to the right of the decimal point and limits the result, so the answer has four significant figures.
c.   The answer will have two significant figures because the number 2.3 has only two significant figures (3.14 has three).
d.   The answer will have three significant figures. The limiting factor is 2.50 because 3 (boxes of candy) is an exact number.

EXAMPLE 2.5   Calculations Using Significant Figures

Carry out the following mathematical operations and give each result to the correct number of significant figures.

EXAMPLE 2.5, CONTINUED

a. 5.18×0.0208
b. $(3.60 \times 10^{-3}) \times (8.123) \div 4.3$
c. $21 + 13.8 + 130.36$

d. $116.8 - 0.33$
e. $(1.33 \times 2.8) + 8.41$

SOLUTION

    Limiting terms   Round to this digit.

a. $5.18 \times 0.0208 = 0.107744 \rightarrow 0.108$

The answer should contain three significant figures because each number being multiplied has three significant figures (Rule 1). The 7 is rounded to 8 because the following digit is greater than 5.

                        Round to this digit.

b. $\dfrac{(3.60 \times 10^{-3})(8.123)}{4.3} = 6.8006 \times 10^{-3} \rightarrow 6.8 \times 10^{-3}$

              Limiting
              term

Because 4.3 has the least number of significant figures (two), the result should have two significant figures (Rule 1).

c.
$$
\begin{array}{r}
21 \\
13.8 \\
130.36 \\
\hline
165.16 \rightarrow \quad 165
\end{array}
$$

In this case 21 is limiting (there are no digits after the decimal point). Thus the answer must have no digits after the decimal point, in accordance with the rule for addition (Rule 2).

d.
$$
\begin{array}{r}
116.8 \\
- \quad 0.33 \\
\hline
116.47 \rightarrow 116.5
\end{array}
$$

Because 116.8 has only one decimal place, the answer must have only one decimal place (Rule 2). The 4 is rounded up to 5 because the digit to the right (7) is greater than 5.

        Limiting term

e. $1.33 \times 2.8 = 3.724$ Round off   3.7

$$
\begin{array}{l}
3.7 \quad \longleftarrow \text{ Limiting term} \\
+ \; 8.41 \\
\hline
12.11 \rightarrow 12.1
\end{array}
$$

Note that in this case we multiplied and then rounded the result to the correct number of significant figures before we performed the addition so that we would know the correct number of decimal places.

SELF-CHECK EXERCISE 2.3

Give the answer for each calculation to the correct number of significant figures.

a. 12.6×0.53
b. $(12.6 \times 0.53) - 4.59$

c. $(25.36 - 4.15) \div 2.317$

Dimensional analysis is introduced very early so that students can practice solving problems using conversion factors as they learn the SI and metric systems. This is a good opportunity to encourage class participation. Students can tell us which conversion factors to use, and what should go "on top" so that the units will cancel. Even if your students are sure they know the metric system, encourage them to work the assigned problems so that they are familiar with how you expect them to set up and solve problems during the rest of the course.

Most students believe that they already know how to solve dimensional-analysis problems. They have usually been exposed to this method for solving problems in a previous course. However, most students usually need more practice in using this method. To help students learn, I begin by insisting that everyone use the steps on page 35 and show units for all problems. Unitless problems are discounted somewhat on the written-out portions of hour exams.

2.6 Problem Solving and Dimensional Analysis

AIM: To show how dimensional analysis can be used to solve various types of problems.

Suppose that the boss at the store where you work on weekends asks you to pick up 2 dozen doughnuts on the way to work. However, you find that the doughnut shop sells by the doughnut. How many doughnuts do you need?

This "problem" is an example of something you encounter all the time: converting from one unit of measurement to another. Examples of this occur in cooking (The recipe calls for 3 cups of cream, which is sold in pints. How many pints do I buy?); traveling (The purse costs 250 pesos. How much is that in dollars?); sports (A recent Tour de France bicycle race was 3215 kilometers long. How many miles is that?); and many other areas.

How do we convert from one unit of measurement to another? Let's explore this process by using the doughnut problem.

$$2 \text{ dozen doughnuts} = ? \text{ individual doughnuts}$$

where ? represents a number you don't know yet. The essential information you must have is the definition of a dozen:

$$1 \text{ dozen} = 12$$

You can use this information to make the needed conversion as follows:

$$2 \text{ \sout{dozen} doughnuts} \times \frac{12}{1 \text{ \sout{dozen}}} = 24 \text{ doughnuts}$$

You need to buy 24 doughnuts.
Note two important things about this process.

1. The factor $\frac{12}{1 \text{ dozen}}$ is a conversion factor based on the definition of the term *dozen.* This conversion factor is a ratio of the two parts of the definition of a dozen given above.
2. The unit dozen itself cancels.

Now let's generalize a bit. To change from one unit to another we will use a conversion factor.

$$\text{Unit}_1 \times \text{conversion factor} = \text{unit}_2$$

The **conversion factor** is a ratio of the two parts of the statement that relates the two units. We will see this in more detail on the following pages.

Earlier in this chapter we considered a pin that measured 2.85 cm in length. What is the length of the pin in inches? We can represent this problem as

$$2.85 \text{ cm} \rightarrow ? \text{ in.}$$

The question mark stands for the number we want to find. To solve this problem, we must know the relationship between inches and centimeters. In Table 2.7, which gives several equivalents between the English and metric systems, we find the relationship

$$\boxed{2.54 \text{ cm} = 1 \text{ in.}}$$

This is called an **equivalence statement.** In other words, 2.54 cm and 1 in. stand for *exactly the same distance.* (See Figure 2.1.) The respective numbers are different because they refer to different *scales (units)* of distance.

The equivalence statement 2.54 cm = 1 in. can lead to either of two conversion factors:

$$\frac{2.54 \text{ cm}}{1 \text{ in.}} \quad \text{or} \quad \frac{1 \text{ in.}}{2.54 \text{ cm}}$$

Note that these *conversion factors* are *ratios of the two parts of the equivalence statement* that relates the two units. Which of the two possible conversion factors do we need? Recall our problem:

$$2.85 \text{ cm} = ? \text{ in.}$$

That is, we want to convert from units of centimeters to inches:

$$2.85 \text{ cm} \times \text{ conversion factor} = ? \text{ in.}$$

We choose the conversion factor that cancels the units we want to get rid of and leaves the units we want in the result. Thus we do the conversion as follows:

$$2.85 \text{ cm} \times \frac{1 \text{ in.}}{2.54 \text{ cm}} = \frac{2.85 \text{ in.}}{2.54} = 1.12 \text{ in.}$$

Note two important facts about this conversion.

1. The centimeter units cancel to give inches for the result. This is exactly what we wanted to accomplish. Using the other conversion factor $\left(2.85 \text{ cm} \times \dfrac{2.54 \text{ cm}}{1 \text{ in.}}\right)$ would not work because the units would not cancel to give inches in the result.
2. As the units changed from centimeters to inches, the number changed from 2.85 to 1.12. Thus 2.85 cm has exactly the same value (is the same length) as 1.12 in. Notice that in this conversion, the number decreased from 2.85 to 1.12. This makes sense because the inch is a larger unit of length than the centimeter is. That is, it takes fewer inches to make the same length in centimeters.

Table 2.7 English–Metric Equivalents

Length
 1 m = 1.094 yd
 2.54 cm = 1 in.
 1 mi = 5280 ft
 1 mi = 1760 yd

Mass
 1 kg = 2.205 lb
 453.6 g = 1 lb

Volume
 1 L = 1.06 qt
 1 ft^3 = 28.32 L

While solving problems in lectures, I often write both of the possible conversion factors on the board (for example, 1/1,000 and 1,000/1) and let students choose which one will produce the desired units. In that way they keep in mind that all conversion factors can be reversed and that they must choose the correct one based on the wording of the problem and the desired units.

Units cancel just as numbers do.

The result in the foregoing conversion has three significant figures as required. Caution: Noting that the term 1 appears in the conversion, you might think that because this number appears to have only one significant figure, the result should only have one significant figure. That is, the answer should be given as 1 in. rather than 1.12 in. However, in the equivalence statement 1 in. = 2.54 cm, the 1 is an exact number (by definition). In other words, exactly 1 in. equals 2.54 cm. Therefore the 1 does not limit the number of significant digits in the result.

When exact numbers are used in a calculation, they never limit the number of significant digits.

We have seen how to convert from centimeters to inches. What about the reverse conversion? For example, if a pencil is 7.00 in. long, what is its length in centimeters? In this case, the conversion we want to make is

$$7.00 \text{ in.} \rightarrow \text{? cm}$$

What conversion factor do we need to make this conversion?

Remember that two conversion factors can be derived from each equivalence statement. In this case, the equivalence statement 2.54 cm = 1 in. gives

$$\frac{2.54 \text{ cm}}{1 \text{ in.}} \quad \text{or} \quad \frac{1 \text{ in.}}{2.54 \text{ cm}}$$

Again, we choose which to use by looking at the *direction* of the required change. For us to change from inches to centimeters, the inches must cancel. Thus the factor

$$\frac{2.54 \text{ cm}}{1 \text{ in.}}$$

is used, and the conversion is done as follows:

$$7.00 \text{ in.} \times \frac{2.54 \text{ cm}}{1 \text{ in.}} = (7.00)(2.54) \text{ cm} = 17.8 \text{ cm}$$

Here the inch units cancel, leaving centimeters as required.

Consider the direction of the required change in order to select the correct conversion factor.

Note that in this conversion, the number increased (from 7.00 to 17.8). This makes sense because the centimeter is a smaller unit of length than the inch. That is, it takes more centimeters to make the same length in inches. *Always take a moment to think about whether your answer makes sense.* This will help you avoid errors.

Changing from one unit to another via conversion factors (based on the equivalence statements between the units) is often called **dimensional analysis.** We will use this method throughout our study of chemistry.

We can now state some general steps for doing conversions by dimensional analysis.

Converting from One Unit to Another

STEP 1
To convert from one unit to another, use the equivalence statement that relates the two units. The conversion factor needed is a ratio of the two parts of the equivalence statement.

STEP 2
Choose the appropriate conversion factor by looking at the direction of the required change (make sure the unwanted units cancel).

STEP 3
Multiply the quantity to be converted by the conversion factor to give the quantity with the desired units.

STEP 4
Check to see that you have the correct number of significant figures.

STEP 5
Ask whether your answer makes sense.

We will now illustrate this procedure in Example 2.6.

EXAMPLE 2.6   Conversion Factors: One-Step Problems

An Italian bicycle has its frame size given as 62 cm. What is the frame size in inches?

SOLUTION

We can represent the problem as

$$62 \text{ cm} = ? \text{ in.}$$

In this problem we want to convert from centimeters to inches.

$$62 \text{ cm} \times \text{conversion factor} = ? \text{ in.}$$

STEP 1
To convert from centimeters to inches, we need the equivalence statement 1 in. = 2.54 cm. This leads to two conversion factors:

$$\frac{1 \text{ in.}}{2.54 \text{ cm}} \quad \text{and} \quad \frac{2.54 \text{ cm}}{1 \text{ in.}}$$

EXAMPLE 2.6, CONTINUED

STEP 2
In this case, the direction we want is

$$\text{Centimeters} \rightarrow \text{inches}$$

so we need the conversion factor $\dfrac{1 \text{ in.}}{2.54 \text{ cm}}$. We know this is the one we want because using it will make the units of centimeters cancel, leaving units of inches.

STEP 3
The conversion is carried out as follows:

$$62 \text{ cm} \times \frac{1 \text{ in.}}{2.54 \text{ cm}} = 24 \text{ in.}$$

STEP 4
The result is limited to two significant figures by the number 62. The centimeters cancel, leaving inches as required.

STEP 5
Note that the number decreased in this conversion. This makes sense; the inch is a larger unit of length than the centimeter.

Solving dimensional-analysis problems in class provides an opportunity to routinely use significant figures and to determine whether or not an answer makes sense. Students often ignore these important steps in their haste to arrive at the "right answer."

SELF-CHECK EXERCISE 2.4

Wine is often bottled in 0.750-L containers. Using the appropriate equivalence statement from Table 2.7, calculate the volume of such a wine bottle in quarts.

I usually work two or three problems on the board if time allows. To save some time in class, I often put the problem to be solved on an overhead transparency so students can copy it in comparative leisure. I then solve the problem on the blackboard so students can think through the steps with me.

Next we will consider a conversion that requires several steps.

EXAMPLE 2.7   **Conversion Factors: Multiple-Step Problems**

The length of the marathon race is approximately 26.2 mi. What is this distance in kilometers?

SOLUTION

The problem before us can be represented as follows:

$$26.2 \text{ mi} = ? \text{ km}$$

EXAMPLE 2.7, CONTINUED

We could accomplish this conversion in several different ways, but because Table 2.7 gives the equivalence statements 1 mi = 1760 yd and 1 m = 1.094 yd, we will proceed as follows:

$$\text{Miles} \rightarrow \text{yards} \rightarrow \text{meters} \rightarrow \text{kilometers}$$

This process will be carried out one conversion at a time to make sure everything is clear.

Miles → Yards

We convert from miles to yards using the conversion factor $\dfrac{1760 \text{ yd}}{1 \text{ mi}}$.

$$26.2 \text{ mi} \times \frac{1760 \text{ yd}}{1 \text{ mi}} = 46{,}112 \text{ yd}$$

Result shown
on calculator

46,112 yd [Round off] 46,100 yd = 4.61 × 10⁴ yd

Yards → Meters

The conversion factor used to convert yards to meters is $\dfrac{1 \text{ m}}{1.094 \text{ yd}}$.

$$4.61 \times 10^4 \text{ yd} \times \frac{1 \text{ m}}{1.094 \text{ yd}} = 4.213894 \times 10^4 \text{ m}$$

Result shown
on calculator

4.213894 × 10⁴ m [Round off] 4.21 × 10⁴ m

Meters → Kilometers

Because 1000 m = 1 km, or 10^3 m = 1 km, we convert from meters to kilometers as follows:

$$4.21 \times 10^4 \text{ m} \times \frac{1 \text{ km}}{10^3 \text{ m}} = 4.21 \times 10^1 \text{ km}$$
$$= 42.1 \text{ km}$$

Thus the marathon (26.2 mi) is 42.1 km.

Remember that we are rounding off at the end of each step to show the correct number of significant figures. However, in doing a multistep calculation, *you* should retain the extra numbers that show on your calculator and round off only at the end of the calculation.

EXAMPLE 2.7, CONTINUED

Once you feel comfortable with the conversion process, you can combine the steps. For the above conversion, the combined expression is

$$\text{miles} \rightarrow \text{yards} \rightarrow \text{meters} \rightarrow \text{kilometers}$$

$$26.2 \text{ mi} \times \frac{1760 \text{ yd}}{1 \text{ mi}} \times \frac{1 \text{ m}}{1.094 \text{ yd}} \times \frac{1 \text{ km}}{10^3 \text{ m}} = 42.1 \text{ km}$$

Note that the units cancel to give the required kilometers and that the result has three significant figures.

SELF-CHECK EXERCISE 2.5

Racing cars at the Indianapolis Motor Speedway now routinely travel around the track at an average speed of 215 mi/h. What is this speed in kilometers per hour?

RECAP: Whenever you work problems, remember the following points:

Units provide a very valuable check on the validity of your solution. Always use them.

1. Always include the units (a measurement always has two parts: a number *and* a unit).
2. Cancel units as you carry out the calculations.
3. Check that your final answer has the correct units. If it doesn't, you have done something wrong.
4. Check that your final answer has the correct number of significant figures.
5. Think about whether your answer makes sense.

Students have no difficulty converting among Fahrenheit, Celsius, and Kelvin temperature scales when the conversion formulas are provided or memorized. Section 2.7 encourages students to think about the relationships among the temperature scales and not to depend solely on memorization. Students should come to understand why memorization is not enough. We are actually trying to develop thought processes, not just trying to produce the right answer by the mechanical use of a formula. Hopefully, they will see that there are reasons why the formulas work, and that, with some thought, we can determine the formulas from the information we know.

2.7 Temperature Conversions: An Approach to Problem Solving

AIM: To define the three temperature scales.
To show how to convert from one scale to another.
To continue to develop problem-solving skills.

When the doctor tells you your temperature is 104 degrees and the weatherperson on TV says it will be 75 degrees tomorrow, they are using the **Fahrenheit scale.** Water boils at 212 °F and freezes at 32 °F, and normal body temperature is 98.6 °F (where °F signifies "Fahrenheit degrees"). This temperature scale is widely used in the United States and Britain, and it is the scale employed in most of the engineering sciences. Another temperature scale used on the European continent and in the physical and life sciences is the **Celsius scale.** In keeping with the metric system, which is based on powers of 10, the freezing and boiling points of water on the Celsius scale are assigned as 0 °C and 100 °C, respectively. On both the Fahrenheit and Celsius scales, the unit of temperature is called a degree, and the symbol for it is followed by the capital letter representing the scale the units are measured on: °C or °F.

Still another temperature scale used in the sciences is the **absolute** or **Kelvin scale.** On this scale water freezes at 273 K and boils at 373 K. On the Kelvin scale, the unit of temperature is called a kelvin and is symbolized by K. Thus on the three scales, the boiling point of water is stated as 212 degrees Fahrenheit (212 °F), 100 degrees Celsius (100 °C), and 373 kelvins (373 K).

The three temperature scales are compared in Figures 2.6 and 2.7. Note several important facts:

1. The size of each temperature unit (each degree) is the same for the Celsius and Kelvin scales. This follows from the fact that the *difference* between the boiling and freezing points of water is 100 units on both of these scales.
2. The Fahrenheit degree is smaller than the Celsius and Kelvin unit. Note that on the Fahrenheit scale there are 180 Fahrenheit degrees between

Although 373 K is often stated as 373 degrees Kelvin, it is more correct to say 373 kelvins.

(a)   (b)

Figure 2.6
Thermometers based on the three temperature scales in (a) ice water and (b) boiling water.

One way to test how well students understand temperature scales is to invent a temperature scale and ask students to convert between the imaginary scale and degrees Celsius. The number of degrees between the boiling and freezing points of water and the offset for the freezing point of water should be different from either the Fahrenheit or the Celsius scales. If students understand the relationship between the imaginary scale and the Celsius scale, they should be able to devise a formula to convert between the two.

Figure 2.7
The three major temperature scales.

the boiling and freezing points of water, as compared to 100 units on the other two scales.

3. The zero points are different on all three scales.

In your study of chemistry, you will sometimes need to convert from one temperature scale to another. We will consider in some detail how this is done. In addition to learning how to change temperature scales, you should also use this section as an opportunity to further develop your skills in problem solving.

Converting Between the Kelvin and Celsius Scales

It is relatively simple to convert between the Celsius and Kelvin scales because the temperature unit is the same size; only the zero points are different. Because 0 °C corresponds to 273 K, converting from Celsius to Kelvin requires that we add 273 to the Celsius temperature. We will illustrate this procedure in Example 2.8.

EXAMPLE 2.8	Temperature Conversion: Celsius to Kelvin

Boiling points will be discussed further in Chapter 14.

Convert the boiling point of water at the top of Mt. Everest (70. °C) to the Kelvin scale. (We add a decimal point after our temperature readings to indicate that the trailing zeros are significant.)

EXAMPLE 2.8, CONTINUED

SOLUTION

This problem asks us to find 70. °C in units of kelvins. We can represent this problem simply as

$$70.\ °C\ =\ ?\ K$$

In doing problems, it is often helpful to draw a diagram in which we try to represent the words in the problem by a picture. This problem can be diagrammed as shown in Figure 2.8(a).

In solving problems, it is often helpful to draw a diagram that depicts what the words are telling you.

In this picture we have shown what we want to find: "What temperature (in kelvins) is the same as 70. °C?" We also know from Figure 2.7 that 0 °C represents the same temperature as 273 K. How many degrees above 0 °C is 70. °C? The answer, of course, is 70. Thus we must add 70. to 0 °C to reach 70.°C. Because degrees are the *same size* on both the Celsius scale and the Kelvin scale (see Figure 2.8b), we must also add 70. to 273 K (same temperature as 0 °C) to reach ? K. That is,

$$?\ K\ =\ 273\ +\ 70.\ =\ 343\ K$$

Thus 70. °C corresponds to 343 K.

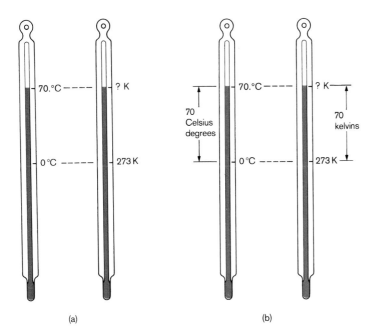

(a)   (b)

Figure 2.8
Converting 70. °C to units measured on the Kelvin scale. (a) We know 0 °C = 273 K. We want to know 70. °C = ? K. (b) There are 70 degrees on the Celsius scale between 0 °C and 70. °C. Because units on these two scales are the same size, there are also 70 kelvins in this same distance on the Kelvin scale.

EXAMPLE 2.8, CONTINUED

Note that to convert from the Celsius to the Kelvin scale, we simply add the temperature in °C to 273. That is,

$$t_C + 273 = t_K$$

Temperature in Celsius degrees   Temperature in kelvins

Using this formula to solve the present problem gives

$$70. + 273 = 343$$

(with units of kelvins, K,) which is the correct answer.

We can summarize what we learned in Example 2.8 as follows: to convert from the Celsius to the Kelvin scale, we can use the formula

$$t_C + 273 = t_K$$

Temperature in Celsius degrees   Temperature in kelvins

EXAMPLE 2.9   **Temperature Conversion: Kelvin to Celsius**

Liquid nitrogen boils at 77 K. What is the boiling point of nitrogen on the Celsius scale?

SOLUTION

The problem to be solved here is 77 K = ? °C. Let's explore this question by examining the picture to the left representing the two temperature scales.

One key point is to recognize that 0 °C = 273 K. Also note that the difference between 273 K and 77 K is 196 kelvins (273 − 77 = 196). That is, 77 K is 196 kelvins below 273 K. The degree size is the same on these two temperature scales, so 77 K must correspond to 196 Celsius degrees below zero or −196 °C. Thus 77 K = ? °C = −196 °C.

We can also solve this problem by using the formula

$$t_C + 273 = t_K$$

However, in this case we want to solve for the Celsius temperature t_C. That is, we want to isolate t_C on one side of the equals sign. To do this we use an

EXAMPLE 2.9, CONTINUED

important general principle: doing *the same thing on both sides of the equals sign* preserves the equality. In other words, it's always okay to perform the same operation on both sides of the equals sign.

To isolate $t_{°C}$ we need to subtract 273 from both sides:

$$t_{°C} + 273 - 273 = t_K - 273$$

Sum is zero

to give

$$t_{°C} = t_K - 273$$

Using this equation to solve the problem, we have

$$t_{°C} = t_K - 273 = 77 - 273 = -196$$

So, as before, we have shown that

$$77 \text{ K} = -196 \text{ °C}$$

SELF-CHECK EXERCISE 2.6

Which temperature is colder, 172 K or -75 °C?

In summary, because the Kelvin and Celsius scales have the same size unit, to switch from one scale to the other we must simply account for the different zero points. We must add 273 to the Celsius temperature to obtain the temperature on the Kelvin scale:

$$t_K = t_{°C} + 273$$

To convert from the Kelvin scale to the Celsius scale, we must subtract 273 from the Kelvin temperature:

$$t_{°C} = t_K - 273$$

Converting Between the Fahrenheit and Celsius Scales

The conversion between the Fahrenheit and Celsius temperature scales requires two adjustments:

1. For the different size units
2. For the different zero points

Figure 2.9
Comparison of the Celsius and Fahrenheit scales.

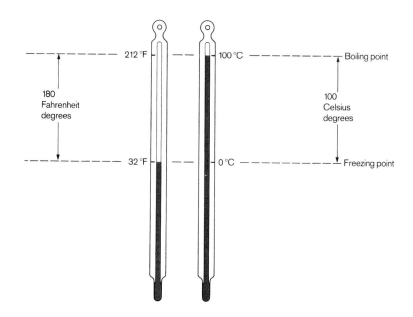

To see how to adjust for the different unit sizes, consider the diagram in Figure 2.9. Note that because 212 °F = 100 °C and 32 °F = 0 °C,

Remember, it's okay to do the same thing to both sides of the equation.

$$212 - 32 = 180 \text{ Fahrenheit degrees} = 100 - 0 = 100 \text{ Celsius degrees}$$

Thus

$$180. \text{ Fahrenheit degrees} = 100. \text{ Celsius degrees}$$

Dividing both sides of this equation by 100. gives

$$\frac{180.}{100.} \text{ Fahrenheit degrees} = \frac{\cancel{100.}}{\cancel{100.}} \text{ Celsius degrees}$$

or

$$1.80 \text{ Fahrenheit degrees} = 1 \text{ Celsius degree}$$

The factor 1.80 is used to convert from one degree size to the other.

Next we have to account for the fact 0 °C is *not* the same as 0 °F. In fact, 32 °F = 0 °C. Although we will not show how to derive it, the equation to convert a temperature in Celsius degrees to the Fahrenheit scale is

$$t_{°F} = 1.80(t_{°C}) + 32$$

Temperature in °F   Temperature in °C

In this equation the term $1.80(t_{°C})$ adjusts for the difference in degree size between the two scales. The 32 in the equation accounts for the different zero points. We will now show how to use this equation.

| EXAMPLE 2.10 | **Temperature Conversion: Celsius to Fahrenheit** |

On a summer day the temperature in the laboratory, as measured on a lab thermometer, is 28 °C. Express this temperature on the Fahrenheit scale.

SOLUTION

This problem can be represented as 28 °C = ? °F. We will solve it using the formula

$$t_F = 1.80(t_C) + 32$$

In this case

$$t_F = ?\ °F = 1.80(28) + 32 = 50.4 + 32$$

Rounds off to 50

$$= 50. + 32 = 82$$

Note that 28 °C is approximately equal to 82 °F. Because the numbers are just reversed, this is an easy reference point to remember for the two scales.

Thus 28 °C = 82 °F.

| EXAMPLE 2.11 | **Temperature Conversion: Celsius to Fahrenheit** |

Express the temperature −40. °C on the Fahrenheit scale.

SOLUTION

We can express this problem as −40. °C = ? °F. To solve it we will use the formula

$$t_F = 1.80(t_C) + 32$$

In this case,

$$t_F = ?\ °F = 1.80(-40.) + 32$$
$$= -72 + 32 = -40$$

So −40 °C = −40 °F. This is a very interesting result and is another useful reference point.

SELF-CHECK EXERCISE 2.7

Hot tubs are often maintained at 41 °C. What is this temperature in Fahrenheit degrees?

An iceberg in Baffin Bay of the Arctic Ocean.

To convert from Celsius to Fahrenheit, we have used the equation

$$t_{\circ F} = 1.80(t_{\circ C}) + 32$$

To convert a Fahrenheit temperature to Celsius, we need to rearrange this equation to isolate Celsius degrees ($t_{\circ C}$). Remember we can always do the same operation to both sides of the equation. First subtract 32 from each side:

$$t_{\circ F} - 32 = 1.80(t_{\circ C}) + \underset{\text{Sum is zero}}{32 - 32}$$

to give

$$t_{\circ F} - 32 = 1.80(t_{\circ C})$$

Next divide both sides by 1.80:

$$\frac{t_{\circ F} - 32}{1.80} = \frac{\cancel{1.80}(t_{\circ C})}{\cancel{1.80}}$$

to give

$$\frac{t_{\circ F} - 32}{1.80} = t_{\circ C}$$

or

$$t_{\circ C} = \underset{\substack{\uparrow \\ \text{Temperature} \\ \text{in } {}^{\circ}C}}{\frac{t_{\circ F} \overset{\substack{\text{Temperature} \\ \text{in } {}^{\circ}F \\ \downarrow}}{} - 32}{1.80}}$$

$$t_{\circ C} = \frac{t_{\circ F} - 32}{1.80}$$

EXAMPLE 2.12   Temperature Conversion: Fahrenheit to Celsius

One of the body's responses to an infection or injury is to elevate its temperature. A certain flu victim has a body temperature of 101 °F. What is this temperature on the Celsius scale?

EXAMPLE 2.12, CONTINUED

SOLUTION

The problem is 101 °F = ? °C.
   Using the formula

$$t_C = \frac{t_F - 32}{1.80}$$

yields

$$t_C = ?\ °C = \frac{\overset{t_F}{\underset{\downarrow}{101}} - 32}{1.80} = \frac{69}{1.80} = 38$$

That is, 101 °F = 38 °C.

SELF-CHECK EXERCISE 2.8

An antifreeze solution in a car's radiator boils at 239°F. What is this temperature on the Celsius scale?

In doing temperature conversions, you will need the following formulas:

Temperature Conversion Formulas

- Celsius to Kelvin

$$t_K = t_C + 273$$

- Kelvin to Celsius

$$t_C = t_K - 273$$

- Celsius to Fahrenheit

$$t_F = 1.80\left(t_C\right) + 32$$

- Fahrenheit to Celsius

$$t_C = \frac{t_F - 32}{1.80}$$

If your class is small enough, you can use a large graduated cylinder filled with colored water to measure the density of some object such as a stone. Measurement of volume and mass provide an opportunity to demonstrate uncertainty in measurements while you are performing density calculations.

Lead has a greater density than feathers.

(a)

(b)

Figure 2.10

(a) Tank of water. (b) Person submerged in the tank, raising the level of the water.

2.8   Density

AIM: To define density and its units.

When you were in elementary school, you may have been embarrassed by your answer to the question "Which is heavier, a pound of lead or a pound of feathers?" If you said lead, you were undoubtedly thinking about density, not mass. **Density** can be defined as the amount of matter present *in a given volume* of substance. That is, density is mass per unit volume, the ratio of the mass of an object to its volume:

$$\text{Density} = \frac{\text{mass}}{\text{volume}}$$

It takes a much bigger volume to make a pound of feathers than to make a pound of lead. This is because lead has a much greater mass per unit volume—a greater density.

The density of a liquid can be determined easily by weighing a known volume of the substance. For example, suppose a student finds that 23.50 mL of a certain liquid weigh 35.062 g. The density of this liquid can be obtained by applying the definition

$$\text{Density} = \frac{\text{mass}}{\text{volume}} = \frac{35.062 \text{ g}}{23.50 \text{ mL}} = 1.492 \text{ g/mL}$$

This result could also be expressed as 1.492 g/cm^3 because 1 mL = 1 cm^3.

The volume of a solid object is often determined indirectly by submerging it in water and measuring the volume of water displaced. In fact, this is the most accurate method for measuring a person's percent body fat. The person is submerged momentarily in a tank of water, and the increase in volume is measured (see Figure 2.10). It is possible to calculate the body density by using the person's weight (mass) and the volume of the person's body determined by submersion. Fat, muscle, and bone have different densities (fat is less dense than muscle tissue, for example), so the fraction of the person's body that is fat can be calculated. The more muscle and the less fat a person has, the higher his or her body density. This is one measure of a person's physical fitness.

EXAMPLE 2.13   Determining Density

At a local pawn shop a student finds a medallion that the shop owner insists is pure platinum. However, the student suspects that the medallion may actually be silver and thus much less valuable. The student buys the medallion only after the shop owner agrees to refund the price if the medallion is returned within two

EXAMPLE 2.13, CONTINUED

days. The student, a chemistry major, then takes the medallion to her lab and measures its density as follows. She first weighs the medallion and finds its mass to be 55.64 g. She then places some water in a graduated cylinder and reads the volume as 75.2 mL. Next she drops the medallion into the cylinder and reads the new volume as 77.8 mL. Is the medallion platinum (density = 21.4 g/cm³) or silver (density = 10.5 g/cm³)?

SOLUTION

The densities of platinum and silver differ so much that the measured density of the medallion will show which metal is present. Because by definition

$$\text{Density} = \frac{\text{mass}}{\text{volume}}$$

to calculate the density of the medallion, we need its mass and its volume. The mass of the medallion is 55.64 g. The volume of the medallion can be obtained by taking the difference between the volume readings of the water in the graduated cylinder before and after the medallion was added.

$$\text{Volume of medallion} = 77.8 \text{ mL} - 75.2 \text{ mL} = 2.6 \text{ mL}$$

The volume appeared to increase by 2.6 mL when the medallion was added, so 2.6 mL represents the volume of the medallion. Now we can use the measured mass and volume of the medallion to determine its density:

$$\text{Density of medallion} = \frac{\text{mass}}{\text{volume}} = \frac{55.64 \text{ g}}{2.6 \text{ mL}} = 21 \text{ g/mL}$$

$$\text{or}$$

$$= 21 \text{ g/cm}^3$$

The medallion is really platinum.

The most common units for density are g/mL = g/cm³.

The measurement of density is easy to do in the classroom. If your students will not be measuring density in the lab, then bring some solid objects and graduated cylinders with water to class to show how the ratio of mass to volume varies. Hydrometers for measuring the density of antifreeze and battery acid are mentioned in the textbook. If you have either of these devices available to you, bring one to class and demonstrate how it is used to measure the density of a solution.

The specific gravity of a substance is the ratio of its density to the density of another substance used as a reference or standard. For a substance liquid, water at 4°C is usually used as the reference substance:

$$\text{specific gravity} = \frac{\text{density of a liquid}}{\text{density of water (4°C)}}$$

SELF-CHECK EXERCISE 2.9

A student wants to identify the main component in a commercial liquid record cleaner. He finds that 35.8 mL of the record cleaner weighs 28.1 g. Of the following possibilities, which is the main component of the cleaner?

Substance	Density, g/cm³
chloroform	1.483
diethyl ether	0.714
isopropyl alcohol	0.785
toluene	0.867

EXAMPLE 2.14   Using Density in Calculations

Mercury has a density of 13.6 g/mL. What volume of mercury must be taken to obtain 225 g of the metal?

Spherical droplets of mercury, a very dense liquid.

SOLUTION

To solve this problem, start with the definition of density:

$$\text{Density} = \frac{\text{mass}}{\text{volume}}$$

and then rearrange this equation to isolate the required quantity. In this case we want to find the volume. Remember that we maintain an equality when we do the same thing to both sides. For example, if we multiply *both sides* of the density definition by volume:

$$\text{Volume} \times \text{density} = \frac{\text{mass}}{\cancel{\text{volume}}} \times \cancel{\text{volume}}$$

volume cancels on the right, leaving

$$\text{Volume} \times \text{density} = \text{mass}$$

We want the volume, so we now divide both sides by density:

$$\frac{\text{Volume} \times \cancel{\text{density}}}{\cancel{\text{Density}}} = \frac{\text{mass}}{\text{density}}$$

to give

$$\text{Volume} = \frac{\text{mass}}{\text{density}}$$

Now we can solve the problem by substituting the given numbers:

$$\text{Volume} = \frac{225 \text{ g}}{13.6 \text{ g/mL}} = 16.5 \text{ mL}$$

We must take 16.5 mL of mercury to obtain an amount that has a mass of 225 g.

The densities of various common substances are given in Table 2.8.

Besides being a tool for the identification of substances, density has many other uses. For example, the liquid in your car's lead storage battery (a solution of sulfuric acid) changes density because the sulfuric acid is consumed as the battery discharges. In a fully charged battery, the density of the solution is about 1.30 g/cm^3. When the density falls below 1.20 g/cm^3, the battery has to be re-

Table 2.8 Densities of Various Common Substances at 20 °C

Substance	Physical state	Density (g/cm^3)
oxygen	gas	0.00133*
hydrogen	gas	0.000084*
ethanol	liquid	0.785
benzene	liquid	0.880
water	liquid	1.000
magnesium	solid	1.74
salt (sodium chloride)	solid	2.16
aluminum	solid	2.70
iron	solid	7.87
copper	solid	8.96
silver	solid	10.5
lead	solid	11.34
mercury	liquid	13.6
gold	solid	19.32

*At I atmosphere pressure

Figure 2.11
A hydrometer being used to determine the density of the antifreeze solution in a car's radiator.

charged. Density measurement is also used to determine the amount of antifreeze, and thus the level of protection against freezing, in the cooling system of a car. Water and antifreeze have different densities, so the measured density of the mixture tells us how much of each is present. The device used to test the density of the solution—a hydrometer—is shown in Figure 2.11.

CHAPTER REVIEW

Key Terms

measurement (p. 16)
scientific notation (p. 16)
units (p. 19)
English system (p. 19)
metric system (p. 19)
SI units (p. 20)
mass (p. 22)
significant figures (p. 25)

rounding off (p. 28)
conversion factor (p. 32)
equivalence statement (p. 33)
dimensional analysis (p. 34)
Fahrenheit scale (p. 38)
Celsius scale (p. 39)
Kelvin (absolute) scale (p. 39)
density (p. 48)

Summary

1. A quantitative observation is called a measurement and always consists of a number and a unit.
2. We can conveniently express very large or very small numbers using scientific notation, which represents the number as a number between 1 and 10 multiplied by 10 raised to a power.
3. Units give a scale on which to represent the results of a measurement. The three systems discussed are the English, metric, and SI systems. The metric and SI systems use prefixes (Table 2.2) to change the size of the units.
4. The mass of an object represents the quantity of matter in that object.
5. All measurements have a degree of uncertainty, which is reflected in the number of significant figures used to express them. Various rules are used to round off to the correct number of significant figures in a calculated result.
6. We can convert from one system of units to another by a method called dimensional analysis, in which conversion factors are used.
7. Temperature can be measured on three different scales: Fahrenheit, Celsius, and Kelvin. We can readily convert among these scales.
8. Density is the amount of matter present in a given volume (mass per unit volume). That is,

$$\text{Density} = \frac{\text{mass}}{\text{volume}}$$

Questions and Problems

All even-numbered exercises have answers in the back of this book and solutions in the Solutions Guide.

2.1 Scientific Notation

QUESTIONS

1. When the number 4,540,000 is expressed in standard scientific notation, it is written as _____ $\times\ 10^6$.
3. A negative exponent in scientific notation indicates a number that is (greater/less) than 1.
2. When 3,314 is written in scientific notation, the exponent indicating the power of ten will be _____.
4. When 0.0021 is written in scientific notation, the exponent is (positive/negative), whereas when 4540 is written in scientific notation, the exponent is (positive/negative).

PROBLEMS

5. For each of the following numbers, if the number is rewritten in standard scientific notation, what will be the value of the exponent (for the power of ten)?
   a. 433
   b. 0.0821
   c. 14.2×10^3
   d. 1,110,000,000
6. For each of the following numbers, if the number is rewritten in standard scientific notation, what will be the value of the exponent (for the power of ten)?
   a. 0.000067
   b. 9,331,442
   c. 1/10,000
   d. 163.1×10^2

7. Express each of the following numbers in scientific (exponential) notation.
    a. 12,500
    b. 37,400,000
    c. 602,300,000,000,000,000,000,000
    d. 375
    e. 0.0202
    f. 0.1550
    g. 0.0000104
    h. 0.000000000000000000129

8. Express each of the following numbers in standard scientific notation.
    a. 0.04731
    b. 4284
    c. 4.201
    d. 0.000000000141
    e. 52.3
    f. 0.04909
    g. 54,331,000
    h. 0.981

9. Express each of the following as an "ordinary" decimal number.
    a. 1.98×10^4
    b. 2.8134×10^6
    c. 7.24×10^{-9}
    d. 1.9444×10^{-6}
    e. 4.921×10^{-3}
    f. 2.90433×10^{-7}
    g. 1.4×10^1
    h. 4.6×10^0
    i. 3.954×10^3
    j. 7.4434×10^5
    k. 2.9×10^2
    l. 1.03×10^9

10. Express each of the following as "ordinary" decimal numbers.
    a. 4.83×10^2
    b. 7.221×10^{-4}
    c. 6.1×10^0
    d. 9.11×10^{-8}
    e. 4.221×10^6
    f. 1.22×10^{-3}
    g. 9.999×10^3
    h. 1.016×10^{-5}
    i. 1.016×10^5
    j. 4.11×10^{-1}
    k. 9.71×10^4
    l. 9.71×10^{-4}

11. Write each of the following numbers in *standard* scientific notation.
    a. 12.3×10^3
    b. 0.0039×10^6
    c. 394.3×10^{-2}
    d. 2232.3×10^{-4}
    e. 832.3×10^2
    f. $143,400 \times 10^3$
    g. 0.000432×10^{-5}
    h. 0.03993×10^{-2}

12. Write each of the following numbers in *standard* scientific notation.
    a. 131.2×10^{-3}
    b. 14.72×10^2
    c. $1,201 \times 10^{-6}$
    d. 44.3×10^4
    e. 0.00721×10^3
    f. 0.0914×10^{-4}
    g. 0.000129×10^5
    h. $0.00001901 \times 10^{-6}$

13. Write each of the following numbers in *standard* scientific notation.
    a. $1/1033$
    b. $1/10^5$
    c. $1/10^{-7}$
    d. $1/0.0002$
    e. $1/3,093,000$
    f. $1/10^{-4}$
    g. $1/10^9$
    h. $1/0.000015$

14. Write each of the following numbers in *standard* scientific notation.
    a. $1/0.00032$
    b. $10^3/10^{-3}$
    c. $10^3/10^3$
    d. $1/55,000$
    e. $(10^5)(10^4)(10^{-4})/(10^{-2})$
    f. $43.2/(4.32 \times 10^{-5})$
    g. $(4.32 \times 10^{-5})/432$
    h. $1/(10^5)(10^{-6})$

2.2 Units

QUESTIONS

15. The international system of measurement that has the meter, kilogram, and second as its fundamental units is called the _____ .

16. Although the standard of mass is, strictly speaking, the kilogram, in chemistry we more commonly make measurements of mass in terms of how many _____ a sample weighs.

17. Indicate the meaning (as a power of 10) for each of the following metric prefixes.
    a. kilo
    b. centi
    c. milli
    d. deci
    e. nano
    f. micro

18. Give the metric prefix that corresponds to each of the following:
    a. 1,000,000
    b. 10^{-3}
    c. 10^{-9}
    d. 10^6
    e. 10^{-2}
    f. 0.000001

2.3 Measurements of Length, Volume, and Mass

QUESTIONS

19. The basic SI unit of length is the _____ , which is equivalent to slightly more than a yard.

20. Which distance is farther, 100 mi or 100 km?

21. As a measure of volume, 1 liter is equivalent to 1 cubic _____ .

22. One liter of volume in the metric system is approximately equivalent to one _____ in the English system.

23. Which is longer, a piece of fabric 100 cm long or a piece 0.50 m long?

24. Which weighs more, a pound of hamburger or a kilogram of hamburger?

25. The length 0.1 m can also be expressed as _____ cm.

26. Which person is taller, a man who is 1.75 m tall or a woman who is 5 ft 2 in. tall?

27. A 1-kg package of hamburger has a mass closest to which of the following?
    a. 8 oz   c. 2 lb
    b. 1 lb   d. 10 lb

28. A 2-L bottle of soda contains a volume closest to which of the following?
    a. 5 gal   c. 2 pt
    b. 1 qt   d. 2 qt

29. A recipe written in metric units calls for 250 mL of milk. Which of the following best approximates this amount?
    a. 1 qt   c. 1 cup
    b. 1 gal   d. 1 pt

30. Which metric system unit is most appropriate for measuring the distance between two cities?
    a. meters   c. centimeters
    b. millimeters   d. kilometers

31–32. Some examples of simple approximate metric–English equivalents are given in Table 2.6.

31. What is the value in dollars of a stack of dimes that is 10 cm high?

32. How many quarters would have to be lined up in a row to reach a length of 1 meter?

2.4 Uncertainty in Measurement

QUESTIONS

33. Estimating a reading between the smallest scale divisions of a measuring device makes the last digit of a measurement _____ .

34. No matter how careful an experimenter may be, a measurement always has some degree of _____ .

35. For the pin shown in Figure 2.5, why is the third figure determined for the length of the pin uncertain? Considering that the third figure is uncertain, explain why the length of the pin is indicated as 2.85 cm rather than, for example, 2.83 or 2.87 cm.

36. Why can the length of the pin shown in Figure 2.5 not be recorded as 2.850 cm?

2.5 Significant Figures

QUESTIONS

37. Indicate the number of significant figures in each of the following:
    a. 12   e. 0.0000101
    b. 1098   f. 1.01×10^{-5}
    c. 2001   g. 1,000.
    d. 2.001×10^3 h. 22.0403

38. Indicate the number of significant figures implied in each of the following statements:
    a. One foot is equivalent to 12 inches.
    b. The continental United States is over 3,000 miles wide.
    c. There are 60 seconds in one minute.
    d. The population of Massachusetts is 5 million.
    e. One liter contains 1000 mL.

Rounding Off Numbers

QUESTIONS

39. When we are rounding a number off, if the number to the right of the digit to be rounded is less than 5, the digit should _____ .

40. When performing a chain of several calculations, we round off the (final/intermediate) answer(s).

PROBLEMS

41. Round off each of the following numbers to four significant digits.
    a. 123.431
    b. 1.23431
    c. 12,343.1
    d. 550,092
    e. 90,221

42. Round off each of the following numbers to three significant digits and write the answer in standard scientific notation.
    a. 312.54
    b. 0.00031254
    c. 31,254,000
    d. 0.31254
    e. 31.254×10^{-3}

43. Round off each of the following numbers to the indicated number of significant digits.
    a. 102.4005 to five digits
    b. 15.9995 to three digits
    c. 1.6385 to four digits
    d. 7.355 to three digits

44. Round off each of the following numbers to the indicated number of significant digits and write the answer in standard scientific notation.
    a. 0.00034159 to three digits
    b. 103.351×10^2 to four digits
    c. 17.9915 to five digits
    d. 3.365×10^5 to three digits

Determining Significant Figures in Calculations

QUESTIONS

45. When numbers are multiplied, the answer should contain the same number of significant figures as the multiplier with the _____ number of significant figures.

46. When numbers are added or subtracted, the limiting term is the one with the smallest number of _____ places.

47. When the calculation (0.0043)(0.0821)(298) is performed, the answer should be reported to _____ significant figures.

48. The quotient $(2.3733 \times 10^2)/(343)$ should be written with _____ significant figures.

49. How many digits after the decimal point should be reported when the calculation (199.0354 + 43.09 + 121.2) is performed?

50. How many digits after the decimal point should be reported when the calculation (10434 − 9.3344) is performed?

PROBLEMS

Note: See Appendix 3 for help in doing mathematical operations with numbers that contain exponents.

51. Evaluate each of the following, and write the answer to the appropriate number of significant figures.
    a. 102.01 + 0.0023 + 0.15
    b. $1.000 \times 10^3 - 1$
    c. 55.0001 + 0.0002 + 0.104
    d. $1.02 \times 10^3 + 1.02 \times 10^2 + 1.02 \times 10^1$

52. Evaluate each of the following and write the answer to the appropriate number of significant figures.
    a. 212.2 + 26.7 + 402.09
    b. 1.0028 + 0.221 + 0.10337
    c. 52.331 + 26.01 − 0.9981
    d. $2.01 \times 10^2 + 3.014 \times 10^3$

53. Evaluate each of the following, and write the answer to the appropriate number of significant figures.
    a. $(0.102)(0.0821)(273)/(1.01)$
    b. $(0.14)(6.022 \times 10^{23})$
    c. $(4.0 \times 10^4)(5.021 \times 10^{-3})(7.34993 \times 10^2)$
    d. $(2.00 \times 10^6)/(3.00 \times 10^{-7})$

54. Evaluate each of the following and write the answer to the appropriate number of significant figures.
    a. $(4.031)(0.08206)(373.1)/(0.995)$
    b. $(12.011)/(6.022 \times 10^{23})$
    c. $(0.500)/(44.02)$
    d. $(0.15)(280.62)$

55. Evaluate each of the following, and write the answer to the appropriate number of significant figures.
    a. $(2.3232 + 0.2034 - 0.16) \times (4.0 \times 10^3)$
    b. $(1.34 \times 10^2 + 3.2 \times 10^1)/(3.32 \times 10^{-6})$
    c. $(4.3 \times 10^6)/(4.334 + 44.0002 - 0.9820)$
    d. $(2.043 \times 10^{-2})^3$

56. Evaluate each of the following and write the answer to the appropriate number of significant figures.
    a. $(2.0944 + 0.0003233 + 12.22)/(7.001)$
    b. $(1.42 \times 10^2 + 1.021 \times 10^3)/(3.1 \times 10^{-1})$
    c. $(9.762 \times 10^{-3})/(1.43 \times 10^2 + 4.51 \times 10^1)$
    d. $(6.1982 \times 10^{-4})^2$

2.6 Problem Solving and Dimensional Analysis

QUESTIONS

57. A _____ represents a ratio based on an equivalence statement between two measurements.

58. How many significant figures are understood for the numbers in the following definition: 1 mi = 5280 ft?

59. Given that 3 ft = 1 yd, determine what conversion factor is appropriate to convert 35 ft to yards; to convert 2.89 yd to feet.

60. Given that 12 in. = 1 ft, determine what conversion factor is appropriate to convert 72 in. to feet; to convert 3.5 ft to inches.

61–62. Apples cost $0.79 per pound.

61. What conversion factor is appropriate to express the cost of 5.3 lb of apples?

62. What conversion factor could be used to determine how many pounds of apples could be bought for $2.00?

PROBLEMS

Note: Appropriate equivalence statements for various units are found inside the back cover of this book.

63. Perform each of the following conversions, being sure to set up clearly the appropriate conversion factor in each case.
    a. 363 ft to inches
    b. 17.4 in. to feet
    c. 2.21 lb to ounces
    d. 26 qt to gallons
    e. 24 ft^2 to square yards
    f. 5 gal to quarts
    g. 5 gal to pints
    h. 25,499 yd to miles

64. Perform each of the following conversions, being sure to set up clearly the appropriate conversion factor in each case.
    a. 2.23 m to yards
    b. 46.2 yd to meters
    c. 292 cm to inches
    d. 881.2 in. to centimeters
    e. 1043 km to miles
    f. 445.5 mi to kilometers
    g. 36.2 m to kilometers
    h. 0.501 km to centimeters

65. Perform each of the following conversions, being sure to set up clearly the appropriate conversion factor in each case.
    a. 62.5 cm to inches
    b. 2.68 in. to centimeters
    c. 3.25 yd to meters
    d. 4.95 m to yards
    e. 62.5 cm to yards
    f. 2.45 mi to kilometers
    g. 4.42 m to inches
    h. 5.01 kg to ounces

66. Perform each of the following conversions, being sure to set up clearly the appropriate conversion factor in each case.
    a. 254.3 g to kilograms
    b. 2.74 kg to grams
    c. 2.74 kg to pounds
    d. 2.74 kg to ounces
    e. 534.1 g to pounds
    f. 1.75 lb to grams
    g. 8.7 oz to grams
    h. 45.9 g to ounces

67. If $1.00 is equivalent to 1.74 German marks, what is $20.00 worth in marks? What is the value in dollars of a 100-mark bill?

68. Boston and New York City are 190 miles apart. What is this distance in kilometers? in meters? in feet?

69. The circumference of the earth is on the order of 25,000 mi. What is this distance in kilometers? in meters?

70. The radius of an atom is on the order of 10^{-10} m. What is this radius in centimeters? in inches? in nanometers?

2.7 Temperature Conversions

QUESTIONS

71. The temperature scale in everyday use in the United States is the _____ scale.

72. The _____ temperature scale uses 0° to represent the normal freezing point of water.

73. The boiling point of water is _____ °F, or _____ °C.

74. The freezing point of water is _____ K.

75. On both the Celsius and Kelvin temperature scales, there are _____ degrees between the normal freezing and boiling points of water.

76. On which temperature scale (°F, °C, or K) does 1 degree represent the smallest change in temperature?

PROBLEMS

77. Convert the following temperatures to kelvins.
    a. −200 °C
    b. 150 °C
    c. −40 °C
    d. 23 °C
    e. −100 °C
    f. −196 °C

78. Convert the following Kelvin temperatures to Celsius degrees.
    a. 275 K
    b. 445 K
    c. 0 K
    d. 77 K
    e. 10,000 K
    f. 2 K

79. Convert the following Fahrenheit temperatures to Celsius degrees.
    a. a beautiful spring day, 68 °F
    b. a hot, humid August day, 86 °F
    c. a warm day in Minnesota in January, −10 °F
    d. the surface of a star, 10,000 °F

80. Convert the following Celsius temperatures to Fahrenheit degrees.
    a. the boiling temperature of ethyl alcohol, 78.1 °C
    b. a hot day at the beach on a Greek isle, 40. °C
    c. the lowest possible temperature, −273 °C
    d. the body temperature of a person with hypothermia, 32 °C

81. Carry out the indicated temperature conversions.
    a. 0 °F to kelvins
    b. 531 K to Celsius degrees
    c. −175 °F to Celsius degrees
    d. 88 °C to Fahrenheit degrees

82. Carry out the indicated temperature conversions.
    a. −185 °F to Celsius degrees
    b. 255 K to Fahrenheit degrees
    c. 400 °C to kelvins
    d. 0 °F to kelvins

2.8 Density

QUESTIONS

83. The ratio of an object's mass to its volume is called the _____ of the object.

84. The most common units for density are _____.

85. A kilogram of lead occupies a much smaller volume than a kilogram of water, because _____ has a much higher density.

86. The _____ of an insoluble object (such as a person, see Figure 2.10) can be determined indirectly by measuring how much water the object displaces.

87. Typically, gases have very (high/low) densities compared to solids and liquids (see Table 2.8).

88. Density may be used as an aid in identifying substances, because every sample of a pure substance always has the _____ density.

89. Referring to Table 2.8, determine whether air, water, alcohol, or aluminum is the most dense.

90. Referring to Table 2.8, determine whether copper, silver, lead, or mercury is the least dense.

PROBLEMS

91. For the masses and volumes indicated, calculate the density in grams per cubic centimeter.
    a. mass = 44.3 g; volume = 22.1 cm^3
    b. mass = 1.23 kg; volume = 0.253 m^3
    c. mass = 4.2 lb; volume = 1.23 ft^3
    d. mass = 234 mg; volume = 2.2 × 10^{-3} cm^3

92. For the masses and volumes indicated, calculate the density in grams per cubic centimeter.
    a. mass = 122.4 g; volume = 4.3 cm^3
    b. mass = 19,302 g; volume = 0.57 m^3
    c. mass = 0.0175 kg; volume = 18.2 mL
    d. mass = 2.49 g; volume = 0.12 m^3

93. 45.0 mL of a liquid weigh 38.2 g. Calculate the density of the liquid.

94. 67.1 mL of a liquid weigh 55.221 g. Calculate the density of the liquid.

95. A rectangular metal bar has a volume of 60. in.^{3}. The bar weighs 1.42 kg. Calculate the density of the metal in grams per cubic centimeter.

96. A material will float on the surface of a liquid if the material has a density less than that of the liquid. Given that the density of water is approximately 1.0 g/mL under many conditions, will a block of material having a volume of 1.2×10^4 in.^{3} and weighing 3.5 lbs float or sink when placed in a reservoir of water?

97. 20.0 g of metal pellets are poured into a graduated cylinder containing 15.6 mL of water, causing the water level to rise to 21.9 mL. Calculate the density of the metal pellets.

98. The density of pure silver is 10.5 g/cm^3 at 20 °C. If 5.25 g of pure silver pellets are added to a graduated cylinder containing 11.2 mL of water, to what volume level will the water in the cylinder rise?

99. Use the information given in Table 2.8 to calculate the volume of 1.00 kg of each of the following substances.
a. ethanol   c. gold
b. copper   d. magnesium

100. Use the information given in Table 2.8 to calculate the mass of 1.00×10^3 cm^3 and 1.00 m^3 of each of the following substances.
a. lead   c. benzene
b. sodium chloride   d. iron

Additional Problems

101. Write each of the following numbers in standard scientific notation.
a. 522.3×10^3 d. 0.08210
b. 0.000003491   e. $444,000 \times 10^{-6}$
c. 102,300   f. 9.234×10^{-7}

102. Express each of the following as an "ordinary" decimal number.
a. 3.011×10^{23} f. 3.001×10^{-2}
b. 5.091×10^9 g. 2.9901×10^{-7}
c. 7.2×10^2 h. 4.2×10^{-1}
d. 1.234×10^5
e. 4.32002×10^{-4}

103. Write each of the following numbers in standard scientific notation, rounding the numbers to three significant digits.
a. 34.000434   d. 0.07305
b. 0.00098012   e. 552.043×10^4
c. 29,000,000

104. Which unit of length in the metric system would be most appropriate in size for measuring each of the following items?
a. this book's thickness
b. the distance from New York to Los Angeles
c. the length of a bacterium
d. the diameter of a piece of thread

105. Make the following conversions.
a. 1.25 in. to feet and to centimeters
b. 2.12 qt to gallons and to liters
c. 2640 ft to miles and to kilometers
d. 1.254 kg lead to its volume in cubic centimeters
e. 250. mL ethanol to its mass in grams
f. 3.5 in.^{3} of mercury to its volume in milliliters and its mass in kilograms

106. On the planet Xgnu, the most common units of length are the blim (for long distances) and the kryll (for shorter distances). Because the Xgnuese have 14 fingers, it is not perhaps surprising that 1400 kryll = 1 blim.
a. Two cities on Xgnu are 36.2 blim apart. What is this distance in kryll?
b. The average Xgnuese is 170 kryll tall. What is this height in blims?
c. This book is presently being used at Xgnu University. The area of the cover of this book is 72.5 square krylls. What is its area in square blims?

107. You pass a road sign saying "New York 110 km." If you drive at a constant speed of 100. km/h, how long should it take you to reach New York?

108. At the mall, you decide to try on a pair of French jeans. Naturally, the waist size of the jeans is given in centimeters. What does a waist measurement of 52 cm correspond to in inches?

109. Suppose your car is rated at 45 mi/gal for highway use and 38 mi/gal for city driving. If you wanted to write your friend in Spain about your car's mileage, what ratings in kilometers per liter would you report?

110. You are in Paris, and you want to buy some peaches for lunch. The sign in the fruit stand indicates that peaches are 11.5 francs per kilogram. Given that there are approximately 5 francs to the dollar, calculate what a pound of peaches will cost in dollars.

111. For a pharmacist dispensing pills or capsules, it is often easier to weigh the medication to be dispensed rather than to count the individual pills. If a single antibiotic capsule weighs 0.65 g, and a pharmacist weighs out 15.6 g of capsules, how many capsules have been dispensed?

112. On the planet Xgnu, the natives have 14 fingers. On the official Xgnuese temperature scale (°X), the boiling point of water (under an atmospheric pressure similar to earth's) is 140 °X, whereas it freezes at 14 °X. Derive the relationship between °X and °C.

113. For a material to float on the surface of water, the material must have a density less than that of water (1.0 g/mL) and must not react with the water or dissolve in it. A spherical ball has a radius of 0.50 cm and weighs 2.0 g. Will this ball float or sink when placed in water? (Note: Volume of a sphere $= \frac{4}{3}\pi r^3$.)

114. A gas cylinder having a volume of 10.5 L contains 36.8 g of gas. What is the density of the gas?

115. Using Table 2.8, calculate the volume of 100. g of each of the following:
    a. oxygen gas (at 1 atmosphere pressure)
    b. benzene
    c. magnesium
    d. copper

116. Ethanol and benzene dissolve in each other. When 100. mL of ethanol is dissolved in 1.00 L of benzene, what is the mass of the mixture? (See Table 2.8.)

117. When 2891 is written in scientific notation, the exponent indicating the power of 10 is _____.

118. For each of the following numbers, if the number is rewritten in scientific notation, will the exponent be positive, negative, or zero?
    a. 0.0421
    b. 2.04
    c. 389
    d. 0.00912

119. For each of the following numbers, if the number is rewritten in scientific notation, will the exponent be positive, negative, or zero?
    a. 4,915,442
    b. 1/1000
    c. 0.001
    d. 3.75

120. For each of the following numbers, by how many places does the decimal point have to be moved to express the number in standard scientific notation? In each case, is the exponent positive or negative?
    a. 102
    b. 0.00000000003489
    c. 2500
    d. 0.00003489
    e. 398,000
    f. 1
    g. 0.3489
    h. 0.0000003489

121. For each of the following numbers, by how many places does the decimal point have to be moved to express the number in standard scientific (exponential) notation? In each case, is the exponent positive or negative?
    a. 0.003901
    b. 0.0034002
    c. 199,000,000
    d. 0.000001024
    e. 652
    f. 92,033
    g. 53
    h. 0.000193

122. For each of the following numbers, by how many places does the decimal point have to be moved to express the number in standard scientific (exponential) notation? In each case, is the exponent positive or negative?
    a. 341
    b. 0.00341
    c. 34.1
    d. 0.0000000341
    e. 42,250
    f. 42.25
    g. 0.000004225
    h. 0.4225

123. Express each of the following numbers in scientific (exponential) notation.
    a. 529
    b. 240,000,000
    c. 301,000,000,000,000,000
    d. 78,444
    e. 0.0003442
    f. 0.000000000902
    g. 0.043
    h. 0.0821

124. Express each of the following as an "ordinary" decimal number.
    a. 2.98×10^{-5}
    b. 4.358×10^{9}
    c. 1.9928×10^{-6}
    d. 6.02×10^{23}
    e. 1.01×10^{-1}
    f. 7.87×10^{-3}
    g. 9.87×10^{7}
    h. 3.7899×10^{2}
    i. 1.093×10^{-1}
    j. 2.9004×10^{0}
    k. 3.9×10^{-4}
    l. 1.904×10^{-8}

125. Write each of the following numbers in *standard* scientific notation.
    a. 102.3×10^{-5}
    b. 32.03×10^{-3}
    c. 59933×10^{2}
    d. 599.33×10^{4}
    e. 5993.3×10^{3}
    f. 2054×10^{-1}
    g. $32,000,000 \times 10^{-6}$
    h. 59.933×10^{5}

126. Write each of the following numbers in *standard* scientific notation. See Appendix 3 if you need help multiplying or dividing numbers with exponents.
    a. $1/10^2$
    b. $1/10^{-2}$
    c. $55/10^3$
    d. $(3.1 \times 10^6)/10^{-3}$
    e. $(10^6)^{1/2}$
    f. $(10^6)(10^4)/(10^2)$
    g. $1/0.0034$
    h. $3.453/10^{-4}$

127. The fundamental unit of length or distance in the metric system is the _____.
128. The unit of temperature in the metric system is the _____.
129. Which distance is farther, 100 km or 50 mi?
130. The unit of volume corresponding to 1/1000 of a liter is referred to as 1 milliliter, or 1 cubic _____.
131. The volume 0.250 L could also be expressed as _____ mL.
132. The distance 10.5 cm could also be expressed as _____ m.
133. Would an automobile moving at a constant speed of 100 km/hr violate a 65-mph speed limit?
134. Which weighs more, 100 g of water or 1 kg of water?
135. Which weighs more, 4.25 grams of gold or 425 milligrams of gold?
136. The length 100 mm can also be expressed as _____ cm.
137. When a measurement is made, the certain numbers plus the first uncertain number are called the _____ of the measurement.
138. In the measurement of the length of the pin indicated in Figure 2.5, what are the *certain* numbers in the measurement shown?
139. Indicate the number of significant figures in each of the following:
     a. This book contains over 500 pages.
     b. A mile is just over 5000 ft.
     c. A liter is equivalent to 1.059 qt.
     d. The population of the United States is approaching 250 million.
     e. A kilogram is 1000 m.
     f. The Boeing 747 cruises at around 600 mi/h.
140. Round off each of the following numbers to three significant digits.
     a. 0.000032421   d. 550,092
     b. 7,212,992   e. 199.99
     c. 2.098988×10^{-7}
141. Round off each of the following numbers to the indicated number of significant digits.
     a. 0.50045 to four digits   c. 15.395 to four digits
     b. 126.5 to three digits   d. 23.0975 to five digits
142. Evaluate each of the following, and write the answer to the appropriate number of significant figures.
     a. $149.2 + 0.034 + 2000.34$
     b. $1.0322 \times 10^3 + 4.34 \times 10^3$
     c. $4.03 \times 10^{-2} - 2.044 \times 10^{-3}$
     d. $2.094 \times 10^5 - 1.073 \times 10^6$
143. Evaluate each of the following, and write the answer to the appropriate number of significant figures.

a. $(0.0432)(2.909)(4.43 \times 10^8)$
b. $(0.8922)/[(0.00932)(4.03 \times 10^2)]$
c. $(3.923 \times 10^2)(2.94)(4.093 \times 10^{-3})$
d. $(4.9211)(0.04434)/[(0.000934)(2.892 \times 10^{-7})]$

144. Evaluate each of the following, and write the answer to the appropriate number of significant figures.
     a. $(2.9932 \times 10^4)[2.4443 \times 10^2 + 1.0032 \times 10^1]$
     b. $[2.34 \times 10^2 + 2.443 \times 10^{-1}]/(0.0323)$
     c. $(4.38 \times 10^{-3})^2$
     d. $(5.9938 \times 10^{-6})^{1/2}$
145. Given that 1 L = 1000 cm^3, determine what conversion factor is appropriate to convert 350 cm^3 to liters; to convert 0.200 L to cubic centimeters.
146. Given that 12 months = 1 year, determine what conversion factor is appropriate to convert 72 months to years; to convert 3.5 years to months.
147. Perform each of the following conversions, being sure to set up clearly the appropriate conversion factor in each case.
     a. 8.43 cm to millimeters
     b. 2.41×10^2 cm to meters
     c. 294.5 nm to centimeters
     d. 404.5 m to kilometers
     e. 1.445×10^4 m to kilometers
     f. 42.2 mm to centimeters
     g. 235.3 m to millimeters
     h. 903.3 nm to micrometers
148. Perform each of the following conversions, being sure to set up clearly the appropriate conversion factor(s) in each case.
     a. 908 oz to kilograms   d. 2.89 gal to milliliters
     b. 12.8 L to gallons   e. 4.48 lb to grams
     c. 125 mL to quarts   f. 550 mL to quarts
149. The mean distance from the earth to the sun is 9.3×10^7 mi. What is this distance in kilometers? in centimeters?
150. Given that one gross = 144 items, how many pencils are contained in 6 gross?
151. Convert the following temperatures to kelvins.
     a. 0 °C   d. 100 °C
     b. 25 °C   e. −175 °C
     c. 37 °C   f. 212 °C
152. Carry out the indicated temperature conversions.
     a. 175 °F to kelvins
     b. 255 K to Celsius degrees
     c. −45 °F to Celsius degrees
     d. 125 °C to Fahrenheit degrees
153. For the masses and volumes indicated, calculate the density in grams per cubic centimeter.

a. mass = 234 g; volume = 2.2 cm^3
b. mass = 2.34 kg; volume = 2.2 m^3
c. mass = 1.2 lb; volume = 2.1 ft^3
d. mass = 4.3 tons; volume = 54.2 yd^3

154. A sample of alcohol has density 0.82 g/mL. What do 55 mL of the alcohol weigh?

155. A sample of organic liquid has density 1.54 g/mL. What do 75.0 mL of the liquid weigh?

156. A solid metal sphere has a volume of 4.2 ft^3. The mass of the sphere is 155 lb. Find the density of the metal sphere in grams per cubic centimeter.

157. A sample containing 33.42 g of metal pellets is poured into a graduated cylinder initially containing 12.7 mL of water, causing the water level in the cylinder to rise to 21.6 mL. Calculate the density of the metal.

158. Convert the following temperatures to Fahrenheit degrees.
a. −5 °C
b. 273 K
c. −196 °C
d. 0 K
e. 86 °C
f. −273 °C

Solutions to Self-Check Exercises

SELF-CHECK EXERCISE 2.1

$357 = 3.57 \times 10^2$
$0.0055 = 5.5 \times 10^{-3}$

SELF-CHECK EXERCISE 2.2

a. Three significant figures. The leading zeros (to the left of the 1) do not count, but the trailing zeros do.

b. Five significant figures. The one captive zero and the two trailing zeros all count.

c. This is an exact number obtained by counting the cars. It has an unlimited number of significant figures.

SELF-CHECK EXERCISE 2.3

a. $12.6 \times 0.53 = 6.678 = 6.7$
↑
Limiting

b. $12.6 \times 0.53 = 6.7$;   6.7 ← Limiting
↑   −4.59
Limiting   2.11 = 2.1

c.
$$\begin{array}{c} 25.36 \\ -\ 4.15 \\ \hline 21.21 \end{array}$$
$\dfrac{21.21}{2.317} = 9.15408 = 9.154$

SELF-CHECK EXERCISE 2.4

$0.750 \, \cancel{L} \times \dfrac{1.06 \text{ qt}}{1 \, \cancel{L}} = 0.795 \text{ qt}$

SELF-CHECK EXERCISE 2.5

$$215 \ \frac{\text{mi}}{\text{h}} \times \frac{1760 \ \text{yd}}{1 \ \text{mi}} \times \frac{1 \ \text{m}}{1.094 \ \text{yd}} \times \frac{1 \ \text{km}}{1000 \ \text{m}} = 346 \ \frac{\text{km}}{\text{h}}$$

SELF-CHECK EXERCISE 2.6

The best way to solve this problem is to convert 172 K to Celsius degrees. To do this we will use the formula $t_C = t_K - 273$

In this case

$$t_C = t_K - 273 = 172 - 273 = -101$$

So 172 K $= -101$ °C

which is a lower temperature than -75 °C. Thus 172 K is colder than -75 °C.

SELF-CHECK EXERCISE 2.7

The problem is 41 °C $= ?$ °F.
    Using the formula $t_F = 1.80(t_C) + 32$

we have

$$t_F = ? \ °F = 1.80(41) + 32 = 74 + 32 = 106$$

That is, 41 °C $= 106$ °F.

SELF-CHECK EXERCISE 2.8

This problem can be stated as 239 °F $= ?$ °C.
    Using the formula

$$t_C = \frac{t_F - 32}{1.80}$$

we have in this case

$$t_C = ? \ °C = \frac{239 - 32}{1.80} = \frac{207}{1.80} = 115$$

That is, 239 °F $= 115$ °C.

SELF-CHECK EXERCISE 2.9

We obtain the density of the record cleaner by dividing its mass by its volume.

$$\text{Density} = \frac{\text{mass}}{\text{volume}} = \frac{28.1 \ \text{g}}{35.8 \ \text{mL}} = 0.785 \ \text{g/mL}$$

This density identifies the liquid as isopropyl alcohol.

3 Matter and Energy

The glowing embers of a wood fire.

CONTENTS

I always start this chapter by asking the question, "Why does anyone study chemistry?" I answer by suggesting that people study chemistry to understand the world around them and to change the world for the better. Examples of interesting phenomena explained by chemistry are the tarnishing of silver, rusting of automobiles, and permanent waving of hair. Two of the many problems chemists can participate in solving that will make the world a better place in which to live include the development of new biodegradable plastics and substitute refrigerants that do not deplete the ozone layer.

This chapter can be very confusing for students because of all the new terms introduced. It is probably a good idea to put Figure 3.10 on page 75 on the board or on an overhead transparency and show the relationships among all the forms of matter mentioned in this chapter.

Why does soda fizz when you open the bottle?

As you look around you, you must wonder about the properties of matter. How do plants grow and why are they green? Why is the sun hot? Why does a hot dog get hot in a microwave oven? Why does wood burn whereas rocks do not? What is a flame? How does soap work? Why does soda fizz when you open it? When iron rusts, what's happening? And why doesn't aluminum rust? How does a cold pack for an athletic injury, which is stored for weeks or months at room temperature, suddenly get cold when you need it? How does a hair permanent work?

The answers to these and endless other questions lie in the domain of chemistry. In this chapter we begin to explore the nature of matter: how it is organized and how and why it changes.

How does this lush vegetation grow in a tropical rain forest, and why is it green?

3.1   Matter

AIM: To define matter and to characterize its three states.

Matter, the "stuff" of which the universe is composed, has two characteristics: it has mass and it occupies space. Matter comes in a great variety of forms: the stars, the air that you are breathing, the gasoline that you put in your car, the chair on which you are sitting, the turkey in the sandwich you may have had for lunch, the tissues in your brain that enable you to read and comprehend this sentence, and so on.

To try to understand the nature of matter, we classify it in various ways. For example, wood, rocks, bone, and steel share certain characteristics. These things are all rigid; they have definite shapes that are difficult to change. On the other hand, water and gasoline, for example, take the shape of any container into which they are poured (see Figure 3.1). Even so, 1 L of water has a volume of 1 L whether it is in a pail or a beaker. In contrast, air takes the shape of its container and fills any container uniformly.

Figure 3.1
Here, colored water takes the shape of various containers.

Table 3.1	The Three States of Matter	
State	Definition	Examples
solid	rigid; has a fixed shape and volume	ice cube, diamond, iron bar
liquid	has a definite volume but takes the shape of its container	gasoline, water, alcohol, blood
gas	has no fixed volume or shape; takes the shape and volume of its container	air, helium, oxygen

The material in Chapter 3 provides a good opportunity for you to bring to class as many different substances as possible. Bring examples of solid, liquid, and gaseous elements. For example, sodium, iodine, aluminum, mercury, sulfur, and chlorine are all interesting and will be used in examples throughout the course. By bringing in salts of different colors and solubilities you can illustrate both compounds and pure substances. Simple bottles of chemicals can show a range of concepts.

It is easy to point out examples of matter without bringing special demonstrations to class. Solids and gases are everywhere, and water can be placed in various containers.

The substances we have just described illustrate the three **states of matter: solid, liquid,** and **gas.** These are defined and illustrated in Table 3.1. The state of a given sample of matter depends on the strength of the forces among the particles contained in the matter; the stronger these forces, the more rigid the matter. We will discuss this in more detail in the next section.

3.2 Physical and Chemical Properties and Changes

AIM: To distinguish between physical and chemical properties.
To distinguish between physical and chemical changes.

When you see a friend, you immediately respond and call him or her by name. We recognize a friend because each person has unique characteristics or properties. The person may be thin and tall, may have blonde hair and blue eyes, and so on. The characteristics just mentioned are examples of **physical properties.** Substances also have physical properties. Typical physical properties of a substance include odor, color, volume, state (gas, liquid, or solid), density, melting point, and boiling point. Another set of properties that we ascribe to a pure substance are its **chemical properties,** which refer to its ability to form new substances.

A mixed crystal of colorless calcite (composed of calcium carbonate) and green malachite.

EXAMPLE 3.1	Identifying Physical and Chemical Properties

Classify each of the following as a physical or a chemical property.

a. The boiling point of a certain alcohol is 78 °C.
b. Diamond is very hard.
c. Sugar ferments to form alcohol.
d. A metal wire conducts an electric current.

SOLUTION

Items (a), (b), and (d) are physical properties; they describe inherent characteristics of each substance, and no change in composition occurs. A metal wire has the same composition after an electric current has passed through it as it did before. Item (c) is a chemical property of sugar. Fermentation involves the formation of a new substance (alcohol).

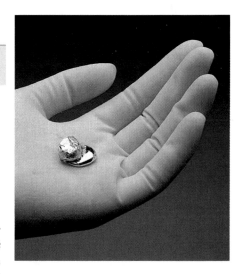

Gallium metal has such a low melting point (30°C) that it melts in the hand.

Students have trouble distinguishing between physical and chemical changes, that is, to know whether the change has occurred in the substance or only in its outward appearance. For that reason, it is better to use examples of physical and chemical changes with which the students are already familiar, such as the burning of wood to produce CO_2, water, and ashes, or the frying of an egg. A good demonstration of chemical change is the electrolytic decomposition of water; you can use the collected O_2 *(g)* and H_2 *(g)* to show the chemical properties of these gases.

The purpose here is to give an overview. Don't worry about the precise definitions of *atom* and *molecule* now.

The letters indicate atoms and the lines indicate attachments (bonds) between atoms.

Point out that physical changes can be of more than one type. For example, some physical changes involve changes of state, whereas others, such as changes in the size of particles, do not. However, chemical changes are all of one type; the chemical composition changes.

When a substance such as sodium chloride or sucrose is dissolved in water, the particles become too small to see. It is a real act of faith on the part of the students to believe that the substance itself has not changed. Unfortunately, it takes quite a while to evaporate to dryness a beaker containing a salt solution. However, you can bring to class a beaker from which you have previously evaporated the water to show the salt crystals recovered unchanged on the bottom of the beaker.

SELF-CHECK EXERCISE 3.1

Which of the following are physical properties and which chemical properties?

a.   Gallium metal melts in your hand.
b.   Platinum does not react with oxygen at room temperature.
c.   This page is white.
d.   The copper sheets that form the "skin" of the Statue of Liberty have acquired a greenish coating over the years.

Matter can undergo changes in both its physical and its chemical properties. To illustrate the fundamental differences between physical and chemical changes, we will consider water. As we will see in much more detail in later chapters, a sample of water contains a very large number of individual units (called molecules), each made up of two atoms of hydrogen and one atom of oxygen—the familiar H_2O. This molecule can be represented as

where the letters stand for atoms and the lines show attachments (called bonds) between atoms, and the molecular model represents water in a more three-dimensional fashion.

What is really occurring when water undergoes the following changes?

We will describe these changes of state precisely in Chapter 14, but you already know something about these processes because you have observed them many times.

When ice melts, the rigid solid is replaced by a mobile liquid that takes the shape of its container. Continued heating brings the liquid to a boil, and the water becomes a gas or vapor that seems to disappear into "thin air." The changes that occur as the substance goes from solid to liquid to gas are represented in Figure 3.2. In ice the water molecules are locked into fixed positions. In the liquid the molecules are still very close together, but some motion is occurring; the positions of the molecules are no longer fixed as they are in ice. In the gaseous state the molecules are much farther apart and move randomly, hitting each other and the walls of the container.

The most important thing about all these changes is that the water molecules are still intact. The motions of individual molecules and the distances between them change, but *H_2O molecules are still present*. These changes of state are **physical changes** because they do not affect the composition of the substance. In each state we still have water (H_2O), not some other substance.

Solid   Liquid   Gas
(Ice)   (Water)   (Steam)

(a)   (b)   (c)

Figure 3.2
The three states of water (where blue spheres represent oxygen atoms and pink spheres represent hydrogen atoms).
(a) Solid: The water molecules are locked into rigid positions and are close together.
(b) Liquid: The water molecules are still close together but can move around to some extent. (c) Gas: The water molecules are far apart and move randomly.

Now suppose we run an electric current through water as illustrated in Figure 3.3. Something very different happens. The water disappears and is replaced by two new gaseous substances, hydrogen and oxygen. An electric current actually causes the water molecules to come apart—the water *decomposes* to hydrogen and oxygen. As we will see in Chapter 18 we can represent this process as follows:

This is a **chemical change** because water (consisting of H_2O molecules) has changed into different substances: hydrogen (containing H_2 molecules) and oxygen (containing O_2 molecules). Thus in this process, the H_2O molecules have been replaced by O_2 and H_2 molecules. Let us summarize

Figure 3.3
Electrolysis, the decomposition of water by an electric current, is a chemical process.

Physical and Chemical Changes

1. A *physical change* involves a change in one or more physical properties, but no change in the fundamental components that make up the substance. The most common physical changes are changes of state: solid ⇔ liquid ⇔ gas.
2. A *chemical change* involves a change in the fundamental components of the substance; a given substance changes into a different substance or substances. Chemical changes are called **reactions:** silver tarnishes by reacting with substances in the air; a plant forms a leaf by combining various substances from the air and soil; and so on.

EXAMPLE 3.2	Identifying Physical and Chemical Changes

Classify each of the following as a physical or a chemical change.

a. Iron metal is melted.
b. Iron combines with oxygen to form rust.
c. Wood burns in air.
d. A rock is broken into small pieces.

Sᴏʟᴜᴛɪᴏɴ

a. Melted iron is just liquid iron and could cool again to the solid state. This is a physical change.
b. When iron combines with oxygen, it forms a different substance (rust) that contains iron and oxygen. This is a chemical change because a different substance forms.
c. Wood burns to form different substances (as we will see later, they include carbon dioxide and water). After the fire, the wood is no longer in its original form. This is a chemical change.
d. When the rock is broken up, all the smaller pieces have the same composition as the whole rock. Each new piece differs from the original only in size and shape. This is a physical change.

SELF-CHECK EXERCISE 3.2

Classify each of the following as a chemical change, a physical change, or a combination of the two.

a. Milk turns sour.
b. Wax is melted over a flame and then catches fire and burns.

Oxygen combines with the chemicals in wood to produce flames. Is a physical or chemical change taking place?

3.3   Elements and Compounds

AIM: To define elements and compounds.

As we examine the chemical changes of matter, we encounter a series of funda-mental substances called **elements.** Elements cannot be broken down into other substances by chemical means. Examples of elements are iron, aluminum, oxy-gen, and hydrogen. All of the matter in the world around us contains elements. The elements sometimes are found in the free state, but more often they are in a combined state. Most substances contain several elements combined together.

The atoms of certain elements have special affinities for each other. They bind together in special ways to form **compounds,** substances that have the same composition no matter where we find them. Because compounds are made of elements, they can be broken down into elements through chemical changes:

Compounds   →| Chemical changes |>   Elements

Water is an example of a compound. Pure water always has the same com-position (the same relative amounts of hydrogen and oxygen) and can be broken down into the elements hydrogen and oxygen by chemical means, such as by the use of an electric current (see Figure 3.3).

As we will discuss in more detail in Chapter 4, each element is made up of a particular kind of atom: a pure sample of the element aluminum contains only aluminum atoms, elemental copper contains only copper atoms, and so on. Thus an element contains only one kind of atom; a sample of iron contains many atoms, but they are all iron atoms. Samples of certain pure elements do contain molecules; for example, hydrogen gas contains H—H (usually written H_2) mole-cules, and oxygen gas contains O—O (O_2) molecules. However, any pure sample of an element contains only atoms of that element, *never* any atoms of any other element.

A compound *always* contains atoms of *different* elements. For example, water contains hydrogen atoms and oxygen atoms, and there are always exactly twice as many hydrogen atoms as oxygen atoms because water consists of H—O—H molecules. A different compound, carbon dioxide, consists of CO_2 molecules and so contains carbon atoms and oxygen atoms (always in the ratio 1:2).

A compound, although it contains more than one type of atom, *always has the same composition*—that is, the same combination of atoms. The properties of a compound are typically very different from those of the elements it contains. For example, the properties of water are quite different from the properties of pure hydrogen and pure oxygen.

CHEMISTRY IN FOCUS

Elements in the Dinosaur Puzzle

*D*inosaurs in amazing variety ruled the earth for over 150 million years! This surely represents one of the most astonishing success stories of any group of organisms to date. However, 65 million years ago the dinosaurs suddenly disappeared, and their demise was almost instantaneous on the geological time scale. What happened? How could a group of animals that dominated the earth for so long so suddenly disappear?

Although many theories have been suggested to explain the extinction of the dinosaurs, none has seemed very convincing until recently. Geologists, examining samples of rocks containing material that was on the earth's surface when the dinosaurs died, have found unusual amounts of the element iridium. Because this metal is often found in relatively large amounts in extraterrestrial objects such as meteorites and comets, it has been suggested that a large object may have collided with the earth. Geologists propose that this happened about 65 million years ago.

It is believed that the tremendous explosion resulting from such an impact would throw millions of tons of fine dust into the atmosphere, blocking much of the sunlight that would normally reach the earth's surface. Presumably this would lead to the disappearance of many plants, because plants depend on the sun's energy to grow. The animals on earth would in turn find their food supply greatly diminished and might well die off.

More recent studies have shown that the element niobium is also present in these core samples in amounts that support the theory that earth was struck by a huge extraterrestrial object. So the presence of the elements iridium and niobium in these soil samples may have given us the answer to a long-standing riddle.

A paleontologist cleans sandstone away from the bones of diplodocus, the longest of the ancient dinosaurs, at Dinosaur National Monument in Colorado.

Packaged food items provide familiar examples of mixtures. For the first time most students will be looking at the list of ingredients as a chemist would.

3.4 Mixtures and Pure Substances

AIM: To distinguish between mixtures and pure substances.

Virtually all of the matter around us consists of mixtures of substances. For example, if you closely observe a sample of soil, you will see that it has many types of components, including tiny grains of sand and remnants of plants. The air we breathe is a complex mixture of such gases as oxygen, nitrogen, carbon dioxide, and water vapor. Even the sparkling water from a drinking fountain contains many substances besides water.

A **mixture** can be defined as something that has variable composition. For example, wood is a mixture (its composition varies greatly depending on the tree from which it originates); wine is a mixture (it can be red or pale yellow, sweet or dry); coffee is a mixture (it can be strong, weak, or bitter); and, although it looks very pure, water pumped from deep in the earth is a mixture (it contains dissolved minerals and gases).

A **pure substance,** on the other hand, always has the same composition. Pure substances are either elements or compounds. For example, pure water is a compound containing individual H_2O molecules. However, as we find it in nature, liquid water always contains other substances in addition to pure water—it is a mixture. This is obvious from the different tastes, smells, and colors of water samples obtained from various locations. However, if we take great pains to purify samples of water from various sources (such as oceans, lakes, rivers, and the earth's interior), we always end up with the same pure substance—water, which is made up only of H_2O molecules. Pure water always has the same physical and chemical properties and is always made of molecules containing hydrogen and oxygen in exactly the same proportions, regardless of the original source of the water. The properties of a pure substance make it possible to identify that substance conclusively.

Mixtures can be separated into pure substances: elements and/or compounds.

For example, the mixture known as air can be separated into oxygen (element), nitrogen (element), water (compound), carbon dioxide (compound), argon (element), and other pure substances.

Soil is a complex mixture containing many chemical compounds.

The earth's atmosphere is a mixture containing many gases and water that condenses to form clouds.

As you might imagine, it is virtually impossible to separate mixtures completely into pure substances. No matter how hard we try, some impurities (components of the original mixture) remain in each of the "pure substances."

Students may have trouble distinguishing among elements, mixtures of elements, and compounds. As an easy demonstration of the difference between a mixture of elements and a compound composed of those elements, you might show a mixture of powdered iron and sulfur in one container and FeS in another. The two substances look different and the iron can be removed from the mixture with a magnet.

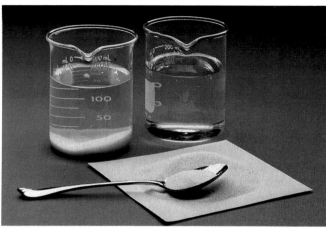

Figure 3.4
When table salt is stirred into water (left), a homogeneous mixture called a solution is formed (right).

Figure 3.5
Sand and water do not mix to form a uniform mixture. After stirring, the sand settles back to the bottom.

A solution is a homogeneous mixture.

Mixtures can be classified as either homogeneous or heterogeneous. A **homogeneous mixture** is *the same throughout.* For example, when we dissolve some salt in water and stir well, all regions of the resulting mixture have the same properties (see Figure 3.4). A homogeneous mixture is also called a **solution.** Of course, different amounts of salt and water can be mixed to form various solutions, but a homogeneous mixture (a solution) does not vary in composition from one region to another.

Coffee is a solution with variable composition. It can be strong or weak.

The air around you is a solution—it is a homogeneous mixture of gases. Solid solutions also exist. Brass is a homogeneous mixture of the metals copper and zinc.

A **heterogeneous mixture** contains regions that have different properties from those of other regions. For example, when we pour sand into water, the resulting mixture has one region containing water and another, very different region containing mostly sand (see Figure 3.5).

Sometimes it is not possible to tell by looking at a substance whether it is a homogeneous mixture or a pure substance. Students find it reassuring when I tell them that I can't tell either.

By mixing fine-grained sand with water, as in Figure 3.5, you can show the difference between a homogeneous and a heterogeneous mixture. When the sand is first mixed with the water and stirred, it looks like a homogeneous mixture. After you stop stirring, the larger grains of sand quickly settle to the bottom of the beaker, demonstrating that a heterogeneous mixture does not have the same composition throughout. You can save the mixture to demonstrate separation by filtration for Section 3.5.

EXAMPLE 3.3   **Distinguishing Between Mixtures and Pure Substances**

Identify each of the following as a pure substance, a homogeneous mixture, or a heterogeneous mixture.

a.   gasoline
b.   a stream with gravel at the bottom
c.   air

d.   brass
e.   copper metal

EXAMPLE 3.3, CONTINUED

SOLUTION

a. Gasoline is a homogeneous mixture containing many compounds.
b. A stream with gravel on the bottom is a heterogeneous mixture.
c. Air is a homogeneous mixture of elements and compounds.
d. Brass is a homogeneous mixture containing the elements copper and zinc. Brass is not a pure substance because the relative amounts of copper and zinc are different in different brass samples.
e. Copper metal is a pure substance (an element).

SELF-CHECK EXERCISE 3.3

Classify each of the following as a pure substance, a homogeneous mixture (solution), or a heterogeneous mixture.
a. wine
b. the oxygen and helium in a scuba tank
c. oil and vinegar salad dressing
d. common salt (sodium chloride)

3.5 Separation of Mixtures

AIM: To describe two methods of separating mixtures.

We have seen that the matter found in nature is typically a mixture of pure substances. For example, seawater is water containing dissolved minerals. We can separate the water from the minerals by boiling, which changes the water to steam (gaseous water) and leaves the minerals behind as solids. If we collect and cool the steam, it condenses to pure water. This separation process, called **distillation,** is shown in Figure 3.6 below.

I like to distill a sample of red wine in lecture to show the separation of colorless ethyl alcohol from the other components of the red wine. Because the color of the residue in the boiling flask is red, the separation is easy to see. You can saturate a cotton swab with the liquid from the receiving flask and ignite it to show that it is ethyl alcohol.

(a)

(b)

Figure 3.6

Distillation of a solution consisting of salt dissolved in water. (a) When the solution is boiled, steam (gaseous water) is driven off. If this steam is collected and cooled, it condenses to form pure water, which drips into the collection flask as shown. (b) After all of the water has been boiled off, the salt remains in the original flask and the water is in the collection flask.

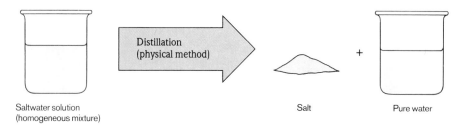

Saltwater solution
(homogeneous mixture)

Salt

Pure water

Figure 3.7
No chemical change occurs when salt water is distilled.

The separation of a mixture sometimes occurs in the natural environment and can be to our benefit (see photo, opposite page).

If you used a sand–water mixture to demonstrate heterogeneous mixtures, you can use it again to demonstrate filtration. Because filtration is time-consuming, you can set up the apparatus and let it filter while you continue to lecture.

When we carry out the distillation of salt water, water is changed from the liquid state to the gaseous state and then back to the liquid state. These changes of state are examples of physical changes. We are separating a mixture of substances, but we are not changing the composition of the individual substances. We can represent this as shown in Figure 3.7.

Suppose we scooped up some sand with our sample of seawater. This sample is a heterogeneous mixture, because it contains an undissolved solid as well

Stirring rod

Mixture
of solid
and liquid

Funnel

Filter paper
traps solid

Figure 3.8
Filtration separates a liquid from a solid.
The liquid passes through the filter paper,
but the solid particles are trapped.

Filtrate (liquid
component of
the mixture)

Figure 3.9
Separation of a sand–saltwater mixture.

When water from the Great Salt Lake evaporates (changes to a gas and escapes), the salt is left behind. This is one commercial source of salt.

as the saltwater solution. We can separate out the sand by simple **filtration.** We pour the mixture onto a mesh, such as a filter paper, which allows the liquid to pass through and leaves the solid behind (see Figure 3.8). The salt can then be separated from the water by distillation. The total separation process is represented in Figure 3.9. All the changes involved are physical changes.

We can summarize the description of matter given in this chapter with the diagram shown in Figure 3.10. Note that a given sample of matter can be a pure substance (either an element or a compound) or, more commonly, a mixture (homogeneous or heterogeneous). We have seen that all matter exists as elements or can be broken down into elements, the most fundamental substances we have encountered up to this point. We will have more to say about the nature of elements in the next chapter.

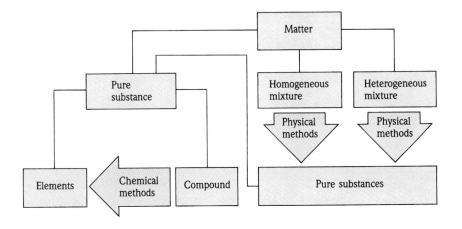

Figure 3.10
The organization of matter.

Section 3.6 provides an introduction to energy. This section can be covered now or with material from Sections 1 and 2 of Chapter 14. There are advantages to each arrangement. If the material is covered now, students receive an introduction to energy, which can be integrated with other basic information. They also have the opportunity to sharpen dimensional-analysis skills. If you follow the order of chapters as written, the next exposure to dimensional analysis is in Chapter 9. However, if you cover Section 3.6 now, it will probably be necessary to review it when you cover Chapter 14 because students often forget that they have been exposed to this material. If you choose not to cover Section 3.6 now, then the energy changes associated with the heating curve for water can be covered all at once, in Chapter 14.

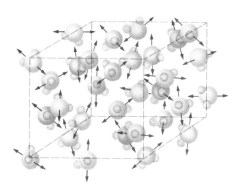

Figure 3.11
In ice, the water molecules vibrate randomly about their positions in the solid. Their motions are represented by arrows.

Gases will be discussed fully in Chapter 13.

calorie: The energy (heat) required to raise the temperature of 1 g of water by 1 °C.

The abbreviations for calorie and joule are cal and J. It is straightforward to convert from one energy unit to another.

3.6   Energy and Energy Changes

AIM: To discuss energy and its effect on matter.

Energy is a familiar term. We speak of solar energy, nuclear energy, and energy from coal and gasoline, and when we're tired, we say that we have run out of energy. Energy allows us to "do things"—to work, to drive to school, and to cook eggs. A common definition of **energy** is the capacity to do work.

One way we use energy is to change the temperature of a substance. For example, we often heat water using the energy provided by a stove or Bunsen burner. As we will see in more detail later, the temperature of a substance reflects the random motions of the components of that substance. For example, in ice the components are water molecules that vibrate randomly about their fixed positions in the solid, as represented in Figure 3.11. When the solid is heated to higher temperatures, the random vibrations become more energetic. Finally, at the melting point of ice, the molecules are vibrating energetically enough to break loose from their positions, and rigid ice changes to liquid water.

We have seen that the motions of molecules in a substance increase as we raise the temperature of that substance. *The amount of energy (heat) required to raise the temperature of one gram of water by one Celsius degree* is called a **calorie** in the metric system of units. The unit of energy in the SI system is called the **joule** (pronounced "jewel"). We can convert between joules and calories by using the definition that exactly 1 calorie = 4.184 joules, which leads to the equivalence statement

$$1 \text{ cal} = 4.184 \text{ J}$$

EXAMPLE 3.4   Converting Calories to Joules

Express 60.1 cal of energy in units of joules.

SOLUTION

By definition 1 cal = 4.184 J, so the conversion factor needed is $\dfrac{4.184 \text{ J}}{1 \text{ cal}}$, and the result is

$$60.1 \ \cancel{\text{cal}} \times \frac{4.184 \text{ J}}{1 \ \cancel{\text{cal}}} = 251 \text{ J}$$

Note that the 1 in the denominator is an exact number by definition and so does not limit the number of significant figures.

EXAMPLE 3.4, CONTINUED

SELF-CHECK EXERCISE 3.4

How many calories of energy correspond to 28.4 J?

Now think about heating a substance from one temperature to another. How does the amount of substance heated affect the energy required? In 2 g of water there are twice as many molecules as in 1 g of water. It takes twice as much energy to change the temperature of 2 g of water by 1 °C, because we must change the motions of twice as many molecules in a 2-g sample as in a 1-g sample. Also, as we would expect, it takes twice as much energy to raise the temperature of a given sample of water by 2 degrees as it does to raise the temperature by 1 degree.

EXAMPLE 3.5	**Calculating Energy Requirements**

Determine the amount of energy (heat) in joules required to raise the temperature of 7.40 g water from 29.0 °C to 46.0 °C.

SOLUTION

In solving any kind of problem, it is often useful to draw a diagram that represents the situation. In this case, we have 7.40 g of water that are to be heated from 29.0 °C to 46.0 °C.

7.40 g water $T = 29.0\ °C$	? energy	7.40 g water $T = 46.0\ °C$

Our task is to determine how much energy is required to accomplish this.

From the discussion in the text, we know that 4.184 J of energy are required to raise the temperature of *one* gram of water by *one* Celsius degree.

1.00 g water $T = 29.0\ °C$	4.184 J	1.00 g water $T = 30.0\ °C$

Because in our case we have 7.40 g of water instead of 1.00 g, it will take 7.40×4.184 J to raise the temperature by one degree.

Relating the gain or loss of a quantity of energy to familiar occurrences provides students with a good common-sense feeling for the amount of energy used. For example, you can calculate the amount of energy required to heat a quart of water for coffee from room temperature to the boiling point. You might want to calculate the amount of energy required and, using an energy bill, the amount of money required to wash dishes or to take a shower.

However, we want to raise the temperature of our sample of water by more than 1 °C. In fact, the temperature change required is from 29.0 °C to 46.0 °C. This is a change of 17.0 °C (46.0 °C − 29.0 °C = 17.0 °C). Thus we will have to supply 17.0 times the energy necessary to raise the temperature of 7.40 g of water by 1 °C.

7.40 g water T = 29.0 °C	17.0 × 7.40 × 4.184 J	7.40 g water T = 46.0 °C

This calculation is summarized as follows:

$4.184\ \dfrac{J}{g\ °C}$	×	7.40 g	×	17.0 °C	=	526 J
Energy per gram of water per degree of temperature	×	Actual grams of water	×	Actual temperature change	=	Energy required

We have shown that 526 J of energy (as heat) are required to raise the temperature of 7.40 g of water from 29.0 °C to 46.0 °C. Note that because 4.184 J of energy are required to heat 1 g of water by 1 °C, the units are J/g °C (joules per gram per Celsius degree).

The result you will get on your calculator is 4.184 × 7.40 × 17.0 = 526.3472, which rounds off to 526.

SELF-CHECK EXERCISE 3.5

Calculate the joules of energy required to heat 454 g of water from 5.4 °C to 98.6 °C.

So far we have seen that the energy (heat) required to change the temperature of a substance depends on

1. The amount of substance being heated (number of grams)
2. The temperature change (number of degrees)

There is another important factor: the identity of the substance.

Different substances respond differently to being heated. We have seen that 4.184 J of energy raises the temperature of 1 g of water 1 °C. In contrast, this same amount of energy applied to 1 g of gold raises its temperature by approximately 32 °C! The point is that some substances require relatively large amounts of energy to change their temperatures, whereas others require relatively little. Chemists describe this difference by saying that substances have different heat capacities. *The amount of energy required to change the temperature of one gram of a substance by one Celsius degree* is called its **specific heat capacity** or, more commonly, its **specific heat.** The specific heat capacities for several substances

Table 3.2	The Specific Heat Capacities of Some Common Substances	
Substance	Specific Heat Capacity (J/g °C)
water (*l*)* (liquid)	4.184
water (*s*) (ice)	2.03
water (*g*) (steam)	2.0
aluminum (*s*)	0.89
iron (*s*)	0.45
mercury (*l*)	0.14
carbon (*s*)	0.71
silver (*s*)	0.24
gold (*s*)	0.13

*The symbols (*s*), (*l*), and (*g*) indicate the solid, liquid, and gaseous states, respectively.

Because water is such a familiar and convenient substance to use when solving specific heat problems, students sometimes believe that these calculations are for water only. Table 3.2 gives the specific heat capacities for a variety of substances. It's probably a good idea to use these in examples in addition to using water.

are listed in Table 3.2. You can see from the table that the specific heat capacity for water is very high compared to those of the other substances listed. This is why lakes and seas are much slower to respond to cooling or heating than are the surrounding land masses.

EXAMPLE 3.6   Calculations Involving Specific Heat Capacity

a.   What quantity of energy (in joules) is required to heat a piece of iron weighing 1.3 g from 25 °C to 46 °C?
b.   What is the answer in calories?

SOLUTION

a.   It is helpful to draw the following diagram to represent the problem.

1.3 g iron $T = 25$ °C	? joules →	1.3 g iron $T = 46$ °C

From Table 3.2 we see that the specific heat capacity of iron is 0.45 J/g °C. That is, it takes 0.45 J to raise the temperature of a 1-g piece of iron by 1 °C.

1.0 g iron $T = 25$ °C	0.45 J →	1.0 g iron $T = 26$ °C

EXAMPLE 3.6, CONTINUED

In this case our sample is 1.3 g, so 1.3 × 0.45 J is required for *each* degree of temperature increase.

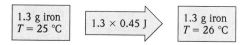

The result you will get on your calculator is: 04.5 × 1.3 × 21 = 12.285, which rounds off to 12.

Because the temperature increase is 21 °C (46 °C − 25 °C = 21 °C), the total amount of energy required is

$$0.45\frac{J}{\cancel{g}\ \cancel{°C}} \times 1.3\,\cancel{g} \times 21\ \cancel{°C} = 12\ J$$

| 1.3 g iron $T = 25$ °C | 21 × 1.3 × 0.45 J | 1.3 g iron $T = 46$ °C |

Note that the final units are joules, as they should be.

b. To calculate this energy in calories, we can use the definition 1 cal = 4.184 J to construct the appropriate conversion factor. We want to change from joules to calories, so cal must be in the numerator and J in the denominator where it cancels:

$$12\,\cancel{J} \times \frac{1\ cal}{4.184\,\cancel{J}} = 2.9\ cal$$

Remember that 1 in this case is an exact number by definition and therefore does not limit the number of significant figures (the number 12 is limiting here).

SELF-CHECK EXERCISE 3.6

A 5.63-g sample of solid gold is heated from 21 °C to 32 °C. How much energy (in joules and calories) is required?

Note that in Example 3.6, to calculate the energy (heat) required, we took the product of the specific heat capacity, the sample size in grams, and the change in temperature in Celsius degrees.

| Energy (heat) required (Q) | = | Specific heat capacity (s) | × | Mass (m) in grams of sample | × | Change in temperature (ΔT) in °C |

We can represent this by the following equation:

$$Q = s \times m \times \Delta T$$

where

The symbol Δ (the Greek letter delta) is shorthand for "change in."

$Q =$ energy (heat) required
$s =$ specific heat capacity
$m =$ mass of the sample in grams
$\Delta T =$ change in temperature in Celsius degrees

This equation always applies when a substance is being heated (or cooled) and no change of state occurs. Before you begin to use this equation, however, make sure you understand where it comes from.

EXAMPLE 3.7   Specific Heat Capacity Calculations: Using the Equation

A 1.6-g sample of a metal that has the appearance of gold requires 5.8 J of energy to change its temperature from 23 °C to 41 °C. Is the metal pure gold?

SOLUTION

We can represent the data given in this problem by the following diagram:

| 1.6 g metal $T = 23$ °C | 5.8 J | 1.6 g metal $T = 41$ °C |

$$\Delta T = 41 \text{ °C} - 23 \text{ °C} = 18 \text{ °C}$$

Using the data given, we can calculate the value of the specific heat capacity for the metal and compare this value to the one for gold given in Table 3.2. We know that

$$Q = s \times m \times \Delta T$$

or, pictorially,

| 1.6 g metal $T = 23$ °C | 5.8 J $= ? \times 1.6 \times 18$ | 1.6 g metal $T = 41$ °C |

When we divide both sides of the equation

$$Q = s \times m \times \Delta T$$

by $m \times \Delta T$, we get

$$\frac{Q}{m \times \Delta T} = s$$

Firewalking: Magic or Science?

*F*or millennia people have been amazed at the ability of Eastern mystics to walk across beds of glowing coals without any apparent discomfort. Even in the United States, thousands of people have performed feats of firewalking as part of motivational seminars. How can this be possible? Do firewalkers have supernatural powers?

Actually, there are good scientific explanations of why firewalking is possible. First, human tissue is mainly composed of water, which has a relatively large specific heat capacity. This means that a large amount of energy must be transferred from the coals to change significantly the temperature of the feet. During the brief contact between feet and coals involved in firewalking, there is relatively little time for energy flow, so the feet do not reach a high enough temperature to cause damage.

Second, although the surface of the coals has a very high temperature, the red-hot layer is very thin. Therefore, the quantity of energy available to heat the feet is smaller than might be expected.

A third factor that aids firewalkers is the so-called Leidenfrost effect,

A Hindu firewalking ceremony in the Fiji Islands.

the phenomenon that allows water droplets to skate around on a hot griddle for a surprisingly long time. The part of the droplet in contact with the hot surface vaporizes first, providing a gaseous layer that both allows the droplet to skate around and acts as a barrier through which energy does not readily flow to the rest of the droplet. The perspiration on the feet of the presumably tense firewalker would have the same effect. And, because firewalking is often done at night when moist grass surrounds the bed of coals, the firewalker's feet are probably damp before the walk, providing ample moisture for the Leidenfrost effect to occur.

Thus, although firewalking is impressive, there are several scientific reasons why anyone with the proper training should be able to do it on a properly prepared bed of coals. (Don't try this on your own!)

EXAMPLE 3.7, CONTINUED

Thus using the data given, we can calculate the value of *s.* In this case,

Q = energy (heat) required = 5.8 J
m = mass of the sample = 1.6 g
ΔT = change in temperature = 18 °C (41 °C − 23 °C = 18 °C)

Thus

$$s = \frac{Q}{m \times \Delta T} = \frac{5.8\ \text{J}}{(1.6\ \text{g})\,(18\ °\text{C})} = 0.20\ \text{J/g} °\text{C}$$

From Table 3.2, the specific heat capacity for gold is 0.13 J/g °C. Thus the metal must not be pure gold.

The result you will get on your calculator is: 5.8/(1.6)(18) = 0.2013889, which rounds off to 0.20.

SELF-CHECK EXERCISE 3.7

A 2.8-g sample of pure metal requires 10.1 J of energy to change its temperature from 21 °C to 36 °C. What is this metal? (Use Table 3.2.)

CHAPTER REVIEW

Key Terms

matter (p. 64)
states of matter (p. 65)
solid (p. 65)
liquid (p. 65)
gas (p. 65)
physical properties (p. 65)
chemical properties (p. 65)
physical change (p. 66)
chemical change (p. 67)
element (p. 69)
compound (p. 69)

mixture (p. 71)
pure substance (p. 71)
homogeneous mixture (p. 72)
solution (p. 72)
heterogeneous mixture (p. 72)
distillation (p. 73)
filtration (p. 75)
energy (p. 76)
calorie (p. 76)
joule (p. 76)
specific heat capacity (p. 78)

Summary

1. Matter can exist in three states—solid, liquid, and gas—and can be described in terms of its physical and chemical properties. Chemical properties describe a substance's ability to undergo a change to a different substance. Physical properties are the characteristics a substance exhibits as long as no chemical change occurs.

2. A physical change involves a change in one or more physical properties, but no change in composition. A chemical change transforms a substance into a new substance or substances.
3. A mixture has variable composition. A homogeneous mixture has the same properties throughout; a heterogeneous mixture does not. A pure substance always has the same composition. We can physically separate mixtures of pure substances by distillation and filtration.
4. Pure substances are of two types: elements, which cannot be broken down chemically into simpler substances, and compounds, which can be broken down chemically into elements.
5. Energy can be defined as the capacity to do work.
6. Specific heat capacity is the amount of energy required to change the temperature of one gram of a substance by one Celsius degree. Each substance has a characteristic specific heat capacity. The energy (heat) required to change the temperature of a substance depends on three factors: the amount of substance, the temperature change, and the specific heat capacity of the substance.

Questions and Problems

All even-numbered exercises have answers in the back of this book and solutions in the Solutions Guide.

3.1 Matter

QUESTIONS

1. Matter is anything that has _____ and occupies space.

2. The three physical _____ of matter are solid, liquid, and gas.

3. Solids and liquids are virtually incompressible; they have _____ volumes.

4. Liquids have definite volumes but are able to take on the shape of their _____.

5. In liquid substances, the molecules are very close to each other but are still able to _____ fairly freely.

6. Matter in the _____ state has no shape and fills completely whatever container holds it.

7. Discuss the similarities and differences between a liquid and a solid.

8. The _____ the forces among the particles in a sample of matter, the more rigid the matter will be.

9. Consider three 10-g samples of water: one as ice, one as a liquid, and one as vapor. How do the volumes of these three samples compare to one another? How is this difference in volume related to the physical state involved?

10. Automobile engines derive their power from the compression and expansion of gases in the cylinder. Why is a gas, as opposed to a liquid or solid, compressible?

3.2 Physical and Chemical Properties and Changes

QUESTIONS

11. Acetone, a solvent widely used in industry, boils at 65 °C; this is an example of a _____ property of acetone.

12. Acetone is highly flammable, because it reacts easily with oxygen gas in the atmosphere; this is an example of a _____ property of acetone.

13–14. Aqueous solutions containing nickel(II) ions are bright green in color. When concentrated ammonia is added to such a solution, the solution turns dark blue and a precipitate forms if the solution is allowed to stand.

13. The fact that nickel(II) solutions are bright green is a _____ property.

14. The change that occurs when ammonia is added to a nickel(II) solution is a _____ property.

15. Describe, on a microscopic basis, the physical changes that take place when an ice cube is heated strongly until it melts and the water vaporizes. Why are these changes *physical* and not chemical in nature?

16. Describe, on a microscopic basis, what happens when a strong electric current is passed through water, and why the resulting changes are *chemical* changes.

17. Classify the following as *physical* or *chemical* properties/changes.
   a. Dry ice (frozen dry ice) gradually vaporizes.
   b. Hair curls when the humidity increases.
   c. Meat blackens when cooked too long on a barbecue.
   d. Salt water boils at a higher temperature than fresh water.
   e. Champagne bubbles when the cork is removed.
   f. Solid sugar forms when a sugar solution is cooled.
   g. Coffee turns a lighter color when cream is added.
   h. An iron poker gets hot when heated in a fireplace.
   i. A light bulb glows.
   j. Sugar dissolves in water.
   k. Ice on a sidewalk melts.

18. Classify the following as *physical* or *chemical* changes/properties.
   a. A copper pan acquires a blue/green coating (patina) with use.
   b. Drain cleaner dissolves a hair clog in a bathroom drain.
   c. Hydrogen peroxide fizzes when applied to a wound.
   d. Acids produced by bacteria in plaque cause teeth to decay.
   e. Marble statues deteriorate when attacked by acid rain.
   f. Grape juice ferments when yeast is added.
   g. Mothballs vaporize in a cedar closet.
   h. Alcohol feels cool as it evaporates from the skin.
   i. Your car battery will run down if you leave the lights on.
   j. Soap lathers when you wash your hands.
   k. The sulfur in eggs will turn a silver spoon black.

3.3 Elements and Compounds

QUESTIONS

19. _____ cannot be broken down into simpler substances by chemical means.

20. A pure sample of a(n) _____ contains only one kind of atom.

21. Certain elements have special affinities for other elements. This causes them to bind together in special ways to form _____ .

22. Compounds can be broken down into _____ by chemical changes.

23. A compound always has (a variable/the same) composition, (a variable/the same) combination of atoms.

24. The properties of a compound are usually very (different from/similar to) those of the elements it contains.

3.4 Mixtures and Pure Substances

QUESTIONS

25. A pure substance always has (a variable/the same) composition.

26. A mixture can be defined as something that has (a variable/the same) composition.

27. A homogeneous mixture is more commonly referred to as a _____ .

28. A mixture of salt and sand would be an example of a _____ mixture.

29. Identify the following as *mixtures* or as *pure substances.*
   a. milk
   b. the paper this book is printed on
   c. a teaspoon of sugar
   d. a teaspoon of sugar dissolved in a glass of water
   e. steel

30. Classify the following as *mixtures* or as *pure substances.*
   a. a white cotton handkerchief
   b. distilled water
   c. water scooped from a pond
   d. the mercury used in a thermometer

31. Classify the following mixtures as *homogeneous* or *heterogeneous*.
    a. a bag of various colored marbles
    b. beach sand
    c. a sample of sodium chloride dissolved in water
    d. a sample of table salt dissolved in water
    e. air

32. Classify the following mixtures as *homogeneous* or *heterogeneous*.
    a. gasoline
    b. a jar of jelly beans
    c. chunky peanut butter
    d. margarine
    e. the paper this question is printed on

3.5 Separation of Mixtures

QUESTIONS

33. Water from the sea can be made fit for drinking by boiling the water and then condensing the vapor back to the liquid state. This process is known as _____.

34. An insoluble contaminant may be removed from a liquid medium most simply by the process of _____.

35. In a common laboratory experiment in general chemistry, students are asked to determine the relative amounts of benzoic acid and charcoal in a solid mixture. Benzoic acid is relatively soluble in hot water, but charcoal is not. Devise a method for separating the two components of this mixture.

36. Describe the process of distillation depicted in Figure 3.6. Does the separation of the components of a mixture by distillation represent a chemical or a physical change?

3.6 Energy and Energy Changes

QUESTIONS

37. _____ is defined as the capacity to do work.

38. _____ is defined as the amount of energy required to raise the temperature of one gram of water by one degree Celsius.

39. Describe what happens to the molecules in a sample of ice as the sample is slowly heated until it liquefies and then vaporizes.

40. Describe what happens to the molecules in liquid water as it is heated and begins to boil.

41. Metallic substances tend to have (higher/lower) specific heat capacities than nonmetallic substances.

42. The quantity of energy required to change the temperature of a sample is calculated by taking the product of the mass of the sample, the specific heat capacity of the sample, and the _____ change undergone by the sample.

43. If it takes 526 J of energy to warm 7.40 g of water by 17 °C, it will take _____ J to warm 3.20 g of water by the same amount.

44. If it takes 526 J of energy to warm 7.40 g of water by 17 °C, it will take _____ J to warm the same amount of water by 34 °C.

45. Convert the following numbers of calories into joules.
    a. 100.0 cal
    b. 5,000.0 cal
    c. 1.00 cal
    d. 1.00×10^3 cal

46. Convert the following numbers of joules into calories.
    a. 1,000.0 J
    b. 550.0 J
    c. 2.45×10^6 J
    d. 1.00 J

47. Convert the following numbers of calories into kilocalories. (Remember: Kilo means 1000.)
    a. 850 cal
    b. 1530 cal
    c. 227,200 cal
    d. 0.00289 cal

48. Convert the following numbers of kilocalories into calories.
    a. 12.30 kcal
    b. 290.4 kcal
    c. 940,000 kcal
    d. 4201 kcal

49. Convert the following numbers of joules (J) into kilojoules (kJ). (Remember: Kilo means 1000.)
    a. 243,000 J   c. 0.251 J
    b. 4.184 J   d. 450.3 J

50. Convert the following numbers of kilojoules (kJ) into joules (J).
    a. 189,900 kJ   c. 2.39 kJ
    b. 24,480 kJ   d. 19.75 kJ

51. Calculate the amount of energy required (in calories) to heat 145 g of water from 22.3 °C to 75.0 °C.

52. Calculate the quantity of heat, in kJ, required to heat 1.00 kg of iron from 25.0 °C to 49.2 °C.

53. Calculate the energy required (in joules) to heat 25.0 g of gold from 120. °C to 155 °C. (See Table 3.2.)

54. It takes 1.25 kJ of energy to heat a certain sample of pure silver from 12.0 °C to 15.2 °C. Calculate the mass of the sample of silver.

55. If 50. J of heat are applied to 10. g of iron, by how much will the temperature of the iron increase? (See Table 3.2.)

56. If 1.34×10^2 kJ of heat are applied to a 50.1 g sample of aluminum, by how much will the temperature of the aluminum sample increase?

57. The specific heat capacity of silver is 0.24 J/g °C. Express this in terms of calories per gram per Celsius degree.

58. The specific heat capacity of the liquid refrigerant Freon-12 is 0.232 cal/g °C. Express this in terms of joules per gram per Celsius degree.

59. Three separate 50.0-g samples of aluminum, iron, and silver are available. Each of these samples is initially at 20.0 °C, and then 1.00 kJ of heat is applied to each sample. What final temperature will each of the samples reach?

60. If the temperatures of separate 25.0-g samples of gold, mercury, and carbon are to be raised by 20. °C, how much heat (in joules) must be applied to each substance?

61. A 22.5-g sample of metal X requires 540. J of energy to heat it from 10. °C to 92 °C. Calculate the specific heat capacity of metal X.

62. A 35.2 g sample of metal Z requires 1,251 J of energy to heat the sample by 25.0 °C. Calculate the specific heat capacity of metal Z.

Additional Problems

63. If solid iron pellets and sulfur powder are poured into a container at room temperature, a simple _____ has been made. If the iron and sulfur are heated until a chemical reaction takes place between them, a(n) _____ will form.

64. Pure substance X is melted, and the liquid is placed in an electrolysis apparatus such as that shown in Figure 3.3. When an electrical current is passed through the liquid, a brown solid forms in one chamber and a white solid forms in the other chamber. Is substance X a compound or an element?

65. When a 1.00-g sample of baking soda is heated, a gas is formed. The solid that remains weighs less than 1.00 g and does not have the same properties as the original sample. Is baking soda an element or a compound?

66. During a very cold winter, the temperature may remain below freezing for extended periods. However, fallen snow can still disappear, even though it cannot melt. This is possible because a solid can vaporize directly, without passing through the liquid state. Is this process (sublimation) a physical or a chemical change?

67. Perform the indicated conversions.
    a. 750,900 cal to kilojoules
    b. 9.985 kJ to kilocalories
    c. 7.899 kcal to joules
    d. 900.4 kcal to kilojoules
    e. 89,930 J to kilocalories
    f. 5.901 kJ to calories

68. Calculate the amount of energy required (in joules) to heat 2.5 kg of water from 18.5 °C to 55.0 °C.

69. If 10. J of heat is applied to 5.0-g samples of each of the substances listed in Table 3.2, which substance's temperature will increase the most? Which substance's temperature will increase the least?

70. A 5-g sample of aluminum and a 5-g sample of iron are heated in a boiling water bath in separate test tubes. The test tubes are then placed together into a beaker containing ice. Which metal will lose the most heat in cooling down?

71. Hydrogen gives off 120. J/g of energy when burned in oxygen, and methane gives off 50. J/g under the same circumstances. If a mixture of 5.0 g of hydrogen and 10. g of methane is burned, and the heat released is transferred to 500. g of water at 25 °C, what final temperature will be reached by the water?

72. A 5.00-g sample of aluminum pellets and a 10.00-g sample of iron pellets are placed together in a dry test tube, and the test tube is heated in a boiling water bath to 100. °C. The

mixture of hot iron and aluminum is then poured into 97.3 g of water at 22.5 °C. To what final temperature is the water heated by the metals?

73. A 50.0-g sample of water at 100. °C is poured into a 50.0-g sample of water at 25 °C. What will be the final temperature of the water?

74. A 25.0-g sample of pure iron at 85 °C is dropped into 75 g of water at 20. °C. What is the final temperature of the water–iron mixture?

75. Discuss the similarities and differences between a liquid and a gas.

76. In gaseous substances, the individual molecules are relatively (close/far apart) and are moving freely, rapidly, and randomly.

77. The fact that the chemical substance sodium chromate is bright yellow is an example of a _____ property.

78. The fact that the chemical substance sodium chromate reacts in solution with lead compounds is an example of a _____ property.

79–80. Solutions containing copper(II) ions are bright blue in color. When sodium hydroxide is added to such a solution, a precipitate is formed that is a much lighter blue.

79. The fact that a solution containing copper(II) ions is bright blue is a _____ property.

80. The fact that a reaction takes place when sodium hydroxide is added to a solution of copper(II) ions is a _____ property.

81. The processes of melting and evaporation involve changes in the _____ of a substance.

82. _____ is the process of making a chemical reaction take place by passage of an electrical current through a substance or solution.

83. Classify the following as *physical* or *chemical* properties/changes.
    a. Milk curdles if a few drops of vinegar are added.
    b. Butter turns rancid if left exposed at room temperature.
    c. Salad dressing separates into layers after standing.
    d. Milk of magnesia neutralizes stomach acid.
    e. The steel in a car has rust spots.
    f. A person is asphyxiated by breathing carbon monoxide.
    g. Spilled acid burns a hole in cotton jeans.
    h. Sweat cools the body as it evaporates from the skin.
    i. Aspirin reduces fever.
    j. Oil feels slippery.
    k. Alcohol burns, forming carbon dioxide and water.

84. Identify the following as *mixtures* or as *pure substances.*
    a. wood   d. wine
    b. aluminum foil   e. a vitamin tablet
    c. salad dressing

85. Classify the following mixtures as *homogeneous* or *heterogeneous.*
    a. blood
    b. rust as scraped from an automobile bumper
    c. salad dressing
    d. window glass
    e. pond water

86. If it takes 4.5 J of energy to warm 5.0 g of aluminum from 25 °C to a certain higher temperature, then it will take _____ J to warm 10. g of aluminum over the same temperature interval.

87. If it takes 103 J of energy to warm a certain mass of iron from 25 °C to 50. °C, then it will take _____ J to warm the same mass of iron from 25 °C to 75 °C.

88. Convert the following numbers of calories into joules (J).
    a. 150.0 cal   c. 10.0 cal
    b. 4462 cal   d. 4.184 cal

89. Convert the following numbers of joules (J) into calories.
    a. 4.184 J   c. 8.02 J
    b. 1520 J   d. 23.29 J

90. Perform the following conversions.
    a. 1.25×10^4 J to kilojoules
    b. 1.25×10^4 J to kilocalories
    c. 512.2 cal to kilojoules
    d. 14.2 kJ to kilocalories

91. Calculate the energy required (in calories) to heat 10.4 g of mercury from 37.0 °C to 42.0 °C.

92. Calculate the energy required (in joules) to heat 75 g of water from 25 °C to 39 °C.

93. If 50. J of heat are applied to 10. g of aluminum, by how much will the temperature of the aluminum increase? (See Table 3.2.)

94. For each of the substances listed in Table 3.2, calculate the quantity of heat required to heat 150. g of the substance by 11.2 °C.

95. Suppose you had 10.0 g samples of each of the substances listed in Table 3.2 and that 1.00 kJ of heat is applied to each of these samples. By what amount would the temperature of each sample be raised?

96. A 55.0-g sample of metal Z requires 675 J of energy to heat it from 25 °C to 118 °C. Calculate the specific heat capacity of metal Z.

Solutions to Self-Check Exercises

SELF-CHECK EXERCISE 3.1

Items (a) and (c) are physical properties. When the solid gallium melts, it forms liquid gallium. There is no change in composition. Items (b) and (d) reflect the ability to change composition and are thus chemical properties. Statement (b) means that platinum does not react with oxygen to form some new substance. Statement (d) means that copper does react in the air to form a new substance, which is green.

SELF-CHECK EXERCISE 3.2

a. Milk turns sour because new substances are formed. This is a chemical change.
b. Melting the wax is a physical change (a change of state). When the wax burns, new substances are formed. This is a chemical change.

SELF-CHECK EXERCISE 3.3

a. Wine is a homogeneous mixture of alcohol and other dissolved substances dispersed uniformly in water.
b. Helium and oxygen form a homogeneous mixture.
c. Oil and vinegar salad dressing is a heterogeneous mixture. (Note the two distinct layers the next time you look at a bottle of dressing.)
d. Common salt is a pure substance (sodium chloride), so it always has the same composition. (Note that other substances such as iodine are often added to commercial preparations of table salt, which is mostly sodium chloride. Thus commercial table salt is a homogeneous mixture.)

SELF-CHECK EXERCISE 3.4

The conversion factor needed is $\dfrac{1 \text{ cal}}{4.184 \text{ J}}$, and the conversion is

$$28.4 \text{ J} \times \frac{1 \text{ cal}}{4.184 \text{ J}} = 6.79 \text{ cal}$$

SELF-CHECK EXERCISE 3.5

We know that it takes 4.184 J of energy to change the temperature of each gram of water by 1 °C, so we must multiply 4.184 by the mass of water (454 g) and the temperature change (98.6 °C − 5.4 °C = 93.2 °C).

$$4.184 \frac{\text{J}}{\text{g} \cdot {}^\circ\text{C}} \times 454 \text{ g} \times 93.2 \text{ }^\circ\text{C} = 1.77 \times 10^5 \text{ J}$$

SELF-CHECK EXERCISE 3.6

From Table 3.2, the specific heat capacity for solid gold is 0.13 J/g °C. Because it takes 0.13 J to change the temperature of *one* gram of gold by *one* Celsius degree, we must multiply 0.13 by the sample size (5.63 g) and the change in temperature (32 °C − 21 °C = 11 °C).

$$0.13 \ \frac{J}{g\,{}^{\circ}C} \times 5.63\ g \times 11\ {}^{\circ}C = 8.1\ J$$

We can change this energy to units of calories as follows:

$$8.1\ J \times \frac{1\ cal}{4.184\ J} = 1.9\ cal$$

SELF-CHECK EXERCISE 3.7

Table 3.2 lists the specific heat capacities of several metals. We want to calculate the specific heat capacity (s) for this metal and then use Table 3.2 to identify the metal. Using the equation

$$Q = s \times m \times \Delta T$$

we can solve for s by dividing both sides by m (the mass of the sample) and by ΔT:

$$\frac{Q}{m \times \Delta T} = s$$

In this case,

$$Q = \text{energy (heat) required} = 10.1\ J$$
$$m = 2.8\ g$$
$$\Delta T = \text{temperature change} = 36\ {}^{\circ}C - 21\ {}^{\circ}C = 15\ {}^{\circ}C$$

so

$$s = \frac{Q}{m \times \Delta T} = \frac{10.1\ J}{(2.8\ g)(15\ {}^{\circ}C)} = 0.24\ J/g\,{}^{\circ}C$$

Table 3.2 shows that silver has a specific heat capacity of 0.24 J/g °C. The metal is silver.

CHAPTER

4 Chemical Foundations: Elements and Atoms

A micrograph of crystals of bismith metal.

CONTENTS

Robert Boyle at 62 years of age.

Now is a good time to bring samples of more elements to class. Your discussion of elements and the introduction to the periodic table will be more interesting if amply illustrated. It is much easier for students to remember that at room temperature chlorine is a yellowish–green gas if they have actually seen a sample of the gas.

Since ancient times, humans have used chemical changes to their advantage. The processing of ores to produce metals for ornaments and tools and the use of embalming fluids are two applications of chemistry that were used before 1000 B.C.

The Greeks were the first to try to explain why chemical changes occur. By about 400 B.C. they had proposed that all matter was composed of four fundamental substances: fire, earth, water, and air.

The next 2000 years of chemical history were dominated by alchemy. Alchemists were often mystics and fakes who were obsessed with the idea of turning cheap metals into gold. However, this period also saw important events: the elements mercury, sulfur, and antimony were discovered, and alchemists learned how to prepare acids.

The first scientist to recognize the importance of careful measurements was the Irishman Robert Boyle (1627–1691). Boyle is best known for his pioneering work on the properties of gases, but his most important contribution to science was probably his insistence that science should be firmly grounded in experiments. For example, Boyle held no preconceived notions about how many elements there might be. His definition of the term *element* was based on experiments: a substance was an element unless it could be broken down into two or more simpler substances. For example, air could not be an element as the Greeks believed, because it could be broken down into many pure substances.

As Boyle's experimental definition of an element became generally accepted, the list of known elements grew, and the Greek system of four elements died. But although Boyle was an excellent scientist, he was not always right. For some reason he ignored his own definition of an element and clung to the alchemists' views that metals were not true elements and that a way would be found eventually to change one metal into another.

4.1   The Elements

AIM: To discuss the relative abundances of the elements.
        To introduce the names of some elements.

In studying the materials of the earth (and other parts of the universe), scientists have found that all matter can be broken down chemically into about one hundred different elements. At first it might seem amazing that the millions of known substances are composed of so few fundamental elements. Fortunately for those trying to understand and systematize it, nature often uses a relatively small number of fundamental units to assemble even extremely complex materials. For example, proteins, a group of substances that serve the human body in almost

Table 4.1   Distribution (Mass Percent) of the 18 Most Abundant
Elements in the Earth's Crust, Oceans, and Atmosphere

Element	Mass Percent	Element	Mass Percent
oxygen	49.2	titanium	0.58
silicon	25.7	chlorine	0.19
aluminum	7.50	phosphorus	0.11
iron	4.71	manganese	0.09
calcium	3.39	carbon	0.08
sodium	2.63	sulfur	0.06
potassium	2.40	barium	0.04
magnesium	1.93	nitrogen	0.03
hydrogen	0.87	fluorine	0.03
		all others	0.49

Tables 4.1 and 4.2 show the distribution of elements on earth and in the human body. These provide an interesting discussion when shown as overhead transparencies in class. Students will already be familiar with many of the element names and are often surprised to find out how much, or how little, of each element is found on earth. It is probably not important that students memorize all the information found in these tables, however.

The sand in Great Sand Dunes National Monument in Colorado is composed of compounds containing the elements silicon and oxygen.

Element 109 has been synthesized, but element 108 has not been reported. So there are presently 108 elements.

uncountable ways, are all made by linking together a few fundamental units to form huge molecules. A nonchemical example is the English language, where hundreds of thousands of words are constructed from only 26 letters. If you take apart the thousands of words in an English dictionary, you will find only these 26 fundamental components. In much the same way, when we take apart all of the substances in the world around us, we find only about one hundred fundamental building blocks—the elements. Compounds are made by combining atoms of the various elements, just as words are constructed from the 26 letters of the alphabet. And just as you had to learn the letters of the alphabet before you learned to read and write, you need to learn the names and symbols of the chemical elements before you can read and write chemistry.

Presently 108 different elements are known, 88 of which occur naturally. (The rest have been made in laboratories.) The elements vary tremendously in abundance. In fact, only 9 elements account for most of the compounds found in the earth's crust. In Table 4.1, the elements are listed in order of abundance (mass percent) in the earth's crust, oceans, and atmosphere. Note that nearly half of the mass is accounted for by oxygen alone. Also note that the 9 most abundant elements account for over 98% of the total mass.

Oxygen, in addition to accounting for about 20% of the earth's atmosphere (where it occurs as O_2 molecules), is also found in virtually all the rocks, sand, and soil on the earth's crust. In these latter materials, oxygen is not present as O_2 molecules but exists in compounds that usually contain silicon and aluminum atoms. The familiar substances of the geological world, such as rocks and sand, contain large groups of silicon and oxygen atoms bound together to form huge clusters.

The comparison of elements forming chemical compounds much as the alphabet forms English words is very powerful. Students can easily understand the analogy, and it provides an explanation for the names of different chemical compounds, many of which are composed of the same elements.

The list of elements found in living matter is very different from that for the earth's crust. Table 4.2 shows the distribution of elements in the human body. Oxygen, carbon, hydrogen, and nitrogen form the basis for all biologically important molecules. Some elements found in the body (called trace elements) are crucial for life, even though they are present in relatively small amounts. For example, chromium helps the body use sugars to provide energy.

One more general comment is important at this point. As we have seen, elements are fundamental to understanding chemistry. However, students are often confused by the many different ways that chemists use the term *element.* Sometimes when we say *element,* we mean a single atom of that element. We might call this the microscopic form of an element. Other times when we use the term *element,* we mean a sample of the element large enough to weigh on a balance. Such a sample contains many, many atoms of the element, and we might call this the macroscopic form of the element. There is yet a further complication. As we will see in more detail in Chapter 5, the macroscopic forms of several elements contain molecules rather than individual atoms as the fundamental components. For example, chemists know that oxygen gas consists of molecules with two oxygen atoms connected together (represented as O—O or more commonly as O_2). Thus when we refer to the element oxygen we might mean a single atom of oxygen, a single O_2 molecule, or a macroscopic sample containing many O_2 molecules. In this text we will try to make clear what we mean when we use the term *element* in a particular case.

Table 4.2   Abundance of Elements in the Human Body

Major Elements	Mass Percent	Trace Elements (in alphabetical order)
oxygen	65.0	arsenic
carbon	18.0	chromium
hydrogen	10.0	cobalt
nitrogen	3.0	copper
calcium	1.4	fluorine
phosphorus	1.0	iodine
magnesium	0.50	manganese
potassium	0.34	molybdenum
sulfur	0.26	nickel
sodium	0.14	selenium
chlorine	0.14	silicon
iron	0.004	vanadium
zinc	0.003	

4.2 Symbols for the Elements

AIM: To introduce the symbols of some elements and suggest how to remember them.

The names of the chemical elements have come from many sources. Often an element's name is derived from a Greek, Latin, or German word that describes some property of the element. For example, gold was originally called *aurum,* a Latin word meaning "shining dawn," and lead was known as *plumbum,* which means "heavy." The names for chlorine and iodine come from Greek words describing their colors, and the name for bromine comes from a Greek word meaning "stench." In addition, it is very common for an element to be named for the place where it was discovered. You can guess where the elements francium, germanium, californium*, and americium* were first found.

We often use abbreviations to simplify the written word. For example, it is much easier to put MA on an envelope than to write out Massachusetts, and we often write USA instead of United States of America. Likewise, chemists have invented a set of abbreviations or **element symbols** for the chemical elements. These symbols usually consist of the first letter or the first two letters of the element names. The first letter is always capitalized, and the second is not. Examples include:

Various forms of the element gold.

fluorine	F	neon	Ne
oxygen	O	silicon	Si
carbon	C		

Sometimes, however, the two letters used are not the first two letters in the name. For example,

| zinc | Zn | cadmium | Cd |
| chlorine | Cl | platinum | Pt |

The symbols for some other elements are based on the original Latin or Greek name.

Current Name	Original Name	Symbol
gold	aurum	Au
lead	plumbum	Pb
sodium	natrium	Na
iron	ferrum	Fe

In the symbol for an element, only the first letter is capitalized.

The only way for students to learn the symbols for the elements is to memorize them. Some symbol pairs, such as Au/Ag, Na/S, K/P, and Mg/Mn are easily confused. Because some students have trouble memorizing, you can suggest they use flash cards with the element name on one side and the symbol on the other.

---

*These elements are made artificially. They do not occur naturally.

When presenting symbols for the elements on overhead transparencies, I provide some alternative groupings for the elements. For example, I make a group of all the elements that have symbols dissimilar to their names, a group of all the elements whose names begin with C, and a group of all elements that have single letters. The alternative groupings break the large list of names and symbols into smaller ones that are easier to memorize.

Figure 4.1
John Dalton (1766–1844) was an English scientist who made his living as a teacher in Manchester. Although Dalton is best known for his atomic theory, he made contributions in many other areas, including meteorology (he recorded daily weather conditions for 46 years, producing a total of 200,000 data entries). A rather shy man, Dalton was colorblind to red (a special handicap for a chemist) and suffered from lead poisoning contracted from drinking stout (strong beer or ale) that had been drawn through lead pipes.

Table 4.3   The Names and Symbols of the Most Common Elements

Element	Symbol	Element	Symbol
aluminum	Al	lithium	Li
antimony (stibium)*	Sb	magnesium	Mg
argon	Ar	manganese	Mn
arsenic	As	mercury (hydragyrum)	Hg
barium	Ba	neon	Ne
bismuth	Bi	nickel	Ni
boron	B	nitrogen	N
bromine	Br	oxygen	O
cadmium	Cd	phosphorus	P
calcium	Ca	platinum	Pt
carbon	C	potassium (kalium)	K
chlorine	Cl	radium	Ra
chromium	Cr	silicon	Si
cobalt	Co	silver (argentium)	Ag
copper (cuprum)	Cu	sodium (natrium)	Na
fluorine	F	strontium	Sr
gold (aurum)	Au	sulfur	S
helium	He	tin (stannum)	Sn
hydrogen	H	titanium	Ti
iodine	I	tungsten (wolfram)	W
iron (ferrum)	Fe	uranium	U
lead (plumbum)	Pb	zinc	Zn

*Where appropriate, the original name is shown in parentheses so that you can see where some of the symbols came from.

A list of the most common elements and their symbols is given in Table 4.3. You can also see the elements represented on a table in the inside front cover of this text. We will explain the form of this table (which is called the periodic table) in later chapters.

4.3   Dalton's Atomic Theory

AIM: To discuss Dalton's theory of atoms.
         To describe and illustrate the law of constant composition.

As scientists of the eighteenth century studied the nature of materials, several things became clear:

1. Most natural materials are mixtures of pure substances.

2. Pure substances are either elements or combinations of elements called compounds.
3. A given compound always contains the same proportions (by mass) of the elements. For example, water *always* contains 8 g of oxygen for every 1 g of hydrogen, and carbon dioxide *always* contains 2.7 g of oxygen for every 1 g of carbon. This principle became known as the **law of constant composition.** It means that a given compound always has the same composition, regardless of where it comes from.

John Dalton (Figure 4.1), an English scientist and teacher, was aware of these observations, and in about 1808 he offered an explanation for them that became known as **Dalton's atomic theory.** The main ideas of this theory (model) can be stated as follows:

Dalton's atomic theory is important and allows the subsequent discussion of compounds and the law of constant composition. The law of constant composition usually needs thorough explanation. It is difficult for students at this stage to understand the relationship between constant masses of each element in a compound and constant numbers of atoms in a compound.

Dalton's Atomic Theory

1. Elements are made of tiny particles called **atoms.**
2. All atoms of a given element are identical.
3. The atoms of a given element are different from those of any other element.
4. Atoms of one element can combine with atoms of other elements to form compounds. A given compound always has the same relative numbers and types of atoms.
5. Atoms are indivisible in chemical processes. That is, atoms are not created or destroyed in chemical reactions. A chemical reaction simply changes the way the atoms are grouped together.

Dalton's model successfully explained important observations such as the law of constant composition. This law makes sense because if a compound always contains the same relative numbers of atoms, it will always contain the same proportions by mass of the various elements.

Like most new ideas, Dalton's model was not accepted immediately. However, Dalton was convinced he was right and *used his model to predict* how a given pair of elements might combine to form more than one compound. For example, nitrogen and oxygen might form a compound containing one atom of nitrogen and one atom of oxygen (written NO), a compound containing two atoms of nitrogen and one atom of oxygen (written N_2O), a compound containing one atom of nitrogen and two atoms of oxygen (written NO_2), and so on (Figure 4.2). When the existence of these substances was verified, it was a triumph for Dalton's model. The fact that Dalton was able to predict correctly the formation of multiple compounds between two elements led to the widespread acceptance of his atomic theory.

NO

NO_2

N_2O

Figure 4.2
Dalton pictured compounds as collections of atoms. Here NO, NO_2, and N_2O are represented. Note that the number of atoms of each type in a molecule is given by a subscript, except that the number 1 is always assumed and never written.

I use the element copper as an example of how Dalton's atomic theory applies to real elements. It is most effective when I bring some pure copper and some simple copper compounds to class.

4.4   Formulas of Compounds

AIM: To explain how a formula describes a compound's composition.

A **compound** is a distinct substance that is composed of the atoms of two or more elements and always contains exactly the same relative masses of those elements. In light of Dalton's atomic theory, this simply means that a compound always contains the same relative *numbers* of atoms of each element. For example, water always contains two hydrogen atoms for each oxygen atom.

The types of atoms and the number of each type in each unit (molecule) of a given compound are conveniently expressed by a **chemical formula.** In a chemical formula the atoms are indicated by the element symbols, and the number of each type of atom is indicated by a subscript, a number that appears to the right of and below the symbol for the element. The formula for water is written H_2O, indicating that each molecule of water contains two atoms of hydrogen and one atom of oxygen (the subscript 1 is always understood and not written). Following are some general principles for writing formulas:

- Each atom present is represented by its element symbol.
- The number of each type of atom is indicated by a subscript written to the right of the element symbol.
- When only one atom of a given type is present, the subscript 1 is not written.

When you introduce chemical formulas, you can remind students that they will not be able to write formulas for compounds unless they have memorized the symbols for the common elements.

EXAMPLE 4.1   Writing Formulas of Compounds

Write the formula for each of the following compounds, listing the elements in the order given.
a.   Each molecule of a compound that has been implicated in the formation of acid rain contains one atom of sulfur and three atoms of oxygen.
b.   Each molecule of a certain compound contains two atoms of nitrogen and five atoms of oxygen.
c.   Each molecule of glucose, a type of sugar, contains six atoms of carbon, twelve atoms of hydrogen, and six atoms of oxygen.

SOLUTION

a.

EXAMPLE 4.1, CONTINUED

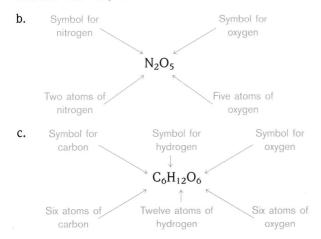

b.
Symbol for nitrogen

Symbol for oxygen

N_2O_5

Two atoms of nitrogen

Five atoms of oxygen

c.
Symbol for carbon

Symbol for hydrogen

Symbol for oxygen

$C_6H_{12}O_6$

Six atoms of carbon

Twelve atoms of hydrogen

Six atoms of oxygen

SELF-CHECK EXERCISE 4.1

Write the formula for each of the following compounds, listing the elements in the order given.

a.   A molecule contains four phosphorus atoms and ten oxygen atoms.
b.   A molecule contains one uranium atom and six fluorine atoms.
c.   A molecule contains one aluminum atom and three chlorine atoms.

4.5   The Structure of the Atom

AIM: To introduce the internal parts of an atom.
To describe Rutherford's experiment to characterize the atom's structure.

Dalton's atomic theory, proposed in about 1808, provided such a convincing explanation for the composition of compounds that it became generally accepted. Scientists came to believe that *elements consist of atoms* and that *compounds are a specific collection of atoms* bound together in some way. But what is an atom like? It might be a tiny ball of matter that is the same throughout with no internal structure—like a ball bearing. Or the atom might be composed of parts—it might be made up of a number of subatomic particles. But if the atom contains parts, there should be some way to break up the atom into its components.

Many scientists pondered the nature of the atom during the 1800s, but it was not until almost 1900 that convincing evidence became available that the atom has a number of different parts.

I introduce this topic by discussing the tremendous scientific feat of determining the structure of atoms that are too small to actually be seen. Students are often amazed that chemists were able to learn so much without aids they take for granted, such as computers.

Glowing Tubes for Signs, Television Sets, and Computers

J.J. Thomson discovered that atoms contain electrons by using a device called a cathode ray tube (often abbreviated CRT today). When he did these experiments, he could not have imagined that he was making television sets and computer monitors possible. A cathode ray tube is a sealed glass tube that contains a gas and has separated metal plates connected to external wires (Figure 4.3). When a source of electrical energy is applied to the metal plates, a glowing beam is produced (Figure 4.4). Thomson became convinced that the glowing gas was caused by a stream of negatively charged particles coming from the metal plate. In addition, because Thomson always got the same kind of negative particles no matter what metal he used, he concluded that all types of atoms must contain these same negative particles (we now call them electrons).

Thomson's cathode ray tube has many modern applications. For example, "neon" signs consist of small-diameter cathode ray tubes containing different kinds of gases to produce various colors. For example, if the gas in the tube is neon, the tube glows with a red–orange color; if argon is present, a blue glow appears. The presence of krypton gives an intense white light.

A television picture tube or computer monitor is also fundamentally a cathode ray tube. In this case the electrons are directed onto a screen containing chemical compounds that glow when struck by fast-moving electrons. The use of various compounds that emit different colors when they are struck by the electrons makes color pictures possible on the screens of these CRTs.

Figure 4.3
Schematic of a cathode ray tube. A stream of electrons passes between the electrodes. The fast-moving particles excite the gas in the tube, causing a glow between the plates.

Diagram labels:
Source of electrical potential
Glass tube
Stream of negative particles (electrons)
(−)   (+)
Gas-filled tube
Metal electrodes

Figure 4.4
A CRT being used to display computer graphics.

Figure 4.5
One of the early models of the atom was the plum pudding model, in which the electrons were pictured as embedded in a positively charged spherical cloud, much as raisins are distributed in an old-fashioned plum pudding.

Figure 4.6
Ernest Rutherford (1871–1937) was born on a farm in New Zealand. In 1895 he placed second in a scholarship competition to attend Cambridge University but was awarded the scholarship when the winner decided to stay home and get married. Rutherford was an intense, hard-driving person who became a master at designing just the right experiment to test a given idea. He was awarded the Nobel Prize in chemistry in 1908.

A physicist in England named J. J. Thomson showed in the late 1890s that the atoms of any element can be made to emit tiny negative particles. (He knew they had a negative charge because he could show that they were repelled by the negative part of an electric field.) Thus he concluded that all types of atoms must contain these negative particles, which are now called **electrons.**

On the basis of his results, Thomson wondered what an atom must be like. Although atoms contain these tiny negative particles, he also knew that whole atoms are not negatively *or* positively charged. Thus he concluded that the atom must also contain positive particles that balance exactly the negative charge carried by the electrons, giving the atom a zero overall charge.

Another scientist pondering the structure of the atom was William Thomson (better known as Lord Kelvin and no relation to J. J. Thomson). Lord Kelvin got the idea (which might have occurred to him during dinner) that the atom might be something like plum pudding (a pudding with raisins randomly distributed throughout). Kelvin reasoned that the atom might be thought of as a uniform "pudding" of positive charge with enough negative electrons scattered within to counterbalance that positive charge (see Figure 4.5). Thus the plum pudding model of the atom came into being.

If you had taken this course in 1910, the plum pudding model would have been the only picture of the atom described. However, our ideas about the atom were changed dramatically in 1911 by a physicist named Ernest Rutherford (Figure 4.6), who learned his physics in J. J. Thomson's laboratory in the late 1890s. By 1911 Rutherford had become a distinguished scientist with many important discoveries to his credit. One of his main areas of interest involved alpha particles (α particles), positively charged particles with a mass approximately 7500 times that of an electron. In studying the flight of these particles through air, Rutherford

It is interesting to show a cathode ray tube in class, if you have one available. You can show how the beam of particles is deflected by a magnet.

Some historians credit J. J. Thomson for the plum pudding model.

Since Plum pudding is not an item in the diet of most American students, it might be better to use the analogy of a muffin with raisins, or a chocolate-chip cookie.

One of Rutherford's co-workers in this experiment was an undergraduate named Ernest Marsden who, like Rutherford, was from New Zealand.

Students are usually impressed whenever I tell them about the role Ernest Marsden played in the Rutherford gold-foil experiment. They often think of chemical discoveries coming only from the "elderly," so that this is an opportune way to show that young people can contribute to the advancement of scientific ideas.

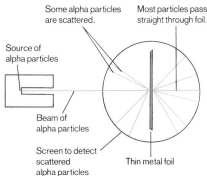

Figure 4.7
Rutherford's experiment on α-particle bombardment of metal foil.

Rutherford's gold-foil experiment is one that students seem to easily understand. It is helpful to organize Rutherford's observations and conclusions into a table so that students appreciate the significance of this experiment.

At this point students do not have enough knowledge to understand how the alpha particles used in the gold-foil experiment are produced. You can either ignore the source of the alpha particles, or if you have the time, you can provide some background information on radioactive decay.

found that some of the α particles were deflected by something in the air. Puzzled by this, he designed an experiment that involved directing α particles toward a thin metal foil. Surrounding the foil was a detector coated with a substance that produced tiny flashes wherever it was hit by an α particle (Figure 4.7). The results of the experiment were very different from those Rutherford anticipated. Although most of the α particles passed straight through the foil, some of the particles were deflected at large angles, as shown in Figure 4.7, and some were reflected backward.

This outcome was a great surprise to Rutherford. (He described this result as comparable to shooting a gun at a piece of paper and having the bullet bounce back.) Rutherford knew that if the plum pudding model of the atom was correct, the massive α particles would crash through the thin foil like cannonballs through paper (as shown in Figure 4.8a). So he expected the α particles to travel through the foil experiencing, at most, very minor deflections of their paths.

Rutherford concluded from these results that the plum pudding model for the atom could not be correct. The large deflections of the α particles could be caused only by a center of concentrated positive charge that would repel the positively charged α particles, as illustrated in Figure 4.8(b). Most of the α particles passed directly through the foil because the atom is mostly open space. The deflected α particles were those that had a "close encounter" with the positive center of the atom, and the few reflected α particles were those that scored a "direct hit" on the positive center. In Rutherford's mind these results could be explained only in terms of a **nuclear atom**—an atom with a dense center of positive charge (the **nucleus**) around which tiny electrons moved in a space that was otherwise empty.

He concluded that the nucleus must have a positive charge to balance the negative charge of the electrons and that it must be small and dense. What was it made of? By 1919 Rutherford concluded that the nucleus of an atom contained what he called protons. A **proton** has the same magnitude (size) of charge as the electron, but its charge is *positive.* We say that the proton has a charge of 1+ and the electron a charge of 1−.

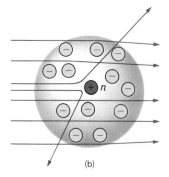

Figure 4.8
(a) The results that the metal foil experiment would have yielded if the plum pudding model had been correct.
(b) Actual results.

(a)

(b)

Rutherford reasoned that the hydrogen atom has a single proton at its center and one electron moving through space at relatively large distance from the proton (the hydrogen nucleus). He also reasoned that other atoms must have nuclei (the plural of *nucleus*) composed of many protons bound together somehow. In addition, Rutherford and a co-worker, James Chadwick, were able to show in 1932 that most nuclei also contain a neutral particle that they named the **neutron.** A neutron is just slightly more massive than a proton but has no charge.

If the atom were expanded to the size of a huge stadium like the Astrodome, the nucleus would be only about as big as a fly at the center.

4.6   Introduction to the Modern Concept of Atomic Structure

AIM: To describe features of subatomic particles.

In the years since Thomson and Rutherford, a great deal has been learned about atomic structure. The simplest view of the atom is that it consists of a tiny nucleus (about 10^{-13} cm in diameter) and electrons that move about the nucleus at an average distance of about 10^{-8} cm from it (Figure 4.9). To visualize how small the nucleus is compared to the size of the atom, consider that if the nucleus were the size of a grape, the electrons would be about one *mile* away. The nucleus contains protons, which have a positive charge equal in magnitude to the electron's negative charge, and neutrons, which have almost the same mass as a proton but no charge. The neutrons' function in the nucleus is not obvious. They may help hold the protons (which repel each other) together to form the nucleus, but we will not be concerned with that here. The relative masses and charges of the electron, proton, and neutron are shown in Table 4.4.

An important question arises at this point: *"If all atoms are composed of these same components, why do different atoms have different chemical properties?"* The answer lies in the number and arrangement of the electrons. The space in which the electrons move accounts for most of the atomic volume. The

I always repeat the information from the text that electrons would be one mile from the nucleus if the nucleus were the size of a grape. Without a point of reference it is difficult for students to appreciate the amount of empty space an atom really contains.

In this model the atom is called a nuclear atom because the positive charge is localized in a small, compact structure (the nucleus) and not spread out uniformly, as in the plum pudding view.

Figure 4.9
A nuclear atom viewed in cross section. (The symbol $\sim$ means approximately.) This drawing does not show the actual scale. The nucleus is actually *much* smaller compared to the size of an atom.

Table 4.4 The Mass and Charge of the Electron, Proton, and Neutron		
Particle	Relative Mass*	Relative Charge
electron	1	1−
proton	1836	1+
neutron	1839	none

*The electron is arbitrarily assigned a mass of 1 for comparison.

Students should memorize the relative masses and charges for electrons, protons, and neutrons in Table 4.4

The difference between absolute and relative masses and charges is difficult for students. Table 4.4 presents numerical values for only relative masses and charges. You might not want to provide absolute values unless your curriculum requires it.

electrons are the parts of atoms that "intermingle" when atoms combine to form molecules. Therefore, the number of electrons a given atom possesses greatly affects the way it can interact with other atoms. As a result, atoms of different elements, which have different numbers of electrons, show different chemical behavior. Although the atoms of different elements also differ in their numbers of protons, it is the number of electrons that really determines chemical behavior. We will discuss how this happens in later chapters.

The *chemistry* of an atom arises from its electrons.

4.7   Isotopes

AIM: To define the terms *isotope, atomic number,* and *mass number.*
To explain the use of the symbol $_Z^A X$ to describe a given atom.

Students learned in Section 4.5 that the overall charge on an atom is zero, but not until now do they learn that the charge is zero because the number of negatively charged electrons always equals the number of positively charged protons.

We have seen that an atom has a nucleus with a positive charge due to its protons and has electrons in the space surrounding the nucleus at relatively large distances from it.

As an example, consider a sodium atom, which has 11 protons in its nucleus. Because an atom has no overall charge, the number of electrons must equal the number of protons. Therefore, a sodium atom has 11 electrons in the space around its nucleus. It is *always* true that a sodium atom has 11 protons and 11 electrons. However, each sodium atom also has neutrons in its nucleus, and different types of sodium atoms exist that have different numbers of neutrons.

All atoms of the same element have the same number of protons (the element's atomic number) and the same number of electrons.

In a free atom, the positive and negative charges always balance to yield a net zero charge.

When Dalton stated his atomic theory in the early 1800s, he assumed all of the atoms of a given element were identical. This idea persisted for over a hundred years, until James Chadwick discovered that the nuclei of most atoms contain neutrons as well as protons. (This is a good example of how a theory changes as new observations are made.) After the discovery of the neutron, Dalton's statement that all atoms of a given element are identical had to be changed to "All atoms of the same element contain the same number of protons and electrons, but atoms of a given element may have different numbers of neutrons."

Now is a good time to point out that at least one part of Dalton's atomic theory was incorrect. All atoms of an element are not exactly alike (as Dalton believed). He had no knowledge of isotopes.

To illustrate this idea, consider the sodium atoms represented in Figure 4.10. These atoms are **isotopes,** or *atoms with the same number of protons but different numbers of neutrons.* The number of protons in a nucleus is called the atom's **atomic number.** The *sum* of the number of neutrons and the number of protons in a given nucleus is called the atom's **mass number.** To specify which of the isotopes of an element we are talking about, we use the symbol

Atomic number: The number of protons.
Mass number: The sum of protons and neutrons.

$$_Z^A X$$

Mass number represents the sum of the numbers of protons and neutrons in an atom. Students often believe that mass number represents the sum of the relative masses of protons and neutrons.

When working with $^A_Z X$ notation, students have a difficult time remembering whether A is the mass or atomic number. You can tell them that isotopes are "top heavy"; that is, the mass number is on the top and the atomic number is on the bottom. For most isotopes, the value of A is greater than the value of Z.

where

   X represents the symbol of the element
   A represents the mass number (number of protons and neutrons)
   Z represents the atomic number (number of protons)

For example, the symbol for one particular type of sodium atom is written

The particular atom represented here is called sodium-23, because it has a mass number of 23. Let's specify the number of each type of subatomic particle. From the atomic number 11 we know that the nucleus contains 11 protons. And because the number of electrons is equal to the number of protons, we know that this atom contains 11 electrons. How many neutrons are present? We can calculate the number of neutrons from the definition of the mass number

$$\text{Mass number} = \text{number of protons} + \text{number of neutrons}$$

or, in symbols,

$$A = Z + \text{number of neutrons}$$

We can isolate (solve for) the number of neutrons by subtracting Z from both sides of the equation

$$A - Z = Z - Z + \text{number of neutrons}$$
$$A - Z = \text{number of neutrons}$$

This is a general result. You can always determine the number of neutrons present in a given atom by subtracting the atomic number from the mass number. In this case ($^{23}_{11}$Na), we know that $A = 23$ and $Z = 11$. Thus

$$A - Z = 23 - 11 = 12 = \text{number of neutrons}$$

In summary, sodium-23 has 11 electrons, 11 protons, and 12 neutrons.

Nucleus
11 protons
12 neutrons

11 electrons

$^{23}_{11}$Na

Nucleus
11 protons
13 neutrons

11 electrons

$^{24}_{11}$Na

Figure 4.10
Two isotopes of sodium. Both have 11 protons and 11 electrons, but they differ in the number of neutrons in their nuclei.

EXAMPLE 4.2   **Interpreting Symbols for Isotopes**

In nature, elements are usually found as a mixture of isotopes. Three isotopes of elemental carbon are $^{12}_6$C (carbon-12), $^{13}_6$C (carbon-13), and $^{14}_6$C (carbon-14). Determine the number of each of the three types of subatomic particles in each of these carbon atoms.

If you do not plan to cover Chapter 19, Radioactivity and Nuclear Energy, you might want to take a few minutes to discuss unstable isotopes and radioactive decay now. The use of carbon-14 to date archaeological objects can be discussed in general terms without mentioning half-life. The fact that carbon-14 decays spontaneously whereas carbon-12 is stable shows a clear difference in behavior among isotopes of a given element.

EXAMPLE 4.2, CONTINUED

SOLUTION

The number of protons and electrons is the same in each of the isotopes and is given by the atomic number of carbon, 6. The number of neutrons can be determined by subtracting the atomic number (Z) from the mass number:

$$A - Z = \text{number of neutrons}$$

The numbers of neutrons in the three isotopes of carbon are

$^{12}_{6}$C: number of neutrons $= A - Z = 12 - 6 = 6$
$^{13}_{6}$C: number of neutrons $= 13 - 6 = 7$
$^{14}_{6}$C: number of neutrons $= 14 - 6 = 8$

In summary,

Symbol	Number of Protons	Number of Electrons	Number of Neutrons
$^{12}_{6}$C	6	6	6
$^{13}_{6}$C	6	6	7
$^{14}_{6}$C	6	6	8

It is interesting to illustrate $^{A}_{Z}$X notation by showing the relative abundance of naturally occurring isotopes of elements such as hydrogen or nitrogen. Students can determine the numbers of electrons, protons, and neutrons in each of the isotopes.

SELF-CHECK EXERCISE 4.2

Give the number of protons, neutrons, and electrons in the atom symbolized by $^{90}_{38}$Sr. Strontium-90 occurs in fallout from nuclear testing. It can accumulate in bone marrow and may cause leukemia and bone cancer.

SELF-CHECK EXERCISE 4.3

Give the number of protons, neutrons, and electrons in the atom symbolized by $^{201}_{80}$Hg.

EXAMPLE 4.3   Writing Symbols for Isotopes

Write the symbol for the magnesium atom (atomic number of 12) with a mass number of 24. How many electrons and how many neutrons does this atom have?

SOLUTION

The atomic number 12 means the atom has 12 protons. The element magnesium is symbolized by Mg. The atom is represented as

$$^{24}_{12}\text{Mg}$$

EXAMPLE 4.3, CONTINUED

and is called magnesium-24. Because the atom has 12 protons, it must also have 12 electrons. The mass number gives the total number of protons and neutrons, which means that this atom has 12 neutrons (24 − 12 = 12).

EXAMPLE 4.4   Calculating Mass Number

Write the symbol for the silver atom ($Z = 47$) that has 61 neutrons.

SOLUTION

The element symbol is ^A_ZAg, where we know that $Z = 47$. We can find A from its definition, $A = Z +$ number of neutrons. In this case,

$$A = 47 + 61 = 108$$

The complete symbol for this atom is $^{108}_{47}$Ag.

SELF-CHECK EXERCISE 4.4

Give the symbol for the phosphorus atom ($Z = 15$) that contains 17 neutrons.

Magnesium burns in oxygen to give a bright, white flame.

4.8 Introduction to the Periodic Table

AIM: To introduce various features of the periodic table.
To describe briefly metals, nonmetals, and metalloids.

In any room where chemistry is taught or practiced, you are almost certain to find a chart called the **periodic table** hanging on the wall. This chart shows all of the known elements and gives a good deal of information about each. As our study of chemistry progresses, the usefulness of the periodic table will become more obvious. This section will simply introduce it.

A simple version of the periodic table is shown in Figure 4.11 (p. 108). Note that each box of this table contains a number written over one or two letters. The letters are the symbols for the elements. The number shown above each symbol is the atomic number (the number of protons and also the number of electrons) for that element. For example, carbon (C) has atomic number 6:

Although the reasons why elements in the periodic table are arranged as they are are not introduced until Chapter 11, students need to use the periodic table to locate groups of elements and to separate metals from nonmetals now.

```
6
C
```

Stress that the number printed above the element symbol in the simple periodic table in this section is the atomic number, not the mass number.

Figure 4.11
The periodic table.

and lead (Pb) has atomic number 82:

82
Pb

Mendeleev actually arranged the elements in order of increasing atomic weight rather than atomic number.

Note that the elements are listed on the periodic table in order of increasing atomic number. They are also arranged in specific horizontal rows and vertical columns. The elements were first arranged in this way in 1869 by Dimitri Mendeleev, a Russian scientist. Mendeleev arranged the elements in this way because of similarities in the chemical properties of various "families" of elements. chlorine are reactive gases that form similar compounds. It was also known that sodium and potassium behave very similarly. Thus the name *periodic table* refers to the fact that as we increase the atomic numbers, every so often an element occurs with properties similar to those of an earlier (lower-atomic-number) element. For example, the elements

9 F
17 Cl
35 Br
53 I
85 At

all show similar chemical behavior and so are listed vertically, as a "family" of elements.

These families of elements with similar chemical properties that lie in the same vertical column on the periodic table are called **groups.** Groups are often referred to by the number over the column (see Figure 4.11). Many of the groups also have names. For example, the first column of elements (Group 1) has the name **alkali metals.** The Group 2 elements are called **alkaline earth metals,** the Group 7 elements are the **halogens,** and the elements in Group 8 are called the **noble gases.** A large collection of elements that spans many vertical columns consists of the **transition metals.**

Most of the elements are **metals.** Metals have the following characteristic physical properties:

Physical Properties of Metals

1. Efficient conduction of heat and electricity
2. Malleability (they can be hammered into thin sheets)
3. Ductility (they can be pulled into wires)
4. A lustrous (shiny) appearance

Students should memorize the names and locations of the common groups in the periodic table. I tell students they need not memorize the names of the elements in each group. They can look up that information in the periodic table itself.

Throughout the text, we will highlight the location of various elements by presenting a small version of the periodic table.

Samples of the halogens make an interesting display when you are introducing the groups of the periodic table.

Rolls of copper pipe for use in plumbing systems. Copper is a transition metal.

Aluminum, which has metallic properties, is widely used for aircraft construction.

Nonmetals sometimes have one or more metallic properties. For example, solid iodine is lustrous, and graphite (a form of pure carbon) is a conductor of electricity.

For example, copper is a typical metal. It is lustrous (although it tarnishes readily); it is an excellent conductor of electricity (it is widely used in electrical wires); and it is readily formed into various shapes, such as pipes for water systems. Copper is one of the transition metals—the metals shown in the center of the periodic table. Iron, aluminum, and gold are other familiar elements that have metallic properties. All of the elements shown to the left and below of the heavy "stair-step" black line in Figure 4.11 are classified as metals, except for hydrogen (Figure 4.12).

The relatively small number of elements that appear in the upper right-hand corner of the periodic table (to the right of the heavy line in Figures 4.11 and 4.12) are called **nonmetals.** Nonmetals generally lack those properties that characterize metals and show much more variation in their properties than metals do. Whereas almost all metals are solids at normal temperatures, many nonmetals

Figure 4.12
The elements classified as metals and as nonmetals.

(such as nitrogen, oxygen, chlorine, and neon) are gaseous and one (bromine) is a liquid. Several nonmetals (such as carbon, phosphorus, and sulfur) are also solids.

The elements that lie close to the "stair-step" line in Figure 4.11 often show a mixture of metallic and nonmetallic properties. These elements, which are called **metalloids** or **semimetals,** include boron, silicon, germanium, arsenic, antimony, and tellurium.

As we continue our study of chemistry, we will see that the periodic table is a valuable tool for organizing accumulated knowledge and that it helps us predict the properties we expect a given element to exhibit. We will also come to understand why there are groups of elements with similar chemical properties.

Elemental sulfur occurs in large deposits in the southern United States.

EXAMPLE 4.5   Interpreting the Periodic Table

For each of the following elements, use the periodic table in the front of the book to give the symbol and atomic number and to specify whether the element is a metal or a nonmetal. Also give the named family to which the element belongs (if any).

a.   iodine
b.   magnesium

c.   gold
d.   lithium

SOLUTION

a.   Iodine (symbol I) is element 53 (its atomic number is 53). Iodine lies to the right of the stair-step line in Figure 4.12 and thus is a nonmetal. Iodine is a member of Group 7, the family of halogens.
b.   Magnesium (symbol Mg) is element 12 (atomic number 12). Magnesium is a metal and is a member of the alkaline earth metal family (Group 2).
c.   Gold (symbol Au) is element 79 (atomic number 79). Gold is a metal and is not a member of a named vertical family. It is classed as a transition metal.
d.   Lithium (symbol Li) is element 3 (atomic number 3). Lithium is a metal in the alkali metal family (Group 1).

Students are almost always confused about the identity of the metalloids. The elements to the tops and sides of the three highest "steps," with the exception of aluminum, are usually considered metalloids.

If you have a piece of silicon or some other metalloid, bring it to class. Silicon looks very much like a metal, yet is brittle. By this time, students have seen samples of both metals and nonmetals, but have probably not had much exposure to the metalloids.

SELF-CHECK EXERCISE 4.5

Give the symbol and atomic number for each of the following elements. Also indicate whether each element is a metal or a nonmetal and whether it is a member of a named family.

a.   argon
b.   chlorine

c.   barium
d.   cesium

CHAPTER REVIEW

Key Terms

element symbols (p. 95)
law of constant composition (p. 97)
Dalton's atomic theory (p. 97)
atom (p. 97)
compound (p. 98)
chemical formula (p. 98)
electron (p. 101)
nuclear atom (p. 102)
nucleus (p. 102)
proton (p. 102)
neutron (p. 103)
isotopes (p. 104)

atomic number, Z (p. 104)
mass number, A (p. 104)
periodic table (p. 107)
groups (p. 109)
alkali metals (p. 109)
alkaline earth metals (p. 109)
halogens (p. 109)
noble gases (p. 109)
transition metals (p. 109)
metals (p. 109)
nonmetals (p. 110)
metalloids (semimetals) (p. 111)

Summary

1. Of the 108 different elements now known, only 9 account for about 98% of the total mass of the earth's crust, oceans, and atmosphere. In the human body, oxygen, carbon, hydrogen, and nitrogen are the most abundant elements.

2. Elements are represented by symbols that usually consist of the first one or two letters of the element's name. Sometimes, however, the symbol is taken from the element's original Latin or Greek name.

3. The law of constant composition states that a given compound always contains the same proportions by mass of the elements of which it is composed.

4. Dalton accounted for this law with his atomic theory. He postulated that all elements are composed of atoms; that all atoms of a given element are identical, but that atoms of different elements are different; that chemical compounds are formed when atoms combine; and that atoms are not created or destroyed in chemical reactions.

5. A compound can be represented by a chemical formula that uses the symbol for each type of atom and gives the number of each type of atom that appears in a molecule of the compound.

6. Atoms consist of a nucleus containing protons and neutrons, surrounded by electrons that occupy a large volume relative to the size of the nucleus. Electrons have a relatively small mass (1/1840 of the proton mass) and a negative charge. Protons have a positive charge equal in magnitude (but opposite in sign) to that of the electron. A neutron has a slightly greater mass than the proton but no charge.

7. Isotopes are atoms with the same number of protons but different numbers of neutrons.

8. The periodic table displays the elements in rows and columns in order of increasing atomic number. Elements that have similar chemical properties fall into vertical columns called groups. Most of the elements are metals. These occur on the left-hand side of the periodic table; the nonmetals appear on the right-hand side.

Questions and Problems

All even-numbered exercises have answers in the back of this book and solutions in the Solutions Guide.

4.1 The Elements

QUESTIONS

1. The ancient Greeks believed that all matter was composed of four fundamental substances: earth, air, fire, and water. How does this early conception of matter compare with our modern theories about matter?

2. Although they were not able to transform base metals into gold, what contributions did the alchemists make to the development of Chemistry?

3. In addition to his important work on the properties of gases, what other valuable contributions did Robert Boyle make to the development of the study of chemistry?

4. How many elements are presently known? How many of these elements occur naturally, and how many are synthesized artificially? What are the most common elements present on the earth?

5. Oxygen, the most abundant element on the earth, makes up a large percentage of the atmosphere. Where else is oxygen found? Is oxygen found more commonly as an element or in compounds?

6. What are the most abundant elements found in living creatures? Are these elements also the most abundant elements found in the nonliving world?

4.2 Symbols for the Elements

Note: Refer to the tables on the inside front cover when appropriate.

QUESTIONS

7. Give the symbols and names for the elements whose chemical symbols consist of only one letter.

8. In some cases, the symbol of an element does not seem to bear any relationship to the name we use for the element. Generally, the symbol for such an element is based on its name in another language. Give the symbols and names for five examples of such elements.

9. Give the chemical symbol for each of the following elements.
   a. tin
   b. lead
   c. carbon
   d. lithium
   e. cobalt
   f. copper

10. Give the chemical symbol for each of the following elements.
   a. hydrogen
   b. iron
   c. magnesium
   d. calcium
   e. gold
   f. helium

11. Give the chemical symbol for each of the following elements.
   a. aluminum
   b. arsenic
   c. argon
   d. antimony
   e. fluorine

12. Give the chemical symbol for each of the following elements.
   a. nitrogen
   b. oxygen
   c. manganese
   d. mercury
   e. neon
   f. nickel

13. For each of the following chemical symbols, give the name of the corresponding element.
    a. U
    b. Fe
    c. Au
    d. Sn
    e. Sb
    f. Si
    g. Sr
    h. S

14. For each of the following chemical symbols, give the name of the corresponding element.
    a. Co
    b. Ag
    c. Cl
    d. Al
    e. Zn
    f. Pt
    g. Cr
    h. Na

4.3 Dalton's Atomic Theory

QUESTIONS

15. Indicate whether each of the following statements is true or false. If a statement is false, correct the statement so that it becomes true.
    a. Most materials occur in nature as pure substances.
    b. A given compound usually contains the same relative number of atoms of its various elements.
    c. Atoms are made up of tiny particles called molecules.

16. Indicate whether each of the following statements is true or false. If a statement is false, correct the statement so that it becomes true.
    a. According to Dalton, the atoms of a given element are identical to the atoms of other related elements.
    b. According to Dalton, atoms may divide into smaller particles during some chemical reaction processes.
    c. Dalton's atomic theory was immediately accepted by scientists worldwide.

4.4 Formulas of Compounds

QUESTIONS

17. What is a compound?

18. A given compound always contains the same relative masses of its constituent elements. How is this related to the relative numbers of each kind of atom present?

19. Write the formula for each of the following substances, listing the elements in the order given.
    a. a molecule containing one phosphorus atom and three chlorine atoms
    b. a molecule containing two boron atoms and six hydrogen atoms
    c. a compound containing one calcium atom for every two chlorine atoms
    d. a molecule containing one carbon atom and four bromine atoms
    e. a compound containing two iron atoms for every three oxygen atoms
    f. a molecule containing three hydrogen atoms, one phosphorus atom, and four oxygen atoms

20. Write the formula for each of the following substances, listing the elements in the order given.
    a. a molecule containing one sulfur atom and two oxygen atoms
    b. a compound containing two iron atoms for every three sulfur atoms
    c. a compound containing an equal number of iron and oxygen atoms
    d. a molecule containing two carbon atoms, four hydrogen atoms, and two chlorine atoms
    e. a compound containing three calcium atoms for every two nitrogen atoms
    f. a benzene molecule, which contains six carbon atoms and six hydrogen atoms

4.5 The Structure of the Atom

QUESTIONS

21. Indicate whether each of the following statements is true or false. If a statement is false, correct the statement so that it becomes true.
    a. In his cathode ray tube experiments, J. J. Thomson obtained beams of different types of particles whose nature depended on which gas was contained in the tube.
    b. Thomson assumed that there must be positively charged particles in the atom, since isolated atoms have no overall charge.
    c. In the plum pudding model of the atom, the atom was envisioned as a sphere of negative charge in which positively charged electrons were randomly distributed.

22. Indicate whether each of the following statements is true or false. If the statement is false, correct the statement so that it becomes true.
    a. Rutherford's bombardment experiments with metal foil suggested that the alpha particles were being deflected by coming near a large, negatively charged atomic nucleus.
    b. The proton and the electron have similar masses but opposite electrical charges.
    c. Some atoms also contain neutrons, which are slightly heavier than protons but carry no charge.

4.6 Introduction to the Modern Concept of Atomic Structure

QUESTIONS

23. The nucleus of an atom contains _____, which are positively charged.
24. The nucleus of an atom may also contain _____, which are uncharged.
25. Compared to the diameter of the atomic nucleus, the average distance of the electrons from the nucleus is relatively (small/large).
26. The proton and the (electron/neutron) have almost equal masses. The proton and the (electron/neutron) have charges that are equal in magnitude but opposite in nature.
27. An average atomic nucleus has a diameter of about _____ m.
28. Although the nucleus of an atom is very important, it is the _____ of the atom that determine its chemical properties.

4.7 Isotopes

QUESTIONS

29. True or false? Atoms that have the same number of neutrons but different numbers of protons are called isotopes.
30. True or false? The mass number of a nucleus represents the number of protons in the nucleus.
31. For an atom, the number of protons and electrons is (different/the same).
32. The _____ number represents the sum of the number of protons and neutrons in a nucleus.
33. How did Dalton's atomic theory have to be modified after the discovery that several isotopes of an element may exist?
34. Are all atoms of the same element identical? If not, how can they differ?
35. For each of the following atomic numbers, write the name and chemical symbol of the corresponding element. (Refer to Figure 4.11).
    a. 46   e. 5
    b. 78   f. 4
    c. 15   g. 56
    d. 94   h. 83
36. For each of the following atomic numbers, write the name and chemical symbol of the corresponding element. (Refer to Figure 4.11)
    a. 38   e. 48
    b. 19   f. 13
    c. 92   g. 27
    d. 3   h. 53
37. Write the atomic symbol ($_Z^A$X) for each of the isotopes described below.
    a. $Z = 8$, number of neutrons = 9
    b. the isotope of chlorine in which $A = 37$
    c. $Z = 27$, $A = 60$
    d. number of protons = 26, number of neutrons = 31
    e. the isotope of I with a mass number of 131
    f. $Z = 3$, number of neutrons = 4
38. Write the atomic symbol ($_Z^A$X) for each of the isotopes described below.
    a. $Z = 13$, number of neutrons = 14
    b. $Z = 13$, $A = 27$
    c. the isotope of aluminum with 14 neutrons in its nucleus
    d. the isotope with 16 protons and 16 neutrons
    e. the isotope of sulfur with mass number 32
    f. $Z = 16$, $A = 16$
39. How many protons and neutrons are contained in the nucleus of each of the following atoms? In an uncharged atom of each element, how many electrons are present?
    a. $_{12}^{24}Mg$ c. $_{21}^{45}Sc$ e. $_{24}^{53}Cr$
    b. $_{16}^{34}O$ d. $_{24}^{52}Cr$ f. $_{24}^{54}Cr$
40. How many protons and neutrons are contained in the nucleus of each of the following atoms? In an uncharged atom of each element, how many electrons are present?
    a. $_{27}^{60}Co$ c. $_{4}^{10}Be$ e. $_{11}^{23}Na$
    b. $_{16}^{33}S$ d. $_{18}^{40}Ar$ f. $_{36}^{84}Kr$

41. Complete the following table.

Name	Symbol	Atomic Number	Mass Number	Neutrons
sodium	___	11	23	___
nitrogen	$^{15}_{7}N$	___	___	___
___	$^{136}_{56}Ba$	___	___	___
lithium	___	___	___	6
boron	___	5	11	___

42. Complete the following table.

Name	Neutrons	Atomic Number	Mass Number	Symbol
chlorine	18	___	___	___
neon	10	___	___	___
helium	2	___	___	___
sodium	12	___	___	___
calcium	21	___	___	___

4.8 Introduction to the Periodic Table

QUESTIONS

43. How are the elements ordered in the periodic table?

44. True or false? The horizontal rows in the periodic table are referred to as "groups" or "families" because these elements have similar chemical properties.

45. List the characteristic physical properties that distinguish the metallic elements from the nonmetallic elements.

46. Where are the metallic elements found on the periodic table? Are there more metallic elements or nonmetallic elements?

47. Under ordinary conditions, what physical state is most common for the metallic elements? Are there any common exceptions?

48. Give several examples of nonmetallic elements that occur in the gaseous state under ordinary conditions.

49. Under ordinary conditions, only a few pure elements occur as liquids. Give an example of a metallic and a nonmetallic element that ordinarily occur as liquids.

50. Some elements show both metallic and nonmetallic properties; these elements are called _____.

51. Write the number and name (if any) of the group (family) to which each of the following elements belongs.
a. cesium
b. Ra
c. Rn
d. chlorine
e. strontium
f. Xe
g. Rb

52. Write the number and name (if any) of the group (family) to which each of the following elements belongs.
a. krypton
b. bromine
c. K
d. aluminum
e. Na
f. barium
g. Ne
h. fluorine

53. For each of the following elements, use the table on the inside front cover of the book to give the chemical symbol and atomic number and to specify whether the element is a metal or a nonmetal. Also give the named family to which the element belongs (if any).
a. rubidium
b. magnesium
c. tellurium
d. astatine

54. For each of the following elements, use the table on the inside front cover of the book to give the chemical symbol and atomic number, and to specify whether the element is a metal or a nonmetal. Also give the named family to which the element belongs (if any).
a. cesium
b. iodine
c. radium
d. xenon

Additional Problems

55. For each of the following elements, give the chemical symbol and the atomic number.
    a. sulfur
    b. sodium
    c. silver
    d. strontium
    e. zinc
    f. mercury
    g. neon
    h. radium

56. Give the group number (if any) in the periodic table for the elements listed in problem 55. If the group has a family name, give that name.

57. List the names, symbols, and atomic numbers of the top four elements in Groups 1, 2, 6, and 7.

58. List the names, symbols, and atomic numbers of the top four elements in Groups 3, 5, and 8.

59. What is the difference between the atomic number and the mass number of an element? Can atoms of two different elements have the same atomic number? Could they have the same mass number? Why or why not?

60. Which subatomic particles contribute most to the atom's mass? Which subatomic particles determine the atom's chemical properties?

61. Is it possible for the same two elements to form more than one compound? Is this consistent with Dalton's atomic theory? Give an example.

62. Carbohydrates, a class of compounds containing the elements carbon, hydrogen, and oxygen, were originally thought to contain one water molecule (H_2O) for each carbon atom present. The carbohydrate glucose contains six carbon atoms. Write a general formula showing the relative numbers of each type of atom present in glucose.

63. When iron rusts in moist air, the product is typically a mixture of two iron–oxygen compounds. In one compound, there is an equal number of iron and oxygen atoms. In the other compound, there are three oxygen atoms for every two iron atoms. Write the formulas for the two iron oxides.

64. How many protons and neutrons are contained in the nucleus of each of the following atoms? For an atom of the element, how many electrons are present?
    a. $^{63}_{29}Cu$
    b. $^{80}_{35}Br$
    c. $^{24}_{12}Mg$

65. Though the common isotope of aluminum has a mass number of 27, isotopes of aluminum have been isolated (or prepared in nuclear reactors) with mass numbers of 24, 25, 26, 28, 29, and 30. How many neutrons are present in each of these isotopes? Why are they all considered aluminum atoms, even though they differ greatly in mass? Write the atomic symbol for each isotope.

66. The principal goal of alchemists was to convert cheaper, more common metals into gold. Considering that gold had no particular practical uses (for example, it was too soft to be used for weapons), why do you think early civilizations placed such emphasis on the value of gold?

67. How did Robert Boyle define an element?

68. Give the chemical symbol for each of the following elements.
    a. barium
    b. bromine
    c. bismuth
    d. boron
    e. potassium
    f. phosphorus

69. Give the chemical symbol for each of the following elements.
    a. chromium
    b. cadmium
    c. iodine
    d. chlorine
    e. platinum

70. Give the chemical symbol for each of the following elements.
    a. zinc
    b. uranium
    c. tungsten
    d. strontium
    e. sodium
    f. silver

71. Give the chemical symbol for each of the following elements.
    a. strontium
    b. sodium
    c. silver

72. For each of the following chemical symbols, give the name of the corresponding element.
    a. B
    b. Ra
    c. As
    d. Ca
    e. Hg
    f. Br
    g. F
    h. P

73. For each of the following chemical symbols, give the name of the corresponding element.
    a. Ba
    b. K
    c. Cd
    d. Li
    e. I
    f. N
    g. Cu
    h. O

74. Write the simplest formula for each of the following substances, listing the elements in the order given.
    a. a molecule containing one carbon atom and two oxygen atoms
    b. a compound containing one aluminum atom for every three chlorine atoms
    c. perchloric acid, which contains one hydrogen atom, one chlorine atom, and four oxygen atoms
    d. a molecule containing one sulfur atom and six chlorine atoms

75. For each of the following atomic numbers, write the name and chemical symbol of the corresponding element. (Refer to Figure 4.11).

a. 7   e. 22
b. 10   f. 18
c. 11   g. 36
d. 28   h. 54

76. Write the atomic symbol ($^A_Z X$) for each of the isotopes described below.
   a. $Z = 6$, number of neutrons $= 7$
   b. the isotope of carbon with a mass number of 13
   c. $Z = 6$, $A = 13$
   d. $Z = 19$, $A = 44$
   e. the isotope of calcium with a mass number of 41
   f. the isotope with 19 protons and 16 neutrons

77. How many protons and neutrons are contained in the nucleus of each of the following atoms? In an atom of each element, how many electrons are present?

a. $^{41}_{22}Ti$ c. $^{76}_{32}Ge$ e. $^{75}_{33}As$
b. $^{64}_{30}Zn$ d. $^{86}_{36}Kr$ f. $^{41}_{19}K$

78. Complete the following table.

Symbol	Protons	Neutrons	Mass Number
$^{41}_{20}Ca$	___	___	___
___	25	30	___
___	47	___	109
$^{45}_{21}Sc$	___	___	___

79. For each of the following elements, use the table on the inside front cover of the book to give the chemical symbol and atomic number and to specify whether the element is a metal or a nonmetal. Also give the named family to which the element belongs (if any).
   a. carbon   c. radon
   b. selenium   d. beryllium

Solutions to Self-Check Exercises

SELF-CHECK EXERCISE 4.1

a. P_4O_{10} b. UF_6 c. $AlCl_3$

SELF-CHECK EXERCISE 4.2

In the symbol $^{90}_{38}Sr$, the number 38 is the atomic number, which represents the number of protons in the nucleus of a strontium atom. Because the atom is neutral overall, it must also have 38 electrons. The number 90 (the mass number) represents the number of protons plus the number of neutrons. Thus the number of neutrons is $A - Z = 90 - 38 = 52$.

SELF-CHECK EXERCISE 4.3

The atom $^{201}_{80}Hg$ has 80 protons, 80 electrons, and $201 - 80 = 121$ neutrons.

SELF-CHECK EXERCISE 4.4

The atomic number for phosphorus is 15 and the mass number is $15 + 17 = 32$. Thus the symbol for the atom is $^{32}_{15}P$.

SELF-CHECK EXERCISE 4.5

Element	Symbol	Atomic Number	Metal or Nonmetal	Family Name
a. argon	Ar	18	nonmetal	noble gas
b. chlorine	Cl	17	nonmetal	halogen
c. barium	Ba	56	metal	alkaline earth metals
d. cesium	Cs	55	metal	alkali metals

CHAPTER

5 Elements, Ions, and Nomenclature

A natural salt flat in Ethiopia.

CONTENTS

There are a many items in this chapter that must be memorized. These facts are part of the basic chemistry vocabulary students will build on during the rest of the course.

Two natural forms of silver: Leaf silver and Silver shot.

A gold nugget weighing 13 lb, 7 oz, that came to be called Tom's Baby, was found by Tom Grove near Breckenridge, Colorado, on July 23, 1887.

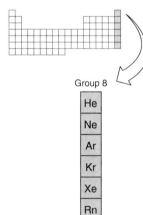

Group 8

| He |
| Ne |
| Ar |
| Kr |
| Xe |
| Rn |

Recall that a molecule is a collection of atoms that behaves as a unit. Molecules are always electrically neutral (zero charge).

The only elemental hydrogen found naturally on earth occurs in the exhaust gases of volcanoes (see photo, p. 122).

In the first four chapters of this text, we spent most of our time discussing the nature of matter, how we learn about it, and how we try to construct models to explain why it behaves the way it does. In Chapter 3, we began learning the language of chemistry. In this chapter we will continue to learn about the chemical nature of elements and compounds and about the language we use to describe chemical compounds. This should help you understand more clearly the things you may be doing and seeing in the laboratory.

5.1   Natural States of the Elements

AIM: To describe the natures of the common elements.

As we have noted, the matter around us consists mainly of mixtures. Most often these mixtures contain compounds, in which atoms from different elements are bound together. Most elements are quite reactive: their atoms tend to combine with those of other elements to form compounds. Thus we do not often find elements in nature in pure form—uncombined with other elements. However, there are notable exceptions. The gold nuggets found at Sutter's Mill in California that launched the Gold Rush in 1849 are virtually pure elemental gold. And platinum and silver are often found in nearly pure form.

Gold, silver, and platinum are members of a class of metals called *noble metals* because they are relatively unreactive. (The term *noble* implies a class set apart.)

Other elements that appear in nature in the uncombined state are the elements in Group 8: helium, neon, argon, krypton, xenon, and radon. Because the atoms of these elements do not combine readily with those of other elements, we call them the **noble gases.** For example, helium gas is found in uncombined form in underground deposits with natural gas.

When we take a sample of air (the mixture of gases that constitute the earth's atmosphere) and separate it into its components, we find several pure elements present. One of these is argon. Argon gas consists of a collection of separate argon atoms, as shown in Figure 5.1.

Air also contains nitrogen gas and oxygen gas. When we examine these two gases, however, we find that they do not contain single atoms, as argon does, but instead contain **diatomic molecules:** molecules made up of *two atoms,* as represented in Figure 5.2. In fact, any sample of elemental oxygen gas at normal temperatures contains O_2 molecules. Likewise, nitrogen gas always contains N_2 molecules.

Hydrogen is another element that forms diatomic molecules. Though virtually all of the hydrogen found on earth is present in compounds with other elements (such as with oxygen in water), when hydrogen is prepared as a free element it contains diatomic H_2 molecules. For example, an electric current can

When diatomic molecules react to form compounds with other atoms, oxygen in the compounds often has a subscript other than two. It is hard for students to accept this because we have not yet discussed bonding. You can tell them that they will understand later why these atoms exist in pairs in the free state, yet can form other combinations with atoms of different elements.

Figure 5.1
Argon gas consists of a collection of separate argon atoms.

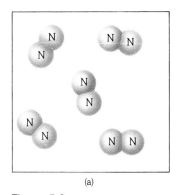

(a)   (b)

Figure 5.2
Gaseous nitrogen and oxygen contain diatomic (two-atom) molecules. (a) Nitrogen gas contains N—N (N_2) molecules. (b) Oxygen gas contains O—O (O_2) molecules.

be used to decompose water (see Figure 5.3 below and Figure 3.3 on p. 67) into elemental hydrogen and oxygen containing H_2 and O_2 molecules, respectively.

Several other elements, in addition to hydrogen, nitrogen, and oxygen, exist as diatomic molecules. For example, when sodium chloride is melted and subjected to an electric current, chlorine gas is produced (along with sodium metal). This chemical change is represented in Figure 5.4 (p. 122). Chlorine gas is a pale green gas that contains Cl_2 molecules.

Chlorine is a member of Group 7, the halogen family. All the elemental forms of the Group 7 elements contain diatomic molecules. Fluorine is a pale yellow gas containing F_2 molecules. Bromine is a brown liquid made up of Br_2 molecules. Iodine is a lustrous, purple solid that contains I_2 molecules.

Table 5.1 on the following page lists the elements that contain diatomic molecules in their pure, elemental forms.

So far we have seen that several elements are gaseous in their elemental forms at normal temperatures (~25 °C). The noble gases (the Group 8 elements)

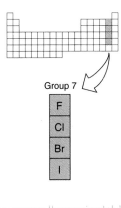

Group 7

F
Cl
Br
I

~ means "approximately"

Water molecules   Electric current   Diatomic hydrogen molecules + Diatomic oxygen molecule

Figure 5.3
The decomposition of two water molecules (H_2O) to form two hydrogen molecules (H_2) and an oxygen molecule (O_2). Note that only the grouping of the atoms changes in this process; no atoms are created or destroyed. There must be the same number of H atoms and O atoms before and after the process. Thus the decomposition of two H_2O molecules (containing four H atoms and two O atoms) yields one O_2 molecule (containing two O atoms) and two H_2 molecules (containing a total of four H atoms).

It is best for students to memorize the short list of diatomic molecules. Later, they will learn why some elements are diatomic.

Platinum is a noble metal used in jewelry and in many industrial processes.

Table 5.1 Elements That Exist as Diatomic Molecules in Their Elemental Forms

Element	Elemental State at 25 °C	Molecule Present
hydrogen	colorless gas	H_2
nitrogen	colorless gas	N_2
oxygen	pale blue gas	O_2
fluorine	pale yellow gas	F_2
chlorine	yellow-green gas	Cl_2
bromine	reddish-brown liquid	Br_2
iodine	lustrous, dark purple solid	I_2

(a)

(b)

(c)

Figure 5.4

(a) Sodium chloride (common table salt) can be decomposed by an electric current to give (b) chlorine gas and (c) sodium metal, shown here being cut by a knife.

A major volcano eruption, such as that of Mt. Pinetubo in the Philippines in 1991, sends huge quantities of ash and gaseous substances into the atmosphere.

contain individual atoms, whereas several other gaseous elements contain diatomic molecules (H_2, N_2, O_2, F_2, and Cl_2).

Only two elements are liquids in their elemental forms at 25 °C: the nonmetal bromine (containing Br_2 molecules) and the metal mercury. The metals gallium and cesium almost qualify in this category; they are solids at 25 °C, but both melt at ~30 °C.

The other elements are solids in their elemental forms at 25 °C. For metals these solids contain large numbers of atoms packed together much like marbles in a jar (see Figure 5.5).

The structures of solid nonmetallic elements are more varied than those of metals. In fact, different forms of the same element often occur. For example, solid carbon occurs in three forms. Different forms of a given element are called *allotropes.* The three allotropes of carbon are the familiar diamond and graphite forms plus a form that has only recently been discovered called *buckminster-fullerene.* These elemental forms have very different properties because of their different structures (see Figure 5.6). Diamond is the hardest natural substance

We write the formulas of the diatomic molecules and the noble gases just as they exist in nature. But we write the formulas of metals as individual atoms even though in the free state they exist as many atoms packed together.

Figure 5.5
In solid metals, the spherical atoms are packed closely together.

known and is often used for industrial cutting tools. Diamonds are also valued as gemstones. Graphite, on the other hand, is a rather soft material useful for writing (pencil "lead" is really graphite) and (in the form of a powder) for lubricating locks. The rather odd name given to buckminsterfullerene comes from the structure of the C_{60} molecules that comprise it. The soccer-ball like structure contains 5- and 6-member rings reminiscent of the structure of geodesic domes suggested by the late industrial designer, Buckminster Fuller. Other "fullerenes" containing molecules with more than 60 carbon atoms have also recently been discovered, leading to a new area of chemistry.

Bromine (Br_2) is a reddish brown liquid.

Figure 5.6
The three solid elemental (allotropes) forms of carbon: (a) diamond, (b) graphite, and (c) buckminsterfullerene. The representation of diamond and graphite are just fragments of much larger structures that extend in all directions from the parts shown here. Buckminsterfullerene contains C_{60} molecules, of which one is shown.

Graphite and diamond, two forms of carbon.

(a)
Diamond

(b)
Graphite

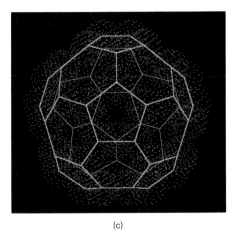

(c)

By showing models of graphite and diamond, you can help students see why graphite is slippery and why diamonds are hard.

CHEMISTRY IN FOCUS

Miracle Coatings

*I*magine a pair of plastic-lens sunglasses that are unscratchable, even if you drop them on concrete or rub them with sandpaper. Research may make such glasses possible, along with cutting tools that never need sharpening, special glass for windshields and buildings that cannot be scratched by wind-blown sand, and speakers that reproduce sound with a crispness unimagined until now. The secret of all these marvels is a thin diamond coating. Diamond is so hard that virtually nothing can scratch it. A thin diamond coating on a speaker cone limits resonance and gives a remarkably pure tone.

But how do you coat something with a diamond? It is nearly impossible to melt diamond (melting point,

Sunglasses would be unscratchable with a diamond coating.

3500 °C). And even if diamond were melted, the object being coated would itself melt immediately at this temperature. Surprisingly, a diamond coating can be applied quite easily to something even as fragile as plastic. First the surface is bathed with a mixture of gaseous methane (CH_4) and hydrogen (H_2). Next the methane is broken apart into its component elements by an energy source similar to that used in microwave ovens. The carbon atoms freed from the methane then form a thin diamond coating on the surface being treated.

The coating of soft, scratchable materials with a super-tough diamond layer should improve many types of consumer products in the near future.

5.2   Ions

AIM: To describe the formation of ions from their parent atoms, and to name them.
To show how the periodic table can help predict which ion a given element forms.

We have seen that an atom has a certain number of protons in its nucleus and an equal number of electrons in the space around the nucleus. This results in an exact balance of positive and negative charges. We say that an atom is a neutral entity—it has *zero net charge.*

We can produce a charged entity, called an **ion,** by taking a neutral atom and adding or removing one or more electrons. For example, a sodium atom ($Z =$ 11) has eleven protons in its nucleus and eleven electrons outside its nucleus.

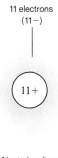

11 electrons
(11−)

11+

Neutral sodium
atom (Na)

If one of the electrons is lost, there will be eleven positive charges but only ten negative charges. This gives an ion with a net positive one (1+) charge: (11+) + (10−) = 1+. We can represent this process as follows:

An ion has a net positive or negative charge.

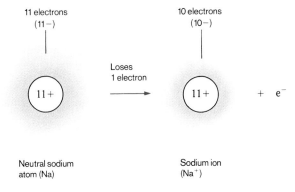

11 electrons
(11−)

10 electrons
(10−)

Loses
1 electron

11+

11+   + e⁻

Neutral sodium
atom (Na)

Sodium ion
(Na⁺)

or, in shorthand form, as

$$Na \rightarrow Na^+ + e^-$$

where Na represents the neutral sodium atom, Na^+ represents the 1+ ion formed, and e^- represents an electron.

A positive ion, called a **cation** (pronounced *cat' eye on*), is produced when one or more electrons are *lost* from a neutral atom. We have seen that sodium loses one electron to become a 1+ cation. Some atoms lose more than one electron. For example, a magnesium atom typically loses two electrons to form a 2+ cation as illustrated on the following page.

Now is a good time for students to learn the notation $Na \rightarrow Na^+ + e^-$, showing how neutral atoms gain or lose electrons to form ions and how ions gain or lose electrons to form neutron atoms. When writing equations of this kind, students often have difficulty placing electrons on the correct side. However, if they understand how neutral atoms lose electrons to form ions and can correctly write equations to represent this process, those skills will help them later in the course.

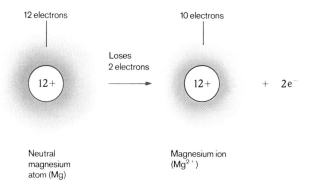

12 electrons   10 electrons

Loses
2 electrons

12+   12+   + 2e⁻

Neutral   Magnesium ion
magnesium   (Mg²⁺)
atom (Mg)

We usually represent this process as follows:

$$Mg \rightarrow Mg^{2+} + 2e^-$$

Aluminum forms a 3+ cation by losing three electrons:

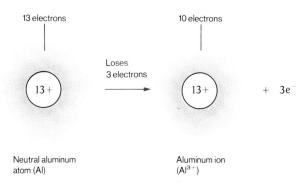

13 electrons   10 electrons

Loses
3 electrons

13+   13+   + 3e⁻

Neutral aluminum   Aluminum ion
atom (Al)   (Al³⁺)

or

$$Al \rightarrow Al^{3+} + 3e^-$$

When first introducing the formation of anions and cations, I usually don't mention where the electrons came from or where they go when they are lost. The simultaneous gain and loss of electrons is covered in Section 8.1.

A cation is named using the name of the parent atom. Thus Na^+ is called the sodium ion (or sodium cation), Mg^{2+} is called the magnesium ion (or magnesium cation), and Al^{3+} is called the aluminum ion (or aluminum cation).

When electrons are *gained* by a neutral atom, an ion with a negative charge is formed. A negatively charged ion is called an **anion** (pronounced *an' ion*). An atom that gains one extra electron forms an anion with a 1− charge. An example of an atom that forms a 1− anion is the chlorine atom, which has seventeen protons and seventeen electrons:

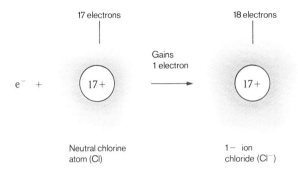

17 electrons   18 electrons

Gains
1 electron

e^- +

17+   17+

Neutral chlorine   1− ion
atom (Cl)   chloride (Cl⁻)

or

$$Cl + e^- \rightarrow Cl^-$$

Note that the anion formed by chlorine has eighteen electrons but only seventeen protons, so the net charge is $(18-) + (17+) = 1-$. Unlike a cation, which is named for the parent atom, an anion is named by taking the root name of the atom and changing the ending. For example, the Cl^- anion produced from the Cl (chlorine) atom is called the *chloride* ion (or chloride anion). Notice that the word *chloride* is obtained from the root of the atom name (*chlor-*) plus the suffix *-ide.* Other atoms that add one electron to form $1-$ ions include

> fluorine
> $$F + e^- \rightarrow F^- \qquad (fluoride \text{ ion})$$
> bromine
> $$Br + e^- \rightarrow Br^- \qquad (bromide \text{ ion})$$
> and iodine
> $$I + e^- \rightarrow I^- \qquad (iodide \text{ ion})$$

Note that the name of each of these anions is obtained by adding *-ide* to the root of the atom name.

Some atoms can add two electrons to form $2-$ anions. Examples include

> oxygen
> $$O + 2e^- \rightarrow O^{2-} \qquad (oxide \text{ ion})$$
> and sulfur
> $$S + 2e^- \rightarrow S^{2-} \qquad (sulfide \text{ ion})$$

Note that the names for these anions are derived in the same way as those for the $1-$ anions.

It is important to recognize that ions are always formed by removing electrons from an atom (to form cations) or adding electrons to an atom (to form anions). *Ions are never formed by changing the number of protons* in an atom's nucleus.

The name of an anion is obtained by adding *-ide* to the root of the atom name.

It is important to understand that isolated atoms do not form ions on their own. Most commonly, ions are formed when metallic elements combine with nometallic elements. As we will discuss in detail in Chapter 8, when metals and nonmetals react, the metal atoms tend to lose one or more electrons, which are in turn gained by the atoms of the nonmetal. Thus reactions between metals and nonmetals tend to form compounds that contain metal cations and nonmetal anions. We will have more to say about these compounds in Section 5.3.

Ion Charges and the Periodic Table

We find the periodic table very useful when we want to know what type of ion is formed by a given atom. Figure 5.7 shows the types of ions formed by atoms in several of the groups on the periodic table. Note that the Group 1 metals all form $1+$ ions (M^+), the Group 2 metals all form 2^+ ions (M^{2+}), and the Group 3 metals form 3^+ ions (M^{3+}). For Groups 1–3, then, the charges of the cations formed are identical to the group numbers.

In contrast to the Group 1, 2, and 3 metals, all of the many *transition metals* form cations with various positive charges. For these elements there is no easy way to predict the charge of the cation that will be formed.

It is important to note that metals always form positive ions. This tendency to lose electrons is a fundamental characteristic of metals. Nonmetals, on the other hand, form negative ions by gaining electrons. Note that the Group 7 atoms all gain one electron to form $1-$ ions and that all the nonmetals in Group 6 gain two electrons to form $2-$ ions.

At this point you should memorize the relationships between the group number and the type of ion formed, as shown in Figure 5.7. You will understand why these relationships exist after we further discuss the theory of the atom in Chapter 11.

For Groups 1, 2, and 3, the charges of the cations equal the group numbers.

Students should memorize the charge of ions in Groups 1, 2, 3, 6, and 7 of the periodic table. Their efforts now will later help them write correct formulas for compounds and help with naming ions.

In the process of naming ions and writing their charges, it is possible students will notice for themselves that all nonmetals form ions with negative charges and all cations form ions with positive charges. This is a good opportunity for students to make a discovery on their own.

Figure 5.7

The ions formed by selected members of Groups 1, 2, 3, 6, and 7.

5.3   Compounds That Contain Ions

AIM: To show how ions combine to form neutral compounds.

Chemists have good reasons to believe that many chemical compounds contain ions. For instance, consider some of the properties of common table salt, sodium chloride (NaCl). It must be heated to about 800 °C to melt and to almost 1500 °C to boil (compare to water, which boils at 100 °C). As a solid, salt will not conduct an electric current, but when melted it is a very good conductor. Pure water will not conduct electricity (will not allow an electric current to flow), but when salt is dissolved in water, the resulting solution readily conducts electricity (see Figure 5.8).

Chemists have come to realize that we can best explain these properties of sodium chloride (NaCl) by picturing it as containing Na^+ ions and Cl^- ions

Melting means that the solid, where the ions are locked into place, is changed to a liquid, where the ions can move.

You can demonstrate that some compounds are composed of ions by performing the light-bulb demonstration with various solutions and with solid sodium chloride. The demonstration is much more vivid in the minds of students than a photograph or a drawing. A 24-volt dry cell battery will provide enough current to light a 25-watt light bulb. If you have time, try solutions containing soluble ionic salts such as KCl and Na_2SO_4.

Pure water

(a)

Source of electric power

Salt dissolved in water

(b)

Figure 5.8
(a) Pure water does not conduct a current, so the circuit is not complete and the bulb does not light. (b) Water containing a dissolved salt conducts electricity and the bulb lights.

Figure 5.9
(a) The arrangement of sodium ions (Na$^+$) and chloride ions (Cl$^-$) in the ionic compound sodium chloride. (b) Solid sodium chloride highly magnified.

packed together as shown in Figure 5.9. Because the positive and negative charges attract each other very strongly, it must be heated to a very high temperature (800 °C) before it melts.

To explore further the significance of the electrical conductivity results, we need to discuss briefly the nature of electric currents. An electric current can travel along a metal wire because *electrons are free to move* through the wire; the moving electrons carry the current. Substances that contain ions can conduct an electric current only *if the ions can move.* The current travels by the movement of the charged ions. In solid NaCl the ions are tightly held and cannot move about, but when the solid is melted and changed to a liquid, the structure is disrupted and the ions can move. As a result, an electric current can travel through the melted salt.

The same reasoning applies to NaCl dissolved in water. When the solid dissolves, the ions are dispersed throughout the water and can move around in the water, allowing it to conduct a current.

We have seen that we recognize substances that contain ions by their characteristic properties. They often have very high melting points, and they conduct an electric current when melted or when dissolved in water.

Many substances contain ions. In fact, whenever a compound forms between a metal and a nonmetal, it can be expected to contain ions. We call these substances **ionic compounds.**

One fact very important to remember is that *a chemical compound must have a net charge of zero.* This means that if a compound contains ions, then

1. There must be both positive ions (cations) and negative ions (anions) present.
2. The numbers of cations and anions must be such that the net charge is zero.

Before students can understand the significance of a solution that carries an electrical current, they must understand what an electrical current is. It helps to mention the role electrons play in sustaining a current.
A substance containing ions that can move can conduct an electric current.

Dissolving NaCl causes the ions to be randomly dispersed in the water, allowing them to move freely. Dissolving is not the same as melting, but both processes free the ions to move.

An ionic compound cannot contain only anions or only cations, because the net charge of a compound must be zero.

The net charge of a compound (zero) is the sum of the positive and negative charges.

For example, note that the formula for sodium chloride is written NaCl, indicating one of each type of these elements. This makes sense because sodium chloride contains Na^+ ions and Cl^- ions. Each sodium ion has a $1+$ charge and each chloride ion has a $1-$ charge, so they must occur in equal numbers to give a net charge of zero.

Writing the formulas of ionic compounds from ions is usually not too difficult. These explanations and Example 5.1 on page 132 can help students algebraically determine what subscripts to use so that the net charge on the compound is zero.

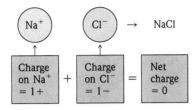

And for *any* ionic compound,

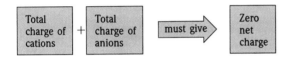

Consider an ionic compound that contains the ions Mg^{2+} and Cl^-. What combination of these ions will give a net charge of zero? To balance the $2+$ charge on Mg^{2+}, we will need two Cl^- ions to give a net charge of zero.

This means that the formula of the compound must be $MgCl_2$. Remember that subscripts are used to give the relative numbers of atoms (or ions).

The subscript 1 in a formula is not written.

Now consider an ionic compound that contains the ions Ba^{2+} and O^{2-}. What is the correct formula? These ions have charges of the same size (but opposite sign), so they must occur in equal numbers to give a net charge of zero. The formula of the compound is BaO, because $(2+) + (2-) = 0$.

Similarly, the formula of a compound that contains the ions Li^+ and N^{3-} is Li_3N, because three Li^+ cations are needed to balance the charge of the N^{3-} anion.

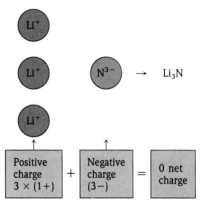

EXAMPLE 5.1   Writing Formulas for Ionic Compounds

The pairs of ions contained in several ionic compounds are listed below. Give the formula for each compound.

a. Ca^{2+} and Cl^- b. Na^+ and S^{2-} c. Ca^{2+} and P^{3-}

SOLUTION

a. Ca^{2+} has a 2+ charge, so two Cl^- ions (each with the charge 1−) will be needed.

The formula is $CaCl_2$.

b.   In this case S^{2-}, with its 2− charge, requires two Na^+ ions to produce a zero net charge.

The formula is Na_2S.

EXAMPLE 5.1, CONTINUED

c.   We have the ions Ca^{2+} (charge 2+) and P^{3-} (charge 3−). We must
figure out how many of each are needed to balance exactly the positive
and negative charges. Let's try two Ca^{2+} and one P^{3-}.

The resulting net charge is $2(2+) + (3-) = (4+) + (3-) = 1-$. This
doesn't work because the net charge is not zero. We can obtain the
same total positive and total negative charges by having three Ca^{2+} ions
and two P^{3-} ions.

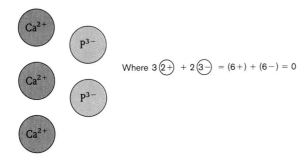

Where $3\,(2+) + 2\,(3-) = (6+) + (6-) = 0$

Thus the formula must be Ca_3P_2.

SELF-CHECK EXERCISE 5.1

Give the formulas for the compounds that contain the following pairs of ions.

a. K^+ and I^- b. Mg^{2+} and N^{3-} c. Al^{3+} and O^{2-}

5.4   Naming Compounds

AIM: To show why it is necessary to have a system for naming
compounds.

When chemistry was an infant science, there was no system for naming compounds. Names such as sugar of lead, blue vitriol, quicklime, Epsom salts, milk of

CHEMISTRY IN FOCUS

Sugar of Lead

*I*n ancient Roman society it was common to boil wine in a lead vessel, driving off much of the water to produce a very sweet, viscous syrup called *sapa.* This syrup was commonly used as a sweetener for many types of food and drink.

We now realize that a major component of this syrup was lead acetate, $Pb(C_2H_3O_2)_2$. This compound has a very sweet taste—hence its original name, sugar of lead.

Many historians believe that the fall of the Roman Empire was due at least in part to lead poisoning, which causes lethargy and mental malfunctions. One major source of this lead was the sapa syrup. In addition, the Romans' highly advanced plumbing system employed lead water pipes, which allowed lead to be leached into their drinking water.

Sadly, this story is more relevant to today's society than you might think. Lead-based solder has been widely used for many years to connect the copper pipes in water systems in homes and commercial buildings. There is evidence that dangerous amounts of lead can be leached from these soldered joints into drinking water. In fact, large quantities of lead have been found in the water that some drinking fountains and water coolers dispense. In response to these problems, the U.S. Congress has passed a law banning lead from the solder used in plumbing systems for drinking water.

An ancient painting showing Romans drinking wine.

To illustrate the need for a systematic naming system, you might provide some colorful examples of common names. Sugars of lead, blue vitriol, and barbituric acid (named after a woman Barbara) are a few you might want to use. Point out that many common names do not provide any information about the kinds or numbers of elements that the compound contains.

Naming is one of the fundamental skills students will need to master to be successful in chemistry. Stress that effort they put into learning naming now will pay off during the remainder of the course.

magnesia, gypsum, and laughing gas were coined by early chemists. Such names are called *common names.* As chemistry grew, it became clear that using common names for compounds would lead to great confusion. More than four million chemical compounds are currently known. Memorizing common names for all these compounds would be impossible.

The solution, of course, was a *system* for naming compounds in which the name tells something about the composition of the compound. After learning the system, you should be able to name a compound when you are given its formula. And you should be able to construct a compound's formula, given its name. In the next few sections we will specify the most important rules for naming compounds other than organic compounds (those based on chains of carbon atoms).

We will begin with the system for naming **binary compounds**—compounds composed of two elements. We can divide binary compounds into two broad classes:

1. Compounds that contain a metal and a nonmetal
2. Compounds that contain two nonmetals

We will describe how to name compounds in each of these classes in the next several sections. Then, in succeeding sections, we will describe the systems used for naming more complex compounds.

5.5 Naming Compounds That Contain a Metal and a Nonmetal

AIM: To show how to name binary compounds of a metal and a nonmetal.

As we saw in Section 5.3, when a metal such as sodium combines with a nonmetal such as chlorine, the resulting compound contains ions. The metal loses one or more electrons to become a cation, and the nonmetal gains one or more electrons to form an anion. The resulting substance is called a **binary ionic compound.** Binary ionic compounds contain a positive ion (cation), which is always written first in the formula, and a negative ion (anion). *To name these compounds we need simply to name the ions.*

In this section we will consider binary ionic compounds of two types.

Type I: The metal present forms only one type of cation.
Type II: The metal present can form two (or more) cations that have different charges.

Type I Binary Ionic Compounds

The following rules apply for Type I ionic compounds:

Rules for Naming Type I Ionic Compounds

1. The cation is always named first and the anion second.
2. A simple cation (obtained from a single atom) takes its name from the name of the element. For example, Na^+ is called sodium in the names of compounds containing this ion.
3. A simple anion (obtained from a single atom) is named by taking the first part of the element name (the root) and adding *-ide.* Thus the Cl^- ion is called chloride.

When you choose examples of compounds to illustrate the naming schemes, try to bring examples of those compounds to class. Pointing them out provides some relief from the monotony of the naming rules for both you and your students.

Covering the naming of compounds in the classroom can be difficult. Writing the rules on the blackboard is time-consuming and students quickly become bored. If you can use an overhead transparency to show the naming rules, you can save time in class. Tell students that all the rules are in the textbook and they don't need to write them. They might spend more time listening to what you say instead of writing frantically.

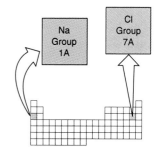

When lecturing I usually reduce the rules for naming Type I ionic compounds to that of "Name the ions." Then I work examples to show what I mean.

A simple cation has the same name as its parent element.

Table 5.2 lists the name and charges for Type I cations and for some monatomic anions. The charges for most of them can be determined from the group numbers. The charge of the silver cation must be memorized.

The charges on Type II cations given in Table 5.3 can be determined from the charge on the anion and don't necessarily need to be memorized, assuming students have memorized the anion charges. It is probably best if students do memorize this short list of common Type II cations, however.

Table 5.2   Common Simple Cations and Anions

Cation	Name	Anion*	Name
H^+	hydrogen	H^-	hydride
Li^+	lithium	F^-	fluoride
Na^+	sodium	Cl^-	chloride
K^+	potassium	Br^-	bromide
Cs^+	cesium	I^-	iodide
Be^{2+}	beryllium	O^{2-}	oxide
Mg^{2+}	magnesium	S^{2-}	sulfide
Ca^{2+}	calcium		
Ba^{2+}	barium		
Al^{3+}	aluminum		
Ag^+	silver		

*The root is given in color.

Naming Type I compounds is relatively easy because, in the beginning, there is only one type of compound. There are no other possibilities to choose among. After Type II and Type III compounds have been introduced, there are many decisions to be made before the correct name can be assigned. Learning how to name compounds is difficult to do in one chapter; you will need to refresh students memories from time to time for the remainder of the course.

Some common cations and anions and their names are listed in Table 5.2. You should memorize these. They are an essential part of your chemical vocabulary.

We will illustrate these rules by naming a few compounds. For example, the compound NaI is called sodium iodide. It contains Na^+ (the sodium cation, named for the parent metal) and I^- (given the name iodide: the root of iodine plus -*ide*). Similarly, the compound CaO is called calcium oxide because it contains Ca^{2+} (the calcium cation) and O^{2-} (the oxide anion).

The rules for naming binary compounds are also illustrated by the following examples:

Compound	Ions Present	Name
NaCl	Na^+, Cl^-	sodium chloride
KI	K^+, I^-	potassium iodide
CaS	Ca^{2+}, S^{2-}	calcium sulfide
CsBr	Cs^+, Br^-	cesium bromide
MgO	Mg^{2+}, O^{2-}	magnesium oxide

It is important to note that in *formulas* of ionic compounds, simple ions are represented by the element symbol: Cl means Cl^-, Na means Na^+, etc. However, when individual *ions* are shown, the charge is always included. Thus the formula of potassium bromide is written KBr, but when the potassium and bromide ions are shown individually, they are written K^+ and Br^-.

EXAMPLE 5.2   Naming Type I Binary Compounds

Name each binary compound.

a.   CsF

b. $AlCl_3$

c. MgI_2

SOLUTION

We will name these compounds by systematically following the rules given above.

a.   CsF

STEP 1
Identify the cation and anion. Cs is in Group 1, so we know it will form the 1+ ion Cs^+. Because F is in Group 7, it forms the 1− ion F^-.

STEP 2
Name the cation. Cs^+ is simply called cesium, the same as the element name.

STEP 3
Name the anion. F^- is called fluoride: we use the root name of the element plus -ide.

STEP 4
Name the compound by combining the names of the individual ions. The name for CsF is cesium fluoride. (Remember that the name of the cation is always given first.)

b.

Compound	Ions Present	Ion Names	Comments
$AlCl_3$	Al^{3+} (Cation)	aluminum	Al (Group 3) always forms Al^{3+}.
	Cl^- (Anion)	chloride	Cl (Group 7) always forms Cl^-.

The name of $AlCl_3$ is aluminum chloride.

c.
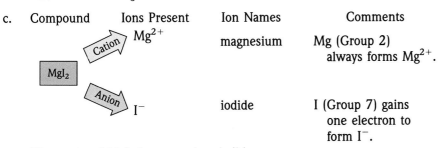

Compound	Ions Present	Ion Names	Comments
MgI_2	Mg^{2+} (Cation)	magnesium	Mg (Group 2) always forms Mg^{2+}.
	I^- (Anion)	iodide	I (Group 7) gains one electron to form I^-.

The name of MgI_2 is magnesium iodide.

EXAMPLE 5.2, CONTINUED

SELF-CHECK EXERCISE 5.2

Name the following compounds.

a. Rb_2O b. SrI_2 c. K_2S

Example 5.2 reminds us of three things:

1. Compounds formed from metals and nonmetals are ionic.
2. In an ionic compound the cation is always named first.
3. The *net* charge on an ionic compound is always zero. Thus in CsF, one of each type of ion (Cs^+ and F^-) is required: (1+) + (1-) = 0 charge. In $AlCl_3$, however, three Cl^- ions are needed to balance the charge of Al^{3+}: (3+) + 3(1-) = 0 charge. In MgI_2, two I^- ions are needed for each Mg^{2+} ion: (2+) + 2(1-) = 0 charge.

Type II Binary Ionic Compounds

So far we have considered binary ionic compounds (Type I) containing metals that always give the same cation. For example, sodium always forms the Na^+ ion, calcium always forms the Ca^{2+} ion, and aluminum always forms the Al^{3+} ion. As we said in the last section, we can predict with certainty that each Group 1 metal will give a 1+ cation and each Group 2 metal will give a 2+ cation. And aluminum always forms Al^{3+}.

However, there are many metals that can form more than one type of cation. For example, lead (Pb) can form Pb^{2+} or Pb^{4+} in ionic compounds. Also, iron (Fe) can produce Fe^{2+} or Fe^{3+}, chromium (Cr) can produce Cr^{2+} or Cr^{3+}, gold (Au) can produce Au^+ or Au^{3+}, and so on. This means that if we saw the name gold chloride, we wouldn't know whether it referred to the compound AuCl (containing Au^+ and Cl^-) or the compound $AuCl_3$ (containing Au^{3+} and three Cl^- ions). Therefore, we need a way of specifying which cation is present in compounds containing metals that can form more than one type of cation.

Chemists have decided to deal with this situation by using a Roman numeral to specify the charge on the cation. To see how this works, consider the compound $FeCl_2$. Iron can form Fe^{2+} or Fe^{3+}, so we must first decide which of these cations is present. We can determine the charge on the iron cation, because we know it must just balance the charge on the two 1− anions (the chloride ions). Thus if we represent the charges as

I introduce Type II compounds by repeating that the charge on Group 1, 2, and 3 cations never changes. I reassure students that these ions, at least, don't have variable charges.

Type II binary ionic compounds contain a metal that can form more than one type of cation.

To emphasize that there is a real difference between compounds such as copper (I) chloride and copper (II) chloride, I provide some characteristics such as melting points and water solubilities.

Table 5.3 Common Type II Cations

Ion	Systematic Name	Older Name
Fe^{3+}	iron(III)	ferric
Fe^{2+}	iron(II)	ferrous
Cu^{2+}	copper(II)	cupric
Cu^{+}	copper(I)	cuprous
Co^{3+}	cobalt(III)	cobaltic
Co^{2+}	cobalt(II)	cobaltous
Sn^{4+}	tin(IV)	stannic
Sn^{2+}	tin(II)	stannous
Pb^{4+}	lead(IV)	plumbic
Pb^{2+}	lead(II)	plumbous
Hg^{2+}	mercury(II)	mercuric
Hg_2^{2+} *	mercury(I)	mercurous

*Mercury(I) ions always occur bound together in pairs to form Hg_2^{2+}.

A dish of copper(II) sulfate.

There are other common Type II cations, such as those of Ni and Cr, whose charges are not given in Table 5.3. Students can usually determine the charge on those cations from the charge on the anion, but they will have difficulty naming compounds with these ions if they are not aware that they are also Type II cations.

we know that ? must represent 2 because

$$(2+) + 2(1-) = 0$$

$FeCl_2$, then, contains one Fe^{2+} ion and two Cl^- ions. We call this compound iron(II) chloride, where the II tells the charge of the iron cation. That is, Fe^{2+} is called iron(II). Likewise, Fe^{3+} is called iron(III). And $FeCl_3$, which contains one Fe^{3+} ion and three Cl^- ions, is called iron(III) chloride.

There is another system for naming ionic compounds containing metals that form two cations. *The ion with the higher charge has a name ending in -ic, and the one with the lower charge has a name ending in -ous.* In this system, for example, Fe^{3+} is called the ferric ion, and Fe^{2+} is called the ferrous ion. The names for $FeCl_3$ and $FeCl_2$, in this system, are ferric chloride and ferrous chloride, respectively.

Table 5.3 gives both names for many Type II cations. We will use the system of Roman numerals exclusively in this text; the other system is not used much today by chemists.

$FeCl_3$ must contain Fe^{3+} to balance the charge of three Cl^- ions.

The old naming system for Type II cations using *-ic* and *-ous* is briefly presented in the text. Students who don't plan to have a career in chemistry will probably have little need to know the old system.

EXAMPLE 5.3   Naming Type II Binary Compounds

Give the systematic name of each of the following compounds.

a. CuCl
b. HgO
c. Fe_2O_3
d. MnO_2
e. $PbCl_4$

EXAMPLE 5.3, CONTINUED

SOLUTION

I tell students to use the periodic table when they cannot remember whether a cation has more than one possible charge. Cations from Groups I, II, or III have only one possible charge and that charge is the same as the group number. Cations formed from transition metals may have more than one possible charge.

All these compounds include a metal that can form more than one type of cation; thus we must first determine the charge on each cation. We do this by recognizing that a compound must be electrically neutral; that is, the positive and negative charges must balance exactly. We will use the known charge on the anion to determine the charge of the cation.

a. In CuCl we recognize the anion as Cl^-. To determine the charge on the copper cation, we invoke the principle of charge balance.

In this case, ?+ must be 1+ because $(1+) + (1-) = 0$. Thus the copper cation must be Cu^+. Now we can name the compound by using the regular steps.

Compound	Ions Present	Ion Names	Comments
CuCl	Cu^+	copper(I)	Copper forms other cations (it is a transition metal), so we must include the I to specify its charge.
	Cl^-	chloride	

The name of CuCl is copper(I) chloride.

b. In HgO we recognize the O^{2-} anion. To yield zero net charge, the cation must be Hg^{2+}.

Compound	Ions Present	Ion Names	Comments
HgO	Hg^{2+}	mercury(II)	The II is necessary to specify the charge.
	O^{2-}	oxide	

The name of HgO is mercury(II) oxide.

c. Because Fe_2O_3 contains three O^{2-} anions, the charge on the iron cation must be 3+.

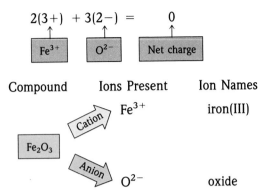

$$2(3+) + 3(2-) = 0$$

Fe^{3+} O^{2-} Net charge

Compound	Ions Present	Ion Names
	Fe^{3+}	iron(III)
Fe$_2$O$_3$		
	O^{2-}	oxide

The name of Fe$_2$O$_3$ is iron(III) oxide.

d.   MnO$_2$ contains two O^{2-} anions, so the charge on the manganese cation is 4+.

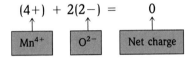

$$(4+) + 2(2-) = 0$$

Mn^{4+} O^{2-} Net charge

Compound	Ions Present	Ion Names
	Mn^{4+}	manganese(IV)
MnO$_2$		
	O^{2-}	oxide

The name of MnO$_2$ is manganese(IV) oxide.

e.   Because PbCl$_4$ contains four Cl$^-$ anions, the charge on the lead cation is 4+.

$$(4+) + 4(1-) = 0$$

Sn^{4+} Cl$^-$ Net charge

Compound	Ions Present	Ion Names
	Pb^{4+}	lead(IV)
SnCl$_4$		
	Cl$^-$	chloride

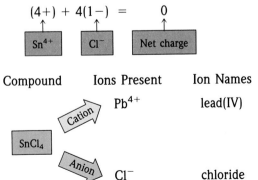

The name for PbCl$_4$ is lead(IV) chloride.

Sometimes transition metals form only one ion, such as silver, which forms Ag^+; zinc, which forms Zn^{2+}; and cadmium, which forms Cd^{2+}. In these cases, chemists do not use a Roman numeral, although it is not "wrong" to do so.

The use of a Roman numeral in a systematic name for a compound is required only in cases where more than one ionic compound forms between a given pair of elements. This occurs most often for compounds that contain transition metals, which frequently form more than one cation. *Metals that form only one cation do not need to be identified by a Roman numeral.* Common metals that do not require Roman numerals are the Group 1 elements, which form only 1+ ions; the Group 2 elements, which form only 2+ ions; and such Group 3 metals as aluminum and gallium, which form only 3+ ions.

As shown in Example 5.3, when a metal ion that forms more than one type of cation is present, the charge on the metal ion must be determined by balancing the positive and negative charges of the compound. To do this, you must be able to recognize the common anions and you must know their charges (see Table 5.2).

EXAMPLE 5.4	Naming Binary Ionic Compounds: Summary

Give the systematic name of each of the following compounds.

a. $CoBr_2$ b. $CaCl_2$ c. Al_2O_3 d. $CrCl_3$

SOLUTION

Compound	Ions and Names	Compound Name	Comments
a. $CoBr_2$	Co^{2+} cobalt(II) Br^- bromide	cobalt(II) bromide	Cobalt is a transition metal; the name of the compound must have a Roman numeral. The two Br^- ions must be balanced by a Co^{2+} cation.
b. $CaCl_2$	Ca^{2+} calcium Cl^- chloride	calcium chloride	Calcium, an alkaline earth metal, forms only the Ca^{2+} ion. A Roman numeral is not necessary.
c. Al_2O_3	Al^{3+} aluminum O^{2-} oxide	aluminum oxide	Aluminum forms only Al^{3+}. A Roman numeral is not necessary.
d. $CrCl_3$	Cr^{3+} chromium(III) Cl^- chloride	chromium(III) chloride	Chromium is a transition metal. The name of the compound must have a Roman numeral. $CrCl_3$ contains Cr^{3+}.

EXAMPLE 5.4, CONTINUED

SELF-CHECK EXERCISE 5.3

Give the names of the following compounds.

a. $PbBr_2$ and $PbBr_4$
b.   FeS and Fe_2S_3
c. $AlBr_3$

d. Na_2S
e. $CoCl_3$

The following flow chart is useful when you are naming binary ionic compounds:

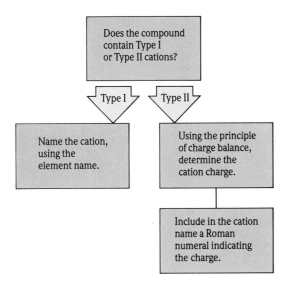

5.6   Naming Binary Compounds That Contain Only Nonmetals (Type III)

AIM: To show how to name binary compounds containing only nonmetals.

Binary compounds that contain only nonmetals are named in accordance with a system similar in some ways to the rules for naming binary ionic compounds, but there are important differences. *Type III binary compounds contain only nonmetals.* The following rules cover the naming of these compounds.

Before presenting the rules for naming Type III compounds, I provide an example compound such as IF_5. The application of each of the rules is illustrated with the example compound. Students have the opportunity to see what each rule means before the next one is presented.

Table 5.4 Prefixes Used to Indicate Numbers in Chemical Names

Prefix	Number Indicated
mono-	1
di-	2
tri-	3
tetra-	4
penta-	5
hexa-	6
hepta-	7
octa-	8

Rules for Naming Type III Binary Compounds

1. The first element in the formula is named first, and the full element name is used.
2. The second element is named as though it were an anion.
3. Prefixes are used to denote the numbers of atoms present. These prefixes are given in Table 5.4.
4. The prefix *mono-* is never used for naming the first element. For example, CO is called carbon monoxide, *not* monocarbon monoxide.

We will illustrate the application of these rules in Example 5.5

A complication to naming many Type III compounds occurs when the combination of prefix and root name results in two consecutive vowels with awkward pronunciations. It seems natural for us to contract tetraoxide to tetroxide, but students find the extra step hard. They are also unsure what letter to omit. The element that causes the most problems is oxygen. For example, monooxide should be monoxide, and pentaoxide should be pentoxide.

EXAMPLE 5.5   Naming Type III Binary Compounds

Name the following binary compounds, which contain two nonmetals (Type III).

a. BF_3 b. NO c. N_2O_5

SOLUTION

a. BF_3

RULE 1
Name the first element, using the full element name: boron.

RULE 2
Name the second element as though it were an anion: fluoride.

RULES 3 AND 4
Use prefixes to denote numbers of atoms.
One boron atom; do not use *mono-* in first position. Three fluorine atoms: use the prefix *tri-*.

The name of BF_3 is boron trifluoride.

b.

Compound	Individual Names	Prefixes	Comments
NO	nitrogen	none	*Mono-* is not used for
	oxide	*mono-*	the first element.

The name for NO is nitrogen monoxide. Note that the *o* in *mono-* has been dropped for easier pronunciation. The *common* name for NO, which is often used by chemists, is nitric oxide.

EXAMPLE 5.5, CONTINUED

c.
Compound	Individual Names	Prefixes	Comments
N_2O_5	nitrogen	*di-*	two N atoms
	oxide	*penta-*	five O atoms

The name for N_2O_5 is dinitrogen pentoxide. The *a* in *penta-* has been dropped for easier pronunciation.

SELF-CHECK EXERCISE 5.4

Name the following compounds.

a. CCl_4 b. NO_2 c. IF_5

Copper reacts with nitric acid to produce colorless NO, which immediately reacts with oxygen in the air to form brown NO_2 gas.

Water and ammonia are always referred to by their common names.

The previous examples illustrate that, to avoid awkward pronunciation, we often drop the final *o* or *a* of the prefix when the second element is oxygen. For example, N_2O_4 is called dinitrogen tetroxide, not dinitrogen tetraoxide, and CO is called carbon monoxide, *not* carbon monooxide.

Some compounds are always referred to by their common names. The two best examples are water and ammonia. The systematic names for H_2O and NH_3 are never used.

To make sure you understand the procedures for naming binary nonmetallic compounds (Type III), do Example 5.6 and Self-Check Exercise 5.5.

EXAMPLE 5.6	Naming Type III Binary Compounds: Summary

Name each of the following compounds:

a. PCl_5 c. SF_6 e. SO_2
b. P_4O_6 d. SO_3 f. N_2O_3

SOLUTION

Compound	Name
a. PCl_5	phosphorus pentachloride
b. P_4O_6	tetraphosphorus hexoxide
c. SF_6	sulfur hexafluoride
d. SO_3	sulfur trioxide
e. SO_2	sulfur dioxide
f. N_2O_3	dinitrogen trioxide

EXAMPLE 5.6, CONTINUED

SELF-CHECK EXERCISE 5.5

Name the following compounds.

a. SiO_2
b. O_2F_2
c. XeF_6

5.7 Naming Binary Compounds: A Review

AIM: To summarize how to name Type I, Type II, and Type III binary compounds.

Because different rules apply for naming various types of binary compounds, we will now consider an overall strategy to use for these compounds. We have considered three types of binary compounds, and naming each of them requires different procedures.

Type I: Ionic compounds with metals that always form a cation with the same charge
Type II: Ionic compounds with metals (usually transition metals) that form cations with various charges
Type III: Compounds that contain only nonmetals

In trying to determine which type of compound you are naming, use the periodic table to help you identify metals and nonmetals and to determine which elements are transition metals.

The flow chart given in Figure 5.10 should help you as you name binary compounds of the various types.

EXAMPLE 5.7   **Naming Binary Compounds: Summary**

Name the following binary compounds.

a.   CuO
b.   SrO
c. B_2O_3
d. $TiCl_4$

e. K_2S
f. OF_2
g. NH_3

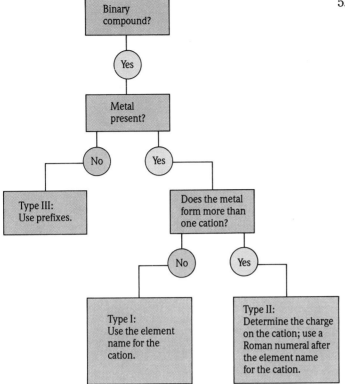

Figure 5.10 is a convenient flow chart to help students systematically name compounds. Students are reluctant to take time to use schemes such as these. You can tell them that if they become stuck on a homework problem, this scheme can help them arrive at the correct answer.

Figure 5.10
A flow chart for naming binary compounds.

EXAMPLE 5.7, CONTINUED

SOLUTION

a.

b.

EXAMPLE 5.7, CONTINUED

c.

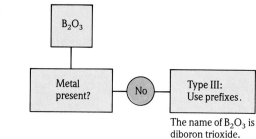

The name of B_2O_3 is
diboron trioxide.

d.

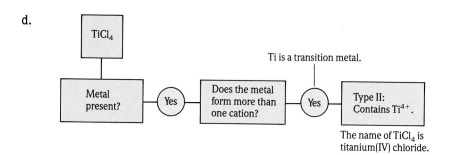

The name of $TiCl_4$ is
titanium(IV) chloride.

e.

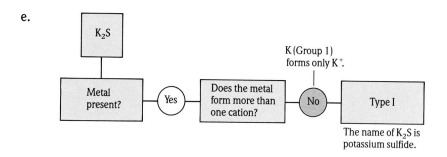

The name of K_2S is
potassium sulfide.

f.

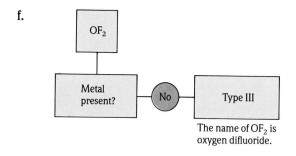

The name of OF_2 is
oxygen difluoride.

EXAMPLE 5.7, CONTINUED

g.

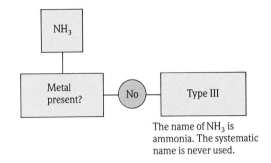

The name of NH_3 is ammonia. The systematic name is never used.

SELF-CHECK EXERCISE 5.6

Name the following binary compounds.

a. ClF_3 d. MnO_2
b. VF_5 e. MgO
c. CuCl   f. H_2O

5.8 Naming Compounds That Contain Polyatomic Ions

AIM: To give the names of common polyatomic ions and explain how to use them in naming compounds.

A type of ionic compound that we have not yet considered is exemplified by ammonium nitrate, NH_4NO_3, which contains the **polyatomic ions** NH_4^+ and NO_3^-. As their name suggests, polyatomic ions are charged entities composed of several atoms bound together. Polyatomic ions are assigned special names that you *must memorize* in order to name the compounds containing them. The most important polyatomic ions and their names are listed in Table 5.5 (p. 150).

Note in Table 5.5 that several series of polyatomic anions contain an atom of a given element and different numbers of oxygen atoms. These anions are called **oxyanions.** When there are two members in such a series, the name of the one with the smaller number of oxygen atoms ends in *-ite,* and the name of the one with the larger number ends in *-ate.* For example, SO_3^{2-} is sulfite and SO_4^{2-} is sulfate. When more than two oxyanions make up a series, *hypo-* (less than) and *per-* (more than) are used as prefixes to name the members of the series with the

There are a lot of polyatomic ions for students to memorize in this section. To students, the names and structures are hard to distinguish from each other. They will require lots of practice before each name and structure begins to look unique.

Ionic compounds containing polyatomic ions are not binary compounds, because they contain more than two elements.

The names and charges of polyatomic ions must be memorized. They are an important part of the vocabulary of chemistry.

Note that the SO_3^{2-} anion has very different properties from SO_3 (sulfur trioxide), a pungent, toxic gas.

Table 5.5 is easier to memorize than students first think. Some of the ions are related, or are part of a series that makes remembering their names and charges easier. Show students how, for example, nitrate and nitrite differ only by an oxygen atom and that their charges are the same. Other pairs or groups of ions are similarly related.

In Table 5.5, chlorine is used to illustrate the series of ions formed between halogens and oxygen. You may wish to show how other halogens such as bromine also form a series of anions with oxygen, and then present their names.

Table 5.5   Names of Common Polyatomic Ions

Ion	Name	Ion	Name
NH_4^+	ammonium	CO_3^{2-}	carbonate
NO_2^-	nitrite	HCO_3^-	hydrogen carbonate
NO_3^-	nitrate		(bicarbonate is a widely
SO_3^{2-}	sulfite		used common name)
SO_4^{2-}	sulfate	ClO^-	hypochlorite
HSO_4^-	hydrogen sulfate	ClO_2^-	chlorite
	(bisulfate is a widely	ClO_3^-	chlorate
	used common name)	ClO_4^-	perchlorate
OH^-	hydroxide	$C_2H_3O_2^-$	acetate
CN^-	cyanide	MnO_4^-	permanganate
PO_4^{3-}	phosphate	$Cr_2O_7^{2-}$	dichromate
HPO_4^{2-}	hydrogen phosphate	CrO_4^{2-}	chromate
$H_2PO_4^-$	dihydrogen phosphate	O_2^{2-}	peroxide

fewest and the most oxygen atoms, respectively. The best example involves the oxyanions containing chlorine:

$$ClO^- \qquad \textit{hypochlorite}$$
$$ClO_2^- \qquad \textit{chlorite}$$
$$ClO_3^- \qquad \textit{chlorate}$$
$$ClO_4^- \qquad \textit{perchlorate}$$

Naming ionic compounds that contain polyatomic ions is very similar to naming binary ionic compounds. For example, the compound NaOH is called sodium hydroxide, because it contains the Na^+ (sodium) cation and the OH^- (hydroxide) anion. To name these compounds, *you must learn to recognize the common polyatomic ions* (that is, you must learn the *composition* and *charge* of each of the ions in Table 5.5). Then when you see the formula $NH_4C_2H_3O_2$, you should immediately recognize its two "parts":

Except for hydroxide and cyanide, the names of polyatomic ions do not have an -*ide* ending.

The correct name is ammonium acetate.

Remember that when transition metals are present, a Roman numeral is required to specify the cation charge, just as in naming Type II binary ionic compounds. For example, the compound $FeSO_4$ is called iron(II) sulfate, because it contains Fe^{2+} to balance the 2− charge on SO_4^{2-}. Note that to determine the charge on the iron cation, you must know that sulfate has a 2− charge.

EXAMPLE 5.8	Naming Compounds That Contain Polyatomic Ions

Give the systematic name of each of the following compounds.

a. Na_2SO_4 c. $Fe(NO_3)_3$ e. Na_2SO_3
b. KH_2PO_4 d. $Mn(OH)_2$

SOLUTION

	Compound	Ions Present	Ion Names	Compound Name
a.	Na_2SO_4	two Na^+ SO_4^{2-}	sodium sulfate	sodium sulfate

b.	KH_2PO_4	K^+
		$H_2PO_4^-$
			potassium
			dihydrogen
			  phosphate
				potassium
				dihydrogen
				  phosphate

c.	$Fe(NO_3)_3$	Fe^{3+} three NO_3^-	iron(III) nitrate	iron(III) nitrate
d.	$Mn(OH)_2$	Mn^{2+} two OH^-	manganese(II) hydroxide	manganese(II) hydroxide
e.	Na_2SO_3	two Na^+ SO_3^{2-}	sodium sulfite	sodium sulfite

Students often name ions such as SO_4^{2-} as sulfide oxide. They do not yet recognize that sulfur and oxygen act as a group and form the sulfate ion.

SELF-CHECK EXERCISE 5.7

Name each of the following compounds.

a. $Ca(OH)_2$ d. $(NH_4)_2Cr_2O_7$ g. $Cu(NO_2)_2$
b. Na_3PO_4 e. $Co(ClO_4)_2$
c. $KMnO_4$ f. $KClO_3$

Example 5.8 illustrates that when more than one polyatomic ion appears in a chemical formula, parentheses are used to enclose the ion and a subscript is written after the closing parenthesis. This also occurs in $(NH_4)_2SO_4$ and $Fe_3(PO_4)_2$.

In naming chemical compounds, use the strategy summarized in Figure 5.11 (p. 152). If the compound being considered is binary, use the procedure summarized in Figure 5.10. If the compound has more than two elements, ask yourself whether it has any polyatomic ions. Use Table 5.5 to help you recognize these ions until you have committed them to memory. If a polyatomic ion is present, name the compound using procedures very similar to those for naming binary ionic compounds.

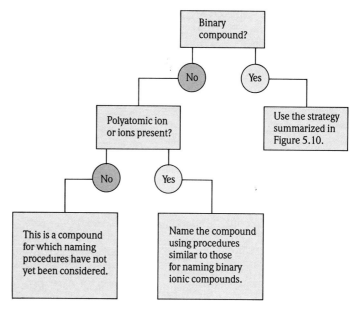

Figure 5.11
Overall strategy for naming chemical compounds.

EXAMPLE 5.9   **Naming Binary Compounds and Compounds That Contain Polyatomic Ions: Summary**

Name the following compounds.

a. Na_2CO_3 d. PCl_3
b. $FeBr_3$ e. $CuSO_4$
c. $CsClO_4$

SOLUTION

	Compound	Name	Comments
a.	Na_2CO_3	sodium carbonate	Contains $2Na^+$ and CO_3^{2-}.
b.	$FeBr_3$	iron(III) bromide	Contains Fe^{3+} and $3Br^-$.
c.	$CsClO_4$	cesium perchlorate	Contains Cs^+ and ClO_4^-.
d.	PCl_3	phosphorus trichloride	Type III binary compound (P and Cl are both nonmetals).
e.	$CuSO_4$	copper(II) sulfate	Contains Cu^{2+} and SO_4^{2-}.

EXAMPLE 5.9, CONTINUED

SELF-CHECK EXERCISE 5.8

Name the following compounds.

a. $NaHCO_3$
b. $BaSO_4$
c. $CsClO_4$
d. BrF_5

e. NaBr
f. KOCl
g. $Zn_3(PO_4)_2$

5.9   Naming Acids

AIM: To show how the anion composition determines the acid's name.
To give names for common acids.

Students are briefly introduced to acids as producers of hydrogen ions in this chapter. Acids are reactants in many reactions in upcoming chapters, so this brief introduction is important.

Acids

When dissolved in water, certain molecules produce H^+ ions (protons). These substances, which are called **acids,** were first recognized by the sour taste of their solutions. For example, citric acid is responsible for the tartness of lemons and limes. Acids will be discussed in detail later. Here we simply present the rules for naming acids.

An acid can be viewed as a molecule with one or more H^+ ions attached to an anion. The rules for naming acids depend on whether the anion contains oxygen.

Table 5.6 Names of Acids That Do Not Contain Oxygen	
Acid	Name
HF	hydrofluoric acid
HCl	hydrochloric acid
HBr	hydrobromic acid
HI	hydroiodic acid
HCN	hydrocyanic acid
H_2S	hydrosulfuric acid

If students learn the names of the acids in Tables 5.6 and 5.7 along with the acids formed from the oxyanions of chlorine, they will know the names of most of the acids they will encounter.

The rules for naming acids are convenient if you plan to introduce other acids not found in the tables in this section.

Rules for Naming Acids

1. If the *anion does not contain oxygen,* the acid is named with the prefix *hydro-* and the suffix *-ic* attached to the root name for the element. For example, when gaseous HCl (hydrogen chloride) is dissolved in water, it forms hydrochloric acid. Similarly, hydrogen cyanide (HCN) and dihydrogen sulfide (H_2S) dissolved in water are called hydrocyanic acid and hydrosulfuric acid, respectively.

2. When the *anion contains oxygen,* the acid name is formed from the root name of the central element of the anion or the anion name, with a suffix of *-ic* or *-ous.* When the anion name ends in *-ate,* the suffix *-ic* (or sometimes *-ric*) is used. For example, H_2SO_4 contains the sulfate anion (SO_4^{2-}) and is called sulfuric acid; H_3PO_4 contains the phosphate anion (PO_4^{3-}) and is called phosphoric acid; and $HC_2H_3O_2$ contains the acetate ion ($C_2H_3O_2^-$) and is called acetic acid. When the anion has an *-ite* ending, the suffix *-ous* is used. For example, H_2SO_3, which contains sulfite (SO_3^{2-}), is called sulfurous acid; and HNO_2, which contains nitrite (NO_2^-), is called nitrous acid.

Students should probably be able to generalize from the names of the oxyacids of chlorine to those of bromine.

The application of rule 2 can be seen in the names of the acids of the oxyanions of chlorine, as shown below. The rules for naming acids are given in schematic form in Figure 5.12. The names of the most important acids are given in Table 5.6 and Table 5.7. These should be memorized.

Acid	Anion	Name
$HClO_4$	perchlor*ate*	perchlor*ic* acid
$HClO_3$	chlor*ate*	chlor*ic* acid
$HClO_2$	chlor*ite*	chlor*ous* acid
$HClO$	hypochlor*ite*	hypochlor*ous* acid

Table 5.7   Names of Some Oxygen-Containing Acids

Acid	Name
HNO_3	nitric acid
HNO_2	nitrous acid
H_2SO_4	sulfuric acid
H_2SO_3	sulfurous acid
H_3PO_4	phosphoric acid
$HC_2H_3O_2$	acetic acid

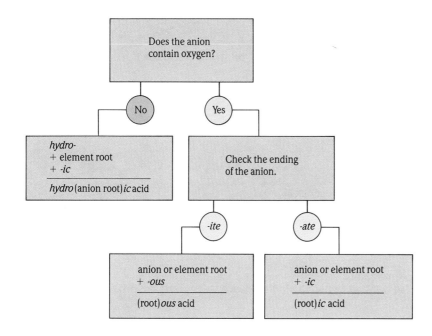

Figure 5.12
A flow chart for naming acids. The acid is considered as one or more H^+ ions attached to an anion.

5.10 Writing Formulas from Names

AIM: To write the formula of a compound, given its name.

So far we have started with the chemical formula of a compound and decided on its systematic name. Being able to reverse the process is also important. Often a laboratory procedure describes a compound by name, but the label on the bottle in the lab shows only the formula of the chemical it contains. It is essential that you be able to get the formula of a compound from its name. In fact, you already know enough about compounds to do this. For example, given the name calcium hydroxide, you can write the formula as $Ca(OH)_2$ because you know that calcium forms only Ca^{2+} ions and that, hydroxide being OH^-, two of these anions will be required in order to give a neutral compound. Similarly, the name iron(II) oxide implies the formula FeO, because the Roman numeral II indicates the presence of the cation Fe^{2+} and the oxide ion is O^{2-}.

We emphasize at this point that it is essential to learn the name, composition, and charge of each of the common polyatomic anions (and the NH_4^+ cation). If you do not recognize these ions by formula and by name, you will not be able to write the compound's name given its formula or the compound's formula given its name. You must also learn the names of the common acids.

Writing formulas from names is usually easier for students than naming formulas because Type I and Type II cations are identified in the name. They must, however, know the structures and charges of the polyatomic ions to be able to write formulas.

I usually don't spend much time in lecture writing formulas from names. I use homework assignments instead to sharpen students' skills.

EXAMPLE 5.10   Writing Formulas from Names

Give the formula for each of the following compounds.

a.   potassium hydroxide
b.   sodium carbonate
c.   nitric acid
d.   cobalt(III) nitrate
e.   calcium chloride
f.   lead(IV) oxide
g.   dinitrogen pentoxide
h.   ammonium perchlorate

SOLUTION

	Name	Formula	Comments
a.	potassium hydroxide	KOH	Contains K^+ and OH^-.
b.	sodium carbonate	Na_2CO_3	We need two Na^+ to balance $CO_3{}^{2-}$.
c.	nitric acid	HNO_3	Common strong acid; memorize.
d.	cobalt(III) nitrate	$Co(NO_3)_3$	Cobalt(III) means Co^{3+}; we need three $NO_3{}^-$ to balance Co^{3+}.
e.	calcium chloride	$CaCl_2$	We need two Cl^- to balance Ca^{2+}; Ca (Group 2) always forms Ca^{2+}.
f.	lead(IV) oxide	PbO_2	Lead(IV) means Pb^{4+}; we need two O^{2-} to balance Pb^{4+}.
g.	dinitrogen pentoxide	N_2O_5	*di-* means two; *pent(a)-* means five.
h.	ammonium perchlorate	NH_4ClO_4	Contains $NH_4{}^+$ and $ClO_4{}^-$.

SELF-CHECK EXERCISE 5.9

Write the formula for each of the compounds.

a.   ammonium sulfate
b.   vanadium(V) fluoride
c.   disulfur dichloride
d.   rubidium peroxide
e.   aluminum oxide

CHAPTER REVIEW

Key Terms

noble gas (p. 120)
diatomic molecule (p. 120)
ion (p. 125)
cation (p. 125)
anion (p. 126)
ionic compound (p. 130)
binary compound (p. 134)
binary ionic compound (p. 135)
polyatomic ion (p. 149)

Summary

1. Each chemical element is composed of a given type of atom. These atoms may exist as individual atoms or as groups of like atoms. For example, the noble gases contain single, separated atoms. However, elements such as oxygen, nitrogen, and chlorine exist as diatomic (two-atom) molecules.

2. When an atom loses one or more electrons, it forms a positive ion called a cation. This behavior is characteristic of metals. When an atom gains one or more electrons, it becomes a negatively charged ion called an anion. This behavior is characteristic of nonmetals. Oppositely charged ions form ionic compounds. A compound is always neutral overall—it has zero net charge.

3. The elements in Groups 1 and 2 on the periodic table form 1+ and 2+ cations, respectively. Group 7 atoms can gain one electron to form 1− ions. Group 6 atoms form 2− ions.

4. Binary compounds can be named systematically by following a set of relatively simple rules. For compounds containing both a metal and a nonmetal, the metal is always named first, followed by a name derived from the root name for the nonmetal. For compounds containing a metal that can form more than one cation (Type II), we use a Roman numeral to specify the cation's charge. In binary compounds containing only nonmetals (Type III), prefixes are used to specify the numbers of atoms.

5. Polyatomic ions are charged entities composed of several atoms bound together. These have special names that must be memorized. Naming ionic compounds that contain polyatomic ions is very similar to naming binary ionic compounds.

6. The names of acids (molecules with one or more H^+ ions attached to an anion) depend on whether or not the anion contains oxygen.

Questions and Problems

All even-numbered exercises have answers in the back of this book and solutions in the Solutions Guide.

5.1 Natural States of the Elements

QUESTIONS

1. Most substances contain _____ rather than elemental substances.

2. Most of the chemical elements are quite _____, and tend to be found in nature only in combined form (in compounds).

3. The noble gas present in relatively large concentrations in the atmosphere is _____.

4. The gaseous elements helium, neon, argon, krypton, and xenon are called the _____ gases because they do not readily react with other elements.

5. Molecules of nitrogen gas and oxygen gas are said to be _____, which means they consist of pairs of atoms.

6. The gases helium, neon, and argon contain (diatomic molecules/single atoms).

7. A simple way to generate elemental hydrogen gas is to pass _____ through water.

8. If sodium chloride (table salt) is melted and then subjected to an electrical current, elemental _____ gas is produced, along with sodium metal.

9. Only four elements are liquids at or just above room temperature: bromine, gallium, cesium, and _____.

10. The two elemental forms of carbon are graphite and _____.

5.2 Ions

QUESTIONS

11. An isolated atom has a net charge of _____.

12. Ions are formed when an atom gains or loses _____; they never involve a change in the atom's nucleus.

13. When an atom loses one electron, the ion formed has a charge of _____.

14. An ion with two more protons than electrons has a charge of _____.

15. Positive ions are called _____, whereas negative ions are called _____.

16. Simple negative ions formed from single atoms are given names that end in _____.

17. The tendency to *lose* electrons is a fundamental property of the _____ elements.

18. The tendency to *gain* electrons is a fundamental property of the _____ elements.

19. For each of the positive ions listed in column 1, use the periodic table to find in column 2 the total number of electrons that ion contains. The same answer may be used more than once.

Column 1	Column 2
[1] Al^{3+}	[a] 2
[2] Fe^{3+}	[b] 10
[3] Mg^{2+}	[c] 21
[4] Sn^{2+}	[d] 23
[5] Co^{2+}	[e] 24
[6] Co^{3+}	[f] 25
[7] Li^{+}	[g] 36
[8] Cr^{3+}	[h] 48
[9] Rb^{+}	[i] 76
[10] Pt^{2+}	[j] 81

20. For each of the following ions, indicate the total number of electrons and protons present in the ion:
    a. Na^{+}
    b. As^{3-}
    c. Fe^{2+}
    d. Ca^{2+}
    e. N^{3-}
    f. Sc^{+}
    g. Sr^{2+}
    h. K^{+}

21. For the following processes that show the formation of ions, use the periodic table to indicate the number of electrons and protons present in both the *ion* and the *neutral atom* from which it is made.

   a. $K \rightarrow K^+ + e^-$ d. $Sr \rightarrow Sr^{2+} + 2e^-$
   b. $Ba \rightarrow Ba^{2+} + 2e^-$ e. $O + 2e^- \rightarrow O^{2-}$
   c. $C + 4e^- \rightarrow C^{4-}$ f. $I + e^- \rightarrow I^-$

22. For the following processes that show the formation of ions, use the periodic table to indicate the number of electrons and protons present in both the *ion* and the *neutral atom* from which it is made.

   a. $Al \rightarrow Al^{3+} + 3e^-$ d. $Cl + e^- \rightarrow Cl^-$
   b. $S + 2e^- \rightarrow S^{2-}$ e. $Na \rightarrow Na^+ + e^-$
   c. $Fe \rightarrow Fe^{3+} + 3e^-$ f. $N + 3e^- \rightarrow N^{3-}$

23. For each of the following atomic numbers, use the periodic table to write the formula (including the charge) for the simple *ion* that the element is most likely to form.

   a. 85   d. 56
   b. 88   e. 13
   c. 34   f. 38

24. For the atom corresponding to each of the following atomic numbers, use the periodic table to write the formula (including the charge) for the simple *ion* that the atom is most likely to form.

   a. 55   d. 35
   b. 20   e. 37
   c. 16   f. 53

5.3 Compounds That Contain Ions

QUESTIONS

25. List some properties of a substance that would lead you to believe it consists of ions. How do these properties differ from those of nonionic compounds?

26. Why does a solution of sodium chloride in water conduct an electric current, whereas a solution of sugar in water does not?

27. Why does an ionic compound conduct an electric current when the compound is melted but not when it is in the solid state?

28. Why must the total number of positive charges in an ionic compound equal the total number of negative charges?

29. For the following pairs of ions, use the principle of electrical neutrality to predict the formula of the binary compound that the ions are most likely to form.

   a. Mg^{2+} and I^- e. Fe^{3+} and O^{2-}
   b. Ti^{4+} and O^{2-} f. Fe^{2+} and O^{2-}
   c. Cr^{3+} and O^{2-} g. Ba^{2+} and Se^{2-}
   d. Rb^+ and N^{3-} h. K^+ and P^{3-}

30. For the following pairs of ions, use the principle of electrical neutrality to predict the formula of the binary compound that the ions will form.

   a. Fe^{2+} and S^{2-} e. Ca^{2+} and N^{3-}
   b. Rb^+ and N^{3-} f. Fe^{3+} and I^-
   c. Ba^{2+} and O^{2-} g. Pb^{4+} and O^{2-}
   d. Al^{3+} and S^{2-} h. Al^{3+} and Br^-

5.4 Naming Compounds

QUESTION

31. A compound containing only two elements is called a _____ compound.

5.5 Naming Compounds That Contain a Metal and a Nonmetal

QUESTIONS

32. In naming ionic compounds, we always name the _____ first.

33. A simple _____ ion has the same name as its parent element.

34. Although the formula of sodium chloride is written simply as NaCl, the compound actually contains _____ and _____ ions.

35. For a metallic element that forms two stable cations, the ending _____ is sometimes used for the cation with the lower charge.

36. We indicate the charge of a metallic element that forms more than one cation by adding a _____ after the name of the cation.

37. Give the name of each of the following simple binary ionic compounds.
    a. Na_2O
    b. K_2S
    c. $MgCl_2$
    d. $CaBr_2$
    e. BaI_2
    f. Al_2S_3
    g. $CsBr$
    h. AgF

38. Give the name of each of the following simple binary ionic compounds.
    a. $AlCl_3$
    b. Na_2S
    c. MgO
    d. $BaCl_2$
    e. LiI
    f. Ag_2O
    g. RaF_2
    h. SrS

39. In which of the following pairs is the name incorrect?
    a. PBr_3, potassium bromide
    b. Na_3P, trisodium phosphate
    c. $CaCl_2$, calcium dichloride
    d. Al_2O_3, dialuminum trioxide
    e. $SnCl_4$, tin(II) chloride

40. In which of the following pairs is the name incorrect?
    a. silver oxide, S_2O
    b. aluminum sulfide, Al_2S_3
    c. trisodium nitride, Na_3N
    d. barium dichloride, $BaCl_2$
    e. strontium hydride, SrH_2

41. Write the name of each of the following ionic substances, using the system that includes a Roman numeral to specify the charge of the cation.
    a. $CrCl_3$
    b. SnI_4
    c. Cu_2O
    d. Fe_2S_3
    e. CuO
    f. $AuCl_3$

42. Write the name of each of the following ionic substances, using the system that includes a Roman numeral to specify the charge of the cation.
    a. $PbCl_2$
    b. Fe_2O_3
    c. SnI_2
    d. Hg_2O
    e. HgS
    f. CuI

43. Write the name of each of the following ionic substances, using *-ous* or *-ic* to indicate the charge of the cation.
    a. PbO
    b. $HgCl_2$
    c. $CuCl_2$
    d. Hg_2Cl_2
    e. $CoCl_2$
    f. $SnCl_2$

44. Write the name for each of the following substances, in which the charge of the cation is indicated using the *-ous/-ic* notation.
    a. PbO_2
    b. $SnBr_2$
    c. Cu_2S
    d. CuI
    e. Hg_2I_2
    f. CrF_3

5.6 Naming Binary Compounds That Contain Only Nonmetals (Type III)

QUESTIONS

45. Name each of the following binary compounds of nonmetallic elements.
    a. CBr_4
    b. N_2O_3
    c. PCl_3
    d. ICl
    e. NCl_3
    f. SiF_4

46. Write the formula for each of the following binary compounds of nonmetallic elements.
    a. CO
    b. SO_3
    c. N_2Cl_4
    d. CI_4
    e. PF_5
    f. P_2O_5

5.7 Naming Binary Compounds: A Review

QUESTIONS

47. Name each of the following binary compounds, being careful to determine from the periodic table whether the compound is likely to be ionic (containing a metal and a nonmetal) or nonionic (containing only nonmetals).
    a. $CoCl_3$
    b. P_2Cl_4
    c. BCl_3
    d. Hg_2O
    e. $SnCl_4$
    f. $SiCl_4$

48. Name each of the following binary compounds, being careful to determine from the periodic table whether the compound is likely to be ionic (containing a metal and a nonmetal) or nonionic (containing only nonmetals).
    a. B_2H_6
    b. Ca_3N_2
    c. CBr_4
    d. Ag_2S
    e. $CuCl_2$
    f. ClF

49. Name each of the following binary compounds, being careful to determine from the periodic table whether the compound is likely to be ionic (containing a metal and a nonmetal) or nonionic (containing only nonmetals).
   a. Mg_2S_3 d. $ClBr$
   b. $AlCl_3$ e. Li_2O
   c. PH_3 f. P_4O_{10}

50. Name each of the following binary compounds, being careful to determine from the periodic table whether the compound is likely to be ionic (containing a metal and a nonmetal) or nonionic (containing only nonmetals).
   a. Al_2O_3 d. Co_2S_3
   b. B_2O_3 e. N_2O_5
   c. N_2O_4 f. Al_2S_3

5.8 Naming Compounds That Contain Polyatomic Ions

QUESTIONS

51. A charged entity composed of several atoms bound together is called a _____ ion.

52. An anion containing an element and one or more _____ atoms is called an oxyanion.

53. For the oxyanions of sulfur, the ending -ite is used for SO_3^{2-} to indicate that it contains _____ than does SO_4^{2-}.

54. Which oxyanion of chlorine contains the most oxygen atoms: hypochlorite, chlorite, or perchlorate?

55. Complete the following list by filling in the missing names or formulas of the oxyanions of chlorine.

   ClO_4^- _____

   _____   hypochlorite

   ClO_3^- _____

   _____   chlorite

56. Complete the following list by filling in the missing names or formulas of the indicated oxyanions.

   BrO^- _____

   _____   iodate

   IO_4^- _____

   _____   hypoiodide

57. Write the formula for each of the following sulfur-containing polyatomic ions, including the overall charge of the ion.
   a. sulfate   c. bisulfate
   b. sulfite   d. hydrogen sulfate

58. Write the formula for each of the following nitrogen-containing polyatomic ions, including the overall charge of the ion.
   a. nitrate   c. ammonium
   b. nitrite   d. cyanide

59. Write the formula for each of the following chlorine-containing ions, including the overall charge of the ion.
   a. chloride
   b. hypochlorite
   c. chlorate
   d. perchlorate

60. Write the formula for each of the following phosphorus-containing anions, including the overall charge of the ion.
   a. phosphate
   b. phosphide
   c. dihydrogen phosphate
   d. hydrogen phosphate

61. Give the name of each of the following polyatomic anions.
   a. MnO_4^- d. $Cr_2O_7^{2-}$
   b. O_2^{2-} e. NO_3^-
   c. CrO_4^{2-} f. SO_3^{2-}

62. Give the name of each of the following polyatomic ions.
   a. HCO_3^- d. $H_2PO_4^-$
   b. NO_2^- e. $C_2H_3O_2^-$
   c. HSO_4^- f. ClO_2^-

63. Name each of the following compounds, which contain polyatomic ions.
   a. $FeCO_3$ d. $KC_2H_3O_2$
   b. $AgCN$ e. $(NH_4)_2Cr_2O_7$
   c. Na_3PO_4 f. $(NH_4)_2CrO_4$

64. Name each of the following compounds, which contain polyatomic ions.
   a. NH_4NO_3 d. $Ca_3(PO_4)_2$
   b. $KClO_3$ e. $NaClO_4$
   c. $PbSO_4$ f. $Cu(OH)_2$

5.9 Naming Acids

QUESTIONS

65. Give a simple definition of an *acid*.

66. Many acids contain the element _____ in addition to hydrogen.

67. Name each of the following acids.
    a. HCl
    b. H_2SO_4
    c. HNO_3
    d. HI
    e. HNO_2
    f. $HClO_3$
    g. HBr
    h. HF
    i. $HC_2H_3O_2$

68. Name each of the following acids.
    a. $HClO_4$
    b. HIO_3
    c. $HBrO_2$
    d. HOCl
    e. H_2SO_3
    f. HCN
    g. H_2S
    h. H_3PO_4

5.10 Writing Formulas from Names

PROBLEMS

69. Write the formula for each of the following simple binary ionic compounds.
    a. lithium bromide
    b. sodium iodide
    c. silver(I) sulfide (usually called silver sulfide)
    d. cesium oxide
    e. beryllium iodide
    f. barium hydride
    g. aluminum fluoride
    h. potassium oxide

70. Write the formula for each of the following ionic substances.
    a. plumbic oxide
    b. stannous bromide
    c. cupric sulfide
    d. cuprous iodide
    e. mercurous chloride
    f. chromic fluoride

71. Write the formula for each of the following binary compounds of nonmetallic elements.
    a. phosphorus triiodide
    b. silicon tetrachloride
    c. dinitrogen pentoxide
    d. iodine monobromide
    e. diboron trioxide
    f. dinitrogen tetroxide
    g. carbon monoxide

72. Write the formula for each of the following binary compounds of nonmetallic elements.
    a. carbon dioxide
    b. sulfur trioxide
    c. dinitrogen tetrachloride
    d. carbon tetraiodide
    e. phosphorus pentafluoride
    f. diphosphorus pentaoxide

73. Write the formula for each of the following compounds, which contain polyatomic ions. Be sure to enclose the polyatomic ion in parentheses if more than one such ion is needed to balance the oppositely charged ion(s).
    a. tin(IV) acetate
    b. sodium peroxide
    c. ammonium hydrogen sulfate
    d. potassium sulfite
    e. mercury(II) sulfate
    f. potassium dihydrogen phosphate
    g. sodium hydrogen sulfite
    h. ammonium phosphate

74. Write the formula for each of the following compounds that contain polyatomic ions. Be sure to enclose the polyatomic ion in parentheses if more than one such ion is needed.
    a. calcium phosphate
    b. ammonium nitrate
    c. aluminum hydrogen sulfate
    d. barium sulfate
    e. ferric nitrate
    f. copper(I) hydroxide

75. Write the formula for each of the following acids.
    a. nitrous acid
    b. acetic acid
    c. hydroiodic (hydriodic) acid
    d. perbromic acid
    e. periodic acid
    f. hydrosulfuric acid
    g. chloric acid
    h. sulfurous acid

76. Write the formula for each of the following acids.
    a. hydrocyanic acid
    b. nitric acid
    c. sulfuric acid
    d. phosphoric acid
    e. hypochlorous acid
    f. hydrofluoric acid
    g. bromous acid
    h. hydrobromic acid

Additional Problems

77. What does an atom lose or gain to become a cation? What does an atom lose or gain to become an anion? Could the same element form an anion under some conditions and a cation under others? Give an example.

78. For Groups 1, 2, 6, and 7 of the periodic table, write representative equations showing how the elements of those groups lose or gain electrons to become simple ions.

79. Although sucrose (table sugar) and sodium chloride (table salt) are very similar in appearance, sodium chloride melts at over 800 °C, whereas table sugar melts at around 185 °C. What sorts of forces give an ionic compound such a high resistance to melting?

80. Before an electrocardiogram (ECG) is recorded for a cardiac patient, the ECG leads are usually coated with a moist paste containing sodium chloride. What property of an ionic substance such as NaCl is being made use of here?

81. What is a *binary* compound? What is a *polyatomic* anion? What is an *oxyanion*?

82. On some periodic tables, hydrogen is listed both as a member of Group 1 and as a member of Group 7. Write an equation showing the formation of H^+ ion and an equation showing the formation of H^- ion.

83. In general, oxyacids contain a nonmetal, oxygen, and what other element?

84. Complete the following list by filling in the missing oxyanion or oxyacid for each pair.

ClO_4^- _____

_____ HIO_3

ClO^- _____

BrO_2^- _____

_____ $HClO_2$

85. Name the following compounds.
    a. $Mg(OH)_2$
    b. Cr_2O_3
    c. P_4O_{10}
    d. $Cr_2(CO_3)_3$
    e. $Fe(NH_4)_2(SO_4)_2$
    f. $KHCO_3$
    g. $CuCN$

86. Name the following compounds.
    a. $AuBr_3$
    b. $Co(CN)_3$
    c. $MgHPO_4$
    d. B_2H_6
    e. NH_3
    f. Ag_2SO_4
    g. $Be(OH)_2$

87. Name the following compounds.
    a. $HClO_3$
    b. $CoCl_3$
    c. B_2O_3
    d. H_2O
    e. $HC_2H_3O_2$
    f. $Fe(NO_3)_3$
    g. $CuSO_4$

88. Name the following compounds.
    a. $(NH_4)_2CO_3$
    b. NH_4HCO_3
    c. $Ca_3(PO_4)_2$
    d. H_2SO_3
    e. MnO_2
    f. HIO_3
    g. KH

89. Most metallic elements form *oxides,* and often the oxide is the most common compound of the element that is found in the earth's crust. Write the formulas for the oxides of the following metallic elements.
    a. potassium
    b. magnesium
    c. iron(II)
    d. iron(III)
    e. zinc(II)
    f. lead(II)
    g. aluminum

90. Consider a hypothetical simple ion M^{4+}. Determine the formula of the compound this ion would form with each of the following anions.
    a. chloride
    b. nitrate
    c. oxide
    d. phosphate
    e. cyanide
    f. sulfate
    g. dichromate

91. Consider a hypothetical element M, which is capable of forming stable simple cations that have charges of $1+, 2+,$ and $3+$, respectively. Write the formulas of the compounds formed by the various M cations with each of the following anions.
    a. chromate
    b. dichromate
    c. sulfide
    d. bromide
    e. bicarbonate
    f. hydrogen phosphate

92. Consider the hypothetical metallic element M, which is capable of forming stable simple cations that have charges of $1+, 2+,$ and $3+$, respectively. Consider also the nonmetallic elements D, E, and F, which form anions that have charges of $1-, 2-,$ and $3-$, respectively. Write the formulas of all possible compounds between metal M and nonmetals D, E, and F.

93. Complete Table 5.A (on page 164) by writing the names and formulas for the ionic compounds formed when the cations listed across the top combine with the anions shown in the left-hand column.

94. Complete Table 5.B (on page 165) by writing the formulas for the ionic compounds formed when the anions listed across the top combine with the cations shown in the left-hand column.

95. The noble metals gold, silver, and platinum are often used in fashioning jewelry because they are relatively _____.

96. The noble gas _____ is frequently found in underground deposits of natural gas.

97. The elementary substances of Group 7 (fluorine, chlorine, bromine, and iodine) consist of molecules containing _____ atom(s).

98. The halogen element that exists under normal conditions in the solid state is _____.

99. When an atom gains two electrons, the ion formed has a charge of _____.

100. An ion with one more electron than protons has a _____ charge.

101. An atom that has lost two electrons has a _____ charge.

102. An atom that has gained one electron has a charge of _____.

103. For each of the negative ions listed in column 1, use the periodic table to find in column 2 the total number of electrons the ion contains. The same answer may be used more than once.

Column 1	Column 2
[1] Se^{2-}	[a] 18
[2] S^{2-}	[b] 35
[3] P^{3-}	[c] 52
[4] O^{2-}	[d] 34
[5] N^{3-}	[e] 36
[6] I^-	[f] 54
[7] F^-	[g] 10
[8] Cl^-	[h] 9
[9] Br^-	[i] 53
[10] At^-	[j] 86

104. For the following processes that show the formation of ions, use the periodic table to indicate the number of electrons and protons present in both the *ion* and the *neutral atom* from which it is made.

Table 5.A

Ions	Fe^{2+}	Al^{3+}	Na^+	Ca^{2+}	$NH_4{}^+$	Fe^{3+}	Ni^{2+}	$Hg_2{}^{2+}$	Hg^{2+}
$CO_3{}^{2-}$									
$BrO_3{}^-$									
$C_2H_3O_2{}^-$									
OH^-									
$HCO_3{}^-$									
$PO_4{}^{3-}$									
$SO_3{}^{2-}$									
$ClO_4{}^-$									
$SO_4{}^{2-}$									
O^{2-}									
Cl^-									

Table 5.B

Ions	bromide	hydrogen carbonate	hydride	acetate	hydrogen sulfate	phosphate
calcium						
strontium						
ammonium						
aluminum						
iron(III)						
nickel(II)						
silver(I)						
gold(III)						
potassium						
mercury(II)						
barium						

a. $Mn \rightarrow Mn^{2+} + 2e^-$ d. $Co \rightarrow Co^{3+} + 3e^-$
b. $Ni \rightarrow Ni^{2+} + 2e^-$ e. $Fe \rightarrow Fe^{2+} + 2e^-$
c. $N + 3e^- \rightarrow N^{3-}$ f. $P + 3e^- \rightarrow P^{3-}$

105. For each of the following atomic numbers, use the periodic table to write the formula (including the charge) for the simple *ion* that the element is most likely to form.
  a. 36   d. 81
  b. 31   e. 35
  c. 52   f. 87

106. For the following pairs of ions, use the principle of electrical neutrality to predict the formula of the binary compound that the ions are most likely to form.
  a. Na^+ and S^{2-} e. Cu^{2+} and Br^-
  b. K^+ and Cl^- f. Al^{3+} and I^-
  c. Ba^{2+} and O^{2-} g. Al^{3+} and O^{2-}
  d. Mg^{2+} and Se^{2-} h. Ca^{2+} and N^{3-}

107. Give the name of each of the following simple binary ionic compounds.
  a. BeO   c. Na_2S e. HCl   g. Ag_2S
  b. MgI_2 d. Al_2O_3 f. LiF   h. CaH_2

108. In which of the following pairs is the name incorrect?
  a. $CaCl_2$, calcium chloride
  b. AlH_3, aluminum trihydride
  c. K_2O, potassium oxide
  d. $Fe(OH)_2$, iron(III) hydroxide
  e. $CoCl_3$, cobalt(II) chloride

109. Write the name of each of the following ionic substances, using the system that includes a Roman numeral to specify the charge of the cation.
  a. $FeBr_2$ c. Co_2S_3 e. Hg_2Cl_2
  b. CoS d. SnO_2 f. $HgCl_2$

110. Write the name of each of the following ionic substances, using -*ous* or -*ic* to indicate the charge of the cation.
  a. $CoBr_3$ c. Fe_2O_3 e. $SnCl_4$
  b. PbI_4 d. FeS f. SnO

111. Name each of the following binary compounds of nonmetallic elements.
  a. XeF_6 d. N_2O_4
  b. OF_2 e. Cl_2O
  c. AsI_3 f. SF_6

112. Name each of the following binary compounds, being careful to determine from the periodic table whether the compound is likely to be ionic (containing a metal and a nonmetal) or nonionic (containing only nonmetals).
    a. Fe_2S_3 c. AsH_3 e. K_2O
    b. $AuCl_3$ d. ClF   f. CO_2
113. Which oxyanion of nitrogen contains a larger number of oxygen atoms, the nitrate ion or the nitrite ion?
114. Write the formula for each of the following carbon-containing polyatomic ions, including the overall charge of the ion.
    a. carbonate   c. acetate
    b. hydrogen carbonate   d. cyanide
115. Write the formula for each of the following chromium-containing ions, including the overall charge of the ion.
    a. chromous   c. chromic
    b. chromate   d. dichromate
116. Give the name of each of the following polyatomic anions.
    a. CO_3^{2-} c. SO_4^{2-} e. ClO_4^{-}
    b. ClO_3^{-} d. PO_4^{3-} f. MnO_4^{-}
117. Name each of the following compounds, which contain polyatomic ions:
    a. LiH_2PO_4 c. $Pb(NO_3)_2$ e. $NaClO_2$
    b. $Cu(CN)_2$ d. Na_2HPO_4 f. $Co_2(SO_4)_3$
118. Write the formula for each of the following simple binary ionic compounds.
    a. calcium chloride
    b. silver(I) oxide (usually called silver oxide)
    c. aluminum sulfide
    d. beryllium bromide
    e. hydrogen sulfide

    f. potassium hydride
    g. magnesium iodide
    h. cesium fluoride
119. Write the formula for each of the following binary compounds of nonmetallic elements.
    a. sulfur dioxide
    b. dinitrogen monoxide
    c. xenon tetrafluoride
    d. tetraphosphorus decoxide
    e. phosphorus pentachloride
    f. sulfur hexafluoride
    g. nitrogen dioxide
120. Write the formula for each of the following ionic substances.
    a. magnesium phosphide   d. ferrous iodide
    b. calcium fluoride   e. barium oxide
    c. cobalt(III) bromide   f. potassium sulfide
121. Write the formula for each of the following compounds, which contain polyatomic ions. Be sure to enclose the polyatomic ion in parentheses if more than one such ion is needed to balance the oppositely charged ion(s).
    a. silver(I) perchlorate (usually called silver perchlorate)
    b. cobalt(III) hydroxide
    c. sodium hypochlorite
    d. potassium dichromate
    e. ammonium nitrite
    f. ferric hydroxide
    g. ammonium hydrogen carbonate
    h. potassium perbromate

Solutions to Self-Check Exercises

SELF-CHECK EXERCISE 5.1

a. KI
    $(1+) + (1-) = 0$

b. Mg_3N_2
    $3(2+) + 2(3-) = (6+) + (6-) = 0$

c. Al_2O_3
    $2(3+) + 3(2-) = 0$

SELF-CHECK EXERCISE 5.2

a. rubidium oxide   b. strontium iodide   c. potassium sulfide

SELF-CHECK EXERCISE 5.3

a. The compound $PbBr_2$ must contain Pb^{2+}—named lead(II)—to balance the charges of the two Br^- ions. Thus the name is lead(II) bromide. The compound $PbBr_4$ must contain Pb^{4+}—named lead(IV)—to balance the charges of the four Br^- ions. The name is therefore lead(IV) bromide.

b. The compound FeS contains the S^{2-} ion (sulfide) and thus the iron cation present must be Fe^{2+}, iron(II). The name is iron(II) sulfide. The compound Fe_2S_3 contains three S^{2-} ions and two iron cations of unknown charge. We can determine the iron charge from the following:

$$2(?+) \quad + \; 3(2-) \quad = 0$$

In this case, ? must represent 3 because

$$2(3+) \; + \; 3(2-) \; = \; 0$$

Thus Fe_2S_3 contains Fe^{3+} and S^{2-}, and its name is iron(III) sulfide.

c. The compound $AlBr_3$ contains Al^{3+} and Br^-. Because aluminum forms only one ion (Al^{3+}), no Roman numeral is required. The name is aluminum bromide.

d. The compound Na_2S contains Na^+ and S^{2-} ions. The name is sodium sulfide. (Because sodium forms only Na^+, no Roman numeral is needed.)

e. The compound $CoCl_3$ contains three Cl^- ions. Thus the cobalt cation must be Co^{3+}, which is named cobalt(III) because cobalt is a transition metal and can form more than one type of cation. Thus the name of $CoCl_3$ is cobalt(III) chloride.

SELF-CHECK EXERCISE 5.4

	Compound	Individual Names	Prefixes	Name
a.	CCl_4	carbon chloride	none *tetra-*	carbon tetrachloride
b.	NO_2	nitrogen oxide	none *di-*	nitrogen dioxide
c.	IF_5	iodine fluoride	none *penta-*	iodine pentafluoride

SELF-CHECK EXERCISE 5.5

a. silicon dioxide   b.   dioxygen difluoride   c.   xenon hexafluoride

---

SELF-CHECK EXERCISE 5.6

a. chlorine trifluoride   d.   manganese(IV) oxide
b. vanadium(V) fluoride   e.   magnesium oxide
c. copper(I) chloride   f.   water

SELF-CHECK EXERCISE 5.7

a. calcium hydroxide
b. sodium phosphate
c. potassium permanganate
d. ammonium dichromate

e. cobalt(II) perchlorate   (Perchlorate has a $1-$ charge, so the cation must be Co^{2+} to balance the two ClO_4^- ions.)
f. potassium chlorate
g. copper(II) nitrite   (This compound contains two NO_2^- (nitrite) ions and thus must contain a Cu^{2+} cation.)

SELF-CHECK EXERCISE 5.8

	Compound	Name	Comments
a.	$NaHCO_3$	sodium hydrogen carbonate	Contains Na^+ and HCO_3^-; often called sodium bicarbonate (common name).
b.	$BaSO_4$	barium sulfate	Contains Ba^{2+} and SO_4^{2-}.
c.	$CsClO_4$	cesium perchlorate	Contains Cs^+ and ClO_4^-.
d.	BrF_5	bromine pentafluoride	Both nonmetals (Type III binary).
e.	NaBr	sodium bromide	Contains Na^+ and Br^- (Type I binary).
f.	KOCl	potassium hypochlorite	Contains K^+ and OCl^-.
g.	$Zn_3(PO_4)_2$	zinc(II) phosphate	Contains Zn^{2+} and PO_4^{3-}; Zn is a transition metal and requires a Roman numeral.

SELF-CHECK EXERCISE 5.9

	Name	Chemical Formula	Comments
a.	ammonium sulfate	$(NH_4)_2SO_4$	Two ammonium ions (NH_4^+) are required for each sulfate ion (SO_4^{2-}) to achieve charge balance.
b.	vanadium(V) fluoride	VF_5	The compound contains V^{5+} ions and requires five F^- ions for charge balance.
c.	disulfur dichloride	S_2Cl_2	The prefix *di-* indicates two of each atom.
d.	rubidium peroxide	Rb_2O_2	Because rubidium is in Group 1, it forms only $1+$ ions. Thus two Rb^+ ions are needed to balance the $2-$ charge on the peroxide ion (O_2^{2-}).
e.	aluminum oxide	Al_2O_3	Aluminum forms only $3+$ ions. Two Al^{3+} ions are required to balance the charge on three O^{2-} ions.

CHAPTER

6 Chemical Reactions: An Introduction

CONTENTS

Writing the formulas of products of reactions and balancing equations depends upon knowing the material in Chapter 5. Most students will not yet have mastered names, formulas, and charges. As you work examples, let students supply the needed formulas and charges to help reinforce the need for continued practice with naming.

Magnesium reacts with oxygen to give an intense white flame.

Young corn plants grow as a result of chemical changes.

Chemistry is about change.

> Plants grow.
> Steel rusts.
> Hair is bleached, dyed, "permed", or straightened.
> Natural gas burns to heat houses.
> Nylon is produced for jackets, swimsuits, and pantyhose.
> Water is decomposed to hydrogen and oxygen gas by an electric current.
> Grape juice ferments to produce wine.

These are just a few examples of chemical changes that affect each of us. Chemical reactions are the heart and soul of chemistry, and in this chapter we will discuss the fundamental ideas about chemical reactions.

6.1   Evidence for a Chemical Reaction

AIM: To discuss the signals that show a chemical reaction has occurred.

How do we know when a chemical reaction has occurred? That is, what are the clues that a chemical change has taken place? A glance back at the processes in the above list suggests that *chemical reactions often give a visual signal.* Steel changes from a smooth, shiny material to a reddish-brown, flaky substance when

Production of plastic film for use in containers such as soft drink bottles.

Nylon being drawn from the boundary between two solutions containing different reactants.

The cold packs and hand warmers mentioned in the textbook are interesting and practical examples of the production and absorption of heat. Hand warmers can often be purchased at outdoor shops, and cold packs at large drugstores.

(a)

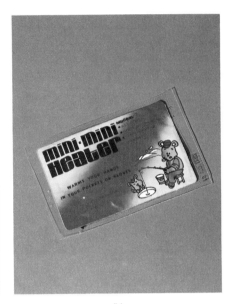

(b)

Figure 6.1
Bubbles of hydrogen and oxygen gas form when an electric current is used to decompose water.

Figure 6.2
(a) An athlete wears a cold pack to help prevent swelling of an injury. The pack is activated by breaking an ampule; this initiates a chemical reaction that absorbs heat rapidly, lowering the temperature of the area to which the pack is applied. (b) A hot pack used to warm hands and feet in winter. When the package is opened, oxygen from the air penetrates a bag containing solid chemicals. The resulting reaction produces heat for several hours.

it rusts. Hair changes color when it is bleached. Solid nylon is formed when two liquid solutions are brought into contact. A blue flame appears when natural gas reacts with oxygen. Chemical reactions, then, often give *visual* clues: a color changes, a solid forms, bubbles are produced (see Figure 6.1), a flame occurs, and so on. However, reactions are not always visible. Sometimes the only signal that a reaction is occurring is a change in temperature as heat is produced or absorbed (see Figure 6.2).

Table 6.1 summarizes common clues to the occurrence of a chemical reaction, and Figure 6.3 (p. 172) gives some examples of reactions that show these clues.

Energy and chemical reactions will be discussed in more detail in Chapter 8.

Show as many kinds of reactions as possible. Table 6.1 presents a list of ''clues'' that a reaction has occurred. Show students examples of each clue. For color change, add hydrochloric acid to K_2CrO_4. For the formation of a solid, mix $BaCl_2$ with Na_2CO_3. For bubbles, add some acetic acid to sodium bicarbonate, or dilute HCl to zinc metal. For heat or flame, drop a tiny piece of sodium into water. The formulas for the reactants and products of these reactions are all included on the lists of monatomic and polyatomic ions to be memorized.

Table 6.1   Some Clues That a Chemical Reaction Has Occurred

1. The color changes.
2. A solid forms.
3. Bubbles form.
4. Heat and/or a flame is produced, or heat is absorbed.

(a)   (b)   (c)   (d)

Figure 6.3

(a) When colorless hydrochloric acid is added to a red solution of cobalt(II) nitrate, the solution turns blue, a sign that a chemical reaction has taken place. (b) A solid forms when a solution of sodium dichromate is added to a solution of lead nitrate. (c) Bubbles of hydrogen gas form when calcium metal reacts with water. (d) Methane gas reacts with oxygen to produce a flame in a bunsen burner.

This section shows how elements rearrange to form new compounds during a chemical reaction. Ball and stick models can be very beneficial when introducing reactions. You can show that there are no balls left over after the reaction has occurred. All of the atoms in the reactants are found in the products.

This is a good place to point out that conditions under which a reaction will occur are sometimes printed above or below the arrow.

6.2   Chemical Equations

AIM: To identify the characteristics of a chemical reaction and the information given by a chemical equation.

Chemists have learned that a chemical change always involves a rearrangement of the ways in which the atoms are grouped. For example, when the methane, CH_4, in natural gas combines with oxygen, O_2, in the air and burns, carbon dioxide, CO_2, and water, H_2O, are formed. A chemical change such as this is called a **chemical reaction.** We represent a chemical reaction by writing a **chemical equation** in which the chemicals present before the reaction (the **reactants**) are shown to the left of an arrow and the chemicals formed by the reaction (the **products**) are shown to the right of an arrow. The arrow indicates the direction of the change and is read as "yields" or "produces":

$$\text{Reactants} \rightarrow \text{products}$$

For the reaction of methane with oxygen, we have

Methane   Oxygen   Carbon   Water
dioxide

$$CH_4 \;+\; O_2 \;\rightarrow\; CO_2 \;+\; H_2O$$

Reactants   Products

Note from this equation that the products contain the same atoms as the reactants but that the atoms are associated in different ways. That is, a *chemical reaction involves changing the ways the atoms are grouped.*

It is important to recognize that **in a chemical reaction, atoms are neither created nor destroyed.** *All atoms present in the reactants must be ac-*

counted for among the products. In other words, there must be the same number of each type of atom on the product side as on the reactant side of the arrow. Making sure that the equation for a reaction obeys this rule is called **balancing the chemical equation** for a reaction.

The equation that we have shown for the reaction between CH_4 and O_2 is not balanced. We can see that it is not balanced by taking the reactants and products apart.

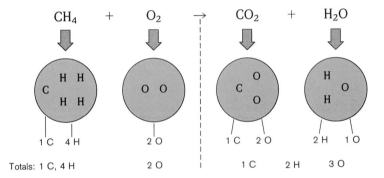

The reaction cannot happen this way because, as it stands, this equation states that one oxygen atom is created and two hydrogen atoms are destroyed. A reaction is only a rearrangement of the way the atoms are grouped; atoms are not created or destroyed. The total number of each type of atom must be the same on both sides of the arrow. We can fix the imbalance in this equation by involving one more O_2 molecule on the left and by showing the production of one more H_2O molecule on the right.

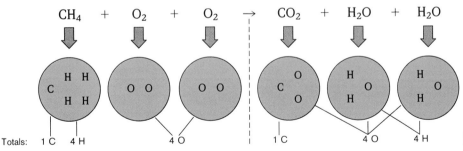

This *balanced chemical equation* shows the actual numbers of molecules involved in this reaction (see Figure 6.4).

When we write the balanced equation for a reaction, we group like molecules together. Thus

$$CH_4 + \boxed{O_2 + O_2} \rightarrow CO_2 + \boxed{H_2O + H_2O}$$

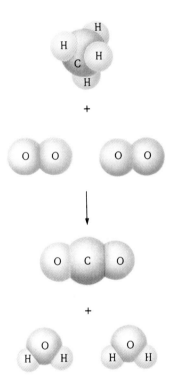

Figure 6.4
The reaction between methane and oxygen to give water and carbon dioxide. Note that there are four oxygen atoms in the products *and* in the reactants; none have been gained or lost in the reaction. Similarly, there are four hydrogen atoms and one carbon atom in the reactants *and* in the products. The reaction simply changes the way the atoms are grouped.

is written

$$CH_4 + \boxed{2O_2} \rightarrow CO_2 + \boxed{2H_2O}$$

The chemical equation for a reaction provides us with two important types of information:

1. The identities of the reactants and products
2. The relative numbers of each

Physical States

Besides specifying the compounds involved in the reaction, we often indicate in the equation the *physical states* of the reactants and products by using the following symbols:

Students often have trouble distinguishing between liquid water and an aqueous solution. This is especially troublesome when water is a product of the reaction.

State	Symbol
solid	(s)
liquid	(l)
gas	(g)
dissolved in water (in aqueous solution)	(aq)

For example, when solid potassium reacts with liquid water, the products are hydrogen gas and potassium hydroxide; the latter remains dissolved in the water. From this information about the reactants and products, we can write the equation for the reaction. Solid potassium is represented by $K(s)$; liquid water is written as $H_2O(l)$; hydrogen gas contains diatomic molecules and is represented as $H_2(g)$; potassium hydroxide dissolved in water is written as $KOH(aq)$. So the *unbalanced* equation for the reaction is

Solid Water Hydrogen Potassium hydroxide
potassium   gas   dissolved in water
$K(s) \quad + H_2O(l) \rightarrow \quad H_2(g) \quad + \quad KOH(aq)$

This reaction is shown in Figure 6.5.

The hydrogen gas produced in this reaction then reacts with the oxygen gas in the air, producing gaseous water and a flame. The *unbalanced* equation for this second reaction is

$$H_2(g) + O_2(g) \rightarrow H_2O(g)$$

Both of these reactions produce a great deal of heat. In Example 6.1 we will practice writing the unbalanced equations for reactions. Then, in the next section, we will discuss systematic procedures for balancing equations.

(a)

(b)

(c)

Figure 6.5
The reactants (a) solid potassium and (b) water. (c) The reaction of potassium with water. The flame occurs because the hydrogen gas, $H_2(g)$, formed burns (reacts with O_2 in the air) at the high temperatures produced by the reaction.

EXAMPLE 6.1	Chemical Equations: Recognizing Reactants and Products

Write the *unbalanced* chemical equation for each of the following reactions.

a. Solid mercury(II) oxide decomposes to produce liquid mercury metal and gaseous oxygen.
b. Solid carbon reacts with gaseous oxygen to form gaseous carbon dioxide.
c. Solid zinc is added to an aqueous solution containing dissolved hydrogen chloride to produce gaseous hydrogen that bubbles out of the solution and zinc(II) chloride that remains dissolved in the water.

SOLUTION

a. In this case we have only one reactant, mercury(II) oxide. The name mercury(II) oxide means that the Hg^{2+} cation is present, so one O^{2-} ion is required for a zero net charge. Thus the formula is HgO, which will be written HgO(s) in this case because it is given as a solid. The products are liquid mercury, written Hg(l), and gaseous oxygen, written $O_2(g)$. (Remember that oxygen exists as a diatomic molecule under normal conditions.) The unbalanced equation is

Reactant   Products

$$HgO(s) \rightarrow Hg(l) + O_2(g)$$

Zinc reacts with hydrochloric acid to produce bubbles of hydrogen gas.

EXAMPLE 6.1, CONTINUED

b. In this case solid carbon, written C(s), reacts with oxygen gas, $O_2(g)$, to form gaseous carbon dioxide, which is written $CO_2(g)$. The equation (which happens to be balanced) is

$$\underset{\text{Reactants}}{C(s) + O_2(g)} \rightarrow \underset{\text{Product}}{CO_2(g)}$$

c. In this reaction solid zinc, Zn(s), is added to an aqueous solution of hydrogen chloride, which is written HCl(aq) and usually called hydrochloric acid. These are the reactants. The products of the reaction are gaseous hydrogen, $H_2(g)$, and aqueous zinc(II) chloride. The name zinc(II) chloride means that the Zn^{2+} ion is present, so two Cl^- ions are needed to achieve a zero net charge. Thus zinc(II) chloride dissolved in water is written $ZnCl_2(aq)$. The unbalanced equation for the reaction is

$$\underset{\text{Reactants}}{Zn(s) + HCl(aq)} \rightarrow \underset{\text{Products}}{H_2(g) + ZnCl_2(aq)}$$

SELF-CHECK EXERCISE 6.1

Identify the reactants and products and write the *unbalanced* equation (including symbols for states) for each of the following chemical reactions.

a. Solid magnesium metal reacts with liquid water to form solid magnesium hydroxide and hydrogen gas.
b. Solid ammonium dichromate (review Table 5.5 if this compound is unfamiliar) decomposes to solid chromium(III) oxide, gaseous nitrogen, and gaseous water.
c. Gaseous ammonia reacts with gaseous oxygen to form gaseous nitrogen monoxide and gaseous water.

6.3 Balancing Chemical Equations

AIM: To show how to write a balanced equation for a chemical reaction.

Now is a good time to remind students about the formulas of elements in their natural states. They should review Section 5.1 and memorize the list of elements that are found as diatomic molecules.

As we saw in the previous section, an unbalanced chemical equation is not an accurate representation of the reaction that occurs. Whenever you see an equation for a reaction, you should ask yourself whether it is balanced. The principle that lies at the heart of the balancing process is that **atoms are conserved in a chemical reaction.** That is, atoms are neither created nor destroyed. They are just grouped differently. The same number of each type of atom is found among the reactants and among the products.

Chemists determine the identity of the reactants and products of a reaction by experimental observation. For example, when methane (natural gas) is burned in the presence of sufficient oxygen gas, the products are always carbon dioxide and water. **The identities (formulas) of the compounds must never be changed in balancing a chemical equation.** In other words, the subscripts in a formula cannot be changed, nor can atoms be added to or subtracted from a formula.

Most chemical equations can be balanced by trial and error—that is, by inspection. Keep trying until you find the numbers of reactants and products that give the same number of each type of atom on both sides of the arrow. For example, consider the reaction of hydrogen gas and oxygen gas to form liquid water. First, we write the unbalanced equation from the description of the reaction.

$$H_2(g) + O_2(g) \rightarrow H_2O(l)$$

We can see that this equation is unbalanced by counting the atoms on both sides of the arrow.

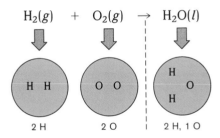

We have one more oxygen atom in the reactants than in the products. Because we cannot create or destroy atoms and because we *cannot change the formulas* of the reactants or products, we must balance the equation by adding more molecules of reactants and/or products. In this case we need one more oxygen atom on the right, so we add another water molecule (which contains one O atom). Then we count all of the atoms again.

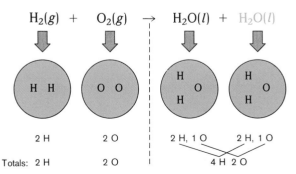

It can be hard to convince students that in order to balance an equation they cannot change the subscripts of products or reactants, but only the coefficients. An analogy to cooking may be helpful. If 1 cup of milk + 1 tbsp of cocoa + 1 tbsp sugar make 1 cup of cocoa, then three cups of milk + 3 tbsp of cocoa + 3 tbsp of sugar make 3 cups of hot cocoa, not strawberry ice cream. Adjusting the amounts of reactants alters the amount of product. It does not produce a different product.

Trial and error is often useful for solving problems. It's ok to make a few wrong turns before you get to the right answer.

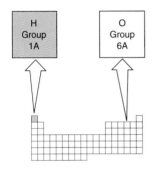

Reactants	Products
2 H	2 H
2 O	1 O

Reactants	Products
2 H	4 H
2 O	2 O

We have balanced the oxygen atoms, but now the hydrogen atoms have become unbalanced. There are more hydrogen atoms on the right than on the left. We can solve this problem by adding another hydrogen molecule (H_2) to the reactant side.

$$H_2(g) \;+\; H_2(g) \;+\; O_2(g) \;\rightarrow\; H_2O(l) \;+\; H_2O(l)$$

Reactants	Products
4 H	4 H
2 O	2 O

Totals: 4 H   2 O   4 H   2 O

The equation is now balanced. We have the same numbers of hydrogen and oxygen atoms represented on both sides of the arrow. Collecting like molecules, we write the balanced equation as

$$2H_2(g) \;+\; O_2(g) \rightarrow 2H_2O(l)$$

Consider next what happens if we multiply every part of this balanced equation by 2:

$$2 \times [2H_2(g) \;+\; O_2(g) \rightarrow 2H_2O(l)]$$

to give

$$4H_2(g) \;+\; 2O_2(g) \rightarrow 4H_2O(l)$$

This equation is balanced (count the atoms to verify this). In fact, we can multiply or divide *all parts* of the original balanced equation by any number to give a new balanced equation. Thus each chemical reaction has many possible balanced equations. Is one of the many possibilities preferred over the others? Yes.

The accepted convention is that the "best" balanced equation is the one with the *smallest integers (whole numbers)*. These integers are called the **coefficients** for the balanced equation. Therefore, for the reaction of hydrogen and oxygen to form water, the "correct" balanced equation is

$$2H_2(g) \;+\; O_2(g) \rightarrow 2H_2O(l)$$

The coefficients 2, 1 (never written), and 2, respectively, are the smallest *integers* that give a balanced equation for this reaction.

Next we will balance the equation for the reaction of liquid ethanol, C_2H_5OH, with oxygen gas to form gaseous carbon dioxide and water. This reaction, among many others, occurs in engines that burn a gasoline–ethanol mixture called gasohol.

The first step in obtaining the balanced equation for a reaction is always to identify the reactants and products from the description given for the reaction. In

Some students may have seen equations balanced using partial molecules, such as $\frac{1}{2}O_2$. Students should be encouraged to use only whole numbers of molecules.

Students need to be exposed to as many different equations as possible. When you work examples, try to choose ones that require different strategies for successful balancing.

this case, we are told that liquid ethanol, $C_2H_5OH(l)$, reacts with gaseous oxygen, $O_2(g)$, to produce gaseous carbon dioxide, $CO_2(g)$, and gaseous water, $H_2O(g)$. Therefore, the unbalanced equation is

$$C_2H_5OH(l) \;+\; O_2(g) \;\rightarrow\; CO_2(g) \;+\; H_2O(g)$$

Liquid   Gaseous   Gaseous   Gaseous
ethanol   oxygen   carbon   water
                           dioxide

When one molecule in an equation is more complicated (contains more elements) than the others, it is best to start with that molecule. The most complicated molecule here is C_2H_5OH, so we begin by considering the products that contain the atoms in C_2H_5OH. We start with carbon. The only product that contains carbon is CO_2. Because C_2H_5OH contains two carbon atoms, we place a 2 before the CO_2 to balance the carbon atoms.

$$C_2H_5OH(l) \;+\; O_2(g) \;\rightarrow\; 2CO_2(g) \;+\; H_2O(g)$$

2 C atoms   2 C atoms

Remember, we cannot change the formula of any reactant or product when we balance an equation. We can only place coefficients in front of the formulas.

Next we consider hydrogen. The only product containing hydrogen is H_2O. C_2H_2OH contains six hydrogen atoms, so we need six hydrogen atoms on the right. Because each H_2O contains two hydrogen atoms, we need three H_2O molecules to yield six hydrogen atoms. So we place a 3 before the H_2O.

$$C_2H_5OH(l) \;+\; O_2(g) \;\rightarrow\; 2CO_2(g) \;+\; 3H_2O(g)$$

(5 + 1) H   (3 × 2) H
6 H   6 H

Finally, we count the oxygen atoms. On the left we have three oxygen atoms (one in C_2H_5OH and two in O_2), and on the right we have seven oxygen atoms (four in $2CO_2$ and three in $3H_2O$). We can correct this imbalance if we have three O_2 molecules on the left. That is, we place a coefficient of 3 before the O_2 to produce the balanced equation.

$$C_2H_5OH(l) \;+\; 3O_2(g) \;\rightarrow\; 2CO_2(g) \;+\; 3H_2O(g)$$

1 O   (3 × 2) O   (2 × 2) O   3 O
       7 O   7 O

At this point you may have a question: why did we choose O_2 on the left when we balanced the oxygen atoms? Why not use C_2H_5OH, which has an oxygen atom? The answer is that if we had changed the coefficient in front of C_2H_5OH, we would have unbalanced the hydrogen and carbon atoms. Now we count all of the atoms as a check to make sure the equation is balanced.

I try to indicate the physical states of the reactants and products when balancing equations on the board so that students realize that this identification of state is an important part of an equation.

In balancing equations, start by looking at the most complicated molecule.

C_2H_5OH

2 C, 6 H, 1 O

O—C—O   H—O—H
O—C—O   H—O—H
                     H—O—H
4 O atoms   3 O atoms

While they are balancing equations, students often try to adjust the coefficients of diatomic molecules and pure elements in the first round. Encourage them to leave the simple components of the equation until last.

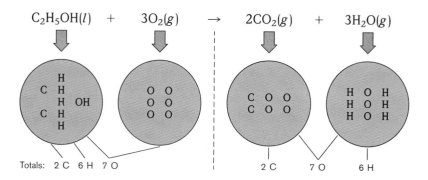

Reactants	Products
2 C	2 C
6 H	6 H
7 O	7 O

The equation is now balanced. We have the same numbers of all types of atoms on both sides of the arrow. Notice that these coefficients are the smallest integers that give a balanced equation.

The process of writing and balancing the equation for a chemical reaction consists of several steps:

How to Write and Balance Equations

STEP 1
Read the description of the chemical reaction. What are the reactants, the products, and their states? Write the appropriate formulas.

STEP 2
Write the *unbalanced* equation that summarizes the information from step 1.

STEP 3
Balance the equation by inspection, starting with the most complicated molecule. Proceed element by element to determine what coefficients are necessary so that the same number of each type of atom appears on both the reactant side and the product side. Do not change the identities (formulas) of any of the reactants or products.

STEP 4
Check to see that the coefficients used give the same number of each type of atom on both sides of the arrow. (Note that an "atom" may be present in an element, compound, or ion.) Also check to see that the coefficients used are the smallest integers that give the balanced equation. This can be done by determining whether all coefficients can be divided by the same integer to give a set of smaller *integer* coefficients.

Simple errors in balancing equations can be caught and corrected if students take the time to check their work and count the number of atoms on both sides of the equation. Encourage students to check their work.

EXAMPLE 6.2   Balancing Chemical Equations I

For the following reaction, write the unbalanced equation and then balance the equation: solid potassium reacts with liquid water to form gaseous hydrogen and potassium hydroxide that dissolves in the water.

SOLUTION

STEP 1
From the description given for the reaction, we know that the reactants are solid potassium K(s), and liquid water, $H_2O(l)$. The products are gaseous hydrogen, $H_2(g)$, and dissolved potassium hydroxide, KOH(aq).

STEP 2
The unbalanced equation for the reaction is

$$K(s) + H_2O(l) \rightarrow H_2(g) + KOH(aq)$$

STEP 3
Although none of the reactants or products is very complicated, we will start with KOH because it contains the most elements (three). We will arbitrarily consider hydrogen first. Note that on the reactant side of the equation in Step 2, there are two hydrogen atoms but on the product side there are three. If we place a coefficient of 2 in front of both H_2O and KOH, we now have four H atoms on each side.

$$K(s) + 2H_2O(l) \rightarrow H_2(g) + 2KOH(aq)$$

| 4 H atoms | 2 H atoms | 2 H atoms |

Also note that the oxygen atoms balance.

$$K(s) + 2H_2O(l) \rightarrow H_2(g) + 2KOH(aq)$$

| 2 O atoms | 2 O atoms |

However, the K atoms do not balance; we have one on the left and two on the right. We can fix this easily by placing a coefficient of 2 in front of K(s) to give the balanced equation:

$$2K(s) + 2H_2O(l) \rightarrow H_2(g) + 2KOH(aq)$$

Reactants	Products
2 K	2 K
4 H	4 H
2 O	2 O

STEP 4

CHECK: There are 2 K, 4 H, and 2 O on both sides of the arrow, and the coefficients are the smallest integers that give a balanced equation. We know this because we cannot divide through by a given integer to give a set

EXAMPLE 6.2, CONTINUED

of smaller *integer* (whole-number) coefficients. For example, if we divide all of the coefficients by 2, we get

$$K(s) + H_2O(l) \rightarrow \tfrac{1}{2}H_2(g) + KOH(aq)$$

This is not acceptable because the coefficient for H_2 is not an integer.

EXAMPLE 6.3   Balancing Chemical Equations II

When I work examples on the board, I use word problems. I then ask students to help me determine what the products and the reactants are, and to help me write correct formulas for each.

Under appropriate conditions at 1000 °C, ammonia gas reacts with oxygen gas to produce gaseous nitrogen monoxide (common name, nitric oxide) and gaseous water. Write the unbalanced and balanced equations for this reaction.

SOLUTION

STEP 1
The reactants are gaseous ammonia, $NH_3(g)$, and gaseous oxygen, $O_2(g)$. The products are gaseous nitrogen monoxide, $NO(g)$, and gaseous water, $H_2O(g)$.

STEP 2
The unbalanced equation for the reaction is

$$NH_3(g) + O_2(g) \rightarrow NO(g) + H_2O(g)$$

STEP 3
In this equation there is no molecule that is obviously the most complicated. Three molecules contain two elements, so we arbitrarily start with NH_3. We arbitrarily begin by looking at hydrogen. A coefficient of 2 for NH_3 and a coefficient of 3 for H_2O give six atoms of hydrogen on both sides.

$$2NH_3(g) + O_2(g) \rightarrow NO(g) + 3H_2O(g)$$
$$\uparrow \qquad\qquad\qquad\qquad \uparrow$$
$$6\ H \qquad\qquad\qquad\qquad 6\ H$$

Reactants	Products
1 N	1 N
3 H	2 H
2 O	1 O

We can balance the nitrogen by giving NO a coefficient of 2.

$$2NH_3(g) + O_2(g) \rightarrow 2NO(g) + 3H_2O(g)$$
$$\uparrow \qquad\qquad\qquad \uparrow$$
$$2\ N \qquad\qquad\qquad 2\ N$$

$\tfrac{5}{2} = 2\tfrac{1}{2}$

$2\tfrac{1}{2}\ O_2$

contains

5 O atoms

Finally, we note that there are two atoms of oxygen on the left and five on the right. The oxygen can be balanced with a coefficient of $\tfrac{5}{2}$ for O_2, because $\tfrac{5}{2} \times O_2$ gives five oxygen atoms.

$$2NH_3(g) + \tfrac{5}{2}O_2(g) \rightarrow 2NO(g) + 3H_2O(g)$$
$$\uparrow \qquad\qquad \uparrow \qquad\qquad \uparrow$$
$$5\ O \qquad\qquad 2\ O \qquad\qquad 3\ O$$

EXAMPLE 6.3, CONTINUED

However, the convention is to have integer (whole-number) coefficients, so we multiply the entire equation by 2.

$$2 \times [2NH_3(g) + \tfrac{5}{2}O_2(g) \rightarrow 2NO(g) + 3H_2O(g)]$$

or

$$2 \times 2NH_3(g) + 2 \times \tfrac{5}{2}O_2(g) \rightarrow 2 \times 2NO(g) + 2 \times 3H_2O(g)$$
$$4NH_3(g) + 5O_2(g) \rightarrow 4NO(g) + 6H_2O(g)$$

Reactants	Products
4 N	4 N
12 H	12 H
10 O	10 O

STEP 4

CHECK: There are 4 N, 12 H, and 10 O atoms on both sides, so the equation is balanced. These coefficients are the smallest integers that give a balanced equation. That is, we cannot divide all coefficients by the same integer and obtain a smaller set of *integers.*

SELF-CHECK EXERCISE 6.2

Propane, C_3H_8, a liquid at 25 °C under high pressure, is often used as a fuel in rural areas where there is no natural gas pipeline. When liquid propane is released from its storage tank, it changes to propane gas that reacts with oxygen gas in a furnace or stove to give gaseous carbon dioxide and gaseous water. Write and balance the equation for this reaction.

HINT: This description of a chemical process contains many words, some of which are crucial to solving the problem and some of which are not. First sort out the important information and use symbols to represent it.

EXAMPLE 6.4	**Balancing Chemical Equations III**

Glass is sometimes decorated by etching patterns on its surface. Etching occurs when hydrofluoric acid (an aqueous solution of HF) reacts with the silicon dioxide in the glass to form gaseous silicon tetrafluoride and liquid water. Write and balance the equation for this reaction.

Decorations on glass are produced by etching with hydrofluoric acid.

SOLUTION

STEP 1

From the description of the reaction we can identify the reactants:

hydrofluoric acid	$HF(aq)$
solid silicon dioxide	$SiO_2(s)$

and the products:

gaseous silicon tetrafluoride	$SiF_4(g)$
liquid water	$H_2O(l)$

EXAMPLE 6.4, CONTINUED

Reactants	Products
1 Si	1 Si
1 H	2 H
1 F	4 F
2 O	1 O

Reactants	Products
1 Si	1 Si
4 H	2 H
4 F	4 F
2 O	1 O

Reactants	Products
1 Si	1 Si
4 H	4 H
4 F	4 F
2 O	2 O

STEP 2

The unbalanced equation is

$$SiO_2(s) \ + \ HF(aq) \rightarrow SiF_4(g) \ + \ H_2O(l)$$

STEP 3

There is no clear choice here for the most complicated molecule. We arbitrarily start with the elements in SiF_4. The silicon is balanced (one atom on each side), but the fluorine is not. To balance the fluorine, we need a coefficient of 4 before the HF.

$$SiO_2(s) \ + \ 4HF(aq) \rightarrow SiF_4(g) \ + \ H_2O(l)$$

Hydrogen and oxygen are not balanced. Because we have four hydrogen atoms on the left and two on the right, we place a 2 before the H_2O:

$$SiO_2(s) \ + \ 4HF(aq) \rightarrow SiF_4(g) \ + \ 2H_2O(l)$$

This balances the hydrogen *and* the oxygen (two atoms on each side).

STEP 4

CHECK: $\underbrace{SiO_2(s) + 4HF(aq)}$ $\rightarrow$ $\underbrace{SiF_4(g) + 2H_2O(l)}$

Totals: 1 Si, 2 O,   4 H, 4 F $\rightarrow$ 1 Si, 4 F,   4 H, 2 O

All atoms check, so the equation is balanced.

SELF-CHECK EXERCISE 6.3

Give the balanced equation for each of the following reactions.

If you are having trouble writing formulas from names, review the appropriate sections of Chapter 5. It is very important that you are able to do this.

a.   When solid ammonium nitrite is heated, it produces nitrogen gas and water vapor.

b.   Gaseous nitrogen monoxide (common name, nitric oxide) decomposes to produce dinitrogen monoxide gas (common name, nitrous oxide) and nitrogen dioxide gas.

c.   Liquid nitric acid decomposes to brown nitrogen dioxide gas, liquid water, and oxygen gas. (This is why bottles of nitric acid become yellow upon standing.)

CHAPTER REVIEW

Key Terms

chemical reaction (p. 172)
chemical equation (p. 172)
reactant (p. 172)

product (p. 172)
balancing a chemical equation (p. 173)
coefficient (p. 178)

Summary

1. Chemical reactions usually give some sort of visual signal—a color changes, a solid forms, bubbles form, heat and/or flame are produced.
2. A chemical equation represents a chemical reaction. Reactants are shown on the left side of an arrow and products on the right. In a chemical reaction, atoms are neither created nor destroyed; they are merely rearranged. A balanced chemical equation gives the relative numbers of reactant and product molecules.
3. A chemical equation for a reaction can be balanced by using a systematic approach. First identify the reactants and products and write the formulas. Next write the unbalanced equation. Then balance by trial and error, starting with the most complicated molecule(s). Finally, check to be sure the equation is balanced.

Questions and Problems

All even-numbered exercises have answers in the back of this book and solutions in the Solutions Guide.

6.1 Evidence for a Chemical Reaction

QUESTIONS

1. List some possible signals that a chemical reaction has occurred. Are these clues always visible? Can you think of any other indications that a chemical reaction has occurred?

2. Commercial drain cleaners are particularly good at dissolving clogs of hair that have collected in bathroom drains. What evidence of the sort described in Table 6.1 is there that such a drain cleaner works by chemical reaction?

3. Sterling silver, as used for fashioning jewelry and dinnerware, tarnishes easily when exposed to air. Is there evidence that products for removing tarnish from silver work by chemical reaction?

4. In northern climates, where salt is used to melt ice on highways, steel automobile bumpers rust more readily. What evidence is there that this is a chemical reaction?

5. Baking powder is used to make certain doughs "rise." Can you think of any evidence that the action of baking powder represents a chemical reaction?

6. When fruit juice is left exposed to the air, it eventually ferments into wine or vinegar. What evidence is there that a chemical reaction is occurring?

6.2 Chemical Equations

QUESTIONS

7. The substances present before a chemical reaction takes place are called the _____, and the substances present after the reaction takes place are called the _____ .

8. In an ordinary chemical reaction, _____ are neither created nor destroyed.

9. The total number of atoms present before and after a chemical reaction takes place must be _____ .

10. Balancing an equation for a reaction ensures that the number of atoms is _____ on both sides of the equation.

11. The notation "(g)" after a substance's formula indicates that the substance exists in the _____ state.

12. In a chemical equation for a reaction, the notation "(aq)" after a substance's formula means that the substance is dissolved in _____ .

PROBLEMS

Note: In some of the following problems you will need to write a chemical formula from the name of the compound. Review Chapter 5 if you are having trouble with this.

13. When sodium hydrogen carbonate (sodium bicarbonate), $NaHCO_3$, is heated strongly in a test tube, carbon dioxide gas, CO_2, and water vapor, H_2O, are evolved from the test tube, leaving a residue of sodium carbonate, Na_2CO_3. Write the unbalanced chemical equation for this process.

14. Hydrogen peroxide, H_2O_2, is a relatively reactive chemical substance. For example, iron in the blood of a wound is able to trigger the decomposition of a hydrogen peroxide solution. When it decomposes, hydrogen peroxide produces water and oxygen gas. Write an unbalanced chemical equation for the breakdown of hydrogen peroxide.

15. When a small piece of sodium, Na, is dropped into a beaker of water, a reaction occurs that produces hydrogen gas, H_2, and dissolved sodium hydroxide, NaOH. Write an unbalanced chemical equation for this process.

16. Solid ammonium carbonate, $(NH_4)_2CO_3$, is used as the active ingredient in "smelling salts." When solid ammonium carbonate is heated, it decomposes into ammonia gas and carbon dioxide gas. Write an unbalanced chemical equation for this process.

17. When turnings of magnesium metal, Mg, are added to a dilute solution of nitric acid, HNO_3, hydrogen gas, H_2, is evolved as the magnesium dissolves in the acid. Upon evaporation of the solution, solid magnesium nitrate, $Mg(NO_3)_2$, is isolated. Write the unbalanced chemical equation for the reaction of magnesium with nitric acid.

18. Sulfur dioxide gas may be generated in small quantities in the laboratory by adding aqueous sulfuric acid to solid sodium sulfite, Na_2SO_3. The other products of the reaction are sodium sulfate and water. Write the unbalanced chemical equation for this process.

19. When a piece of chalk (which consists primarily of calcium carbonate, $CaCO_3$) is treated with sulfuric acid, H_2SO_4, the chalk dissolves in the acid to produce a solution of calcium sulfate, $CaSO_4$, and carbon dioxide gas, CO_2, is evolved. Write an unbalanced chemical equation for this process.

20. Calcium metal is moderately reactive. If turnings of calcium are added to water, the metal begins to bubble as hydrogen gas is formed. The water begins to turn cloudy, as relatively insoluble calcium hydroxide begins to form. Write the unbalanced chemical equation for the reaction of calcium metal with water.

21. When a strip of copper wire is placed in an aqueous solution of silver nitrate, solid silver precipitates out of solution, as the copper wire dissolves to give a solution of copper(II) nitrate. Write an unbalanced chemical equation for this process.

22. If solutions of lead nitrate, $Pb(NO_3)_2$, and potassium iodide, KI, are mixed, a yellow precipitate of lead iodide, PbI_2, forms immediately and settles from the solution. The liquid remaining is a solution of potassium nitrate, KNO_3. Write the unbalanced chemical equation for the reaction of lead nitrate with potassium iodide.

23. Methyl alcohol (methanol), CH_3OH, burns cleanly in air, producing carbon dioxide gas and water vapor. Write an unbalanced chemical equation for this reaction.

24. If bottles of ammonia and hydrochloric acid are opened near each other, the ammonia and hydrogen chloride fumes from the bottles react in the air, forming solid ammonium chloride. Chemistry labs (which usually contain these solutions) often become "dusty" from this process. Write the unbalanced chemical equation for the reaction of gaseous ammonia with gaseous hydrogen chloride.

25. When sulfuric acid is added to sodium chloride and the mixture is heated, hydrogen chloride gas is generated, leaving a solid residue of sodium sulfate. Write an unbalanced chemical equation for this reaction.

26. The marble used in construction projects consists primarily of polished calcium carbonate, $CaCO_3$. The marble of many public monuments and buildings is slowly deteriorating as this substance reacts with the sulfuric acid contained in acid rain. The marble is converted by this reaction into calcium sulfate, $CaSO_4$, and carbon dioxide gas. Although calcium carbonate is insoluble in water, calcium sulfate is somewhat soluble, and as exposure to the acid continues, the monument or building gradually dissolves. Write the unbalanced chemical equation for the reaction of solid calcium carbonate with sulfuric acid solution.

27. When aqueous potassium chromate is added to a solution of lead(II) nitrate, a bright orange precipitate of lead(II) chromate forms, leaving potassium nitrate in solution. Write the unbalanced chemical equation for this process.

28. Ozone gas is a form of elemental oxygen containing molecules with *three* oxygen atoms, O_3. Ozone is produced from atmospheric oxygen gas, O_2, by the high energy outbursts found in lightning storms. Write an unbalanced equation for the formation of ozone gas from oxygen gas.

29. When sulfuric acid is added to ordinary table sugar (sucrose), $C_{12}H_{22}O_{11}$, a long black "snake" of elemental carbon forms, with the release of a cloud of steam (water vapor). Write the unbalanced chemical equation for this process.

30. Sand consists primarily of silicon dioxide, SiO_2. Pure elemental silicon for use in the semiconductor industry can be produced by reaction of sand with elemental carbon at high temperatures. Carbon monoxide gas is also a product of the reaction. Write an unbalanced chemical equation for the reaction of silicon dioxide with carbon.

31. When magnesium metal is heated to several hundred degrees in an atmosphere of pure nitrogen gas, a green deposit of magnesium nitride, Mg_3N_2, forms. Write the unbalanced chemical equation for this process.

32. Although they were formerly called the inert gases, the heavier elements of Group 8 do form relatively stable compounds. For example, at high temperatures in the presence of an appropriate catalyst, xenon gas will combine directly with fluorine gas, to produce solid xenon tetrafluoride. Write the unbalanced chemical equation for this process.

33. When bright red mercuric oxide, HgO, is heated in a test tube, oxygen gas is evolved, and droplets of liquid mercury condense at the cooler top of the test tube. Write the unbalanced chemical equation for this process.

34. If ordinary table sugar (sucrose, $C_{12}H_{22}O_{11}$) is heated strongly in pure oxygen gas, it will ignite and burn, producing carbon dioxide gas and water vapor. Write the unbalanced chemical equation for this process.

6.3 Balancing Chemical Equations

QUESTIONS

35. When balancing a chemical equation, one must never change the _____ of any reactant or product.

36. After balancing a chemical reaction equation, we ordinarily make sure that the coefficients are the smallest _____ possible.

37. Balance each of the following chemical equations.
    a. $Fe(s) + O_2(g) \rightarrow Fe_2O_3(s)$
    b. $Na(s) + Br_2(l) \rightarrow NaBr(s)$
    c. $K(s) + N_2(g) \rightarrow K_3N(s)$
    d. $KHCO_3(s) \rightarrow K_2CO_3(s) + CO_2(g) + H_2O(g)$
    e. $P_4(s) + Cl_2(g) \rightarrow PCl_3(l)$
    f. $Al_2O_3(s) \rightarrow Al(s) + O_2(g)$
    g. $C_3H_8(g) + O_2(g) \rightarrow CO_2(g) + H_2O(g)$
    h. $C_5H_{12}(l) + O_2(g) \rightarrow CO_2(g) + H_2O(g)$

38. Balance each of the following chemical equations.
    a. $Br_2(l) + KI(aq) \rightarrow KBr(aq) + I_2(s)$
    b. $Co(s) + O_2(g) \rightarrow Co_2O_3(s)$
    c. $P_4(s) + O_2(g) \rightarrow P_4O_{10}(s)$
    d. $Al(s) + HNO_3(aq) \rightarrow Al(NO_3)_2(aq) + H_2(g)$
    e. $PBr_3(l) + H_2O(l) \rightarrow H_3PO_3(aq) + HBr(aq)$
    f. $NO(g) + O_2(g) \rightarrow NO_2(g)$
    g. $C_2H_6(g) + O_2(g) \rightarrow CO_2(g) + H_2O(g)$
    h. $CuO(s) + H_2SO_4(aq) \rightarrow CuSO_4(aq) + H_2O(l)$

39. Balance each of the following chemical equations.
    a. $C_4H_{10}(g) + O_2(g) \rightarrow CO_2(g) + H_2O(g)$
    b. $C_6H_{14}(g) + O_2(g) \rightarrow CO_2(g) + H_2O(g)$
    c. $B(s) + Cl_2(g) \rightarrow BCl_3(l)$
    d. $NO(g) + O_2(g) \rightarrow NO_2(g)$
    e. $Tl_2O_3(s) + CO_2(g) \rightarrow Tl_2(CO_3)_3(s)$
    f. $OF_2(g) \rightarrow O_2(g) + F_2(g)$
    g. $NaClO_3(s) \rightarrow NaCl(s) + O_2(g)$
    h. $Al(OH)_3(s) \rightarrow Al_2O_3(s) + H_2O(l)$

41. Balance each of the following chemical equations.
    a. $Li(s) + Cl_2(g) \rightarrow LiCl(s)$
    b. $Ba(s) + N_2(g) \rightarrow Ba_3N_2(s)$
    c. $NaHCO_3(s) \rightarrow Na_2CO_3(s) + CO_2(g) + H_2O(g)$
    d. $Al(s) + HCl(aq) \rightarrow AlCl_3(aq) + H_2(g)$
    e. $NiS(s) + O_2(g) \rightarrow NiO(s) + SO_2(g)$
    f. $CaH_2(s) + H_2O(l) \rightarrow Ca(OH)_2(s) + H_2(g)$
    g. $H_2(g) + CO(g) \rightarrow CH_3OH(l)$
    h. $B_2O_3(s) + C(s) \rightarrow B_4C_3(s) + CO_2(g)$

43. Balance each of the following chemical equations.
    a. $Co(s) + O_2(g) \rightarrow Co_2O_3(s)$
    b. $Li(s) + Br_2(l) \rightarrow LiBr(s)$
    c. $Cs(s) + N_2(g) \rightarrow Cs_3N(s)$
    d. $Co_2S_3(s) + H_2(g) \rightarrow Co(s) + H_2S(g)$
    e. $CoS(s) + O_2(g) \rightarrow CoO(s) + SO_2(g)$
    f. $BaH_2(s) + H_2O(l) \rightarrow Ba(OH)_2(s) + H_2(g)$
    g. $NH_3(g) + Cl_2(g) \rightarrow NH_4Cl(s) + NCl_3(g)$
    h. $MnCl_2(s) + Al(s) \rightarrow Mn(s) + AlCl_3(s)$

40. Balance each of the following chemical equations.
    a. $Ba(NO_3)_2(aq) + KOH(aq) \rightarrow Ba(OH)_2(s) + KNO_3(aq)$
    b. $Cu(s) + HNO_3(aq) \rightarrow Cu(NO_3)_2(aq) + H_2(g)$
    c. $Cr(s) + S(s) \rightarrow Cr_2S_3(s)$
    d. $AgNO_3(aq) + Zn(s) \rightarrow Zn(NO_3)_2(aq) + Ag(s)$
    e. $H_2O(l) + Br_2(l) \rightarrow HBr(aq) + HOBr(aq)$
    f. $SnS(s) + O_2(g) \rightarrow SnO_2(s) + SO_2(g)$
    g. $Pb(NO_3)_2(aq) + KCl(aq) \rightarrow PbCl_2(s) + KNO_3(aq)$
    h. $FeO(s) + C(s) \rightarrow Fe(s) + CO(g)$

42. Balance each of the following chemical equations.
    a. $SiI_4(s) + Mg(s) \rightarrow Si(s) + MgI_2(s)$
    b. $MnO_2(s) + Mg(s) \rightarrow Mn(s) + MgO(s)$
    c. $Ba(s) + S_8(s) \rightarrow BaS(s)$
    d. $NH_3(g) + Cl_2(g) \rightarrow NH_4Cl(s) + NCl_3(g)$
    e. $Cu_2S(s) + S_8(s) \rightarrow CuS(s)$
    f. $Al(s) + H_2SO_4(aq) \rightarrow Al_2(SO_4)_3(aq) + H_2(g)$
    g. $NaCl(s) + H_2SO_4(l) \rightarrow HCl(g) + Na_2SO_4(s)$
    h. $CO(g) + O_2(g) \rightarrow CO_2(g)$

44. Balance each of the following chemical equations.
    a. $Ba(NO_3)_2(aq) + Na_2CrO_4(aq) \rightarrow BaCrO_4(s) + NaNO_3(aq)$
    b. $PbCl_2(aq) + K_2SO_4(aq) \rightarrow PbSO_4(s) + KCl(aq)$
    c. $C_2H_5OH(l) + O_2(g) \rightarrow CO_2(g) + H_2O(l)$
    d. $CaC_2(s) + H_2O(l) \rightarrow Ca(OH)_2(s) + H_2C_2(g)$
    e. $Sr(s) + HNO_3(aq) \rightarrow Sr(NO_3)_2(aq) + H_2(g)$
    f. $BaO_2(s) + H_2SO_4(aq) \rightarrow BaSO_4(s) + H_2O_2(aq)$
    g. $AsI_3(s) \rightarrow As(s) + I_2(s)$
    h. $CuSO_4(aq) + KI(s) \rightarrow CuI(s) + I_2(s) + K_2SO_4(aq)$

Additional Problems

45. Sodium hydrogen carbonate, more commonly known as baking soda, can be produced by bubbling carbon dioxide gas through an ice-cold aqueous solution containing large amounts of sodium chloride and ammonia. Because sodium hydrogen carbonate is not very soluble in cold water, it precipitates, whereas the other product of the reaction, ammonium chloride, remains dissolved. Write an unbalanced chemical equation for the process. *Hint:* water is a reactant.

46. The metallic elements of Group 1 of the periodic table are all very reactive. For example, each of these metals reacts with water, producing hydrogen gas and leaving a solution of the metal's hydroxide compound. Write unbalanced chemical equations for the reactions of the Group 1 metals with water.

47. Crude gunpowders often contain a mixture of potassium nitrate and charcoal (carbon). When such a mixture is heated until reaction occurs, a solid residue of potassium carbonate is produced. The explosive force of the gunpowder comes from the fact that two gases are also produced (carbon monoxide and nitrogen), which increase in volume with great force and speed. Write an unbalanced chemical equation for the process.

48. The sugar sucrose, which is present in many fruits and vegetables, reacts in the presence of certain yeast enzymes to produce ethyl alcohol (ethanol) and carbon dioxide gas. Balance the following equation for this reaction of sucrose.

$$C_{12}H_{22}O_{11}(aq) \rightarrow C_2H_6O(aq) + CO_2(g)$$

49. Methanol (methyl alcohol), CH_3OH, is a very important industrial chemical. Formerly, methanol was prepared by heating wood to high temperatures in the absence of air. The complex compounds present in wood are degraded by this process into a charcoal residue and a volatile portion that is rich in methanol. Today, methanol is instead synthe-

sized from carbon monoxide and elemental hydrogen. Write the balanced chemical equation for this latter process.

50. The Hall process is an important method by which pure aluminum is prepared from its oxide (alumina, Al_2O_3) by indirect reaction with graphite (carbon). Balance the following equation, which is a simplified representation of this process.

$$Al_2O_3(s) + C(s) \rightarrow Al(s) + CO_2(g)$$

51. Iron oxide ores, commonly a mixture of FeO and Fe_2O_3, are given the general formula Fe_3O_4. They yield elemental iron when heated to a very high temperature with either carbon monoxide or elemental hydrogen. Balance the following equations for these processes.

$$Fe_3O_4(s) + H_2(g) \rightarrow Fe(s) + H_2O(g)$$
$$Fe_3O_4(s) + CO_2(g) \rightarrow Fe(s) + CO_2(g)$$

52. The elements of Group 2 all react with elemental fluorine and chlorine gases to produce the metal fluorides or the metal chlorides. For each of the elements of Group 2, write a balanced chemical equation for its reaction with fluorine gas and for its reaction with chlorine gas.

53. When steel wool (iron) is heated in pure oxygen gas, the steel wool bursts into flame and a fine powder consisting of a mixture of iron oxides (FeO and Fe_2O_3) forms. Write *separate* unbalanced equations for the reaction of iron with oxygen to give each of these products.

54. A common demonstration in introductory chemistry courses is the "volcano" reaction of ammonium dichromate, $(NH_4)_2Cr_2O_7$. When a small pile of the bright orange ammonium dichromate is heated strongly, the compound decomposes with a great deal of sparking into a large volume of green chromium(III) oxide, nitrogen gas, and water vapor. Write the balanced chemical equation for this process.

55. When elemental boron, B, is burned in oxygen gas, the product is diboron trioxide. If the diboron trioxide is then reacted with a measured quantity of water, it reacts with the water to form what is commonly known as boric acid, $B(OH)_3$. Write a balanced chemical equation for each of these processes.

56. A common experiment in introductory chemistry courses involves heating a weighed mixture of potassium chlorate, $KClO_3$, and potassium chloride. Potassium chlorate decomposes when heated, producing additional potassium chloride and evolving oxygen gas. By measurement of the volume of oxygen gas produced in this experiment, students can calculate the relative percentage of $KClO_3$ and KCl in the original mixture. Write the balanced chemical equation for this process.

57. A common demonstration in chemistry courses involves adding a tiny speck of manganese(IV) oxide to a concentrated hydrogen peroxide, H_2O_2, solution. Hydrogen peroxide is unstable, and it decomposes quite spectacularly under these conditions to produce oxygen gas and steam (water vapor). Manganese(IV) oxide is a catalyst for the decomposition of hydrogen peroxide and is not consumed in the reaction. Write the balanced equation for the decomposition reaction of hydrogen peroxide.

58. The benches in many undergraduate chemistry laboratories are often covered by a film of white dust. This may be due to poor housekeeping, but the dust is usually ammonium chloride, produced by the reaction in the laboratory air of hydrogen chloride and ammonia; most labs keep handy solutions of these common reagents. Write the balanced chemical equation for the reaction of gaseous ammonia and hydrogen chloride to form solid ammonium chloride.

59. Glass is a mixture of several compounds, but a major constituent of most glass is calcium silicate, $CaSiO_3$. Glass can be etched by treatment with hydrogen fluoride: HF attacks the calcium silicate of the glass, producing gaseous and water-soluble products (which can be removed by washing the glass). For example, the volumetric glassware in chemistry laboratories is often graduated by using this process. Balance the following equation for the reaction of hydrogen fluoride with calcium silicate.

$$CaSiO_3(s) + HF(g) \rightarrow CaF_2(aq) + SiF_4(g) + H_2O(l)$$

60. Suppose that you were about to fry some chicken and that you left the oil on to heat for too long and at too high a temperature. What could happen? Is there evidence that a chemical reaction is occurring?

61. If you had a "sour stomach," you might try an over-the-counter antacid tablet to relieve the problem. Can you think of evidence that the action of such an antacid is a chemical reaction?

62. When iron, Fe, wire is heated in the presence of sulfur, S, the iron soon begins to glow, and a chunky, blue-black mass of iron(II) sulfide, FeS, is formed. Write an unbalanced chemical equation for this reaction.

63. When finely divided solid sodium, Na, is dropped into a flask containing chlorine gas, Cl_2, an explosion occurs and a fine powder of sodium chloride, NaCl, is deposited on the walls of the flask. Write the unbalanced chemical equation for this process.

64. Suburban homes sometimes use bottled propane gas, C_3H_8, for cooking purposes. When propane is burned in air, carbon dioxide, CO_2, and water vapor, H_2O, are produced (the

presence of water vapor may be demonstrated by holding a cool dish above a burning propane flame, allowing liquid water to form on the cool surface). Write the unbalanced chemical equation for the reaction of gaseous propane with oxygen gas, O_2.

65. When hydrogen sulfide, H_2S, gas is bubbled through a solution of lead(II) nitrate, $Pb(NO_3)_2$, a black precipitate of lead(II) sulfide, PbS, forms, and nitric acid, HNO_3, is produced. Write an unbalanced chemical equation for this reaction.

66. When an electric current is passed through an aqueous solution of potassium iodide, hydrogen gas is evolved at one electrode, while elemental iodine, I_2, is produced at the other electrode. If the solution remaining is then evaporated, solid potassium hydroxide is recovered. Write an unbalanced chemical equation for this reaction.

67. When a strip of magnesium metal is heated in oxygen, it bursts into an intensely white flame and produces a finely powdered dust of magnesium oxide. Write an unbalanced chemical equation for this process.

68. Although magnesium metal does not react readily with cold water, magnesium reacts easily with steam, producing magnesium hydroxide and hydrogen gas. Write an unbalanced chemical equation for this process.

69. When solid red phosphorus, P_4, is burned in air, the phosphorus combines with oxygen, producing a choking cloud of tetraphosphorus decoxide. Write the unbalanced chemical equation for this reaction.

70. When copper(II) oxide is boiled in an aqueous solution of sulfuric acid, a strikingly blue solution of copper(II) sulfate forms along with additional water. Write the unbalanced chemical equation for this reaction.

71. When lead(II) sulfide is heated to high temperatures in a stream of pure oxygen gas, solid lead(II) oxide forms with the release of gaseous sulfur dioxide. Write the unbalanced chemical equation for this reaction.

72. When sodium sulfite is boiled with sulfur, the sulfite ions, SO_3^{2-}, are converted to thiosulfate ions, $S_2O_3^{2-}$, resulting in a solution of sodium thiosulfate, $Na_2S_2O_3$. Write the unbalanced chemical equation for this reaction.

73. Balance each of the following chemical equations.
   a. $Cl_2(g) + KBr(aq) \rightarrow Br_2(l) + KCl(aq)$
   b. $Cr(s) + O_2(g) \rightarrow Cr_2O_3(s)$
   c. $P_4(s) + H_2(g) \rightarrow PH_3(g)$
   d. $Al(s) + H_2SO_4(aq) \rightarrow Al_2(SO_4)_3(aq) + H_2(g)$
   e. $PCl_3(l) + H_2O(l) \rightarrow H_3PO_3(aq) + HCl(aq)$
   f. $SO_2(g) + O_2(g) \rightarrow SO_3(g)$
   g. $C_7H_{16}(l) + O_2(g) \rightarrow CO_2(g) + H_2O(g)$
   h. $C_2H_6(g) + O_2(g) \rightarrow CO_2(g) + H_2O(g)$

74. Balance each of the following chemical equations.
   a. $Ba(NO_3)_2(aq) + KF(aq) \rightarrow BaF_2(s) + KNO_3(aq)$
   b. $Zn(s) + HCl(aq) \rightarrow ZnCl_2(aq) + H_2(g)$
   c. $Fe(s) + S(s) \rightarrow Fe_2S_3(s)$
   d. $C_6H_{12}O_6(s) + O_2(g) \rightarrow CO_2(g) + H_2O(g)$
   e. $H_2O(l) + Cl_2(g) \rightarrow HCl(aq) + HOCl(aq)$
   f. $ZnS(s) + O_2(g) \rightarrow ZnO(s) + SO_2(g)$
   g. $PbSO_4(s) + NaCl(aq) \rightarrow Na_2SO_4(aq) + Na_2PbCl_4(aq)$
   h. $Fe_2O_3(s) + C(s) \rightarrow Fe_3O_4(s) + CO(g)$

75. Balance each of the following chemical equations.
   a. $SiCi_4(l) + Mg(s) \rightarrow Si(s) + MgCl_2(s)$
   b. $NO(g) + Cl_2(g) \rightarrow NOCl(g)$
   c. $MnO_2(s) + Al(s) \rightarrow Mn(s) + Al_2O_3(s)$
   d. $Cr(s) + S_8(s) \rightarrow Cr_2S_3(s)$
   e. $NH_3(g) + F_2(g) \rightarrow NH_4F(s) + NF_3(g)$
   f. $Ag_2S(s) + H_2(g) \rightarrow Ag(s) + H_2S(g)$
   g. $O_2(g) \rightarrow O_3(g)$
   h. $Na_2SO_3(aq) + S_8(s) \rightarrow Na_2S_2O_3(aq)$

76. Balance each of the following chemical equations.
   a. $Pb(NO_3)_2(aq) + K_2CrO_4(aq)$
       $\rightarrow PbCrO_4(s) + KNO_3(aq)$
   b. $BaCl_2(aq) + Na_2SO_4(aq) \rightarrow BaSO_4(s) + NaCl(aq)$
   c. $CH_3OH(l) + O_2(g) \rightarrow CO_2(g) + H_2O(g)$
   d. $Na_2CO_3(aq) + S(s) + SO_2(g)$
       $\rightarrow CO_2(g) + Na_2S_2O_3(aq)$
   e. $Cu(s) + H_2SO_4(aq) \rightarrow CuSO_4(aq) + SO_2(g) + H_2O(l)$
   f. $MnO_2(s) + HCl(aq) \rightarrow MnCl_2(aq) + Cl_2(g) + H_2O(l)$
   g. $As_2O_3(s) + KI(aq) + HCl(aq)$
       $\rightarrow AsI_3(s) + KCl(aq) + H_2O(l)$
   h. $Na_2S_2O_3(aq) + I_2(aq) \rightarrow Na_2S_4O_6(aq) + NaI(aq)$

Solutions to Self-Check Exercises

SELF-CHECK EXERCISE 6.1

a. $Mg(s) + H_2O(l) \rightarrow Mg(OH)_2(s) + H_2(g)$
Note that magnesium (which is in Group 2) always forms the Mg^{2+} cation and thus requires two OH^- anions for a zero net charge.

b. Ammonium dichromate contains the polyatomic ions NH_4^+ and $Cr_2O_7^{2-}$ (you should have these memorized). Because NH_4^+ has a 1+ charge, two NH_4^+ cations are required for each $Cr_2O_7^{2-}$, with its 2− charge, to give the formula $(NH_4)_2Cr_2O_7$. Chromium(III) oxide contains the ions Cr^{3+}—signified by chromium(III)—and O^{2-} (the oxide ion). To achieve a net charge of zero, the solid must contain two Cr^{3+} ions for every three O^{2-} ions, so the formula is Cr_2O_3. Nitrogen gas contains diatomic molecules and is written $N_2(g)$, and gaseous water is written $H_2O(g)$. Thus the unbalanced equation for the decomposition of ammonium dichromate is

$$(NH_4)_2Cr_2O_7(s) \rightarrow Cr_2O_3(s) + N_2(g) + H_2O(g)$$

c. Gaseous ammonia, $NH_3(g)$, and gaseous oxygen, $O_2(g)$, react to form nitrogen monoxide gas, $NO(g)$, plus gaseous water, $H_2O(g)$. The unbalanced equation is

$$NH_3(g) + O_2(g) \rightarrow NO(g) + H_2O(g)$$

SELF-CHECK EXERCISE 6.2

STEP 1
The reactants are propane, $C_3H_8(g)$, and oxygen, $O_2(g)$; the products are carbon dioxide, $CO_2(g)$, and water, $H_2O(g)$. All are in the gaseous state.

STEP 2
The unbalanced equation for the reaction is

$$C_3H_8(g) + O_2(g) \rightarrow CO_2(g) + H_2O(g)$$

STEP 3
We start with C_3H_8 because it is the most complicated molecule. C_3H_8 contains three carbon atoms per molecule, so a coefficient of 3 is needed for CO_2.

$$C_3H_8(g) + O_2(g) \rightarrow 3CO_2(g) + H_2O(g)$$

Also, each C_3H_8 molecule contains eight hydrogen atoms, so a coefficient of 4 is required for H_2O.

$$C_3H_8(g) + O_2(g) \rightarrow 3CO_2(g) + 4H_2O(g)$$

The final element to be balanced is oxygen. Note that the left side of the equation now has two oxygen atoms, and the right side has ten. We can balance the oxygen by using a coefficient of 5 for O_2.

$$C_3H_8(g) + 5O_2(g) \rightarrow 3CO_2(g) + 4H_2O(g)$$

STEP 4

CHECK: 3 C, 8 H, 10 O → 3 C, 8 H, 10 O

                      Reactant   Product
                       atoms   atoms

We cannot divide all coefficients by a given integer to give smaller integer coefficients.

SELF-CHECK EXERCISE 6.3

a. $NH_4NO_2(s) \rightarrow N_2(g) + H_2O(g)$ (unbalanced)
     $NH_4NO_2(s) \rightarrow N_2(g) + 2H_2O(g)$ (balanced)

b. $NO(g) \rightarrow N_2O(g) + NO_2(g)$ (unbalanced)
     $3NO(g) \rightarrow N_2O(g) + NO_2(g)$ (balanced)

c. $HNO_3(l) \rightarrow NO_2(g) + H_2O(l) + O_2(g)$ (unbalanced)
     $4HNO_3(l) \rightarrow 4NO_2(g) + 2H_2O(l) + O_2(g)$ (balanced)

CHAPTER

7 Reactions in Aqueous Solutions

CONTENTS

This is really a crucial chapter. So much chemistry occurs in aqueous solution that it's vital for students to have a feeling for what happens in solution. Show them lots of reactions and, as a class, try to figure out the product of each. This may be challenging, but it also can be fun.

Sodium hydrogen carbonate reacts in an acidic solution to give bubbles of carbon dioxide gas.

The chemical reactions that are most important to us occur in water—in aqueous solutions. Virtually all of the chemical reactions that keep each of us alive and well take place in the aqueous medium present in our bodies.

In this chapter we will study some common types of reactions that take place in water, and we will become familiar with some of the driving forces that make these reactions occur. We will also learn how to predict the products for these reactions and how to write various equations to describe them.

7.1   Predicting Whether a Reaction Will Occur

AIM: To learn about some of the factors that cause reactions to occur.

In this text we have already seen many chemical reactions. Now let's consider an important question: why does a chemical reaction occur? What causes reactants to "want" to form products? As chemists have studied reactions, they have recognized several "tendencies" in reactants that drive them to form products. That is, there are several "driving forces" (types of changes) that pull reactants toward products—changes that tend to make reactions go in the direction of the arrow. The most common of these driving forces are

1. Formation of a solid
2. Formation of water
3. Formation of a gas
4. Transfer of electrons

When two or more chemicals are brought together, if any of these things can occur, a chemical change (a reaction) is likely to take place. Accordingly, when we are confronted with a set of reactants and want to predict whether a reaction will occur and what products might form, we will consider these driving forces. They will help us organize our thoughts as we encounter new reactions.

In this chapter we will consider examples of reactions in aqueous solutions where the first two of these driving forces are operating. In Chapter 8 we will consider reactions that involve formation of gases and electron transfer.

I always explain that it can be hard, even for an experienced chemist, to predict whether or not a reaction will occur when two substances are mixed together. Memorizing the four driving forces can help students determine whether or not a reaction will proceed to the right.

Figure 7.1
The precipitation reaction that occurs when yellow potassium chromate, $K_2CrO_4(aq)$, is added to a colorless barium nitrate solution, $Ba(NO_3)_2(aq)$.

7.2 Reactions in Which a Solid Forms

AIM: To learn to identify the solid that forms in a precipitation reaction.

One driving force for a chemical reaction is the formation of a solid, a process called **precipitation.** The solid that forms is called a **precipitate,** and the reaction is known as a **precipitation reaction.** For example, when an aqueous (water) solution of potassium chromate, $K_2CrO_4(aq)$, which is yellow, is added to a colorless aqueous solution containing barium nitrate, $Ba(NO_3)_2(aq)$, a yellow solid forms (see Figure 7.1). The fact that a solid forms tells us that a reaction—a chemical change—has occurred. That is, we have a situation where

$$\text{Reactants} \rightarrow \text{Products}$$

What is the equation that describes this chemical change? To write the equation, we must decipher the identities of the reactants and products. The reactants have already been described: $K_2CrO_4(aq)$ and $Ba(NO_3)_2(aq)$. Is there some way in which we can predict the identities of the products? What is the brownish-yellow solid? The best way to predict the identity of this solid is to first *consider what products are possible.* To do this we need to know what chemical species are present in the solution that results when the reactant solutions are mixed. First, let's think about the nature of each reactant in an aqueous solution.

What Happens When an Ionic Compound Dissolves in Water?

The designation $Ba(NO_3)_2(aq)$ means that barium nitrate (a white solid) has been dissolved in water. Note from its formula that barium nitrate contains the Ba^{2+} and NO_3^- ions. **In virtually every case when a solid containing ions dissolves in water, the ions separate** and move around independently. That is, $Ba(NO_3)_2(aq)$ does not contain $Ba(NO_3)_2$ units. Rather, it contains separated Ba^{2+} and NO_3^- ions. In the solution there are two NO_3^- ions for every Ba^{2+} ion. Chemists know that separated ions are present in this solution because it is an excellent conductor of electricity (see Figure 7.2). Pure water does not conduct an electric current. Ions must be present in water for a current to flow.

When each unit of a substance that dissolves in water produces separated ions, the substance is called a **strong electrolyte.** Barium nitrate is a strong electrolyte in water, because each $Ba(NO_3)_2$ unit produces the separated ions $(Ba^{2+}, NO_3^-, NO_3^-)$.

Similarly, aqueous K_2CrO_4 also behaves as a strong electrolyte. Potassium chromate contains the K^+ and CrO_4^{2-} ions, so an aqueous solution of potassium

Pure water

(a)

Source of electric power

Free ions present in water

(b)

Figure 7.2
Electrical conductivity of aqueous solutions. (a) Pure water does not conduct an electric current. The lamp does not light. (b) When an ionic compound is dissolved in water, current flows and the lamp lights. The result of this experiment is strong evidence that ionic compounds dissolved in water exist in the form of separate ions.

A conductivity apparatus can be used to show the nature of ionic solutions and to show that something new has been formed when precipitation occurs. Demonstrate the presence of ions in solutions of $Ba(NO_3)_2$ and K_2CrO_4. Compare the amount of light produced from the two solutions with the amount of light produced from pure water. Then, mix solutions containing equimolar amounts of Ba^{2+} and CrO_4^{2-}. Have students figure out what solid is formed. Then immerse the conductivity apparatus in the solution with the precipitate to show that ions have been removed from solution.

chromate (which is prepared by dissolving solid K_2CrO_4 in water) contains these separated ions. That is, $K_2CrO_4(aq)$ does not contain K_2CrO_4 units but instead contains K^+ cations and CrO_4^{2-} anions, which move around independently. (There are two K^+ ions for each CrO_4^{2-} ion.)

The idea introduced here is very important: when ionic compounds dissolve, the *resulting solution contains the separated ions.* Therefore we can represent the mixing of $K_2CrO_4(aq)$ and $Ba(NO_3)_2(aq)$ in two ways. We usually write these reactants as

$$K_2CrO_4(aq) \ + \ Ba(NO_3)_2(aq) \rightarrow \text{Products}$$

However, a more accurate representation of the situation is

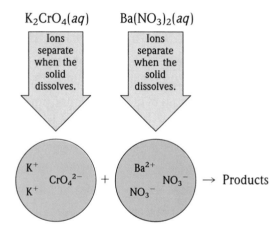

We can express this information in equation form as follows:

$$2K^+(aq) + CrO_4^{2-}(aq) + Ba^{2+}(aq) + 2NO_3^-(aq) \rightarrow \text{Products}$$

If you do not have equipment for performing conductivity demonstrations, draw some beakers on the board. Separate beakers can contain ions from a different compound in solution. When the contents of the two beakers are "mixed," you can show all four ions in solution before the reaction occurs. This type of discussion can be more effective than a live demonstration.

The ions in $K_2CrO_4(aq)$ The ions in $Ba(NO_3)_2(aq)$

Thus the *mixed solution* contains four types of ions: K^+, CrO_4^{2-}, Ba^{2+}, and NO_3^-. Now that we know what the reactants are, we can make some educated guesses about the possible products.

How to Decide What Products Form

Which of these ions combine to form the yellow solid observed when the original solutions are mixed? This is not an easy question to answer. Even an experienced chemist is not sure what will happen in a new reaction. The chemist tries to think of the various possibilities, considers the likelihood of each possibility, and then makes a prediction (an educated guess). Only after identifying each product experimentally can the chemist be sure what reaction actually has taken place.

However, an educated guess is very useful because it indicates what kinds of products are most likely to result. It gives us a place to start. So the best way to proceed is first to think of the various possibilities and then to decide which of them is most likely.

What are the possible products of the reaction between $K_2CrO_4(aq)$ and $Ba(NO_3)_2(aq)$ or, more accurately, what reaction can occur among the ions K^+, CrO_4^{2-}, Ba^{2+}, and NO_3^-? We already know some things that will help us decide. We know that a *solid compound must have a zero net charge.* This means that the product of our reaction must contain *both anions and cations* (negative and positive ions). For example, K^+ and Ba^{2+} could not combine to form the solid because such a solid would have a positive charge. Similarly, CrO_4^{2-} and NO_3^- could not combine to form a solid because that solid would have a negative charge.

Something else that will help us is an observation that chemists have made by examining many compounds: *most ionic materials contain only two types of ions*—one type of cation and one type of anion. This idea is illustrated by the following compounds (among many others):

Compound	Cation	Anion
NaCl	Na^+	Cl^-
KOH	K^+	OH^-
Na_2SO_4	Na^+	SO_4^{2-}
NH_4Cl	NH_4^+	Cl^-
Na_2CO_3	Na^+	CO_3^{2-}

All the possible combinations of a cation and an anion to form uncharged compounds from among the ions K^+, CrO_4^{2-}, Ba^{2+}, and NO_3^-, are shown below:

	NO_3^-	CrO_4^{2-}
K^+	KNO_3	K_2CrO_4
Ba^{2+}	$Ba(NO_3)_2$	$BaCrO_4$

So the list of compounds that *might* make up the solid are

$$K_2CrO_4 \quad KNO_3 \quad BaCrO_4 \quad Ba(NO_3)_2$$

Which of these possibilities is most likely to represent the yellow solid? We know it's not K_2CrO_4 or $Ba(NO_3)_2$; these are the reactants. They were present (dissolved) in the separate solutions that were mixed initially. The only real possibilities are KNO_3 and $BaCrO_4$. To decide which of these is more likely to represent the yellow solid, we need more facts. An experienced chemist, for example,

I work the first example by asking students to suggest possible combinations of ions that could be the product. Only after all possible combinations are listed do we go back and eliminate unlikely combinations. By this time, some students are thoroughly confused and have difficulty understanding why Na^+ and Ca^{2+} cannot form a compound. I usually review the basics of compound formation before we proceed.

After the possible products are narrowed to two, I present the solubility rules found in Table 7.1 on page 199. It helps to put the rules on an overhead transparency and show them for the remainder of lecture. Students can refer to them without opening their texts as you work examples.

knows that KNO_3 is a white solid. On the other hand, the CrO_4^{2-} ion is yellow. Therefore the yellow solid must be $BaCrO_4$.

We have determined that one product of the reaction between $K_2CrO_4(aq)$ and $Ba(NO_3)_2(aq)$ is $BaCrO_4(s)$, but what happened to the K^+ and NO_3^- ions? The answer is that these ions are left dissolved in the solution. That is, KNO_3 does not form a solid when the K^+ and NO_3^- ions are present in water. In other words, if we took the white solid $KNO_3(s)$ and put it in water, it would totally dissolve (the white solid would "disappear" yielding a colorless solution). So when we mix $K_2CrO_4(aq)$ and $Ba(NO_3)_2(aq)$, $BaCrO_4(s)$ forms but KNO_3 is left behind in solution [we write it as $KNO_3(aq)$]. (If we poured the mixture through a filter to remove the solid $BaCrO_4$ and then evaporated all of the water, we would obtain the white solid KNO_3.)

After all this thinking, we can finally write the unbalanced equation for the precipitation reaction.

$$K_2CrO_4(aq) \;+\; Ba(NO_3)_2(aq) \rightarrow BaCrO_4(s) \;+\; KNO_3(aq)$$

We can represent this reaction in pictures as follows:

Note that the K^+ and NO_3^- ions are not involved in the chemical change. They remain dispersed in the water before and after the reaction.

Using Solubility Rules

In the example considered above we were finally able to identify the products of the reaction by using two types of chemical knowledge:

1. Knowledge of facts
2. Knowledge of concepts

For example, knowing the colors of the various compounds proved very helpful. This represents factual knowledge. Awareness of the concept that solids always have a net charge of zero was also essential. These two kinds of knowledge allowed us to make good guesses about the identity of the solid that formed.

As you continue to study chemistry, you will see that a balance of factual and conceptual knowledge is always required. You must both *memorize* important facts and *understand* crucial concepts in order to succeed.

In the present case we are dealing with a reaction in which an ionic solid forms—that is, a process in which ions that are dissolved in water combine to

---

For the first examples of precipitation reactions, I set up the anions and cations in a grid so that the formulas for all possible products can be easily written. After the first few, I use the analogy of switching partners on a double date. This analogy allows students to begin to view these reactions as double-displacement reactions.

I usually don't expect students to memorize the solubility rules; I provide them for use on hourly exams. Students have enough difficulty determining the products with the rules in front of them.

Solids must contain both anions and cations in the relative numbers necessary to produce zero net charge.

not always soluble

give a solid. We know that for a solid to form, both positive and negative ions must be present in relative numbers that give zero net charge. However, oppositely charged ions in water do not always react to form a solid, as we have seen for K^+ and NO_3^-. In addition, Na^+ and Cl^- can coexist in water in very large numbers with no formation of solid NaCl. Another way of saying the same thing is to say that when solid NaCl (common salt) is placed in water, it dissolves—the white solid "disappears" as the Na^+ and Cl^- ions are dispersed throughout the water. (You probably have observed this phenomenon in preparing salt water to cook food.) The following two statements, then, are really saying the same thing.

Solid NaCl is very soluble in water.

Solid NaCl does not form when one solution containing Na^+ is mixed with another solution containing Cl^-.

To predict whether a given pair of dissolved ions will form a solid when mixed, we must know some facts about the solubilities of various types of ionic compounds. In this text we will use the term **soluble solid** to mean a solid that readily dissolves in water; the solid "disappears" as the ions are dispersed in the water. The terms **insoluble solid** and **slightly soluble solid** are taken to mean the same thing: a solid where such a tiny amount dissolves in water that it is undetectable with the naked eye. The solubility information about common solids that is summarized in Table 7.1 is based on observations of the behavior of many compounds. This is factual knowledge that you will need to predict what will happen in chemical reactions where a solid might form.

Table 7.1   General Rules for Solubility of Ionic Compounds (Salts) in Water at 25 °C

---

1. Most nitrate (NO_3^-) salts are soluble.
2. Most salts of Na^+, K^+, and NH_4^+ are soluble.
3. Most chloride salts are soluble. Notable exceptions are AgCl, $PbCl_2$, and Hg_2Cl_2.
4. Most sulfate salts are soluble. Notable exceptions are $BaSO_4$, $PbSO_4$, and $CaSO_4$.
5. Most hydroxide compounds are only slightly soluble.* The important exceptions of soluble hydroxides are NaOH, KOH, and $Ca(OH)_2$.
6. Most sulfide (S^{2-}), carbonate (CO_3^{2-}), and phosphate (PO_4^{3-}) salts are only slightly soluble.*

---

*The terms *insoluble* and *slightly soluble* really mean the same thing: such a tiny amount dissolves that it is not possible to detect it with the naked eye.

One effective demonstration that can be used in class or in the laboratory is the three-bottle experiment. Three solutions are prepared and their identities made known, but not which bottle contains which solution. Portions of the solutions are mixed in pairs. From the results of mixing and the solubility rules, students are asked to determine which bottle contains which solution. One set of solutions that works well is Na_2S, $Ba(NO_3)_2$, and K_2CO_3. You will need to tell your students, or show them in advance, that BaS forms a greyish precipitate, whereas $BaCO_3$ is white. Depending upon what compounds you have on hand, you can devise many different three-bottle experiments, and even some with six bottles. These experiments teach several skills, including making good observations, organizing data, drawing conclusions, and predicting the products of reactions.

We have worked to establish in students' minds that in a precipitation reaction a solid product forms. The precipitate is obviously not water soluble. Now, we introduce the term slightly soluble. Stress to your students that compounds that we call slightly soluble are virtually insoluble in water and that most of the compound remains a solid. Only a few ions are formed.

I always use one example such as Na_2SO_4 plus KCl so that students don't think that a precipitate forms every time solutions of ionic compounds are mixed.

Notice that in Table 7.1 the term *salt* is used to mean *ionic compound.* Many chemists use the terms salt and ionic compound interchangeably. In Example 7.1, we will illustrate how to use the solubility rules to predict the products of reactions among ions.

EXAMPLE 7.1	Identifying Precipitates in Reactions Where a Solid Forms

AgNO$_3$ is usually called silver nitrate rather than silver(I) nitrate.

When an aqueous solution of silver nitrate is added to an aqueous solution of potassium chloride, a white solid forms. Identify the white solid and write the balanced equation for the reaction that occurs.

SOLUTION

First let's use the description of the reaction to represent what we know:

$$AgNO_3(aq) + KCl(aq) \rightarrow \text{White solid}$$

Remember, try to determine the essential facts from the words and represent these facts by symbols or diagrams. To answer the main question (What is the white solid?) we must establish what ions are present in the mixed solution. That is, we must know what the reactants are really like. Remember that *when ionic substances dissolve in water, the ions separate.* So we can write the equation

$$\underbrace{Ag^+(aq) + NO_3^-(aq)}_{\text{Ions in } AgNO_3(aq)} + \underbrace{K^+(aq) + Cl^-(aq)}_{\text{Ions in } KCl(aq)} \rightarrow \text{Products}$$

	NO$_3^-$	Cl$^-$
Ag$^+$	AgNO$_3$	AgCl
K$^+$	KNO$_3$	KCl

or use pictures

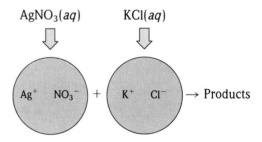

AgNO$_3$(aq)   KCl(aq)

Ag$^+$ NO$_3^-$ + K$^+$ Cl$^-$ $\rightarrow$ Products

to represent the ions present in the mixed solution before any reaction occurs. In summary:

Ag(NO$_3$)(aq) + KCl(aq)   Solution contains   Ag$^+$ NO$_3^-$ K$^+$ Cl$^-$

EXAMPLE 7.1, CONTINUED

Now we will consider what solid *might* form from this collection of ions. Because the solid must contain both positive and negative ions, the possible compounds that can be assembled from this collection of ions are

$$AgNO_3 \quad KCl \quad AgCl \quad KNO_3$$

$AgNO_3$ and KCl are the substances already dissolved in the reactant solutions, so we know that they do not represent the white solid product. We are left with two possibilities:

$$AgCl \quad and \quad KNO_3$$

Another way to obtain these two possibilities is by *ion interchange.* This means that in the reaction of $AgNO_3(aq)$ and $KCl(aq)$, we take the cation from one reactant and combine it with the anion of the other reactant.

$$Ag^+ \quad + \quad NO_3^- \quad + \quad K^+ \quad + \quad Cl^- \rightarrow Products$$

Possible solid products

Ion interchange leads to the following possible solids:

$$\boxed{AgCl \quad or \quad KNO_3}$$

To decide whether AgCl or KNO_3 is the white solid, we need the solubility rules (Table 7.1). Rule 2 states that most salts containing K^+ are quite soluble in water. Rule 1 says that most nitrate salts (those containing NO_3^-) are soluble. So the salt KNO_3 is water-soluble. That is, when K^+ and NO_3^- are mixed in water, a solid (KNO_3) does *not* form.

On the other hand, rule 3 states that, although most chloride salts (salts that contain Cl^-) are soluble, AgCl is an exception. That is, AgCl(s) is insoluble in water. Thus the white solid must be AgCl. Now we can write

$$AgNO_3(aq) + KCl(aq) \rightarrow AgCl(s) + ?$$

What is the other product?

To form AgCl(s), we have used the Ag^+ and Cl^- ions:

$$Ag^+(aq) + NO_3^-(aq) + K^+(aq) + Cl^-(aq) \rightarrow AgCl(s)$$

This leaves the K^+ and NO_3^- ions. What do they do? Nothing. Because KNO_3 is very soluble in water (rules 1 and 2), the K^+ and NO_3^- ions remain separate in the water; the KNO_3 remains dissolved and we represent it as

EXAMPLE 7.1, CONTINUED

$KNO_3(aq)$. We can now write the full equation:

$$AgNO_3(aq) + KCl(aq) \rightarrow AgCl(s) + KNO_3(aq)$$

Figure 7.3 shows the precipitation of $AgCl(s)$ that occurs when this reaction takes place. In graphic form, the reaction is

In predicting what reaction will occur when ions are mixed in water, focus on the ions that are in solution before any reaction occurs.

The following is a useful strategy for predicting what will occur when two solutions containing dissolved salts are mixed.

How to Predict Precipitates When Solutions of Two Ionic Compounds Are Mixed

STEP 1
Write the reactants as they actually exist before any reaction occurs. Remember that when a salt dissolves, its ions separate.

STEP 2
Consider the various solids that could form. To do this, simply *exchange the anions* of the added salts.

STEP 3
Use the solubility rules (Table 7.1) to decide whether a solid forms and, if so, to predict the identity of the solid.

Figure 7.3
Precipitation of silver chloride occurs when solutions of silver nitrate and potassium chloride are mixed. The K^+ and NO_3^- ions remain in solution.

EXAMPLE 7.2   **Using Solubility Rules to Predict Reactions**

Using the solubility rules in Table 7.1, predict what will happen when the following solutions are mixed. Write the balanced equation for any reaction that occurs.

a. $KNO_3(aq)$ and $BaCl_2(aq)$
b. $Na_2SO_4(aq)$ and $Pb(NO_3)_2(aq)$
c. $KOH(aq)$ and $Fe(NO_3)_3(aq)$

EXAMPLE 7.2, CONTINUED

SOLUTION

a. STEP 1
    $KNO_3(aq)$ represents an aqueous solution obtained by dissolving solid
    KNO_3 in water to give the ions $K^+(aq)$ and $NO_3^-(aq)$. Likewise, $BaCl_2(aq)$
    is a solution formed by dissolving solid $BaCl_2$ in water to produce $Ba^{2+}(aq)$
    and $Cl^-(aq)$. When these two solutions are mixed, the following ions will
    be present:

    $$K^+, \quad NO_3^-, \quad\quad Ba^{2+}, \quad Cl^-$$

    From $KNO_3(aq)$ From $BaCl_2(aq)$

    STEP 2
    To get the possible products, we exchange the anions.

    $$K^+ \quad NO_3^- \quad Ba^{2+} \quad Cl^-$$

    This yields the possibilities KCl and $Ba(NO_3)_2$. These are the solids that
    might form. Notice that two NO_3^- ions are needed to balance the 2+
    charge on Ba^{2+}.

    STEP 3
    The rules listed in Table 7.1 indicate that KCl and $Ba(NO_3)_2$ are both soluble
    in water. So no precipitate forms when $KNO_3(aq)$ and $BaCl_2(aq)$ are mixed.
    All of the ions remain dissolved in the solution. This means that no reaction
    takes place. That is, no chemical change occurs.

b. STEP 1
    The following ions are present in the mixed solution before any reaction
    occurs:

    $$Na^+, \quad SO_4^{2-}, \quad\quad Pb^{2+}, \quad NO_3^-$$

    From   From
    $Na_2SO_4(aq)$ $Pb(NO_3)_2(aq)$

EXAMPLE 7.2, CONTINUED

STEP 2
Exchanging anions as follows:

$$Na^+ \quad SO_4^{2-} \quad Pb^{2+} \quad NO_3^-$$

yields the *possible* solid products $PbSO_4$ and $NaNO_3$.

STEP 3
Using Table 7.1, we see that $NaNO_3$ is soluble in water (rules 1 and 2) but that $PbSO_4$ is only slightly soluble (rule 4). Thus when these solutions are mixed, solid $PbSO_4$ forms. The balanced reaction is

$$Na_2SO_4(aq) + Pb(NO_3)_2(aq) \rightarrow PbSO_4(s) + \underbrace{2NaNO_3(aq)}_{\text{Remains dissolved}}$$

which can be represented as

c. STEP 1
The ions present in the mixed solution before any reaction occurs are

$$\underbrace{K^+, \quad OH^-,}_{\text{From KOH}(aq)} \quad \underbrace{Fe^{3+}, \quad NO_3^-}_{\text{From Fe(NO}_3)_3(aq)}$$

STEP 2
Exchanging anions as follows:

$$K^+ \quad OH^- \quad Fe^{3+} \quad NO_3^-$$

yields the possible solid products KNO_3 and $Fe(OH)_3$.

STEP 3
Rules 1 and 2 (Table 7.1) state that KNO_3 is soluble, whereas $Fe(OH)_3$ is only slightly soluble (rule 5). Thus when these solutions are mixed, solid $Fe(OH)_3$ forms. The balanced equation for the reaction is

$$3KOH(aq) + Fe(NO_3)_3(aq) \rightarrow Fe(OH)_3(s) + 3KNO_3(aq)$$

which can be represented as

SELF-CHECK EXERCISE 7.1

Predict whether a solid will form when the following pairs of solutions are mixed. If so, identify the solid and write the balanced equation for the reaction.

a. $Ba(NO_3)_2(aq)$ and $NaCl(aq)$

b. $Na_2S(aq)$ and $Cu(NO_3)_2(aq)$

c. $NH_4Cl(aq)$ and $Pb(NO_3)_2(aq)$

CHEMISTRY IN FOCUS

The Chemistry of Teeth

*I*f dental chemistry continues to progress at its present rate, tooth decay may soon be a thing of the past. Cavities are holes that develop in tooth enamel, which is composed of the mineral hydroxyapatite, $Ca_5(PO_4)_3OH$. Recent studies have shown that there is constant dissolving and reforming of the tooth mineral in the saliva at the tooth's surface. Demineralization (dissolving of tooth enamel) is caused mainly by acids in the saliva that are created by bacteria as they digest food.

In the first stages of tooth decay, parts of the tooth surface become porous and spongy and develop Swiss-cheese-like holes that, if untreated, eventually turn into cavities (see the accompanying photograph). How-

Tooth enamel in early stages of decay.

ever, recent results indicate that if the affected tooth is bathed in a solution containing appropriate amounts of Ca^{2+}, PO_4^{3-}, and F^-, it remineralizes. In this process, F^- replaces some of the OH^- present in the original tooth mineral. That is, some of the $Ca_5(PO_4)_3OH$ is changed to $Ca_5(PO_4)_3F$. The remineralized area is more resistant to decay in the future, because the fluoride mineral is less soluble than the original tooth enamel. In addition, it has been shown that the presence of Sr^{2+} in the remineralizing fluid significantly increases resistance to decay.

If these results hold up under further study, the work of dentists will change dramatically. Dentists will be much more involved in preventing damage to teeth than in repairing damage that has already occurred. One can picture the routine use of a remineralization rinse that will repair enamel in problem areas before they become cavities. Dental drills could join leeches as a medical anachronism (and patients would probably miss them about as much!).

7.3   Describing Reactions in Aqueous Solutions

AIM: To describe reactions in solutions by writing molecular, complete ionic, and net ionic equations.

Much important chemistry, including virtually all of the reactions that make life possible, occurs in aqueous solutions. We will now consider the types of equations used to represent reactions that occur in water. For example, as we saw earlier, when we mix aqueous potassium chromate with aqueous barium nitrate, a reaction occurs to form solid barium chromate and dissolved potassium nitrate. One way to represent this reaction is by the equation

$$K_2CrO_4(aq) + Ba(NO_3)_2(aq) \rightarrow BaCrO_4(s) + 2KNO_3(aq)$$

When you contrast molecular and complete ionic equations, it might be helpful to mention that molecular equations tend to represent substances as we get them from the bottle. For complete ionic equations, show substances as they actually exist in aqueous solution.

This is called the **molecular equation** for the reaction that shows the complete formulas of all reactants and products. However, although this equation shows the reactants and products of the reaction, it does not give a very clear picture of what actually occurs in solution. As we have seen, aqueous solutions of potassium chromate, barium nitrate, and potassium nitrate contain the individual ions, not molecules as is implied by the molecular equation. Thus the **complete ionic equation,**

When we write complete ionic equations, students are often surprised that the solid product is not written as individual ions, as are the spectator ions. It is hard for students to grasp that the ions have combined to form a new compound.

Ions from K_2CrO_4 Ions from $Ba(NO_3)_2$

$$2K^+(aq) + CrO_4{}^{2-}(aq) + Ba^{2+}(aq) + 2NO_3{}^-(aq)$$
$$\rightarrow BaCrO_4(s) + 2K^+(aq) + 2NO_3{}^-(aq)$$

A strong electrolyte is a substance that completely breaks apart into ions when dissolved in water. The resulting solution readily conducts an electric current.

better represents the actual forms of the reactants and products in solution. *In a complete ionic equation, all substances that are strong electrolytes are represented as ions.* Notice that $BaCrO_4$ is not written as the separate ions, because it is present as a solid; it is not dissolved.

The complete ionic equation reveals that only some of the ions participate in the reaction. Notice that the K^+ and $NO_3{}^-$ ions are present in solution both before and after the reaction. Ions such as these, which do not participate directly in a reaction in solution, are called **spectator ions.** The ions that participate in this reaction are the Ba^{2+} and $CrO_4{}^{2-}$ ions, which combine to form solid $BaCrO_4$:

$$Ba^{2+}(aq) + CrO_4{}^{2-}(aq) \rightarrow BaCrO_4(s)$$

The net ionic equation includes only those components that undergo a change in the reaction.

When they are writing net ionic equations, students often forget that the charges and the number of atoms must be balanced. Work an example that requires balancing to draw attention to this step.

This equation, called the **net ionic equation,** includes only those components that are directly involved in the reaction. Chemists usually write the net ionic equation for a reaction in solution, because it gives the actual forms of the reactants and products and includes only the species that undergo a change.

Types of Equations for Reactions in Aqueous Solutions

Three types of equations are used to describe reactions in solutions.

- The *molecular equation* shows the overall reaction but not necessarily the actual forms of the reactants and products in solution.
- The *complete ionic equation* represents all reactants and products that are strong electrolytes as ions. All reactants and products are included.
- The *net ionic equation* includes only those components that undergo a change. Spectator ions are not included.

Before I move on to reactions that form water, I give one last example that ties together Sections 7.1, 7.2, and 7.3. I ask students to tell me the product of the reaction between $Zn(NO_3)_2$ and KOH, ask them to help me balance the equation, and then to help me write complete ionic and net ionic equations.

To make sure these ideas are clear, we will do another example. In the last section we considered the reaction between aqueous solutions of lead nitrate and sodium sulfate. The molecular equation for this reaction is

$$Pb(NO_3)_2(aq) \; + \; Na_2SO_4(aq) \rightarrow PbSO_4(s) \; + \; 2NaNO_3(aq)$$

Because any ionic compound that is dissolved in water is present as the separated ions, we can write the complete ionic equation as follows:

$$Pb^{2+}(aq) \; + \; 2NO_3^-(aq) \; + \; 2Na^+(aq) \; + \; SO_4^{2-}(aq)$$
$$\rightarrow PbSO_4(s) \; + \; 2Na^+(aq) \; + \; 2NO_3^-(aq)$$

The $PbSO_4$ is not written as separate ions because it is present as a solid. The ions that take part in the chemical change are the Pb^{2+} and the SO_4^{2-} ions, which combine to form solid $PbSO_4$. Thus the net ionic equation is

$$Pb^{2+}(aq) \; + \; SO_4^{2-} \rightarrow PbSO_4(s)$$

The Na^+ and NO_3^- ions do not undergo any chemical change; they are spectator ions.

EXAMPLE 7.3	Writing Equations for Reactions in Solution

For each of the following reactions, write the molecular equation, the complete ionic equation, and the net ionic equation.

a. Aqueous sodium chloride is added to aqueous silver nitrate to form solid silver chloride plus aqueous sodium nitrate.

b. Aqueous potassium hydroxide is mixed with aqueous iron(III) nitrate to form solid iron(III) hydroxide and aqueous potassium nitrate.

Because silver is present as Ag^+ in all of its common ionic compounds, we usually delete the (I) when naming silver compounds.

EXAMPLE 7.3, CONTINUED

SOLUTION

a.   *Molecular equation:*

$$NaCl(aq) + AgNO_3(aq) \rightarrow AgCl(s) + NaNO_3(aq)$$

Complete ionic equation:

$$Na^+(aq) + Cl^-(aq) + Ag^+(aq) + NO_3^-(aq)$$
$$\rightarrow AgCl(s) + Na^+(aq) + NO_3^-(aq)$$

Net ionic equation:

$$Cl^-(aq) + Ag^+(aq) \rightarrow AgCl(s)$$

b.   *Molecular equation:*

$$3KOH(aq) + Fe(NO_3)_3(aq) \rightarrow Fe(OH)_3(s) + 3KNO_3(aq)$$

Complete ionic equation:

$$3K^+(aq) + 3OH^-(aq) + Fe^{3+}(aq) + 3NO_3^-(aq)$$
$$\rightarrow Fe(OH)_3(s) + 3K^+(aq) + 3NO_3^-(aq)$$

Net ionic equation:

$$3OH^-(aq) + Fe^{3+}(aq) \rightarrow Fe(OH)_3(s)$$

SELF-CHECK EXERCISE 7.2

For each of the following reactions, write the molecular equation, the complete ionic equation, and the net ionic equation.

a.   Aqueous sodium sulfide is mixed with aqueous copper(II) nitrate to produce solid copper(II) sulfide and aqueous sodium nitrate.

b.   Aqueous ammonium chloride and aqueous lead(II) nitrate react to form solid lead(II) chloride and aqueous ammonium nitrate.

A lemon tastes sour because it contains citric acid.

7.4 Reactions That Form Water: Acids and Bases

AIM: To describe the key characteristics of the reactions between strong acids and strong bases.

In this section we encounter two very important classes of compounds: acids and bases. Acids were first associated with the sour taste of citrus fruits. In fact, the word *acid* comes from the Latin word *acidus,* which means "sour." Vinegar tastes sour because it is a dilute solution of acetic acid; citric acid is responsible for the sour taste of a lemon. Bases, sometimes called *alkalis,* are characterized by their bitter taste and slippery feel, like wet soap. Most commercial preparations for unclogging drains are highly basic.

Acids have been known for hundreds of years. For example, the *mineral acids* sulfuric acid, H_2SO_4, and nitric acid, HNO_3, so named because they were originally obtained by the treatment of minerals, were discovered around 1300. However, it was not until the late 1800s that the essential nature of acids was discovered by Svante Arrhenius, then a Swedish graduate student in physics.

Arrhenius, who was trying to discover why only certain solutions could conduct an electric current, found that conductivity arose from the presence of ions. In his studies of solutions, Arrhenius observed that when the substances HCl, HNO_3, and H_2SO_4 were dissolved in water, they behaved as strong electrolytes. He suggested that this was the result of ionization reactions in water.

$$HCl \xrightarrow{H_2O} H^+(aq) + Cl^-(aq)$$

$$HNO_3 \xrightarrow{H_2O} H^+(aq) + NO_3^-(aq)$$

$$H_2SO_4 \xrightarrow{H_2O} H^+(aq) + HSO_4^-(aq)$$

Arrhenius proposed that an **acid** *is a substance that produces H^+ ions (protons) when it is dissolved in water.*

Studies show that when HCl, HNO_3, and H_2SO_4 are placed in water, *virtually every molecule* dissociates to give ions. This means that when 100 molecules of HCl are dissolved in water, 100 H^+ ions and 100 Cl^- ions are produced. Virtually no HCl molecules exist in aqueous solution (see Figure 7.4). Because these substances are strong electrolytes that produce H^+ ions, they are called **strong acids.**

Arrhenius also found that *aqueous solutions that exhibit basic behavior always contain hydroxide ions.* He defined a **base** as a *substance that produces hydroxide ions in water.* The base most commonly used in the chemical laboratory is sodium hydroxide, NaOH, which contains Na^+ and OH^- ions and is very

Students have already received a brief introduction to acids in Section 5.9, Naming Acids.

Don't taste chemicals!

The Nobel Prize in chemistry was awarded to Arrhenius in 1903 for his studies of solution conductivity.

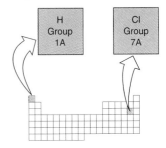

The Arrhenius definition of an acid: a substance that produces H^+ ions in aqueous solution.

I don't talk about the existence of weak acids in this section. Students are introduced to weak acids in Section 17.2

Each HCl molecule dissociates when it dissolves in water.

Figure 7.4

When gaseous HCl is dissolved in water, each molecule dissociates to produce H^+ and Cl^- ions. That is, HCl behaves as a strong electrolyte.

If you have equipment for a conductivity demonstration, it can be effective to show that strong acids and bases conduct electricity just as ionic salts do.

soluble in water. Sodium hydroxide, like all ionic substances, produces separated cations and anions when it is dissolved in water.

$$NaOH(s) \xrightarrow{H_2O} Na^+(aq) + OH^-(aq)$$

Although dissolved sodium hydroxide is usually represented as NaOH(aq), you should remember that the solution really contains separated Na^+ and OH^- ions. In fact, for every 100 units of NaOH dissolved in water, 100 Na^+ and 100 OH^- ions are produced.

Potassium hydroxide (KOH) has properties markedly similar to those of sodium hydroxide. It is very soluble in water and produces separated ions.

$$KOH(s) \xrightarrow{H_2O} K^+(aq) + OH^-(aq)$$

Because these hydroxide compounds are strong electrolytes that contain OH^- ions, they are called **strong bases.**

When strong acids and strong bases (hydroxides) are mixed, the fundamental chemical change that always occurs is that *H^+ ions react with OH^- ions to form water.*

$$H^+(aq) + OH^-(aq) \rightarrow H_2O(l)$$

You might want to remind students that the formation of water is one of the driving forces that causes reactions to proceed toward the right.

Students have difficulty understanding how water can be the product of a reaction when all the solutions are already aqueous. I always tell them that we are producing new water molecules from the ions that are mixed together.

Hydrochloric acid is an aqueous solution that contains dissolved hydrogen chloride. It is a strong electrolyte.

You can compare the net ionic equations of precipitation reactions and neutralization reactions. Most precipitation reactions have net ionic equations with different ions, but neutralization reactions all have the same net ionic equation.

Water is a very stable compound, as evidenced by the abundance of it on the earth's surface. Therefore, when substances that can form water are mixed, there is a strong tendency for the reaction to occur. In particular, the hydroxide ion OH^- has a high affinity for H^+ ions, because water is produced in the reaction between these ions.

The tendency to form water is the second of the driving forces for reactions that we mentioned in Section 7.1. Any compound that produces OH^- ions in water reacts vigorously to form H_2O with any compound that can furnish H^+ ions. For example, the reaction between hydrochloric acid and aqueous sodium hydroxide is represented by the following molecular equation:

$$HCl(aq) + NaOH(aq) \rightarrow H_2O(l) + NaCl(aq)$$

Because HCl, NaOH, and NaCl exist as completely separated ions in water, the complete ionic equation for this reaction is

$$H^+(aq) + Cl^-(aq) + Na^+(aq) + OH^-(aq) \rightarrow$$
$$H_2O(l) + Na^+(aq) + Cl^-(aq)$$

Notice that the Cl^- and Na^+ are spectator ions (they undergo no changes), so the net ionic equation is

$$H^+(aq) + OH^-(aq) \rightarrow H_2O(l).$$

Thus the only chemical change that occurs when these solutions are mixed is that water is formed from H^+ and OH^- ions.

EXAMPLE 7.4	Writing Equations for Acid–Base Reactions

Nitric acid is a strong acid. Write the molecular, complete ionic, and net ionic equations for the reaction of aqueous nitric acid and aqueous potassium hydroxide.

SOLUTION

Molecular equation:

$$HNO_3(aq) \ + \ KOH(aq) \rightarrow H_2O(l) \ + \ KNO_3(aq)$$

Complete ionic equation:

$$H^+(aq) \ + \ NO_3^-(aq) \ + \ K^+(aq) \ + \ OH^-(aq) \rightarrow$$
$$H_2O(l) \ + \ K^+(aq) \ + \ NO_3^-(aq)$$

Net ionic equation:

$$H^+(aq) \ + \ OH^-(aq) \rightarrow H_2O(l)$$

Note that K^+ and NO_3^- are spectator ions and that the formation of water is the driving force for this reaction.

There are two important things to note as we examine the reaction of hydrochloric acid with aqueous sodium hydroxide and the reaction of nitric acid with aqueous potassium hydroxide.

Hydrochloric acid is an aqueous solution of HCl.

1. The net ionic equation is the same in both cases; water is formed.

$$H^+(aq) \ + \ OH^-(aq) \rightarrow H_2O(l)$$

2. Besides water, which is *always a product* of the reaction of an acid with OH^-, the second product is an ionic compound, which might precipitate or remain dissolved, depending on its solubility.

$$HCl(aq) + NaOH(aq) \rightarrow H_2O(l) + NaCl(aq)$$
$$HNO_3(aq) + KOH(aq) \rightarrow H_2O(l) + KNO_3(aq)$$

Dissolved ionic compounds

This ionic compound is called a **salt.** In the first case the salt is sodium chloride, and in the second case the salt is potassium nitrate. We can obtain these soluble salts in solid form (both are white solids) by evaporating the water.

The following points about strong acids and bases are particularly important.

1. The common strong acids are aqueous solutions of HCl, HNO_3, and H_2SO_4.

2. A strong acid is a substance that completely dissociates in water. (Each molecule breaks up into an H^+ ion plus an anion.)

3. A strong base is a metal hydroxide compound that is very soluble in water. The most common strong bases are NaOH and KOH, which completely break up into separated ions (Na^+ and OH^- or K^+ and OH^-) when they are dissolved in water.

4. The net ionic equation for the reaction of a strong acid and a strong base is always the same: it shows the production of water.

$$H^+(aq) + OH^-(aq) \rightarrow H_2O(l)$$

5. In the reaction of a strong acid and a strong base, one product is always water and the other is always an ionic compound called a salt, which remains dissolved in the water. This salt can be obtained as a solid by evaporating the water.

6. The reaction of H^+ and OH^- is often called an acid–base reaction, where H^+ is the acidic ion and OH^- is the basic ion.

CHAPTER REVIEW

Key Terms

precipitation reaction (p. 195)
strong electrolyte (p. 195)
soluble solid (p. 199)
insoluble (slightly soluble) solid (p. 199)
molecular equation (p. 206)
complete ionic equation (p. 206)
spectator ions (p. 206)

net ionic equation (p. 206)
acid (p. 209)
strong acid (p. 209)
base (p. 209)
strong base (p. 210)
salt (p. 211)

Summary

1. Four driving forces that favor chemical change (chemical reaction) are formation of a solid, formation of water, formation of a gas, and transfer of electrons.

2. A reaction where a solid forms is called a precipitation reaction. General rules on solubility help predict whether a solid—and what solid—will form when two solutions are mixed.

3. Three types of equations are used to describe reactions in solution: (1) the molecular equation, which shows the complete formulas of all reactants and products; (2) the complete ionic equation, in which all reactants and products that are strong electrolytes are shown as ions; and (3) the net ionic equation, which includes only those components of the solution that undergo a change. Spectator ions (those ions that remain unchanged in a reaction) are not included in a net ionic equation.

4. A strong acid is a compound in which virtually every molecule dissociates in water to give an H^+ ion and an anion. Similarly, a strong base is a metal hydroxide compound that is soluble in water, giving OH^- ions and cations. The products of the reaction of a strong acid and a strong base are water and a salt.

Questions and Problems

All even-numbered exercises have answers in the back of this book and solutions in the Solutions Guide.

7.1 Predicting Whether a Reaction Will Occur

QUESTIONS

1. Why is water an important solvent? Although you have not yet studied water in detail, can you think of some properties of water that make it so important?

2. What is a "driving force"? What are some of the driving forces discussed in this section that tend to make reactions likely to occur? Can you think of any other possible driving forces?

7.2 Reactions in Which a Solid Forms

QUESTIONS

3. When two solutions are mixed, and a solid forms and settles out of the mixture, the solid is commonly referred to as a _____.

4. When two solutions of ionic substances are mixed and a precipitate forms, what is the net charge of the precipitate? Why?

5. When a soluble ionic substance such as sodium chloride is dissolved in water, the _____ that are released behave independently of each other.

6. When an ionic substance dissolves, the resulting solution contains the separated _____.

7. A substance that completely breaks up into ions when dissolved in water is called a (strong/weak) electrolyte.

8. What information is used to determine what products are formed when two solutions of ionic substances are mixed to form a precipitate? Is there a way to determine, merely by inspection, what the products are?

9. When aqueous solutions of potassium chloride, KCl, and silver nitrate, AgNO₃, are mixed, a precipitate forms, but the precipitate is *not* potassium chloride. What does this tell you about the solubility of KCl in water?

10. What do we mean when we say that a solid is "slightly" soluble in water? Is there a practical difference between *slightly* soluble and *in*soluble?

11. On the basis of the general solubility rules given in Table 7.1, predict which of the following substances are likely to be soluble in water.
    a. aluminum chloride
    b. nickel(II) sulfide
    c. sodium carbonate
    d. silver carbonate
    e. ammonium dichromate
    f. ammonium bromide
    g. iron(III) hydroxide

12. On the basis of the general solubility rules given in Table 7.1, predict which of the following substances are likely to be soluble in water.
    a. barium nitrate
    b. potassium carbonate
    c. sodium sulfate
    d. copper(II) hydroxide
    e. mercury(I) chloride
    f. ammonium phosphate
    g. chromium(III) sulfide
    h. lead sulfate

13. On the basis of the general solubility rules given in Table 7.1, explain *why* the following compounds are *not* likely to be soluble in water. That is, indicate *which* of the solubility rules covers each substance's particular situation.
    a. cobalt(II) sulfide, CoS
    b. iron(III) hydroxide, $Fe(OH)_3$
    c. calcium phosphate, $Ca_3(PO_4)_2$
    d. barium carbonate, $BaCO_3$

14. On the basis of the general solubility rules given in Table 7.1, for each of the following compounds indicate *which* of the solubility rules reveals that the substance is expected to be insoluble.
    a. mercuric sulfide, HgS
    b. nickel(II) hydroxide, $Ni(OH)_2$
    c. barium sulfate, $BaSO_4$
    d. zirconium carbonate, $ZrCO_3$

15. On the basis of the general solubility rules given in Table 7.1, predict the identity of the precipitate that forms when aqueous solutions of the following substances are mixed. If no precipitate is likely, indicate why (which rules apply).
    a. silver nitrate, $AgNO_3$, and sodium iodide, NaI
    b. sodium sulfate, Na_2SO_4, and barium chloride, $BaCl_2$
    c. sodium phosphate, Na_3PO_4, and calcium chloride, $CaCl_2$
    d. ammonium fluoride, NH_4F, and potassium sulfide, K_2S
    e. calcium chloride, $CaCl_2$, and sodium sulfate, Na_2SO_4
    f. lead(II) nitrate, $Pb(NO_3)_2$, and barium chloride, $BaCl_2$

16. On the basis of the general solubility rules given in Table 7.1, predict the identity of the precipitate that forms when aqueous solutions of the following substances are mixed. If no precipitate is likely, indicate which rule(s) applies.
    a. mercuric nitrate, $Hg(NO_3)_2$, and sodium sulfide, Na_2S
    b. sulfuric acid, H_2SO_4, and calcium nitrate, $Ca(NO_3)_2$
    c. ammonium chloride, NH_4Cl, and silver nitrate, $AgNO_3$
    d. barium nitrate, $Ba(NO_3)_2$, and sodium hydroxide, $NaOH$
    e. phosphoric acid, H_3PO_4, and nickel(II) chloride, $NiCl_2$
    f. potassium carbonate, K_2CO_3, and copper(II) sulfate, $CuSO_4$

PROBLEMS

17. On the basis of the general solubility rules given in Table 7.1, write a balanced molecular equation for the precipitation reactions that take place when the following aqueous solutions are mixed. Circle the formula of the precipitate (solid) that forms. If no precipitation reaction is likely for the reactants given, so indicate.
    a. hydrochloric acid, HCl, and silver nitrate, $AgNO_3$
    b. nitric acid, HNO_3, and ammonium chloride, NH_4Cl
    c. sulfuric acid, H_2SO_4, and barium nitrate, $Ba(NO_3)_2$
    d. hydrochloric acid, HCl, and lead acetate, $Pb(C_2H_3O_2)_2$
    e. lead acetate, $Pb(C_2H_3O_2)_2$, and sodium chloride, $NaCl$
    f. barium chloride, $BaCl_2$, and sodium carbonate, Na_2CO_3

18. On the basis of the general solubility rules given in Table 7.1, write a balanced molecular equation for the precipitation reactions that take place when the following aqueous solutions are mixed. Circle the formula of the precipitate (solid) that forms. If no precipitation reaction is likely for the solutes given, so indicate.
    a. nitric acid, HNO_3, and barium chloride, $BaCl_2$
    b. ammonium sulfide, $(NH_4)_2S$, and cobalt(II) chloride, $CoCl_2$
    c. sulfuric acid, H_2SO_4, and lead(II) nitrate, $Pb(NO_3)_2$
    d. calcium chloride, $CaCl_2$, and carbonic acid, H_2CO_3
    e. sodium acetate, $NaC_2H_3O_2$, and ammonium nitrate, NH_4NO_3
    f. sodium phosphate, Na_3PO_4, and chromium(III) chloride, $CrCl_3$

19. Balance each of the following equations that describe precipitation reactions.
    a. $Pb(NO_3)_2(aq) + H_2SO_4(aq) \rightarrow PbSO_4(s) + HNO_3(aq)$
    b. $BaCl_2(aq) + Na_2SO_4(aq) \rightarrow BaSO_4(s) + NaCl(aq)$
    c. $HCl(aq) + Pb(C_2H_3O_2)_2(aq)$
    $$\rightarrow PbCl_2(s) + HC_2H_3O_2(aq)$$

21. For each of the following precipitation reactions, complete and balance the equation, indicating clearly which product is the precipitate.
    a. $Pb(NO_3)_2(aq) + CoCl_2(aq) \rightarrow$
    b. $Ca(C_2H_3O_2)_2(aq) + K_2SO_4(aq) \rightarrow$
    c. $FeCl_3(aq) + NaOH(aq) \rightarrow$

20. Balance each of the following equations that describe precipitation reactions.
    a. $AgNO_3(aq) + H_2SO_4(aq) \rightarrow Ag_2SO_4(s) + HNO_3(aq)$
    b. $Ca(NO_3)_2(aq) + H_2SO_4(aq) \rightarrow CaSO_4(s) + HNO_3(aq)$
    c. $Pb(NO_3)_2(aq) + H_2SO_4(aq) \rightarrow PbSO_4(s) + HNO_3(aq)$

22. For each of the following precipitation reactions, complete and balance the equation, indicating clearly which product is the precipitate.
    a. $NiCl_2(aq) + H_2S(aq) \rightarrow$
    b. $CuSO_4(aq) + NaOH(aq) \rightarrow$
    c. $Ba(NO_3)_2(aq) + Na_2CO_3(aq) \rightarrow$

7.3 Describing Reactions in Aqueous Solutions

QUESTIONS

23. An equation for a precipitation reaction that shows only those ions that actually form the precipitate is called the _____ for the reaction.

24. Ions that do not directly participate in a reaction in solution are called _____ ions.

PROBLEMS

25. Write balanced net ionic equations for the reactions that occur when the following aqueous solutions are mixed. If no reaction is likely to occur, so indicate.
    a. silver nitrate, $AgNO_3$, and potassium chloride, KCl
    b. nickel(II) sulfate, $NiSO_4$, and barium chloride, $BaCl_2$
    c. ammonium phosphate, $(NH_4)_3PO_4$, and calcium chloride, $CaCl_2$
    d. hydrofluoric acid, HF, and potassium sulfate, K_2SO_4
    e. calcium chloride, $CaCl_2$, and ammonium sulfate, $(NH_4)_2SO_4$
    f. lead(II) nitrate, $Pb(NO_3)_2$, and barium chloride, $BaCl_2$

26. Write balanced net ionic equations for the reactions that occur when the following aqueous solutions are mixed. If no reaction is likely to occur, so indicate.
    a. iron(III) nitrate and potassium hydroxide
    b. nickel(II) chloride and hydrosulfuric acid
    c. silver nitrate and ammonium chloride
    d. sodium sulfate and barium chloride
    e. potassium bromide and mercury(I) nitrate
    f. barium nitrate and potassium sulfate

27. A common analysis for the quantity of halide ions (Cl^-, Br^-, and I^-) in a sample is to precipitate and weigh the halide ions as their silver, Ag^+, salts. For example, a given sample of seawater can be treated with dilute silver nitrate, $AgNO_3$, solution to precipitate the halides. The mixture of precipitated silver halides can then be filtered from the solution, dried, and weighed as an indication of the halide content of the original sample. Write the net ionic equations showing the precipitation of halide ions from seawater with silver nitrate.

28. Barium sulfate, $BaSO_4$, is very insoluble in water. In analytical chemistry, samples containing barium ion, Ba^{2+}, may be analyzed by treating the sample with dilute sulfuric acid, H_2SO_4. Write the net ionic equation for the precipitation of barium ions from a sample by the use of sulfuric acid.

29. Many plants are poisonous because their stems and leaves contain oxalic acid, $H_2C_2O_4$, or sodium oxalate, $Na_2C_2O_4$; when ingested, these substances cause swelling of the respiratory tract and suffocation. A standard analysis for determining the amount of oxalate ion, $C_2O_4{}^{2-}$, in a sample is to

30. As discussed in the solubility rules in this chapter, most carbonate ($CO_3{}^{2-}$) salts are only slightly soluble. In the northeastern United States, mountain ranges tend to contain a lot of limestone, whose major constituents are salts containing the carbonate ion. Rain and snow tend to partially

precipitate this species as calcium oxalate, which is insoluble in water. Write the net ionic equation for the reaction between sodium oxalate and calcium chloride, $CaCl_2$.

dissolve these carbonate salts, carrying them off into rivers and streams. When water from these sources is used in industry and the home, carbonate salts tend to precipitate on the inside walls of pipes and boilers. Write balanced net ionic equations for the precipitation reactions of carbonate ion with Fe^{2+}, Pb^{2+}, Al^{2+}, and Zn^{2+} ions.

7.4 Reactions That Form Water: Acids and Bases

QUESTIONS

31. When an acid like HCl is placed in water, virtually _____percent of the molecules ionize.

32. According to the simple Arrhenius theory, a(n) _____ is a substance that produces hydrogen ions, H^+, when dissolved in water.

33. The fundamental net ionic process that occurs when a strong acid reacts with a strong base is_____.

34. The common strong acids are H_2SO_4, HNO_3, and _____.

35. If 1,000 NaOH units were dissolved in a sample of water, the NaOH would ionize to produce _____Na^+ ions and _____OH^- ions.

36. The ionic compound produced when a strong acid and a strong base react is called a(n) _____.

PROBLEMS

37. In addition to the three strong acids emphasized in the chapter (HCl, HNO_3, and H_2SO_4), hydrobromic acid, HBr, and perchloric acid, $HClO_4$, are also strong acids. Write equations for the dissociation of each of these additional strong acids in water.

38. In addition to the strong bases NaOH and KOH discussed in this chapter, the hydroxide compounds of other Group 1 elements also behave as strong bases when dissolved in water. Write equations for RbOH and CsOH that show what ions they form when they dissolve in water.

39. Complete and balance each of the following molecular equations for strong acid/strong base reactions; circle the formula of the salt produced in each.
    a. $HCl(aq) + RbOH(aq) \rightarrow$
    b. $HClO_4(aq) + NaOH(aq) \rightarrow$
    c. $HBr(aq) + NaOH(aq) \rightarrow$
    d. $H_2SO_4(aq) + CsOH(aq) \rightarrow$

40. Below are indicated the *products* of acid–base reactions. Complete (and balance) the equations to show which strong acids/strong bases have reacted to form these products.
    a. _____ $\rightarrow KCl(aq) + H_2O(l)$
    b. _____ $\rightarrow Na_2SO_4(aq) + H_2O(l)$
    c. _____ $\rightarrow CsClO_4(aq) + H_2O(l)$
    d. _____ $\rightarrow KNO_3(aq) + H_2O(l)$

Additional Problems

41. Complete and balance each of the following precipitation reactions; circle the formula of the solid that precipitates.
    a. $CaCl_2(aq) + K_2SO_4(aq) \rightarrow$
    b. $KOH(aq) + Fe(NO_3)_3(aq) \rightarrow$
    c. $Na_2SO_4(aq) + Pb(C_2H_3O_2)_2(aq) \rightarrow$
    d. $FeCl_3(aq) + H_2S(aq) \rightarrow$
    e. $CuSO_4(aq) + NaOH(aq) \rightarrow$

42. Distinguish between the *molecular* equation, the *complete ionic* equation, and the *net ionic* equation for a reaction in solution. Which type of equation most clearly shows the species that actually react with one another?

43. Write balanced net ionic equations for the reactions that occur when the following aqueous solutions are mixed. If no reaction is likely to occur, so indicate.
    a. silver nitrate and calcium nitrate
    b. nickel(II) chloride and lead(II) acetate
    c. silver nitrate and ammonium carbonate
    d. sodium sulfate and calcium chloride
    e. mercury(I) nitrate and copper(II) chloride
    f. calcium nitrate and sulfuric acid

44. In the subfield of chemistry called *qualitative analysis,* precipitation reactions are used as a means of identifying the

ions present in a sample. For example, lead ions in solution precipitate when hydrochloric acid is added. If the precipitate of lead chloride is dissolved by heating it in boiling water, the lead may be reprecipitated as a bright yellow solid with potassium chromate. Lead ion is the only metal ion that behaves this way. Write net ionic equations for both of these precipitation reactions.

45. Using the general solubility rules given in Table 7.1, name three reactants that would form precipitates with each of the following ions in aqueous solution. Write the net ionic equation for each of your suggestions.
   a. chloride ion
   b. calcium ion
   c. iron(III) ion
   d. sulfate ion
   e. mercury(I) ion, Hg_2^{2+}
   f. silver ion
46. Without first writing a full molecular or ionic equation, write the net ionic equations for any precipitation reactions that occur when aqueous solutions of the following compounds are mixed. If no reaction occurs, so indicate.
   a. cobalt(III) chloride and sodium hydroxide
   b. silver nitrate and ammonium carbonate
   c. barium nitrate and hydrochloric acid
   d. barium chloride and sulfuric acid
   e. copper(II) sulfate and nickel(II) chloride
   f. calcium chloride and phosphoric acid
   g. aluminum chloride and potassium hydroxide
47. List the formulas of four strong acids and four strong bases. What do we mean when we say that these acids and bases are *strong?* Write a full molecular equation for the reaction of each of the four acids you have chosen with each of the four bases.
48. The hydroxide compounds of the common Group 2 elements—$Ca(OH)_2$, $Mg(OH)_2$, $Sr(OH)_2$, and $Ba(OH)_2$—are not very soluble in water. However, the small amount of the compound that does dissolve produces ions. Therefore, these hydroxide compounds are sometimes considered strong bases. Write an equation for each of these bases, showing the ions produced when they dissolve in water.
49. Complete and balance each of the following molecular equations for strong acid/strong base reactions; circle the formula of the salt produced in each.
   a. $HCl(aq) + Ca(OH)_2(aq) \rightarrow$
   b. $H_2SO_4(aq) + Sr(OH)_2(aq) \rightarrow$
   c. $HNO_3(aq) + Ba(OH)_2(aq) \rightarrow$
   d. $HClO_4(aq) + Mg(OH)_2(aq) \rightarrow$
50. One important measurement for a chemical reaction is the quantity of heat energy liberated or absorbed when the reaction takes place. It has been found that when any strong acid is reacted with an excess amount of any strong base, exactly the same amount of heat energy is liberated per unit amount of strong acid. Explain this observation. *Hint:* think of the net ionic equation.

51. For the cations listed in the left-hand column, give the formulas of the precipitates that would form with each of the anions in the right-hand column. If no precipitate is expected for a particular combination, so indicate.

Cations	Anions
Ag^+	$C_2H_3O_2^-$
Ba^{2+}	Cl^-
Ca^{2+}	CO_3^{2-}
Fe^{3+}	NO_3^-
Hg_2^{2+}	OH^-
Na^+	PO_4^{3-}
Ni^{2+}	S^{2-}
Pb^{2+}	SO_4^{2-}

52. On the basis of the general solubility rules given in Table 7.1, predict which of the following substances are likely to be soluble in water.
   a. potassium hexacyanoferrate(III), $K_3Fe(CN)_6$
   b. ammonium molybdate, $(NH_4)_2MoO_4$
   c. osmium(II) carbonate, $OsCO_3$
   d. gold phosphate, $AuPO_4$
   e. sodium hexanitrocobaltate(III), $Na_3Co(NO_2)_6$
   f. barium hydroxide, $Ba(OH)_2$
   g. iron(III) chloride, $FeCl_3$
53. On the basis of the general solubility rules given in Table 7.1, explain *why* the following compounds are *not* likely to be soluble in water. That is, indicate *which* of the solubility rules covers each substance's particular situation.
   a. iron(III) sulfide
   b. chromium(III) hydroxide
   c. nickel(II) carbonate
   d. aluminum phosphate
54. On the basis of the general solubility rules given in Table 7.1, predict the identity of the precipitate that forms when aqueous solutions of the following substances are mixed. If no precipitate is likely, indicate why (which rules apply).
   a. iron(III) chloride and sodium hydroxide
   b. nickel(II) nitrate and ammonium sulfide
   c. silver nitrate and potassium chloride
   d. sodium carbonate and barium nitrate
   e. potassium chloride and mercury(I) nitrate
   f. barium nitrate and sulfuric acid
55. On the basis of the general solubility rules given in Table 7.1, write a balanced molecular equation for the precipitation reactions that take place when the following aqueous solutions are mixed. Circle the formula of the precipitate

(solid) that forms. If no precipitation reaction is likely for the reactants given, so indicate.
  a. silver nitrate and barium nitrate
  b. nickel(II) chloride and silver acetate
  c. lead nitrate and ammonium carbonate
  d. sodium sulfate and calcium acetate
  e. nickel(II) chloride and mercury(I) nitrate
  f. calcium nitrate and sulfuric acid

56. Balance each of the following equations that describe precipitation reactions.
  a. $Co(NO_3)_2(aq) + (NH_4)_2S(aq) \rightarrow CoS(s) + NH_4NO_3(aq)$
  b. $CaCl_2(aq) + H_2SO_4(aq) \rightarrow CaSO_4(s) + HCl(aq)$
  c. $FeCl_3(aq) + Na_3PO_4(aq) \rightarrow FePO_4(s) + NaCl(aq)$

57. For each of the following precipitation reactions, complete and balance the equation, indicating clearly which product is the precipitate.
  a. $Pb(NO_3)_2(aq) + Na_2CO_3(aq) \rightarrow$
  b. $K_2SO_4(aq) + CaCl_2(aq) \rightarrow$
  c. $Ni(C_2H_3O_2)_2(aq) + Na_2S(aq) \rightarrow$

58. For each of the following *unbalanced* molecular equations, write the corresponding *balanced net ionic equation* for the reaction.
  a. $HCl(aq) + AgNO_3(aq) \rightarrow AgCl(s) + HNO_3(aq)$
  b. $CaCl_2(aq) + Na_3PO_4(aq) \rightarrow Ca_3(PO_4)_2(s) + NaCl(aq)$
  c. $Pb(NO_3)_2(aq) + BaCl_2(aq) \rightarrow PbCl_2(s) + Ba(NO_3)_2(aq)$
  d. $FeCl_3(aq) + NaOH(aq) \rightarrow Fe(OH)_3(s) + NaCl(aq)$

59. For each of the following *unbalanced* molecular equations, write the corresponding *balanced net ionic equation* for the reaction.
  a. $Ca(NO_3)_2(aq) + H_2SO_4(aq) \rightarrow CaSO_4(s) + HNO_3(aq)$
  b. $Na_2CO_3(aq) + CoCl_2(aq) \rightarrow CoCO_3(s) + NaCl(aq)$
  c. $NiCl_2(aq) + Na_2S(aq) \rightarrow NiS(s) + NaCl(aq)$
  d. $Hg_2(NO_3)_2(aq) + NaCl(aq) \rightarrow Hg_2Cl_2(s) + NaNO_3(aq)$

60. Most sulfide compounds of the transition metals are insoluble in water. Many of these metal sulfides have striking and characteristic colors by which we can identify them. Therefore, in the analysis of mixtures of metal ions, it is very common to precipitate the metal ions by using hydrogen sulfide, H_2S. Suppose you had a mixture of Fe^{2+}, Cr^{3+}, and Ni^{2+}. Write net ionic equations for the precipitation of these metal ions by the use of H_2S.

61. _____ and nitric acids are sometimes referred to as the "mineral" acids.

62. The strong bases considered in this chapter are the _____ compounds of the Group 1 elements.

63. When substances that can react to form water are mixed, there is a (strong/weak) tendency for the reaction to occur readily.

64. Complete and balance each of the following molecular equations for strong acid/strong base reactions; circle the formula of the salt produced in each.
  a. $HClO_4(aq) + RbOH(aq) \rightarrow$
  b. $HNO_3(aq) + KOH(aq) \rightarrow$
  c. $H_2SO_4(aq) + NaOH(aq) \rightarrow$
  d. $HBr(aq) + CsOH(aq) \rightarrow$

65. What strong acid and what strong base would react in aqueous solution to produce the salts listed below?
  a. potassium sulfate, K_2SO_4
  b. sodium nitrate, $NaNO_3$
  c. cesium bromide, $CsBr$
  d. sodium perchlorate, $NaClO_4$

66. Using the general solubility rules given in Table 7.1, name three reactants that would form precipitates with each of the following ions in aqueous solutions. Write the balanced molecular equation for each of your suggested reactants.
  a. sulfide ion
  b. carbonate ion
  c. hydroxide ion
  d. phosphate ion

Solutions to Self-Check Exercises

SELF-CHECK EXERCISE 7.1

a. The ions present are

$$Ba^{2+}(aq) + 2NO_3^-(aq) + Na^+(aq) + Cl^-(aq) \rightarrow$$

Ions in $Ba(NO_3)_2(aq)$ Ions in $NaCl(aq)$

Exchanging the anions gives the possible solid products $BaCl_2$ and $NaNO_3$. Using Table 7.1, we see that both substances are very soluble (rules 1, 2, and 3). Thus no solid forms.

b.   The ions present in the mixed solution before any reaction occurs are

$$\underbrace{2Na^+(aq) + S^{2-}(aq)}_{\substack{\text{Ions in} \\ Na_2S(aq)}} + \underbrace{Cu^{2+}(aq) + 2NO_3^-\,(aq)}_{\substack{\text{Ions in} \\ Cu(NO_3)_2(aq)}} \rightarrow$$

Exchanging the anions gives the possible solid products CuS and $NaNO_3$. By rules 1 and 2 in Table 7.1, $NaNO_3$ is soluble, and by rule 6, CuS should be insoluble. Thus CuS will precipitate. The balanced equation is

$$Na_2S(aq) + Cu(NO_3)_2(aq) \rightarrow CuS(s) + 2NaNO_3(aq)$$

c.   The ions present are

$$\underbrace{NH_4^+(aq) + Cl^-(aq)}_{\substack{\text{Ions in} \\ NH_4Cl(aq)}} + \underbrace{Pb^{2+}(aq) + 2NO_3^-(aq)}_{\substack{\text{Ions in} \\ Pb(NO_3)_2(aq)}} \rightarrow$$

Exchanging the anions gives the possible solid products NH_4NO_3 and $PbCl_2$. NH_4NO_3 is soluble (rules 1 and 2) and $PbCl_2$ is insoluble (rule 3). Thus $PbCl_2$ will precipitate. The balanced equation is

$$2NH_4Cl(aq) + Pb(NO_3)_2(aq) \rightarrow PbCl_2(s) + 2NH_4NO_3(aq)$$

SELF-CHECK EXERCISE 7.2

a.   *Molecular equation:*

$$Na_2S(aq) + Cu(NO_3)_2(aq) \rightarrow CuS(s) + 2NaNO_3(aq)$$

Complete ionic equation:

$$2Na^+(aq) + S^{2-}(aq) + Cu^{2+}(aq) + 2NO_3^-(aq)$$
$$\rightarrow CuS(s) + 2Na^+(aq) + 2NO_3^-(aq)$$

Net ionic equation:

$$S^{2-}(aq) + Cu^{2+}(aq) \rightarrow CuS(s)$$

b.   *Molecular equation:*

$$2NH_4Cl(aq) + Pb(NO_3)_2(aq) \rightarrow PbCl_2(s) + 2NH_4NO_3(aq)$$

Complete ionic equation:

$$2NH_4^+(aq) + 2Cl^-(aq) + Pb^{2+}(aq) + 2NO_3^-(aq)$$
$$\rightarrow PbCl_2(s) + 2NH_4^+(aq) + 2NO_3^-(aq)$$

Net ionic equation:

$$2Cl^-(aq) + Pb^{2+}(aq) \rightarrow PbCl_2(s)$$

8 Classifying Chemical Reactions

CONTENTS

Potassium reacts with water to yield hydrogen gas which catches
fire due to the energy (heat) produced by the reaction.

In Chapter 7 we considered two important types of reactions that occur in aqueous solutions. Precipitation reactions are driven by the formation of an insoluble solid, whereas acid–base reactions are driven by the formation of that very stable substance, water.

A third driving force for chemical reactions is the transfer of electrons from one substance to another. A good example of this type of reaction occurs when a metal, which tends to lose one or more electrons, reacts with a nonmetal, which tends to gain one or more electrons. This is the type of reaction that occurs when steel rusts in the air. We will consider these important reactions in this chapter.

Because there are so many different chemical reactions, it is helpful to classify these reactions in various ways. This makes it easier to remember reactions that we have studied and helps us understand the new reactions that we encounter. The following diagram gives an overview of the classes of reactions we have mentioned up to this point. The driving force for each type of reaction is given in parentheses.

It is important to stress that there are many ways to classify chemical reactions. The system presented in this chapter is not the only one. Some reactions fall into more than one category. For example, a combustion reaction is also an oxidation–reduction reaction. The different categories of oxidation–reduction reactions can be very confusing to students.

As we explore oxidation–reduction reactions in this chapter, we will see that there are several subclasses of this type of reaction.

8.1   Reactions of Metals with Nonmetals (Oxidation–Reduction)

Discuss the importance of oxidation–reduction reactions. Battery power, corrosion of metals, and combustion of fossil fuels are all oxidation–reduction reactions that cost consumers money.

AIM: To describe the general characteristics of a reaction between a metal and a nonmetal.
To introduce electron transfer as a driving force for a chemical reaction.

In Chapter 5 we spent considerable time discussing the reaction of a metal and a nonmetal to form an ionic compound. A typical example is the formation of

I like to demonstrate to students that the loss of electrons occurs only when there is an element present that can gain electrons. You can use a small piece of sodium metal to illustrate that sodium, by itself, does not form sodium ions. Similarly, chlorine gas in a cylinder does not spontaneously gain electrons. But if you heat the sodium in a glass deflagrating spoon and place it in the container of chlorine, the two elements react to form ions.

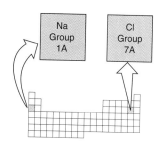

sodium chloride from sodium metal and chlorine gas:

$$2Na(s) + Cl_2(g) \rightarrow 2NaCl(s)$$

Let's examine what happens in this reaction. Sodium metal is composed of sodium atoms, each of which has a net charge of zero. (The positive charges of the eleven protons in its nucleus are exactly balanced by the negative charges on the eleven electrons.) Similarly, the chlorine molecule consists of two uncharged chlorine atoms (each has seventeen protons and seventeen electrons). However, in the product (sodium chloride), the sodium is present as Na^+ and the chlorine as Cl^-. By what process do the neutral atoms become ions? The answer is that one electron is transferred from each sodium atom to each chlorine atom.

$$Na + Cl \rightarrow Na^+ + Cl^-$$

After the electron transfer, each sodium has ten electrons and eleven protons (a net charge of $1+$), and each chlorine has eighteen electrons and seventeen protons (a net charge of $1-$).

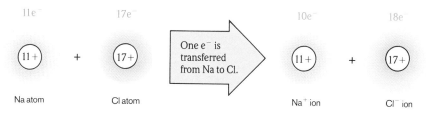

When introducing gain and loss of electrons, I find it helpful to use drawings of atoms similar to those in this section of the text. Students can see both the number of protons and the number of electrons in the ion. This makes it easy for them to determine algebraically the charge on the ion.

Thus the reaction of a metal with a nonmetal to form an ionic compound involves the transfer of one or more electrons from the metal (which forms a cation) to the nonmetal (which forms an anion). This tendency to transfer electrons from metals to nonmetals is the fourth driving force for reactions that we listed in Section 7.1. Such a reaction that *involves a transfer of electrons* is called an **oxidation–reduction reaction.** These reactions are often called **redox** reactions by chemists.

There are many examples of oxidation–reduction reactions in which a metal reacts with a nonmetal to form an ionic compound. Consider the reaction of magnesium metal with oxygen,

$$2Mg(s) + O_2(g) \rightarrow 2MgO(s)$$

which produces a bright, white light useful in camera flash units. Note that the reactants contain uncharged atoms, but the product contains ions.

MgO

Contains Mg^{2+}, O^{2-}

Therefore, in this reaction, each magnesium atom loses two electrons ($Mg \rightarrow Mg^{2+} + 2e^-$) and each oxygen atom gains two electrons ($O + 2e^- \rightarrow O^{2-}$).

The rusting of metal occurs by an oxidation–reduction reaction.

The first oxidation–reduction reactions I discuss are ones that occur between two pure elements. Only after students understand the basic concepts do I present a reaction between a pure element and a compound, such as that between NaBr and Cl_2. Often, students are not sure which two of the three elements participate in the reaction. I provide them with both reactants and products so that they need only determine the species that is oxidized and the one that is reduced.

We might represent this reaction as follows:

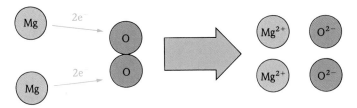

Another example is

$$2Al(s) + Fe_2O_3(s) \rightarrow 2Fe(s) + Al_2O_3(s)$$

which is a reaction (called the thermite reaction) that produces so much energy (heat) that the iron is initially formed as a liquid (see Figure 8.1). In this case the aluminum is originally present as the elemental metal (which contains uncharged Al atoms) and ends up in Al_2O_3, where it is present as Al^{3+} cations (the $2Al^{3+}$ ions just balance the charge of the $3O^{2-}$ ions). Therefore, in the reaction each aluminum atom loses three electrons.

$$Al \rightarrow Al^{3+} + 3e^-$$

The opposite process occurs with the iron, which is initially present as Fe^{3+} ions in Fe_2O_3 and ends up as uncharged atoms in the elemental iron. Thus each iron cation gains three electrons to form an uncharged atom:

$$Fe^{3+} + 3e^- \rightarrow Fe$$

We can represent this reaction in schematic form as follows:

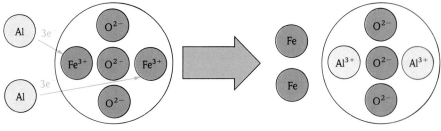

This equation is read, "An aluminum atom yields an aluminum ion with a 3+ charge and three electrons."

Figure 8.1
The thermite reaction gives off so much heat that the iron formed is molten.

EXAMPLE 8.1 **Identifying Electron Transfer in Oxidation–Reduction Reactions**

For each of the following reactions, show how electrons are gained and lost.

a.　$2Al(s) + 3I_2(s) \rightarrow 2AlI_3$
(This reaction is shown in Figure 8.2. Note the flame plus the purple "smoke," which is due to excess I_2 being driven off by the heat.)

b.　$2Cs(s) + F_2(g) \rightarrow 2CsF(s)$

EXAMPLE 8.I, CONTINUED

SOLUTION

a. In AlI_3 the ions are Al^{3+} and I^- (aluminum always forms Al^{3+}, and iodine always forms I^-). In $Al(s)$ the aluminum is present as uncharged atoms. Thus aluminum goes from Al to Al^{3+} by losing three electrons $(Al \rightarrow Al^{3+} + 3e^-)$. In I_2 each iodine atom is uncharged. Thus each iodine atom goes from I to I^- by gaining one electron $(I + e^- \rightarrow I^-)$. A schematic for this reaction is

Figure 8.2
When powdered aluminum and iodine react, the heat given off changes some of the solid iodine to a purple cloud of vapor.

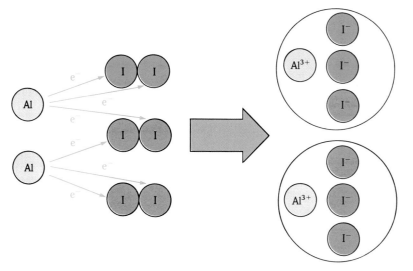

You can illustrate your lectures with some of the more spectacular oxidation–reduction demonstrations. Thermite, $KClO_3$-sugar and Al-I_2 are commonly used. Both thermite and Al-I_2 are discussed in the text. The actual demonstrations are much more dramatic than the still photographs indicate. Detailed instructions and precautions for preparing thermite, $KClO_3$-sugar, and other oxidation–reduction demonstrations can be found in *Chemical Demonstrations* by Bassam Z. Shakshiri (Madison, Wisconsin: The University of Wisconsin Press, 1983).

b. In CsF the ions present are Cs^+ and F^-. Cesium metal, $Cs(s)$, contains uncharged cesium atoms; and fluorine gas, $F_2(g)$, contains uncharged fluorine atoms. Thus in the reaction each cesium atom loses one electron $(Cs \rightarrow Cs^+ + e^-)$ and each fluorine atom gains one electron $(F + e^- \rightarrow F^-)$. The schematic for this reaction is

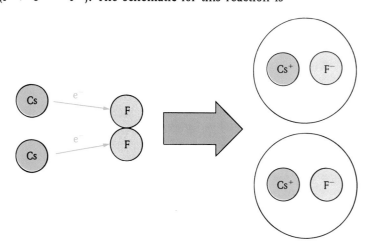

EXAMPLE 8.1, CONTINUED

SELF-CHECK EXERCISE 8.1

For each reaction, show how electrons are gained and lost by the elements.

a. $2Na(s) + Br_2(l) \rightarrow 2NaBr(s)$

b. $2Ca(s) + O_2(g) \rightarrow 2CaO(s)$

So far we have discussed only electron transfer (oxidation–reduction) reactions that involve a metal and a nonmetal. Electron transfer reactions can also take place between two nonmetals. We will not discuss these reactions in detail here. All we will say at this point is that one sure sign of an oxidation–reduction reaction between nonmetals is the presence of oxygen, $O_2(g)$, as a reactant or product. In fact, oxidation got its name from oxygen. Thus the reactions

$$CH_4(g) + 2O_2(g) \rightarrow CO_2(g) + 2H_2O(g)$$

and

$$2SO_2(g) + O_2(g) \rightarrow 2SO_3(g)$$

are electron transfer reactions, even though it is not obvious at this point.

Make a special note of the following points:

1. When a metal reacts with a nonmetal, an ionic compound is formed. The ions are formed when the metal transfers one or more electrons to the nonmetal, the metal atom becoming a cation and the nonmetal atom becoming an anion. *Therefore a metal–nonmetal reaction can always be assumed to be an oxidation–reduction reaction, which involves electron transfer.*
2. Two nonmetals can also undergo an oxidation–reduction reaction. At this point we can recognize these cases only by looking for O_2 as a reactant or product. When two nonmetals react, the compound formed is not ionic. More will be said about these reactions in Chapter 18.

8.2   Ways to Classify Reactions

AIM: To learn various classification schemes for reactions.

So far in our study of chemistry we have seen many, many chemical reactions—and this is just Chapter 8. In the world around us and in our bodies, literally millions of chemical reactions are taking place. Obviously, we need a system for putting reactions into meaningful classes that will make them easier to remember and easier to understand.

In Chapters 7 and 8 we have so far considered the following "driving forces" for chemical reactions:

1. Formation of a solid
2. Formation of water
3. Transfer of electrons

We will now discuss how to classify reactions involving these processes. For example, in the reaction

$$K_2CrO_4(aq) + Ba(NO_3)_2(aq) \rightarrow BaCrO_4(s) + 2KNO_3(aq)$$

Solution   Solution   Solid   Solution
formed

solid $BaCrO_4$ (a precipitate) is formed. Because the *formation of a solid when two solutions are mixed* is called *precipitation,* we call this a **precipitation reaction.** In fact, any reaction in which a solid is formed when two solutions are mixed can be classified as a precipitation reaction.

Notice in this reaction that two anions (NO_3^- and CrO_4^{2-}) are simply exchanged. Note that CrO_4^{2-} was originally associated with K^+ in K_2CrO_4 and that NO_3^- was associated with Ba^{2+} in $Ba(NO_3)_2$. In the products these associations are reversed. Because of this double exchange, we sometimes call this reaction a double-exchange reaction or **double-displacement reaction.** We might represent such a reaction as

$$AB + CD \rightarrow AC + BD$$

So we can classify a reaction such as the one above as a precipitation reaction or as a double-displacement reaction. Either name is correct, but the former is more commonly used by chemists.

In Chapter 7 we also considered reactions in which water is formed when a strong acid is mixed with a strong base. All of these reactions had the same net ionic equation:

$$H^+(aq) + OH^-(aq) \rightarrow H_2O(l)$$

Go over the different terms and definitions used to describe reactions in Chapters 7 and 8. It is hard for students to remember that a precipitation reaction can also be called a double-displacement reaction.

I present to students lists of criteria and clues they can use to help them recognize the types of reactions. The characteristics of a precipitation reaction, for example, are the production of a solid product and the presence of ionic reactants and products.

After students write correct formulas for the products of double-displacement reactions, the equations are not necessarily balanced. Students forget to check the coefficients. Work an example that is not balanced after the products are written to remind students to balance the equation after they have written the products.

My list of criteria that students can use to help identify strong acid—strong base reactions includes the following: H^+ appears on the reactant side combined with an anion. OH^- appears on the reactant side combined with a metal ion. The product is water. The net ionic equation is always $H^+ (aq) + OH^- (aq) \rightarrow H_2O(l)$.

CHEMISTRY IN FOCUS

Oxidation–Reduction Reactions Launch the Space Shuttle

*L*aunching into space a vehicle that weighs millions of pounds requires unimaginable quantities of energy—all furnished by oxidation–reduction reactions.

Notice from Figure 8.3 that three cylindrical objects are attached to the shuttle orbiter. In the center is a tank 28 feet in diameter and 154 feet long that contains liquid oxygen and liquid hydrogen (in separate compartments). These fuels are fed to the orbiter's rocket engines, where they react to form water and release a huge quantity of energy.

$$2H_2 + O_2 \rightarrow 2H_2O + energy$$

Note that we can recognize this reaction as an oxidation–reduction reaction because O_2 is a reactant.

Two solid-fuel rockets 12 feet in diameter and 150 feet long are also attached to the orbiter. Each rocket contains 1.1 million pounds fuel: ammonium perchlorate (NH_4ClO_4) and powdered aluminum mixed with a binder ("glue"). Because the rock-

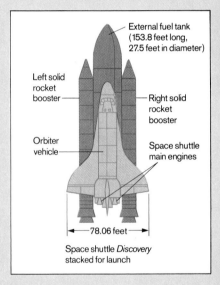

Figure 8.3
For launch, the space shuttle orbiter is attached to two solid-fuel rockets (left and right) and a fuel tank (center) that supplies hydrogen and oxygen to the orbiter's engines.

Adapted with permission from *Chemical and Engineering News*, September 19, 1988, Volume 66, Number 33, p. 9. Copyright 1988, American Chemical Society.

ets are so large, they are built in segments and assembled at the launch site as shown in Figure 8.4. Each segment is filled with the syrupy propellant (Figure 8.5), which then solidifies to a consistency much like that of a hard rubber eraser.

The oxidation–reduction reaction between the ammonium perchlorate and the aluminum is represented as follows:

$$3NH_4ClO_4(s) + 3Al(s)$$
$$\rightarrow Al_2O_3(s) + AlCl_3(s)$$
$$+ 3NO(g) + 6H_2O(g) + energy$$

It produces temperatures of about 5700 °F and 3.3 million pounds of thrust in each rocket.

Thus we can see that oxidation–reduction reactions furnish the energy to launch the space shuttle.

Solid booster

Solid propellant

Aft field joint (point of failure in *Challenger*'s right booster)

149.16 feet long, 12.17 feet in diameter

Figure 8.4
The solid-fuel rockets are assembled from segments to make loading the fuel more convenient.

Figure 8.5
A rocket segment being filled with the propellant mixture.

Adapted with permission from *Chemical and Engineering News,* September 19, 1988, Volume 66, Number 33, p. 9. Copyright 1988, American Chemical Society.

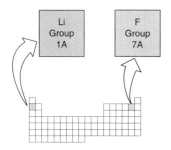

The list of criteria for identifying oxidation–reduction reactions is more complicated than those for precipitation and acid–base reactions because there are many types of oxidation–reduction reactions. I usually provide criteria only for subcategories of oxidation–reduction reactions.

The origin of the H^+ ion is a strong acid, such as $HCl(aq)$ or $HNO_3(aq)$, and the origin of the OH^- ion is a strong base, such as $NaOH(aq)$ and $KOH(aq)$. An example is

$$HCl(aq) + KOH(aq) \rightarrow H_2O(l) + KCl(aq)$$

We classify these reactions as **acid–base reactions.** You can recognize this as an acid–base reaction because it *involves an H^+ ion that ends up in the product water.*

The third driving force—the one we have considered in the present chapter—is electron transfer. We see evidence of this driving force particularly in the "desire" of a metal to donate electrons to nonmetals. An example is

$$2Li(s) + F_2(g) \rightarrow 2LiF(s)$$

where each lithium atom loses one electron to form Li^+, and each fluorine atom gains one electron to form the F^- ion. The process of electron transfer is also called oxidation–reduction. Thus we classify the above reaction as an **oxidation–reduction reaction.**

An additional driving force for chemical reactions that we have not yet discussed is *formation of a gas.* A reaction in aqueous solution that forms a gas (which escapes as bubbles) is pulled toward products by this event. An example is the reaction

$$2HCl(aq) + Na_2CO_3(aq) \rightarrow CO_2(g) + H_2O(l) + NaCl(aq)$$

for which the net ionic equation is

$$2H^+(aq) + CO_3^{2-}(aq) \rightarrow CO_2(g) + H_2O(l)$$

Note that this reaction forms carbon dioxide gas as well as water, so it illustrates two of the driving forces that we have considered. Because this reaction involves H^+ that ends up in the product water, we classify it as an acid–base reaction.

Consider another reaction that forms a gas:

$$Zn(s) + 2HCl(aq) \rightarrow H_2(g) + ZnCl_2(aq)$$

How might we classify this reaction? A careful look at the reactants and products shows the following:

$$Zn(s) \quad + \quad 2HCl(aq) \quad \rightarrow \quad H_2(g) \quad + \quad ZnCl_2(aq)$$

| Contains uncharged Zn atoms | Really $2H^+(aq) + 2Cl^-(aq)$ | Contains uncharged H atoms | Really $Zn^{2+}(aq) + 2Cl^-(aq)$ |

Note that in the reactant zinc metal, Zn exists as uncharged atoms, whereas in the product it exists as Zn^{2+}. Thus each Zn loses two electrons. Where have these electrons gone? They have been transferred to two H^+ ions to form H_2. The

schematic for this reaction is

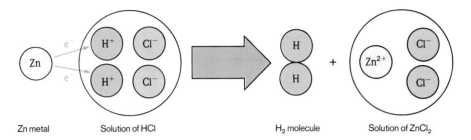

Zn metal   Solution of HCl   H₂ molecule   Solution of ZnCl₂

This is an electron transfer process, so the reaction can be classified as an oxidation—reduction reaction.

Another way this reaction is sometimes classified is based on the fact that a *single* type of anion (Cl^-) has been exchanged between H^+ and Zn^{2+}. That is, Cl^- is originally associated with H^+ in HCl and ends up associated with Zn^{2+} in the product $ZnCl_2$. We can call this a *single-replacement reaction* in contrast to double-displacement reactions, in which two types of anions are exchanged. We can represent a single replacement as

$$A + BC \rightarrow B + AC$$

After presenting the three ways to classify reactions, I present a variety of reactions on an overhead transparency and ask students to help me classify them. We go over the criteria for each type of reaction so that students learn to locate the clues that distinguish each.

8.3   Other Ways to Classify Reactions

AIM: To consider additional classes of chemical reactions.

So far in this chapter we have classified chemical reactions in several ways. The most commonly used of these classifications are

1. Precipitation reactions
2. Acid—base reactions
3. Oxidation—reduction reactions

However, there are still other ways to classify reactions that you may encounter in your future studies of chemistry. We will consider several of these in this section.

I usually tell students in advance that all the new reaction types in Section 8.3 are subcategories of oxidation—reduction reactions.

Combustion Reactions

Many chemical reactions that involve oxygen produce energy (heat) so rapidly that a flame results. Such reactions are called **combustion reactions.** We have considered some of these reactions previously. For example, the methane in

My criteria for identifying combustion reactions are two: the presence of molecular oxygen and a carbon-containing molecule as reactants, and the presence of product molecules CO_2 and H_2O.

Figure 8.6
Classes of reactions. Combustion reactions are a special type of oxidation–reduction reaction.

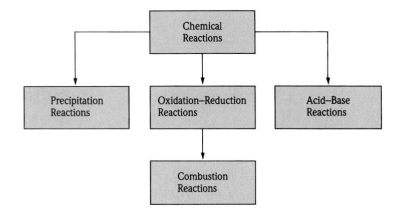

natural gas reacts with oxygen according to the following balanced equation:

$$CH_4(g) + 2O_2(g) \rightarrow CO_2(g) + 2H_2O(g)$$

This reaction produces the flame of the common laboratory burner and is used to heat most homes in the United States. Recall that we originally classified this reaction as an oxidation–reduction reaction in Section 8.1. Thus we can say that the reaction of methane with oxygen is both an oxidation–reduction reaction and a combustion reaction. Combustion reactions, in fact, are a special class of oxidation–reduction reactions (see Figure 8.6).

There are many combustion reactions, most of which are used to provide heat or electricity for homes or businesses or energy for transportation. Some examples are

Combustion of propane (used to heat some rural homes)

$$C_3H_8(g) + 5O_2(g) \rightarrow 3CO_2(g) + 4H_2O(g)$$

Combustion of gasoline* (used to power cars and trucks)

$$2C_8H_{18}(l) + 25O_2(g) \rightarrow 16CO_2(g) + 18H_2O(g)$$

Combustion of coal* (used to generate electricity)

$$C(s) + O_2(g) \rightarrow CO_2(g)$$

*This substance is really a complex mixture of compounds, but the reaction shown is representative of what takes place.

Synthesis (Combination) Reactions

One of the most important activities in chemistry is the synthesis of new compounds. Each of our lives has been greatly affected by synthetic compounds such as plastic, polyester, and aspirin. When a given compound is formed from simpler materials, we call this a **synthesis** (or **combination**) **reaction.**

In many cases synthesis reactions start with elements, as shown by the following examples:

Synthesis of water

$$2H_2(g) + O_2(g) \rightarrow 2H_2O(l)$$

Synthesis of carbon dioxide

$$C(s) + O_2(g) \rightarrow CO_2(g)$$

Synthesis of nitrogen monoxide

$$N_2(g) + O_2(g) \rightarrow 2NO(g)$$

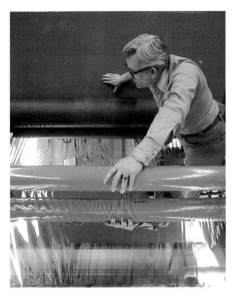

Production of Mylar film is an example of a synthesis reaction.

Notice that each of these reactions involves oxygen, so each can be classified as an oxidation–reduction reaction. The first two reactions are also commonly called combustion reactions because they produce flames. The reaction of hydrogen with oxygen to produce water, then, can be classified three ways: as an oxidation–reduction reaction, as a combustion reaction, and as a synthesis reaction.

There are also many synthesis reactions that do not involve oxygen:

Synthesis of sodium chloride

$$2Na(s) + Cl_2(g) \rightarrow 2NaCl(s)$$

Synthesis of magnesium fluoride

$$Mg(s) + F_2(g) \rightarrow MgF_2(s)$$

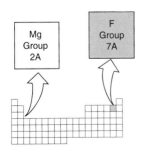

We have discussed the formation of sodium chloride before and have noted that it is an oxidation–reduction reaction; uncharged sodium atoms lose electrons to form Na^+ ions, and uncharged chlorine atoms gain electrons to form Cl^- ions. The synthesis of magnesium fluoride is also an oxidation–reduction reaction because Mg^{2+} and F^- ions are produced from the uncharged atoms.

We have seen that synthesis reactions in which the reactants are elements are oxidation–reduction reactions as well. In fact, we can think of these synthesis reactions as another subclass of the oxidation–reduction class of reactions.

Students can identify synthesis reactions by looking for two pure elements or small compounds on the reactant side and a larger compound on the product side.

Students should be able to predict the products for double–displacement reactions. However, most students are not able to write correct products for decomposition reactions without more experience.

My list of criteria for identifying decomposition reactions includes the presence of a larger compound on the reactant side and pure elements or smaller compounds on the product side.

Now is a good time to present as many different reactions as you can from Sections 8.2 and 8.3. Ask students to help you classify them. Try to encourage students to learn all possible names for each type of reaction. Even non-redox reactions can have more than one name.

Decomposition Reactions

In many cases a compound can be broken down into simpler compounds or all the way to the component elements. This is usually accomplished by heating or by the application of an electric current. Such reactions are called **decomposition reactions.** We have discussed decomposition reactions before, including

Decomposition of water

$$2H_2O(l) \xrightarrow[\text{current}]{\text{Electric}} 2H_2(g) + O_2(g)$$

Decomposition of mercury(II) oxide

$$2HgO(s) \rightarrow 2Hg(l) + O_2(g)$$

Because O_2 is involved in the first reaction, we recognize it as an oxidation–reduction reaction. In the second reaction, HgO, which contains Hg^{2+} and O^{2-} ions, is decomposed to the elements, which contain uncharged atoms. In this process each Hg^{2+} gains two electrons and each O^{2-} loses two electrons, so this is both a decomposition reaction and an oxidation–reduction reaction.

A decomposition reaction, in which a compound is broken down into its elements, is just the opposite of the synthesis (combination) reaction where elements combine to form the compound. For example, we have just discussed the synthesis of sodium chloride from its elements. Sodium chloride can be decomposed into its elements by melting it and passing an electric current through it:

$$2NaCl(l) \xrightarrow[\text{current}]{\text{Electric}} 2Na(l) + Cl_2(g)$$

There are other schemes for classifying reactions that we have not considered. However, we have covered many of the classifications that are commonly used by chemists as they pursue their science in laboratories and industrial plants.

It should be apparent that many important reactions can be classified as oxidation–reduction reactions. As shown in Figure 8.7, various types of reactions can be viewed as subclasses of the overall oxidation–reduction category.

EXAMPLE 8.2   Classifying Reactions

Classify each of the following reactions in as many ways as possible.

a. $2K(s) + Cl_2(g) \rightarrow 2KCl(s)$
b. $Fe_2O_3(s) + 2Al(s) \rightarrow Al_2O_3(s) + 2Fe(s)$
c. $2Mg(s) + O_2(g) \rightarrow 2MgO(s)$
d. $HNO_3(aq) + NaOH(aq) \rightarrow H_2O(l) + NaNO_3(aq)$
e. $KBr(aq) + AgNO_3(aq) \rightarrow AgBr(s) + KNO_3(aq)$
f. $PbO_2(s) \rightarrow Pb(s) + O_2(g)$

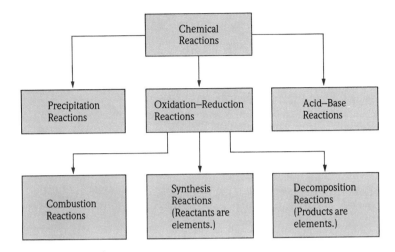

Figure 8.7
Summary of classes of reactions.

EXAMPLE 8.2, CONTINUED

SOLUTION

a. This is both a synthesis reaction (elements combine to form a compound) and an oxidation–reduction reaction (uncharged potassium and chlorine atoms are changed to K^+ and Cl^- ions in KCl).

b. This is an oxidation–reduction reaction. Iron is present in $Fe_2O_3(s)$ as Fe^{3+} ions and in elemental iron, Fe(s), as uncharged atoms. So each Fe^{3+} must gain three electrons to form Fe. The reverse happens to aluminum, which is present initially as uncharged aluminum atoms, each of which loses three electrons to give Al^{3+} ions in Al_2O_3. Note that this reaction might also be called a single-replacement reaction because O is switched from Fe to Al.

c. This is both a synthesis reaction (elements combine to form a compound) and an oxidation–reduction reaction (each magnesium atom loses two electrons to give Mg^{2+} ions in MgO, and each oxygen atom gains two electrons to give O^{2-} in MgO).

d. This is an acid–base reaction. It might also be called a double-displacement reaction because NO_3^- and OH^- "switch partners."

e. This is a precipitation reaction that might also be called a double-displacement reaction.

f. This is a decomposition reaction (a compound breaks down into elements). It also is an oxidation–reduction reaction, because the ions in PbO_2 (Pb^{4+} and O^{2-}) are changed to uncharged atoms in the elements Pb(s) and $O_2(g)$. That is, electrons are transferred from O^{2-} to Pb^{4+} in the reaction.

EXAMPLE 8.2, CONTINUED

SELF-CHECK EXERCISE 8.2

Classify each of the following reactions in as many ways as possible.

a. $4NH_3(g) + 5O_2(g) \rightarrow 4NO(g) + 6H_2O(g)$
b. $S_8(s) + 8O_2(g) \rightarrow 8SO_2(g)$
c. $2Al(s) + 3Cl_2(g) \rightarrow 2AlCl_3(s)$
d. $2AlN \rightarrow 2Al(s) + N_2(g)$
e. $BaCl_2(aq) + Na_2SO_4(aq) \rightarrow BaSO_4(s) + 2NaCl(aq)$
f. $2Cs(s) + Br_2(l) \rightarrow 2CsBr(s)$
g. $KOH(aq) + HCl(aq) \rightarrow H_2O(l) + KCl(aq)$
h. $2C_2H_2(g) + 5O_2(g) \rightarrow 4CO_2(g) + 2H_2O(l)$

CHAPTER REVIEW

Key Terms

oxidation–reduction reaction (p. 223)
combustion reaction (p. 231)
synthesis (combination) reaction (p. 233)
decomposition reaction (p. 234)

Summary

1. Reactions of metals and nonmetals involve a transfer of electrons and are called oxidation–reduction reactions. A reaction between a nonmetal and oxygen is also an oxidation–reduction reaction. Combustion reactions involve oxygen and are a subgroup of oxidation–reduction reactions.

2. When a given compound is formed from simpler materials, such as elements, the reaction is called a synthesis or combination reaction. The reverse process, which occurs when a compound is broken down into its component elements, is called a decomposition reaction. These reactions are also subgroups of oxidation–reduction reactions.

Questions and Problems

All even-numbered exercises have answers in the back of this book and solutions in the Solutions Guide.

8.1 Reactions of Metals with Nonmetals (Oxidation–Reduction)

QUESTIONS

1. A chemical reaction in which electrons are transferred from one species to another is called a(n) ———— reaction.

2. The reaction $2Na + Cl_2 \rightarrow 2NaCl$, like any reaction between a metal and a nonmetal, involves the ———— of electrons.

3. What do we mean when we say that the transfer of electrons can be the "driving force" for a reaction? Give an example of a reaction where this happens.

4. When a metal reacts with a nonmetal, which atoms form the cations of the ionic product? Which atoms form the negative ions? Which atoms lose electrons, and which gain them?

5. If sodium atoms were to react with a nonmetal such as chlorine, how many electrons would each sodium atom lose? What charge would the resulting sodium ions have? If magnesium atoms were to react with chlorine, how many electrons would each magnesium atom lose? What charge would the resulting magnesium ions have?

6. If a bromine molecule, Br_2, were to react with a metal such as sodium, what charge would the resulting bromide ions have? How many electrons would be gained by each bromine atom? How many electrons would be gained by each Br_2 molecule?

7. For the reaction $Ba(s) + Cl_2(g) \rightarrow BaCl_2(s)$, show how electrons are gained and lost by the atoms.

8. For the reaction $2Al(s) + 3Br_2(l) \rightarrow 2AlBr_3(s)$, show how electrons are gained and lost by the atoms.

9. Balance the equation for each of the following oxidation–reduction chemical reactions.
   a. $Fe(s) + O_2(g) \rightarrow Fe_2O_3(s)$
   b. $Al(s) + Cl_2(g) \rightarrow AlCl_3(s)$
   c. $Mg(s) + P_4(s) \rightarrow Mg_3P_2(s)$
   d. $Mg(s) + O_2(g) \rightarrow MgO(s)$
   e. $Mg(s) + Cu(NO_3)_2(aq) \rightarrow Mg(NO_3)_2(aq) + Cu(s)$

10. Balance each of the following oxidation–reduction chemical reactions.
    a. $Na(s) + S(s) \rightarrow Na_2S(s)$
    b. $Cu(s) + AgNO_3(aq) \rightarrow Cu(NO_3)_2(aq) + Ag(s)$
    c. $Mg(s) + HCl(aq) \rightarrow MgCl_2(aq) + H_2(g)$
    d. $Pb(s) + S(s) \rightarrow PbS_2(s)$
    e. $Li(s) + H_2O(l) + LiOH(aq) + H_2(g)$

8.2 Ways to Classify Reactions

QUESTIONS

11. What is a *double*-displacement reaction? What is a *single*-displacement reaction? Write balanced chemical equations for two examples of each type.

12. Two "driving forces" for reactions discussed in this section are the formation of water in an acid–base reaction and the formation of a gaseous product. Write balanced chemical equations for two examples of each type.

13. Identify each of the following unbalanced reaction equations as belonging to one or more of the following categories: precipitation, acid–base, or oxidation–reduction.
    a. $Pb(NO_3)_2(aq) + H_2SO_4(aq) \rightarrow PbSO_4(s) + HNO_3(aq)$
    b. $HCl(aq) + RbOH(aq) \rightarrow H_2O(l) + RbCl(aq)$
    c. $H_2(g) + O_2(g) \rightarrow H_2O(l)$
    d. $Co(NO_3)_2(aq) + (NH_4)_2S(aq)$
       $\rightarrow CoS(s) + NH_4NO_3(aq)$
    e. $HClO_4(aq) + NaOH(aq) \rightarrow NaClO_4(aq) + H_2O(l)$
    f. $HCl(aq) + Pb(C_2H_3O_2)_2(aq)$
       $\rightarrow PbCl_2(s) + HC_2H_3O_2(aq)$
    g. $HBr(aq) + NaOH(aq) \rightarrow H_2O(l) + NaBr(aq)$
    h. $BaCl_2(aq) + Na_2SO_4(aq) \rightarrow BaSO_4(s) + NaCl(aq)$
    i. $Mg(s) + HCl(aq) \rightarrow MgCl_2(aq) + H_2(g)$

14. Identify each of the following unbalanced reaction equations as belonging to one or more of the following categories: precipitation, acid–base, or oxidation–reduction.
    a. $H_2O_2(aq) \rightarrow H_2O(l) + O_2(g)$
    b. $H_2SO_4(aq) + Cu(s) \rightarrow CuSO_4(aq) + H_2(g)$
    c. $H_2SO_4(aq) + NaOH(aq) \rightarrow Na_2SO_4(aq) + H_2O(l)$
    d. $H_2SO_4(aq) + Ba(OH)_2(aq) \rightarrow BaSO_4(s) + H_2O(l)$
    e. $AgNO_3(aq) + CuCl_2(aq) \rightarrow Cu(NO_3)_2(aq) + AgCl(s)$
    f. $KOH(aq) + CuSO_4(aq) \rightarrow Cu(OH)_2(s) + K_2SO_4(aq)$
    g. $Cl_2(g) + F_2(g) \rightarrow ClF(g)$
    h. $NO(g) + O_2(g) \rightarrow NO_2(g)$
    i. $Ca(OH)_2(s) + HNO_3(aq) \rightarrow Ca(NO_3)_2(aq) + H_2O(l)$

8.3 Other Ways to Classify Reactions

QUESTIONS

15. A reaction in which a compound reacts rapidly with elemental oxygen, usually with the release of heat or light, is referred to as a _____ reaction.

16. Reactions involving the combustion of fuel substances make up a subclass of _____ reactions.

17. The formation of a compound from the direct combination of two or more elemental substances, or from simpler compounds, is referred to as a(n) _____ reaction.

18. A reaction such as $Mg_3N_2(s) \rightarrow 3Mg(s) + N_2(g)$, in which a compound is broken down into its elements, is called a(n) _____ reaction.

19. Balance each of the following equations that describe combustion reactions.
   a. $C_2H_6(g) + O_2(g) \rightarrow CO_2(g) + H_2O(g)$
   b. $C_4H_{10}(g) + O_2(g) \rightarrow CO_2(g) + H_2O(g)$
   c. $C_6H_{14}(g) + O_2(g) \rightarrow CO_2(g) + H_2O(g)$

20. Balance each of the following equations that describe combustion reactions.
   a. $C_3H_8(l) + O_2(g) \rightarrow CO_2(g) + H_2O(g)$
   b. $C_4H_{10}(l) + O_2(g) \rightarrow CO_2(g) + H_2O(g)$
   c. $C_5H_{12}(l) + O_2(g) \rightarrow CO_2(g) + H_2O(g)$

21. Complete and balance each of the following equations that describe combustion reactions.
   a. $C_2H_2(g) + O_2(g) \rightarrow$
   b. $C_3H_8(g) + O_2(g) \rightarrow$
   c. $C_2H_4O_2(l) + O_2(g) \rightarrow$

22. Complete and balance each of the following equations that describe combustion reactions.
   a. $C_{19}H_{40}(s) + O_2(g) \rightarrow$
   b. $C_6H_{12}O_6(s) + O_2(g) \rightarrow$
   c. $C_{12}H_{22}O_{11}(s) + O_2(g) \rightarrow$

23. Balance each of the following equations that describe synthesis reactions.
   a. $B(s) + Cl_2(s) \rightarrow BCl_3(l)$
   b. $NO(g) + O_2(g) \rightarrow NO_2(g)$
   c. $Tl_2O_3(s) + CO_2(g) \rightarrow Tl_2(CO_3)_3(s)$
   d. $Fe(s) + O_2(g) \rightarrow Fe_2O_3(s)$
   e. $NaOH(s) + CO_2(g) \rightarrow Na_2CO_3(s) + H_2O(l)$

24. Balance each of the following equations that describe synthesis reactions.
   a. $Co(s) + S(s) \rightarrow Co_2S_3(s)$
   b. $NO(g) + O_2(g) \rightarrow NO_2(g)$
   c. $FeO(s) + CO_2(g) \rightarrow FeCO_3(s)$
   d. $Al(s) + F_2(g) \rightarrow AlF_3(s)$
   e. $NH_3(g) + H_2CO_3(aq) \rightarrow (NH_4)_2CO_3(s)$

25. Balance each of the following equations that describe decomposition reactions.
   a. $OF_2(g) \rightarrow O_2(g) + F_2(g)$
   b. $Al_2(CO_3)_3(s) \rightarrow Al_2O_3(s) + CO_2(g)$
   c. $NaClO_3(s) \rightarrow NaCl(s) + O_2(g)$
   d. $Al(OH)_3(s) \rightarrow Al_2O_3(s) + H_2O(g)$
   e. $(NH_4)_2Cr_2O_7(s) \rightarrow N_2(g) + Cr_2O_3(s) + H_2O(g)$

26. Balance each of the following equations that describe decomposition reactions.
   a. $NI_3(s) \rightarrow N_2(g) + I_2(s)$
   b. $BaCO_3(s) \rightarrow BaO(s) + CO_2(g)$
   c. $C_6H_{12}O_6(s) \rightarrow C(s) + H_2O(g)$
   d. $Cu(NH_3)_4SO_4(s) \rightarrow CuSO_4(s) + NH_3(g)$
   e. $NaN_3(s) \rightarrow Na_3N(s) + N_2(g)$

Additional Problems

27. For each of the following metals, how many electrons will the metal atoms lose when the metal reacts with a nonmetal?
   a. sodium
   b. potassium
   c. magnesium
   d. barium
   e. aluminum

28. For each of the following nonmetals, how many electrons will each atom of the nonmetal gain in reacting with a metal?
   a. oxygen
   b. fluorine
   c. nitrogen
   d. chlorine
   e. sulfur
   f. bromine
   g. iodine

29. There is much overlapping of the classification schemes for reactions discussed in this chapter. Give an example of a reaction that is, at the same time, an oxidation–reduction reaction, a combustion reaction, and a synthesis reaction.

30. Most substances used as fuels are compounds that contain carbon and hydrogen; oxygen also occurs in some of these compounds. When these compounds react with oxygen, what are the most common products of the reaction?

31. Classify the reactions represented by the following unbalanced equations by as many methods as possible. Use both the classifications developed in this chapter and those discussed in Chapter 7. Balance the equations.
   a. $I_4O_9(s) \rightarrow I_2O_6(s) + I_2(s) + O_2(g)$

b. $Mg(s) + AgNO_3(aq) \rightarrow Mg(NO_3)_2(aq) + Ag(s)$
c. $SiCl_4(l) + Mg(s) \rightarrow MgCl_2(s) + Si(s)$
d. $CuCl_2(aq) + AgNO_3(aq) \rightarrow Cu(NO_3)_2(aq) + AgCl(s)$
e. $Al(s) + Br_2(l) \rightarrow AlBr_3(s)$

32. Classify the reactions represented by the following unbalanced equations by as many methods as possible. Use both the classifications developed in this chapter and those discussed in Chapter 7. Balance the equations.
a. $C_3H_8O(l) + O_2(g) \rightarrow CO_2(g) + H_2O(g)$
b. $HCl(aq) + AgC_2H_3O_2(aq) \rightarrow AgCl(s) + HC_2H_3O_2(aq)$
c. $HCl(aq) + Al(OH)_3(s) \rightarrow AlCl_3(aq) + H_2O(l)$
d. $H_2O_2(aq) \rightarrow H_2O(l) + O_2(g)$
e. $N_2H_4(l) + O_2(g) \rightarrow N_2(g) + H_2O(g)$

33. The metals of Group 1 of the periodic table are very reactive. When reacted with virtually any nonmetallic element, these metals all easily lose one electron to form unipositive (1+) ions. Write and balance the reactions the Group 1 metals sodium, potassium, and cesium would be expected to undergo with the nonmetallic elements chlorine, sulfur, bromine, and oxygen.

34. Corrosion of metals costs us many billions of dollars annually, slowly destroying cars, bridges, and buildings. Corrosion of a metal involves the oxidation of the metal by the oxygen in the air, typically in the presence of moisture. Write a balanced equation for the reaction of each of the following metals with O_2: Zn, Al, Fe, Cr, and Ni.

35. Elemental chlorine, Cl_2, is very reactive, combining with most metallic substances. Write a balanced equation for the reaction of each of the following metals with Cl_2: Na, Al, Zn, Ca, and Fe.

36. For the elements in the left-hand column, choose from the right-hand column the *charge* of the ion each element would form in undergoing a simple oxidation–reduction reaction.

Elements	Charges
Al	1+
Ba	1−
Br	2+
Ca	2−
Cl	3+
Cs	3−
I	
K	
Li	
Mg	
Na	
O	
Rb	
S	
Sr	

37. If a chlorine molecule, Cl_2, were to react with a metal such as sodium, what charge would the resulting chloride ions have? If a sulfur molecule were to react with sodium, what charge would the resulting sulfide ions have?

38. If a potassium atom were to react with a nonmetal, what charge would the resulting potassium ion have? If a calcium atom were to react with the same nonmetal, what charge would the resulting calcium ion have?

39. For the reaction $16Fe(s) + 3S_8(s) \rightarrow 8Fe_2S_3(s)$, show how electrons are gained and lost by the atoms.

40. Balance the equation for each of the following oxidation–reduction chemical reactions.
a. $Na(s) + O_2(g) \rightarrow Na_2O_2(s)$
b. $Fe(s) + H_2SO_4(aq) \rightarrow FeSO_4(aq) + H_2(g)$
c. $Al_2O_3(s) \rightarrow Al(s) + O_2(g)$
d. $Fe(s) + Br_2(l) \rightarrow FeBr_3(s)$
e. $Zn(s) + HNO_3(aq) \rightarrow Zn(NO_3)_2(aq) + H_2(g)$

41. Identify each of the following unbalanced reaction equations as belonging to one or more of the following categories: precipitation, acid–base, or oxidation–reduction.
a. $Fe(s) + H_2SO_4(aq) \rightarrow Fe_3(SO_4)_2(aq) + H_2(g)$
b. $HClO_4(aq) + RbOH(aq) \rightarrow RbClO_4(aq) + H_2O(l)$
c. $Ca(s) + O_2(g) \rightarrow CaO(s)$
d. $H_2SO_4(aq) + NaOH(aq) \rightarrow Na_2SO_4(aq) + H_2O(l)$
e. $Pb(NO_3)_2(aq) + Na_2CO_3(aq)$
$\rightarrow PbCO_3(s) + NaNO_3(aq)$
f. $K_2SO_4(aq) + CaCl_2(aq) \rightarrow KCl(aq) + CaSO_4(s)$
g. $HNO_3(aq) + KOH(aq) \rightarrow KNO_3(aq) + H_2O(l)$
h. $Ni(C_2H_3O_2)_2(aq) + Na_2S(aq)$
$\rightarrow NiS(s) + NaC_2H_3O_2(aq)$
i. $Ni(s) + Cl_2(g) \rightarrow NiCl_2(s)$

42. Balance each of the following equations that describe combustion reactions.
a. $C_5H_{12}(l) + O_2(g) \rightarrow CO_2(g) + H_2O(g)$
b. $C_2H_6O(l) + O_2(g) \rightarrow CO_2(g) + H_2O(g)$
c. $C_6H_6(l) + O_2(g) \rightarrow CO_2(g) + H_2O(g)$

43. Complete and balance each of the following equations that describe combustion reactions.
a. $C_2H_4(g) + O_2(g) \rightarrow$
b. $C_8H_{18}(l) + O_2(g) \rightarrow$
c. $C_{30}H_{62}(s) + O_2(g) \rightarrow$

44. Balance each of the following equations that describe synthesis reactions.
a. $FeO(s) + O_2(g) \rightarrow Fe_2O_3(s)$
b. $CO(g) + O_2(g) \rightarrow CO_2(g)$
c. $H_2(g) + Cl_2(g) \rightarrow HCl(g)$
d. $K(s) + S_8(s) \rightarrow K_2S(s)$
e. $Na(s) + N_2(g) \rightarrow Na_3N(s)$

45. Balance each of the following equations that describe de-composition reactions.
    a. $NaHCO_3(s) \rightarrow Na_2CO_3(s) + H_2O(g) + CO_2(g)$
    b. $NaClO_3(s) \rightarrow NaCl(s) + O_2(g)$
    c. $HgO(s) \rightarrow Hg(l) + O_2(g)$
    d. $C_{12}H_{22}O_{11}(s) \rightarrow C(s) + H_2O(g)$
    e. $H_2O_2(l) \rightarrow H_2O(l) + O_2(g)$
46. Write a balanced oxidation–reduction equation for the reaction of each of the metals in the left-hand column with each of the nonmetals in the right-hand column.

Ba	O_2
K	S
Mg	Cl_2
Rb	N_2
Ca	Br_2
Li	

47. Fluorine gas, F_2, is so reactive that it attacks the other halo-gen elements (Cl_2, Br_2, I_2) forming *inter*halogen compounds (for example, ClF, BrF, IF). Write the balanced oxidation–reduction equations for the reactions of fluorine gas with each of the other elemental halogens to form ClF, BrF and IF.
48. Sulfuric acid, H_2SO_4, oxidizes many metallic elements. One of the effects of acid rain is that it produces sulfuric acid in the atmosphere that reacts with metals used in construction. Write balanced oxidation–reduction equations for the reaction of sulfuric acid with Fe, Zn, Cu, Co, and Ni.
49. Although the metals of Group 2 of the periodic table are not nearly as reactive as those of Group 1, many of the Group 2 metals will combine with common nonmetals, especially at elevated temperatures. Write balanced chemical equations for the reactions of Mg, Ca, Sr, and Ba with Cl_2, Br_2, and O_2.
50. Elemental sulfur (S_8) combines with many metals, especially at higher temperatures. Write balanced chemical equations for the reaction of sulfur (S_8) with each of the following metals: Na, Ba, Al, Fe, K, Ca, Mg.

Solutions to Self-Check Exercises

SELF-CHECK EXERCISE 8.1

a. The compound NaBr contains the ions Na^+ and Br^-. Thus each sodium atom loses one electron ($Na \rightarrow Na^+ + e^-$), and each bromine atom gains one electron ($Br + e^- \rightarrow Br^-$).

$$Na + Na + Br - Br \rightarrow (Na^+Br^-) + (Na^+Br^-)$$

b. The compound CaO contains the Ca^{2+} and O^{2-} ions. Thus each calcium atom loses two electrons ($Ca \rightarrow Ca^{2+} + 2e^-$), and each oxygen atom gains two electrons ($O + 2e^- \rightarrow O^{2-}$).

$$Ca + Ca + O - O \rightarrow (Ca^{2+}O^{2-}) + (Ca^{2+}O^{2-})$$

SELF-CHECK EXERCISE 8.2

a. oxidation–reduction reaction
    combustion reaction
b. synthesis reaction
    oxidation–reduction reaction
    combustion reaction
c. synthesis reaction
    oxidation–reduction reaction
d. decomposition reaction
    oxidation–reduction reaction
e. precipitation reaction
    (and double displacement)
f. synthesis reaction
    oxidation–reduction reaction
g. acid–base reaction
    (and double displacement)
h. combustion reaction
    oxidation–reduction reaction

9 Chemical Composition

Samples of mints and jellybeans, both having a mass of 50 grams.

CONTENTS

Chapter 9 covers one of the truly fundamental operations of chemistry: counting atoms and molecules by weighing samples of them. Understanding chemical stoichiometry is a universal goal of an introductory chemistry course. To help reach this important goal, I give considerable space at the beginning of this chapter to nonchemical analogies that clarify the connection between the mass of a sample and the number of components present in the sample. It seems worthwhile to spend some time in class on these analogies before getting to the chemical examples.

Mylar film, an important synthetic polymer.

A synthetic polymer leg brace.

One very important chemical activity is the synthesis of new compounds. Nylon, the artificial sweetener aspartame (Nutra-Sweet™), synthetic rubber, the polymer Kevlar used in bulletproof vests, polyvinyl chloride (PVC) for plastic water pipes, Teflon, and so many other materials that make our lives easier—all originated in some chemist's laboratory. When a chemist makes a new substance, the first order of business is to identify it. What is its composition? What is its chemical formula?

In this chapter we will learn to determine a compound's formula. Before we can do that, however, we need to think about counting atoms. How do we determine the number of each type of atom in a substance so that we can write its formula? Of course atoms are too small to count individually. As we will see in this chapter, we typically count atoms by weighing them. So let us first consider the general principle of counting by weighing.

9.1 Counting by Weighing

AIM: To discuss the concept of average mass and explore how counting can be done by weighing.

Suppose you work in a candy store that sells gourmet jelly beans by the bean. People come in and ask for 50 beans, 100 beans, 1000 beans, and so on, and you have to count them out—a tedious process at best. As a good problem solver, you try to come up with a better system. It occurs to you that it might be far more efficient to buy a scale and count the jelly beans by weighing them. How can you count jelly beans by weighing them? What information about the individual beans do you need to know?

Assume that all of the jelly beans are identical and that each has a mass of 5 g. If a customer asks for 1000 jelly beans, what mass of jelly beans would be required? Each bean has a mass of 5 g, so you would need 1000 beans × 5 g/bean, or 5000 g (5 kg). It takes just a few seconds to weigh out 5 kg of jelly beans. It would take much longer to count out 1000 of them.

In reality, jelly beans are not identical. For example, let's assume that you weigh 10 beans individually and get the following results:

Bean	Mass
1	5.1 g
2	5.2 g
3	5.0 g
4	4.8 g
5	4.9 g
6	5.0 g
7	5.0 g

Bean	Mass
8	5.1 g
9	4.9 g
10	5.0 g

Can we count these nonidentical beans by weighing? Yes. The key piece of information we need is the *average mass* of the jelly beans. Let's compute the average mass for our 10-bean sample.

The reasons for learning this material can be overshadowed by the mechanics of getting the right answer. We are learning the material in Chapters 9 and 10 so we can answer the important questions "How much can we make?" and "How much should we add?" The same questions are important when cooking. The analogy between chemical equations and a recipe can help explain this material.

$$\text{Average mass} = \frac{\text{total mass of beans}}{\text{number of beans}}$$

$$= \frac{5.1 \text{ g} + 5.2 \text{ g} + 5.0 \text{ g} + 4.8 \text{ g} + 4.9 \text{ g} + 5.0 \text{ g} + 5.0 \text{ g} + 5.1 \text{ g} + 4.9 \text{ g} + 5.0 \text{ g}}{10}$$

$$= \frac{50.0}{10} = 5.0 \text{ g}$$

The average mass is 5.0 g. Thus to count out 1000 beans, we need to weigh out 5000 g of beans. This sample of beans, in which the beans have an average mass of 5.0 g, can be treated exactly like a sample where all of the beans are identical. Objects do not need to have identical masses in order to be counted by weighing. We simply need to know the average mass of the objects. For purposes of counting, the objects *behave as though they were all identical,* as though they each actually had the average mass.

We resort to counting atoms by weighing because they are too small to count individually. I tell students that the small size of an individual atom means that we must find another way to determine how many atoms there are in a given sample.

Suppose a customer comes into the store and says, "I want to buy a bag of candy for each of my kids. One of them likes jelly beans and the other one likes mints. Please put a scoopful of jelly beans in a bag and a scoopful of mints in another bag." Then the customer recognizes a problem. "Wait! My kids will fight unless I bring home exactly the same number of candies for each one. Both bags must have the same number of pieces because they'll definitely count them and compare. But I'm really in a hurry, so we don't have time to count them here. Is there a simple way you can be sure the bags will contain the same number of candies?"

You need to solve this problem quickly. Suppose you know the average masses of the two kinds of candy:

Jelly beans: average mass = 5 g
Mints: average mass = 15 g

You fill the scoop with jelly beans and dump them onto the scale, which reads 500 g. Now the key question: what mass of mints do you need to give the same number of mints as there are jelly beans in 500 g of jelly beans? Comparing the average masses of the jelly beans (5 g) and mints (15 g), you realize that each mint has three times the mass of each jelly bean:

You could demonstrate the jelly-bean concept of using a centigram balance and several bundles of jelly beans, M&Ms, or some other regularly shaped candy. The number of candies determined by weighing could then be verified by counting.

$$\frac{15 \text{ g}}{5 \text{ g}} = 3$$

Students are able to follow discussions with jelly beans and mints very well until I say that samples of jelly beans and mints with identical mass ratios contain the same number of candies. It is important to go over this idea very slowly. Using candies in a nontechnical analogy helps to explain this concept.

This means that you must weigh out an amount of mints that is three times the mass of the jelly beans:

$$3 \times 500 \text{ g} = 1500 \text{ g}$$

You weigh out 1500 g of mints and put them in a bag. The customer leaves with your assurance that the bag containing 500 g of jelly beans and the bag containing 1500 g of mints both contain the same number of candies.

In solving this problem, you have discovered a principle that is very important in chemistry: two samples containing different types of components, A and B, both contain the same number of components if the ratio of the sample masses is the same as the ratio of the masses of the individual components of A and B.

Let's illustrate this rather intimidating statement by using the example we just discussed. The individual components have the masses 5 g (jelly beans) and 15 g (mints). Consider several cases.

Each sample contains 1 component:

$$\text{Mass of mint} = 15 \text{ g}$$
$$\text{Mass of jelly bean} = 5 \text{ g}$$

Each sample contains 10 components:

$$10 \text{ mints} \times \frac{15 \text{ g}}{\text{mint}} = 150 \text{ g of mints}$$

$$10 \text{ jelly beans} \times \frac{5 \text{ g}}{\text{jelly bean}} = 50 \text{ g of jelly beans}$$

Each sample contains 100 components:

$$100 \text{ mints} \times \frac{15 \text{ g}}{\text{mint}} = 1500 \text{ g of mints}$$

$$100 \text{ jelly beans} \times \frac{5 \text{ g}}{\text{jelly bean}} = 500 \text{ g of jelly beans}$$

Note in each case that the ratio of the masses is always 3 to 1:

$$\frac{1500}{500} = \frac{150}{50} = \frac{15}{5} = \frac{3}{1}$$

which is the ratio of the masses of the individual components:

$$\frac{\text{Mass of mint}}{\text{Mass of jelly bean}} = \frac{15}{5} = \frac{3}{1}$$

Each child's bag of candy contained 100 pieces of candy.

Some hardware stores still sell nails by weight. Banks and mints "count" coins into bundles or rolls by weighing. Both nails and coins make good examples to use in class if you prefer to use something other than jelly beans.

Any two samples of mints and jelly beans that have a *mass ratio* of $15/5 = 3/1$ will contain the same number of components. And these same ideas apply also to atoms, as we will see in the next section.

9.2 Atomic Masses: Counting Atoms by Weighing

Use elements such as aluminum or magnesium as examples to illustrate this section. They have low atomic masses and are not diatomic. You can use the same elements as demonstrations to show the mass of a mole of elements in Section 9.3.

AIM: To discuss atomic mass and its experimental determination.

In Chapter 6 we considered the balanced equation for the reaction of solid carbon and gaseous oxygen to form gaseous carbon dioxide:

$$C(s) + O_2(g) \rightarrow CO_2(g)$$

Now suppose you have a small pile of solid carbon and want to know how many oxygen molecules are required to convert all of the carbon into carbon dioxide. The balanced equation tells us that one oxygen molecule is required for each carbon atom.

$$C(s) \quad + \quad O_2(g) \quad \rightarrow \quad CO_2(g)$$
1 atom reacts with 1 molecule to yield 1 molecule

To determine the number of oxygen molecules required, we must know how many carbon atoms are present in the pile of carbon. But individual atoms are far too small to see. We must learn to count atoms by weighing samples containing large numbers of them.

In the last section we saw that we can easily count things like jelly beans and mints by weighing. Exactly the same principles can be applied to counting atoms.

Because atoms are so tiny, the normal units of mass—the gram and the kilogram—are much too large to be convenient. For example, the mass of a single carbon atom is 1.99×10^{-23} g. To avoid using terms like 10^{-23} when describing the mass of an atom, scientists have defined a much smaller unit of mass called the **atomic mass unit,** which is abbreviated **amu.** In terms of grams,

$$1 \text{ amu} = 1.66 \times 10^{-24} \text{ g}$$

Now let's return to our problem of counting carbon atoms. To count carbon atoms by weighing, we need to know the mass of individual atoms, just as we needed to know the mass of the individual jelly beans. Recall from Chapter 4 that the atoms of a given element exist as isotopes. The isotopes of carbon are $^{12}_6C$, $^{13}_6C$, and $^{14}_6C$. Any sample of carbon contains a mixture of these isotopes, always in the same proportions. Each of these isotopes has a slightly different mass. Therefore, just like the nonidentical jelly beans, we need to use an average mass for the carbon atoms. The **average atomic mass** for carbon atoms is 12.01 amu. This means that any sample of carbon from nature *can be treated as though it*

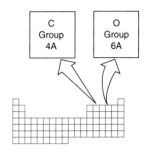

Even though students understand counting jelly beans by weighing, it is hard for them to transfer the idea to the abstract concept of counting atoms or molecules by weighing. We first introduce counting atoms by weighing with atomic mass units, which are not units of measurement students are accustomed to. The use of the amu as the unit of measure also increases the abstractness of the ideas. You might reassure students that they will soon see how counting by weighing can actually be used with familiar units such as grams.

I tell students that the average mass of jelly beans does not represent the mass of any single jelly bean because individual jelly beans differ slightly in mass. Likewise, the average mass of an atom does not represent the mass of any individual atom. By comparing atoms to jelly beans it is possible to tell students that atoms are slightly different in mass without ˙ introducing isotopes.

were composed of identical carbon atoms, each with a mass of 12.01 amu. Now that we know the average mass of the carbon atom, we can count carbon atoms by weighing samples of natural carbon. For example, what mass of natural carbon must we take to have 1000 carbon atoms present? Because 12.01 amu is the average mass,

Remember that 1000 is an exact number here.

$$\text{Mass of 1000 natural carbon atoms} = (1000 \text{ atoms}) (12.01 \frac{\text{amu}}{\text{atom}})$$

$$= 12{,}010 \text{ amu} = 1.201 \times 10^4 \text{ amu}$$

When you first work examples in which numbers of atoms are converted to amu, remember that students have not solved numerical problems for several chapters. It can be beneficial to review how to construct a conversion factor and how to cancel units.

Now let's assume that when we weigh the pile of natural carbon mentioned earlier, the result is 3.00×10^{20} amu. How many carbon atoms are present in this sample? We know that an average carbon atom has the mass 12.01 amu, so we can compute the number of carbon atoms by using the equivalence statement

$$1 \text{ carbon atom} = 12.01 \text{ amu}$$

to construct the appropriate conversion factor,

$$\frac{1 \text{ carbon atom}}{12.01 \text{ amu}}$$

The calculation is carried out as follows:

$$3.00 \times 10^{20} \text{ amu} \times \frac{1 \text{ carbon atom}}{12.01 \text{ amu}} = 2.50 \times 10^{19} \text{ carbon atoms}$$

The principles we have just discussed for carbon apply to all the other elements as well. All the elements as found in nature typically consist of a mixture of various isotopes. So to count the atoms in a sample of a given element by weighing, we must know the mass of the sample and the average mass for that element. Some average masses for common elements are listed in Table 9.1.

Table 9.1   Average atomic mass values for some common elements.

Element	Average Atomic Mass (amu)
Hydrogen	1.008
Carbon	12.01
Nitrogen	14.01
Oxygen	16.00
Sodium	22.99
Aluminum	26.98

EXAMPLE 9.1 — Calculating Mass Using Atomic Mass Units (amu)

Calculate the mass, in amu, of a sample of aluminum that contains 75 atoms.

SOLUTION

To solve this problem we use the average mass for an aluminum atom: 26.98 amu. We set up the equivalence statement:

$$1 \text{ Al atom} = 26.98 \text{ amu}$$

which gives the conversion factor we need:

$$75 \text{ Al atoms} \times \frac{26.98 \text{ amu}}{\text{Al atom}} = 2024 \text{ amu}$$

The 75 in this problem is an exact number—the number of atoms.

SELF-CHECK EXERCISE 9.1

Calculate the mass of a sample that contains 23 nitrogen atoms.

The opposite calculation can also be carried out. That is, if we know the mass of a sample, we can determine the number of atoms present. This procedure is illustrated in Example 9.2.

EXAMPLE 9.2 — Calculating the Number of Atoms from the Mass

Calculate the number of sodium atoms present in a sample that has a mass of 1172.49 amu.

SOLUTION

We can solve this problem by using the average atomic mass for sodium (see Table 9.1) of 22.99 amu. The appropriate equivalence statement is

$$1 \text{ Na atom} = 22.99 \text{ amu}$$

which gives the conversion factor we need:

$$1172.49 \text{ amu} \times \frac{1 \text{ Na atom}}{22.99 \text{ amu}} = 51.00 \text{ Na atoms}$$

EXAMPLE 9.2, CONTINUED

SELF-CHECK EXERCISE 9.2

Calculate the number of oxygen atoms in a sample that has a mass of 288 amu.

To summarize, we have seen that we can count atoms by weighing if we know the average atomic mass for that type of atom. This is one of the fundamental operations in chemistry, as we will see in the next section.

The average atomic mass for each element is listed in tables found inside the front cover of this book. Chemists often call these values the *atomic weights* for the elements, although this terminology is slowly passing out of use.

9.3   The Mole

Section 9.3 explains how quantities of atoms expressed in amu can be expressed in grams, which students can measure in the laboratory.

AIM: To explain the mole concept and Avogadro's number.
To show how to convert among moles, mass, and number of atoms in a given sample.

In the last section we used atomic mass units for mass, but these are extremely small units. In the laboratory a much larger unit, the gram, is the convenient unit for mass. In this section we will learn to count atoms in samples with masses given in grams.

Let's assume we have a sample of aluminum that has a mass of 26.98 g. What mass of copper contains exactly the same number of atoms as this sample of aluminum?

To answer this question, we need to know the average atomic masses for aluminum (26.98 amu) and copper (63.55 amu). Which atom has the greater atomic mass, aluminum or copper? The answer is copper. If we have 26.98 g of aluminum, do we need more or less than 26.98 g of copper to have the same number of copper atoms as aluminum atoms? We need more than 26.98 g of copper because each copper atom has a greater mass than each aluminum atom. Therefore, a given number of copper atoms will weigh more than an equal number of aluminum atoms. How much copper do we need? Because the average masses of aluminum and copper atoms are 26.98 amu and 63.55 amu, respectively, 26.98 g of aluminum and 63.55 g of copper contain exactly the same number of atoms. So we need 63.55 g of copper. As we saw in the first section when we were discussing candy, *samples where the ratio of the masses is the same as the ratio of the masses of the individual atoms always contain the same number of atoms.* In the case just considered, the ratios are

$$\underbrace{\frac{26.98\ \text{g}}{63.55\ \text{g}}}_{\substack{\text{Ratio of}\\ \text{sample}\\ \text{masses}}} = \underbrace{\frac{26.98\ \text{amu}}{63.55\ \text{amu}}}_{\substack{\text{Ratio of}\\ \text{atomic}\\ \text{masses}}}$$

Therefore 26.98 g of aluminum contains the same number of aluminum atoms as 63.55 g of copper contains copper atoms.

Now compare carbon (average atomic mass, 12.01 amu) and helium (average atomic mass, 4.003 amu). A sample of 12.01 g of carbon contains the same number of atoms as 4.003 g of helium. In fact, if we weigh out samples of all the elements such that each sample has a mass equal to that element's average atomic mass in grams, these samples all contain the same number of atoms (Figure 9.1). This number (the number of atoms present in all of these samples) assumes special importance in chemistry. It is called the mole, the unit all chemists use in describing numbers of atoms. The **mole** (abbreviated mol) can be defined as *the number equal to the number of carbon atoms in 12.01 grams of carbon.* Techniques for counting atoms very precisely have been used to determine this number to be 6.022×10^{23}. This number is called **Avogadro's number.** *One mole of something consists of 6.022×10^{23} units of that substance.* Just as a dozen eggs is 12 eggs, a mole of eggs is 6.022×10^{23} eggs. And a mole of water contains 6.022×10^{23} H_2O molecules.

The magnitude of the number 6.022×10^{23} is very difficult to imagine. To give you some idea, 1 mol of seconds represents a span of time 4 million times as long as the earth has already existed! One mole of marbles is enough to cover the

Lead bar

207.2 g

Silver bars

107.9 g

Pile of copper

63.55 g

Figure 9.1
All these samples of pure elements contain the *same number* (a mole) of atoms: 6.022×10^{23} atoms.

You can point out that students won't need to memorize average atomic masses for common elements. Average atomic masses are given in the periodic table.

Table 9.2 Comparison of 1-Mol Samples of Various Elements

Element	Number of Atoms Present	Mass of Sample (g)
Aluminum	6.022×10^{23}	26.98
Gold	6.022×10^{23}	196.97
Iron	6.022×10^{23}	55.85
Sulfur	6.022×10^{23}	32.06
Boron	6.022×10^{23}	10.81
Xenon	6.022×10^{23}	131.30

Figure 9.2
Samples of 1 mol of sulfur, iron, iodine, and mercury. Each of these samples contains 1 mol of atoms.

The mass of 1 mol of an element is equal to its average atomic mass (the atomic weight) in grams.

A 1 mol sample of graphite (a form of carbon) weighs 12.01 g.

If you used elements such as aluminum and magnesium to illustrate atomic mass and atomic mass units, you can use the same elements in this section to illustrate the mole.

entire earth to a depth of 50 miles! However, because atoms are so tiny, a mole of atoms or molecules is a perfectly manageable quantity to use in a reaction (Figure 9.2).

How do we use the mole in chemical calculations? Recall that Avogadro's number is defined such that a 12.01-g sample of carbon contains 6.022×10^{23} atoms. By the same token, because the average atomic mass of hydrogen is 1.008 amu (Table 9.1), 1.008 g of hydrogen contains 6.022×10^{23} hydrogen atoms. Similarly, 26.98 g of aluminum contains 6.022×10^{23} aluminum atoms. The point is that a sample of *any* element that weighs a number of grams equal to the average atomic mass of that element contains 6.022×10^{23} atoms (1 mol) of that element.

Table 9.2 shows the masses of several elements that contain 1 mol of atoms.

In summary, *a sample of an element with a mass equal to that element's average atomic mass expressed in grams contains 1 mol of atoms.*

To do chemical calculations, you *must* understand what the mole means and how to determine the number of moles in a given mass of a substance. However, before we do any calculations, let's be sure that the process of counting by weighing is clear. Consider the following "bag" of H atoms (symbolized by dots), which contains 1 mol (6.022×10^{23}) of H atoms and has a mass of 1.008 g. Assume the bag itself has no mass.

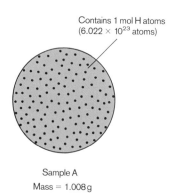

Contains 1 mol H atoms
(6.022×10^{23} atoms)

Sample A
Mass = 1.008 g

Now consider another "bag" of hydrogen atoms in which the number of hydrogen atoms is unknown.

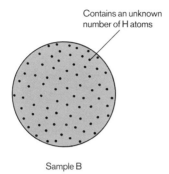

Contains an unknown number of H atoms

Sample B

We want to find out how many H atoms are present in sample ("bag") B. How can we do that? We can do it by weighing the sample. We find the mass of sample B to be 0.500 g.

How does this measured mass help us determine the number of atoms in sample B? We know that 1 mol of H atoms has a mass of 1.008 g. Sample B has a mass of 0.500 g, which is approximately half the mass of a mole of H atoms.

Sample A
Mass = 1.008 g

Contains 1 mol
of H atoms

Because the mass of B
is about half the
mass of A

Sample B
Mass = 0.500 g

Must contain
about 1/2 mol
of H atoms

We carry out the actual calculation by using the equivalence statement

$$1 \text{ mol H atoms} = 1.008 \text{ g H}$$

to construct the conversion factor we need:

$$0.500 \text{ g H} \times \frac{1 \text{ mol H}}{1.008 \text{ g H}} = 0.496 \text{ mol H in sample B}$$

Let's summarize. We know the mass of 1 mol of H atoms, so we can determine the number of moles of H atoms in any other sample of pure hydrogen by weighing the sample and *comparing* its mass to 1.008 g (the mass of 1 mol of H atoms). We can follow this same process for any element, because we know the mass of 1 mol for each of the elements.

Also, because we know that 1 mol is 6.022×10^{23} units, once we know the *moles* of atoms present, we can easily determine the *number* of atoms present. In the case considered above, we have approximately 0.5 mol of H atoms in sample B. This means that about 1/2 of 6×10^{23}, or 3×10^{23}, H atoms are

Solving mole problems can seem more practical and less theoretical if you can weigh a sample of some element in class and then ask how many atoms or how many moles are in the pile. Let students help you set up appropriate conversion factors to solve the problem. The pencil and paper exercise then becomes a real problem.

present. We carry out the actual calculation by using the equivalence statement

$$1 \text{ mol} = 6.022 \times 10^{23}$$

to determine the conversion factor we need:

$$0.496 \text{ mol H atoms} \times \frac{6.022 \times 10^{23} \text{ H atoms}}{1 \text{ mol H atoms}} = 2.99 \times 10^{23} \text{ H atoms in sample B}$$

These procedures are illustrated in Example 9.3.

EXAMPLE 9.3   **Calculating Moles and Number of Atoms**

Aluminum (Al), a metal with a high strength-to-weight ratio and a high resistance to corrosion, is often used for structural purposes such as in high-quality bicycle frames. Compute both the number of moles of atoms and the number of atoms in a 10.0-g sample of aluminum.

SOLUTION

In this case we want to change from mass to moles of atoms:

The mass of 1 mol (6.022 × 10²³ atoms) of aluminum is 26.98 g. The sample we are considering has a mass of 10.0 g. Its mass is less than 26.98 g, so this sample contains less than 1 mol of aluminum atoms. We calculate the number of moles of aluminum atoms in 10.0 g by using the equivalence statement

$$1 \text{ mol Al} = 26.98 \text{ g Al}$$

to construct the appropriate conversion factor:

$$10.0 \text{ g Al} \times \frac{1 \text{ mol Al}}{26.98 \text{ g Al}} = 0.371 \text{ mol Al}$$

Next we convert from moles of atoms to the number of atoms, using the equivalence statement

$$6.022 \times 10^{23} \text{ Al atoms} = 1 \text{ mol Al atoms}$$

We have

$$0.371 \text{ mol Al} \times \frac{6.022 \times 10^{23} \text{ Al atoms}}{1 \text{ mol Al}} = 2.23 \times 10^{23} \text{ Al atoms}$$

EXAMPLE 9.3, CONTINUED

We can summarize this calculation as follows:

EXAMPLE 9.4   **Calculating the Number of Atoms**

A silicon chip used in an integrated circuit of a microcomputer has a mass of 5.68 mg. How many silicon (Si) atoms are present in this chip? The average atomic mass for silicon is 28.08 amu.

SOLUTION

Our strategy for doing this problem is to convert from milligrams of silicon to grams of silicon, then to moles of silicon, and finally to atoms of silicon:

A silicon chip of the type used in electronic equipment.

where each arrow in the schematic represents a conversion factor. Because 1 g = 1000 mg, we have

$$5.68 \ \cancel{\text{mg Si}} \times \frac{1 \ \text{g Si}}{1000 \ \cancel{\text{mg Si}}} = 5.68 \times 10^{-3} \ \text{g Si}$$

Next, because the average mass of silicon is 28.08 amu, we know that

$$1 \ \text{mol Si atoms} = 28.08 \ \text{g Si}$$

$$5.68 \times 10^{-3} \ \cancel{\text{g Si}} \times \frac{1 \ \text{mol Si}}{28.08 \ \cancel{\text{g Si}}} = 2.02 \times 10^{-4} \ \text{mol Si}$$

Using the definition of a mole (1 mol = 6.022×10^{23}), we have

$$2.02 \times 10^{-4} \ \cancel{\text{mol Si}} \times \frac{6.022 \times 10^{23} \ \text{atoms}}{1 \ \cancel{\text{mol Si}}} = 1.22 \times 10^{20} \ \text{Si atoms}$$

EXAMPLE 9.4, CONTINUED

We can summarize this calculation as follows:

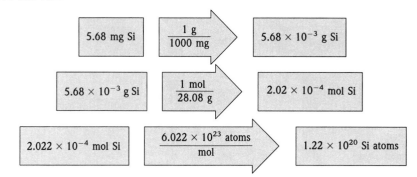

SELF-CHECK EXERCISE 9.3

Cobalt (Co) is a metal that is added to steel to improve its resistance to corrosion (for example, to make stainless steel). Calculate both the number of moles in a sample of cobalt containing 5.00×10^{20} atoms and the mass of the sample.

Problem Solving: Does the Answer Make Sense?

When you finish a problem, always think about the "reasonableness" of your answers. In Example 9.4, 5.68 mg of silicon is clearly much less than 1 mol of silicon (which has a mass of 28.08 g), so the final answer of 1.22×10^{20} atoms (compared to 6.022×10^{23} atoms in a mole) at least lies in the right direction. That is, 1.22×10^{20} atoms is a smaller number than 6.022×10^{23}. Paying careful attention to units and making this type of general check can help you detect errors such as an inverted conversion factor or a number that was incorrectly entered into your calculator.

9.4 Molar Mass

AIM: To define molar mass.
To show how to convert between moles and mass of a given sample of a chemical compound.

When you explain how to compute molar masses, emphasize to students that atomic masses must be used rather than atomic numbers.

A chemical compound is, fundamentally, a collection of atoms. For example, methane (the major component of natural gas) consists of molecules each of which contains one carbon and four hydrogen atoms (CH_4). How can we calculate the mass of 1 mol of methane; that is, what is the mass of 6.022×10^{23} CH_4 molecules? Because each CH_4 molecule contains one carbon atom and four hydrogen atoms, 1 mol of CH_4 molecules consists of 1 mol of carbon atoms and 4 mol of hydrogen atoms (Figure 9.3). The mass of 1 mol of methane can be found by summing the masses of carbon and hydrogen present:

Note that when we say 1 mol of methane, we mean 1 mol of methane *molecules*.

$$
\begin{array}{ll}
\text{Mass of 1 mol of C} = 1 \times 12.01 \text{ g} = 12.01 \text{ g} \\
\text{Mass of 4 mol of H} = 4 \times 1.008 \text{ g} = \underline{4.032 \text{ g}} \\
\text{Mass of 1 mol of } CH_4 \phantom{= 4 \times 1.008 \text{ g}} = 16.04 \text{ g}
\end{array}
$$

Remember that the least number of decimal places limits the number of significant figures in addition.

The quantity 16.04 g is called the molar mass for methane: the mass of 1 mol of CH_4 molecules. The **molar mass*** of any substance is the *mass (in grams) of 1 mol of the substance.* The molar mass is obtained by summing the masses of the component atoms.

A substance's molar mass (in grams) is the mass of 1 mol of that substance.

EXAMPLE 9.5   Calculating Molar Mass

Calculate the molar mass of sulfur dioxide.

SOLUTION

The formula for sulfur dioxide is SO_2. We need to compute the mass of 1 mol of SO_2 molecules—the molar mass for sulfur dioxide. We know that 1 mol of SO_2 molecules contains 1 mol of sulfur atoms and 2 mol of oxygen atoms:

$$
\begin{array}{ll}
\text{Mass of 1 mol of S} = 1 \times 32.06 = 32.06 \text{ g} \\
\text{Mass of 2 mol of O} = 2 \times 16.00 = \underline{32.00 \text{ g}} \\
\text{Mass of 1 mol of } SO_2 = 64.06 \text{ g}
\end{array}
$$

The molar mass of SO_2 is 64.06 g. It represents the mass of 1 mol of SO_2 molecules.

1 mol CH_4 molecules

Figure 9.3
Various numbers of methane molecules showing their constituent atoms.

---

*The term *molecular weight* was traditionally used instead of molar mass. The terms *molecular weight* and *molar mass* mean exactly the same thing. Because the term *molar mass* more accurately describes the concept, it will be used in this text.

This chapter and Chapter 10 introduce concepts that are illustrated with many calculations, so it may be impossible to work as many problems as you would like. You might want to choose just a few problems to discuss in detail. It can be helpful to work through the problems on the blackboard and to put the worked-out solutions on an overhead transparency so that students can follow your discussion without missing any details.

The mass of the electron is so small that Na^+ and Na have the same mass for our purposes, even though Na^+ has one electron less than Na. Also the mass of Cl virtually equals the mass of Cl^- even though it has one more electron than Cl.

Students often use atomic masses expressed to one or two significant figures to make calculating molar mass simple. For example, hydrogen is rounded to 1 and carbon to 12. Students who develop this habit now generally will have difficulty obtaining the correct answer to stoichiometry problems in the next chapter.

EXAMPLE 9.5, CONTINUED

SELF-CHECK EXERCISE 9.4

Polyvinyl chloride (called PVC), which is widely used for floor coverings ("vinyl") and for plastic pipes in plumbing systems, is made from a molecule with the formula C_2H_3Cl. Calculate the molar mass of this substance.

Some substances exist as a collection of ions rather than as separate molecules. An example is ordinary table salt, sodium chloride (NaCl), which is composed of an array of Na^+ and Cl^- ions. There are no NaCl molecules present. In some books the term **formula weight** is used instead of molar mass for ionic compounds. However, in this book we will apply the term *molar mass* to both ionic and molecular substances.

To calculate the molar mass for sodium chloride, we must realize that 1 mol of NaCl contains 1 mol of Na^+ ions and 1 mol of Cl^- ions.

Therefore the molar mass (in grams) for sodium chloride represents the sum of the mass of 1 mol of sodium ions and the mass of 1 mol of chloride ions.

$$\text{Mass of 1 mol of } Na^+ = 22.99 \text{ g}$$
$$\text{Mass of 1 mol of } Cl^- = \underline{35.45 \text{ g}}$$
$$\text{Mass of 1 mol of NaCl} = 58.44 \text{ g} = \text{molar mass}$$

The molar mass of NaCl is 58.44 g. It represents the mass of 1 mol of sodium chloride.

EXAMPLE 9.6   Calculating Mass from Moles

Calcium carbonate, $CaCO_3$, (also called calcite) is the principal mineral found in limestone, marble, chalk, pearls, and the shells of marine animals such as clams.

a.   Calculate the molar mass of calcium carbonate.
b.   A certain sample of calcium carbonate contains 4.86 mol. What is the mass in grams of this sample?

EXAMPLE 9.6, CONTINUED

SOLUTION

a.   Calcium carbonate is an ionic compound composed of Ca^{2+} and CO_3^{2-} ions. One mole of calcium carbonate contains 1 mol of Ca^{2+} and 1 mol of CO_3^{2-} ions. We calculate the molar mass by summing the masses of the components.

An abalone shell consists mainly of the mineral calcium carbonate, $CaCO_3$.

$$\text{Mass of 1 mol of } Ca^{2+} = 1 \times 40.08 \text{ g} = 40.08 \text{ g}$$

Mass of 1 mol of CO_3^{2-} (contains 1 mol of C and 3 mol of O):

$$1 \text{ mol of C} \quad = 1 \times 12.01 \text{ g} = 12.01 \text{ g}$$
$$3 \text{ mol of O} \quad = 3 \times 16.00 \text{ g} = \underline{48.00 \text{ g}}$$
$$\text{Mass of 1 mol of } CaCO_3 = 100.09 \text{ g} = \text{molar mass}$$

b.   We determine the mass of 4.86 mol of $CaCO_3$ by using the molar mass.

$$4.86 \text{ mol } CaCO_3 \times \frac{100.09 \text{ g } CaCO_3}{1 \text{ mol } CaCO_3} = 486 \text{ g } CaCO_3$$

which can be diagrammed as follows:

Note that the sample under consideration contains nearly 5 mol and thus should have a mass of nearly 500 g, so our answer makes sense.

SELF-CHECK EXERCISE 9.5

Calculate the molar mass for sodium sulfate, Na_2SO_4. A sample of sodium sulfate with a mass of 300.0 g represents what number of moles of sodium sulfate?

For average atomic masses, look inside the front cover of the book.

    In summary, the molar mass of a substance can be obtained by adding together the masses of the component atoms. The molar mass (in grams) represents the mass of 1 mol of the substance. Once we know the molar mass of a compound, we can compute the number of moles present in a sample of known mass. The reverse, of course, is also true.

EXAMPLE 9.7   Calculating Moles from Mass

Juglone, a dye known for centuries, is produced from the husks of black walnuts. It is also a natural herbicide (weed killer) that kills off competitive plants around the black walnut tree but does not affect grass and other noncompetitive plants. The formula for juglone is $C_{10}H_6O_3$.

EXAMPLE 9.7, CONTINUED

a.   Calculate the molar mass of juglone.
b.   A sample of 1.56 g of pure juglone was extracted from black walnut husks. How many moles of juglone does this sample represent?

SOLUTION

a.   The molar mass is obtained by summing the masses of the component atoms. In 1 mol of juglone there are 10 mol of carbon atoms, 6 mol of hydrogen atoms, and 3 mol of oxygen atoms.

$$
\begin{aligned}
\text{Mass of 10 mol of C} &= 10 \times 12.01 \text{ g} = 120.1 \text{ g} \\
\text{Mass of } 6 \text{ mol of H} &= 6 \times 1.008 \text{ g} = 6.048 \text{ g} \\
\text{Mass of } 3 \text{ mol of O} &= 3 \times 16.00 \text{ g} = \underline{48.00 \text{ g}} \\
\text{Mass of 1 mol of } C_{10}H_6O_3 &= 174.1 \text{ g} = \text{molar mass}
\end{aligned}
$$

b.   The mass of 1 mol of this compound is 174.1 g, so 1.56 g is much less than a mole. We can determine the exact fraction of a mole by using the equivalence statement

$$1 \text{ mol} = 174.1 \text{ g juglone}$$

to derive the appropriate conversion factor:

$$1.56 \text{ g juglone} \times \frac{1 \text{ mol juglone}}{174.1 \text{ g juglone}} = 0.00896 \text{ mol juglone}$$

$$= 8.96 \times 10^{-3} \text{ mol juglone}$$

| 1.56 g juglone | $\dfrac{1 \text{ mol}}{174.1 \text{ g}}$ | 8.96×10^{-3} mol juglone |

EXAMPLE 9.8   **Calculating Number of Molecules**

Isopentyl acetate, $C_7H_{14}O_2$, the compound responsible for the scent of bananas, can be produced commercially. Interestingly, bees release about 1 μg (1 × 10^{-6} g) of this compound when they sting. This attracts other bees, which then join the attack. How many moles and how many molecules of isopentyl acetate are released in a typical bee sting?

SOLUTION

We are given a mass of isopentyl acetate and want the number of molecules, so we must first compute the molar mass.

Close-up of bee's stinging apparatus.
Vespula maculata

EXAMPLE 9.8, CONTINUED

$$7 \text{ mol C} \times 12.01 \ \frac{g}{\text{mol}} = 84.07 \text{ g C}$$

$$14 \text{ mol H} \times 1.008 \ \frac{g}{\text{mol}} = 14.112 \text{ g H}$$

$$2 \text{ mol O} \times 16.00 \ \frac{g}{\text{mol}} = \underline{32.00 \text{ g O}}$$
$$130.18 \text{ g}$$

This means that 1 mol of isopentyl acetate (6.022×10^{23} molecules) has a mass of 130.18 g.

Next we determine the number of moles of isopentyl acetate in 1 μg, which is 1×10^{-6} g. To do this, we use the equivalence statement

$$1 \text{ mol isopentyl acetate} = 130.18 \text{ g isopentyl acetate}$$

which yields the conversion factor we need:

$$1 \times 10^{-6} \text{ g C}_7\text{H}_{14}\text{O}_2 \times \frac{1 \text{ mol C}_7\text{H}_{14}\text{O}_2}{130.18 \text{ g C}_7\text{H}_{14}\text{O}_2} = 8 \times 10^{-9} \text{ mol C}_7\text{H}_{14}\text{O}_2$$

Using the equivalence statement 1 mol $= 6.022 \times 10^{23}$ units, we can determine the number of molecules:

$$8 \times 10^{-9} \text{ mol C}_7\text{H}_{14}\text{O}_2 \times \frac{6.022 \times 10^{23} \text{ molecules}}{1 \text{ mol C}_7\text{H}_{14}\text{O}_2} = 5 \times 10^{15} \text{ molecules}$$

This very large number of molecules is released in each bee sting.

SELF-CHECK EXERCISE 9.6

The substance Teflon, the slippery coating on many frying pans, is made from the C_2F_4 molecule. Calculate the number of C_2F_4 units present in 135 g of Teflon.

9.5 Percent Composition of Compounds

AIM: To show how to find the mass percent of an element in a given compound.

So far we have discussed the composition of compounds in terms of the numbers of constituent atoms. It is often useful to know a compound's composition in

If you have a compound that students have recently used in lab or if there is some other compound with which they are familiar, bring it to class as an illustration, and then calculate its molar mass. I use $Cu(NO_3)_2$ for this purpose.

The formula for ethanol is written C_2H_5OH, although you might expect it to be written simply as C_2H_6O.

terms of the *masses* of its elements. We can obtain this information from the formula of the compound by comparing the mass of each element present in 1 mol of the compound to the total mass of 1 mol of the compound. The mass fraction for each element is calculated as follows:

$$\text{Mass fraction for a given element} = \frac{\text{mass of the element present in 1 mol of compound}}{\text{mass of 1 mol of compound}}$$

The mass fraction is converted to *mass percent* by multiplying by 100.

We will illustrate this concept using the compound ethanol, an alcohol obtained by fermenting the sugar in grapes, corn, and other fruits and grains. Ethanol is often added to gasoline as an octane enhancer to form a fuel called gasohol.

Note from its formula that each molecule of ethanol contains two carbon atoms, six hydrogen atoms, and one oxygen atom. This means that each mole of ethanol contains 2 mol of carbon atoms, 6 mol of hydrogen atoms, and 1 mol of oxygen atoms. We calculate the mass of each element present and the molecular weight for ethanol as follows:

$$\text{Mass of C} = 2 \text{ mol} \times 12.01 \ \frac{g}{mol} = 24.02 \text{ g}$$

$$\text{Mass of H} = 6 \text{ mol} \times 1.008 \ \frac{g}{mol} = 6.048 \text{ g}$$

$$\text{Mass of O} = 1 \text{ mol} \times 16.00 \ \frac{g}{mol} = \underline{16.00 \text{ g}}$$

$$\text{Mass of 1 mol of } C_2H_5OH = 46.07 \text{ g} = \text{molar mass}$$

The **mass percent** (sometimes called the weight percent) of carbon in ethanol can be computed by comparing the mass of carbon in 1 mol of ethanol to the total mass of 1 mol of ethanol and multiplying the result by 100.

$$\text{Mass percent of C} = \frac{\text{mass of C in 1 mol } C_2H_5OH}{\text{mass of 1 mol } C_2H_5OH} \times 100$$

$$= \frac{24.02 \text{ g}}{46.07 \text{ g}} \times 100 = 52.14\%$$

That is, ethanol contains 52.14% by mass of carbon. The mass percents of hydrogen and oxygen in ethanol are obtained in a similar manner.

$$\text{Mass percent of H} = \frac{\text{mass of H in 1 mol } C_2H_5OH}{\text{mass of 1 mol } C_2H_5OH} \times 100$$

$$= \frac{6.048 \text{ g}}{46.07 \text{ g}} \times 100 = 13.13\%$$

Wine fermentation tanks at a winery in Napa Valley, California.

$$\text{Mass percent of O} = \frac{\text{mass of O in 1 mol C}_2\text{H}_5\text{OH}}{\text{mass of 1 mol C}_2\text{H}_5\text{OH}} \times 100$$

$$= \frac{16.00 \text{ g}}{46.07 \text{ g}} \times 100 = 34.73\%$$

The mass percentages of all the elements in a compound add up to 100%, although rounding-off effects may produce a small deviation. Adding up the percentages is a good way to check the calculations. In this case, the sum of the mass percents is 52.14% + 13.13% + 34.73% = 100.00%.

Sometimes, because of rounding-off effects, the sum of the mass percents in a compound is not exactly 100%.

EXAMPLE 9.9 Calculating Mass Percent

Carvone is a substance that occurs in two forms, both of which have the same molecular formula ($C_{10}H_{14}O$) and molar mass. One type of carvone gives caraway seeds their characteristic smell; the other is responsible for the smell of spearmint oil. Compute the mass percent of each element in carvone.

SOLUTION

Because the formula for carvone is $C_{10}H_{14}O$, the masses of the various elements in 1 mol of carvone are

$$\text{Mass of C in 1 mol} = 10 \text{ mol} \times 12.01 \ \frac{\text{g}}{\text{mol}} = 120.1 \text{ g}$$

$$\text{Mass of H in 1 mol} = 14 \text{ mol} \times 1.008 \ \frac{\text{g}}{\text{mol}} = 14.11 \text{ g}$$

$$\text{Mass of O in 1 mol} = 1 \text{ mol} \times 16.00 \ \frac{\text{g}}{\text{mol}} = \underline{16.00 \text{ g}}$$

$$\text{Mass of 1 mol of C}_{10}\text{H}_{14}\text{O} = 150.21 \text{ g}$$

$$\text{Molar mass} = 150.2 \text{ g} \quad \text{(rounding to the correct number of significant figures)}$$

The 120.1 limits the sum to one decimal place.

Next we find the fraction of the total mass contributed by each element and convert it to a percentage.

$$\text{Mass percent of C} = \frac{120.1 \text{ g C}}{150.2 \text{ g C}_{10}\text{H}_{14}\text{O}} \times 100 = 79.96\%$$

$$\text{Mass percent of H} = \frac{14.11 \text{ g H}}{150.2 \text{ g C}_{10}\text{H}_{14}\text{O}} \times 100 = 9.394\%$$

$$\text{Mass percent of O} = \frac{16.00 \text{ g O}}{150.2 \text{ g C}_{10}\text{H}_{14}\text{O}} \times 100 = 10.65\%$$

EXAMPLE 9.9, CONTINUED

CHECK: Add the individual mass percent values—they should total 100% within a small range due to rounding off. In this case, the percentages add up to 100.00%.

SELF-CHECK EXERCISE 9.7

Penicillin, an important antibiotic (antibacterial agent), was discovered accidentally by the Scottish bacteriologist Alexander Fleming in 1928, although he was never able to isolate it as a pure compound. This and similar antibiotics have saved millions of lives that would otherwise have been lost to infections. Penicillin, like many of the molecules produced by living systems, is a large molecule containing many atoms. One type of penicillin, penicillin F, has the formula $C_{14}H_{20}N_2SO_4$. Compute the mass percent of each element in this compound.

$\underline{9.6}$ Formulas of Compounds

AIM: To describe empirical formulas of compounds.

The sections on formulas follow naturally from moles and numbers of atoms. However, the progression doesn't seem as natural to students as it does to us.

Assume that you have mixed two solutions and a solid product (a precipitate) forms. How can you find out what the solid is? What is its formula? There are several possible approaches you can take to answering these questions. For example, we saw in Chapter 7 that we can usually predict the identity of a precipitate formed when two solutions are mixed in a reaction of this type if we know some facts about the solubilities of ionic compounds.

However, although an experienced chemist can often predict the product expected in a chemical reaction, the only sure way to identify the product is to perform experiments. Usually we compare the physical properties of the product to the properties of known compounds.

Sometimes a chemical reaction gives a product that has never been obtained before. In such a case, a chemist determines what compound has been formed by determining which elements are present and how much of each. These data can be used to obtain the formula of the compound. In Section 9.5 we used the formula of the compound to determine the mass of each element present in a mole of the compound. To obtain the formula of an unknown compound, we do the opposite. That is, we use the measured masses of the elements present to determine the formula.

Recall that the formula of a compound gives the relative numbers of various types of atoms present. For example, the formula CO_2 tells us that for each carbon atom there are two oxygen atoms in each molecule of carbon dioxide. So to determine the formula of a substance we need to count the atoms. As we have seen in this chapter, we can do this by weighing. Suppose we know that a compound contains only the elements carbon, hydrogen, and oxygen, and we weigh out a 0.2015-g sample for analysis. Using methods we will not discuss here, we find that this 0.2015-g sample of compound contains 0.0806 g of carbon, 0.01353 g of hydrogen, and 0.1074 g of oxygen. We have just learned how to convert these masses to numbers of atoms by using the atomic mass of each element. We begin by converting to moles.

Carbon

$$(0.0806 \text{ g C}) \times \frac{1 \text{ mol C atoms}}{12.01 \text{ g C}} = 0.00671 \text{ mol C atoms}$$

Hydrogen

$$(0.01353 \text{ g H}) \times \frac{1 \text{ mol H atoms}}{1.008 \text{ g H}} = 0.01342 \text{ mol H atoms}$$

Oxygen

$$(0.1074 \text{ g O}) \times \frac{1 \text{ mol O atoms}}{16.00 \text{ g O}} = 0.00671 \text{ mol O atoms}$$

Let's review what we have established. We now know that 0.2015 g of the compound contains 0.00671 mol of C atoms, 0.01342 mol of H atoms, and 0.00671 mol of O atoms. Because 1 mol is 6.022×10^{23}, these quantities can be converted to actual numbers of atoms.

Carbon

$$(0.00671 \text{ mol C atoms})\frac{(6.022 \times 10^{23} \text{ C atoms})}{1 \text{ mol C atoms}} = 4.04 \times 10^{21} \text{ C atoms}$$

Hydrogen

$$(0.01342 \text{ mol H atoms})\frac{(6.022 \times 10^{23} \text{ H atoms})}{1 \text{ mol H atoms}} = 8.08 \times 10^{21} \text{ H atoms}$$

Oxygen

$$(0.00671 \text{ mol O atoms})\frac{(6.022 \times 10^{23} \text{ O atoms})}{1 \text{ mol O atoms}} = 4.04 \times 10^{21} \text{ O atoms}$$

These are the numbers of the various types of atoms *in 0.2015 g of compound.* What do these numbers tell us about the formula of the compound? Note the following:

1. The compound contains the same number of C and O atoms.
2. There are twice as many H atoms as C atoms or O atoms.

We can represent this information by the formula CH_2O, which expresses the *relative* numbers of C, H, and O atoms present. Is this the true formula for the

Figure 9.4
The glucose molecule. The molecular formula is $C_6H_{12}O_6$, as can be verified by counting the atoms. The empirical formula for glucose is CH_2O.

Show several examples of molecules with identical empirical formulas but different molecular formulas. Formaldehyde, acetic acid, and glucose all have the empirical formula CH_2O, but are different molecules, with different characteristics. You might bring to class samples of each of these and demonstrate a few of their properties and highlight their differences.

compound? In other words, is the compound made up of CH_2O molecules? It may be. However, it might also be made up of $C_2H_4O_2$ molecules, $C_3H_6O_3$ molecules, $C_4H_8O_4$ molecules, $C_5H_{10}O_5$ molecules, $C_6H_{12}O_6$ molecules, and so on. Note that each of these molecules has the required $1:2:1$ ratio of carbon to hydrogen to oxygen atoms (the ratio shown by experiment to be present in the compound).

When we break a compound down into its separate elements and "count" the atoms present, we learn only the ratio of atoms—we get only the *relative* numbers of atoms. The formula of a compound that expresses the smallest whole-number ratio of the atoms present is called the **empirical formula** or *simplest formula*. A compound that contains the molecules $C_4H_8O_4$ has the same empirical formulas as a compound that contains $C_6H_{12}O_6$ molecules. The empirical formula for both is CH_2O. The actual formula of a compound—the one that gives the composition of the molecules that are present—is called the **molecular formula**. The sugar called glucose is made of molecules with the molecular formula $C_6H_{12}O_6$ (Figure 9.4). Note from the molecular formula for glucose that the empirical formula is CH_2O. We can represent the molecular formula as a multiple (by 6) of the empirical formula:

$$C_6H_{12}O_6 = (CH_2O)_6$$

In the next section, we will explore in more detail how to calculate the empirical formula for a compound from the relative masses of the elements present. As we will see in Sections 9.8 and 9.9, we must know the molar mass of a compound to determine its molecular formula.

EXAMPLE 9.10 **Determining Empirical Formulas**

In each case below, the molecular formula for a compound is given. Determine the empirical formula for each compound.

a. C_6H_6. This is the molecular formula for benzene, a liquid commonly used in industry as a starting material for many important products.
b. $C_{12}H_4Cl_4O_2$. This is the molecular formula for a substance commonly called dioxin, a powerful poison that sometimes occurs as a by-product in the production of other chemicals.
c. $C_6H_{16}N_2$. This is the molecular formula for one of the reactants used to produce nylon.

Students often forget the effect that subscripts outside of parentheses have on the number of atoms of each element in a compound. For example, it is very easy to forget that $Ca_3(PO_4)_2$ has two phosphorus atoms. Students cannot remember whether the subscripts are added or multiplied and often mistakenly report the number of oxygen atoms as six.

SOLUTION

a. $C_6H_6 = (CH)_6$; CH is the empirical formula. Each subscript in the empirical formula is multiplied by 6 to obtain the molecular formula.
b. $C_{12}H_4Cl_4O_2$; $C_{12}H_4Cl_4O_2 = (C_6H_2Cl_2O)_2$; $C_6H_2Cl_2O$ is the empirical formula. Each subscript in the empirical formula is multiplied by 2 to obtain the molecular formula.

EXAMPLE 9.10, CONTINUED

c. $C_6H_{16}N_2 = (C_3H_8N)_2$; C_3H_8N is the empirical formula. Each subscript in the empirical formula is multiplied by 2 to obtain the molecular formula.

9.7   Calculation of Empirical Formulas

AIM: To show how to calculate empirical formulas.

As we said in the previous section, one of the most important things we can learn about a new compound is its chemical formula. To calculate the empirical formula of a compound, we first determine the relative masses of the various elements that are present.

One way to do this is to measure the masses of elements that react to form the compound. For example, suppose we weigh out 0.2636 g of pure nickel metal into a crucible and heat this metal in the air so that the nickel can react with oxygen to form a nickel oxide compound. After the sample has cooled, we weigh it again and find its mass to be 0.3354 g. The gain in mass is due to the oxygen that reacts with the nickel to form the oxide. Therefore, the mass of oxygen present in the compound is the total mass of the product minus the mass of the nickel:

Total mass of nickel oxide	−	Mass of nickel originally present	=	Mass of oxygen that reacted with the nickel

or

$$0.3354 \text{ g} - 0.2636 \text{ g} = 0.0718 \text{ g}$$

Note that the mass of nickel present in the compound is the nickel metal originally weighed out. So we know that the nickel oxide contains 0.2636 g of nickel and 0.0718 g of oxygen. What is the empirical formula of this compound?

To answer this question we must convert the masses to numbers of atoms, using atomic masses:

$$0.2636 \text{ g Ni} \times \frac{1 \text{ mol Ni atoms}}{58.69 \text{ g Ni}} = 0.004491 \text{ mol Ni atoms}$$

$$0.0718 \text{ g O} \times \frac{1 \text{ mol O atoms}}{16.00 \text{ g O}} = 0.00449 \text{ mol O atoms}$$

The key task for students in this section is to learn how grams of each element in a compound can provide information about the number of atoms of each element in a compound.

The most straightforward kind of compound to use initially is one with two types of atoms that has an empirical formula of one for each atom.

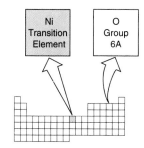

Students should know that when the data are given in terms of grams or as mass percents, the only kind of formula that can be determined is the empirical formula.

Four significant figures allowed.

Three significant figures allowed.

There are several ways to explain to students why calculating moles of each element in a compound leads to the correct empirical formula for that compound. After calculating moles, I approach empirical-formula problems by asking students to name the element that occurs ''the most'' in a given compound. For a compound that contains 0.3354 mol C, 1.174 mol H, and 0.1677 mol N, most students can answer that the compound has more H than C or N. I then ask, ''How much more H is there than N?'' If we divide 1.174 mol H by 0.1677 mol N, we see that the answer is seven times as much H as N and two times as much C as N. Because this is really what an empirical formula tells us, we have solved the problem. This approach might work for you as a way to explain how moles really represent the relative number of atoms.

When the resulting ratio of atoms is not an integer, students begin to worry. They are not sure that their calculations are correct. Warn them to be especially careful to use the correct number of significant figures. I always tell students that, for exams and practice problems, the number of atoms, if not an integer, will be an easily recognizable fraction of an integer.

These mole quantities represent numbers of atoms (remember that a mole of atoms is 6.022×10^{23} atoms). It is clear from the moles of atoms that the compound contains an equal number of Ni and O atoms, so the formula is NiO. This is the *empirical formula;* it expresses the smallest whole-number (integer) ratio of atoms:

$$\frac{0.004491 \text{ mol Ni atoms}}{0.00449 \text{ mol O atoms}} = \frac{1 \text{ Ni}}{1 \text{ O}}$$

That is, this compound contains equal numbers of nickel atoms and oxygen atoms. We say the ratio of nickel atoms to oxygen atoms is $1:1$ (1 to 1).

EXAMPLE 9.11   Calculating Empirical Formulas

An oxide of aluminum is formed by the reaction of 4.151 g of aluminum with 3.692 g of oxygen. Calculate the empirical formula for this compound.

SOLUTION

We know that the compound contains 4.151 g of aluminum and 3.692 g of oxygen. But we need to know the relative number of each type of atom to write the formula, so we must convert these masses to moles of atoms to get the empirical formula. We carry out the conversion by using the atomic masses of the elements.

$$4.151 \text{ g Al} \times \frac{1 \text{ mol Al}}{26.98 \text{ g Al}} = 0.1539 \text{ mol Al atoms}$$

$$3.692 \text{ g O} \times \frac{1 \text{ mol O}}{16.00 \text{ g O}} = 0.2308 \text{ mol O atoms}$$

Because chemical formulas use only whole numbers, we next find the integer (whole-number) ratio of the atoms. To do this we start by dividing both numbers by the smallest of the two. This converts the smallest number to 1.

$$\frac{0.1539 \text{ mol Al}}{0.1539} = 1.000 \text{ mol Al atoms}$$

$$\frac{0.2308 \text{ mol O}}{0.1539} = 1.500 \text{ mol O atoms}$$

Note that dividing both numbers of moles of atoms by the *same* number does not change the *relative* numbers of oxygen and aluminum atoms. That is,

$$\frac{0.2308 \text{ mol O}}{0.1539 \text{ mol Al}} = \frac{1.500 \text{ mol O}}{1.000 \text{ mol Al}}$$

Thus we know that the compound contains 1.500 mol of O atoms for every 1.000 mol of Al atoms, or, in terms of individual atoms, we could say that the compound

EXAMPLE 9.II, CONTINUED

contains 1.500 O atoms for every 1.000 Al atom. However, because only *whole* atoms combine to form compounds, we must find a set of *whole numbers* to express the empirical formula. When we multiply both 1.000 and 1.500 by 2, we get the integers we need.

$$1.500 \text{ O} \times 2 = 3.000 = 3 \text{ O atoms}$$
$$1.000 \text{ Al} \times 2 = 2.000 = 2 \text{ Al atoms}$$

Therefore this compound contains two Al atoms for every three O atoms, and the empirical formula is Al_2O_3. Note that the *ratio* of atoms in this compound is given by each of the following fractions:

$$\frac{0.2308 \text{ O}}{0.1539 \text{ Al}} = \frac{1.500 \text{ O}}{1.000 \text{ Al}} = \frac{3 \text{ O}}{2 \text{ Al}}$$

The smallest whole-number ratio corresponds to the subscripts of the empirical formula, Al_2O_3.

Sometimes the relative numbers of moles you get when you calculate an empirical formula will turn out to be nonintegers, as was the case in Example 9.11. When this happens, you must convert to the appropriate whole numbers. This is done by multiplying all the numbers by the same small integer, which can be found by trial and error. The multiplier needed is almost always between 1 and 6. We will now summarize what we have learned about calculating empirical formulas.

Steps for Determining the Empirical Formula of a Compound

- **STEP I**
  Obtain the mass of each element present (in grams).

- **STEP 2**
  Determine the number of moles of each type of atom present.

- **STEP 3**
  Divide the number of moles of each element by the smallest number of moles to convert the smallest number to 1. If all of the numbers so obtained are integers (whole numbers), these are the subscripts in the empirical formula. If one or more of these numbers are not integers, go on to step 4.

- **STEP 4**
  Multiply the numbers you derived in step 3 by the smallest integer that will convert all of them to whole numbers. This set of whole numbers represents the subscripts in the empirical formula.

EXAMPLE 9.12   Calculating Empirical Formulas for Binary Compounds

When a 0.3546-g sample of vanadium metal is heated in air, it reacts with oxygen to achieve a final mass of 0.6330 g. Calculate the empirical formula of this vanadium oxide.

SOLUTION

STEP I

All the vanadium that was originally present will be found in the final compound, so we can calculate the mass of oxygen that reacted by taking the following difference:

$$0.6330 \text{ g} \quad - \quad 0.3546 \text{ g} \quad = \quad 0.2784 \text{ g}$$

STEP 2

Using the atomic masses (50.94 for V and 16.00 for O), we obtain

$$0.3546 \text{ g V} \times \frac{1 \text{ mol V atoms}}{50.94 \text{ g V}} = 0.006961 \text{ mol V atoms}$$

$$0.2784 \text{ g O} \times \frac{1 \text{ mol O atoms}}{16.00 \text{ g O}} = 0.01740 \text{ mol O atoms}$$

STEP 3

Then we divide both numbers of moles by the smaller, 0.006961.

$$\frac{0.006961 \text{ mol V atoms}}{0.006961} = 1.000 \text{ mol V atoms}$$

$$\frac{0.01740 \text{ mol O atoms}}{0.006961} = 2.500 \text{ mol O atoms}$$

Because one of these numbers (2.500) is not an integer, we go on to step 4.

STEP 4

We note that $2 \times 2.500 = 5.000$ and $2 \times 1.000 = 2.000$, so we multiply both numbers by 2 to get integers.

$$2 \times 1.000 \text{ V} = 2.000 \text{ V} = 2 \text{ V}$$
$$2 \times 2.500 \text{ O} = 5.000 \text{ O} = 5 \text{ O}$$

EXAMPLE 9.12, CONTINUED

This compound contains 2 V atoms for every 5 O atoms, and the empirical formula is V_2O_5.

SELF-CHECK EXERCISE 9.8

In a lab experiment it was observed that 0.6884 g of lead combines with 0.2356 g of chlorine to form a binary compound. Calculate the empirical formula of this compound.

The same procedures we have used for binary compounds also apply to compounds containing three or more elements, as Example 9.13 illustrates.

EXAMPLE 9.13　**Calculating Empirical Formulas for Compounds Containing Three or More Elements**

A sample of lead arsenate, an insecticide used against the potato beetle, contains 1.3813 g of lead, 0.00672 g of hydrogen, 0.4995 g of arsenic, and 0.4267 g of oxygen. Calculate the empirical formula for lead arsenate.

SOLUTION

STEP 1
The compound contains 1.3813 g Pb, 0.00672 g H, 0.4995 g As, and 0.4267 g O.

STEP 2
We use the atomic masses of the elements present to calculate the moles of each.

$$1.3813 \text{ g Pb} \times \frac{1 \text{ mol Pb}}{207.2 \text{ g Pb}} = 0.006667 \text{ mol Pb}$$

$$0.00672 \text{ g H} \times \frac{1 \text{ mol H}}{1.008 \text{ g H}} = 0.00667 \text{ mol H}$$

Only three significant figures allowed.

$$0.4995 \text{ g As} \times \frac{1 \text{ mol As}}{74.92 \text{ g As}} = 0.006667 \text{ mol As}$$

$$0.4267 \text{ g O} \times \frac{1 \text{ mol O}}{16.00 \text{ g O}} = 0.02667 \text{ mol O}$$

EXAMPLE 9.13, CONTINUED

STEP 3
Now we divide by the smallest number of moles.

$$\frac{0.006667 \text{ mol Pb}}{0.006667} = 1.000 \text{ mol Pb}$$

$$\frac{0.00667 \text{ mol H}}{0.006667} = 1.00 \text{ mol H}$$

$$\frac{0.006667 \text{ mol As}}{0.006667} = 1.000 \text{ mol As}$$

$$\frac{0.02667 \text{ mol O}}{0.006667} = 4.000 \text{ mol O}$$

The numbers of moles are all whole numbers, so the empirical formula is $PbHAsO_4$.

SELF-CHECK EXERCISE 9.9

Sevin, the commercial name for an insecticide used to protect crops such as cotton, vegetables, and fruit is made from carbamic acid. A chemist analyzing a sample of carbamic acid finds 0.8007 g of carbon, 0.9333 g of nitrogen, 0.2016 g of hydrogen, and 2.133 g of oxygen. Determine the empirical formula for carbamic acid.

When a compound is analyzed to determine the relative amounts of the elements present, the results are usually given in terms of percentages by masses of the various elements. In Section 9.5 we learned to calculate the percent composition of a compound from its formula. Now we will do the opposite. Given the percent composition, we will calculate the empirical formula.

Percent by mass for a given element means the grams of that element in 100 g of the compound.

To understand this procedure, you must understand the meaning of *percent.* Remember that percent means parts of a given component per 100 parts of the total mixture. For example, if a given compound is 15% carbon (by mass), the compound contains 15 g of carbon per 100 g of compound.

Students are confused at first when confronted with percent compositions after they have learned to determine empirical formulas with grams of elements in a compound. But most students quickly understand that a percentage can be converted to a number of grams out of 100 grams and they then can continue to solve problems.

Calculation of the empirical formula of a compound when one is given its percent composition is illustrated in Example 9.14.

EXAMPLE 9.14   **Calculating Empirical Formulas from Percent Composition**

Cisplatin, the common name for a platinum compound that is used to treat cancerous tumors, has the composition (mass percent) 65.02% platinum, 9.34%

EXAMPLE 9.14, CONTINUED

nitrogen, 2.02% hydrogen, and 23.63% chlorine. Calculate the empirical formula for cisplatin.

SOLUTION

STEP 1

Determine how many grams of each element are present in 100 g of compound. Cisplatin is 65.02% platinum (by mass), which means there are 65.02 g of platinum (Pt) per 100.00 g of compound. Similarly, a 100.00-g sample of cisplatin contains 9.34 g of nitrogen (N), 2.02 g of hydrogen (H), and 26.63 g of chlorine (Cl).

If we have a 100.00-g sample of cisplatin, we have 65.02 g Pt, 9.34 g N, 2.02 g H, and 23.63 g Cl.

STEP 2

Determine the number of moles of each type of atom. We use the atomic masses to calculate moles.

$$65.02 \text{ g Pt} \times \frac{1 \text{ mol Pt}}{195.1 \text{ g Pt}} = 0.3333 \text{ mol Pt}$$

$$9.34 \text{ g N} \times \frac{1 \text{ mol N}}{14.01 \text{ g N}} = 0.667 \text{ mol N}$$

$$2.02 \text{ g H} \times \frac{1 \text{ mol H}}{1.008 \text{ g H}} = 2.00 \text{ mol H}$$

$$23.63 \text{ g Cl} \times \frac{1 \text{ mol Cl}}{35.45 \text{ g Cl}} = 0.6666 \text{ mol Cl}$$

STEP 3

Divide through by the smallest number of moles.

$$\frac{0.3333 \text{ mol Pt}}{0.3333} = 1.000 \text{ mol Pt}$$

$$\frac{0.667 \text{ mol N}}{0.3333} = 2.00 \text{ mol N}$$

$$\frac{2.00 \text{ mol H}}{0.3333} = 6.01 \text{ mol H}$$

$$\frac{0.6666 \text{ mol Cl}}{0.3333} = 2.000 \text{ mol Cl}$$

The empirical formula for cisplatin is $PtN_2H_6Cl_2$. Note that the number for hydrogen is slightly greater than 6 because of rounding-off effects.

EXAMPLE 9.14, CONTINUED

SELF-CHECK EXERCISE 9.10

The most common form of nylon (Nylon-6) is 63.68% carbon, 12.38% nitrogen, 9.80% hydrogen, and 14.14% oxygen. Calculate the empirical formula for Nylon-6.

Note from Example 9.14 that once the percentages are converted to masses, this example is the same as earlier examples in which the masses were given directly.

Solving empirical and molecular formula problems in class is very time-consuming. I usually move from empirical-formula problems to molecular-formula problems quickly, because finding the molecular formula also requires finding the empirical formula. One type of problem provides practice with both.

Remind students that, unless the molar mass of the compound is given in the problem, they can only calculate an empirical formula for the compound.

$\underline{9.8}$ Calculation of Molecular Formulas

AIM: To calculate the molecular formula of a compound, given its empirical formula and molar mass.

If we know the composition of a compound in terms of the masses (or mass percentages) of the elements present, we can calculate the empirical formula but not the molecular formula. For reasons that will become clear as we consider Example 9.15, to obtain the molecular formula we must know the molar mass. In this section we will consider compounds where both the percent composition and the molar mass are known.

EXAMPLE 9.15   Calculating Molecular Formulas

A white powder is analyzed and found to have an empirical formula of P_2O_5. The compound has a molar mass of 283.88. What is the compound's molecular formula?

SOLUTION

To obtain the molecular formula, we must compare the empirical formula mass to the molar mass. The empirical formula mass for P_2O_5 is the mass of 1 mol of P_2O_5 units.

EXAMPLE 9.15, CONTINUED

Recall that the molecular formula contains a whole number of empirical formula units. That is,

$$\text{Molecular formula} = (\text{empirical formula})_n$$

where n is a small whole number. Now, because

$$\text{Molecular formula} = n \times \text{empirical formula}$$

then

$$\text{Molar mass} = n \times \text{empirical formula mass}$$

Solving for n gives

$$n = \frac{\text{molar mass}}{\text{empirical formula mass}}$$

Thus to determine the molecular formula, we first divide the molar mass by the empirical formula mass. This tells us how many empirical formula masses there are in one molar mass.

$$\frac{\text{Molar mass}}{\text{Empirical formula mass}} = \frac{283.88 \text{ g}}{141.94 \text{ g}} = 2$$

This result means that $n = 2$ for this compound, so the molecular formula consists of two empirical formula units, and the molecular formula is $(P_2O_5)_2$, or P_4O_{10}. The structure of this interesting compound is shown in Figure 9.5.

SELF-CHECK EXERCISE 9.11

A compound used as an additive for gasoline to help prevent engine knock shows the following percentage composition:

71.65% Cl   24.27% C   4.07% H

The molar mass is known to be 98.96g. Determine the empirical formula and the molecular formula for this compound.

Most students realize that they first need to determine the empirical formula before they can determine the molecular formula. Students first must focus on the deciding how and why the molar mass of the empirical formula is important; this abstract understanding is a prerequisite to mastering the skill of finding the molecular formula.

Figure 9.5
The structure of P_4O_{10} as a "ball and stick" model. This compound has a great affinity for water and is often used as a desiccant, or drying agent.

It is important to realize that the molecular formula is always an integer multiple of the empirical formula. For example, the sugar glucose (see Figure 9.4) has the empirical formula CH_2O and the molecular formula $C_6H_{12}O_6$. In this case there are six empirical formula units in each glucose molecule:

$$(CH_2O)_6 = C_6H_{12}O_6$$

In general, we can represent the molecular formula in terms of the empirical formula as follows:

Molecular formula = (empirical formula)$_n$, where n is an integer.

$$(Empirical formula)_n = molecular formula$$

where n is an integer. If $n = 1$, the molecular formula is the same as the empirical formula. For example, for carbon dioxide the empirical formula (CO_2) and the molecular formula (CO_2) are the same, so $n = 1$. On the other hand, for tetraphosphorus decoxide the empirical formula is P_2O_5 and the molecular formula is $P_4O_{10} = (P_2O_5)_2$. In this case $n = 2$.

CHAPTER REVIEW

Key Terms

atomic mass unit (p. 245)
average atomic mass (p. 245)
mole (p. 249)
Avogadro's number (p. 249)

molar mass (p. 255)
mass percent (p. 260)
empirical formula (p. 264)
molecular formula (p. 264)

Summary

1. We can count individual units by weighing if we know the average mass of the units. Thus when we know the average mass of the atoms of an element as that element occurs in nature, we can calculate the number of atoms in any given sample of that element by weighing the sample.
2. A mole is a unit of measure equal to 6.022×10^{23}, which is called Avogadro's number. One mole of any substance contains 6.022×10^{23} units.
3. One mole of an element has a mass equal to the element's atomic mass expressed in grams. The molar mass of any compound is the mass (in grams) of 1 mol of the compound and is the sum of the masses of the component atoms.
4. Percent composition consists of the mass percent of each element in a compound:

$$Mass\ percent = \frac{mass\ of\ a\ given\ element\ in\ 1\ mol\ of\ compound}{mass\ of\ 1\ mol\ of\ compound} \times 100$$

5. The empirical formula of a compound is the simplest whole-number ratio of the atoms present in the compound; it can be derived from the percent composition of the compound. The molecular formula is the exact formula of the molecules present; it is always an integer multiple of the empirical formula. The following diagram shows the relationship between these different ways of expressing the same information.

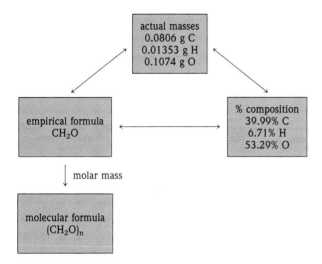

Questions and Problems*

All even-numbered exercises have answers in the back of this book and solutions in the Solutions Guide.

9.1 Counting by Weighing

QUESTIONS

1. Merchants usually sell small nuts, washers, and bolts by weight (like jelly beans!) rather than by individually counting the items. Suppose a particular type of washer weighs 0.110 g on the average. What would 100 such washers weigh? How many washers would there be in 100. g of washers?

2. A particular small laboratory cork weighs 1.63 g, whereas a rubber lab stopper of the same size weighs 4.31 g. How many corks would there be in 500. g of such corks? How many rubber stoppers would there be in 500. g of similar stoppers? How many grams of rubber stoppers would be needed to contain the same number of stoppers as there are corks in 1 kg of corks?

*The element symbols and formulas are given in some problems but not in others to help you learn this necessary "vocabulary."

9.2 Atomic Masses: Counting Atoms by Weighing

QUESTIONS

3. The *average* atomic mass of an element represents the weighted average mass of all the _____ of the element.

4. When using the average atomic mass of an element for calculations, we pretend that all atoms of the element are _____.

PROBLEMS

5. Using the average atomic masses given in Table 9.1, calculate the masses of each of the following samples.
   a. 125 atoms of sodium
   b. 1.0×10^6 atoms of hydrogen
   c. 10 atoms of carbon
   d. 1.9×10^{19} atoms of aluminum
   e. 10,000 atoms of nitrogen

6. Using the average atomic masses given in Table 9.1, calculate the number of atoms present in each of the following samples.
   a. 10.81 amu of boron
   b. 320.6 amu of sulfur
   c. 19,691 amu of gold
   d. 19,695 amu of xenon
   e. 3588.3 amu of aluminum

7. If an average oxygen atom weighs 16.0 amu, how many oxygen atoms are contained in 6.40×10^6 amu of oxygen? What will 3.41×10^{19} oxygen atoms weigh?

8. If an average gold atom weighs 196.97 amu, what does a sample containing 1.00×10^4 gold atoms weigh? How many gold atoms are contained in a sample of gold that has a mass of 2.955×10^5 amu?

9.3 The Mole

QUESTIONS

9. 6.022×10^{23} atoms of carbon weigh _____ grams.

10. The atomic mass of an element in grams contains _____ atoms.

PROBLEMS

11. What mass of oxygen contains the same number of atoms as 14.01 g of nitrogen?

12. What mass of carbon contains the same number of atoms as 1.008 g of hydrogen?

13. What mass of hydrogen contains the same number of atoms as 7.00 g of nitrogen?

14. What mass of nitrogen contains the same number of atoms as 48 g of oxygen?

15. Calculate the average mass in grams of 1 atom of carbon.

16. Calculate the average mass in grams of 1 atom of oxygen.

17. Which weighs more, 1 mol of hydrogen atoms or 0.5 mol of oxygen atoms?

18. Which weighs more, 0.50 mol of oxygen atoms or 4 mol of hydrogen atoms?

19. Using the average masses given inside the front cover, calculate how many *moles* of each element the following *masses* represent.
   a. 19.5 g of platinum, Pt
   b. 11.78 g of cobalt, Co
   c. 2.395 g of titanium, Ti
   d. 1.00 k of barium, Ba
   e. 1.00 lb of magnesium, Mg
   f. 86.2 g of sodium, Na
   g. 91.4 g of strontium, Sr

20. Using the average atomic masses given inside the front cover, calculate the number of *moles* of each element in samples with the following masses.
   a. 26.2 g of gold
   b. 41.5 g of calcium
   c. 335 mg of barium
   d. 1.42×10^{-3} g of palladium
   e. 3.05×10^{-5} μg of nickel
   f. 1.00 lb of iron
   g. 12.01 g of carbon

21. Using the average atomic masses given inside the front cover, calculate the *mass in grams* of each of the following samples.
   a. 1.25 mol of nickel, Ni
   b. 0.00255 mol of zirconium, Zr

22. Using the average atomic masses given inside the front cover, calculate the *mass in grams* of each of the following samples.
   a. 2.00 mol of iron
   b. 0.521 mol of nickel

c. 1.89×10^{-4} mol of lead, Pb
d. 55.56 mol of beryllium, Be
e. 2.6×10^{7} mol of aluminum, Al
f. 0.45 mol of barium, Ba
g. 0.00115 mol of chromium, Cr

c. 1.23×10^{-3} mol of platinum
d. 72.5 mol of lead
e. 0.00102 mol of magnesium
f. 4.87×10^{3} mol of aluminum
g. 211.5 mol of lithium
h. 1.72×10^{-6} mol of sodium

23. Using the average atomic masses given inside the front cover, calculate the number of *atoms* present in each of the following samples.
    a. 1.50 g of silver, Ag
    b. 0.0015 mol of copper, Cu
    c. 0.0015 g of copper, Cu
    d. 2.00 kg of magnesium, Mg
    e. 2.34 oz of calcium, Ca
    f. 2.34 g of calcium, Ca
    g. 2.34 mol of calcium, Ca

24. Using the average atomic masses given inside the front cover, calculate the indicated quantity.
    a. the number of atoms of Fe in 1.00 g Fe
    b. the number of moles of Fe in 1.00 g of Fe
    c. the number of moles of Fe that 1.62×10^{24} atoms represents
    d. the mass of a sample of Fe containing 1.62×10^{24} atoms
    e. the number of atoms of Fe in 0.343 mol of Fe
    f. the mass of 0.343 mol of Fe
    g. the mass of 6.022×10^{23} Fe atoms
    h. the number of moles of Fe represented by 6.022×10^{23} atoms of Fe

9.4 Molar Mass

QUESTIONS

25. The _____ of a substance is the mass (in grams) of 1 mol of the substance.

26. The molar mass of a substance can be obtained by _____ the atomic weights of the component atoms.

PROBLEMS

27. Calculate the molar mass for each of the following substances.
    a. methane, CH_4
    b. sodium nitrate, $NaNO_3$
    c. carbon monoxide, CO
    d. carbon dioxide, CO_2
    e. ammonium fluoride, NH_4F
    f. potassium chlorate, $KClO_3$

28. Calculate the molar mass of each of the following substances.
    a. butane, C_4H_{10}
    b. sodium perchlorate, $NaClO_4$
    c. aluminum iodide, AlI_3
    d. dimethyl ether, C_2H_6O
    e. benzene, C_6H_6
    f. nitrogen triiodide, NI_3

29. Calculate the molar mass for each of the following substances.
    a. silver(I) oxide
    b. iron(II) sulfate
    c. ammonium nitrate
    d. ethanol, C_2H_5OH
    e. chlorine dioxide
    f. strontium chloride

30. Calculate the molar mass for each of the following substances.
    a. calcium phosphate, $Ca_3(PO_4)_2$
    b. lithium chlorate, $LiClO_3$
    c. sulfur hexafluoride, SF_6
    d. tetrammine copper(II) sulfate, $Cu(NH_3)_4SO_4$
    e. sodium hydrogen carbonate, $NaHCO_3$
    f. lithium aluminum hydride, $LiAlH_4$

31. Calculate the number of *moles* of the indicated substance present in each of the following samples.
    a. 10.0 g of nitrogen dioxide, NO_2
    b. 22.5 g of potassium chlorate, $KClO_3$
    c. 100.0 g of potassium nitrate, KNO_3
    d. 1.0 kg of sodium chloride, NaCl
    e. 5.0 mg of sodium phosphate, Na_3PO_4

32. Calculate the number of *moles* of the indicated substance present in each of the following samples.
    a. 4.15 g of chlorine dioxide
    b. 56.1 g of potassium fluoride
    c. 2.91×10^{-4} g of ammonium nitrate
    d. 2.21 kg of iron(II) oxide
    e. 4.12 mg of cobalt(II) chloride

33. Calculate the number of *moles* of the indicated substance present in each of the following samples.
    a. 18.0 g of dextrose, $C_6H_{12}O_6$
    b. 21.94 g of nitrous oxide, N_2O
    c. 21.94 g of nitric oxide, NO
    d. 1.24 oz of gold acetate, $Au(C_2H_3O_2)_3$
    e. 44.2 g of ammonium dichromate, $(NH_4)_2Cr_2O_7$

34. Calculate the number of *moles* of the indicated substance present in each of the following samples.
    a. 0.000314 g of aluminum chloride
    b. 1.00 lb of sodium hydroxide
    c. 7.4 μg of arsenic triiodide
    d. 44.2 kg of sodium hydrogen carbonate
    e. 6.21 g of potassium hydrogen phosphate

35. Calculate the mass in grams of each of the following samples.
    a. 0.10 mol of potassium bromide
    b. 0.35 mol of magnesium chloride
    c. 1.49 mol of silicon tetrafluoride
    d. 2.75 millimol (1 millimol = 0.001 mol) of zinc sulfide
    e. 10.2 millimol of sodium nitrate

36. Calculate the mass in grams of each of the following samples.
    a. 1.50 mol aluminum iodide
    b. 1.91×10^{-3} mol of benzene, C_6H_6
    c. 4.00 mol of glucose, $C_6H_{12}O_6$
    d. 4.56×10^5 mol of ethanol, C_2H_5OH
    e. 2.27 mol of calcium nitrate

37. Calculate the mass in grams of each of the following samples.
    a. 0.25 millimol of sodium dihydrogen phosphate, NaH_2PO_4
    b. 1.40 mol of lithium carbonate, Li_2CO_3
    c. 1.02×10^{-5} mol of calcium nitrate, $Ca(NO_3)_2$
    d. 0.0045 mol of gold(III) chloride, $AuCl_3$
    e. 5.029 mol of potassium chromate, $KCrO_4$

38. Calculate the mass in grams of each of the following samples.
    a. 1.27 millimol of carbon dioxide
    b. 4.12×10^3 mol of nitrogen trichloride
    c. 0.00451 mol of ammonium nitrate
    d. 18.0 mol of water
    e. 62.7 mol of copper(II) sulfate

39. Calculate the number of *molecules* present in each of the following samples.
    a. 14.2 g of P_4O_{10}
    b. 0.225 mol of XeF_4O_2
    c. 5.75 g of $TeBr_2$
    d. 10.0 g of $Pb(C_2H_5)_4$
    e. 4.26 mol of $Sr(NO_3)_2$

40. Calculate the *number of molecules* present in each of the following samples.
    a. 4.29 mol of nitrogen dioxide
    b. 4.29 g of nitrogen dioxide
    c. 1.95×10^{-10} mol of hydrogen fluoride
    d. 1.95×10^{-10} g of hydrogen fluoride
    e. 4.61 g of ammonia

41. Calculate the number of moles of carbon atoms present in each of the following samples.
    a. 1.271 g of ethanol, C_2H_6O
    b. 3.982 g of 1,4-dichlorobenzene, $C_6H_4Cl_2$
    c. 0.4438 g of carbon suboxide, C_3O_2
    d. 2.910 g of methylene chloride, CH_2Cl_2

42. Calculate the number of moles of sulfur atoms present in each of the following samples.
    a. 2.01 g of sodium sulfate
    b. 2.01 g of sodium sulfite
    c. 2.01 g of sodium sulfide
    d. 2.01 g of sodium thiosulfate, $Na_2S_2O_3$

9.5 Percent Composition of Compounds

QUESTIONS

43. The mass fraction of an element present in a compound can be obtained by comparing the mass of the particular element present in 1 mol of the compound to the _____ mass of the compound.

44. The mass percentage of a given element in a compound must always be (greater/less) than 100%.

PROBLEMS

45. Calculate the percent by mass of each element in the following compounds.
    a. benzene, C_6H_6

46. Calculate the percent by mass of each element in the following compounds.
    a. sodium sulfate

b.   magnesium nitrate, $Mg(NO_3)_2$
c.   formaldehyde, CH_2O
d.   calcium hypochlorite, $Ca(OCl)_2$
e.   calcium carbonate, $CaCO_3$
f.   sodium phosphate, Na_3PO_4
g.   calcium nitrate, $Ca(NO_3)_2$
h.   aluminum chloride, $AlCl_3$

47.   Calculate the percent by mass of the element mentioned *first* in the formula for each of the following compounds.
   a.   methane, CH_4
   b.   sodium nitrate
   c.   carbon monoxide, CO
   d.   nitrogen dioxide
   e.   1-octanol, $C_8H_{18}O$
   f.   calcium phosphate
   g.   3-phenylphenol, $C_{12}H_{10}O$
   h.   aluminum acetate, $Al(C_2H_3O_2)_3$

49.   Calculate the percent by mass of the element mentioned *first* in the formulas for each of the following compounds.
   a.   adipic acid, $C_6H_{10}O_4$
   b.   ammonium nitrate, NH_4NO_3
   c.   caffeine, $C_8H_{10}N_4O_2$
   d.   chlorine dioxide, ClO_2
   e.   cyclohexanol, $C_6H_{11}OH$
   f.   dextrose, $C_6H_{12}O_6$
   g.   eicosane, $C_{20}H_{42}$
   h.   ethanol, C_2H_5OH

51.   For each of the following samples of ionic substances, calculate the number of moles and masses of the *negative ions* present in the sample.
   a.   10.0 g of calcium carbonate, $CaCO_3$
   b.   26.2 g of sodium phosphate, Na_3PO_4
   c.   0.0250 g of calcium nitrate, $Ca(NO_3)_2$
   d.   0.200 mol of aluminum chloride, $AlCl_3$

b.   sodium sulfite
c.   sodium sulfide
d.   sodium thiosulfate, $Na_2S_2O_3$
e.   potassium phosphate
f.   potassium hydrogen phosphate
g.   potassium dihydrogen phosphate
h.   potassium phosphide

48.   Calculate the percent by mass of the element mentioned *first* in the formulas for each of the following compounds.
   a.   sodium fluoride
   b.   titanium(IV) chloride
   c.   iron(III) sulfate
   d.   cobalt(III) nitrate
   e.   hydrogen peroxide, H_2O_2
   f.   aluminum phosphide
   g.   barium carbonate
   h.   lithium perchlorate

50.   Calculate the percent by mass of the element mentioned *first* in the formulas for each of the following compounds.
   a.   iron(III) chloride
   b.   oxygen difluoride, OF_2
   c.   benzene, C_6H_6
   d.   ammonium perchlorate, NH_4ClO_4
   e.   silver oxide
   f.   cobalt(II) chloride
   g.   dinitrogen tetroxide
   h.   manganese(II) chloride

52.   For each of the following ionic substances, calculate the *percentage* of the overall molar mass of the compound that is represented by the *negative ions* the compound contains.
   a.   calcium phosphate
   b.   cadmium sulfate
   c.   iron(III) sulfate
   d.   manganese(II) chloride

9.6 Formulas of Compounds

QUESTIONS

53.   What experimental evidence about a new compound must be known before its formula can be determined?

55.   Give the empirical formula that corresponds to each of the following molecular formulas.
   a.   sodium peroxide, Na_2O_2
   b.   terephthalic acid, $C_8H_6O_4$
   c.   phenobarbital, $C_{12}H_{12}N_2O_3$
   d.   1,4-dichloro-2-butene, $C_4H_6Cl_2$

54.   What does the *empirical* formula of a compound represent? How does the *molecular* formula differ from the empirical formula?

56.   Which of the following pairs of compounds have the same *empirical* formula?
   a.   acetylene, C_2H_2, and benzene, C_6H_6
   b.   ethane, C_2H_6, and butane, C_4H_{10}
   c.   nitrogen dioxide, NO_2, and dinitrogen tetroxide, N_2O_4
   d.   diphenyl ether, $C_{12}H_{10}O$, and phenol, C_6H_5OH

9.7 Calculation of Empirical Formulas

PROBLEMS

57. A 0.3037-g sample of a new compound has been analyzed and found to contain the following masses of elements: carbon, 0.1656 g; hydrogen, 0.02779 g; oxygen, 0.1103 g. Calculate the empirical formula of the compound.

58. If 1.000 g of barium metal is heated in a stream of pure oxygen gas, 1.117 g of an oxide is produced. Calculate the empirical formula of the oxide.

59. A 0.5998-g sample of a new compound has been analyzed and found to contain the following masses of elements: carbon, 0.2322 g; hydrogen, 0.05848 g; oxygen, 0.3091 g. Calculate the empirical formula of the compound.

60. A compound has the following percentages by mass: barium, 58.84%; sulfur, 13.74%; oxygen, 27.43%. Determine the empirical formula of the compound.

61. When 4.008 g of calcium metal is heated in air, the sample gains 1.600 g of oxygen in forming calcium oxide. Calculate the empirical formula of calcium oxide.

62. Analysis of a certain compound yielded the following percentages of the elements by mass: nitrogen, 29.16%; hydrogen, 8.392%; carbon, 12.50%; oxygen, 49.95%. Determine the empirical formula of the compound.

63. When 3.269 g of zinc is heated in pure oxygen, the sample gains 0.800 g of oxygen in forming the oxide. Calculate the empirical formula of zinc oxide.

64. If cobalt metal is mixed with excess sulfur and heated strongly, a sulfide is produced that contains 55.06% cobalt by mass. Calculate the empirical formula of the sulfide.

65. When 1.916 g of titanium is strongly heated in a stream of pure oxygen, an oxide sample weighing 3.196 g results. Calculate the empirical formula of the oxide.

66. If 10.00 g of copper metal is heated strongly in the air, the sample gains 2.52 g of oxygen in forming an oxide. Determine the empirical formula of this oxide.

67. A compound has been analyzed and found to have the following percentage composition: oxygen, 25.81%; sodium, 74.19%. Calculate the empirical formula of the compound.

68. A compound has the following percentages by mass: aluminum, 32.13%; fluorine, 67.87%. Calculate the empirical formula of the compound.

69. A compound was analyzed and found to have the following percentage composition: aluminum, 35.93%; sulfur, 64.06%. Calculate the empirical formula of the compound.

70. When lithium metal is heated strongly in an atmosphere of pure nitrogen, the product contains 59.78% Li and 40.22% N on a mass basis. Determine the empirical formula of the compound.

71. A compound was analyzed and found to have the following percentage composition: hydrogen, 2.056%; sulfur, 32.69%; oxygen 65.26%. Calculate the empirical formula of the compound.

72. A compound was analyzed and found to have the following percentage composition: aluminum, 15.77%; sulfur, 28.11%; oxygen 56.12%. Calculate the empirical formula of the compound.

73. A compound consists of 40.00% C, 6.713% H, and 53.28% O on a mass basis. Determine the empirical formula of the compound.

74. A compound consists of 21.60% Na, 33.31% Cl, and 45.09% O on a mass basis. Determine the empirical formula of the compound.

9.8 Calculation of Molecular Formulas

QUESTIONS

75. How does the *molecular* formula of a compound differ from the empirical formula? Can a compound's empirical and molecular formulas be the same? Explain.

76. What information do we need to determine the molecular formula of a compound if we know only the empirical formula?

PROBLEMS

77. A compound with the empirical formula CH_2O was found to have a molar mass between 89 and 91 g. What is the molecular formula of the compound?

78. A compound with empirical formula CH was found by experiment to have a molar mass of approximately 78. What is the molecular formula of the compound?

79. A compound with the empirical formula CH_2 was found to have a molar mass of approximately 84 g. What is the molecular formula of the compound?

80. A compound with the empirical formula NO_2 was found to have a molar mass of 92 g. What is the molecular formula of the compound?

81. A compound consists of 40.00% C, 6.713% H, and 53.28% O on a mass basis and has a molar mass of approximately 180 g. Determine the molecular formula of the compound.

82. A compound consists of 65.45% C, 5.492% H, and 29.06% O on a mass basis and has a molar mass of approximately 110. Determine the molecular formula of the compound.

Additional Problems

83. Use the periodic table inside the front cover of this book to determine the atomic mass (per mole) or molar mass of each of the substances in column 1, and find that mass in column 2.

Column 1	Column 2
[1] molybdenum	[a] 33.99 g
[2] lanthanum	[b] 79.9 g
[3] carbon tetrabromide	[c] 95.94 g
[4] mercury(II) oxide	[d] 125.84 g
[5] titanium(IV) oxide	[e] 138.9 g
[6] manganese(II) chloride	[f] 143.1 g
[7] phosphine, PH_3	[g] 156.7 g
[8] tin(II) fluoride	[h] 216.6 g
[9] lead(II) sulfide	[i] 239.3 g
[10] copper(I) oxide	[j] 331.6 g

84. Complete the following table.

Mass of sample	Moles of sample	Atoms in sample
5.00 g Al		
	0.00250 mol Fe	
		2.6×10^{24} atoms Cu
0.00250 g Mg		
	2.7×10^{-3} mol Na	
		1.00×10^4 atoms U

85. Complete the following table.

Mass of sample	Moles of sample	Molecules in sample
2.98 g N_2O_4		
	0.050 mol CO_2	
		4.8×10^{25} molecules $SiCl_4$
0.3049 g $C_6H_5O_4$		
	2.1×10^{-2} mol C_2F_4	
		1 million molecules C_2H_5OH

86. Consider a hypothetical compound composed of elements X, Y, and Z with the empirical formula X_2YZ_3. Given that the atomic masses of X, Y, and Z are 41.2, 57.7, and 63.9, respectively, calculate the percentage composition by mass of the compound. If the molecular formula of the compound is found by molar mass determination to be actually $X_4Y_2Z_6$, what is the percentage of each element present? Explain your results.

87. A binary compound of magnesium and nitrogen is analyzed, and 1.2791 g of the compound is found to contain 0.9240 g of magnesium. When a second sample of this compound is treated with water and heated, the nitrogen is driven off as ammonia, leaving a compound that contains 60.31% magnesium and 39.69% oxygen by mass. Calculate the empirical formulas of the two magnesium compounds.

88. When a 2.118-g sample of copper is heated in an atmosphere in which the amount of oxygen present is restricted, the sample gains 0.2666 g of oxygen in forming a reddish brown oxide. However, when 2.118 g of copper is heated in a stream of pure oxygen, the sample gains 0.5332 g of oxygen. Calculate the empirical formulas of the two oxides of copper.

89. Nitrogen and oxygen combine under different conditions to give a series of binary compounds: N_2O, NO, NO_2, N_2O_4. Calculate the percent by mass of nitrogen in each.

90. Calculate the number of atoms of each element present in each of the following samples.
    a. 2.37 g of lead(II) nitrate
    b. 22.4 g of elemental oxygen

c. 101.9 mg of elemental sulfur, S_8
d. 43.7 μg of uranium(VI) fluoride

91. Calculate the mass in grams of each of the following samples.
    a. 10,000,000,000 nitrogen molecules
    b. 2.49×10^{20} carbon dioxide molecules
    c. 7.0983 mol of sodium chloride
    d. 9.012×10^{-6} mol of 1,2-dichloroethane, $C_2H_4Cl_2$

92. Calculate the mass of carbon in grams, the percent carbon by mass, and the number of individual carbon atoms present in each of the following samples.
    a. 7.819 g of carbon suboxide, C_3O_2
    b. 1.53×10^{21} molecules of carbon monoxide
    c. 0.200 mol of phenol, C_6H_6O

93. Find the item in column 2 that best explains or completes the statement or question in column 1.

Column 1

[1] 1 amu
[2] 1008 amu
[3] mass of the "average" atom of an element
[4] number of carbon atoms in 12.01 g of carbon
[5] 6.022×10^{23} molecules
[6] total mass of all atoms in 1 mol of a compound
[7] smallest whole-number ratio of atoms present in a molecule
[8] formula showing actual number of atoms present in a molecule
[9] product formed when any carbon-containing compound is burned in O_2
[10] have the same empirical formulas, but different molecular formulas

Column 2

[a] 6.02×10^{23}
[b] atomic mass
[c] mass of 1000 hydrogen atoms
[d] benzene, C_6H_6, and acetylene, C_2H_2
[e] carbon dioxide
[f] empirical formula
[g] 1.66×10^{-24} g
[h] molecular formula
[i] molar mass
[j] 1 mol

94. Calculate the number of grams of lead that contain the same number of atoms as 1.00 g of neon.

95. Calculate the number of grams of uranium that contain the same number of atoms as 1.00 g of sodium.

96. Calculate the number of grams of mercury that contain the same number of atoms as 5.00 g of tellurium.

97. Calculate the number of grams of lithium that contain the same number of atoms as 1.00 kg of zirconium.

98. Given that the molar mass of carbon tetrachloride, CCl_4, is 153.8 g, calculate the mass in grams of 1 molecule of CCl_4.

99. Calculate the mass in grams of oxygen present in 1.000 g of each of the following compounds.
    a. formaldehyde, CH_2O c. calcium carbonate
    b. barium peroxide   d. serine, $C_3H_7O_3N$

100. Calculate the mass in grams of nitrogen present in 5.000 g of each of the following compounds.

a. glycine, $C_2H_5O_2N$ c. calcium nitrate
b. magnesium nitride, Mg_3N_2 d. dinitrogen tetroxide

101. A strikingly beautiful copper compound with the common name "blue vitriol" has the following elemental composition: 25.45% Cu, 12.84% S, 4.036% H, 57.67% O. Determine the empirical formula of the compound.

102. A magnesium salt has the following elemental composition: 16.39% Mg, 18.89% N, 64.72% O. Determine the empirical formula of the salt.

103. The mass 1.66×10^{-24} g is equivalent to 1 _____.

104. Even though most elements consist of several isotopes, we use the _____ mass of the isotopes in most chemical calculations.

105. Using the average atomic masses given in Table 9.1, calculate the number of atoms present in each of the following samples.

a. 160,000 amu of oxygen
b. 8139.81 amu of nitrogen
c. 13,490 amu of aluminum
d. 5040 amu of hydrogen
e. 367,495.15 amu of sodium

106. If an average sodium atom weighs 22.99 amu, how many sodium atoms are contained in 1.98×10^{13} amu of sodium? What will 3.01×10^{23} sodium atoms weigh?

107. Using the average atomic masses given inside the front cover, calculate how many *moles* of each element the following *masses* represent.
a. 1.5 mg of chromium
b. 2.0×10^{-3} g of strontium
c. 4.84×10^{4} g of boron
d. 3.6×10^{-6} μg of californium
e. 1.0 ton (2000 lb) of iron
f. 20.4 g of barium
g. 62.8 g of cobalt

108. Using the average atomic masses given inside the front cover, calculate the *mass in grams* of each of the following samples.
a. 5.0 mol of potassium
b. 0.000305 mol of mercury
c. 2.31×10^{-5} mol of manganese
d. 10.5 mol of phosphorus
e. 4.9×10^{4} mol of iron
f. 125 mol of lithium
g. 0.01205 mol of fluorine

109. Using the average atomic masses given inside the front cover, calculate the number of *atoms* present in each of the following samples.
a. 2.89 g of gold
b. 0.000259 mol of platinum
c. 0.000259 g of platinum
d. 2.0 lb of magnesium
e. 1.90 mL of liquid mercury (density = 13.6 g/mL)
f. 4.30 mol of tungsten
g. 4.30 g of tungsten

110. Calculate the molar mass for each of the following substances.
a. propane, C_3H_8
b. iron(III) sulfate
c. dinitrogen pentoxide
d. sucrose, $C_{12}H_{22}O_{11}$
e. ammonium carbonate
f. barium chloride

111. Calculate the molar mass for each of the following substances.
a. adipic acid, $C_6H_{10}O_4$

b. caffeine, $C_8H_{10}N_4O_2$
c. eicosane, $C_{20}H_{42}$
d. cyclohexanol, $C_6H_{11}OH$
e. vinyl acetate, $C_4H_6O_2$
f. dextrose, $C_6H_{12}O_6$

112. Calculate the number of *moles* of the indicated substance present in each of the following samples.
a. 21.2 g of ammonium sulfide
b. 44.3 g of calcium nitrate
c. 4.35 g of dichlorine monoxide
d. 1.0 lb of ferric chloride
e. 1.0 kg of ferric chloride

113. Calculate the number of *moles* of the indicated substance present in each of the following samples.
a. 0.00056 g of potassium cyanide
b. 1.90 mg of mescaline, $C_{11}H_{17}O_3N$
c. 23.09 g of calcium phosphate
d. 2.6×10^{-3} mg of rubidium oxide
e. 6.0 μg of ammonium nitrate

114. Calculate the mass in grams of each of the following samples.
a. 2.6×10^{-2} mol of copper(II) sulfate, $CuSO_4$
b. 3.05×10^{3} mol of tetrafluoroethylene, C_2F_4
c. 7.83 millimol (1 millimol = 0.001 mol) of 1,4-pentadiene, C_5H_8
d. 6.30 mol of bismuth trichloride, $BiCl_3$
e. 12.2 mol of sucrose, $C_{12}H_{22}O_{11}$

115. Calculate the mass in grams of each of the following samples.
a. 3.09 mol of ammonium carbonate
b. 4.01×10^{-6} mol of sodium hydrogen carbonate
c. 88.02 mol of carbon dioxide
d. 1.29 millimol of silver(I) nitrate
e. 0.0024 mol of chromium(III) chloride

116. Calculate the number of *molecules* present in each of the following samples.
a. 3.45 g of $C_6H_{12}O_6$
b. 3.45 mol of $C_6H_{12}O_6$
c. 25.0 g of ICl_5
d. 1.00 g of B_2H_6
e. 1.05 millimol of $Al(NO_3)_3$

117. Calculate the number of moles of oxygen atoms present in each of the following samples.
a. 1.002 g of phenol, C_6H_6O
b. 4.901 g of sodium peroxide
c. 124.5 mg of *p*-aminobenzoic acid, $C_7H_7O_2N$
d. 9.821 g of aluminum nitrate

118. Calculate the percent by mass of each element in the following compounds.

a. calcium phosphate
b. cadmium sulfate
c. iron(III) sulfate
d. manganese(II) chloride
e. ammonium carbonate
f. sodium hydrogen carbonate
g. carbon dioxide
h. silver(I) nitrate

119. Calculate the percent by mass of the element mentioned *first* in the formulas for each of the following compounds.
    a. propane, C_3H_8
    b. potassium chlorate, $KClO_3$
    c. dinitrogen pentoxide, N_2O_5
    d. sucrose, $C_{12}H_{22}O_{11}$
    e. ammonium fluoride, NH_4F
    f. sodium perchlorate, $NaClO_4$
    g. ammonium carbonate, $(NH_4)_2CO_3$
    h. barium chloride, $BaCl_2$

120. Calculate the percent by mass of the element mentioned *first* in the formula for each of the following compounds.
    a. iron(II) sulfate
    b. silver(I) oxide
    c. strontium chloride
    d. vinyl acetate, $C_4H_6O_2$
    e. methanol, CH_3OH

f. aluminum oxide
g. potassium chlorite
h. potassium chloride

121. A 1.2569-g sample of a new compound has been analyzed and found to contain the following masses of elements: carbon, 0.7238 g; hydrogen, 0.07088 g; nitrogen, 0.1407 g; oxygen, 0.3214 g. Calculate the empirical formula of the compound.

122. A 0.7221-g sample of a new compound has been analyzed and found to contain the following masses of elements: carbon, 0.2990 g; hydrogen, 0.05849 g; nitrogen, 0.2318 g; oxygen, 0.1328 g. Calculate the empirical formula of the compound.

123. When 2.004 g of calcium is heated in pure nitrogen gas, the sample gains 0.4670 g of nitrogen. Calculate the empirical formula of the calcium nitride formed.

124. When 4.01 g of mercury is strongly heated in air, the resulting oxide weighs 4.33 g. Calculate the empirical formula of the oxide.

125. When 1.00 g of metallic chromium is heated with elemental chlorine gas, 3.045 g of a chromium chloride salt results. Calculate the empirical formula of the compound.

126. When barium metal is heated in chlorine gas, a binary compound forms that consists of 65.95% Ba and 34.05% Cl by mass. Calculate the empirical formula of the compound.

Solutions to Self-Check Exercises

SELF-CHECK EXERCISE 9.1

The average mass of nitrogen is 14.01 amu. The appropriate equivalence statement is 1 N atom = 14.01 amu, which yields the conversion factor we need:

$$23 \text{ N atoms} \times \frac{14.01 \text{ amu}}{\text{N atom}} = 322.2 \text{ amu}$$

↑
(exact)

SELF-CHECK EXERCISE 9.2

The average mass of oxygen is 16.00 amu, which gives the equivalence statement 1 O atom = 16.00 amu. The number of oxygen atoms present is

$$288 \text{ amu} \times \frac{1 \text{ O atom}}{16.00 \text{ amu}} = 18.0 \text{ O atoms}$$

SELF-CHECK EXERCISE 9.3

Note that the sample of 5.00×10^{20} atoms of cobalt is less than 1 mol (6.022×10^{23} atoms) of cobalt. What fraction of a mole it represents can be determined as follows:

$$5.00 \times 10^{20} \text{ atoms Co} \times \frac{1 \text{ mol Co}}{6.022 \times 10^{23} \text{ atoms Co}} = 8.30 \times 10^{-4} \text{ mol Co}$$

Because the mass of 1 mol of cobalt atoms is 58.93 g, the mass of 5.00×10^{20} atoms can be determined as follows:

$$8.30 \times 10^{-4} \text{ mol Co} \times \frac{58.93 \text{ g Co}}{1 \text{ mol Co}} = 4.89 \times 10^{-2} \text{ g Co}$$

SELF-CHECK EXERCISE 9.4

Each molecule of C_2H_3Cl contains two carbon atoms, three hydrogen atoms, and one chlorine atom, so 1 mol of C_2H_3Cl molecules contains 2 mol of C atoms, 3 mol of H atoms, and 1 mol of Cl atoms.

$$
\begin{aligned}
\text{Mass of 2 mol of C atoms: } & 2 \times 12.01 = 24.02 \text{ g} \\
\text{Mass of 3 mol of H atoms: } & 3 \times 1.008 = 3.024 \text{ g} \\
\text{Mass of 1 mol of Cl atoms: } & 1 \times 35.45 = \underline{35.45 \text{ g}} \\
& 62.494 \text{ g}
\end{aligned}
$$

The molar mass of C_2H_3Cl is 62.49 g (rounding to the correct number of significant figures).

SELF-CHECK EXERCISE 9.5

The formula for sodium sulfate is Na_2SO_4. One mole of Na_2SO_4 contains 2 mol of sodium ions and 1 mol of sulfate ions.

1 mol of $Na_2SO_4 \rightarrow$ 1 mol of

$$
\begin{aligned}
\text{Mass of 2 mol of Na}^+ &= 2 \times 22.99 &&= 45.98 \text{ g} \\
\text{Mass of 1 mol of SO}_4^{2-} &= 32.06 + 4(16.00) &&= \underline{96.06 \text{ g}} \\
\text{Mass of 1 mol of Na}_2\text{SO}_4 & &&= 142.04 \text{ g}
\end{aligned}
$$

The molar mass for sodium sulfate is 142.04 g.

A sample of sodium sulfate with a mass of 300.0 g represents more than 1 mol. (Compare 300.0 g to the molar mass of Na_2SO_4.) We calculate the number of moles of Na_2SO_4 present in 300.0 g as follows:

$$300.0 \text{ g Na}_2\text{SO}_4 \times \frac{1 \text{ mol Na}_2\text{SO}_4}{142.04 \text{ g Na}_2\text{SO}_4} = 2.112 \text{ mol Na}_2\text{SO}_4$$

SELF-CHECK EXERCISE 9.6

First we must compute the mass of 1 mol of C_2F_4 molecules (the molar mass). Because 1 mol of C_2F_4 contains 2 mol of C atoms and 4 mol of F atoms, we have:

$$2 \text{ mol C} \times \frac{12.01 \text{ g}}{\text{mol}} = 24.02 \text{ g C}$$

$$4 \text{ mol F} \times \frac{19.00 \text{ g}}{\text{mol}} = \underline{76.00 \text{ g F}}$$

Mass of 1 mol of C_2F_4:   100.02 g = molar mass

Using the equivalence statement 100.02 g C_2F_4 = 1 mol C_2F_4, we calculate the moles of C_2F_4 units in 135 g of Teflon.

$$135 \text{ g } C_2F_4 \text{ units} \times \frac{1 \text{ mol } C_2F_4}{100.02 \text{ g } C_2F_4} = 1.35 \text{ mol } C_2F_4 \text{ units}$$

Next, using the equivalence statement 1 mol = 6.022×10^{23} units, we calculate the number in C_2F_4 units in 135 g of Teflon.

$$1.35 \text{ mol } C_2F_4 \times \frac{6.022 \times 10^{23} \text{ units}}{1 \text{ mol}} = 8.13 \times 10^{23} \text{ } C_2F_4 \text{ units}$$

SELF-CHECK EXERCISE 9.7

The molar mass of penicillin F is computed as follows:

$$\text{C: } 14 \text{ mol} \times 12.01 \text{ } \frac{g}{\text{mol}} = 168.1 \text{ g}$$

$$\text{H: } 20 \text{ mol} \times 1.008 \text{ } \frac{g}{\text{mol}} = 20.16 \text{ g}$$

$$\text{N: } 2 \text{ mol} \times 14.01 \text{ } \frac{g}{\text{mol}} = 28.02 \text{ g}$$

$$\text{S: } 1 \text{ mol} \times 32.06 \text{ } \frac{g}{\text{mol}} = 32.06 \text{ g}$$

$$\text{O: } 4 \text{ mol} \times 16.00 \text{ } \frac{g}{\text{mol}} = \underline{64.00 \text{ g}}$$

Mass of 1 mol of $C_{14}H_{20}N_2SO_4$ = 312.38 g = 312.4 g

$$\text{Mass percent of C} = \frac{168.1 \text{ g C}}{312.4 \text{ g } C_{14}H_{20}N_2SO_4} \times 100 = 53.81\%$$

$$\text{Mass percent of H} = \frac{20.16 \text{ g H}}{312.4 \text{ g } C_{14}H_{20}N_2SO_4} \times 100 = 6.453\%$$

$$\text{Mass percent of N} = \frac{28.02 \text{ g N}}{312.4 \text{ g } C_{14}H_{20}N_2SO_4} \times 100 = 8.969\%$$

$$\text{Mass percent of S} = \frac{32.06 \text{ g S}}{312.4 \text{ g } C_{14}H_{20}N_2SO_4} \times 100 = 10.26\%$$

$$\text{Mass percent of O} = \frac{64.00 \text{ g O}}{312.4 \text{ g C}_{14}\text{H}_{20}\text{N}_2\text{SO}_4} \times 100 = 20.49\%$$

CHECK: The percentages add up to 99.98%.

SELF-CHECK EXERCISE 9.8

STEP I

0.6884 g lead and 0.2356 g chlorine

STEP 2

$$0.6884 \text{ g Pb} \times \frac{1 \text{ mol Pb}}{207.2 \text{ g Pb}} = 0.003322 \text{ mol Pb}$$

$$0.2356 \text{ g Cl} \times \frac{1 \text{ mol Cl}}{35.45 \text{ g Cl}} = 0.006646 \text{ mol Cl}$$

STEP 3

$$\frac{0.003322 \text{ mol Pb}}{0.003322} = 1.000 \text{ mol Pb}$$

$$\frac{0.006646 \text{ mol Cl}}{0.003322} = 2.000 \text{ mol Cl}$$

These numbers are integers, so step 4 is unnecessary. The empirical formula is $PbCl_2$.

SELF-CHECK EXERCISE 9.9

STEP I

0.8007 g C, 0.9333 g N, 0.2016 g H, and 2.133 g O

STEP 2

$$0.8007 \text{ g C} \times \frac{1 \text{ mol C}}{12.01 \text{ g C}} = 0.06667 \text{ mol C}$$

$$0.9333 \text{ g N} \times \frac{1 \text{ mol N}}{14.01 \text{ g N}} = 0.06667 \text{ mol N}$$

$$0.2016 \text{ g H} \times \frac{1 \text{ mol H}}{1.008 \text{ g H}} = 0.2000 \text{ mol H}$$

$$2.133 \text{ g O} \times \frac{1 \text{ mol O}}{16.00 \text{ g O}} = 0.1333 \text{ mol O}$$

STEP 3

$$\frac{0.06667 \text{ mol C}}{0.06667} = 1.000 \text{ mol C}$$

$$\frac{0.06667 \text{ mol N}}{0.06667} = 1.000 \text{ mol N}$$

$$\frac{0.2000 \text{ mol H}}{0.06667} = 3.000 \text{ mol H}$$

$$\frac{0.1333 \text{ mol O}}{0.06667} = 2.000 \text{ mol O}$$

The empirical formula is CNH_3O_2.

SELF-CHECK EXERCISE 9.10

STEP 1

In 100.00 g of Nylon-6 the masses of elements present are 63.68 g C, 12.38 g N, 9.80 g H, and 14.14 g O.

STEP 2

$$63.68 \text{ g C} \times \frac{1 \text{ mol C}}{12.01 \text{ g C}} = 5.302 \text{ mol C}$$

$$12.38 \text{ g N} \times \frac{1 \text{ mol N}}{14.01 \text{ g N}} = 0.8836 \text{ mol N}$$

$$9.80 \text{ g H} \times \frac{1 \text{ mol H}}{1.008 \text{ g H}} = 9.72 \text{ mol H}$$

$$14.14 \text{ g O} \times \frac{1 \text{ mol O}}{16.00 \text{ g O}} = 0.8837 \text{ mol O}$$

STEP 3

$$\frac{5.302 \text{ mol C}}{0.8836} = 6.000 \text{ mol C}$$

$$\frac{0.8836 \text{ mol N}}{0.8836} = 1.000 \text{ mol N}$$

$$\frac{9.72 \text{ mol H}}{0.8836} = 11.0 \text{ mol H}$$

$$\frac{0.8837 \text{ mol O}}{0.8836} = 1.000 \text{ mol O}$$

The empirical formula for Nylon-6 is $C_6NH_{11}O$.

SELF-CHECK EXERCISE 9.11

STEP 1

First we convert the mass percents to mass in grams. In 100.0 g of the compound, there are 71.65 g of chlorine, 24.27 g of carbon, and 4.07 g of hydrogen.

STEP 2
We use these masses to compute the moles of atoms present.

$$71.65 \ \text{g Cl} \times \frac{1 \ \text{mol Cl}}{35.45 \ \text{g Cl}} = 2.021 \ \text{mol Cl}$$

$$24.27 \ \text{g C} \times \frac{1 \ \text{mol C}}{12.01 \ \text{g C}} = 2.021 \ \text{mol C}$$

$$4.07 \ \text{g H} \times \frac{1 \ \text{mol H}}{1.008 \ \text{g H}} = 4.04 \ \text{mol H}$$

STEP 3
Dividing each mole value by 2.02 (the smallest number of moles present), we obtain the empirical formula $ClCH_2$.

To determine the molecular formula, we must compare the empirical formula mass to the molar mass. The empirical formula mass is 49.48.

Cl:	35.45
C:	12.01
2 H: $2 \times ($	1.008)
$ClCH_2$:	49.48 = empirical formula mass

The molar mass is known to be 98.96. We know that

$$\text{Molar mass} = n \times (\text{empirical formula mass})$$

So we can obtain the value of n as follows:

$$\frac{\text{Molar mass}}{\text{Empirical formula mass}} = \frac{98.96}{49.48} = 2$$

$$\text{Molecular formula} = (ClCH_2)_2 = Cl_2C_2H_4$$

This substance is composed of molecules with the formula $Cl_2C_2H_4$.

CHAPTER

10 Chemical
Quantities

CONTENTS

This chapter involves the application of many fundamentals: the use of names to write formulas, balancing equations, the conversion of mass to moles, and the use of the balanced equation to determine the moles of other reactants or products. If students have not mastered material covered earlier in the course, they will have difficulty with this chapter.

Dry ice (solid CO_2) forms CO_2 gas bubbles when placed in water (in this case water containing food coloring). The clouds form due to the cold temperature of the dry ice causing water vapor to condense from the air.

A chemist monitoring air quality in a mobile environmental lab.

Suppose you work for a consumer advocate organization and you want to test a company's advertising claims about the effectiveness of its antacid. The company claims that its product neutralizes 10 times as much stomach acid per tablet as its nearest competitor. How would you test the validity of this claim?

Or suppose that after graduation you go to work for a chemical company that makes methanol (methyl alcohol, or "wood" alcohol), a substance used as a starting material for the manufacture of products such as antifreeze and aviation fuels. Pure methanol is also used as a fuel in the cars that race in the Indianapolis 500. About 10 *billion* pounds of methanol are produced annually. You are working with an experienced chemist who is trying to improve the company's process for making methanol from the reaction of gaseous hydrogen with carbon monoxide gas. The first day on the job, you are instructed to order enough hydrogen and carbon monoxide to produce 6.0 kg of methanol in a test run. How would you determine how much carbon monoxide and hydrogen you should order?

After you study this chapter, you will be able to answer questions such as these.

10.1   Information Given by Chemical Equations

AIM: To identify the molecular and mass information given in a balanced equation.

Reactions are what chemistry is really all about. Recall that chemical changes are really rearrangements of atom groupings that can be described by chemical equations. In this section we will review the meaning and usefulness of chemical equations by considering one of the processes mentioned in the introduction: the reaction between gaseous carbon monoxide and hydrogen to produce liquid methanol, $CH_3OH(l)$. The reactants and products are

$$\text{Unbalanced: } CO(g) + H_2(g) \rightarrow CH_3OH(l)$$

Reactants   Product

Many of the problems in this chapter are word problems that must first be converted to chemical equations before they can be solved. Students often have some difficulty writing correct equations, and many forget to balance them. With each example that I work, I ask students whether or not the equation is balanced and ask them to help me balance it, if it is not. In this way, I try to make balancing the equation a routine part of solving the problem.

Because atoms are just rearranged (not gained or lost) in a chemical reaction, we must always balance a chemical equation. That is, we must choose coefficients that give the same number of each type of atom on both sides. Using the smallest set of integers that satisfies this condition gives the balanced equation

$$\text{Balanced: } CO(g) + 2H_2(g) \rightarrow CH_3OH(l)$$

CHECK: Reactants: 1 C, 1 O, 4 H;   Products: 1 C, 1 O, 4 H

It is important to recognize that the coefficients in a balanced equation give the *relative* numbers of molecules. That is, we could multiply this balanced equation by any number and still have a balanced equation. For example, we could multiply by 12:

$$12[CO(g) + 2H_2(g) \rightarrow CH_3OH(l)]$$

to obtain

$$12CO(g) + 24H_2(g) \rightarrow 12CH_3OH(l)$$

This is still a balanced equation (check to be sure). Because 12 represents a dozen, we could even describe the reaction in terms of dozens:

$$1 \text{ dozen } CO(g) + 2 \text{ dozen } H_2(g) \rightarrow 1 \text{ dozen } CH_3OH(l)$$

We could also multiply the original equation by a very large number, such as 6.022×10^{23}:

$$6.022 \times 10^{23}[CO(g) + 2H_2(g) \rightarrow CH_3OH(l)]$$

which leads to the equation

$$6.022 \times 10^{23} CO(g) + 2(6.022 \times 10^{23})H_2(g) \rightarrow 6.022 \times 10^{23} CH_3OH(l)$$

Just as 12 is called a dozen, chemists call 6.022×10^{23} a *mole* (abbreviated mol). Our equation, then, can be written in terms of moles:

$$1 \text{ mol } CO(g) + 2 \text{ mol } H_2(g) \rightarrow 1 \text{ mol } CH_3OH(l)$$

Various ways of interpreting this balanced chemical equation are given in Table 10.1.

When students are solving stoichiometry problems, a recipe can be used as a good analogy for a balanced equation. One needs to know what the ingredients are and what relative amounts of ingredients are needed for both recipes and chemical equations. An unbalanced chemical equation is like a list of ingredients for a recipe without the amounts specified. Chemical equations—after the formulas for products and reactants are known and the equation is balanced—can be manipulated like a recipe. We can halve the chemical recipe to make less product or triple the chemical recipe to make more product.

One mole is 6.022×10^{23} units.

Table 10.1 Information Conveyed by the Balanced Equation for the Production of Methanol

$CO(g)$	+	$2H_2(g)$	$\rightarrow$	$CH_3OH(l)$
1 molecule CO	+	2 molecules H_2	$\rightarrow$	1 molecule CH_3OH
1 dozen CO molecules	+	2 dozen H_2 molecules	$\rightarrow$	1 dozen CH_3OH molecules
6.022×10^{23} CO molecules	+	$2(6.022 \times 10^{23})$ H_2 molecules	$\rightarrow$	6.022×10^{23} CH_3OH molecules
1 mol CO molecules	+	2 mol H_2 molecules	$\rightarrow$	1 mol CH_3OH molecules

Table 10.1 shows some information provided by the coefficients in a balanced equation. I usually go over the information in this table in class because an understanding of what the coefficients represent is essential. In addition to pointing out all the information that coefficients convey, I like to show that they do not represent grams of compound. Students are frequently confused by the distinction between moles, molecules, and grams.

Tanks for storing propane, a fuel often used when natural gas is unavailable.

EXAMPLE 10.1   **Relating Moles to Molecules in Chemical Equations**

Propane, C_3H_8, is a fuel commonly used in rural areas where natural gas is unavailable. Propane reacts with oxygen gas to produce heat and the products carbon dioxide and water. This combustion reaction is represented by the unbalanced equation

$$C_3H_8(g) + O_2(g) \rightarrow CO_2(g) + H_2O(g)$$

Give the balanced equation for this reaction, and state the meaning of the equation in terms of numbers of molecules and moles of molecules.

SOLUTION

Using the techniques explained in Chapter 6, we can balance the equation.

$$C_3H_8(g) + 5O_2(g) \rightarrow 3CO_2(g) + 4H_2O(g)$$

CHECK: 3 C, 8 H,   10 O → 3 C,   8 H, 10 O

This equation can be interpreted in terms of molecules as follows:

1 molecule of C_3H_8 reacts with 5 molecules of O_2 to give 3 molecules of CO_2 plus 4 molecules of H_2O

or as follows in terms of moles (of molecules):

1 mol of C_3H_8 reacts with 5 mol of O_2 to give 3 mol of CO_2 plus 4 mol of H_2O

10.2   Mole–Mole Relationships

AIM: To use a balanced equation to determine relationships between moles of reactants and moles of products.

Now that we have discussed the meaning of a balanced chemical equation in terms of moles of reactants and products, we can use an equation to predict the moles of products that a given number of moles of reactants will yield. For example, consider the decomposition of water to give hydrogen and oxygen, which is represented by the following balanced equation:

$$2H_2O(l) \rightarrow 2H_2(g) + O_2(g)$$

This equation tells us that 2 mol of H_2O yields 2 mol of H_2 and 1 mol of O_2.

Now suppose that we have 4 mol of water. If we decompose 4 mol of water, how many moles of products do we get?

One way to answer this question is to multiply the entire equation by 2 (that will give us 4 mol of H_2O).

$$2[2H_2O(l) \rightarrow 2H_2(g) + O_2(g)]$$
$$4H_2O(l) \rightarrow 4H_2(g) + 2O_2(g)$$

Now we can state that

4 mol of H_2O yields 4 mol of H_2 plus 2 mol of O_2

which answers the question of how many moles of products we get with 4 mol of H_2O.

Next, suppose we decompose 5.8 mol of water. What numbers of moles of products are formed in this process? We could answer this question by rebalancing the chemical equation as follows: First, we divide *all coefficients* of the balanced equation

$$2H_2O(l) \rightarrow 2H_2(g) + O_2(g)$$

by 2, to give

$$H_2O(l) \rightarrow H_2(g) + \frac{1}{2}O_2(g)$$

Now, because we have 5.8 mol of H_2O, we multiply this equation by 5.8.

$$5.8[H_2O(l) \rightarrow H_2(g) + \frac{1}{2}O_2(g)]$$

This gives

$$5.8H_2O(l) \rightarrow 5.8H_2(g) + 5.8(\frac{1}{2})O_2(g)$$
$$5.8H_2O(l) \rightarrow 5.8H_2(g) + 2.9O_2(g)$$

(Verify that this is a balanced equation.) Now we can state that

5.8 mol of H_2O yields 5.8 mol of H_2 plus 2.9 mol of O_2

This procedure of rebalancing the equation to obtain the number of moles involved in a particular situation always works, but it can be cumbersome. In Example 10.2 we will develop a more convenient procedure, which uses conversion factors, or **mole ratios,** based on the balanced chemical equation.

This equation with noninteger coefficients makes sense only if the equation means moles (of molecules) of the various reactants and products.

Students can have trouble with mole ratios as conversion factors. Conversion factors used until now have been equalities. However, a ratio such as 1 in./2.54 cm does not express the same relationship as 2 mol H/1 mol H_2O. Students know that 2 mol H does not *equal* 1 mol H_2O, but you might point out the different meanings of the conversion factors here. You can read this conversion as 2 mol H *per* 1 mol H_2O, or 2 mol H *yields* 1 mol H_2O, but not as 2 mol H *equals* 1 mol H_2O.

EXAMPLE 10.2   Determining Mole Ratios

What number of moles of O_2 will be produced by the decomposition of 5.8 mol of water?

SOLUTION

Our problem can be diagrammed as follows:

| 5.8 mol H_2O | yields ⟹ | ? mol of O_2 |

When I introduce mole ratios, I show all the mole–mole relationships that exist for a particular, balanced equation. I like to establish in students' minds that valid relationships can exist between a reactant and a product, between two reactants, or between two products.

The statement 2 mol H_2O = 1 mol O_2 is not true in a literal sense, but it correctly expresses the chemical equivalence between H_2O and O_2.

EXAMPLE 10.2, CONTINUED

To answer this question, we need to know the relationship between moles of H_2O and moles of O_2 in the balanced equation (conventional form):

$$2H_2O(l) \rightarrow 2H_2(g) + O_2(g)$$

From this equation we can state that

which can be represented by the following equivalence statement:

$$2 \text{ mol } H_2O = 1 \text{ mol } O_2$$

We now want to use this equivalence statement to obtain the conversion factor (mole ratio) that we need. Because we want to go from moles of H_2O to moles of O_2, we need the mole ratio

$$\frac{1 \text{ mol } O_2}{2 \text{ mol } H_2O}$$

so that mol H_2O will cancel in the conversion from moles of H_2O to moles of O_2.

$$5.8 \text{ mol } H_2O \times \frac{1 \text{ mol } O_2}{2 \text{ mol } H_2O} = 2.9 \text{ mol } O_2$$

So if we decompose 5.8 mol of H_2O, we will get 2.9 mol of O_2. Note that this is the same answer we obtained above when we rebalanced the equation to give

$$5.8H_2O(l) \rightarrow 5.8H_2(g) + 2.9O_2(g)$$

We saw in Example 10.2 that to determine the moles of a product that can be formed from a specified number of moles of a reactant, we can use the balanced equation to obtain the appropriate mole ratio. We will now extend these ideas in Example 10.3.

EXAMPLE 10.3   **Using Mole Ratios in Calculations**

Calculate the number of moles of oxygen required to react exactly with 4.30 mol of propane, C_3H_8, in the reaction described by the following balanced equation:

$$C_3H_8(g) + 5O_2(g) \rightarrow 3CO_2(g) + 4H_2O(g)$$

SOLUTION

In this case the problem can be stated as follows:

EXAMPLE 10.3, CONTINUED

To solve this problem, we need to consider the relationship between the reactants C_3H_8 and O_2. Using the balanced equation, we find that

$$1 \text{ mol of } C_3H_8 \text{ requires } 5 \text{ mol of } O_2$$

which can be represented by the equivalence statement

$$1 \text{ mol } C_3H_8 = 5 \text{ mol } O_2$$

This leads to the required mole ratio

$$\frac{5 \text{ mol } O_2}{1 \text{ mol } C_3H_8}$$

for converting from moles of C_3H_8 to moles of O_2. We construct the conversion ratio this way so that mol C_3H_8 cancels:

$$4.30 \text{ mol } C_3H_8 \times \frac{5 \text{ mol } O_2}{1 \text{ mol } C_3H_8} = 21.5 \text{ mol } O_2$$

We can now answer the original question:

$$4.30 \text{ mol of } C_3H_8 \text{ requires } 21.5 \text{ mol of } O_2$$

SELF-CHECK EXERCISE 10.1

Calculate the moles of CO_2 formed when 4.30 mol of C_3H_8 react with the required 21.5 mol of O_2.

HINT: Use the moles of C_3H_8, and obtain the mole ratio between C_3H_8 and CO_2 from the balanced equation.

10.3 Mass Calculations

AIM: To learn to relate masses of reactants and products in a chemical reaction.

In the last section we saw how to use the balanced equation for a reaction to calculate the numbers of moles of reactants and products for a particular case. However, moles represent numbers of molecules, and we cannot count molecules directly. In chemistry we count by weighing. Therefore, in this section we

This section develops most of the common stoichiometry-problem variations. There is never enough time to work as many examples as I would like. The only way students can be exposed to all the variations is for them to work lots of homework problems.

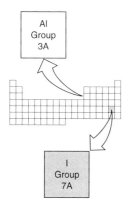

will review the procedures for converting between moles and masses and will see how these procedures are applied to chemical calculations.

To develop these procedures we will consider the reaction between pow-dered aluminum metal and finely ground iodine to produce aluminum iodide. The balanced equation for this vigorous chemical reaction is

$$2Al(s) \; + \; 3I_2(s) \rightarrow 2AlI_3(s)$$

Suppose we have 35.0 g of aluminum. What mass of I_2 should we weigh out to react exactly with this amount of aluminum?

To answer this question we need to think about what the balanced equation tells us. The equation states that

$$2 \text{ mol of Al requires } 3 \text{ mol of } I_2$$

which leads to the mole ratio

$$\frac{3 \text{ mol } I_2}{2 \text{ mol Al}}$$

We can use this ratio to calculate the moles of I_2 needed:

$$\text{Moles of Al present} \times \frac{3 \text{ mol } I_2}{2 \text{ mol Al}} = \text{moles of } I_2 \text{ required}$$

It is helpful to organize the data in mass-to-mol conversions by asking the two questions: "What do we know?" and "What do we want to calculate?" Students always want to skip these steps, but when they review their notes and work problems on their own, this organization may help them to solve similar problems.

This leads us to the question: what number of moles of Al are present? The problem states that we have 35.0 g of aluminum, so we must convert from grams to moles of aluminum. This is something we already know how to do. Using the table of average atomic masses inside the front cover of the book, we find the atomic mass of aluminum to be 26.98. This means that 1 mol of aluminum has a mass of 26.98 g. We can use the equivalence statement

$$1 \text{ mol Al} = 26.98 \text{ g}$$

to find the moles of Al in 35.0 g.

$$35.0 \text{ g Al} \times \frac{1 \text{ mol Al}}{26.98 \text{ g Al}} = 1.30 \text{ mol Al}$$

Now that we have moles of Al, we can find the moles of I_2 required.

$$1.30 \text{ mol Al} \times \frac{3 \text{ mol } I_2}{2 \text{ mol Al}} = 1.95 \text{ mol } I_2$$

Remember that to show the correct significant figures in each step, we are rounding off after each calculation. In doing problems, you should carry extra numbers, rounding off only at the end.

We now know the *moles* of I_2 required to react with the 1.30 mol of Al (35.0 g). The next step is to convert 1.95 mol of I_2 to grams so we will know how much to weigh out. We do this by using the molar mass of I_2. The atomic mass of iodine is 126.9 g (for 1 mol of I atoms), so the molar mass of I_2 is

$$2 \times 126.9 \text{ g/mol} = 253.8 \text{ g/mol} = \text{mass of 1 mol of } I_2$$

Now we convert the 1.95 mol of I_2 to grams of I_2.

$$1.95 \; \cancel{mol \; I_2} \times \frac{253.8 \text{ g } I_2}{\cancel{mol \; I_2}} = 495 \text{ g } I_2$$

We have solved the problem. We need to weigh out 495 g of iodine (contains I_2 molecules) to react exactly with the 35.0 g of aluminum. We will further develop procedures for dealing with masses of reactants and products in Example 10.4.

Students who have made a habit of using unitless numbers and who have a tendency to ignore significant figures will be at a disadvantage when solving these problems. For full credit on hour exams I insist that students set up their problems showing all units that cancel and that their answers contain the correct number of significant figures.

EXAMPLE 10.4   Using Mass–Mole Conversions with Mole Ratios

Propane, C_3H_8, when used as a fuel, reacts with oxygen to produce carbon dioxide and water according to the following unbalanced equation:

$$C_3H_8(g) + O_2(g) \rightarrow CO_2(g) + H_2O(g)$$

What mass of oxygen will be required to react exactly with 96.1 g of propane?

SOLUTION

To deal with the amounts of reactants and products, we first need the balanced equation for this reaction:

$$C_3H_8(g) + 5O_2(g) \rightarrow 3CO_2(g) + 4H_2O(g)$$

Always balance the equation for the reaction first.

Next, let's summarize what we know and what we want to find.

What we know:
- The balanced equation for the reaction
- The mass of propane available (96.1 g)

What we want to calculate:
- The mass of oxygen (O_2) required to react exactly with all the propane

Before you begin working problems, you might want to examine a balanced equation and review with students the many different kinds of questions about reactants and products that can be answered, depending on the mole ratio.

Our problem, in schematic form, is

Using the ideas we developed when we discussed the aluminum–iodine reaction, we will proceed as follows:

1. We are given the number of grams of propane, so we must convert to moles of propane (C_3H_8).
2. Then we can use the coefficients in the balanced equation to determine the moles of oxygen (O_2) required.
3. Finally, we will use the molar mass of O_2 to calculate grams of oxygen.

EXAMPLE 10.4, CONTINUED

We can sketch out this strategy as follows:

$$C_3H_8(g) \quad + \quad 5O_2(g) \quad \rightarrow \quad 3CO_2(g) \quad + \quad 4H_2O(g)$$

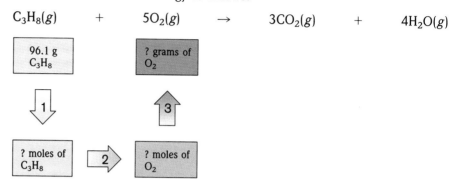

Thus the first question we must answer is *How many moles of propane are present in 96.1 g of propane?* The molar mass of propane is 44.09 g (3 × 12.01 + 8 × 1.008). The moles of propane present can be calculated as follows:

$$96.1 \text{ g } C_3H_8 \times \frac{1 \text{ mol } C_3H_8}{44.09 \text{ g } C_3H_8} = 2.18 \text{ mol } C_3H_8$$

Next we recognize that each mole of propane reacts with 5 mol of oxygen. This gives us the equivalence statement

$$1 \text{ mol } C_3H_8 = 5 \text{ mol } O_2$$

from which we construct the mole ratio

$$\frac{5 \text{ mol } O_2}{1 \text{ mol } C_3H_8}$$

that we need to convert from moles of propane molecules to moles of oxygen molecules.

$$2.18 \text{ mol } C_3H_8 \times \frac{5 \text{ mol } O_2}{1 \text{ mol } C_3H_8} = 10.9 \text{ mol } O_2$$

Notice that the mole ratio is set up so that the moles of C_3H_8 cancel and the resulting units are moles of O_2.

Because the original question asked for the *mass* of oxygen needed to react with 96.1 g of propane, we must convert the 10.9 mol of O_2 to grams, using the molar mass of O_2 (32.00 = 2 × 16.00).

$$10.9 \text{ mol } O_2 \times \frac{32.0 \text{ g } O_2}{1 \text{ mol } O_2} = 349 \text{ g } O_2$$

EXAMPLE 10.4, CONTINUED

Therefore, 349 g of oxygen is required to burn 96.1 g of propane. We can summarize this problem by writing out a "conversion string" that shows how the problem was done.

$$96.1 \text{ g } C_3H_8 \times \frac{1 \text{ mol } C_3H_8}{44.09 \text{ g } C_3H_8} \times \frac{5 \text{ mol } O_2}{1 \text{ mol } C_3H_8} \times \frac{32.0 \text{ g } O_2}{1 \text{ mol } O_2} = 349 \text{ g } O_2$$

This is a convenient way to make sure the final units are correct. The procedure we have followed is summarized in Figure 10.1.

Use units as a check to see that you have used the correct conversion factors (mole ratios).

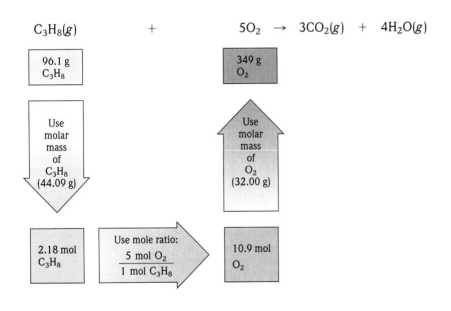

$$C_3H_8(g) \quad + \quad 5O_2 \rightarrow 3CO_2(g) + 4H_2O(g)$$

Figure 10.1
Summary of the steps required to solve Example 10.4.

An aid to solving stoichiometry problems is a diagram similar to the one in Figure 10.1. If you consistently use the same kind of diagram each time you solve a problem and encourage students to do the same, students who would otherwise be unable to solve stoichiometry problems can learn how.

SELF-CHECK EXERCISE 10.2

What mass of carbon dioxide is produced when 96.1 g of propane react with sufficient oxygen?

SELF-CHECK EXERCISE 10.3

Calculate the mass of water formed by the complete reaction of 96.1 g of propane with oxygen.

So far in this chapter, we have spent considerable time "thinking through" the procedures for calculating the masses of reactants and products in chemical reactions. We can summarize these procedures in the following steps:

Steps for Calculating the Masses of Reactants and Products in Chemical Reactions

STEP 1
Balance the equation for the reaction.

STEP 2
If amounts are given as masses, convert the known masses of reactants or products to moles.

STEP 3
Use the balanced equation to set up the appropriate mole ratio(s).

STEP 4
Use the mole ratio(s) to calculate the number of moles of the desired reactant or product.

STEP 5
Convert from moles back to mass, if this is required by the problem.

The process of using a chemical equation to calculate the relative masses of reactants and products involved in a reaction is called **stoichiometry** (pronounced stoý·kē·oḿ·ĕtry). Chemists say that the balanced equation for a chemical reaction describes the stoichiometry of the reaction.

We will now consider a few more examples that involve chemical stoichiometry. Because real-world examples often involve very large or very small masses of chemicals that are most conveniently expressed by using scientific notation, we will deal with such a case in Example 10.5.

EXAMPLE 10.5   **Stoichiometric Calculations: Using Scientific Notation**

Solid lithium hydroxide is used in space vehicles to remove exhaled carbon dioxide from the living environment. The products are solid lithium carbonate and liquid water. What mass of gaseous carbon dioxide can 1.00×10^3 g of lithium hydroxide absorb?

SOLUTION

STEP 1
Using the description of the reaction, we can write the unbalanced equation

$$LiOH(s) + CO_2(g) \rightarrow Li_2CO_3(s) + H_2O(l)$$

If you need a review of writing formulas of ionic compounds, see Chapter 5.

EXAMPLE 10.5, CONTINUED

The balanced equation is

$$2LiOH(s) + CO_2(g) \rightarrow Li_2CO_3(s) + H_2O(l)$$

Check this for yourself.

STEP 2
We convert the given mass of LiOH to moles, using the molar mass of LiOH, which is $6.941 + 16.00 + 1.008 = 23.95$ g.

$$1.00 \times 10^3 \text{ g LiOH} \times \frac{1 \text{ mol LiOH}}{23.95 \text{ g LiOH}} = 41.8 \text{ mol LiOH}$$

STEP 3
The appropriate mole ratio is $\dfrac{1 \text{ mol CO}_2}{2 \text{ mol LiOH}}$.

STEP 4
Using this mole ratio, we calculate the moles of CO_2 needed to react with the given mass of LiOH.

$$41.8 \text{ mol LiOH} \times \frac{1 \text{ mol CO}_2}{2 \text{ mol LiOH}} = 20.9 \text{ mol CO}_2$$

STEP 5
We calculate the mass of CO_2 by using its molar mass (44.01 g).

$$20.9 \text{ mol CO}_2 \times \frac{44.01 \text{ g CO}_2}{1 \text{ mol CO}_2} = 920. \text{ g CO}_2 = 9.20 \times 10^2 \text{ g CO}_2$$

Carrying extra significant figures and rounding off only at the end gives an answer of 919 g CO_2.

Thus 1.00×10^3 g of LiOH(s) can absorb 920. g of $CO_2(g)$.

We can summarize this problem as follows:

Astronaut Sidney M. Gutierrez changes the lithium hydroxide cannisters on Space Shuttle Columbia. The lithium hydroxide is used to purge carbon dioxide from the air in the shuttle's cabin.

EXAMPLE 10.5, CONTINUED

The conversion string is

$$1.00 \times 10^3 \text{ g LiOH} \times \frac{1 \text{ mol LiOH}}{23.95 \text{ g LiOH}} \times \frac{1 \text{ mol CO}_2}{2 \text{ mol LiOH}} \times \frac{44.01 \text{ g CO}_2}{1 \text{ mol CO}_2}$$

$$= 9.20 \times 10^2 \text{ g CO}_2$$

SELF-CHECK EXERCISE 10.4

Hydrofluoric acid, an aqueous solution containing dissolved hydrogen fluoride, is used to etch glass by reacting with the silica, SiO_2, in the glass to produce gaseous silicon tetrafluoride and liquid water. The unbalanced equation is

$$HF(aq) + SiO_2(s) \rightarrow SiF_4(g) + H_2O(l)$$

a. Calculate the mass of hydrogen fluoride needed to react with 5.68 g of silica. *Hint:* Think carefully about this problem. What is the balanced equation for the reaction? What is given? What do you need to calculate? Sketch a map of the problem before you do the calculations.
b. Calculate the mass of water produced in the reaction described in part a.

EXAMPLE 10.6 **Stoichiometric Calculations: Comparing Two Reactions**

When students read word problems describing two separate reactions, they can be easily confused. Unless they have carefully read the problem, their first inclination is to combine all the substances mentioned in the problem into one chemical equation. You can use a problem that describes two different reactions to emphasize to students that they must determine what the problem asks before they try to solve it.

Baking soda, $NaHCO_3$, is often used as an antacid. It neutralizes excess hydrochloric acid secreted by the stomach. The balanced equation for the reaction is

$$NaHCO_3(s) + HCl(aq) \rightarrow NaCl(aq) + H_2O(l) + CO_2(g)$$

Milk of magnesia, which is an aqueous suspension of magnesium hydroxide, $Mg(OH)_2$, is also used as an antacid. The balanced equation for the reaction is

$$Mg(OH)_2(s) + 2HCl(aq) \rightarrow 2H_2O(l) + MgCl_2(aq)$$

Which is the more effective antacid, 1.00 g of $NaHCO_3$ or 1.00 g of $Mg(OH)_2$?

SOLUTION

Before we begin, let's think about the problem to be solved. The question we must ask for each antacid is *How many moles of HCl will react with 1.00 g of each antacid?* The antacid that reacts with the larger number of moles of HCl is more effective because it will neutralize more moles of acid. A schematic for this

EXAMPLE IO.6, CONTINUED

procedure is

Antacid   +   HCl   →   Products

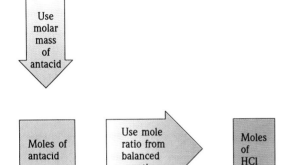

Notice that in this case we do not need to calculate how many grams of HCl react; we can answer the question with moles of HCl. We will now solve this problem for each antacid. Both of the equations are balanced, so we can proceed with the calculations.

Using the molar mass of $NaHCO_3$, which is $22.99 + 1.008 + 12.01 + 3(16.00) = 84.01$ g, we determine the moles of $NaHCO_3$ in 1.00 g of $NaHCO_3$.

$$1.00 \text{ g } NaHCO_3 \times \frac{1 \text{ mol } NaHCO_3}{84.01 \text{ g } NaHCO_3} = 0.0119 \text{ mol } NaHCO_3$$

$$= 1.19 \times 10^{-2} \text{ mol } NaHCO_3$$

Next we determine the moles of HCl, using the mole ratio $\dfrac{1 \text{ mol HCl}}{1 \text{ mol } NaHCO_3}$.

$$1.19 \times 10^{-2} \text{ mol } NaHCO_3 \times \frac{1 \text{ mol HCl}}{1 \text{ mol } NaHCO_3} = 1.19 \times 10^{-2} \text{ mol HCl}$$

Thus 1.00 g of NaHCO neutralizes 1.19×10^{-2} mol of HCl. We need to compare this to the number of moles of HCl that 1.00 g of $Mg(OH)_2$ neutralizes.

Using the molar mass of $Mg(OH)_2$, which is $24.31 + 2(16.00) + 2(1.008) = 58.33$ g, we determine the moles of $Mg(OH)_2$ in 1.00 g of $Mg(OH)_2$.

Methyl Alcohol: Fuel with a Future?

Southern California is famous for many things, and among them, unfortunately, is smog. Smog is produced when pollutants in the air are trapped near the ground and are caused to react by sunlight. One step being considered by the state of California to help solve the smog problem is to replace gasoline with methyl alcohol (usually called methanol). One advantage of methanol is that it reacts more nearly completely than gasoline with oxygen in a car's engine, thus releasing lower amounts of unburned fuel into the atmosphere. Methanol also produces less carbon monoxide (CO) in the exhaust than does gasoline. Carbon monoxide is not only toxic itself, but it also encourages the formation of nitrogen dioxide by the reaction

$$CO(g) + O_2(g) + NO(g) \rightarrow CO_2(g) + NO_2(g)$$

Nitrogen dioxide is a brown gas that leads to ozone formation and acid rain.

Using methanol as a fuel is not a new idea. For example, it is the only fuel allowed in the open-wheeled race cars used in the Indianapolis 500 and in similar races. Methanol works very well in racing engines because it has outstanding antiknock character-

A crewman adds methanol fuel to a race car in the Indianapolis 500 during a pit stop.

istics, even at the tremendous speeds at which these engines operate (around 11,000 revolutions per minute).

The news about methanol is not all good, however. One problem is lower fuel mileage. Because it takes about twice as many gallons of methanol than of gasoline to travel a given distance, a methanol-powered car's fuel tank must be twice the usual size. However, although costs vary greatly depending on market conditions, the cost of methanol averages

about half that of gasoline, so the net cost is about the same for both fuels.

A second disadvantage of methanol is that its high affinity for water causes condensation from the air, which leads to increased corrosion of the fuel tank and fuel lines. This problem can be solved by using more expensive stainless steel for these parts.

The most serious problem with methanol may be its tendency to form formaldehyde, HCHO, when it is combusted. Formaldehyde has been implicated as a carcinogen (a substance that causes cancer). Formaldehyde can also lead to ozone formation in the air, which causes even more severe smog. Researchers are now working on catalytic converters for exhaust systems to help decompose the formaldehyde.

To test the feasibility of methanol as a motor fuel, California has operated more than 600 vehicles on methanol since 1980. Because accessibility to methanol is limited, cars are now being prepared that can run on methanol or gasoline. These vehicles will soon be tested on a large scale in California. So if you live in southern California, in a few years your neighborhood "gas station" may actually be pumping methanol.

EXAMPLE 10.6, CONTINUED

$$1.00 \text{ g Mg(OH)}_2 \times \frac{1 \text{ mol Mg(OH)}_2}{58.3 \text{ g Mg(OH)}_2} = 0.0171 \text{ mol Mg(OH)}_2$$

$$= 1.71 \times 10^{-2} \text{ mol Mg(OH)}_2$$

To determine the moles of HCl that react with this amount of $Mg(OH)_2$, we use the mole ratio $\dfrac{2 \text{ mol HCl}}{1 \text{ mol Mg(OH)}_2}$.

$$1.71 \times 10^{-2} \text{ mol Mg(OH)}_2 \times \frac{2 \text{ mol HCl}}{1 \text{ mol Mg(OH)}_2} = 3.42 \times 10^{-2} \text{ mol HCl}$$

Therefore 1.00 g of $Mg(OH)_2$ neutralizes 3.42×10^{-2} mol of HCl. We have already calculated that 1.00 g of $NaHCO_3$ neutralizes only 1.19×10^{-2} mol of HCl. Therefore $Mg(OH)_2$ is a more effective antacid than $NaHCO_3$ on a mass basis.

Although the steps we took in solving the problem were not labeled in this example, they were followed. Identify the steps to make sure you understand them.

SELF-CHECK EXERCISE 10.5

In Example 10.6 we answered one of the questions we posed in the introduction to this chapter. Now let's see if you can answer the other question posed there. Determine what mass of carbon monoxide and what mass of hydrogen are required to form 6.0 kg of methanol by the reaction

$$CO(g) + 2H_2(g) \rightarrow CH_3OH(l)$$

10.4 Calculations Involving a Limiting Reactant

AIM: To learn to recognize the limiting reactant in a reaction.
To learn to use the limiting reactant to do stoichiometric calculations.

When chemicals are mixed together so that they can undergo a reaction, they are often mixed in **stoichiometric quantities**—that is, in exactly the correct amounts so that all reactants "run out" (are used up) at the same time. To clarify

I begin this section by presenting a simple balanced equation, such as the one that describes the reaction between sodium and chlorine, $2Na + Cl_2 \rightarrow 2NaCl$. If we add 2 mol Na to 1 mol Cl_2, the reaction yields 0 mol Na, 0 mol Cl_2, and 2 mol NaCl. We used up all reactants because we added just enough chlorine to react with all the sodium. I then ask "What happens if we add 0.5 mol Cl_2 to 2 mol Na?" You can help students better understand the concept of limiting reactants if you use a simple equation and molar quantities that are easy to manipulate than if you start with a more complex system.

Ammonia being dissolved in irrigation water to provide fertilizer for a field of corn.

Only after developing the concept of limiting reactants with moles do I extend these problems to include grams.

this concept, we will consider the production of hydrogen for use in the manufacture of ammonia. Ammonia, a very important fertilizer itself and a starting material for other fertilizers, is made by combining nitrogen from the air with hydrogen. The hydrogen for this process is produced by the reaction of methane with water according to the balanced equation

$$CH_4(g) + H_2O(g) \rightarrow 3H_2(g) + CO(g)$$

Let's consider the question *What mass of water is required to react exactly with 249 g of methane?* That is, how much water will just use up all of the 249 g of methane, leaving no methane or water remaining?

This problem requires the same strategies we developed in the last section. Again, drawing a map of the problem is helpful.

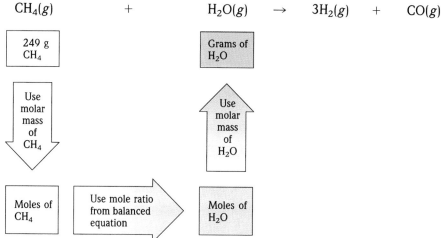

We first convert the mass of CH_4 to moles, using the molar mass of CH_4 (16.0 g/mol).

$$249 \text{ g } CH_4 \times \frac{1 \text{ mol } CH_4}{16.04 \text{ g } CH_4} = 15.5 \text{ mol } CH_4$$

Because in the balanced equation 1 mol of CH_4 reacts with 1 mol of H_2O, we have

$$15.5 \text{ mol } CH_4 \times \frac{1 \text{ mol } H_2O}{1 \text{ mol } CH_4} = 15.5 \text{ mol } H_2O$$

Therefore 15.5 mol of H_2O will react exactly with the given mass of CH_4. Converting 15.5 mol of H_2O to grams of H_2O (molar mass = 18.02 g/mol) gives

$$15.5 \text{ mol } H_2O \times \frac{18.02 \text{ g } H_2O}{1 \text{ mol } H_2O} = 279 \text{ g } H_2O$$

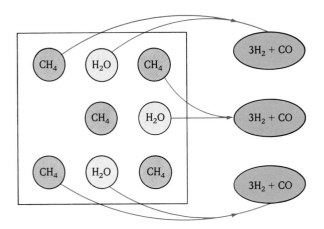

Figure 10.2
A mixture of $5CH_4$ and $3H_2O$ molecules undergoes the reaction $CH_4(g)$ + $H_2O(g) \rightarrow 3H_2(g)$ + $CO(g)$. Note that the H_2O molecules are used up first, leaving two CH_4 molecules unreacted.

This result means that if 249 g of methane is mixed with 279 g of water, both reactants will "run out" at the same time. The reactants have been mixed in stoichiometric quantities.

If, on the other hand, 249 g of methane is mixed with 300 g of water, the methane will be consumed before the water runs out. The water will be in *excess.* In this case, the quantity of products formed will be determined by the quantity of methane present. Once the methane is consumed, no more products can be formed, even though some water still remains. In this situation, because the amount of methane *limits* the amount of products that can be formed, it is called the **limiting reactant,** or **limiting reagent.** In any stoichiometry problem, where reactants are not mixed in stoichiometric quantities, it is essential to determine which reactant is limiting in order to calculate correctly the amounts of products that will be formed. This concept is illustrated in Figure 10.2. Note from this figure that because there are fewer water molecules than CH_4 molecules, the water is consumed first. After the water molecules are gone, no more products can form. So in this case water is the limiting reactant.

You probably have been dealing with limiting-reactant problems for most of your life. For example, suppose a lemonade recipe calls for 1 cup of sugar for every 6 lemons. You have 12 lemons and 3 cups of sugar. Which ingredient is limiting, the lemons or the sugar?*

---

*The ratio of lemons to sugar that the recipe calls for is 6 lemons to 1 cup of sugar. We can calculate the number of lemons required to "react with" the 3 cups of sugar as follows:

$$3 \; \cancel{\text{cups sugar}} \times \frac{6 \text{ lemons}}{1 \; \cancel{\text{cup sugar}}} = 18 \text{ lemons}$$

Thus 18 lemons would be required to use up 3 cups of sugar. However, we have only 12 lemons, so the lemons are limiting.

You cannot remind students often enough that it is not possible to tell which of two reactants will run out first just by comparing the number of grams of each reactant.

The reactant that is consumed first limits the amounts of products that can form.

The concept of limiting reactant as applied to chemical reactions is very hard for students to grasp. They can appreciate examples such as the number of lemons limiting the amount of lemonade one can make, but they find it difficult to transfer this same reasoning to chemical reactions. At first, they determine which reactant is limiting based on the number of grams of each reactant. Even after you work some examples, it may be hard for students to remember to compare the number of moles of reactants and not the number of grams.

You might point out that there are some clues within problems that can help students to know when they do or do not need to be concerned with determining limiting reactants. For example, when one reactant is in excess, then the second reactant will be the limiting reactant. When grams of two or more reactants are given, then the limiting reactant must be determined before the problem can be solved.

EXAMPLE 10.7   Stoichiometric Calculations: Identifying the Limiting Reactant

Suppose 25.0 kg (2.50 × 10^4 g) of nitrogen gas and 5.00 kg (5.00 × 10^3 g) of hydrogen gas are mixed and reacted to form ammonia. Calculate the mass of ammonia produced when this reaction is run to completion.

SOLUTION

The unbalanced equation for this reaction is

$$N_2(g) \, + \, H_2(g) \rightarrow NH_3(g)$$

which leads to the balanced equation

$$N_2(g) \, + \, 3H_2(g) \rightarrow 2NH_3(g)$$

This problem is different from the others we have done so far in that we are mixing *specified amounts of two reactants* together. To know how much product forms, we must determine which reactant is consumed first. That is, we must determine which is the limiting reactant in this experiment. To do so we must add a step to our normal procedure. We can map this process as follows:

Problems that ask students to find the limiting reactant are much more straightforward than problems that provide masses of two reactants and ask students to determine the amount of product formed. In the first kind of problem, students know they must determine the limiting reactant. In the second kind, they must look at the data and determine for themselves that they need to find the limiting reactant.

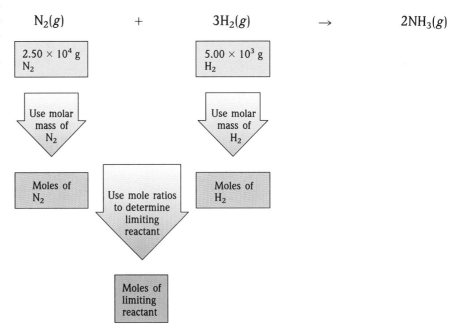

We will use the moles of the limiting reactant to calculate the moles and then the grams of the product.

EXAMPLE 10.7, CONTINUED

$$N_2(g) \quad + \quad 3H_2(g) \quad \rightarrow \quad 2NH_3(g)$$

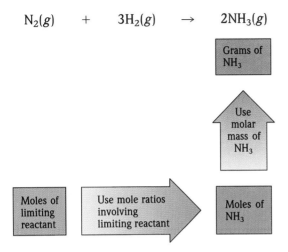

We first calculate the moles of the two reactants present:

$$2.50 \times 10^4 \text{ g } N_2 \times \frac{1 \text{ mol } N_2}{28.02 \text{ g } N_2} = 8.92 \times 10^2 \text{ mol } N_2$$

$$5.00 \times 10^3 \text{ g } H_2 \times \frac{1 \text{ mol } H_2}{2.016 \text{ g } H_2} = 2.48 \times 10^3 \text{ mol } H_2$$

Now we must determine which reactant is limiting (will be consumed first). We have 8.92×10^2 mol of N_2. Let's determine *how many moles of H_2 are required to react with this much N_2.* Because 1 mol of N_2 reacts with 3 mol of H_2, the number of moles of H_2 we need to react completely with 8.92×10^2 mol of N_2 is determined as follows:

$$8.92 \times 10^2 \text{ mol } N_2 \times \frac{3 \text{ mol } H_2}{1 \text{ mol } N_2} = 2.68 \times 10^3 \text{ mol } H_2$$

Is N_2 or H_2 the limiting reactant? The answer comes from the comparison

Moles of H_2 available	smaller than	Moles of H_2 required
2.48×10^3		2.68×10^3

EXAMPLE 10.7, CONTINUED

We see that 8.92×10^2 mol of N_2 require 2.68×10^3 mol of H_2 to react completely. However, only 2.48×10^3 mol of H_2 are present. This means that the hydrogen will be consumed before the nitrogen runs out, so hydrogen is the *limiting reactant* in this particular situation.

Note that in our effort to determine the limiting reactant, we could have started instead with the given amount of hydrogen and calculated the moles of nitrogen required.

$$2.48 \times 10^3 \text{ mol H}_2 \times \frac{1 \text{ mol N}_2}{3 \text{ mol H}_2} = 8.27 \times 10^2 \text{ mol N}_2$$

Thus 2.48×10^3 mol of H_2 require 8.27×10^2 mol of N_2. Because 8.93×10^2 mol of N_2 are actually present, the nitrogen is in excess.

If nitrogen is in excess, hydrogen will "run out" first; again we find that hydrogen limits the amount of ammonia formed.

Because the moles of H_2 present are limiting, we must use this quantity to determine the moles of NH_3 that can form.

$$2.48 \times 10^3 \text{ mol H}_2 \times \frac{2 \text{ mol NH}_3}{3 \text{ mol H}_2} = 1.65 \times 10^3 \text{ mol NH}_3$$

Next we convert moles of NH_3 to mass of NH_3.

$$1.65 \times 10^3 \text{ mol NH}_3 \times \frac{17.0 \text{ g NH}_3}{1 \text{ mol NH}_3} = 2.80 \times 10^4 \text{ g NH}_3 = 28.0 \text{ kg NH}_3$$

Therefore 25.0 kg of N_2 and 5.00 kg of H_2 can form 28.0 kg of NH_3.

The strategy used in Example 10.7 is summarized in Figure 10.3.

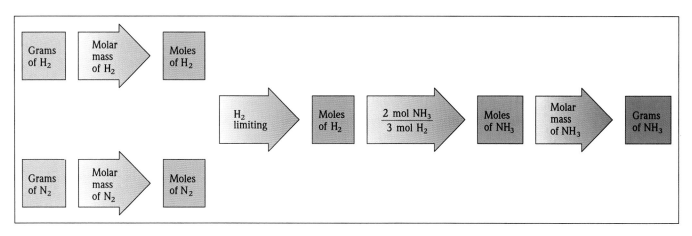

Figure 10.3
A map of the procedure used in Example 10.7.

The following list summarizes the steps to take in solving stoichiometry problems in which the amounts of two (or more) reactants are given.

Steps for Solving Stoichiometric Problems Involving Limiting Reactants

STEP 1
Write and balance the equation for the reaction.

STEP 2
Convert known masses of reactants to moles.

STEP 3
Using the numbers of moles of reactants and the appropriate mole ratios, determine which reactant is limiting.

STEP 4
Using the amount of the limiting reactant and the appropriate mole ratios, compute the number of moles of the desired product.

STEP 5
Convert from moles of product to grams of product, using the molar mass (if this is required by the problem).

Copper oxide(II) reacting with ammonia in a heated tube.

EXAMPLE 10.8   Stoichiometric Calculations: Reactions Involving the Masses of Two Reactants

Nitrogen gas can be prepared by passing gaseous ammonia over solid copper(II) oxide at high temperatures. The other products of the reaction are solid copper and water vapor. How many grams of N_2 are formed when 18.1 g of NH_3 is reacted with 90.4 g of CuO?

SOLUTION

STEP 1
From the description of the problem, we obtain the following balanced equation:

$$2NH_3(g) \ + \ 3CuO(s) \rightarrow N_2(g) \ + \ 3Cu(s) \ + \ 3H_2O(g)$$

STEP 2
Next from the masses of reactants available we must compute the moles of NH_3 (molar mass = 17.0) and of CuO (molar mass = 79.5).

$$18.1 \ g \ NH_3 \ \times \ \frac{1 \ mol \ NH_3}{17.03 \ g \ NH_3} = 1.06 \ mol \ NH_3$$

$$90.4 \ g \ CuO \ \times \ \frac{1 \ mol \ CuO}{79.55 \ g \ CuO} = 1.14 \ mol \ CuO$$

STEP 3
To determine which reactant is limiting, we use the mole ratio between CuO and NH_3.

$$1.06 \ mol \ NH_3 \ \times \ \frac{3 \ mol \ CuO}{2 \ mol \ NH_3} = 1.59 \ mol \ CuO$$

Then we compare how much CuO we have with how much of it we need.

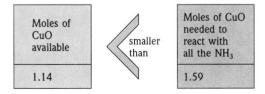

Moles of CuO available	smaller than	Moles of CuO needed to react with all the NH_3
1.14		1.59

Therefore, 1.59 mol of CuO is required to react with 1.06 mol of NH_3, but only 1.14 mol of CuO is actually present. So the amount of CuO is limiting; CuO will run out before NH_3 does.

EXAMPLE 10.8, CONTINUED

STEP 4

CuO is the limiting reactant, so we must use the amount of CuO in calculating the amount of N_2 formed. Using the mole ratio between CuO and N_2 from the balanced equation, we have

$$1.14 \text{ mol CuO} \times \frac{1 \text{ mol } N_2}{3 \text{ mol CuO}} = 0.380 \text{ mol } N_2$$

STEP 5

Using the molar mass of N_2 (28.02), we can now calculate the mass of N_2 produced.

$$0.380 \text{ mol } N_2 \times \frac{28.02 \text{ g } N_2}{1 \text{ mol } N_2} = 10.6 \text{ g } N_2$$

SELF-CHECK EXERCISE 10.6

Lithium nitride, an ionic compound containing the Li^+ and N^{3-} ions, is prepared by the reaction of lithium metal and nitrogen gas. Calculate the mass of lithium nitride formed from 56.0 g of nitrogen gas and 56.0 g of lithium in the unbalanced reaction

$$Li(s) + N_2(g) \rightarrow Li_3N(s)$$

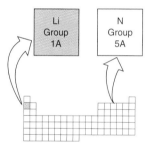

10.5 Percent Yield

AIM: To calculate actual yield as a percentage of theoretical yield.

In the previous section we learned how to calculate the amount of product that forms when specified amounts of reactants are mixed together. In doing these calculations, we used the fact that the amount of product is controlled by the limiting reactant. Products stop forming when one reactant runs out.

The amount of product calculated in this way is called the **theoretical yield** of that product. In Example 10.8, 10.6 g of nitrogen represents the theoretical yield. This is the *maximum amount* of nitrogen that can be produced from the quantities of reactants used. Actually, however, the amount of product predicted (the theoretical yield) is seldom obtained. One reason for this is the presence of side reactions (other reactions that consume one or more of the reactants or products).

It is hard for students to accept that the answers to mass-to-mass stoichiometry problems are actually the theoretical yield. Learning the definitions of theoretical and actual yield is the hardest part of this section. You might want to remind students that actual yields are determined by someone in the laboratory. Unless they have the opportunity to produce a compound in the lab and find the actual yield, the difference between the theoretical and actual yields cannot be determined.

The *actual yield* of product, which is the amount of product *actually obtained,* is often compared to the theoretical yield. This comparison, usually expressed as a percent, is called the **percent yield.**

$$\frac{\text{Actual yield}}{\text{Theoretical yield}} \times 100 = \text{percent yield}$$

For example, *if* the reaction considered in Example 10.8 *actually* gave 6.63 g of nitrogen instead of the *predicted* 10.6 g, the percent yield of nitrogen would be

$$\frac{6.63 \text{ g N}_2}{10.6 \text{ g N}_2} \times 100 = 62.5\%$$

EXAMPLE 10.9   Stoichiometric Calculations: Determining Percent Yield

Earlier, we stated that methanol can be produced by the reaction between carbon monoxide and hydrogen. Let's consider this process again. Suppose 68.5 kg (6.85×10^4 g) of CO(g) is reacted with 8.60 kg (8.60×10^3 g) of H$_2$(g).

a.   Calculate the theoretical yield of methanol.
b.   If 3.57×10^4 g of CH$_3$OH is actually produced, what is the percent yield of methanol?

SOLUTION

a.   STEP 1
      The balanced equation is $2H_2(g) + CO(g) \rightarrow CH_3OH(l)$

STEP 2
Next we calculate the moles of reactants.

$$6.85 \times 10^4 \text{ g CO} \times \frac{1 \text{ mol CO}}{28.01 \text{ g CO}} = 2.45 \times 10^3 \text{ mol CO}$$

$$8.60 \times 10^3 \text{ g H}_2 \times \frac{1 \text{ mol H}_2}{2.016 \text{ g H}_2} = 4.27 \times 10^3 \text{ mol H}_2$$

STEP 3
Then we determine which reactant is limiting. Using the mole ratio between CO and H$_2$ from the balanced equation, we have

$$2.45 \times 10^3 \text{ mol CO} \times \frac{2 \text{ mol H}_2}{1 \text{ mol CO}} = 4.90 \times 10^3 \text{ mol H}_2$$

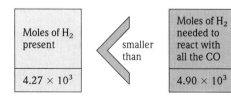

Moles of H$_2$ present	smaller than	Moles of H$_2$ needed to react with all the CO
4.27×10^3		4.90×10^3

EXAMPLE 10.9, CONTINUED

We see that 2.45×10^3 mol of CO requires 4.90×10^3 mol of H_2. Because only 4.27×10^3 mol of H_2 is actually present, H_2 *is limiting.*

STEP 4
We must therefore use the amount of H_2 and the mole ratio between H_2 and CH_3OH to determine the maximum amount of methanol that can be produced in the reaction.

$$4.27 \times 10^3 \text{ mol } H_2 \times \frac{1 \text{ mol } CH_3OH}{2 \text{ mol } H_2} = 2.14 \times 10^3 \text{ mol } CH_3OH$$

This represents the theoretical yield in moles.

STEP 5
Using the molar mass of CH_3OH (32.00 g), we can calculate the theoretical yield in grams.

$$2.14 \times 10^3 \text{ mol } CH_3OH \times \frac{32.00 \text{ g } CH_3OH}{1 \text{ mol } CH_3OH} = 6.85 \times 10^4 \text{ g } CH_3OH$$

So, from the amounts of reactants given, the maximum amount of CH_3OH that can be formed is 6.85×10^4 g. This is the *theoretical yield.*

b. The percent yield is

$$\frac{\text{Actual yield (grams)}}{\text{Theoretical yield (grams)}} \times 100 = \frac{3.57 \times 10^4 \text{ g } CH_3OH}{6.85 \times 10^4 \text{ g } CH_3OH} \times 100$$
$$= 52.1\%$$

SELF-CHECK EXERCISE 10.7

Titanium(IV) oxide is a white compound used as a coloring pigment. In fact, the page you are now reading is white because of the presence of this compound in the paper. Solid titanium(IV) oxide can be prepared by reacting gaseous titanium(IV) chloride with oxygen gas. A second product of this reaction is chlorine gas.

$$TiCl_4(g) + O_2(g) \rightarrow TiO_2(s) + Cl_2(g)$$

a. Suppose 6.71×10^3 g of titanium(IV) chloride is reacted with 2.45×10^3 g of oxygen. Calculate the maximum mass of titanium(IV) oxide that can form.
b. If the percent yield of TiO_2 is 75%, what mass is actually formed?

CHAPTER REVIEW

Key Terms

mole ratio (p. 295)
stoichiometry (p. 302)
stoichiometric quantities (p. 307)
limiting reactant (limiting reagent) (p. 309)
theoretical yield (p. 315)
percent yield (p. 316)

Summary

1. A balanced equation relates the numbers of molecules of reactants and products. It can also be expressed in terms of the numbers of moles of reactants and products.
2. The process of using a chemical equation to calculate the relative amounts of reactants and products involved in the reaction is called doing stoichiometric calculations. To convert between moles of reactants and moles of products, we use mole ratios derived from the balanced equation.
3. Often reactants are not mixed in stoichiometric quantities (so they do not "run out" at the same time). In that case, we must use the limiting reactant to calculate the amounts of products formed.
4. The actual yield of a reaction is usually less than its theoretical yield. The actual yield is often expressed as a percentage of the theoretical yield called the percent yield.

Questions and Problems

All even-numbered exercises have answers in the back of this book and solutions in the Solutions Guide.

10.1 Information Given by Chemical Equations

QUESTIONS

1. What do the coefficients of a balanced chemical equation tell us about the proportions in which atoms and molecules react on an individual (microscopic) basis?

2. What do the coefficients of a balanced chemical equation tell us about the proportions in which substances react on a macroscopic (mole) basis?

3. Explain why the coefficients of a balanced chemical equation are interpreted on a mole basis, rather than directly on the basis of mass.

4. Explain why, in the balanced chemical equation $C + O_2 \rightarrow CO_2$, we know that 1 g of C will *not* react exactly with 1 g of O_2.

PROBLEMS

5. For each of the following reactions, give the balanced equation for the reaction and state the meaning of the equation in terms of numbers of *individual molecules* and in terms of *moles* of molecules.
   a. $NO(g) + O_2(g) \rightarrow NO_2(g)$
   b. $AgC_2H_3O_2(aq) + CuSO_4(aq) \rightarrow Ag_2SO_4(s) + Cu(C_2H_3O_2)_2(aq)$
   c. $PCl_3(l) + H_2O(l) \rightarrow H_3PO_3(l) + HCl(g)$
   d. $C_2H_6(g) + Cl_2(g) \rightarrow C_2H_5Cl(g) + HCl(g)$

6. For each of the following reactions, give the balanced equation for the reaction and state the meaning of the equation in terms of numbers of *individual molecules* and in terms of *moles* of molecules.
   a. $HC_2H_3O_2(aq) + Ca(OH)_2(aq) \rightarrow Ca(C_2H_3O_2)_2(aq) + H_2O(l)$
   b. $Ba(OH)_2 \rightarrow BaO(s) + H_2O(g)$
   c. $P_4(s) + H_2(g) \rightarrow PH_3(g)$
   d. $Al(s) + H_2SO_4(aq) \rightarrow Al_2(SO_4)_3(s) + H_2(g)$

10.2 Mole–Mole Relationships

QUESTIONS

7. True or false? For the reaction represented by the chemical equation

$$2H_2(g) + O_2(g) \rightarrow 2H_2O(l)$$

the mole ratio that would be used to convert moles of O_2 reacting to moles of H_2O formed is 2 mol H_2O/2 mol H_2.

9. Consider the balanced equation

$$CH_4(g) + 2O_2(g) \rightarrow CO_2(g) + 2H_2O(g)$$

What is the mole ratio that would enable you to calculate the number of moles of oxygen needed to react exactly with a given number of moles of $CH_4(g)$? What mole ratios would you use to calculate how many moles of each product forms from a given number of moles of CH_4?

8. True or false? For the reaction represented by the balanced chemical equation

$$N_2(g) + 3I_2(s) \rightarrow 2NI_3(s)$$

if you want to prepare 2 mol of $NI_3(s)$, then 1 g of $N_2(g)$ and 3 g of $I_2(s)$ will be needed.

10. Consider the unbalanced chemical equation

$$CH_3CH_2OH(l) + O_2(g) \rightarrow CO_2(g) + H_2O(g)$$

Balance the equation. Then write the mole ratios that would enable you to calculate the number of moles of each product that would form for a given number of moles of $CH_3CH_2OH(l)$ reacting in air.

PROBLEMS

11. For each of the following balanced reactions, calculate how many *moles of product* would be produced by complete conversion of 0.15 mol of the reactant indicated in boldface. Write clearly the mole ratio used for the conversion.
    a. **2Mg**(s) $+ O_2(g) \rightarrow 2MgO(s)$
    b. 2Mg(s) $+$ **O_2**(g) $\rightarrow 2MgO(s)$
    c. **4Fe**(s) $+ 3O_2(g) \rightarrow 2Fe_2O_3(s)$
    d. 4Fe(s) $+$ **$3O_2$**(g) $\rightarrow 2Fe_2O_3(s)$

12. For each of the following balanced reactions, calculate how many *moles of each product* would be produced by the complete conversion of 0.20 mol of the reactant indicated in boldface. State clearly the mole ratio used for the conversion.
    a. $2AgNO_3(aq) +$ **$NiSO_4$**$(aq) \rightarrow Ag_2SO_4(aq) + Ni(NO_3)_2(aq)$
    b. **2Al**(s) $+ 3H_2SO_4(aq) \rightarrow Al_2(SO_4)_3(aq) + 3H_2(aq)$
    c. **$2NI_3$**(s) $\rightarrow N_2(g) + 3I_2(s)$
    d. $H_3PO_4(aq) +$ **3NaOH**$(aq) \rightarrow Na_3PO_4(aq) + 3H_2O(l)$

13. For each of the balanced reactions below, calculate how many *moles of each product* would be produced by complete conversion of 1.25 mol of the reactant indicated in boldface. Write clearly the mole ratio used for the conversion.
    a. **$2C_2H_5OH$**$(l) + 7O_2(g) \rightarrow 4CO_2(g) + 6H_2O(g)$
    b. **N_2**$(g) + O_2(g) \rightarrow 2NO(g)$
    c. **$2NaClO_2$**$(s) + Cl_2(g) \rightarrow 2ClO_2(g) + 2NaCl(s)$
    d. **$3H_2$**$(g) + N_2(g) \rightarrow 2NH_3(g)$

14. For each of the balanced reactions in Question 12, calculate how many moles of each product would be produced by complete conversion of 0.20 g of the reactant indicated in boldface.

15. For each of the following *unbalanced* equations, indicate how many *moles* of the *second reactant* would be required to react exactly with *0.25 mol* of the *first reactant*. Write clearly the mole ratio used for the conversion.
    a. $CO(g) + O_2(g) \rightarrow CO_2(g)$
    b. $AgNO_3(aq) + CuSO_4(aq) \rightarrow Ag_2SO_4(s) + Cu(NO_3)_2(aq)$
    c. $PCl_3(l) + H_2O(l) \rightarrow H_3PO_3(l) + HCl(g)$
    d. $CH_4(g) + Cl_2(g) \rightarrow CCl_4(l) + HCl(g)$

16. For each of the following *unbalanced* equations, indicate how many *moles* of the *second reactant* would be required to react exactly with *3.125 mol* of the *first reactant*. Write clearly the mole ratio used for the conversion.
    a. $N_2(g) + H_2(g) \rightarrow NH_3(g)$
    b. $Al(s) + O_2(g) \rightarrow Al_2O_3(s)$
    c. $Ag^+(aq) + CO_3^{2-}(aq) \rightarrow Ag_2CO_3(s)$
    d. $C_5H_{12}(l) + O_2(g) \rightarrow CO_2(g) + H_2O(g)$

10.3 Mass Calculations

QUESTIONS

17. What quantity serves as the conversion factor between the mass of a sample and how many moles the sample contains?

18. What does it mean to say that the balanced chemical equation for a reaction describes the *stoichiometry* of the reaction?

PROBLEMS

19. Using the average atomic masses given inside the front cover, calculate how many *moles* of each element the following *masses* represent.
    a. 19.5 g of gold
    b. 11.78 g of iron
    c. 2.395 g of carbon dioxide
    d. 1.00 kg of barium chloride
    e. 1.00 mg of magnesium
    f. 86.2 g of sodium acetate
    g. 91.4 g of beryllium

20. Using the average atomic masses given inside the front cover, calculate the number of *moles* of each substance the following *masses* represent.
    a. 14.15 g of carbon
    b. 4.01×10^{-3} g of diboron trioxide
    c. 2.5 kg of tin(IV) chloride
    d. 1.13×10^{-3} g of sodium fluoride
    e. 40.5 g of sucrose, $C_{12}H_{22}O_{11}$
    f. 4.7 μg of uranium(IV) oxide
    g. 62.1 mg lithium carbonate

21. Using the average atomic masses given inside the front cover, calculate the *mass in grams* of each of the following samples.
    a. 1.25 mol of platinum(IV) chloride
    b. 0.00255 mol of copper(II) oxide
    c. 1.89×10^{-4} mol of ethane, C_2H_6
    d. 55.56 mol of beryllium
    e. 2.6×10^7 mol of diboron trioxide
    f. 0.45 mol of sodium fluoride
    g. 0.00115 mol of calcium nitrate

22. Using the average atomic masses given inside the front cover, calculate the *mass in grams* of each of the following samples.
    a. 0.101 mol of hydrogen chloride
    b. 2.00 mol of nitrogen triiodide
    c. 0.253 mol of calcium carbonate
    d. 125 mol of carbon dioxide
    e. 1.51×10^{-3} mol of gold(III) chloride
    f. 7.42×10^2 mol of sulfur dioxide
    g. 0.000315 mol of chloroplatinic acid, H_2PtCl_4

23. For each of the following *unbalanced* equations, indicate how many *moles of each product* could be produced by complete reaction of *1.00 g* of the reactant indicated in boldface. Write clearly the mole ratio used for the conversion.
    a. $LiOH(s) + \mathbf{CO_2}(g) \rightarrow Li_2CO_3(s) + H_2O(l)$
    b. $\mathbf{Ba(OH)_2}(s) \rightarrow BaO(s) + H_2O(g)$
    c. $C_2H_4(g) + \mathbf{Cl_2}(g) \rightarrow C_2H_4Cl_2(l)$
    d. $H_2SO_4(aq) + \mathbf{NaOH}(aq) \rightarrow Na_2SO_4(aq) + H_2O(l)$

24. For each of the following *unbalanced* equations, indicate how many *moles* of the *second reactant* would be required for complete reaction of 1.00 g of the *first reactant*. State clearly the mole ratio used for the conversion.
    a. $SO_2(g) + NaOH(s) \rightarrow Na_2SO_3(s) + H_2O(l)$
    b. $Cl_2(g) + Br_2(l) \rightarrow BrCl(l)$
    c. $PbO(s) + HCl(aq) \rightarrow PbCl_2(s) + H_2O(l)$
    d. $HCO_2H(aq) + KOH(aq) \rightarrow KHCO_2(aq) + H_2O(l)$

25. For each of the following *unbalanced* equations, calculate how many *grams of (each) product* would be produced by complete reaction of *10.0 g* of the reactant indicated in boldface. Write clearly the mole ratio used for the conversion.
    a. **Na**(s) + N_2(g) → Na_3N(s)
    b. BaO(s) + **HCl**(aq) → $BaCl_2$(aq) + H_2O(l)
    c. **Cl_2**(g) + KI(s) → KCl(s) + I_2(s)
    d. **HgO**(s) → Hg(l) + O_2(g)

26. For each of the following *unbalanced* equations, calculate how many *milligrams of (each) product* would be produced by complete reaction of *10.0 mg* of the reactant indicated in boldface. Write clearly the mole ratio used for the conversion.
    a. **$FeSO_4$**(aq) + K_2CO_3(aq) → $FeCO_3$(s) + K_2SO_4(aq)
    b. **Cr**(s) + $SnCl_4$(l) → $CrCl_3$(s) + Sn(s)
    c. Fe(s) + **S_8**(s) → Fe_2S_3(s)
    d. Ag(s) + **HNO_3**(aq) → $AgNO_3$(aq) + H_2O(l) + NO(g)

27. Although mixtures of hydrogen and oxygen are highly explosive, pure elemental hydrogen gas itself burns quietly in air with a pale blue flame, producing water vapor.

$$2H_2(g) + O_2(g) → 2H_2O(g)$$

Calculate the mass (in grams) of water vapor produced when 56.0 g of pure hydrogen gas burns.

28. Given the information in Problem 27, calculate the mass of oxygen gas that would be needed to burn 1.00 g of hydrogen gas.

29. When elemental carbon is burned in the open atmosphere, with plenty of oxygen gas present, the product is carbon dioxide.

$$C(s) + O_2(g) → CO_2(g)$$

However, when the amount of oxygen present during the burning of the carbon is restricted, carbon monoxide is more likely to result.

$$2C(s) + O_2(g) → 2CO(g)$$

What mass of each product is expected when a 5.00-g sample of pure carbon is burned under each of these conditions?

30. If a shiny strip of zinc metal is placed in a blue copper sulfate solution, the zinc dissolves slowly, copper metal forms, and the blue color of the solution fades gradually. The unbalanced equation is

$$Zn(s) + CuSO_4(aq) → ZnSO_4(aq) + Cu(s)$$

If 1.00 g of zinc metal is reacted with a large excess of copper sulfate solution, how many grams of copper metal will be produced?

31. Although we usually think of substances as "burning" only in oxygen gas, the process of rapid oxidation to produce a flame may also take place in other strongly oxidizing gases. For example, when iron is heated and placed in pure chlorine gas, the iron "burns" according to the following (unbalanced) reaction.

$$Fe(s) + Cl_2(g) → FeCl_3(s)$$

How many milligrams of iron(III) chloride result when 15.5 mg of iron is reacted with an excess of chlorine gas?

32. Hydrochloric acid will attack and corrode many of the metallic elements, resulting in the generation of hydrogen gas and producing a solution of the metal chloride. Consider the following unbalanced reactions:

$$Zn(s) + HCl(aq) → ZnCl_2(aq) + H_2(g)$$
$$Al(s) + HCl(aq) → AlCl_3(aq) + H_2(g)$$
$$Mg(s) + HCl(aq) → MgCl_2(aq) + H_2(g)$$

Calculate the mass of each metal required to produce 1.00 g of hydrogen by its respective reaction.

33. When very small samples of oxygen gas are needed in the laboratory, the gas may be generated by any number of simple chemical reactions, such as

$$2KClO_3(s) → 2KCl(s) + 3O_2(g)$$

What mass of oxygen gas should be produced when 5.00 g of $KClO_3$ is heated?

34. Elemental fluorine and chlorine gases are very reactive. For example, they react with each other to produce an *inter*halogen compound, chlorine monofluoride.

$$Cl_2(g) + F_2(g) → 2ClF(g)$$

Calculate how many grams of chlorine gas are required to react completely with 5.00 mg of fluorine gas.

35. With the news that calcium deficiencies in many women's diets may contribute to the development of osteoporosis (bone weakening), the use of dietary calcium supplements has become increasingly common. Many of these calcium supplements consist of nothing more than calcium carbonate, $CaCO_3$. When a calcium carbonate tablet is ingested, it dissolves by reaction with stomach acid, which contains hydrochloric acid, HCl. The unbalanced equation is

$$CaCO_3(s) + HCl(aq) \rightarrow CaCl_2(aq) + H_2O(l) + CO_2(g)$$

What mass of HCl is required to react with a tablet containing 500. mg of calcium carbonate?

36. Sulfur is sometimes used to contain mercury spills in the lab. Sulfur reacts with mercury according to the balanced chemical equation:

$$S(s) + Hg(l) \rightarrow HgS(s)$$

Calculate how many grams of sulfur are required to react exactly with 0.10 g of mercury.

37. Passage of an electric current ("electrolysis") through molten sodium chloride can be used as a means of preparing pure metallic sodium and chlorine gas.

$$NaCl(l) \xrightarrow[\text{current}]{\text{Electric}} Na(s) + Cl_2(g)$$

How many grams of pure sodium and how many grams of pure chlorine should be produced when 5.00 g of sodium chloride is completely reacted by this process?

38. Solutions of sodium hydroxide cannot be kept for very long because they absorb carbon dioxide from the air, forming sodium carbonate. The unbalanced equation is

$$NaOH(aq) + CO_2(g) \rightarrow Na_2CO_3(aq) + H_2O(l)$$

Calculate the number of grams of carbon dioxide that can be absorbed by complete reaction with a solution that contains 5.00 g of sodium hydroxide.

39. Ethanol, C_2H_5OH, has been used for some time as a source of heat in such commercial products as Sterno. Ethanol burns in air to produce gaseous carbon dioxide and water vapor. Write the balanced chemical equation for this process. What mass of carbon dioxide and what mass of water vapor are produced when 10.0 g of pure ethanol is burned? What mass of oxygen is needed for the complete burning of the 10.0 g of ethanol?

40. Powdered magnesium metal burns in oxygen with an intensely bright white flame, a fact that has been made use of in photographic flash units. The balanced equation for this reaction is:

$$2Mg(s) + O_2(g) \rightarrow 2MgO(s)$$

How many grams of $MgO(s)$ are produced by complete reaction of 1.25 g of magnesium metal?

10.4 Calculations Involving a Limiting Reactant

QUESTIONS

41. What is the *limiting reactant* for a process? Why does a reaction stop when the limiting reactant is consumed, even though there may be plenty of the other reactants present?

42. Explain how one determines which reactant in a process is the limiting reactant? Does this depend only on the masses of the reactant present? Is the mole ratio in which the reactants combine involved?

43. What is the *theoretical yield* for a reaction, and how does this quantity depend on the limiting reactant?

44. Can a reactant be present "in excess" in a reaction? Does the presence of an excess of one or more reactants affect the theoretical yield of the reaction?

PROBLEMS

45. For each of the following reactions, suppose exactly 10.0 g of each reactant is taken. Determine which reactant is limiting, and calculate what mass of product is expected, assuming that the limiting reactant is completely consumed.
    a. $N_2(g) + O_2(g) \rightarrow 2NO(g)$
    b. $N_2(g) + 3H_2(g) \rightarrow 2NH_3(g)$
    c. $2CO(g) + O_2(g) \rightarrow 2CO_2(g)$
    d. $PCl_3(l) + Cl_2(g) \rightarrow PCl_5(s)$

46. For each of the following *unbalanced* chemical equations, suppose that exactly 15.0 g of each reactant is taken. Determine which reactant is limiting, and calculate what mass of each product is expected. (Assume that the limiting reactant is completely consumed.)
    a. $Al(s) + HCl(aq) \rightarrow AlCl_3(aq) + H_2(g)$
    b. $NaOH(aq) + CO_2(g) \rightarrow Na_2CO_3(aq) + H_2O(l)$
    c. $Pb(NO_3)_2(aq) + HCl(aq) \rightarrow PbCl_2(s) + HNO_3(aq)$
    d. $K(s) + I_2(s) \rightarrow KI(s)$

47. For each of the following *unbalanced* chemical equations, suppose 10.0 g of *each* reactant is taken. Show by calculation which reactant is the limiting reagent. Calculate the mass of each product that is expected.
    a. $C_3H_8(g) + O_2(g) \rightarrow CO_2(g) + H_2O(g)$
    b. $Al(s) + Cl_2(g) \rightarrow AlCl_3(s)$
    c. $NaOH(s) + CO_2(g) \rightarrow Na_2CO_3(s) + H_2O(l)$
    d. $NaHCO_3(s) + HCl(aq) \rightarrow NaCl(aq) + H_2O(l) + CO_2(g)$

48. For each of the following *unbalanced* chemical equations, suppose that exactly 50.0 g of each reactant is taken. Determine which reactant is limiting, and calculate what mass of the product in boldface is expected. (Assume that the limiting reactant is completely consumed.)
    a. $NH_3(g) + Na(s) \rightarrow NaNH_2(s) + H_2(g)$
    b. $BaCl_2(aq) + Na_2SO_4(aq) \rightarrow BaSO_4(s) + NaCl(aq)$
    c. $SO_2(g) + NaOH(s) \rightarrow Na_2SO_3(s) + H_2O(l)$
    d. $Al(s) + H_2SO_4(l) \rightarrow Al_2(SO_4)_3(s) + H_2(g)$

49. For each of the following *unbalanced* chemical equations, suppose 1.00 g of *each* reactant is taken. Show by calculation which reactant is the limiting reagent. Calculate the mass of each product that is expected.
    a. $UO_2(s) + HF(aq) \rightarrow UF_4(aq) + H_2O(l)$
    b. $NaNO_3(aq) + H_2SO_4(aq) \rightarrow Na_2SO_4(aq) + HNO_3(aq)$
    c. $Zn(s) + HCl(aq) \rightarrow ZnCl_2(aq) + H_2(g)$
    d. $B(OH)_3(s) + CH_3OH(l) \rightarrow B(OCH_3)_3(s) + H_2O(l)$

50. For each of the following *unbalanced* chemical equations, suppose 10.0 mg of *each* reactant is taken. Show by calculation which reactant is the limiting reagent. Calculate the mass of each product that is expected.
    a. $CO(g) + H_2(g) \rightarrow CH_3OH(l)$
    b. $Al(s) + I_2(s) \rightarrow AlI_3(s)$
    c. $Ca(OH)_2(aq) + HBr(aq) \rightarrow CaBr_2(aq) + H_2O(l)$
    d. $Cr(s) + H_3PO_4(aq) \rightarrow CrPO_4(s) + H_2(g)$

51. Ordinarily, elemental nitrogen gas is not very reactive. However, when one of the more reactive metallic elements is heated in pure nitrogen, nitrogen is reduced to the nitride ion, N^{3-}. For example, when magnesium metal is heated in pure nitrogen, magnesium nitride results.

    $$3Mg(s) + N_2(g) \rightarrow Mg_3N_2(s)$$

    What is the theoretical yield of magnesium nitride when 5.00 g of magnesium is heated in 25.0 grams of nitrogen gas? How many grams of the excess reactant will remain after the reaction is complete?

52. An experiment that led to the foundation of a whole new field of chemistry involved the formation of urea, CN_2H_4O, by the controlled reaction of ammonia and carbon dioxide.

    $$2NH_3(g) + CO_2(g) \rightarrow CN_2H_4O(s) + H_2O(l)$$
    Urea

    What is the theoretical yield of urea when 100. g of ammonia is reacted with 100. g of carbon dioxide?

53. Most general chemistry laboratories contain bottles of several common reagents at each student's desk for ready use. Among these reagents are solutions of ammonia ("ammonium hydroxide") and hydrogen chloride (hydrochloric acid). Because the substances dissolved in these two solutions are normally gaseous, they tend to leave the solution relatively easily. Then they react with each other, depositing a fine dust of ammonium chloride on the lab benches.

    $$NH_3(g) + HCl(g) \rightarrow NH_4Cl(s)$$

    What is the maximum amount of ammonium chloride that can be deposited when 125 mg of ammonia is evolved from one bottle and 190. mg of HCl is evolved from another?

54. Powdered iron metal reacts spectacularly with elemental chlorine gas, accompanied by much heat and flame.

    $$Fe(s) + Cl_2(g) \rightarrow FeCl_3(s)$$

    Suppose 5.00 g of iron is added to 10.0 g of chlorine gas. Show which reactant is the limiting reactant. Calculate the mass of product produced and the mass of unreacted starting material remaining.

55. During World War I, the substance phosphine, PH_3, was used as a poisonous gas against allied troops in their trenches (phosphine is heavier than air). Phosphine may be produced by the reaction

    $$Na_3P(s) + H_2O(l) \rightarrow PH_3(g) + NaOH(aq) \text{ (unbalanced)}$$

    What is the theoretical yield of phosphine when 150 g of sodium phosphide, Na_3P, is dissolved in 250 mL of water (density 1.0 g/mL)?

56. The copper(II) ion in a copper(II) sulfate solution reacts with potassium iodide to produce the triiodite ion, I_3^-. This reaction is commonly used to determine how much copper is present in a given sample.

    $$CuSO_4(aq) + KI(aq) \rightarrow CuI(s) + KI_3(aq) + K_2SO_4(aq)$$

    If 2.00 g of KI is added to a solution containing 0.525 g of $CuSO_4$, calculate the mass of each product produced.

57. Hydrogen peroxide is used as a cleaning agent in the treatment of cuts and abrasions for several reasons. It is an oxidizing agent that can directly kill many microorganisms; it decomposes upon contact with blood, releasing elemental oxygen gas (which inhibits the growth of anaerobic microorganisms); and it foams upon contact with blood, which provides a cleansing action. In the laboratory, small quantities of hydrogen peroxide can be prepared by the action of an acid on an alkaline earth metal peroxide, such as barium peroxide.

$$BaO_2(s) + 2HCl(aq) \rightarrow H_2O_2(aq) + BaCl_2(aq)$$

What amount of hydrogen peroxide should result when 1.50 g of barium peroxide is treated with 25.0 mL of hydrochloric acid solution containing 0.0272 g of HCl per mL?

58. Silicon carbide, SiC, is one of the hardest materials known. Surpassed in hardness only by diamond, it is sometimes known commercially as carborundum. Silicon carbide is used primarily as an abrasive for sandpapers and is manufactured by heating common sand (silicon dioxide, SiO_2) with carbon in a furnace.

$$SiO_2(s) + C(s) \rightarrow CO(g) + SiC(s) \quad \text{(unbalanced)}$$

What mass of silicon carbide should result when 1.0 kg of pure sand is heated with an excess of carbon?

10.5 Percent Yield

QUESTIONS

59. What is the *actual yield* of a reaction? What is the *percent yield* of a reaction? How do the actual yield and the percent yield differ from the theoretical yield?

60. The text explains that one reason why the actual yield for a reaction may be less than the theoretical yield is side reactions. Suggest some other reasons why the percent yield for a reaction might not be 100%.

61. According to her pre-lab stoichiometric calculation, a student's lab experiment should have produced 5.51 g of NaCl. While actually doing the experiment, this student obtained only 4.32 g of NaCl. What was her percent yield?

62. A student calculated the theoretical yield of $BaSO_4(s)$ in a precipitation experiment to be 1.352 g. When she filtered, dried, and weighed her precipitate, however, her yield was only 1.279 g. Calculate the student's percent yield.

PROBLEMS

63. The compound sodium thiosulfate pentahydrate, $Na_2S_2O_3 \cdot 5H_2O$, is important commercially to the photography business as "hypo," because it has the ability to dissolve unreacted silver salts from photographic film during development. Sodium thiosulfate pentahydrate can be produced by boiling elemental sulfur in an aqueous solution of sodium sulfite.

$$S_8(s) + Na_2SO_3(aq) + H_2O(l)$$
$$\rightarrow Na_2S_2O_3 \cdot 5H_2O(s) \quad \text{(unbalanced)}$$

What is the theoretical yield of sodium thiosulfate pentahydrate when 3.25 g of sulfur is boiled with 13.1 g of sodium sulfite? Sodium thiosulfate pentahydrate is very soluble in water. What is the percent yield of the synthesis if a student doing this experiment is able to isolate (collect) only 5.26 g of the product?

64. Silicon is an element that is found in many rocks ("silicates") and is very much in demand in the pure elemental form for the semiconductor industry. Relatively pure silicon can be made from ordinary sand, by reduction with carbon at high temperatures.

$$SiO_2(s) + C(s) \rightarrow Si(s) + CO(g) \quad \text{(unbalanced)}$$

What mass of pure elemental silicon should result when 100. kg of sand (SiO_2) is heated in a suitable furnace with 100. kg of carbon? What is the percent yield if only 17.2 kg of silicon actually results from the process?

65. Although they were formerly called the inert gases, at least the heavier elements of Group 8 do form relatively stable compounds. For example, xenon combines directly with elemental fluorine at elevated temperatures in the presence of a nickel catalyst.

$$Xe(g) + 2F_2(g) \rightarrow XeF_4(s)$$

What is the theoretical mass of xenon tetrafluoride that should form when 130. g of xenon is reacted with 100. g of F_2? What is the percent yield if only 145 g of XeF_4 is actually isolated?

Additional Problems

67. Natural waters often contain relatively high levels of calcium ion, Ca^{2+}, and hydrogen carbonate ion (bicarbonate), HCO_3^-, from the leaching of minerals into the water. When such water is used commercially or in the home, heating of the water leads to the formation of solid calcium carbonate, $CaCO_3$, which forms a deposit (or "scale") on the interior of boilers, pipes, and other plumbing fixtures.

$$Ca(HCO_3)_2(aq) \rightarrow CaCO_3(s) + CO_2(g) + H_2O(l)$$

If a sample of well water contains 2.0×10^{-3} mg of $Ca(HCO_3)_2$ per milliliter, what mass of $CaCO_3$ scale would 1.0 mL of this water be capable of depositing?

68. One process for the commercial production of baking soda (sodium hydrogen carbonate) involves the following reaction, in which the carbon dioxide is used in its solid form ("dry ice") both to serve as a source of reactant and to cool the reaction system to a temperature low enough for the sodium hydrogen carbonate to precipitate:

$$NaCl(aq) + NH_3(aq) + H_2O(l) + CO_2(s)$$
$$\rightarrow NH_4Cl(aq) + NaHCO_3(s)$$

Because they are relatively cheap, sodium chloride and water are typically present in excess. What is the expected yield of $NaHCO_3$ when one performs such a synthesis using 10.0 g of ammonia and 15.0 g of dry ice, with an excess of NaCl and water?

69. When sulfur is burned in air, the noxious, toxic gas sulfur dioxide results.

$$S(s) + O_2(g) \rightarrow SO_2(g)$$

66. Anhydrous calcium chloride, $CaCl_2$, is frequently used in the laboratory as a drying agent for solvents, because it absorbs 6 mol of water molecules for every mole of $CaCl_2$ used (forming a stable solid hydrated salt, $CaCl_2 \cdot 6H_2O$). Calcium chloride is typically prepared by treating calcium carbonate with hydrogen chloride gas.

$$CaCO_3(s) + 2HCl \rightarrow CaCl_2(s) + CO_2(g) + H_2O(g)$$

A large amount of heat is generated by this reaction, so the water produced from the reaction is usually driven off as steam. Some liquid water may remain, however, and it may dissolve some of the desired calcium chloride. What is the percent yield if 155 g of calcium carbonate is treated with 250. g of anhydrous hydrogen chloride and only 142 g of $CaCl_2$ is obtained?

Until a few years ago, many hydrocarbon fuels that were burned for heat and power were "high-sulfur" fuels, which contributed greatly to air pollution in the United States and other countries. When such fuels are burned, the sulfur they contain is converted to sulfur dioxide. Suppose such a fuel is 1.5% sulfur by mass and has density of 0.82 g/mL. What mass of sulfur dioxide is released when 1000 gal of this fuel is burned?

70. When the sugar glucose, $C_6H_{12}O_6$, is burned in air, carbon dioxide and water vapor are produced. Write the balanced chemical equation for this process, and calculate the theoretical yield of carbon dioxide when 1.00 g of glucose is burned completely.

71. When elemental copper is strongly heated with sulfur, a mixture of CuS and Cu_2S is produced, with CuS predominating.

$$Cu(s) + S(s) \rightarrow CuS(s) \qquad 2Cu(s) + S(s) \rightarrow Cu_2S(s)$$

What is the theoretical yield of CuS when 31.8 g of Cu(s) is heated with 50.0 g of S? (Assume only CuS is produced in the reaction.) What is the percent yield of CuS if only 40.0 g of CuS can be isolated from the mixture?

72. Potassium chromate is used in chemical analysis to precipitate quantitatively lead ion from solution (lead chromate is very insoluble).

$$Pb^{2+}(aq) + CrO_4^{2-}(aq) \rightarrow PbCrO_4(s)$$

Suppose a solution is known to contain no more than 15 mg of lead ion. What mass of potassium chromate should be added to guarantee precipitation of all the lead ion?

73. The traditional method of analysis for the amount of chloride ion present in a sample was to dissolve the sample in water and then slowly to add a solution of silver nitrate. Silver chloride is very insoluble in water, and by adding a slight excess of silver nitrate, it is possible effectively to remove all chloride ion from the sample.

$$Ag^+(aq) + Cl^-(aq) \rightarrow AgCl(s)$$

Suppose a 1.054-g sample is known to contain 10.3% chloride ion by mass. What weight of silver nitrate must be used to completely precipitate the chloride ion from the sample? What mass of silver chloride will be obtained?

74. For each of the following reactions, give the balanced equation for the reaction and state the meaning of the equation in terms of numbers of *individual molecules* and in terms of *moles* of molecules.
    a. $UO_2(s) + HF(aq) \rightarrow UF_4(aq) + H_2O(l)$
    b. $NaC_2H_3O_2(aq) + H_2SO_4(aq) \rightarrow Na_2SO_4(aq) + HC_2H_3O_2(aq)$
    c. $Mg(s) + HCl(aq) \rightarrow MgCl_2(aq) + H_2(g)$
    d. $B_2O_3(s) + H_2O(l) \rightarrow B(OH)_3(aq)$

75. True or false? For the reaction represented by the balanced chemical equation

$$Mg(OH)_2(aq) + 2HCl(aq) \rightarrow 2H_2O(l) + MgCl_2(aq)$$

if 0.40 mol of $Mg(OH)_2$ is to react, then 0.20 mol of HCl will be needed.

76. Consider the balanced equation

$$C_3H_8(g) + 5O_2(g) \rightarrow 3CO_2(g) + 4H_2O(g)$$

What is the mole ratio that would enable you to calculate the number of moles of oxygen needed to react exactly with a given number of moles of $C_3H_8(g)$? What are the mole ratios that would enable you to calculate how many moles of each product form from a given number of moles of C_3H_8 reacting?

77. For each of the following balanced reactions, calculate how many *moles of each product* would be produced by complete conversion of 0.50 mol of the reactant indicated in boldface. Write clearly the mole ratio used for the conversion.
    a. $\mathbf{2H_2O_2}(l) \rightarrow 2H_2O(l) + O_2(g)$
    b. $\mathbf{2KClO_3}(s) \rightarrow 2KCl(s) + 3O_2(g)$
    c. $\mathbf{2Al}(s) + 6HCl(aq) \rightarrow 2AlCl_3(aq) + 3H_2(g)$
    d. $\mathbf{C_3H_8}(g) + 5O_2(g) \rightarrow 3CO_2(g) + 4H_2O(g)$

78. For each of the following balanced equations, indicate how many *moles of the product* could be produced by complete

reaction of *1.00 g* of the reactant indicated in boldface. Write clearly the mole ratio used for the conversion.
    a. $\mathbf{NH_3}(g) + HCl(g) \rightarrow NH_4Cl(s)$
    b. $\mathbf{CaO}(s) + CO_2(g) \rightarrow CaCO_3(s)$
    c. $\mathbf{4Na}(s) + O_2(g) \rightarrow 2Na_2O(s)$
    d. $\mathbf{2P}(s) + 3Cl_2(g) \rightarrow 2PCl_3(l)$

79. Using the average atomic masses given inside the front cover, calculate how many *moles* of each element the following *masses* represent.
    a. 1.5 mg of lithium
    b. 2.0×10^{-3} g of dinitrogen monoxide
    c. 4.84×10^4 g of phosphorus trichloride
    d. 3.6×10^{-2} µg of uranium(VI) fluoride
    e. 1.0 kg of lead(II) sulfide
    f. 20.4 g of sulfuric acid
    g. 62.8 g of glucose, $C_6H_{12}O_6$

80. Using the average atomic masses given inside the front cover, calculate the *mass in grams* of each of the following samples.
    a. 5.0 mol of nitric acid
    b. 0.000305 mol of mercury
    c. 2.31×10^{-5} mol of potassium chromate
    d. 10.5 mol of aluminum chloride
    e. 4.9×10^4 mol of sulfur hexafluoride
    f. 125 mol of ammonia
    g. 0.01205 mol of sodium peroxide

81. For each of the following *unbalanced* equations, indicate how many *moles* of the *second reactant* would be required for complete reaction of *1.00 g* of the *first reactant*. Write clearly the mole ratio used for the conversion.
    a. $CO(g) + O_2(g) \rightarrow CO_2(g)$
    b. $AgNO_3(aq) + CuSO_4(aq) \rightarrow Ag_2SO_4(s) + Cu(NO_3)_2(aq)$
    c. $Al(s) + HCl(g) \rightarrow AlCl_3(s) + H_2(g)$
    d. $C_3H_8(g) + O_2(g) \rightarrow CO_2(g) + H_2O(g)$

82. One step in the commercial production of sulfuric acid, H_2SO_4, involves the conversion of sulfur dioxide, SO_2, into sulfur trioxide, SO_3.

$$2SO_2(g) + O_2(g) \rightarrow 2SO_3(g)$$

If 150 kg of SO_2 reacts completely, what mass of SO_3 should result?

83. Many metals occur naturally as sulfide compounds; examples include ZnS and CoS. Air pollution often accompanies the processing of these ores, because toxic sulfur dioxide is released as the ore is converted from the sulfide to the oxide by roasting (smelting). For example, consider the unbalanced equation for the roasting reaction for zinc.

$$ZnS(s) + O_2(g) \rightarrow ZnO(s) + SO_2(g)$$

How many kilograms of sulfur dioxide are produced when 1.0×10^2 kg of ZnS is roasted in excess oxygen by this process?

84. Elemental chlorine oxidizes the bromide ion of sodium bromide as follows:

$$Cl_2(g) + 2NaBr(aq) \rightarrow 2NaCl(aq) + Br_2(l)$$

How many grams of elemental bromine are produced when 25.0 g of elemental chlorine gas is pumped slowly into a large excess of sodium bromide solution?

85. When elemental copper is placed in a solution of silver(I) nitrate, the following oxidation–reduction reaction takes place, forming elemental silver.

$$Cu(s) + 2AgNO_3(aq) \rightarrow Cu(NO_3)_2(aq) + Ag(s)$$

What mass of copper is required to remove all the silver from a silver nitrate solution containing 1.95 mg of silver nitrate?

86. When small quantities of elemental hydrogen gas are needed for laboratory work, the hydrogen is often generated by chemical reaction of a metal with acid. For example, zinc reacts with hydrochloric acid, releasing gaseous elemental hydrogen:

$$Zn(s) + 2HCl(aq) \rightarrow ZnCl_2(aq) + H_2(g)$$

What mass of hydrogen gas is produced when 2.50 g of zinc is reacted with excess aqueous hydrochloric acid?

87. The gaseous hydrocarbon acetylene, C_2H_2, is used in welders' torches because of the large amount of heat released when acetylene burns with oxygen.

$$2C_2H_2(g) + 5O_2(g) \rightarrow 4CO_2(g) + 2H_2O(g)$$

How many grams of oxygen gas are needed for the complete combustion of 150 g of acetylene?

88. For each of the following *unbalanced* chemical equations, suppose exactly 5.0 g of each reactant is taken. Determine which reactant is limiting, and calculate what mass of each product is expected, assuming that the limiting reactant is completely consumed.

a. $Na(s) + Br_2(l) \rightarrow NaBr(s)$
b. $Zn(s) + CuSO_4(aq) \rightarrow ZnSO_4(aq) + Cu(s)$
c. $NH_4Cl(aq) + NaOH(aq) \rightarrow NH_3(g) + H_2O(l) + NaCl(aq)$
d. $Fe_2O_3(s) + CO(g) \rightarrow Fe(s) + CO_2(g)$

89. For each of the following *unbalanced* chemical equations, suppose 25.0 g of *each* reactant is taken. Show by calculation which reactant is the limiting reagent. Calculate the theoretical yield in grams of the product in boldface.

a. $C_2H_5OH(l) + O_2(g) \rightarrow \mathbf{CO_2}(g) + H_2O(l)$
b. $N_2(g) + O_2(g) \rightarrow \mathbf{NO}(g)$
c. $NaClO_2(aq) + Cl_2(g) \rightarrow ClO_2(g) + \mathbf{NaCl}(aq)$
d. $H_2(g) + N_2(g) \rightarrow \mathbf{NH_3}(g)$

90. Hydrazine, N_2H_4, emits a large quantity of energy when it reacts with oxygen, which has led to hydrazine's use as a fuel for rockets:

$$N_2H_4(l) + O_2(g) \rightarrow N_2(g) + 2H_2O(g)$$

How many moles of each of the gaseous products are produced when 20.0 g of pure hydrazine is ignited in the presence of 20.0 g of pure oxygen? How many grams of each product are produced?

91. Although elemental chlorine, Cl_2, is added to drinking water supplies primarily to kill microorganisms, another beneficial reaction that also takes place removes sulfides (which would impart unpleasant odors or tastes to the water). For example, the noxious-smelling gas hydrogen sulfide (its odor resembles that of rotten eggs) is removed from water by chlorine by the reaction:

$$H_2S(aq) + Cl_2(aq) \rightarrow HCl(aq) + S_8(s) \quad \text{(unbalanced)}$$

What mass of sulfur is removed from the water when 50. L of water containing 1.5×10^{-5} g of H_2S per liter is treated with 1.0 g of $Cl_2(g)$?

92. Before going to lab, a student read in his lab manual that the percent yield for a difficult reaction to be studied was likely to be only 40% of the theoretical yield. The student's pre-lab stoichiometric calculations predict that the theoretical yield should be 12.5 g. What is the student's actual yield likely to be?

Solutions to Self-Check Exercises

SELF-CHECK EXERCISE 10.1

The problem can be stated as follows:

$$4.30 \text{ mol } C_3H_8 \quad \boxed{\text{yields}} \quad ? \text{ mol } CO_2$$

From the balanced equation

$$C_3H_8(g) \ + \ 5O_2(g) \to 3CO_2(g) \ + \ 4H_2O(g)$$

we derive the equivalence statement

$$1 \text{ mol } C_3H_8 \ = \ 3 \text{ mol } CO_2$$

The appropriate conversion factor (moles of C_3H_8 must cancel) is 3 mol CO_2/1 mol C_3H_8, and the calculation is

$$4.30 \text{ mol } C_3H_8 \ \times \ \frac{3 \text{ mol } CO_2}{1 \text{ mol } C_3H_8} \ = \ 12.9 \text{ mol } CO_2$$

Thus we can say

$$4.30 \text{ mol } C_3H_8 \text{ yields } 12.9 \text{ mol } CO_2$$

SELF-CHECK EXERCISE 10.2

The problem can be sketched as follows:

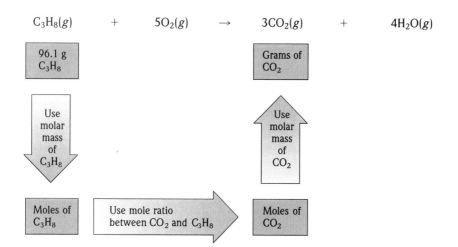

We have already done the first step in Example 10.4.

To find out how many moles of CO_2 can be produced from 2.18 mol of C_3H_8, we see from the balanced equation that 3 mol of CO_2 is produced for each mole of C_3H_8 reacted. The mole ratio we need is 3 mol CO_2/1 mol C_3H_8. The conversion is therefore

$$2.18 \text{ mol } C_3H_8 \times \frac{3 \text{ mol } CO_2}{1 \text{ mol } C_3H_8} = 6.54 \text{ mol } CO_2$$

Next, using the molar mass of CO_2, which is $12.01 + 32.00 = 44.01$ g, we calculate the mass of CO_2 produced.

$$6.54 \text{ mol } CO_2 \times \frac{44.01 \text{ g } CO_2}{1 \text{ mol } CO_2} = 288 \text{ g } CO_2$$

The sequence of steps we took to find the mass of carbon dioxide produced from 96.1 g of propane is summarized in the following diagram.

SELF-CHECK EXERCISE 10.3

We sketch the problem as follows:

$$C_3H_8(g) + 5O_2(g) \rightarrow 3CO_2(g) + 4H_2O(g)$$

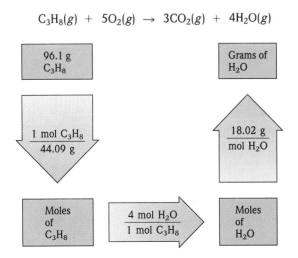

Then we do the calculations.

Therefore 157 g of H_2O is produced from 96.1 g C_3H_8.

SELF-CHECK EXERCISE 10.4

a.   We first write the balanced equation.

$$SiO_2(s) + 4HF(aq) \rightarrow SiF_4(g) + 2H_2O(l)$$

The map of the steps required is

$SiO_2(s)$ $\qquad$ + $\qquad\qquad$ $4HF(aq) \rightarrow SiF_4(g) + 2H_2O(l)$

5.68 g SiO₂		Grams of HF

Use molar mass of SiO₂ ↓ $\qquad\qquad\qquad$ ↑ Use molar mass of HF

Moles of SiO₂	Use mole ratio between HF and SiO₂ ⟹	Moles of HF

We convert 5.68 g of SiO_2 to moles as follows:

$$5.68 \text{ g } SiO_2 \times \frac{1 \text{ mol } SiO_2}{60.09 \text{ g } SiO_2} = 9.45 \times 10^{-2} \text{ mol } SiO_2$$

Using the balanced equation, we obtain the appropriate mole ratio and convert to moles of HF.

$$9.45 \times 10^{-2} \; \cancel{\text{mol SiO}_2} \times \frac{4 \text{ mol HF}}{1 \; \cancel{\text{mol SiO}_2}} = 3.78 \times 10^{-1} \text{ mol HF}$$

Finally, we calculate the mass of HF by using its molar mass.

$$3.78 \times 10^{-1} \; \cancel{\text{mol HF}} \times \frac{20.01 \text{ g HF}}{\cancel{\text{mol HF}}} = 7.56 \text{ g HF}$$

b.   The map for this problem is

$$\text{SiO}_2(s) + 4\text{HF}(aq) \rightarrow \text{SiF}_4(g) + 2\text{H}_2\text{O}(l),$$

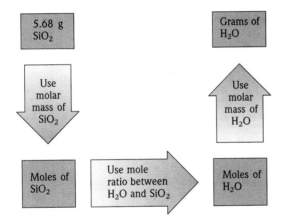

We have already accomplished the first conversion in part a. Using the balanced equation, we obtain moles of H_2O as follows:

$$9.45 \times 10^{-2} \; \cancel{\text{mol SiO}_2} \times \frac{2 \text{ mol H}_2\text{O}}{1 \; \cancel{\text{mol SiO}_2}} = 1.89 \times 10^{-1} \text{ mol H}_2\text{O}$$

The mass of water formed is

$$1.89 \times 10^{-1} \; \cancel{\text{mol H}_2\text{O}} \times \frac{18.02 \text{ g H}_2\text{O}}{\cancel{\text{mol H}_2\text{O}}} = 3.41 \text{ g H}_2\text{O}$$

SELF-CHECK EXERCISE 10.5

In this problem, we know the mass of the product to be formed by the reaction

$$\text{CO}(g) + 2\text{H}_2(g) \rightarrow \text{CH}_3\text{OH}(l)$$

and we want to find the masses of reactants needed. The procedure is the same one we have been following. We must first convert the mass of CH_3OH to moles, then use the balanced equation to obtain moles of H_2 and CO needed, and then convert these moles to masses. Using the molar mass of CH_3OH (32.04 g/mol), we convert to moles of CH_3OH.

First we convert kilograms to grams.

$$6.0 \; \cancel{kg} \; CH_3OH \; \times \; \frac{1000 \; g}{\cancel{kg}} \; = \; 6.0 \; \times \; 10^3 \; g \; CH_3OH$$

Next we convert 6.0×10^3 g CH_3OH to moles of CH_3OH, using the conversion factor 1 mol CH_3OH/32.04 g CH_3OH.

$$6.0 \; \times \; 10^3 \; \cancel{g \; CH_3OH} \; \times \; \frac{1 \; mol \; CH_3OH}{32.04 \; \cancel{g \; CH_3OH}} \; = \; 1.9 \; \times \; 10^2 \; mol \; CH_3OH$$

Then we have two questions to answer:

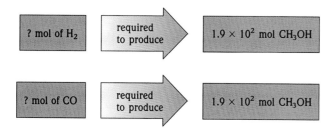

To answer these questions, we use the balanced equation

$$CO(g) \; + \; 2H_2(g) \rightarrow CH_3OH(l)$$

to obtain mole ratios between the reactants and the products. In the balanced equation the coefficients for both CO and CH_3OH are 1, so we can write the equivalence statement

$$1 \; mol \; CO \; = \; 1 \; mol \; CH_3OH$$

Using the mole ratio 1 mol CO/1 mol CH_3OH, we can now convert from moles of CH_3OH to moles of CO.

$$1.9 \; \times \; 10^2 \; \cancel{mol \; CH_3OH} \; \times \; \frac{1 \; mol \; CO}{1 \; \cancel{mol \; CH_3OH}} \; = \; 1.9 \; \times \; 10^2 \; mol \; CO$$

To calculate the moles of H_2 required, we construct the equivalence statement between CH_3OH and H_2, using the coefficients in the balanced equation.

$$2 \; mol \; H_2 \; = \; 1 \; mol \; CH_3OH$$

Using the mole ratio 2 mol H_2/1 mol CH_3OH, we can convert moles of CH_3OH to moles of H_2.

$$1.9 \; \times \; 10^2 \; \cancel{mol \; CH_3OH} \; \times \; \frac{2 \; mol \; H_2}{1 \; \cancel{mol \; CH_3OH}} \; = \; 3.8 \; \times \; 10^2 \; mol \; H_2$$

We now have the moles of reactants required to produce 6.0 kg of CH_3OH. Since we need the masses of reactants, we must use the molar masses to convert from moles to mass.

$$1.9 \times 10^2 \text{ mol CO} \times \frac{28.01 \text{ g CO}}{1 \text{ mol CO}} = 5.3 \times 10^3 \text{ g CO}$$

$$3.8 \times 10^2 \text{ mol H}_2 \times \frac{2.016 \text{ g H}_2}{1 \text{ mol H}_2} = 7.7 \times 10^2 \text{ g H}_2$$

Therefore we need 5.3×10^3 g CO to react with 7.7×10^2 g H_2 to form 6.0×10^3 g (6.0 kg) of CH_3OH. This whole process is mapped in the following diagram.

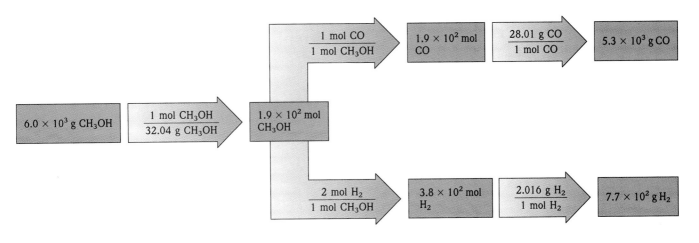

SELF-CHECK EXERCISE 10.6

STEP I
The balanced equation for the reaction is

$$6Li(s) \ + \ N_2(g) \rightarrow 2Li_3N(s)$$

STEP 2
To determine the limiting reactant, we must convert the masses of lithium (atomic mass = 6.941 g) and nitrogen (molar mass = 28.02 g) to moles.

$$56.0 \text{ g Li} \times \frac{1 \text{ mol Li}}{6.941 \text{ g Li}} = 8.07 \text{ mol Li}$$

$$56.0 \text{ g N}_2 \times \frac{1 \text{ mol N}_2}{28.02 \text{ g N}_2} = 2.00 \text{ mol N}_2$$

STEP 3
Using the mole ratio from the balanced equation, we can calculate the moles of lithium required to react with 2.00 mol of nitrogen.

$$2.00 \text{ mol N}_2 \times \frac{6 \text{ mol Li}}{1 \text{ mol N}_2} = 12.0 \text{ mol Li}$$

Therefore 12.0 mol of Li is required to react with 2.00 mol of N_2. However, we only have 8.07 mol of Li, so lithium is limiting. It will be consumed before the nitrogen runs out.

STEP 4
Because lithium is the limiting reactant, we must use the 8.07 mol of Li to determine how many moles of Li_3N can be formed.

$$8.07 \ \text{mol Li} \times \frac{2 \ \text{mol Li}_3\text{N}}{6 \ \text{mol Li}} = 2.69 \ \text{mol Li}_3\text{N}$$

STEP 5
We can now use the molar mass of Li_3N (34.83) to calculate the mass of Li_3N formed.

$$2.69 \ \text{mol Li}_3\text{N} \times \frac{34.83 \ \text{g Li}_3\text{N}}{1 \ \text{mol Li}_3\text{N}} = 93.7 \ \text{g Li}_3\text{N}$$

SELF-CHECK EXERCISE 10.7

a.   **STEP 1**
The balanced equation is

$$TiCl_4(g) + O_2(g) \rightarrow TiO_2(s) + 2Cl_2(g)$$

STEP 2
The numbers of moles of reactants are

$$6.71 \times 10^3 \ \text{g TiCl}_4 \times \frac{1 \ \text{mol TiCl}_4}{189.68 \ \text{g TiCl}_4} = 3.54 \times 10^1 \ \text{mol TiCl}_4$$

$$2.45 \times 10^3 \ \text{g O}_2 \times \frac{1 \ \text{mol O}_2}{32.00 \ \text{g O}_2} = 7.66 \times 10^1 \ \text{mol O}_2$$

STEP 3
In the balanced equation both $TiCl_4$ and O_2 have coefficients of 1, so

$$1 \ \text{mol TiCl}_4 = 1 \ \text{mol O}_2$$

and

$$3.54 \times 10^1 \ \text{mol TiCl}_4 \times \frac{1 \ \text{mol O}_2}{1 \ \text{mol TiCl}_4} = 3.54 \times 10^1 \ \text{mol O}_2 \ \text{required}$$

We have 7.66×10^1 mol of O_2, so the O_2 is in excess and the $TiCl_4$ is limiting. This makes sense. $TiCl_4$ and O_2 react in a 1:1 mole ratio, so the $TiCl_4$ is limiting because fewer moles of $TiCl_4$ are present than moles of O_2.

STEP 4
We will now use the moles of $TiCl_4$ (the limiting reactant) to determine the moles of TiO_2 that would form if the reaction produced 100% of the expected yield (the theoretical yield).

$$3.54 \times 10^1 \; \cancel{\text{mol TiCl}_4} \times \frac{1 \; \text{mol TiO}_2}{1 \; \cancel{\text{mol TiCl}_4}} = 3.54 \times 10^1 \; \text{mol TiO}_2$$

The mass of TiO_2 expected for 100% yield is

$$3.54 \times 10^1 \; \cancel{\text{mol TiO}_2} \times \frac{79.88 \; \text{g TiO}_2}{\cancel{\text{mol TiO}_2}} = 2.83 \times 10^3 \; \text{g TiO}_2$$

This amount represents the theoretical yield.

b.   Because the reaction is said to give only a 75.0% yield of TiO_2, we use the definition of percent yield,

$$\frac{\text{Actual yield}}{\text{Theoretical yield}} \times 100 = \% \; \text{yield}$$

to write the equation

$$\frac{\text{Actual yield}}{2.83 \times 10^3 \; \text{g TiO}_2} \times 100 = 75.0\% \; \text{yield}$$

We now want to solve for the actual yield. First we divide both sides by 100.

$$\frac{\text{Actual yield}}{2.83 \times 10^3 \; \text{g TiO}_2} \times \frac{\cancel{100}}{\cancel{100}} = \frac{75.0}{100} = 0.750$$

Then we multiply both sides by 2.83×10^3 g TiO_2.

$$\cancel{2.83 \times 10^3 \; \text{g TiO}_2} \times \frac{\text{actual yield}}{\cancel{2.83 \times 10^3 \; \text{g TiO}_2}} = 0.750 \times 2.83 \times 10^3 \; \text{g TiO}_2$$

$$\begin{aligned} \text{Actual yield} &= 0.750 \times 2.83 \times 10^3 \; \text{g TiO}_2 \\ &= 2.12 \times 10^3 \; \text{g TiO}_2 \end{aligned}$$

Thus 2.12×10^3 g of $TiO_2(s)$ is actually obtained in this reaction.

11 Modern Atomic Theory

CONTENTS

The brilliant colors in fireworks are due to the light emitted by various types of atoms. Each atom emits a color that is characteristic of that atom.

Although students may not have previously encountered the terms frequency and wavelength, they are familiar with the relative energies associated with X rays, ultraviolet and microwave radiation. So this chapter begins with material with which students are somewhat familiar.

Burning wood embers.

Light from the sun being diffracted by a prism.

The concept of atoms is a very useful one. It explains many important observations, such as why compounds always have the same composition (a specific compound always contains the same types and numbers of atoms) and how chemical reactions occur (they involve a rearrangement of atoms).

Once chemists came to "believe" in atoms, a logical question followed: what are atoms like? What is the structure of an atom? In Chapter 4 we learned to picture the atom with a positively charged nucleus composed of protons and neutrons at its center and electrons moving around the nucleus in a space very large compared to the size of the nucleus.

In this chapter we will look at atomic structure in more detail. In particular, we will develop a picture of the electron arrangements in atoms—a picture that allows us to account for the chemistry of the various elements. Recall from our discussion of the periodic table in Chapter 4 that, although atoms exhibit a great variety of characteristics, certain elements can be grouped together because they behave similarly. For example, fluorine, chlorine, bromine, and iodine (the halogens) show great chemical similarity. Likewise lithium, sodium, potassium, rubidium, and cesium (the alkali metals) exhibit many similar properties, and the noble gases (helium, neon, argon, krypton, xenon, and radon) are all very nonreactive. Although the members of each of these groups of elements show great similarity *within* the group, the differences in behavior *between* groups are striking. In this chapter we will see that it is the way the electrons are arranged in various atoms that accounts for these facts. However, before we examine atomic structure, we must consider the nature of electromagnetic radiation, which plays a central role in the study of the atom's behavior.

11.1 Electromagnetic Radiation and Energy

AIM: To characterize electromagnetic radiation and introduce the concept of quantized energy.

One of the ways in which energy travels through space is by **electromagnetic radiation.** The light from the sun, the energy used to cook food in a microwave oven, the X rays used by dentists, and the radiant heat from a fireplace are all examples of electromagnetic radiation. Although these forms of radiant energy seem quite different, they all exhibit the same type of wave-like behavior and travel at the speed of light in a vacuum.

Waves have three primary characteristics: wavelength, frequency, and speed. **Wavelength** (symbolized by the Greek letter lambda, λ) is the *distance between two consecutive peaks or troughs in a wave,* as shown in Figure 11.1. **Frequency** (symbolized by the Greek letter nu, ν) indicates how many waves

CHEMISTRY IN FOCUS

Solar Polar Bears

*T*he polar bear, a regal beast that has dominated the world's arctic regions for thousands of years, can exist in its incredibly hostile environment partly because its fur is an almost perfect absorber and converter of solar radiation.

What color is a polar bear's fur? White, the obvious answer, is incorrect. The hairs on a polar bear are completely colorless and transparent. The bear *appears* white because of the way the rough inner surfaces of the hollow hairs reflect visible light. The most interesting characteristic of each of these hollow fibers is its ability to act as a solar converter designed to trap ultraviolet light and transmit it to the bear's black skin. In the summer, the sun directly supplies up to

A pair of polar bears in the Arctic.

25% of the bear's total energy requirements. This allows the polar bear to be very active in pursuing prey, while still building up the layers of blubber it needs to survive the winter. This fascinating system also en-

sures that although the skin of the polar bear is very warm, the outer layers of the fur remain at about the same temperature as the bear's surroundings. Because of the small temperature difference between the outer layers of its fur and the air, the bear loses very little energy through heat leakage to the environment.

Human society could benefit greatly from what we know about polar bear fur as we continue to search for more efficient energy supplies. Clearly, covering our roofs with polar bear pelts is not an option, but polar bear hair provides an excellent model for the possible development of synthetic optical fibers to convert the sun's radiant energy to other energy forms.

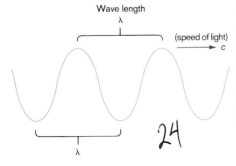

Wave length
λ

(speed of light)
c

24

(a)

(b)

Figure 11.1

(a) A light wave moves through space at a speed of 3×10^8 m/s, which is denoted by the letter c. The distance between successive peaks or troughs is the wavelength, denoted by λ. (b) You can picture the wave traveling through space much as a swell (peak) in the ocean approaches the shore. A surfer is propelled toward the beach at the speed at which the wave is traveling.

CHEMISTRY IN FOCUS

Atmospheric Effects

*T*he gaseous atmosphere of the earth is crucial to life in many different ways. One of the most important characteristics of the atmosphere is the way its molecules absorb radiation from the sun.

If it weren't for the protective nature of the atmosphere, the sun would "fry" us with its high-energy radiation. We are protected by the atmospheric ozone, a form of oxygen consisting of O_3 molecules, which absorbs high-energy radiation and thus prevents it from reaching the earth. This explains why we are so concerned that chemicals released into the atmosphere are destroying this high-altitude ozone.

The atmosphere also plays a central role in controlling the earth's temperature, a phenomenon called the *greenhouse effect*. The atmospheric gases CO_2, H_2O, CH_4, N_2O, and others do not absorb light in the visible region (see Figure 11.3). Therefore the visible light from the sun passes through the atmosphere to warm the earth. In turn, the earth radiates this energy back toward space as infrared radiation. (For example, think of the heat radiated from black asphalt on a hot summer day.) But the gases listed above are strong *absorbers* of *infrared* waves, and they reradiate some of this energy back toward the earth as shown in Figure 11.2. Thus these gases act as an insulating blanket keeping the earth much warmer than it would be without them. (If these gases were not

A composite satellite image of the earth's biomass constructed from the radiation given off by living matter over a multi-year period.

We use the generic term *light* for all forms of electromagnetic radiation.

pass a given point per second. **Speed** indicates how fast a given peak moves through space.

Electromagnetic radiation, which we usually call light, is divided into the various classes according to wavelength, as shown in Figure 11.3 (p. 342). Radiation provides an important means of energy transfer. For example, the energy present, all of the heat the earth radiates would be lost into space.)

However, there is a problem. When we burn fossil fuels (coal, petroleum, and natural gas), one of the products is CO_2. Because we use such huge quantities of fossil fuels, the CO_2 content in the atmosphere is increasing gradually but significantly. This should cause the earth to get warmer, eventually changing the weather patterns on the earth's surface and melting the polar ice caps, which would flood many low-lying areas.

Because the natural forces that control the earth's temperature are not very well understood at this point, it is difficult to decide whether the greenhouse warming has already started. But many scientists think it has. For example, the 1980s were among the warmest years the earth experienced since people started keeping records.

The greenhouse effect is something we must watch closely. Controlling it may mean lowering our dependence on fossil fuel and increasing our reliance on nuclear, solar, or other power sources. In recent years, the trend has been in the opposite direction.

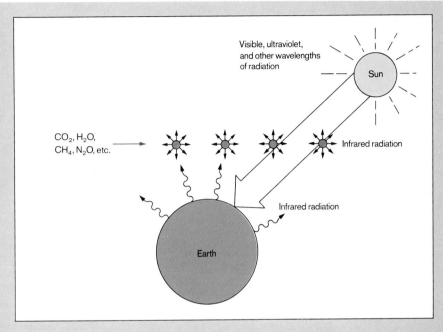

Figure 11.2
Certain of the gases in the earth's atmosphere reflect back some of the infrared (heat) radiation produced by the earth. This keeps the earth warmer than it would be otherwise.

from the sun reaches the earth mainly in the form of visible and ultraviolet radiation, and the glowing coals of a fireplace transmit heat energy by infrared radiation. In a microwave oven, the water molecules in food absorb microwave radiation, which increases their motions. This energy is then transferred to other types of molecules by collisions, causing an increase in the food's temperature.

I ask students to memorize the types of electromagnetic radiation and their frequencies. Even nonscientists will encounter many of these types of radiation in everyday life.

Figure 11.3
The different wavelengths of electromagnetic radiation.

25

Light as a wave

Light as a stream of photons
(packets of energy)

Figure 11.4
Electromagnetic radiation (a beam of light) can be pictured in two ways: as a wave and as a stream of individual packets of energy called photons.

Although we will not go into any details here, scientists have much evidence that a beam of light can also be thought of as a stream of tiny packets of energy called **photons.** So what is the exact nature of light? Does it consist of waves or is it a stream of discrete particles of energy? It seems to be both (Figure 11.4). This situation is often referred to as the wave–particle nature of light.

11.2   The Energy Levels of Hydrogen

AIM: To show how the emission spectrum of hydrogen demonstrates the quantized nature of energy.

An atom can lose energy by *emitting* a photon.

Each photon of blue light carries a larger quantity of energy than a photon of red light.

The atomic spectra of the elements can be demonstrated by using gas discharge tubes or by flaming solutions of salts using cotton swabs or wires. A longer lasting flame-test demonstration can be performed by preparing aqueous slurries of salts containing cations such as Sr^{2+}, Ca^{2+}, Ba^{2+}, K^+, Na^+, and Li^+, adding methanol, and then lighting each solution in a glass dish.

An atom with excess energy is said to be in an *excited state.* An excited atom can release some or all of its excess energy by emitting a photon (a "particle" of electromagnetic radiation) and thus moving to a lower energy state. The lowest possible energy state of an atom is called its *ground state.*

We can learn a great deal about the energy states of hydrogen atoms by observing the photons they emit. To understand the significance of this, you need to know that the *different wavelengths of light carry different amounts of energy per photon.* For example, a beam of red light has lower-energy photons than a beam of blue light.

When a hydrogen atom absorbs energy from some outside source, it uses this energy to enter an excited state. It can release this excess energy (go back to its ground state) by emitting a photon of light (Figure 11.5). We can picture this process in terms of the energy-level diagram shown in Figure 11.6. The impor-

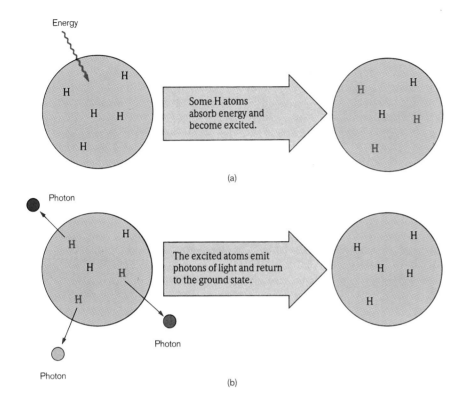

(a)

(b)

Figure 11.5
(a) A sample of H atoms receives energy from an external source, which causes some of the atoms to become excited (to possess excess energy). (b) The excited atoms (H) can release the excess energy by emitting photons. The energy of each emitted photon corresponds exactly to the energy lost by each excited atom.

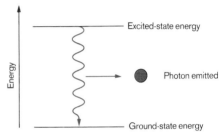

Figure 11.6
When an excited H atom returns to its ground state, it emits a photon that contains the energy released by the atom. Thus the energy of the photon corresponds to the difference in energy between the excited state and the ground state. In this case, red light is emitted.

tant point here is that *the energy contained in the photon corresponds to the change in energy that the atom experiences* in going from the excited state to the ground state.

Consider the following experiment. Suppose we take a sample of H atoms and put lots of energy into the system (as represented in Figure 11.5). When we study the photons of visible light emitted, we see only certain colors (Figure 11.7). That is, *only certain types of photons* are produced. We don't see all colors; we see only selected colors. This is a very significant result. Let's discuss carefully what it means.

A particular color (wavelength) of light carries a particular amount of energy per photon.

26

410 nm   434 nm   486 nm   656 nm

It is interesting at this point to discuss why sodium vapor lamps produce yellow light and why neon lights are red.

Figure 11.7
When excited hydrogen atoms return to their ground states, they emit photons of certain energies, and thus certain colors. Shown here are the colors and wavelengths (in nanometers) of the photons in the visible region that are emitted by excited hydrogen atoms.

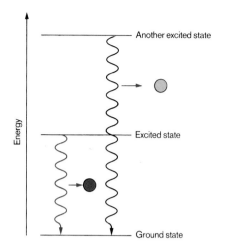

Figure 11.8
Hydrogen atoms have several excited-state energy levels. The color of the photon emitted depends on the energy change that produces it. A larger energy change may correspond to a blue photon, whereas a smaller change may produce a red photon.

The concept of quantized energy is hard for students to grasp. The explanation in the text using stair-steps of energy compared to a continuous ramp of energy is effective, but students will need to read this section carefully.

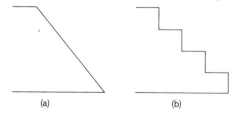

Figure 11.11
The difference between continuous and quantized energy levels can be illustrated by comparing a flight of stairs with a ramp. (a) A ramp varies continuously in elevation. (b) A flight of stairs allows only certain elevations; the elevations are quantized.

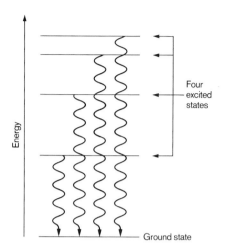

Figure 11.9
Each photon emitted by an excited hydrogen atom corresponds to a particular energy change in the hydrogen atom. In this diagram the horizontal lines represent discrete energy levels present in the hydrogen atom. A given H atom can exist in any of these energy states and can undergo any of the energy changes shown. It is also possible for an atom to change from one excited state to another excited state.

27

Figure 11.10
(a) Continuous energy levels. Any energy value is allowed. (b) Discrete (quantized) energy levels. Only certain energy states are allowed.

Because only certain photons are emitted, we know that only certain energy changes are occurring (Figure 11.8). This means that the hydrogen atom must have *certain discrete energy levels* (Figure 11.9). Excited hydrogen atoms *always* emit photons with the same discrete colors (wavelengths)—those shown in Figure 11.7. They *never* emit photons with energies (colors) in between those shown. So we can conclude that all hydrogen atoms have the same set of discrete energy levels. We say the energy levels of hydrogen are **quantized.** That is, only *certain values are allowed.* Scientists have found that the energy levels of *all* atoms are quantized.

The quantized nature of the energy levels in atoms was a surprise when scientists discovered it. It had been assumed prior to this that an atom could exist at any energy level. That is, everyone had assumed that atoms could have a continuous set of energy levels rather than only certain discrete values (Figure 11.10). A useful analogy here is the contrast between the elevations allowed by a ramp, which vary continuously, and those allowed by a set of steps, which are discrete (Figure 11.11). The discovery of the quantized nature of energy has radically changed our view of the atom, as we will see in the next few sections.

11.3   The Bohr Model of the Atom

AIM: To describe briefly Bohr's model of the hydrogen atom.

In 1911 at the age of twenty-five, Niels Bohr (Figure 11.12) received his Ph.D. in physics. He was convinced that the atom could be pictured as a small positive nucleus with electrons orbiting around it.

Over the next two years, Bohr constructed a model of the hydrogen atom with quantized energy levels that agreed with the hydrogen emission results we have just discussed. Bohr pictured the electron moving in circular orbits corresponding to the various allowed energy levels. He suggested that the electron could jump to a different orbit by absorbing or emitting a photon of light with exactly the correct energy content. Thus in the Bohr atom, the energy levels in the hydrogen atom represented certain allowed circular orbits (Figure 11.13).

At first Bohr's model appeared very promising. It fit the hydrogen atom very well. However, when this model was applied to atoms other than hydrogen, it did not work. In fact, further experiments showed that the Bohr model is fundamentally incorrect. Although the Bohr model paved the way for later theories, it is important to realize that the current theory of atomic structure is not the same as the Bohr model. Electrons do *not* move around the nucleus in circular orbits like planets orbiting the sun. Surprisingly, as we shall see later in this chapter, we do not know exactly how the electrons move in an atom.

Before I explain the Bohr model, I always begin the discussion by stating that Bohr's ideas provided a good starting place, but were incorrect. Even so, some students come away with the idea that electrons travel in circular orbits around the nucleus.

Although Bohr's model is consistent with the energy levels for hydrogen, it is fundamentally incorrect.

Figure 11.12
Niels Hendrik David Bohr (1885–1962) as a boy lived in the shadow of his younger brother Harald, who played on the 1908 Danish Olympic Soccer Team and later became a distinguished mathematician. In school, Bohr received his poorest marks in composition and struggled with writing during his entire life. In fact, he wrote so poorly that he was forced to dictate his Ph.D. thesis to his mother. He is one of the very few people who felt the need to write rough drafts of postcards. Nevertheless, Bohr was a brilliant physicist. After receiving his Ph.D. in Denmark, he constructed a quantum model for the hydrogen atom by the time he was 27. Even though his model later proved to be incorrect, Bohr remained a central figure in the drive to understand the atom. He was awarded the Nobel Prize in physics in 1922.

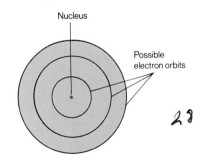

Figure 11.13
The Bohr model of the hydrogen atom represented the electron as restricted to certain circular orbits around the nucleus.

Louis Victor de Broglie

There is now much experimental evidence suggesting that all matter exhibits both wave and particle properties.

Figure 11.14
A representation of the photo of the firefly experiment. Remember that a picture is brightest where the film has been exposed to the most light. Thus the intensity of the color reflects how often the firefly visited a given point in the room. Notice that the brightest area is in the center of the room near the source of the sex attractant.

11.4　The Wave Mechanical Model of the Atom

AIM:　To describe how the electron's position is represented in the wave mechanical model.

By the mid-1920s it had become apparent that the Bohr model was incorrect. Scientists needed to pursue a totally new approach. Two young physicists, Louis Victor de Broglie and Werner Schrödinger, suggested that because light seems to have both wave and particle characteristics (it behaves simultaneously as a wave and as a stream of particles), the electron might also exhibit both of these characteristics. Although everyone had assumed that the electron was a tiny particle, these scientists said it might be useful to find out whether it could be described as a wave.

When Schrödinger carried out a mathematical analysis based on this idea, he found that it led to a new model for the hydrogen atom that seemed to apply equally well to other atoms—something Bohr's model failed to do. We will now explore a general picture of this model, which is called the **wave mechanical model** of the atom.

In the Bohr model, the electron was assumed to move in circular orbits. In the wave mechanical model, on the other hand, the electron states are described by orbitals. *Orbitals are nothing like orbits.* To approximate the idea of an orbital, picture a single male firefly in a room in the center of which an open vial of female sex-attractant hormones is suspended. The room is extremely dark and there is a camera in one corner with its shutter open. Every time the firefly "flashes," the camera records a pinpoint of light and thus the firefly's position in the room at that moment. The firefly senses the sex attractant, and as you can imagine, it spends a lot of time at or close to it. However, now and then the insect flies randomly around the room.

When the film is taken out of the camera and developed, the picture will probably look like Figure 11.14. Because a picture is brightest where the film has been exposed to the most light, the color intensity at any given point tells us how often the firefly visited a given point in the room. Notice that, as we might expect, the firefly spent the most time near the room's center.

Now suppose you are watching the firefly in the dark room. You see it flash at a given point far from the center of the room. Where do you expect to see it next? There is really no way to be sure. The firefly's flight path is not precisely predictable. However, if you had seen the time-exposure picture of the firefly's activities (Figure 11.14), you would have some idea where to look next. Your best chance would be to look more toward the center of the room. Figure 11.14 suggests there is the highest probability (the highest odds, the greatest likelihood)

The probability map for an electron is hard to illustrate in class. Solid models, even if they accurately describe the shape of an orbital, are not sufficient by themselves. Their sharp boundaries often lead students to think that the electron moves along the surface. A solid model cannot accurately depict the differences in probability within the orbital. The use of transparencies that show the fuzzy boundaries of orbitals along with the use of three-dimensional models can help make this concept clear.

of finding the firefly at any particular moment near the center of the room. You *can't be sure* the firefly will fly toward the center of the room, but it *probably* will. So the time-exposure picture is a kind of "probability map" of the firefly's flight pattern.

According to the wave mechanical model, the electron in the hydrogen atom can be pictured as being something like this firefly. Schrödinger found that he could not precisely describe the electron's path. His mathematics enabled him only to predict the probabilities of finding the electron at given points in space around the nucleus. In its ground state the hydrogen electron has a probability map like that shown in Figure 11.15. The more intense the color at a particular point, the more probable it is that the electron will be found at that point at a given instant. The model gives *no information about when* the electron occupies a certain point in space or *how it moves.* In fact, we have good reasons to believe that we can *never know* the details of electron motion, no matter how sophisticated our models may become. But one thing we feel confident about is that the electron *does not* orbit the nucleus in circles as Bohr suggested.

After discussing the wave mechanical model, I have found it helps to reemphasize the important points before moving to Section 11.5

29

Figure 11.15
The orbitals being filled for elements in various parts of the periodic table. Note that in going along a horizontal row (a period), the $(n + 1)s$ orbital fills before the nd orbital. The group label indicates the number of valence electrons (the number of s plus the number of p electrons in the highest occupied principal energy level) for the elements in each group.

11.5 The Hydrogen Orbitals

AIM: To describe the shapes of orbitals designated by *s, p,* and *d.*

The probability map for the hydrogen electron shown in Figure 11.15 is called an **orbital.** Although the probability of finding the electron decreases at greater distances from the nucleus, the probability of finding it even at great distances from the nucleus never becomes exactly zero. A useful analogy might be the lack of a sharp boundary between the earth's atmosphere and "outer space." The atmosphere fades away gradually, but there are always a few molecules present. Because the edge of an orbital is "fuzzy," an orbital does not have an exactly defined size. So chemists arbitrarily define its size as the sphere that contains 90% of the total electron probability (Figure 11.16(a)). This means that the electron spends 90% of the time inside this surface and 10% somewhere outside this surface. (Note that we are *not* saying the electron travels only on the *surface* of the sphere.) The orbital represented in Figure 11.16(b) is named the **1s orbital,** and it describes the hydrogen electron's lowest energy state (the ground state).

In Section 11.2 we saw that the hydrogen atom can absorb energy to transfer the electron to a higher energy state (an excited state). In terms of the obsolete Bohr model, this meant the electron was transferred to an orbit with a larger radius. In the wave mechanical model, these higher energy states correspond to different kinds of orbitals with different shapes.

At this point we need to stop and talk about how the hydrogen atom is organized. Remember, we showed earlier that the hydrogen atom has discrete

Solid models, even though they have limitations, can demonstrate the three-dimensional shapes of orbitals and show that individual p and d orbitals project away from each other as much as possible.

(a)   (b)

Figure 11.16
The hydrogen 1s orbital. (a) The size of the orbital is defined by a sphere that contains 90% of the total electron probability. That is, the electron can be found *inside* this sphere 90% of the time. (b) The 1s orbital is often represented simply as a sphere. However, the most accurate picture of the orbital is the probability map represented in (a).

Students easily confuse the meanings of the terms principal energy level, sublevel, and orbital. The inverted pyramids in Figures 11.18 and 11.19 can help alleviate this confusion.

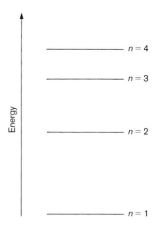

Figure 11.17
The first four principal energy levels in the hydrogen atom. Each level is assigned an integer, n.

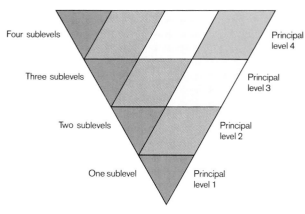

Figure 11.18
An illustration of how principal levels can be divided into sublevels.

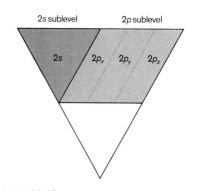

Figure 11.19
Principal level 2 shown as divided into the $2s$ and $2p$ sublevels.

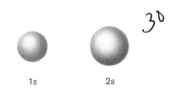

Figure 11.20
The relative sizes of the $1s$ and $2s$ orbitals of hydrogen.

energy levels. We call these levels **principal energy levels** and label them with integers (Figure 11.17). Next we find that each of these levels is subdivided into **sublevels.** The following analogy should help you understand this. Picture an inverted triangle (Figure 11.18). We divide the principal levels into various numbers of sublevels. Principal level 1 consists of one sublevel, principal level 2 has two sublevels, principal level 3 has three sublevels, and principal level 4 has four sublevels.

Like our triangle, the principal energy levels in the hydrogen atom contain sublevels. As we will see presently, these sublevels contain spaces for the electron that we call orbitals. Principal energy level 1 consists of just one sublevel, or one type of orbital. The spherical shape of this orbital is shown in Figure 11.16. We label this orbital $1s$. The number 1 is for the principal energy level, and s is a shorthand way to label a particular sublevel (type of orbital).

Principal energy level 2 has two sublevels. (Note the correspondence between the principal energy level number and the number of sublevels.) These sublevels are labeled $2s$ and $2p$. The $2s$ sublevel consists of one orbital (called the $2s$), and the $2p$ sublevel consists of three orbitals (called $2p_x$, $2p_y$, and $2p_z$). Let's return to the inverted triangle to illustrate this. Figure 11.19 shows principal level 2 divided into the sublevels $2s$ and $2p$ (which is subdivided into $2p_x$, $2p_y$, and $2p_z$). The orbitals have the shapes shown in Figures 11.20 and 11.21. The $2s$

orbital is spherical like the 1s orbital but larger in size (see Figure 11.20). The three 2p orbitals are not spherical but have two "lobes." These orbitals are shown in Figure 11.21 both as electron probability maps and as surfaces that contain 90% of the total electron probability. Notice that the label x, y, or z on a given 2p orbital tells along which axis the lobes of that orbital are directed.

What we have learned so far about the hydrogen atom is summarized in Figure 11.22. Principal energy level 1 has one sublevel, which contains the 1s orbital. Principal energy level 2 contains two sublevels, one of which contains the 2s orbital and one of which contains the 2p orbitals (three of them).

We should also summarize what we have said about orbital labels. The number tells the principal energy level. The letter tells the shape. The letter s means a spherical orbital; the letter p means a two-lobed orbital. The x, y, or z subscript on a p orbital label tells along which of the coordinate axis the two lobes lie.

One important characteristic of orbitals is that as the level number increases, the average distance of the electron in that orbital from the nucleus also increases. That is, when the hydrogen electron is in the 1s orbital (the ground state), it spends most of its time much closer to the nucleus than when it occupies the 2s orbital (an excited state).

You may be wondering at this point why hydrogen, which has only one electron, has more than one orbital. It is best to think of an orbital as a *potential space* for an electron. The hydrogen electron can occupy only a single orbital at a time, but the other orbitals are still available should the electron be transferred into one of them. For example, when a hydrogen atom is in its ground state (lowest possible energy state), the electron is in the 1s orbital. By adding the correct amount of energy (for example, a specific photon of light), we can excite the electron to the 2s orbital or to one of the 2p orbitals.

So far we have discussed only two of hydrogen's energy levels. There are many others. For example, level 3 has three sublevels (see Figure 11.18), which

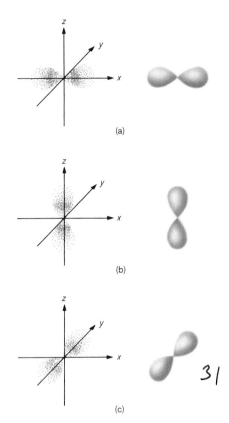

(a)

(b)

(c)

3|

Figure 11.21
The three 2p orbitals: (a) $2p_x$, (b) $2p_z$, (c) $2p_y$. The x, y, or z label indicates along which axis the two lobes are directed. Each orbital is shown both as a probability map and as a surface that encloses 90% of the electron probability.

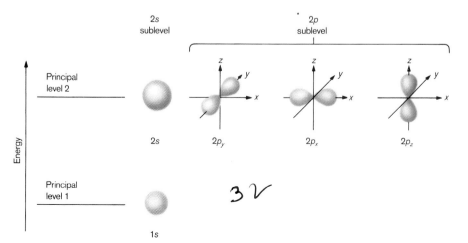

2s sublevel

2p sublevel

Principal level 2

2s $2p_y$ $2p_x$ $2p_z$

Energy

Principal level 1

3✓

1s

Figure 11.22
A diagram of principal energy levels 1 and 2 showing the shapes of orbitals that comprise the sublevels.

1s   2s   **33**   3s

Figure 11.23
The relative sizes of the spherical 1s, 2s, and 3s orbitals of hydrogen.

Think of orbitals as ways of dividing up the space around a nucleus.

d_{yz} d_{xz}

d_{xy} $d_{x^2-y^2}$

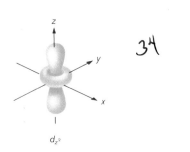

34

d_{z^2}

Figure 11.24
The shapes and labels of the five 3d orbitals.

we label 3s, 3p, and 3d. The 3s sublevel contains a single 3s orbital, a spherical orbital larger than 1s and 2s (Figure 11.23). Sublevel 3p contains three orbitals: $3p_x$, $3p_y$, and $3p_z$, which are shaped like the 2p orbitals except that they are larger. The 3d sublevel contains five 3d orbitals with the shapes and labels shown in Figure 11.24. (You do not need to memorize the 3d orbital shapes and labels. They are shown for completeness.)

Notice as you compare levels 1, 2, and 3 that a new type of orbital (sublevel) is added in each principal energy level. (Recall that the p orbitals are added in level 2 and the d orbitals in level 3.) This makes sense because in going farther out from the nucleus, there is more space available and thus room for more orbitals.

It might help you to understand that the number of orbitals increases with the principal energy level if you think of a theater in the round. Picture a round stage with circular rows of seats surrounding it. The farther from the stage a row of seats is, the more seats it contains because the circle is larger. Orbitals divide up the space around a nucleus somewhat like the seats in this circular theater. The greater the distance from the nucleus, the more space there is and the more orbitals we find.

The pattern of increasing numbers of orbitals continues with level 4. Level 4 has four sublevels labeled 4s, 4p, 4d, and 4f. The 4s sublevel has a single 4s orbital. The 4p sublevel contains three orbitals ($4p_x$, $4p_y$, and $4p_z$). The 4d sublevel has five 4d orbitals. The 4f sublevel has seven 4f orbitals.

The 4s, 4p, and 4d orbitals have the same shapes as the earlier s, p, and d orbitals, respectively, but are larger. We will not be concerned here with the shapes of the f orbitals.

11.6   The Wave Mechanical Model: A Summary

AIM: To summarize the energy levels and orbitals of the wave mechanical model of the atom.
To describe electron spin.

A model for the atom is of little use if it does not apply to all atoms. The Bohr model was discarded because it could be applied only to hydrogen. The wave mechanical model can be applied to all atoms in basically the same form as we have just used it for hydrogen. In fact, the major triumph of this model is its ability to explain the periodic table of the elements. Recall that the elements on the periodic table are arranged in vertical groups, which contain elements that typically show similar chemical properties. The wave mechanical model of the atom allows us to explain, based on electron arrangements, why these similarities occur. We will see in due time how this is done.

Remember that an atom has as many electrons as it has protons to give it a zero overall charge. Therefore, all atoms beyond hydrogen have more than one electron. Before we can consider the atoms beyond hydrogen, we must describe one more property of electrons that determines how they can be arranged in an atom's orbital. This property is spin. Each electron appears to be spinning as a top spins on its axis. Like the top, an electron can spin only in one of two directions. We often represent spin with an arrow: either ↑ or ↓. One arrow represents the electron spinning in the one direction, and the other represents the electron spinning in the opposite direction. For our purposes, what is most important about electron spin is that two electrons must have *opposite* spins to occupy the same orbital. That is, two electrons that have the same spin cannot occupy the same orbital. This leads to the **Pauli exclusion principle:** an atomic orbital can hold a maximum of two electrons, and those two electrons must have opposite spins.

Before we apply the wave mechanical model to atoms beyond hydrogen, we will summarize the model for convenient reference.

The box diagram on this page is a convenient summary of the information presented so far in this chapter. To understand each of the seven points, students must read the appropriate section in the text. I tell them that rereading the most difficult sections will help them understand this abstract material. Students need to understand the material well enough to apply the wave mechanical model to energy levels of atoms other than hydrogen.

Principal Components of the Wave Mechanical Model of the Atom

1. Atoms have a series of energy levels called **principal energy levels,** which are designated by whole numbers symbolized by n; n can equal 1, 2, 3, 4, ... Level 1 corresponds to $n = 1$, level 2 corresponds to $n = 2$, and so on.
2. The energy of the level increases as the value of n increases.
3. Each principal energy level contains one or more *types* of orbitals, called **sublevels.**
4. The number of sublevels present in a given principal energy level equals n. For example, level 1 contains one sublevel ($1s$); level 2 contains two sublevels (two types of orbitals), the $2s$ orbital and the three $2p$ orbitals; and so on. These are summarized in the following table.

n	Sublevels (Types of Orbitals) Present			
1	$1s$ (1)			
2	$2s$ (1)	$2p$ (3)		
3	$3s$ (1)	$3p$ (3)	$3d$ (5)	
4	$4s$ (1)	$4p$ (3)	$4d$ (5)	$4f$ (7)

The number of each type of orbital is shown in parentheses.
5. The n value is always used to label the orbitals of a given principal level and is followed by a letter that indicates the type (shape) of the orbital. For example, the designation $3p$ means an orbital in level 3 that has two lobes (a p orbital always has two lobes).
6. An orbital can be empty or it can contain one or two electrons, but never more than two. If two electrons occupy the same orbital, they must have opposite spins.
7. The shape of an orbital does not indicate the details of electron movement. It indicates the probability distribution for an electron residing in that orbital.

EXAMPLE 11.1	**Understanding the Wave Mechanical Model of the Atom**

Indicate whether each of the following statements about atomic structure is true or false.

a. An *s* orbital is always spherical in shape.
b. The 2*s* orbital is the same size as the 3*s* orbital.
c. The number of lobes on a *p* orbital increases as *n* increases. That is, a 3*p* orbital has more lobes than a 2*p* orbital.
d. Level 1 has one *s* orbital, level 2 has two *s* orbitals, level 3 has three *s* orbitals, and so on.
e. The electron path is indicated by the surface of the orbital.

SOLUTION

a. True. The size of the sphere increases as *n* increases, but the shape is always spherical.
b. False. The 3*s* orbital is larger (the electron is farther from the nucleus on average) than the 2*s* orbital.
c. False. A *p* orbital always has two lobes.
d. False. Each principal energy level has only one *s* orbital.
e. False. The electron is *somewhere inside* the orbital surface 90% of the time. The electron does not move around *on* this surface.

SELF-CHECK EXERCISE 11.1

Define the following terms.

a. Bohr orbits
b. orbitals

c. orbital size
d. sublevel

11.7 Electron Arrangements in the First Eighteen Atoms

AIM: To describe how the principal energy levels fill with electrons in atoms beyond hydrogen.
To introduce valence electrons and core electrons.

We will now describe the electron arrangements in atoms with $Z = 1$ to $Z = 18$ by placing electrons in the various orbitals in the principal energy levels, starting with $n = 1$, and then continuing with $n = 2$, $n = 3$, and so on. For the first

Students readily grasp the electron configurations of the first
18 elements, especially if you have time to cover each element in order.

11.7 Electron Arrangements in the First Eighteen Atoms   **353**

eighteen elements, the individual sublevels fill in the following order: $1s$, then
$2s$, then $2p$, then $3s$, then $3p$.

The most favorable orbital in an atom is always the $1s$, because in this
orbital a negatively charged electron is closer to the positively charged nucleus
than in any other orbital. That is, the $1s$ orbital describes the space around the
nucleus that is closest to the nucleus. As n increases, the orbital becomes larger—
the electron, on average, occupies space farther from the nucleus.

So in its ground state hydrogen has its lone electron in the $1s$ orbital. This is
commonly represented in two ways. First, we say that hydrogen has the electron
arrangement, or **electron configuration,** $1s^1$. This just means there is one electron in the $1s$ orbital. We can also represent this configuration by using an **orbital
diagram,** also called a **box diagram,** in which orbitals are represented by boxes
grouped by sublevel with small arrows indicating the electrons. For *hydrogen,* the
electron configuration and box diagram are

When I work examples, I show the electron
configuration and the box diagram side by
side so that students can compare the
information in each.

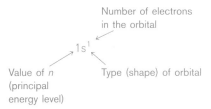

Value of n (principal energy level) — $1s^1$ — Number of electrons in the orbital — Type (shape) of orbital

$$1s$$

H: $1s^1$ $\boxed{\uparrow}$

Configuration   Orbital diagram

The arrow represents an electron spinning in a particular direction. The next
element is *helium, Z = 2.* It has two protons in its nucleus and so has two
electrons. Because the $1s$ orbital is the most desirable, both electrons go there but
with opposite spins. For helium, the electron configuration and box diagram are

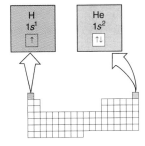

Two electrons in $1s$ orbital

$$1s$$

He: $1s^2$ $\boxed{\uparrow\downarrow}$

The opposite electron spins are shown by the opposing arrows in the box.

Lithium $(Z = 3)$ has three electrons, two of which go into the $1s$ orbital.
That is, two electrons fill that orbital. The $1s$ orbital is the only orbital for $n = 1$,
so the third electron must occupy an orbital with $n = 2$—in this case the $2s$
orbital. This gives a $1s^2 2s^1$ configuration. The electron configuration and box
diagram are

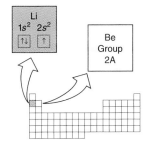

$$1s \quad 2s$$

Li: $1s^2 2s^1$ $\boxed{\uparrow\downarrow}$ $\boxed{\uparrow}$

The next element, *beryllium,* has four electrons, which occupy the $1s$ and
$2s$ orbitals with opposite spins.

$$1s \quad 2s$$

Be: $1s^2 2s^2$ $\boxed{\uparrow\downarrow}$ $\boxed{\uparrow\downarrow}$

Boron has five electrons, four of which occupy the $1s$ and $2s$ orbitals. The
fifth electron goes into the second type of orbital with $n = 2$, one of the $2p$
orbitals.

$$1s \quad 2s \quad 2p$$

B: $1s^2 2s^2 2p^1$ $\boxed{\uparrow\downarrow}$ $\boxed{\uparrow\downarrow}$ $\boxed{\uparrow}\square\square$

When you explain that the last electron in carbon occupies an empty $2p_y$ orbital instead of adding a second electron to $2p_x$, you might use the analogy of a restaurant with three tables for two. When three strangers take seats, they usually occupy tables by themselves. It is not until all three tables have an occupant that the next diner sits at a table that is already occupied.

Because all the $2p$ orbitals have the same energy, it does not matter which $2p$ orbital the electron occupies.

Carbon, the next element, has six electrons: two electrons occupy the $1s$ orbital, two occupy the $2s$ orbital, and two occupy $2p$ orbitals. There are three $2p$ orbitals, so each of the mutually repulsive electrons occupies a different $2p$ orbital. For reasons we will not consider, in the separate $2p$ orbitals the electrons have the same spin.

The configuration for carbon could be written $1s^2 2s^2 2p^1 2p^1$ to indicate that the electrons occupy separate $2p$ orbitals. However, the configuration is usually given as $1s^2 2s^2 2p^2$, and it is understood that the electrons are in different $2p$ orbitals.

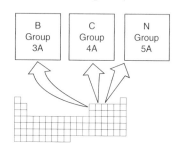

Like charges repel.

C: $1s^2 2s^2 2p^2$

$$\begin{array}{ccc} 1s & 2s & 2p \\ \boxed{\uparrow\downarrow} & \boxed{\uparrow\downarrow} & \boxed{\uparrow\,|\,\uparrow\,|\,} \end{array}$$

Note the like spins for the unpaired electrons in the $2p$ orbitals.

The configuration for *nitrogen,* which has seven electrons, is $1s^2 2s^2 2p^3$. The three electrons in $2p$ orbitals occupy separate orbitals and have like spins.

N: $1s^2 2s^2 2p^3$

$$\begin{array}{ccc} 1s & 2s & 2p \\ \boxed{\uparrow\downarrow} & \boxed{\uparrow\downarrow} & \boxed{\uparrow\,|\,\uparrow\,|\,\uparrow} \end{array}$$

The configuration for *oxygen,* which has eight electrons, is $1s^2 2s^2 2p^4$. One of the $2p$ orbitals is now occupied by a pair of electrons with opposite spins, as required by the Pauli exclusion principle.

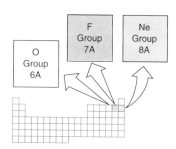

O: $1s^2 2s^2 2p^4$

$$\begin{array}{ccc} 1s & 2s & 2p \\ \boxed{\uparrow\downarrow} & \boxed{\uparrow\downarrow} & \boxed{\uparrow\downarrow\,|\,\uparrow\,|\,\uparrow} \end{array}$$

The electron configurations and orbital diagrams for *fluorine* (nine electrons) and *neon* (ten electrons) are

F: $1s^2 2s^2 2p^5$

$$\begin{array}{ccc} 1s & 2s & 2p \\ \boxed{\uparrow\downarrow} & \boxed{\uparrow\downarrow} & \boxed{\uparrow\downarrow\,|\,\uparrow\downarrow\,|\,\uparrow} \end{array}$$

Ne: $1s^2 2s^2 2p^6$

$$\begin{array}{ccc} 1s & 2s & 2p \\ \boxed{\uparrow\downarrow} & \boxed{\uparrow\downarrow} & \boxed{\uparrow\downarrow\,|\,\uparrow\downarrow\,|\,\uparrow\downarrow} \end{array}$$

With neon, the orbitals with $n = 1$ and $n = 2$ are completely filled.

For *sodium,* which has eleven electrons, the first ten electrons occupy the $1s$, $2s$, and $2p$ orbitals, and the eleventh electron must occupy the first orbital with $n = 3$, the $3s$ orbital. The electron configuration for sodium is $1s^2 2s^2 2p^6 3s^1$. To avoid writing the inner-level electrons, we often abbreviate the configuration $1s^2 2s^2 2p^6 3s^1$ as $[\text{Ne}]3s^1$, where $[\text{Ne}]$ represents the electron configuration of neon, $1s^2 2s^2 2p^6$.

[Ne] is shorthand for $1s^2 2s^2 2p^6$.

The orbital diagram for sodium is

$$\begin{array}{cccc} 1s & 2s & 2p & 3s \\ \boxed{\uparrow\downarrow} & \boxed{\uparrow\downarrow} & \boxed{\uparrow\downarrow\,|\,\uparrow\downarrow\,|\,\uparrow\downarrow} & \boxed{\uparrow} \end{array}$$

H $1s^1$							He $1s^2$
Li $2s^1$	Be $2s^2$	B $2p^1$	C $2p^2$	N $2p^3$	O $2p^4$	F $2p^5$	Ne $2p^6$
Na $3s^1$	Mg $3s^2$	Al $3p^1$	Si $3p^2$	P $3p^3$	S $3p^4$	Cl $3p^5$	Ar $3p^6$

Figure 11.25
The electron configurations in the sublevel last occupied for the first eighteen elements.

The next element, *magnesium, Z* = 12, has the electron configuration $1s^2 2s^2 2p^6 3s^2$, or [Ne]$3s^2$.

The next six elements, *aluminum* through *argon,* have electron configurations obtained by filling the $3p$ orbitals one electron at a time. Figure 11.25 summarizes the electron configurations of the first eighteen elements by giving the number of electrons in the type of orbital (sublevel) occupied last.

EXAMPLE II.2   Writing Orbital Diagrams

Write the orbital diagram for magnesium.

SOLUTION

Magnesium (Z = 12) has twelve electrons that are placed successively in the $1s$, $2s$, $2p$, and $3s$ orbitals to give the electron configuration $1s^2 2s^2 2p^6 3s^2$. The orbital diagram is

$$1s \quad 2s \quad \quad 2p \quad \quad 3s$$

↑↓	↑↓	↑↓	↑↓	↑↓	↑↓

Only occupied orbitals are shown here.

SELF-CHECK EXERCISE II.2

Write the complete electron configuration and the orbital diagram for each of the elements aluminum through argon.

At this point it is useful to introduce the concept of **valence electrons**— that is, *the electrons in the outermost (highest) principal energy level of an atom.* For example, nitrogen, which has the electron configuration $1s^2 2s^2 2p^3$, has electrons in principal levels 1 and 2. Therefore, level 2 (which has $2s$ and $2p$ sublevels) is the valence level of nitrogen, and the $2s$ and $2p$ electrons are the valence electrons. For the sodium atom (electron configuration $1s^2 2s^2 2p^6 3s^1$, or [Ne]$3s^1$) the valence electron is the electron in the $3s$ orbital, because in this case

When I write electron configurations, I underline the valence electrons for each example so that students learn to distinguish them from core electrons.

Students are easily confused when trying to write core notation. They are unsure as to whether to move back to the last noble gas or to move forward to the next noble gas. Remind them that they must go back to the previous noble gas.

principal energy level 3 is the outermost level that contains an electron. The valence electrons are the most important electrons to chemists because, being the outermost electrons, they are the ones involved when atoms attach to each other (form bonds), as we will see in the next chapter. The inner electrons, which are known as **core electrons,** are not involved in bonding atoms to each other.

Note in Figure 11.25 that a very important pattern is developing: except for helium, *the elements in the same group (vertical column of the periodic table) have the same number of electrons in a given type of orbital* (sublevel), except that the orbitals are in different principal energy levels. Remember that the elements were originally organized into groups on the periodic table on the basis of similarities in chemical properties. Now we understand the reason behind these groupings. Elements with the same valence electron arrangement show very similar chemical behavior.

11.8 Electron Configurations and the Periodic Table

AIM: To describe the electron configurations of atoms with Z greater than 18.

In the last section we saw that we can describe the atoms beyond hydrogen by simply filling the atomic orbitals starting with level $n = 1$ and working outward in order. This works fine until we reach the element *potassium* $(Z = 19)$, which is the next element after argon. Because the $3p$ orbitals are fully occupied in argon, we might expect the next electron to go into a $3d$ orbital (recall that for $n = 3$ the sublevels are $3s$, $3p$, and $3d$). However, experiments show that the chemical properties of potassium are very similar to those of lithium and sodium. Because we have learned to associate similar chemical properties with similar valence-electron arrangements, we predict that the valence-electron configuration for potassium is $4s^1$, resembling sodium $(3s^1)$ and lithium $(2s^1)$. That is, we expect the last electron in potassium to occupy the $4s$ orbital instead of one of the $3d$ orbitals. This means that principal energy level 4 begins to fill before level 3 has been completed. This conclusion is confirmed by many types of experiments. So the electron configuration of potassium is

[Ar] is shorthand for $1s^22s^22p^63s^23p^6$.

K: $1s^22s^22p^63s^23p^64s^1$, or $[Ar]4s^1$

The next element is *calcium,* with an additional electron that also occupies the $4s$ orbital.

Ca: $1s^22s^22p^63s^23p^64s^2$, or $[Ar]4s^2$

The $4s$ orbital is now full.

K $4s^1$	Ca $4s^2$	Sc $3d^1$	Ti $3d^2$	V $3d^3$	Cr $4s^13d^5$	Mn $3d^5$	Fe $3d^6$	Co $3d^7$	Ni $3d^8$	Cu $4s^13d^{10}$	Zn $3d^{10}$	Ga $4p^1$	Ge $4p^2$	As $4p^3$	Se $4p^4$	Br $4p^5$	Kr $4p^6$

Figure 11.26

Partial electron configurations for the elements potassium through krypton. The transition metals shown in green (scandium through zinc) have the general configuration $[Ar]4s^23d^n$, except for chromium and copper.

Students need to memorize the valence electron configurations of Cr and Cu, which are exceptions to transition metal electron configurations.

After calcium the next electrons go into the $3d$ orbitals to complete principal energy level 3. The elements that correspond to filling the $3d$ orbitals are called transition metals. Then the $4p$ orbitals fill. Figure 11.26 gives partial electron configurations for the elements potassium through krypton.

Note from Figure 11.26 that all of the transition metals have the general configuration $[Ar]4s^23d^n$ except chromium ($4s^13d^5$) and copper ($4s^13d^{10}$). The reasons for these exceptions are complex and will not be discussed here.

Instead of continuing to consider the elements individually, we will now look at the overall relationship between the periodic table and orbital filling. Figure 11.27 shows which type of orbital is filling in each area of the periodic table. Note the following points:

1. In a principal energy level that has d orbitals, the s orbital from the *next* level fills before the d orbitals in the current level. That is, the $(n + 1)s$ orbitals always fill before the nd orbitals. For example, the $5s$ orbitals fill for rubidium and strontium before the $4d$ orbitals fill for the second row of transition metals (yttrium through cadmium).

The $(n + 1)s$ orbital fills before the nd orbitals fill.

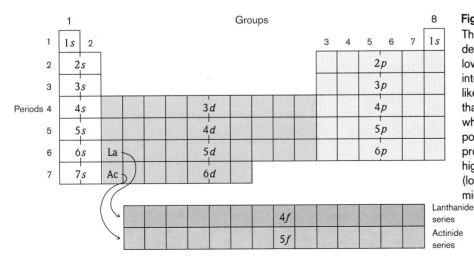

Figure 11.27

The probability map, or orbital, that describes the hydrogen electron in its lowest possible energy state. The more intense the color of a given dot, the more likely it is that the electron will be found at that point. We have no information about when the electron will be at a particular point or about how it moves. Note that the probability of the electron's presence is highest closest to the positive nucleus (located at the center of this diagram), as might be expected.

error

Figures 11.27 and 11.28 illustrate two ways to remember the order of electron filling. It is probably better for students to learn to use the periodic table, as in Figure 11.27. However, many students prefer to memorize the order of filling shown in Figure 11.28.

Lanthanides are elements in which the 4f orbitals are being filled.

Actinides are elements in which the 5f orbitals are being filled.

2. After lanthanum, which has the electron configuration $[Xe]6s^25d^1$, a group of fourteen elements called the **lanthanide series,** or the lanthanides, occurs. This series of elements corresponds to the filling of the seven 4f orbitals.

3. After actinium, which has the configuration $[Rn]7s^26d^1$, a group of fourteen elements called the **actinide series,** or the actinides, occurs. This series corresponds to the filling of the seven 5f orbitals.

4. Except for helium, the group numbers indicate the sum of electrons in the ns and np orbitals in the highest principal energy level that contains electrons (where n is the number that indicates a particular principal energy level). These electrons are the valence electrons, the electrons in the outermost principal energy level of a given atom.

To further help you understand the connection between orbital filling and the periodic table, Figure 11.28 shows the orbitals in the order in which they fill.

A periodic table is almost always available to you. If you understand the relationship between the electron configuration of an element and its position on the periodic table, you can figure out the expected electron configuration of any atom.

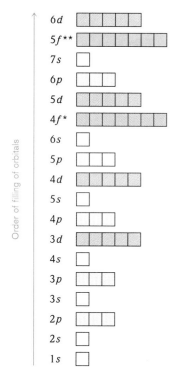

Figure 11.28
A box diagram showing the order in which orbitals fill to produce the atoms in the periodic table. Each box can hold two electrons.

---

*After the 6s orbital is full, one electron goes into a 5d orbital. This corresponds to the element lanthanum ($[Xe]6s^25d^1$). After lanthanum, the 4f orbitals fill with electrons before any other electrons go into 5d.

**After the 7s orbital is full, one electron goes into 6d. This is actinium ($[Rn]7s^26d^1$). The 5f orbitals then fill before any other electrons go into 6d.

EXAMPLE 11.3   Determining Electron Configurations

Using the periodic table inside the front cover, give the electron configurations for sulfur (S), gallium (Ga), hafnium (Hf), and radium (Ra).

SOLUTION

Sulfur is element 16 and resides in Period 3, where the 3p orbitals are being filled (see Figure 11.29). Because sulfur is the fourth among the "3p elements," it must have four 3p electrons. Sulfur's electron configuration is

$$\text{S:} \qquad 1s^22s^22p^63s^23p^4, \text{ or } [Ne]3s^23p^4$$

Gallium is element 31 in Period 4 just after the transition metals (see Figure 11.29). It is the first element in the "4p series" and has a $4p^1$ arrangement. Gallium's electron configuration is

$$\text{Ga:} \qquad 1s^22s^22p^63s^23p^64s^23d^{10}4p^1, \text{ or } [Ar]4s^23d^{10}4p^1$$

Hafnium is element 72 and is found in Period 6, as shown in Figure 11.29. Note that it occurs just after the lanthanide series (see Figure 11.27). Thus the 4f orbitals are already filled. Hafnium is the second member of the 5d transition series and has two 5d electrons. Its electron configuration is

Hf: $1s^22s^22p^63s^23p^64s^23d^{10}4p^65s^24d^{10}5p^66s^24f^{14}5d^2$, or $[Xe]6s^24f^{14}5d^2$

EXAMPLE 11.3, CONTINUED

Figure 11.29
The positions of the elements considered in Example 11.3.

Sulfur is recovered from deposits in the earth by melting it with hot water. Here the water–sulfur mixture is being sprayed into pits to allow the water to evaporate.

Radium is element 88 and is in Period 7 (and Group 2), as shown in Figure 11.29. Thus radium has two electrons in the $7s$ orbital, and its electron configuration is

Ra: $1s^2 2s^2 2p^6 3s^2 3p^6 4s^2 3d^{10} 4p^6 5s^2 4d^{10} 5p^6 6s^2 4f^{14} 5d^{10} 6p^6 7s^2$, or $[Rn]7s^2$

SELF-CHECK EXERCISE 11.3

Using the periodic table inside the front cover, predict the electron configurations for fluorine, silicon, cesium, lead, and iodine. If you have trouble, use Figure 11.27.

Summary: The Wave Mechanical Model and Valence-Electron Configurations

The concepts we have discussed in this chapter are very important. They allow us to make sense of a good deal of chemistry. When it was first observed that elements with similar properties occur periodically as the atomic number increases, chemists wondered why. Now we have an explanation. The wave mechanical model pictures the electrons in an atom as arranged in orbitals, with each orbital capable of holding two electrons. As we build up the atoms, the same types of orbitals recur in going from one principal energy level to another. This means that particular valence-electron configurations recur periodically. For reasons we will explore in the next chapter, elements with a particular type of valence configuration all show very similar chemical behavior. Thus groups of elements, such as the alkali metals, show similar chemistry because all the elements in that group have the same type of valence-electron arrangement. This concept, which explains so much chemistry, is the greatest contribution of the wave mechanical model to modern chemistry.

Molten lead being poured into a mold to make toy soldiers.

Representative elements — Transition elements — Representative elements

Period number, highest occupied electron level

Period	1 (ns^1)	2 (ns^2)	3	4	5	6	7	8	9	10	11	12	3 (ns^2np^1)	4 (ns^2np^2)	5 (ns^2np^3)	6 (ns^2np^4)	7 (ns^2np^5)	8 (ns^2np^6)
1	1 H $1s^1$																	2 He $1s^2$
2	3 Li $2s^1$	4 Be $2s^2$											5 B $2s^22p^1$	6 C $2s^22p^2$	7 N $2s^22p^3$	8 O $2s^22p^4$	9 F $2s^22p^5$	10 Ne $2s^22p^6$
3	11 Na $3s^1$	12 Mg $3s^2$											13 Al $3s^23p^1$	14 Si $3s^23p^2$	15 P $3s^23p^3$	16 S $3s^23p^4$	17 Cl $3s^23p^5$	18 Ar $3s^23p^6$
4	19 K $4s^1$	20 Ca $4s^2$	21 Sc $4s^23d^1$	22 Ti $4s^23d^2$	23 V $4s^23d^3$	24 Cr $4s^13d^5$	25 Mn $4s^23d^5$	26 Fe $4s^23d^6$	27 Co $4s^23d^7$	28 Ni $4s^23d^8$	29 Cu $4s^13d^{10}$	30 Zn $4s^23d^{10}$	31 Ga $4s^24p^1$	32 Ge $4s^24p^2$	33 As $4s^24p^3$	34 Se $4s^24p^4$	35 Br $4s^24p^5$	36 Kr $4s^24p^6$
5	37 Rb $5s^1$	38 Sr $5s^2$	39 Y $5s^24d^1$	40 Zr $5s^24d^2$	41 Nb $5s^14d^4$	42 Mo $5s^14d^5$	43 Tc $5s^14d^6$	44 Ru $5s^14d^7$	45 Rh $5s^14d^8$	46 Pd $5s^04d^{10}$	47 Ag $5s^14d^{10}$	48 Cd $5s^24d^{10}$	49 In $5s^25p^1$	50 Sn $5s^25p^2$	51 Sb $5s^25p^3$	52 Te $5s^25p^4$	53 I $5s^25p^5$	54 Xe $5s^25p^6$
6	55 Cs $6s^1$	56 Ba $6s^2$	57 La $6s^25d^1$	72 Hf $6s^25d^2$	73 Ta $6s^25d^3$	74 W $6s^25d^4$	75 Re $6s^25d^5$	76 Os $6s^25d^6$	77 Ir $6s^25d^7$	78 Pt $6s^15d^9$	79 Au $6s^15d^{10}$	80 Hg $6s^25d^{10}$	81 Tl $6s^26p^1$	82 Pb $6s^26p^2$	83 Bi $6s^26p^3$	84 Po $6s^26p^4$	85 At $6s^26p^5$	86 Rn $6s^26p^6$
7	87 Fr $7s^1$	88 Ra $7s^2$	89 Ac $7s^26d^1$	104 Unq $7s^26d^2$	105 Unp $7s^26d^3$	106 Unh $7s^26d^4$	107 Uns $7s^26d^5$	108 Uno	109 Une $7s^26d^7$									

Group number

Figure 11.30
The periodic table with atomic symbols, atomic numbers, and partial electron configurations. (The configuration for element 108 is not shown because it is not yet known.)

For reference, the valence-electron configurations for all the elements are shown on the periodic table in Figure 11.30. Note the following points:

The group label gives the total number of valence electrons for that group.

1. The group labels for Groups 1, 2, 3, 4, 5, 6, 7, and 8 indicate the *total number* of valence electrons for the atoms in these groups. For example, all the elements in Group 5 have the configuration ns^2np^3. (Any d electrons present are always in the next lower principal energy level than the valence electrons and so are not counted as valence electrons.)

An overhead transparency similar to Figure 11.30 can illustrate the first two points listed in the text and can help show how the numbers and arrangements of electrons are related to periodic trends in Section 11.9.

2. The elements in Groups 1, 2, 3, 4, 5, 6, 7, and 8 are often called the **main-group elements,** or **representative elements.** Remember that every member of a given group (except for helium) has the same valence-electron configuration, except that the electrons are in different principal energy levels.

3. We will not be concerned in this text with the configurations for the f transition elements (lanthanides and actinides), so these are not included in Figure 11.30.

11.9   Atomic Properties and the Periodic Table

AIM: To show the general trends in atomic properties in the periodic table.

With all of this talk about electron probability and orbitals, we must not lose sight of the fact that chemistry is still fundamentally a science based on the observed properties of substances. We know that wood burns, steel rusts, plants grow, sugar tastes sweet, and so on because we *observe* these phenomena. The atomic theory is an attempt to help us understand why these things occur. If we understand why, we can hope to better control the chemical events that are so crucial in our daily lives.

In the next chapter we will see how our ideas about atomic structure help us understand how and why atoms combine to form compounds. As we explore this, and as we use theories to explain other types of chemical behavior later in the text, it is important that we distinguish the observation (steel rusts) from the attempts to explain why the observed event occurs (theories). The observations remain the same over the decades, but the theories (our explanations) change as we gain a clearer understanding of how nature operates. A good example of this is the replacement of the Bohr model for atoms by the wave mechanical model.

Because the observed behavior of matter lies at the heart of chemistry, you need to understand thoroughly the characteristic properties of the various elements and the trends (systematic variations) that occur in those properties. To that end, we will now consider some especially important properties of atoms and see how they vary, horizontally and vertically, on the periodic table.

Metals and Nonmetals

The most fundamental classification of the chemical elements is into metals and nonmetals. **Metals** typically have the following physical properties: a lustrous appearance, the ability to change shape without breaking (they can be pulled into a wire or pounded into a thin sheet), and excellent conductivity of heat and electricity. **Nonmetals** typically do not have these physical properties, although there are some exceptions. (For example, solid iodine is lustrous; the graphite form of carbon is an excellent conductor of electricity; and the diamond form of carbon is an excellent conductor of heat.) However, it is the *chemical* differences between metals and nonmetals that interests us the most: *metals tend to lose electrons to form positive ions, and nonmetals tend to gain electrons to form negative ions.* When a metal and a nonmetal react, a transfer of one or more electrons from the metal to the nonmetal often occurs.

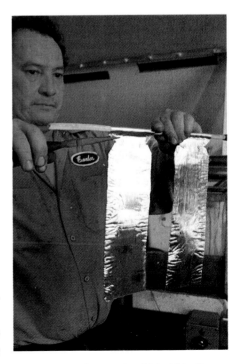

Thin sheets of the metal gold (gold leaf) to be used for decorative purposes.

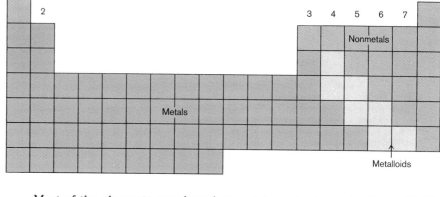

Figure 11.31
The classification of elements as metals, nonmetals, and metalloids.

Most of the elements are classed as metals, as is shown in Figure 11.31. Note that the metals are found on the left-hand side and at the center of the periodic table. The relatively few nonmetals are in the upper right-hand corner of the table. A few elements exhibit both metallic and nonmetallic behavior; they are classed as metalloids or semimetals.

It is important to understand that simply being classified as a metal does not mean that an element behaves exactly like all other metals. For example, some metals can lose one or more electrons much more easily than others. In particular, cesium can give up its outermost electron (a $6s$ electron) more easily than can lithium (a $2s$ electron). In fact, for the alkali metals (Group 1) the ease of giving up an electron varies as follows:

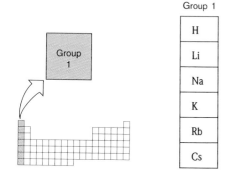

Group 1

| H |
| Li |
| Na |
| K |
| Rb |
| Cs |

$$Cs \ > \ Rb \ > \ K \ > \ Na \ > \ Li$$

Loses an
electron
most easily

Least likely
to lose an
electron

Note that as we go down the group, the metals become more likely to lose an electron. This makes sense because as we go down the group, the electron being removed resides, on average, farther and farther from the nucleus. That is, the $6s$ electron lost from Cs is much farther from the attractive positive nucleus—and so is easier to remove—than the $2s$ electron that must be removed from a lithium atom.

The same trend is also seen in the Group 2 metals (alkaline earth metals): the farther down in the group the metal resides, the more likely it is to lose an electron.

Just as metals vary somewhat in their properties, so do nonmetals. In general, the elements that can most effectively pull electrons from metals occur in the upper right-hand corner of the periodic table.

As a general rule, we can say that the most chemically active metals appear in the lower left-hand region of the periodic table, whereas the most chemically active nonmetals appear in the upper right-hand region. The properties of the semimetals, or metalloids, lie between the metals and the nonmetals, as might be expected.

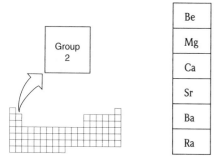

Group 2

| Be |
| Mg |
| Ca |
| Sr |
| Ba |
| Ra |

Ionization Energies

The **ionization energy** of an atom is the energy required to remove an electron from an individual atom in the gas phase:

$$M(g) \xrightarrow{\text{Ionization energy}} M^+(g) + e^-$$

As we have noted, the most characteristic chemical property of a metal atom is losing electrons to nonmetals. Another way of saying this is to say that *metals have relatively low ionization energies*—a relatively small amount of energy is needed to remove an electron from a typical metal.

Recall that metals at the bottom of a group lose electrons more easily than those at the top. In other words, ionization energies tend to decrease in going from the top to the bottom of a group.

Students have difficulty associating high and low ionization energy with the relative ease of removing electrons. You can tell them that high ionization energy means that an electron is *hard* to remove.

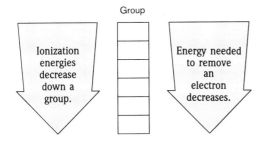

In contrast to metals, nonmetals have relatively large ionization energies. Nonmetals tend to gain, not lose, electrons. Recall that metals appear on the left-hand side of the periodic table and nonmetals appear on the right. Thus it is not surprising that ionization energies tend to increase from left to right across a given period on the periodic table.

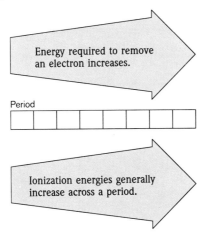

In general, the elements that appear in the lower left-hand region of the periodic table have the lowest ionization energies (and are therefore the most chemically active metals). On the other hand, the elements with the highest ionization energies (the most chemically active nonmetals) occur in the upper right-hand region of the periodic table.

Atomic Size

The sizes of atoms vary as shown in Figure 11.32. Notice that atoms get larger as we go down a group on the periodic table and that they get smaller as we go from left to right across a period.

We can understand the increase in size that we observe as we go down a group by remembering that as the principal energy level increases, the average distance of the electrons from the nucleus also increases. So atoms get bigger as electrons are added to larger principal energy levels.

Explaining the decrease in **atomic size** across a period requires a little thought about the atoms in a given row (period) of the periodic table. Recall that the atoms in a particular period all have their outermost electrons in a given

Students should learn the reasons for periodic trends in ionization energy and atomic size along with learning the periodic trends.

Figure 11.32
Relative atomic sizes for selected atoms. Note that atomic size increases down a group and decreases across a period.

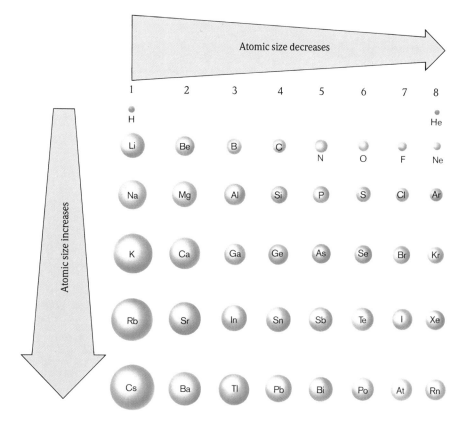

CHEMISTRY IN FOCUS

Fireworks

*T*he art of using mixtures of chemicals to produce explosives is an ancient one. Black powder—a mixture of potassium nitrate, charcoal, and sulfur—was being used in China well before A.D. 1000, and it has been used through the centuries in military explosives, in construction blasting, and for fireworks.

Before the nineteenth century, fireworks were confined mainly to rockets and loud bangs. Orange and yellow colors came from the presence of charcoal and iron filings. However, with the great advances in chemistry in the nineteenth century, new compounds found their way into fireworks. Salts of copper, strontium, and barium added brilliant colors. Magnesium and aluminum metals gave a dazzling white light. Fireworks, in fact, have changed very little in the last century.

How do fireworks produce their brilliant colors and loud bangs? Actually, only a handful of different chemicals are responsible for most of the spectacular effects. To produce the noise and flashes, an oxidizer (some-

Fireworks over New York City.

thing with a strong affinity for electrons) is reacted with a metal such as magnesium or aluminum mixed with sulfur. The resulting reaction produces a brilliant flash, which is due to the aluminum or magnesium burning, and a loud report is produced by the rapidly expanding gases. For a color effect, an element with a colored flame is included.

Yellow colors in fireworks are due to sodium. Strontium salts give the red color familiar from highway safety flares. Barium salts give a green color.

Although you might think that the chemistry of fireworks is simple, achieving the vivid white flashes and the brilliant colors requires complex combinations of chemicals. For example, because the white flashes produce high flame temperatures, the colors tend to wash out. Another problem arises from the use of sodium salts. Because sodium produces an extremely bright yellow color, sodium salts cannot be used when other colors are desired. In short, the manufacture of fireworks that produce the desired effects and are also safe to handle requires very careful selection of chemicals.*

---

*The chemical mixtures in fireworks are very dangerous. *Do not* experiment with chemicals on your own.

principal energy level. That is, the atoms in Period 1 have their outer electrons in the $1s$ orbital (principal energy level 1), the atoms in Period 2 have their outermost electrons in principal energy level 2 ($2s$ and $2p$ orbitals), and so on (see Figure 11.27). Because all the orbitals in a given principal energy level are expected to be the same size, we might expect the atoms in a given period to be the same size. However, remember that the number of protons in the nucleus increases as we move from atom to atom in the period. The resulting increase in positive charge on the nucleus tends to pull the electrons closer to the nucleus. So instead of remaining the same size across a period as electrons are added in a given principal energy level, the atoms get smaller as the electron "cloud" is drawn in by the increasing nuclear charge.

CHAPTER REVIEW

Key Terms

electromagnetic radiation (p. 338)
wavelength (p. 338)
frequency (p. 338)
photon (p. 342)
quantized energy levels (p. 344)
wave mechanical model (p. 346)
orbital (p. 347)
principal energy levels (p. 348)
sublevels (p. 348)
Pauli exclusion principle (p. 351)
electron configuration (p. 353)

orbital (box) diagram (p. 353)
valence electrons (p. 355)
core electrons (p. 356)
lanthanide series (p. 358)
actinide series (p. 358)
main-group (representative)
    elements (p. 360)
metals (p. 361)
nonmetals (p. 361)
ionization energy (p. 363)
atomic size (p. 364)

Summary

1. Energy travels through space by electromagnetic radiation ("light"), which can be characterized by the wavelength and frequency of the waves. Light can also be thought of as packets of energy called photons. Atoms can gain energy by absorbing a photon and can lose energy by emitting a photon.

2. The emissions of energy from hydrogen atoms produce only certain energies as hydrogen changes from a higher to a lower energy. This shows that the energy levels of hydrogen are quantized.

3. The Bohr model of the hydrogen atom postulated that the electron moved in circular orbits corresponding to the various allowed energy levels. Though it worked well for hydrogen, the Bohr model did not work for other atoms.

4. The wave mechanical model explains atoms by postulating that the electron has both wave and particle characteristics. Electron states are described by orbitals, which are probability maps indicating how likely it is to find the electron at a given point in space. The orbital size can be thought of as a surface containing 90% of the total electron probability.

5. According to the Pauli exclusion principle, an atomic orbital can hold a maximum of two electrons, and those electrons must have opposite spins.

6. Atoms have a series of energy levels, called principal energy levels (n), which contain one or more sublevels (types of orbitals). The number of sublevels increases with increasing n.

7. Valence electrons are the s and p electrons in the outermost principal energy level of an atom. Core electrons are the inner electrons of an atom.

8. Metals are found at the left and center of the periodic table. The most chemically active metals are found in the lower left-hand corner of the periodic table. The most chemically active nonmetals are located in the upper right-hand corner.

9. Ionization energy, the energy required to remove an electron from a gaseous atom, decreases going down a group and increases going from left to right across a period.

10. For the representative elements, atomic size increases going down a group but decreases going across a period in the periodic table.

Questions and Problems

All even-numbered exercises have answers in the back of this book and solutions in the Solutions Guide.

11.1 Electromagnetic Radiation and Energy

QUESTIONS

1. Give several examples of different types of electromagnetic radiation. How are these types of radiation similar? How do they differ?

2. What is the *wavelength* of electromagnetic radiation?

3. What is the *frequency* of electromagnetic radiation? How do the wavelength and frequency differ?

4. The text gives several examples of the absorption of electromagnetic radiation by molecules. What happens to a molecule when it absorbs radiation?

5. A "packet" of electromagnetic energy is called a _____.

6. The visible light region of the electromagnetic spectrum corresponds to a wavelength on the order of _____ meters.

11.2 The Energy Levels of Hydrogen

QUESTIONS

7. An atom that possesses excess energy is said to be in a/an _____ state.

8. An atom can release its excess energy by emitting a _____ of electromagnetic radiation.

9. Different _____ of light carry different amounts of energy per photon.

11. Describe briefly why the study of electromagnetic radiation has been important to our understanding of the arrangement of electrons in atoms.

13. Because a given element's atoms emit only certain photons of light, only certain _____ are occurring in those particular atoms.

15. What does it mean to say that the energy levels of hydrogen are *quantized*? What experimental evidence exists for this?

10. A beam of red light has (higher/lower) energy photons than blue light.

12. What do we mean when we say that excited hydrogen atoms always emit radiation at the same discrete wavelengths—that is, that only certain types of photons are emitted when a hydrogen atom releases its excess energy?

14. The energy of an emitted photon corresponds to the difference in energy between the excited state of the emitting atom and its _____ state.

16. What is meant by the *ground state* of an atom?

11.3 The Bohr Model of the Atom

QUESTIONS

17. What are the essential points of Bohr's theory of the structure of the hydrogen atom?

19. How does the Bohr theory account for the observed phenomenon of the emission of discrete wavelengths of light by excited atoms?

18. According to Bohr, what sorts of motions do electrons have in an atom, and what happens when energy is applied to the atom?

20. Why was Bohr's theory for the hydrogen atom initially accepted, and why was it ultimately discarded?

11.4 The Wave Mechanical Model of the Atom

QUESTIONS

21. What major assumption (that was analogous to what had already been demonstrated for electromagnetic radiation) did de Broglie and Schrödinger make about the motion of tiny particles?

23. Why was Schrödinger not able to describe exactly the pathway an electron takes as it moves through the space of an atom?

22. Discuss briefly the difference between an orbit (as described by Bohr for hydrogen) and an orbital (as described by the more modern, wave mechanical picture of the atom).

24. Explain why we cannot *exactly* specify the location of an electron in an atom but can only discuss where an electron is *most likely* to be at any given time.

11.5 The Hydrogen Orbitals

QUESTIONS

25. When we draw a picture of an orbital, we are indicating that the probability of finding the electron within this region of space is greater than 90%. Why is this probability never 100%?

27. What are the differences between the $2s$ orbital and the $1s$ orbital of hydrogen? How are they similar?

29. As an electron moves from one principal energy level to a higher-number level, the distance of the electron from the nucleus (increases/decreases).

31. Although a hydrogen atom has only one electron, the hydrogen atom possesses a complete set of available orbitals. What purpose do these additional orbitals serve?

26. What do the principal energy levels of an atom represent? How are these principal energy levels like the orbits postulated by Bohr, and how do they differ from those orbits?

28. What overall shape do the $2p$ orbitals have? How are the individual $2p$ orbitals alike, and how do they differ?

30. If an electron moves from the $1s$ orbital to the $2s$ orbital, its energy (increases/decreases).

32. Into how many sublevels is the fourth principal energy level of the hydrogen atom divided? What designations are given to these sublevels?

11.6 The Wave Mechanical Model: A Summary

QUESTIONS

33. What do we mean when we say that each electron in an atom has its own intrinsic "spin"?

34. What is meant by the *Pauli exclusion principle?* How many electrons are in an orbital, according to this principle? Why?

35. The energy of a principal energy level (increases/decreases) as n increases.

36. The number of sublevels in a principal energy level (increases/decreases) as n increases.

37. According to the Pauli exclusion principle, a given orbital can never contain more than _____ electrons.

38. The shape of an orbital provides only a _____ map for the electron residing in the orbital; it does not indicate the details of the electron's movement.

39. Sketch the general shape of each of the following types of orbitals.
    a. an s orbital
    b. a p orbital
    c. a d orbital

40. Which of the following orbital designations is(are) not correct?
    a. $1d$ d. $4d$
    b. $2d$ e. $3f$
    c. $3d$ f. $4f$

11.7 Electron Arrangements in the First Eighteen Atoms

QUESTIONS

41. Why is the $1s$ orbital the first orbital to be filled in any atom?

42. In the electron configuration for nitrogen ($Z = 7$), how many of the $2p$ orbitals will be occupied by electrons?

43. Which electrons of an atom are the *valence* electrons? Why are these electrons especially important?

44. What important feature do the elements in a given group (vertical column) of the periodic table have in common?

PROBLEMS

45. Write the full electronic configuration ($1s^2 2s^2$, etc.) for each of the following elements.
    a. silicon, $Z = 14$ c. magnesium, $Z = 12$
    b. argon, $Z = 18$ d. helium, $Z = 2$

46. Write the electronic configuration ($1s^2 2s^2$, and so on) for each of the following elements.
    a. strontium, $Z = 38$ c. helium, $Z = 2$
    b. zinc, $Z = 30$ d. bromine, $Z = 35$

47. Write the full electronic configuration ($1s^2 2s^2$, etc.) for each of the following elements.
    a. sodium, $Z = 11$ c. nitrogen, $Z = 7$
    b. cesium, $Z = 55$ d. beryllium, $Z = 4$

48. Write the full electronic configuration ($1s^2 2s^2$, etc.) for each of the following elements.
    a. carbon, $Z = 6$ c. sulfur, $Z = 16$
    b. phosphorus, $Z = 15$ d. boron, $Z = 5$

49. Write the complete orbital diagram for each of the following elements, using boxes to represent orbitals and arrows to represent electrons.
    a. sodium, $Z = 11$ c. chlorine, $Z = 17$
    b. calcium, $Z = 20$ d. argon, $Z = 18$

50. Write the complete orbital diagram for each of the following elements, using boxes to represent orbitals and arrows to represent electrons.
    a. aluminum, $Z = 13$ c. bromine, $Z = 35$
    b. phosphorus, $Z = 15$ d. argon, $Z = 18$

51. How many valence electrons does each of the following atoms have?
    a. lithium, $Z = 3$ c. argon, $Z = 18$
    b. aluminum, $Z = 13$ d. phosphorus, $Z = 15$

52. How many valence electrons does each of the following atoms possess?
    a. magnesium, $Z = 12$ c. bromine, $Z = 35$
    b. carbon, $Z = 6$ d. cesium, $Z = 55$

11.8 Electron Configurations and the Periodic Table

QUESTIONS

53. Why do we believe that the valence electrons of calcium and potassium reside in the $4s$ orbital rather than in the $3d$ orbital?

54. What do we mean by the *core electrons* of an atom?

PROBLEMS

55. Using the symbol of the previous noble gas to indicate the inner-core electrons, write the valence shell electron configuration for each of the following elements.
    a. rubidium, $Z = 37$ c. phosphorus, $Z = 15$
    b. chlorine, $Z = 17$ d. barium, $Z = 56$

57. Using the symbol of the previous noble gas to indicate the inner-core electrons, write the valence shell electron configuration for each of the following elements.
    a. arsenic, $Z = 33$ c. scandium, $Z = 21$
    b. bromine, $Z = 35$ d. radium, $Z = 88$

59. How many $3d$ electrons are found in each of the following elements?
    a. titanium, $Z = 22$ c. zinc, $Z = 30$
    b. cobalt, $Z = 27$ d. chromium, $Z = 24$

61. For each of the following elements, indicate which set of orbitals is being filled last.
    a. tungsten, $Z = 74$
    b. iron, $Z = 26$
    c. europium, $Z = 63$
    d. cerium, $Z = 58$

63. Write the shorthand valence shell electron configuration of each of the following elements, basing your answer on the element's location on the periodic table.
    a. tungsten, $Z = 74$
    b. cadmium, $Z = 48$
    c. gold, $Z = 79$
    d. platinum, $Z = 78$

56. Using the symbol of the previous noble gas to indicate the inner-core electrons, write the valence shell electron configuration for each of the following elements.
    a. calcium, $Z = 20$ c. yttrium, $Z = 39$
    b. francium, $Z = 87$ d. cerium, $Z = 58$

58. Using the symbol of the previous noble gas to indicate the core electrons, write the valence shell electron configuration for each of the following elements.
    a. silicon, $Z = 14$ c. vanadium, $Z = 23$
    b. gallium, $Z = 31$ d. scandium, $Z = 21$

60. How many $4d$ electrons are found in each of the following elements?
    a. yttrium, $Z = 39$ c. strontium, $Z = 38$
    b. zirconium, $Z = 40$ d. cadmium, $Z = 48$

62. For each of the following elements, indicate which set of orbitals is being filled last.
    a. lanthanum, $Z = 57$
    b. lutetium, $Z = 71$
    c. osmium, $Z = 76$
    d. zinc, $Z = 30$

64. Write the shorthand valence shell electron configuration of each of the following elements, basing your answer on the element's location on the periodic table.
    a. uranium, $Z = 92$
    b. manganese, $Z = 25$
    c. mercury, $Z = 80$
    d. francium, $Z = 87$

11.9 Atomic Properties and the Periodic Table

QUESTIONS

65. What are some of the physical properties that distinguish the metallic elements from the nonmetals? Are these properties absolute, or do some nonmetallic elements exhibit some metallic properties (and vice versa)?

67. Give some similarities that exist among the elements of Group 1.

69. Which metallic elements within a given group (vertical column) of the periodic table lose electrons most easily? Why?

71. Where are the most nonmetallic elements located in the periodic table? Why do these elements pull electrons from metallic elements so effectively during a reaction?

73. Explain why the atoms of the elements at the bottom of a given group (vertical column) of the periodic table are *larger* than the atoms of the elements at the top of the same group.

66. What types of ions do the metals and the nonmetallic elements form? Do the metals lose or gain electrons in doing this? Do the nonmetallic elements gain or lose electrons in doing this?

68. Give some similarities that exist among the elements of Group 7.

70. Which elements in a given period (horizontal row) of the periodic table lose electrons most easily? Why?

72. Why do the metallic elements of a given period (horizontal row) typically have much lower ionization energies than do the nonmetallic elements of the same period?

74. Though all the elements in a given period (horizontal row) of the periodic table have their valence electrons in the same types of orbitals, the sizes of the atoms decrease from left to right within a period. Explain why.

PROBLEMS

75. In each of the following sets of elements, indicate which element shows the most active chemical behavior.
   a. Li, K, Cs
   b. Mg, Sr, Ra
   c. S, Se, Te
77. In each of the following sets of elements, indicate which element has the largest atomic size.
   a. Li, Rb, Cs
   b. B, Ga, Tl
   c. F, Cl, I

76. In each of the following sets of elements, indicate which element would be expected to have the lowest ionization energy.
   a. Li, K, Cs
   b. Li, C, F
   c. F, Cl, I
   d. Rb, Sr, In
78. In each of the following sets of elements, indicate which element has the smallest atomic size.
   a. Na, K, Rb
   b. Na, Si, S
   c. N, P, As
   d. N, O, F

Additional Problems

79. The distance in meters between two consecutive peaks (or troughs) in a wave is called the _____.
80. The speed at which electromagnetic radiation moves through a vacuum is called the _____.
81. The portion of the electromagnetic spectrum between wavelengths of approximately 400 and 700 nanometers is called the _____ light region.
82. A beam of light can be thought of as consisting of a stream of light particles called _____.
83. The lowest possible energy state of an atom is called the _____ state.
84. The energy levels of hydrogen (and other atoms) are _____, which means that only certain values of energy are allowed.
85. According to Bohr, the electron in the hydrogen atom moved around the nucleus in circular paths called _____.
86. In the modern theory of the atom, a(n) _____ represents a region of space in which there is a high probability of finding an electron.
87. Electrons found in the outermost principal energy level of an atom are referred to as _____ electrons.
88. An element with partially filled d-orbitals is called a/an _____.
89. The _____ of electromagnetic radiation represents the number of waves passing a given point in space each second.
90. Only two electrons can occupy a given orbital in an atom, and to be in the same orbital, they must have opposite _____.
91. One bit of evidence that the present theory of atomic structure is "correct" lies in the magnetic properties of matter. Atoms with *unpaired* electrons are attracted by magnetic fields and thus are said to exhibit *paramagnetism.* The de-

gree to which this effect is observed is directly related to the *number* of unpaired electrons present in the atom. On the basis of the electron orbital diagrams for the following elements, indicate which atoms would be expected to be paramagnetic, and tell how many unpaired electrons each atom contains.
   a. phosphorus, $Z = 15$
   b. iodine, $Z = 53$
   c. germanium, $Z = 32$
92. Without referring to your textbook or a periodic table, write the full electron configuration, the orbital box diagram, and the noble gas shorthand configuration for the elements with the following atomic numbers.
   a. $Z = 19$
   b. $Z = 22$
   c. $Z = 14$
   d. $Z = 26$
   e. $Z = 30$
93. Without referring to your textbook or a periodic table, write the full electron configuration, the orbital box diagram, and the noble gas shorthand configuration for the elements with the following atomic numbers.
   a. $Z = 28$
   b. $Z = 18$
   c. $Z = 35$
   d. $Z = 37$
   e. $Z = 23$
94. Write the general valence configuration (for example, ns^1 for Group 1) for the group in which each of the following elements is found.
   a. barium, $Z = 56$
   b. bromine, $Z = 35$
   c. tellurium, $Z = 52$
   d. potassium, $Z = 19$
   e. sulfur, $Z = 16$
95. How many valence electrons does each of the following atoms have?
   a. cesium, $Z = 55$
   b. strontium, $Z = 38$
   c. bromine, $Z = 35$
   d. selenium, $Z = 34$

96. In the text (Section 11.4) it was mentioned that current theories of atomic structure suggest that all matter and all energy demonstrate both particle-like and wave-like properties under the appropriate conditions, although the wave-like nature of matter becomes apparent only in very small and very fast-moving particles. The relationship between wavelength (λ) observed for a particle and the mass and velocity of that particle is called the de Broglie relationship. It is

$$\lambda = h/mv$$

in which h is Planck's constant (6.63×10^{-34} J s*), m represents the mass of the particle in kilograms, and v represents the velocity of the particle in meters per second. Calculate the "de Broglie wavelength" for each of the following, and use your numerical answers to explain why macroscopic (large) objects are not ordinarily discussed in terms of their "wave-like" properties.
   a. an electron moving at 0.90 times the speed of light
   b. a 150-g ball moving at a speed of 10. m/s
   c. a 75-kg person walking at a speed of 2 km/h
97. Light waves move through space at a speed of _____ meters per second.
98. How do we know that the energy levels of the hydrogen atom are not *continuous,* as physicists originally assumed?
99. As an electron moves from one principal energy level to a higher-number level, the attractive force that the nucleus exerts on the electron (increases/decreases).
100. Into how many sublevels is the third principal energy level of hydrogen divided? What are the names of the orbitals that constitute these sublevels? What are the general shapes of these orbitals?
101. What condition related to spin must be fulfilled for two electrons both to occupy a particular orbital?
102. Which of the following orbital designations is(are) *not* correct?
   a. $1p$ d. $2p$
   b. $3d$ e. $5f$
   c. $3f$ f. $6s$
103. Why do the two electrons in the $2p$ sublevel of carbon occupy *different* $2p$ orbitals?

104. Write the full electronic configuration ($1s^2 2s^2$, etc.) for each of the following elements.
   a. chlorine, $Z = 17$ c. aluminum, $Z = 13$
   b. oxygen, $Z = 8$ d. lithium, $Z = 3$
105. Write the complete orbital diagram for each of the following elements, using boxes to represent orbitals and arrows to represent electrons.
   a. scandium, $Z = 21$ c. potassium, $Z = 19$
   b. sulfur, $Z = 16$ d. nitrogen, $Z = 7$
106. How many valence electrons does each of the following atoms have?
   a. nitrogen, $Z = 7$ c. sodium, $Z = 11$
   b. chlorine, $Z = 17$ d. aluminum, $Z = 13$
107. What name is given to the group of ten elements in which the electrons are filling the $3d$ sublevel?
108. Using the symbol of the previous noble gas to indicate the inner-core electrons, write the valence shell electron configuration for each of the following elements.
   a. zirconium, $Z = 40$ c. germanium, $Z = 32$
   b. iodine, $Z = 53$ d. cesium, $Z = 55$
109. Using the symbol of the previous noble gas to indicate inner-core electrons, write the valence shell electron configuration for each of the following elements.
   a. titanium, $Z = 22$ c. antimony, $Z = 51$
   b. selenium, $Z = 34$ d. strontium, $Z = 38$
110. For each of the following elements, indicate which set of orbitals is being filled last.
   a. chromium, $Z = 24$ c. uranium, $Z = 92$
   b. silver, $Z = 47$ d. germanium, $Z = 32$
111. Write the shorthand valence shell electron configuration of each of the following elements, basing your answer on the element's location on the periodic table.
   a. nickel, $Z = 28$ c. hafnium, $Z = 72$
   b. niobium, $Z = 41$ d. astatine, $Z = 85$
112. Metals have relatively (low/high) ionization energies, whereas nonmetals have relatively (high/low) ionization energies.
113. In each of the following sets of elements, indicate which element shows the most active chemical behavior.
   a. B, Al, In
   b. Na, Al, S
   c. B, C, F
114. In each of the following sets of elements, indicate which element has the largest atomic size.
   a. Be, Ca, Ba
   b. Mg, Si, Cl
   c. Be, N, F

---

*Note that s is the abbreviation for "seconds."

Solutions to Self-Check Exercises

SELF-CHECK EXERCISE 11.1

a. Circular pathways for electrons in the Bohr model.
b. Probability maps that give the likelihood that the electron will occupy a given point in space. The details of electron motion are not described by an orbital.
c. The surface that contains 90% of the total electron probability. That is, the electron is found somewhere inside this surface 90% of the time.
d. A sublevel is a given type of orbital within a principal energy level. For example, there are three sublevels in principal level 3; they consist of the $3s$ orbital, the three $3p$ orbitals, and the five $3d$ orbitals.

SELF-CHECK EXERCISE 11.2

Element	Electron Configuration	Orbital Diagram $1s$	$2s$	$2p$	$3s$	$3p$
Al	$1s^2 2s^2 2p^6 3s^2 3p^1$ [Ne]$3s^2 3p^1$	↑↓	↑↓	↑↓ ↑↓ ↑↓	↑↓	↑ □ □
Si	[Ne]$3s^2 3p^2$	↑↓	↑↓	↑↓ ↑↓ ↑↓	↑↓	↑ ↑ □
P	[Ne]$3s^2 3p^3$	↑↓	↑↓	↑↓ ↑↓ ↑↓	↑↓	↑ ↑ ↑
S	[Ne]$3s^2 3p^4$	↑↓	↑↓	↑↓ ↑↓ ↑↓	↑↓	↑↓ ↑ ↑
Cl	[Ne]$3s^2 3p^5$	↑↓	↑↓	↑↓ ↑↓ ↑↓	↑↓	↑↓ ↑↓ ↑
Ar	[Ne]$3s^2 3p^6$	↑↓	↑↓	↑↓ ↑↓ ↑↓	↑↓	↑↓ ↑↓ ↑↓

SELF-CHECK EXERCISE 11.3

Fluorine (F): In Group 7 and Period 2, it is the fifth "$2p$ element." The configuration is $1s^2 2s^2 2p^5$, or [He]$2s^2 2p^5$.

Silicon (Si): In Group 4 and Period 3, it is the second of the "$3p$ elements." The configuration is $1s^2 2s^2 2p^6 3s^2 3p^2$, or [Ne]$3s^2 3p^2$.

Cesium (Cs): In Group 1 and Period 6, it is the first of the "$6s$ elements." The configuration is $1s^2 2s^2 2p^6 3s^2 3p^6 4s^2 3d^{10} 4p^6 5s^2 4d^{10} 5p^6 6s^1$, or [Xe]$6s^1$.

Lead (Pb): In Group 4 and Period 6, it is the second of the "$6p$ elements." The configuration is [Xe]$6s^2 4f^{14} 5d^{10} 6p^2$.

Iodine (I): In Group 7 and Period 5, it is the fifth of the "$5p$ elements." The configuration is [Kr]$5s^2 4d^{10} 5p^5$.

12 Chemical Bonding

Diamond, whose crystals are shown in this photomicrograph, is composed solely of carbon atoms bonded into one of the hardest materials known.

CONTENTS

Students have difficulty with the material in this chapter because it depends on visualizing three-dimensional shapes from two-dimensional drawings and written descriptions. This is hard for the inexperienced to do. Models can help students understand many of the topics presented.

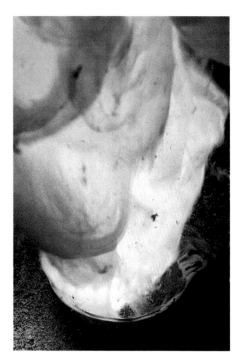

Zinc metal and sulfur react vigorously to form zinc sulfide.

The world around us is composed almost entirely of compounds and mixtures of compounds. Rocks, coal, soil, petroleum, trees, and human beings are all complex mixtures of chemical compounds in which different kinds of atoms are bound together. Most of the pure elements found in the earth's crust also contain many atoms bound together. In a gold nugget each gold atom is bound to many other gold atoms, and in a diamond many carbon atoms are bonded very strongly to each other. Substances composed of unbound atoms do exist in nature, but they are very rare. (Examples include the argon atoms in the atmosphere and the helium atoms found in natural gas reserves.)

The manner in which atoms are bound together has a profound effect on the chemical and physical properties of substances. For example, both graphite and diamond are composed solely of carbon atoms. However, graphite is a soft, slippery material used as a lubricant in locks, and diamond is one of the hardest materials known, valuable both as a gemstone and in industrial cutting tools. Why do these materials, both composed solely of carbon atoms, have such different properties? The answer lies in the different ways in which the carbon atoms are bound to each other in these substances.

Molecular bonding and structure play the central role in determining the course of chemical reactions, many of which are vital to our survival. Most reactions in biological systems are very sensitive to the structures of the participating molecules; in fact, very subtle differences in shape sometimes serve to channel the chemical reaction one way rather than another. Molecules that act as drugs must have exactly the right structure to perform their functions correctly.

To understand the behavior of natural materials, we must understand the nature of chemical bonding and the factors that control the structures of compounds. In this chapter, we will present various classes of compounds that illustrate the different types of bonds. We will then develop models to describe the structure and bonding that characterize the materials found in nature.

12.1   Types of Chemical Bonds

AIM: To describe ionic and covalent bonds and explain how they are formed.

To introduce the polar covalent bond.

A water molecule

What is a chemical bond? Although there are several possible ways to answer this question, we will define a **bond** as a force that holds groups of two or more atoms together and makes them function as a unit. For example, in water the fundamental unit is the H—O—H molecule, which we describe as being held together by the two O—H bonds. We can obtain information about the strength of a bond by measuring the energy required to break the bond, the **bond energy.**

Atoms can interact with one another in several ways to form aggregates. We will consider specific examples to illustrate the various types of chemical bonds.

In Chapter 7 we saw that when solid sodium chloride is dissolved in water, the resulting solution conducts electricity, a fact that convinces chemists that sodium chloride is composed of Na^+ and Cl^- ions. Thus when sodium and chlorine react to form sodium chloride, electrons are transferred from the sodium atoms to the chlorine atoms to form Na^+ and Cl^- ions, which then aggregate to form solid sodium chloride. The resulting solid sodium chloride is a very sturdy material; it has a melting point of approximately 800 °C. The strong bonding forces present in sodium chloride result from the attractions among the closely packed, oppositely charged ions. This is an example of **ionic bonding.** Ionic substances are formed when an atom that loses electrons relatively easily reacts with an atom that has a high affinity for electrons. In other words, an **ionic compound** results when a metal reacts with a nonmetal.

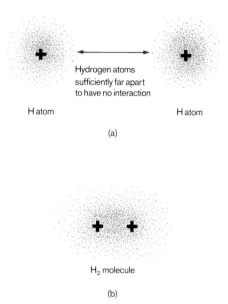

Figure 12.1
The formation of a bond between two hydrogen atoms. (a) Two separate hydrogen atoms. (b) When two hydrogen atoms come close together, the two electrons are attracted simultaneously by both nuclei. This produces the bond. Note the relatively large electron probability between the nuclei indicating sharing of the electrons.

We have seen that a bonding force develops when two very different types of atoms react to form oppositely charged ions. But how does a bonding force develop between two identical atoms? Let's explore this situation by considering what happens when two hydrogen atoms are brought close together, as shown in Figure 12.1. When hydrogen atoms are close together, the two electrons are simultaneously attracted to both nuclei. Note in Figure 12.1(b) how the electron probability increases between the two nuclei indicating that the electrons are shared by the two nuclei.

The type of bonding we encounter in the hydrogen molecule and in many other molecules where *electrons are shared by nuclei* is called **covalent bonding.** Note that in the H_2 molecule the electrons reside primarily in the space between the two nuclei, where they are attracted simultaneously by both protons. Although we will not go into detail about it here, the increased attractive forces in this area lead to the formation of the H_2 molecule from the two separated hydrogen atoms. When we say that a bond is formed between the hydrogen atoms, we mean that the H_2 molecule is more stable than two separated hydrogen atoms by a certain quantity of energy (the bond energy).

So far we have considered two extreme types of bonding. In ionic bonding, the participating atoms are so different that one or more electrons are transferred to form oppositely charged ions. The bonding results from the attractions between these ions. In covalent bonding, two identical atoms share electrons equally. The bonding results from the mutual attraction of the two nuclei for the shared electrons. Between these extremes are intermediate cases in which the atoms are not so different that electrons are completely transferred but are different enough so that unequal sharing of electrons results, forming what is called a

Ionic and covalent bonds are the extreme bond types.

It is easy for students to gain a basic understanding of bonding because the terms sharing, unequal sharing, and transfer can all be related to the human experience.

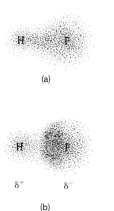

(a)

(b)

Figure 12.2
Probability representations of the electron sharing in HF. (a) What the probability map would look like if the two electrons in the H—F bond were shared equally. (b) The actual situation, where the shared pair spends more time close to the fluorine atom than to the hydrogen atom. This gives fluorine a slight excess of negative charge and the hydrogen a slight deficit of negative charge (a slight positive charge).

We wouldn't ordinarily ask students to memorize electronegativity values. However, students should be aware of periodic trends and be able to compare the electronegativities of two elements in the same group or the same period.

Students should know that fluorine is the element with the highest electronegativity, even if they don't know the numerical value.

polar covalent bond. The hydrogen fluoride (HF) molecule contains this type of bond, which produces the following charge distribution:

$$H - F$$
$$\delta^+ \quad \delta^-$$

where δ (delta) is used to indicate a partial or fractional charge.

The most logical explanation for the development of *bond polarity* (the partial positive and negative charges on the atoms in such molecules as HF) is that the electrons in the bonds are not shared equally. For example, we can account for the polarity of the HF molecule by assuming that the fluorine atom has a stronger attraction than the hydrogen atom for the shared electrons (Figure 12.2). Because bond polarity has important chemical implications, we find it useful to assign a number that indicates an atom's ability to attract shared electrons. In the next section we show how this is done.

12.2   Electronegativity

AIM: To discuss the nature of bonds in terms of electronegativity.

We saw in the last section that when a metal and a nonmetal react, one or more electrons are transferred from the metal to the nonmetal to give ionic bonding. On the other hand, two identical atoms react to form a covalent bond in which electrons are shared equally. When *different* nonmetals react, a bond forms in which electrons are shared *unequally,* giving a polar covalent bond. The unequal sharing of electrons between two atoms is described by a property called **electronegativity:** *the relative ability of an atom in a molecule to attract shared electrons to itself.*

Chemists determine electronegativity values for the elements (Figure 12.3) by measuring the polarities of the bonds between various atoms. Note that electronegativity generally increases going from left to right across a period and decreases going down a group for the representative elements. The range of electronegativity values is from 4.0 for fluorine to 0.7 for cesium and francium. Remember, the higher the atom's electronegativity value, the closer the shared electrons tend to be to that atom when it forms a bond.

The polarity of a bond depends on the *difference* between the electronegativity values of the atoms forming the bond. If the atoms have very similar electronegativities, the electrons are shared almost equally and the bond shows little polarity. If the atoms have very different electronegativity values, a very polar bond is formed. In extreme cases one or more electrons are actually transferred, forming ions and an ionic bond. For example, when an element from Group 1

When two identical atoms with high electronegativity values form a covalent bond (as in the oxygen molecule), students sometimes think that the bond is polar covalent because the atoms have high electronegativity values. They overlook the fact that the difference in electronegativity is zero.

Figure 12.3

Electronegativity values for selected elements. Note that electronegativity generally increases across a period and decreases down a group. Note also that metals have relatively low electronegativity values and that nonmetals have relatively high values.

(electronegativity values of about 0.8) reacts with an element from Group 7 (electronegativity values of about 3), ions are formed and an ionic substance results. In general, if the *difference* between the electronegativities of two elements is about 1.7 or greater, the bond is considered to be ionic.

The relationship between electronegativity and bond type is shown in Table 12.1. The various types of bonds are summarized in Figure 12.4.

Figure 12.4

The three possible types of bonds: (a) a covalent bond formed between identical atoms; (b) a polar covalent bond, with both ionic and covalent components; and (c) an ionic bond, with no electron sharing.

Table 12.1   The Relationship Between Electronegativity and Bond Type

Electronegativity Difference Between the Bonding Atoms	Bond Type	Covalent Character	Ionic Character
Zero ↓ Intermediate ↓ Large	Covalent ↓ Polar covalent ↓ Ionic	Decreases	Increases

EXAMPLE 12.1 **Using Electronegativity to Determine Bond Polarity**

Using the electronegativity values given in Figure 12.3, arrange the following bonds in order of increasing polarity: H—H, O—H, Cl—H, S—H, and F—H.

SOLUTION

The polarity of the bond increases as the difference in electronegativity increases. From the electronegativity values in Figure 12.3, the following variation in bond polarity is expected (the electronegativity value appears in parentheses below each element).

Bond	H—H	S—H	Cl—H	O—H	F—H
Electronegativity values	(2.1)(2.1)	(2.5)(2.1)	(3.0)(2.1)	(3.5)(2.1)	(4.0)(2.1)
Difference in electronegativity values	$2.1 - 2.1 = 0$	$2.5 - 2.1 = 0.4$	$3.0 - 2.1 = 0.9$	$3.5 - 2.1 = 1.4$	$4.0 - 2.1 = 1.9$
Bond type	Covalent	All polar covalent			

Increasing polarity

Therefore, in order of increasing polarity, we have

H—H   S—H   Cl—H   O—H   F—H

Least polar   Most polar

SELF-CHECK EXERCISE 12.1

For each of the following pairs of bonds, choose the bond that will be more polar.

a. H—P, H—C   c. N—O, S—O
b. O—F, O—I   d. N—H, Si—H

You might want to begin your discussion of dipole moments with simple molecules containing only two or three atoms. Whether or not a molecule has a dipole depends in part upon the three-dimensional shape of the molecule, and we have not covered molecular structure yet.

12.3 Bond Polarity and Dipole Moments

AIM: To describe how bond polarity is related to molecular polarity.

We saw in Section 12.1 that hydrogen fluoride has a positive end and a negative end. A molecule such as HF that has a center of positive charge and a center of negative charge is said to have a **dipole moment**. The dipolar character of a molecule is often represented by an arrow. This arrow points toward the negative charge center, and its tail indicates the positive center of charge:

Any diatomic (two-atom) molecule that has a polar bond has a dipole moment. Some polyatomic (more than two atoms) molecules also have dipole moments. For example, because the oxygen atom in the water molecule has a greater electronegativity than the hydrogen atoms, the electrons are not shared equally. This results in a charge distribution (Figure 12.5a) that causes the molecule to behave as though it had two centers of charge—one positive and one negative. So the water molecule has a dipole moment.

The fact that the water molecule is polar (has a dipole moment) has a profound impact on its properties. In fact, it is not overly dramatic to state that the polarity of the water molecule is crucial to life as we know it on earth. Because water molecules are polar, they can surround and attract both positive and negative ions (Figure 12.6). These attractions allow ionic materials to dissolve in

(a)

(b)

Figure 12.5
(a) The charge distribution in the water molecule. The oxygen has a charge of $2\delta^-$ because it pulls δ^- of charge from each hydrogen atom ($\delta^- + \delta^- = 2\delta^-$). (b) The water molecule behaves as if it had a positive end and a negative end, as indicated by the arrow.

Students can usually determine whether or not diatomic molecules will have a dipole and can arrange the molecules in order of increasing polarity using a table of electronegativity values. However, when molecules contain more than two atoms, it is very hard for students to determine whether or not there is a dipole.

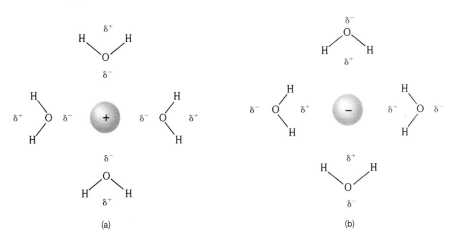

(a)   (b)

Figure 12.6
(a) Polar water molecules are strongly attracted to positive ions by their negative ends. (b) They are also strongly attracted to negative ions by their positive ends.

To show the significance of dipole moment, spend a few minutes discussing the behavior of water, as illustrated in Figures 12.6 and 12.7.

Figure 12.7
Polar water molecules are strongly attracted to each other.

water. Also, the polarity of water molecules causes them to attract each other strongly (Figure 12.7). This means that much energy is required to change water from a liquid to a gas (the molecules must be separated from each other to undergo this change of state). Therefore it is the polarity of the water molecule that causes water to remain a liquid at the temperatures on the earth's surface. If it were nonpolar, water would be a gas and the oceans would be empty.

12.4 Stable Electron Configurations and Charges on Ions

AIM: To define a stable electron configuration.
To show how to predict the formulas of ionic compounds.

We have seen many times that when a metal and a nonmetal react to form an ionic compound, the metal atom loses one or more electrons to the nonmetal. In Chapter 5, where binary ionic compounds were introduced, we saw that in these reactions, Group 1 metals always form 1+ cations, Group 2 metals always form 2+ cations, and aluminum in Group 3 always forms a 3+ cation. For the nonmetals, the Group 7 elements always form 1− anions, and the Group 6 elements always form 2− anions. This is further illustrated in Table 12.2.

Notice something very interesting about the ions in Table 12.2: they all have the electron configuration of neon; a noble gas. That is, sodium loses its one valence electron (the $3s$) to form Na^+, which has a [Ne] electron configuration. Likewise Mg loses its two valence electrons to form Mg^{2+}, which also has a [Ne] electron configuration. On the other hand, the nonmetal atoms gain just the

Table 12.2   The Formation of Ions by Metals and Nonmetals

Group	Ion Formation	Electron Configuration	
		Atom	*Ion*
1	$Na \rightarrow Na^+ + e^-$	$[Ne]3s^1$ — e^- lost →	$[Ne]$
2	$Mg \rightarrow Mg^{2+} + 2e^-$	$[Ne]3s^2$ — $2e^-$ lost →	$[Ne]$
3	$Al \rightarrow Al^{3+} + 3e^-$	$[Ne]3s^23p^1$ — $3e^-$ lost →	$[Ne]$
6	$O + 2e^- \rightarrow O^{2-}$	$[He]2s^22p^4 + 2e^- \rightarrow [He]2s^22p^6 = [Ne]$	
7	$F + e^- \rightarrow F^-$	$[He]2s^22p^5 + e^- \rightarrow [He]2s^22p^6 = [Ne]$	

number of electrons needed for them to achieve the noble gas electron configuration. The O atom gains two electrons and the F atom gains one electron to give O^{2-} and F^-, respectively, both of which have the [Ne] electron configuration. We can summarize these observations as follows:

1. Representative (main group) metals form ions by losing enough electrons to achieve the configuration of the previous noble gas (that is, the noble gas that occurs before the metal in question on the periodic table). For example, note from the periodic table inside the front cover of the text that neon is the noble gas previous to sodium and magnesium. Similarly, helium is the noble gas previous to lithium and beryllium.

2. Nonmetals form ions by gaining enough electrons to achieve the configuration of the next noble gas (that is, the noble gas that follows the element in question on the periodic table). For example, note that neon is the noble gas that follows oxygen and fluorine, and argon is the noble gas that follows sulfur and chlorine.

This brings us to an important general principle. In observing millions of stable compounds, chemists have learned that **in almost all stable chemical compounds of the representative elements, all of the atoms have achieved a noble gas electron configuration.** The importance of this observation cannot be overstated. It forms the basis for all of our fundamental ideas about why and how atoms bond to each other.

We have already seen this principle operating in the formation of ions (see Table 12.2). We can summarize this behavior as follows: when representative metals and nonmetals react, they transfer electrons in such a way that both the cation and the anion have noble gas electron configurations.

On the other hand, when nonmetals react with each other, they share electrons in ways that lead to a noble gas electron configuration for each atom in

Now that we know something about the electron configurations of atoms, we can explain *why* these various ions are formed.

This idea—that there is a strong tendency for atoms to adopt a noble gas electron structure—is central to an understanding of the electronic structure of ions and the Lewis dot structures of molecules. This unifying concept is stressed in this chapter and should be reinforced in classroom discussion.

I begin this discussion by showing the electron configurations of Mg^{2+} and Cl^- in the ionic compound $MgCl_2$. Mg^{2+} has the same electron configuration as Ne, and Cl^- has the same electron configuration as Ar. I then show that covalently bonded molecules such as H_2 achieve the electron configuration of the noble gas He by sharing a pair of electrons. From these two examples it is possible to generalize that all atoms that transfer or share electrons do so to achieve a noble gas configuration.

Box diagrams that show how individual orbitals become full when atoms share electrons can help make it clearer to students that the formation of covalent and ionic bonds usually results in a valence electron configuration equal to that of a noble gas.

Atoms in stable compounds almost always have a noble gas electron configuration.

It is harder for students to understand how metals achieve the previous noble gas configuration by losing electrons than it is for them to understand how nonmetals achieve the next noble gas configuration by gaining electrons.

Students have already learned how to predict the charge on an ion from the group number of the periodic table. Now, they learn another way to predict charge, by examining the valence electron configuration.

Differences in electronegativities greater than 2 lead to ionic bonds.

the resulting molecule. For example, oxygen ($[He]2s^22p^4$), which needs two more electrons to achieve a [Ne] configuration, can get these electrons by combining with two H atoms (each of which has one electron):

O: [He]   2s   2p

to form water, H_2O. This fills the valence orbitals of oxygen.

In addition, each H shares two electrons with the oxygen atom:

which fills the H $1s$ orbital, giving it a $1s^2$ or [He] electron configuration. We will have much more to say about covalent bonding in Section 12.6.

At this point let's summarize the ideas we have introduced so far.

- When a *nonmetal and a Group 1, 2, or 3 metal* react to form a binary ionic compound, the ions form in such a way that the valence electron configuration of the *nonmetal* is *completed* to achieve the configuration of the *next* noble gas, and the valence orbitals of the *metal* are *emptied* to achieve the configuration of the *previous* noble gas. In this way both ions achieve noble gas electron configurations.
- When *two nonmetals* react to form a covalent bond, they share electrons in a way that completes the valence electron configurations of both atoms. That is, both nonmetals attain noble gas electron configurations by sharing electrons.

Predicting Formulas of Ionic Compounds

To show how to predict what ions form when a metal reacts with a nonmetal, we will consider the formation of an ionic compound from calcium and oxygen. We can predict what compound will form by considering the valence electron configurations of the two atoms.

$$Ca: \quad [Ar]4s^2$$
$$O: \quad [He]2s^22p^4$$

From Figure 12.3 we see that the electronegativity of oxygen (3.5) is much greater than that of calcium (1.0), giving a difference of 2.5. Because of this large difference, electrons are transferred from calcium to oxygen to form an oxygen anion and a calcium cation. How many electrons are transferred? We can base our prediction on the observation that noble gas configurations are the most stable. Note that oxygen needs two electrons to fill its valence orbitals ($2s$ and $2p$) and achieve the configuration of neon ($1s^22s^22p^6$), which is the next noble gas.

$$O \quad + 2e^- \rightarrow O^{2-}$$
$$[He]2s^22p^4 + 2e^- \rightarrow [He]2s^22p^6, \text{ or } [Ne]$$

And by losing two electrons, calcium can achieve the configuration of argon (the previous noble gas).

$$Ca \quad \rightarrow Ca^{2+} + 2e^-$$
$$[Ar]4s^2 \rightarrow [Ar] \ + 2e^-$$

Two electrons are therefore transferred as follows:

$$Ca + O \rightarrow Ca^{2+} + O^{2-}$$
$$\underset{2e^-}{\curvearrowright}$$

To predict the formula of the ionic compound, we use the fact that chemical compounds are always electrically neutral—they have the same total quantities of positive and negative charges. In this case we must have equal numbers of Ca^{2+} and O^{2-} ions, and the empirical formula of the compound is CaO.

The same principles can be applied to many other cases. For example, consider the compound formed from aluminum and oxygen. Aluminum has the electron configuration $[Ne]3s^23p^1$. To achieve the neon configuration, aluminum must lose three electrons, forming the Al^{3+} ion.

$$Al \quad \rightarrow Al^{3+} + 3e^-$$
$$[Ne]3s^23p^1 \rightarrow [Ne] \ + 3e^-$$

Therefore the ions will be Al^{3+} and O^{2-}. Because the compound must be electrically neutral, there will be three O^{2-} ions for every two Al^{3+} ions, and the compound has the empirical formula Al_2O_3.

Table 12.3 shows common elements that form ions with noble gas electron configurations in ionic compounds.

Notice that our discussion in this section refers to metals in Groups 1, 2, and 3 (the representative metals). The transition metals exhibit more complicated behavior (they form a variety of ions), which we will not be concerned with in this text.

It helps students to learn to predict the formulas of ionic compounds produced in a reaction between pure elements if you show the core notations of reactants. Students can then see that the valence electron configurations of the elements are not equal to those of noble gases. The core notations of ions in the products will, however, be equal to that of a noble gas.

$3 \times 2-$ balances $2 \times 3+$

Table 12.3	Common Ions with Noble Gas Configurations in Ionic Compounds				
Group 1	Group 2	Group 3	Group 6	Group 7	Electron Configuration
Li^+	Be^{2+}				[He]
Na^+	Mg^{2+}	Al^{3+}	O^{2-}	F^-	[Ne]
K^+	Ca^{2+}		S^{2-}	Cl^-	[Ar]
Rb^+	Sr^{2+}		Se^{2-}	Br^-	[Kr]
Cs^+	Ba^{2+}		Te^{2-}	I^-	[Xe]

Table 12.3 shows groups of ions all with the electron configuration of the same noble gas. It can be useful to choose one of the groups, for example, K^+, Ca^{2+}, S^{2-}, and Cl^-. Show how each element in the group gains or loses electrons to assume the electron configuration of Ar.

A three-dimensional model showing the relative sizes of the anions and cations in an ionic compound helps show the regular arrangement of ions and shows how individual anions and cations are surrounded by ions of opposite charge.

(a)

(b)

Figure 12.8
The structure of lithium fluoride. (a) This structure represents the ions as packed spheres. (b) This structure shows the positions (centers) of the ions. The spherical ions are packed in the way that maximizes the ionic attractions.

12.5 Ionic Bonding and Structures of Ionic Compounds

AIM: To discuss ionic structures.
To discuss factors governing ionic size.

When metals and nonmetals react, the resulting ionic compounds are very stable; large amounts of energy are required to "take them apart." For example, the melting point of sodium chloride is approximately 800 °C. The strong bonding in these ionic compounds results from the attractions between the oppositely charged cations and anions.

We write the formula of an ionic compound such as lithium fluoride simply as LiF, but this is really the empirical, or simplest, formula. The actual solid contains huge and equal numbers of Li^+ and F^- ions packed together in a way that maximizes the attractions of the oppositely charged ions. A representative part of the lithium fluoride structure is shown in Figure 12.8(a). In this structure the larger F^- ions are packed together like hard spheres, and the much smaller Li^+ ions are interspersed regularly among the F^- ions. The structure shown in Figure 12.8(b) represents only a tiny part of the actual structure, which continues in all three dimensions with the same pattern as that shown.

The structures of virtually all binary ionic compounds can be explained by a model that involves packing the ions as though they were hard spheres. The larger spheres (usually the anions) are packed together, and the small ions occupy the interstices (spaces or holes) among them.

To understand the packing of ions it helps to realize that *a cation is always smaller than the parent atom, and an anion is always larger than the parent atom.* This makes sense because when a metal loses all of its valence electrons to form a cation, it gets much smaller. On the other hand, in forming an anion, a nonmetal gains enough electrons to achieve the next noble gas electron configuration and so becomes much larger. The relative sizes of the Group 1 and Group 7 atoms and their ions are shown in Figure 12.9.

Ionic Compounds Containing Polyatomic Ions

So far in this chapter we have discussed only binary ionic compounds, which contain ions derived from single atoms. However, many compounds contain polyatomic ions: charged species composed of several atoms. For example, ammonium nitrate contains the NH_4^+ and NO_3^- ions. These ions with their opposite charges attract each other in the same way as do the simple ions in binary ionic compounds. However, the *individual* polyatomic ions are held together by covalent bonds, with all of the atoms behaving as a unit. For example, in the ammonium ion, NH_4^+, there are four N—H covalent bonds. Likewise the nitrate ion,

When spheres are packed together, they do not fill up all of the space. The spaces (holes) that are left can be occupied by smaller spheres.

It is hard for students to understand the structures of ionic compounds composed of metal cations and polyatomic anions. Because polyatomic ions are covalently bonded, students can't understand how they can be part of an ionic compound. For example, when Na_2SO_4 is dissolved in water, sodium ions and sulfate ions are produced. However, the sulfate ions don't dissociate further because sulfur and oxygen are covalently bonded.

Lewis structures of ionic compounds are different from those of covalent compounds. As students draw Lewis structures of ionic compounds, they must remember to transfer electrons from the cation to the anion. For example, in potassium bromide the potassium ion has no electrons, not one electron, and the bromide ion has eight electrons, not seven.

NO_3^-, contains three covalent N—O bonds. Thus, although ammonium nitrate is an ionic compound because it contains the NH_4^+ and NO_3^- ions, it also contains covalent bonds in the individual polyatomic ions. When ammonium nitrate is dissolved in water, it behaves as a strong electrolyte like the binary ionic compounds sodium chloride and potassium bromide. As we saw in Chapter 7, this occurs because when an ionic solid dissolves, the ions are freed to move independently and can conduct an electric current.

The common polyatomic ions, which are listed in Table 5.5, are all held together by covalent bonds.

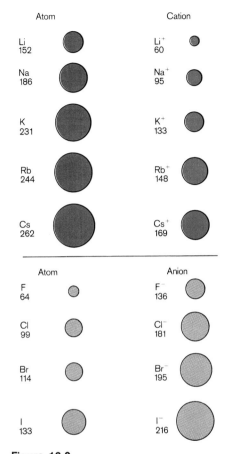

12.6 Lewis Structures

AIM: To show how to write Lewis structures.

Bonding involves just the valence electrons of atoms. Valence electrons are transferred when a metal and a nonmetal react to form an ionic compound. Valence electrons are shared between nonmetals in covalent bonds.

The **Lewis structure** is a representation of a molecule that shows how the valence electrons are arranged among the atoms in the molecule. These representations are named after G. N. Lewis, who conceived the idea while lecturing to a class of general chemistry students in 1902. The rules for writing Lewis structures are based on observations of many molecules from which chemists have learned that the *most important requirement for the formation of a stable compound is that the atoms achieve noble gas electron configurations.*

We have already seen this rule operate in the reaction of metals and nonmetals to form binary ionic compounds. An example is the formation of KBr, where the K^+ ion has the [Ar] electron configuration and the Br^- ion has the [Kr] electron configuration. In writing Lewis structures, *we include only the valence electrons.* Using dots to represent valence electrons, we write the Lewis structure for KBr as follows:

$$K^+ \qquad\qquad [\,:\!\ddot{Br}\!:\,]^-$$

Noble gas configuration [Ar]   Noble gas configuration [Kr]

No dots are shown on the K^+ ion because it has lost its only valence electron (the 4s electron). The Br^- ion is shown with eight electrons because it has a filled valence shell.

Next we will consider Lewis structures for molecules with covalent bonds, involving nonmetals in the first and second periods. The principle of achieving a noble gas electron configuration applies to these elements as follows:

1. Hydrogen forms stable molecules where it shares two electrons. That is, it follows a **duet rule.** For example, when two hydrogen atoms, each

Figure 12.9
Relative sizes of some ions and their parent atoms. Note that cations are smaller and anions are larger than their parent atoms. The sizes (radii) are given in units of picometers (1 pm $=$ 10^{-12} m).

Remember that the electrons in the highest principal energy level of an atom are called the valence electrons.

Lewis structures show only valence electrons.

G. N. Lewis in his lab.

You can effectively show why noble gases usually do not form compounds by drawing Lewis structures for several different noble gases.

Carbon, nitrogen, oxygen, and fluorine almost always obey the octet rule in stable molecules.

with one electron, combine to form the H_2 molecule, we have

By sharing electrons, each hydrogen in H_2 has, in effect, two electrons; that is, each hydrogen has a filled valence shell.

2. Helium does not form bonds because its valence orbitals are already filled; it is a noble gas. Helium has the electron configuration $1s^2$ and can be represented by the Lewis structure

$$He :$$

[He] configuration

3. The second-row nonmetals carbon through fluorine form stable molecules when they are surrounded by enough electrons to fill the valence orbitals—that is, the one $2s$ and the three $2p$ orbitals. Eight electrons are required to fill these orbitals, so these elements typically obey the **octet rule**; they are surrounded by eight electrons. An example is the F_2 molecule, which has the following Lewis structure:

| F atom with seven valence electrons | F_2 molecule | F atom with seven valence electrons |

Note that each fluorine atom in F_2 is, in effect, surrounded by eight valence electrons, two of which are shared with the other atom. This is a **bonding pair** of electrons, as we discussed earlier. Each fluorine atom also has three pairs of electrons that are not involved in bonding. These are called **lone pairs.**

4. Neon does not form bonds because it already has an octet of valence electrons (it is a noble gas). The Lewis structure is

Note that only the valence electrons ($2s^2 2p^6$) of the neon atom are represented by the Lewis structure. The $1s^2$ electrons are core electrons and are not shown.

Next we want to develop some general procedures for writing Lewis structures for molecules. Remember that Lewis structures involve only the valence electrons on atoms, so before we proceed, we will review the relationship of an element's position on the periodic table to the number of valence electrons it has. Recall that the group number gives the total number of valence electrons. For example, all Group 6 elements have six valence electrons (valence configuration ns^2np^4).

Group 6

O $2s^2 2p^4$
S $3s^2 3p^4$
Se $4s^2 4p^4$
Te $5s^2 5p^4$

Similarly, all Group 7 elements have seven valence electrons (valence configuration ns^2np^5).

Group 7

F $2s^2 2p^5$
Cl $3s^2 3p^5$
Br $4s^2 4p^5$
I $5s^2 5p^5$

In writing the Lewis structure for a molecule, we need to keep the following things in mind:

1. We must include all the valence electrons from all atoms. The total number of electrons available is the sum of all the valence electrons from all the atoms in the molecule.
2. Atoms that are bonded to each other share one or more pairs of electrons.
3. The electrons are arranged so that each atom is surrounded by enough electrons to fill the valence orbitals of that atom. This means two electrons for hydrogen and eight electrons for second-row nonmetals.

The best way to make sure we arrive at the correct Lewis structure for a molecule is to use a systematic approach. We will use the approach summarized by the following rules.

Steps for Writing Lewis Structures

The rules for writing Lewis structures will help students write correct Lewis structures every time. However, students often choose not to use formal rules. They are very successful in drawing Lewis structures of simple molecules without using rules but have difficulty subsequently drawing Lewis structures of more complex molecules.

STEP 1

Obtain the sum of the valence electrons from all of the atoms. Do not worry about keeping track of which electrons come from which atoms. It is the *total* number of electrons that is important.

STEP 2

Use one pair of electrons to form a bond between each pair of bound atoms. For convenience, a line (instead of a pair of dots) is generally used to indicate each pair of bonding electrons.

STEP 3

Arrange the remaining electrons to satisfy the duet rule for hydrogen and the octet rule for each second-row element.

When working examples in class it is helpful to show the rules for writing Lewis structures on an overhead transparency.

To see how these rules are applied, we will write the Lewis structures of several molecules.

EXAMPLE 12.2   **Writing Lewis Structures: Simple Molecules**

Write the Lewis structure of the water molecule.

SOLUTION

We will follow the steps listed above.

EXAMPLE 12.2, CONTINUED

STEP 1
We sum the *valence* electrons for H_2O.

$$1 \quad + \quad 1 \quad + \quad 6 \quad = 8 \text{ valence electrons}$$
$$\uparrow \qquad\quad \uparrow \qquad\quad \uparrow$$
$$H \qquad\quad H \qquad\quad O$$
(Group 1)   (Group 1)   (Group 6)

STEP 2
Using a pair of electrons per bond, we draw in the two O—H bonds, using a line to indicate each pair of bonding electrons.

$$H-O-H$$

Note that

$$H-O-H \quad \text{represents} \quad H:O:H$$

STEP 3
We arrange the remaining electrons around the atoms to achieve a noble gas electron configuration for each atom. Four electrons have been used in forming the two bonds, so four electrons (8 − 4) remain to be distributed. Each hydrogen is satisfied with two electrons (duet rule), but oxygen needs eight electrons to have a noble gas electron configuration. So the remaining four electrons are added to oxygen as two lone pairs. Dots are used to represent the lone pairs.

   Lone pairs

This is the correct Lewis structure for the water molecule. Each hydrogen shares two electrons, and the oxygen has four electrons and shares four to give a total of eight.

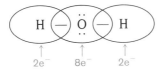

Note that a line is used to represent a shared pair of electrons (bonding electrons) and dots are used to represent unshared pairs.

Students determine the total number of valence electrons in a compound by adding together the number of valence electrons found in each atom. Remind students that this number can be determined from the group number of the periodic table.

$$H-\overset{..}{\underset{..}{O}}-H$$

might also be drawn as

$$H:\overset{..}{O}:H$$

Using a pair of dots instead of a dash to indicate a shared pair of electrons makes it easier for students to correctly determine the number of valence electrons they have used and how many they still need to distribute.

SELF-CHECK EXERCISE 12.2

Write the Lewis structure for HCl.

Writing Lewis structures of more complex molecules or ions can be difficult. Often, students don't know how to correctly arrange the atoms. To them, N—O—O—O⁻ is just as reasonable as any other arrangement of atoms for the nitrate ion. It can help to point out that atoms in molecules are usually arranged symmetrically. You may need to show them the arrangement of the atoms in complex molecules before they begin to write the Lewis structures.

O—C—O

represents

O : C : O

:Ö—C—Ö:

represents

:O : C : O :

O=C=O

represents

O : : C : : O

12.7   Lewis Structures of More Complex Molecules

AIM: To show how to write Lewis structures for molecules with multiple bonds.

Now let's write the Lewis structure for carbon dioxide. Summing the valence electrons gives

$$4 \quad + \quad 6 \quad + \quad 6 \quad = 16$$

$$\begin{array}{ccc} \uparrow & \uparrow & \uparrow \\ C & O & O \\ (Group\ 4) & (Group\ 6) & (Group\ 6) \end{array}$$

After forming a bond between the carbon and each oxygen,

$$O—C—O$$

we distribute the remaining electrons to achieve noble gas electron configurations on each atom. In this case twelve electrons (16 − 4) remain after the bonds are drawn. The distribution of these electrons is determined by a trial-and-error process. We have six pairs of electrons to distribute. Suppose we try three pairs on each oxygen to give

$$:Ö—C—Ö:$$

Is this correct? To answer this question we need to check two things:

1. The total number of electrons. There are sixteen valence electrons in this structure, which is the correct number.
2. The octet rule for each atom. Each oxygen has eight electrons around it, but the carbon only has four. This cannot be the correct Lewis structure.

How can we arrange the sixteen available electrons to achieve an octet for each atom? Suppose there are two shared pairs between the carbon and each oxygen:

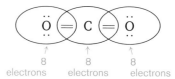

Now each atom is surrounded by eight electrons, and the total number of electrons is sixteen, as required. This is the correct Lewis structure for carbon dioxide, which has two *double* bonds. A **single bond** involves two atoms sharing one electron pair. A **double bond** involves two atoms sharing two pairs of electrons.

In thinking about the Lewis structure for CO_2, you may have come up with

$$:O≡C—\overset{..}{\underset{..}{O}}: \quad \text{or} \quad :\overset{..}{\underset{..}{O}}—C≡O:$$

Note that both of these structures have the required sixteen electrons and that both have octets of electrons around each atom (verify this for yourself). Both of these structures have a **triple bond** in which three electron pairs are shared. Are these valid Lewis structures for CO_2? Yes. So there really are three Lewis structures for CO_2:

$$:\overset{..}{\underset{..}{O}}—C≡O: \quad \overset{..}{\underset{..}{O}}=C=\overset{..}{\underset{..}{O}} \quad :O≡C—\overset{..}{\underset{..}{O}}:$$

This brings us to a new term, **resonance.** A molecule shows resonance when *more than one Lewis structure can be drawn for the molecule.* In such a case we call the various Lewis structures **resonance structures.**

Of the three resonance structures for CO_2 shown above, the one in the center with two double bonds most closely fits our experimental information about the CO_2 molecule. In this text we will not be concerned about how to choose which resonance structure for a molecule gives the "best" description of that molecule's properties.

Next let's consider the Lewis structure of the CN^- (cyanide) ion. Summing the valence electrons, we have

$$\underset{4 + 5 + 1 = 10}{\overset{\displaystyle CN^-}{↗ ↑ ↑}}$$

Note that the negative charge means an extra electron must be added. After drawing a single bond (C—N), we distribute the remaining electrons to achieve a noble gas configuration for each atom. Eight electrons remain to be distributed. We can try various possibilities, such as

$$\overset{..}{\underset{..}{C}}—\overset{..}{N} \quad \text{or} \quad :\overset{..}{C}—N: \quad \text{or} \quad :C—\overset{..}{\underset{..}{N}}:$$

These structures are incorrect. To show why none is a valid Lewis structure, count the electrons around the C and N atoms. In the left structure, neither atom satisfies the octet rule. In the center structure, C has eight electrons but N has only four. In the right structure, the opposite is true. Remember that both atoms must simultaneously satisfy the octet rule. Therefore, the correct arrangement is

$$:C≡N:$$

(Satisfy yourself that both carbon and nitrogen have eight electrons.) In this case we have a triple bond between C and N, in which three electron pairs are shared. Because this is an anion, we indicate the charge by using square brackets around the Lewis structure.

$$[:C≡N:]^-$$

In summary, sometimes we need double or triple bonds to satisfy the octet rule. Writing Lewis structures is a trial-and-error process. Start with single bonds between the bonded atoms and add multiple bonds as needed.

$$:O≡C—\overset{..}{\underset{..}{O}}:$$

represents

$$:O:::C:\overset{..}{\underset{..}{O}}:$$

Students have difficulty understanding the concept of resonance structures. If asked to show all resonance structures for a molecule, they will often draw several completely incorrect Lewis structures, in addition to the correct one.

$$:C≡N:$$

represents

$$:C:::N:$$

Students have more difficulty writing Lewis structures of ions than they do of neutral molecules. They are unsure of the way that the charge on the ion affects the total number of valence electrons to be distributed.

Writing Lewis structures involves some trial and error. The first arrangement of electrons may not provide complete octets for all atoms. Tell students that they may need to try another arrangement if the first attempt is not satisfactory.

We will write the Lewis structure for NO_2^- in Example 12.3 to make sure the procedures for writing Lewis structures are clear.

EXAMPLE 12.3	**Writing Lewis Structures: Resonance Structures**

Write the Lewis structure for the NO_2^- anion.

SOLUTION

STEP 1
Sum the valence electrons for NO_2^-.

$$\text{Valence electrons: } 6 + 5 + 6 + 1 = 18 \text{ electrons}$$

$$\uparrow \quad \uparrow \quad \uparrow \quad \uparrow$$
$$O \quad N \quad O \quad -1$$
$$\text{charge}$$

STEP 2
Put in single bonds.

$$O-N-O$$

STEP 3
Satisfy the octet rule. In placing the electrons, we find there are two Lewis structures that satisfy the octet rule:

$$[\overset{..}{O}=\overset{..}{N}-\overset{..}{\underset{..}{O}}:]^- \quad \text{and} \quad [:\overset{..}{\underset{..}{O}}-\overset{..}{N}=\overset{..}{O}]^-$$

Verify that each atom in these structures is surrounded by an octet of electrons. Try some other arrangements to see whether other structures exist in which the eighteen electrons can be used to satisfy the octet rule. It turns out that these are the only two that work. Note that this is another case where resonance occurs; there are two valid Lewis structures.

Remind students that the elements carbon, nitrogen, oxygen, and fluorine almost always obey the octet rule. Compounds that contain only those elements should have an octet surrounding each atom.

SELF-CHECK EXERCISE 12.3

Ozone is a very important constituent of the atmosphere. At upper levels it protects us by absorbing high-energy radiation from the sun. Near the earth's surface it produces harmful air pollution. Write the Lewis structure for ozone, O_3.

Now let's consider a few more cases in Example 12.4.

EXAMPLE 12.4	**Writing Lewis Structures: Summary**

Give the Lewis structure for each of the following:

a. HF
b. N_2
c. NH_3
d. CH_4
e. CF_4
f. NO^+
g. NO_3^-

EXAMPLE 12.4, CONTINUED

SOLUTION

In each case we apply the three steps for writing Lewis structures. Recall that lines are used to indicate shared electron pairs and that dots are used to indicate nonbonding pairs (lone pairs). The following table summarizes our results.

Molecule or Ion	Total Valence Electrons	Draw Single Bonds	Calculate Number of Electrons Remaining	Use Remaining Electrons to Achieve Noble Gas Configurations	Check Atom	Check Electrons
a. HF	$1 + 7 = 8$	H—F	$8 - 2 = 6$	H—$\ddot{\text{F}}$:	H F	2 8
b. N_2	$5 + 5 = 10$	N—N	$10 - 2 = 8$	:N≡N:	N	8
c. NH_3	$5 + 3(1) = 8$	H—N—H $\quad$ H	$8 - 6 = 2$	H—$\ddot{\text{N}}$—H $\qquad$ H	H N	2 8
d. CH_4	$4 + 4(1) = 8$	$\quad$ H H—C—H $\quad$ H	$8 - 8 = 0$	$\quad$ H H—C—H $\quad$ H	H C	2 8
e. CF_4	$4 + 4(7) = 32$	$\quad$ F F—C—F $\quad$ F	$32 - 8 = 24$	$\quad$:$\ddot{\text{F}}$: :$\ddot{\text{F}}$—C—$\ddot{\text{F}}$: $\quad$:$\ddot{\text{F}}$:	F C	8 8
f. NO^+	$5 + 6 - 1 = 10$	N—O	$10 - 2 = 8$	$[:N≡O:]^+$	N O	8 8
g. NO_3^-	$5 + 3(6) + 1 = 24$	$\begin{bmatrix} \text{O} \\ \text{N} \\ \text{O} \quad \text{O} \end{bmatrix}$	$24 - 6 = 18$	resonance structure 1	N O	8 8
			NO$_3^-$ shows resonance.	resonance structure 2	N O	8 8
				resonance structure 3	N O	8 8

EXAMPLE 12.4, CONTINUED

SELF-CHECK EXERCISE 12.4

You may wonder how to decide which atom is
the central atom in molecules of binary
compounds. In cases where there is one atom
of a given element and several atoms of a
second element, the single atom is almost
always the central atom of the molecule.

Write the Lewis structures for the following molecules:

a. NF_3
b. O_2
c.   CO
d. PH_3
e. H_2S

f. SO_4^{2-}
g. NH_4^+
h. ClO_3^-
i. SO_2

Remember, when writing Lewis structures, you don't have to worry about
which electrons come from which atoms in a molecule. It is best to think of a
molecule as a new entity that uses all the available valence electrons from the
various atoms to achieve the strongest possible bonds. Think of the valence electrons as belonging to the molecule, rather than to the individual atoms. Simply
distribute all the valence electrons so that noble gas electron configurations are
obtained for each atom, without regard to the origin of each particular electron.

Some Exceptions to the Octet Rule

Students are often dismayed to discover that
some elements do not obey the octet rule.
You can ease their concerns by reminding
them that some elements usually do obey
the octet rule. If an atom suspected of
violating the octet rule is bonded to elements
that almost always obey that rule, students
should first make sure that those atoms that
do obey the octet rule have an octet and
then they should distribute the remaining
electrons around the suspect atom.

The idea that covalent bonding can be predicted by achieving noble gas electron
configurations for all atoms is a simple and very successful idea. The rules we
have used for Lewis structures describe correctly the bonding in most molecules.
However, with such a simple model, some exceptions are inevitable. Boron, for
example, tends to form compounds where the boron atom has fewer than eight
electrons around it—that is, it does not have a complete octet. Boron trifluoride,
BF_3, a gas at normal temperatures and pressures, reacts very energetically with
molecules such as water and ammonia that have unshared electron pairs (lone
pairs).

The violent reactivity of BF_3 with electron-rich molecules arises because the
boron atom is electron-deficient. The Lewis structure that seems most consistent
with the properties of BF_3 (twenty-four valence electrons) is

Note that in this structure the boron atom has only six electrons around it. The
octet rule for boron could be satisfied by drawing a structure with a double bond

between the boron and one of the fluorines. However, experiments indicate that each B—F bond is a single bond in accordance with the above Lewis structure. This structure is also consistent with the reactivity of BF$_3$ with electron-rich molecules. For example, BF$_3$ reacts vigorously with NH$_3$ to form H$_3$NBF$_3$.

$$\underset{H}{\overset{H}{H-N:}} + \underset{F:}{\overset{F:}{B-F:}} \rightarrow \underset{H}{\overset{H}{H-N-}}\underset{F:}{\overset{F:}{B-F:}}$$

Note that in the product H$_3$NBF$_3$, which is very stable, boron has an octet of electrons.

It is also characteristic of beryllium to form molecules where the beryllium atom is electron-deficient.

The elements carbon, nitrogen, oxygen, and fluorine obey the octet rule in the vast majority of their compounds. However, even these elements show a few exceptions. One important example is the oxygen molecule, O$_2$. The following Lewis structure that satisfies the octet rule can be drawn for O$_2$ (see Self-Check Exercise 12.4).

$$:\overset{..}{O}=\overset{..}{O}:$$

However, this structure does not agree with the *observed behavior* of oxygen. For example, the photo in Figure 12.10 shows that when liquid oxygen is poured between the poles of a strong magnet, it "sticks" there. This provides clear evidence that oxygen is paramagnetic—that is, it contains unpaired electrons. However, the above Lewis structure shows only pairs of electrons. That is, no unpaired electrons are shown. There is no simple Lewis structure that satisfactorily explains the paramagnetism of the O$_2$ molecule.

Any molecule that contains an odd number of electrons does not conform to our rules for Lewis structures. For example, NO and NO$_2$ have eleven and seventeen valence electrons, respectively, and conventional Lewis structures cannot be drawn for these cases.

Even though there are a few exceptions, most molecules can be described by Lewis structures in which all the atoms have noble gas electron configurations, and this is a very useful model for chemists.

Figure 12.10
When liquid oxygen is poured between the poles of a magnet, it "sticks" until it boils away. This shows that the O$_2$ molecule has unpaired electrons (is paramagnetic).

Paramagnetic substances have unpaired electrons and are drawn toward the space between a magnet's poles.

12.8 Molecular Structure

AIM: To define molecular structure and bond angles.

So far in this chapter we have considered Lewis structures for molecules. These structures represent the arrangement of the *valence electrons* in a molecule. We use the word *structure* in another way when we talk about the **molecular structure** or **geometric structure** of a molecule. These terms refer to the three-dimensional arrangement of the *atoms* in the molecule. For example, the water

Use models to demonstrate the molecular structure of those molecules used as examples in this section. It is very difficult for students to see the difference between trigonal planar and tetrahedral molecules without models.

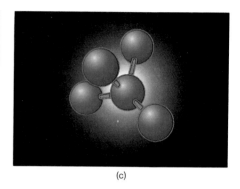

(a)   (b)   (c)

Computer graphics of (a) a linear molecule containing three atoms, (b) a trigonal planar molecule, and (c) a tetrahedral molecule.

molecule is known to have the molecular structure

$$\underset{H}{}\overset{O}{\diagup}\underset{H}{}$$

which is often called "bent" or "V-shaped." To describe the structure more precisely, we often specify the **bond angle.** For the H_2O molecule the bond angle is about 106°.

$$\underset{\sim106°}{\underset{H}{}\overset{O}{\diagup}\underset{H}{}}$$

On the other hand, some molecules exhibit a **linear structure** (all atoms in a line). An example is the CO_2 molecule.

$$\underset{180°}{O-C-O}$$

Note that a linear molecule has a 180° bond angle.

A third type of molecular structure is illustrated by BF_3, which is planar or flat (all four atoms in the same plane) with 120° bond angles.

The name usually given to this structure is **trigonal planar structure,** although triangular might seem to make more sense.

Another type of molecular structure is illustrated by methane, CH_4. This molecule has the molecular structure shown in Figure 12.11, which is called a

Figure 12.11
The tetrahedral molecular structure of methane. This representation is called a ball-and-stick model; the atoms are represented by balls and the bonds by sticks. The dashed line shows the outline of the tetrahedron.

tetrahedral structure or a **tetrahedron.** The dashed lines shown connecting the H atoms define the four identical triangular faces of the tetrahedron.

In the next section we will discuss these various molecular structures in more detail. In that section we will learn how to predict the molecular structure of a molecule by looking at the molecule's Lewis structure.

Toothpicks and styrofoam spheres make inexpensive and effective model kits for students to use in class.

12.9 Molecular Structure: The VSEPR Model

AIM: To describe how molecular geometry can be predicted from the number of electron pairs.

The structures of molecules play a very important role in determining their chemical properties. This is particularly important for biological molecules; a slight change in the structure of a large biomolecule can completely destroy its usefulness to a cell and may even change the cell from a normal one to a cancerous one.

Many experimental methods now exist for determining the molecular structure of a molecule—that is, the three-dimensional arrangement of the atoms. These methods must be used when accurate information about the structure is required. However, it is often useful to be able to predict the *approximate* molecular structure of a molecule. In this section we consider a simple model that allows us to do this. This model, called the **valence shell electron pair repulsion (VSEPR) model,** is useful for predicting the molecular structures of molecules formed from nonmetals. The main idea of this model is that *the structure around a given atom is determined by minimizing repulsions between electron pairs.* This means that the bonding and nonbonding electron pairs (lone pairs) around a given atom are positioned *as far apart as possible.* To see how this model works, we will first consider the molecule $BeCl_2$, which has the following Lewis structure (it is an exception to the octet rule).

$$: \ddot{Cl}—Be—\ddot{Cl} :$$

After learning the five possible shapes of molecules, students need to know how they can assign the appropriate shape to a molecule. I tell students that the VSEPR model can help them predict which molecular structure a molecule will assume.

When you discuss the VSEPR model, stress the difference between the arrangement of the electron pairs and the molecular structure (arrangement of atoms). Students at first can see no difference between the two definitions.

Note that there are two pairs of electrons around the beryllium atom. What arrangement of these electron pairs allows them to be as far apart as possible to minimize the repulsions? The best arrangement places the pairs on opposite sides of the beryllium atom at $180°$ from each other.

$$—Be—$$
$$180°$$

This is the maximum possible separation for two electron pairs. Now that we have determined the optimal arrangement of the electron pairs around the central atom, we can specify the molecular structure of $BeCl_2$—that is, the positions of the atoms. Because each electron pair on beryllium is shared with a chlorine atom, the molecule has a **linear structure** with a 180° bond angle.

$$: \overset{..}{\underset{..}{Cl}} — Be — \overset{..}{\underset{..}{Cl}} :$$

180°

Whenever two pairs of electrons are present around an atom, they should always be placed at an angle of 180° to each other to give a linear arrangement.

Next let's consider BF_3, which has the following Lewis structure (it is another exception to the octet rule).

$$: \overset{..}{\underset{..}{F}} : \\ | \\ : \overset{..}{\underset{..}{F}} — B — \overset{..}{\underset{..}{F}} :$$

Here the boron atom is surrounded by three pairs of electrons. What arrangement minimizes the repulsions among three pairs of electrons? Here the greatest distance between electron pairs is achieved by angles of 120°.

120°   B   120°

120°

Because each of the electron pairs is shared with a fluorine atom, the molecular structure is

F
120°   B   120°   or   F —— B
F   F   F ---- F
120°

This is a planar (flat) molecule with a triangular arrangement of F atoms, commonly described as a trigonal planar structure. *Whenever three pairs of electrons are present around an atom, they should always be placed at the corners of a triangle (in a plane at an angle of 120° to each other).*

Next let's consider the methane molecule, which has the Lewis structure

H   H
| ..
H—C—H   or   H : C : H
| ..
H   H

There are four pairs of electrons around the central carbon atom. What arrangement of these electron pairs best minimizes the repulsions? First we try a square

I show how molecular structure is related to the number of atoms in a compound by using hypothetical atoms X and Y. If X is a central atom with no unshared electrons bonded to two Ys, then we can arrange the two Y atoms around X in many different ways. I then show some possible arrangements, with the two Y atoms at right angles to X or with the two Y atoms forming an acute angle with X. The VSEPR model tells us what the correct structure is likely to be. The electron pairs that form the bonds between X and Y produce fewest repulsive forces if they are on opposite sides of X. Similar forms of reasoning can be used for other possible shapes.

planar arrangement:

The carbon atom and the electron pairs are all in a plane represented by the surface of the paper, and the angles between the pairs are all 90°.

Is there another arrangement with angles greater than 90° that would put the electron pairs even farther away from each other? The answer is yes. We can get larger angles than 90° by using the following three-dimensional structure, which has angles of approximately 109.5°.

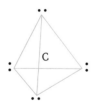

In this drawing the wedge indicates a position above the surface of the paper and the dashed lines indicate positions behind that surface. The regular line indicates a position on the surface of the page. The figure formed by connecting the lines is called a tetrahedron, so we call this arrangement of electron pairs the **tetrahedral arrangement.**

A tetrahedron has four equal triangular faces.

This is the maximum possible separation of four pairs around a given atom. *Whenever four pairs of electrons are present around an atom, they should always be placed at the corners of a tetrahedron (the tetrahedral arrangement).*

Now that we have the arrangement of electron pairs that gives the least repulsion, we can determine the positions of the atoms and thus the molecular structure of CH_4. In methane each of the four electron pairs is shared between the carbon atom and a hydrogen atom. Thus the hydrogen atoms are placed as shown in Figure 12.12, and the molecule has a tetrahedral structure with the carbon atom at the center.

Recall that the main idea of the VSEPR model is to find the arrangement of electron pairs around the central atom which minimizes the repulsions. Then we can determine the *molecular structure* by knowing how the electron pairs are shared with the peripheral atoms. A systematic procedure for using the VSEPR model to predict the structure of a molecule is outlined in the list on the following page.

Figure 12.12

The molecular structure of methane. The tetrahedral arrangement of electron pairs produces a tetrahedral arrangement of hydrogen atoms.

(a)

(b)

(c)

Figure 12.13

(a) The tetrahedral arrangement of electron pairs around the nitrogen atom in the ammonia molecule. (b) Three of the electron pairs around nitrogen are shared with hydrogen atoms as shown, and one is a lone pair. Although the arrangement of *electron pairs* is tetrahedral, as in the methane molecule, the hydrogen atoms in the ammonia molecule occupy only three corners of the tetrahedron. A lone pair occupies the fourth corner. (c) The NH_3 molecule has the trigonal pyramid structure (a pyramid with a triangle as a base).

Balloons tied together can be useful models for the electron arrangements in molecules such as water or ammonia.

These four steps for predicting molecular structure will help students determine the correct molecular structure of most molecules. The most common error students make here is failing to write the Lewis structure first. Unshared pairs of electrons are likely to be forgotten without a correct Lewis structure.

Steps for Predicting Molecular Structure Using the VSEPR Model

STEP 1
Draw the Lewis structure for the molecule.

STEP 2
Count the electron pairs and arrange them in the way that minimizes repulsion (that is, put the pairs as far apart as possible).

STEP 3
Determine the positions of the atoms from the way the electron pairs are shared.

STEP 4
Determine the name of the molecular structure from the positions of the *atoms.*

EXAMPLE 12.5   **Predicting Molecular Structure Using the VSEPR Model, I**

Predict the structure of ammonia, NH_3, using the VSEPR model.

SOLUTION

STEP 1
Draw the Lewis structure.

$$H-\overset{\displaystyle ..}{\underset{\displaystyle |}{N}}-H$$
$$H$$

STEP 2
Count the pairs of electrons and arrange them to minimize repulsions. The NH_3 molecule has four pairs of electrons around the N atom: three bonding pairs and one nonbonding pair. From the discussion of the methane molecule, we know that the best arrangement of four electron pairs is the tetrahedral structure shown in Figure 12.13(a).

STEP 3
Determine the positions of the atoms. The three H atoms share electron pairs as shown in Figure 12.13(b).

EXAMPLE 12.5, CONTINUED

STEP 4
Name the molecular structure. It is very important to recognize that the name of the molecular structure is always based on the *positions of the atoms. The placement of the electron pairs determines the structure, but the name is based on the positions of the atoms.* Thus it is incorrect to say that the NH_3 molecule is tetrahedral. It has a tetrahedral arrangement of electron pairs but *not* a tetrahedral arrangement of atoms. The molecular structure of ammonia is a **trigonal pyramid** (one side is different from the other three) rather than a tetrahedron.

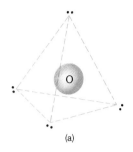
(a)

EXAMPLE 12.6   **Predicting Molecular Structure Using the VSEPR Model, II**

Describe the molecular structure of the water molecule.

SOLUTION

STEP 1
The Lewis structure for water is

$$H-\overset{..}{\underset{..}{O}}-H$$

STEP 2
There are four pairs of electrons: two bonding pairs and two nonbonding pairs. To minimize repulsions, these are best arranged in a tetrahedral structure as shown in Figure 12.14(a).

STEP 3
Although H_2O has a tetrahedral arrangement of *electron pairs,* it is *not a tetrahedral molecule.* The *atoms* in the H_2O molecule form a V shape, as shown in Figure 12.14(b) and (c).

STEP 4
The molecular structure is called V-shaped or bent.

(b)

(c)

Figure 12.14
(a) The tetrahedral arrangement of the four electron pairs around oxygen in the water molecule. (b) Two of the electron pairs are shared between oxygen and the hydrogen atoms, and two are lone pairs. (c) The V-shaped molecular structure of the water molecule.

SELF-CHECK EXERCISE 12.5

Predict the arrangement of electron pairs around the central atom. Then sketch and name the molecular structure for each of the following molecules or ions.

a. NH_4^+ d. H_2S
b. SO_4^{2-} e. ClO_3^-
c. NF_3 f. BeF_2

Table 12.4 Arrangements of Electron Pairs and the Resulting Molecular Structures for Two, Three, and Four Electron Pairs.

Case	Number of Electron Pairs	Bonds	Electron Pair Arrangement	Ball-and-Stick Model	Angle Between Pairs	Molecular Structure	Partial Lewis Structure	Ball-and-Stick Model	Example
1	2	2	Linear		180°	Linear	A — B — A		BeF_2
2	3	3	Trigonal planar (triangular)		120°	Trigonal planar (triangular)	A B A (with central B and A below)		BF_3
3	4	4	Tetrahedral		109.5°	Tetrahedral	A — B — A (with A above and A below)		CH_4
4	4	3	Tetrahedral		109.5°	Trigonal pyramid	A — B — A (with lone pair and A below)		NH_3
5	4	2	Tetrahedral		109.5°	Bent or V-shaped	A — B — A (with two lone pairs)		H_2O

The various cases we have considered are summarized in Table 12.4. Note the following general rules.

Rules for Predicting Molecular Structure Using the VSEPR Model

- Two pairs of electrons on a central atom in a molecule are always placed 180° apart. This is a linear arrangement of pairs.
- Three pairs of electrons on a central atom in a molecule are always placed 120° apart in the same plane as the central atom. This is a trigonal planar (triangular) arrangement of pairs.

- Four pairs of electrons on a central atom in a molecule are always placed 109.5° apart. This is a tetrahedral arrangement of electron pairs.
- When *every pair* of electrons on the central atom is *shared* with another atom, the molecular structure has the same name as the arrangement of electron pairs.

Number of Pairs	Name of Arrangement
2	linear
3	trigonal planar
4	tetrahedral

- When one or more of the electron pairs around a central atom are unshared (lone pairs), the name for the molecular structure is *different* from that for the arrangement of electron pairs (see cases 4 and 5 in Table 12.4).

12.10 Molecular Structure: Molecules with Double Bonds

AIM: To describe how the VSEPR model is applied to molecules with double bonds.

In this section students will see how VSEPR applies to molecules that contain double bonds. Students usually have no difficulty with the concept that the electrons in a double bond act as a single unit of electron density, and are able to predict the molecular structure as long as they have a correct Lewis structure from which to work.

To this point we have applied the VSEPR model only to molecules (and ions) that contain single bonds. In this section we will show that this model applies equally well to species with one or more double bonds. We will develop the procedures for dealing with molecules with double bonds by considering examples whose structures are known.

First we will examine the structure of carbon dioxide, a substance that may be contributing to the warming of the earth. The carbon dioxide molecule has the Lewis structure

$$\ddot{O}=C=\ddot{O}$$

as discussed in Section 12.7. Carbon dioxide is known by experiment to be a linear molecule. That is, it has a 180° bond angle

Recall from Section 12.9 that two electron pairs around a central atom can minimize their mutual repulsions by taking positions on opposite sides of the atom (at 180° from each other). This causes a molecule like $BeCl_2$, which has the

Lewis structure

to have a linear structure (see Section 12.9). Now recall that CO_2 has two double bonds and is known to be linear, so the double bonds must be at 180° from each other. Therefore, we conclude that each double bond in this molecule acts *effectively* as one repulsive unit. This conclusion makes sense if we think of a bond in terms of an electron density "cloud" between two atoms. For example, we can picture the single bonds in $BeCl_2$ as follows:

The minimum repulsion between these two electron density clouds occurs when they are on opposite sides of the Be atom (180° angle between them).

Each double bond in CO_2 involves the sharing of four electrons between the carbon atom and an oxygen atom. Thus we might expect the bonding cloud to be "fatter" than for a single bond:

However, the repulsive effects of these two clouds produce the same result as for single bonds; the bonding clouds have minimum repulsions when they are positioned on opposite sides of the carbon. The bond angle is 180°, and so the molecule is linear:

In summary, examination of CO_2 leads us to the conclusion that in using the VSEPR model for molecules with double bonds, each double bond should be treated the same as a single bond. In other words, although a double bond involves four electrons, these electrons are restricted to the space between a given pair of atoms. Therefore these four electrons do not function as two independent pairs but are "tied together" to form one effective repulsive unit.

Formaldehyde, H_2CO, is a small organic molecule which can be used to illustrate trigonal planar structure. Because it has no resonance structures, its molecular structure might be easier for students to determine than that of molecules or ions with resonance structures.

We reach this same conclusion by considering the known structures of other molecules that contain double bonds. For example, consider the ozone molecule which has 18 valence electrons and exhibits the resonance structures:

$$: \overset{..}{\underset{..}{O}}—\overset{..}{O}=\overset{..}{\underset{.}{O}} \quad \longleftrightarrow \quad \overset{.}{\underset{..}{O}}=\overset{..}{O}—\overset{..}{\underset{..}{O}} :$$

The ozone molecule is known to have a bond angle close to 120°. Recall that 120° angles represent the minimum repulsion for three pairs of electrons

This indicates that the double bond in the ozone molecule is behaving as one effective repulsive unit:

These and other examples lead us to the following rule: *When using the VSEPR model to predict the molecular geometry of a molecule, a double bond is counted the same as a single electron pair.*

Thus CO_2 has two "effective pairs" that lead to its linear structure whereas O_3 has three "effective pairs" that lead to its bent structure with a 120° bond angle. Therefore, to use the VSEPR for molecules (or ions) that have double bonds, we use the same steps as those given in Section 12.9, but we count any double bond the same as a single electron pair. Although we have not shown it here, triple bonds also count as one repulsive unit in applying the VSEPR model.

If you want to illustrate the number of effective pairs in a molecule when it contains a triple bond, acetylene, which has the structure $H—C{\equiv}C—H$, works well.

EXAMPLE 12.7   **Predicting Molecular Structure Using the VSEPR Model, III**

Predict the structure of the nitrate ion.

SOLUTION

STEP 1
The Lewis structures for NO_3^- are

STEP 2
In each resonance structure there are effectively three pairs of electrons: the two single bonds and the double bond (which counts as one pair). These three "effective pairs" will require a trigonal planar arrangement (120° angles).

STEP 3
The atoms are all in a plane, with the nitrogen at the center and the three oxygens at the corners of a triangle (trigonal planar arrangement).

STEP 4
The NO_3^- ion has a trigonal planar structure.

Key Terms

bond (p. 376)
bond energy (p. 376)
ionic bonding (p. 377)
ionic compound (p. 377)
covalent bonding (p. 377)
polar covalent bond (p. 378)
electronegativity (p. 378)
dipole moment (p. 381)
Lewis structure (p. 387)
duet rule (p. 387)
octet rule (p. 388)
bonding pair (p. 388)

lone pair (p. 388)
single bond (p. 392)
double bond (p. 392)
triple bond (p. 393)
resonance (p. 393)
molecular (geometric) structure (p. 397)
linear structure (p. 398)
trigonal planar structure (p. 398)
tetrahedral structure (p. 399)
valence shell electron pair repulsion (VSEPR) model (p. 399)

Summary

1. Chemical bonds hold groups of atoms together. They can be classified into several types. An ionic bond is formed when a transfer of electrons occurs to form ions; in a purely covalent bond, electrons are shared equally between identical atoms. Between these extremes lies the polar covalent bond, in which electrons are shared unequally between atoms with different electronegativities.

2. Electronegativity is defined as the relative ability of an atom in a molecule to attract the electrons shared in a bond. The difference in electronegativity values between the atoms involved in a bond determines the polarity of that bond.

3. In stable chemical compounds, the atoms tend to achieve a noble gas electron configuration. In the formation of a binary ionic compound involving representative elements, the valence-electron configuration of the nonmetal is completed: it achieves the configuration of the next noble gas. The valence orbitals of the metal are emptied to give the electron configuration of the previous noble gas. Two nonmetals share the valence electrons so that both atoms have completed valence-electron configurations (noble gas configurations).

4. Lewis structures are drawn to represent the arrangement of the valence electrons in a molecule. The rules for drawing Lewis structures are based on the observation that nonmetal atoms tend to achieve noble gas electron configurations by sharing electrons. This leads to a duet rule for hydrogen and to an octet rule for other atoms.

5. Some molecules have more than one valid Lewis structure, a situation called resonance. Although Lewis structures in which the atoms have noble gas electron configurations correctly describe most molecules, there are some notable exceptions, including O_2, NO, NO_2, and the molecules that contain Be and B.

6. The molecular structure of a molecule describes how the atoms are arranged in space.

7. The molecular structure of a molecule can be predicted by using the valence shell electron pair repulsion (VSEPR) model. This model bases its prediction on minimum repulsions among the electron pairs around an atom, which means arranging the electron pairs as far apart as possible.

Questions and Problems

All even-numbered exercises have answers in the back of this book and solutions in the Solutions Guide.

12.1 Types of Chemical Bonds

QUESTIONS

1. A chemical _____ represents a force that holds groups of atoms together, making them function as a unit.

2. The quantity of energy necessary to break a chemical bond is referred to as the _____.

3. _____ compounds result when a metallic element reacts with a nonmetal.

4. A(n) _____ bond requires the complete transfer of an electron from one atom to another.

5. A hydrogen molecule is more _____ than two separated hydrogen atoms by an amount of energy called the bond energy.

6. A _____ covalent bond results when a pair of electrons is shared unequally between two bonded atoms.

12.2 Electronegativity

QUESTIONS

7. Two _____ atoms react to form a covalent bond in which the electrons are shared equally.

8. The ability of an atom to attract a shared pair of electrons toward itself is referred to as the atom's _____.

9. A molecule with distinct centers of positive and negative charge is said to possess a _____ moment.

10. The extent of polarity of a bond depends on the _____ between the electronegativity values of the atoms forming the bond.

PROBLEMS

11. For each of the following pairs of elements, identify which element would be expected to be more electronegative. It should not be necessary to look at a table of actual electronegativity values.
   a. Ca or Br
   b. P or S
   c. C or N

12. For each of the following sets of elements, identify which element would be expected to be most electronegative and which would be expected to be least electronegative.
   a. K, Sc, Ca
   b. Br, F, At
   c. C, O, N

13. On the basis of the electronegativity values given in Figure 12.3, indicate whether each of the following bonds would be expected to be ionic, covalent, or polar covalent.
    a. H—F   c. Cl—F
    b. Na—F   d. Ca—F

14. On the basis of the electronegativity values given in Figure 12.3, indicate whether each of the following bonds would be expected to be ionic, covalent, or polar covalent.
    a. N—N   c. N—Cl
    b. N—O   d. N—Na

15. Which of the following molecules contain polar covalent bonds?
    a. dinitrogen, N_2 c. dioxygen, O_2
    b. hydrogen iodide, HI   d. ozone, O_3

16. Which of the following molecules contain polar covalent bonds?
    a. sulfur, S_8 c. hydrogen peroxide, H_2O_2
    b. hydrogen, H_2 d. chlorine monofluoride, ClF

17. On the basis of the electronegativity values given in Figure 12.3, indicate which is the more polar bond in each of the following pairs.
    a. H—F or H—Cl   c. H—Br or H—Cl
    b. H—Cl or H—I   d. H—I or H—Br

18. On the basis of the electronegativity values given in Figure 12.3, indicate which is the more polar bond in each of the following pairs.
    a. H—O or H—N   c. H—O or H—F
    b. H—N or H—F   d. H—O or H—Cl

19. On the basis of the electronegativity values given in Figure 12.3, indicate which is the more polar bond in each of the following pairs.
    a. H—N or H—P   c. H—P or H—S
    b. H—O or H—S   d. H—S or H—I

20. On the basis of the electronegativity values given in Figure 12.3, indicate which bond of the following pairs has a more ionic character.
    a. Na—O or Na—N   c. Na—Cl or K—Cl
    b. K—S or K—P   d. Na—Cl or Mg—Cl

12.3 Bond Polarity and Dipole Moments

QUESTIONS

21. What is a *dipole moment*? Give four examples of molecules that possess dipole moments, drawing the direction of the dipole as shown in this section.

22. Why is the presence of a dipole moment in the water molecule so important? What are some properties of water that are determined by its polarity?

PROBLEMS

23. In each of the following diatomic molecules, which end of the molecule is negative relative to the other end?
    a. hydrogen chloride, HCl
    b. carbon monoxide, CO
    c. bromine monofluoride, BrF

24. In each of the following diatomic molecules, which end of the molecule is positive relative to the other end?
    a. hydrogen fluoride, HF
    b. chlorine monofluoride, ClF
    c. iodine monochloride, ICl

25. For each of the following bonds, draw a figure indicating the direction of the bond dipole and indicating which end of the bond is positive and which is negative.
    a. O—H   c. O—C
    b. O—F   d. O—N

26. For each of the following bonds, draw a figure indicating the direction of the bond dipole. Show which end of the bond is positive and which is negative.
    a. Cl—F   c. Br—Cl
    b. I—Br   d. Br—F

27. For each of the following bonds, draw a figure indicating the direction of the bond dipole and indicating which end of the bond is positive and which is negative.
    a. C—S   c. C—Cl
    b. C—P   d. C—H

28. For each of the following bonds, draw a figure indicating the direction of the bond dipole and indicating which end of the bond is positive and which is negative.
    a. S—P   c. S—N
    b. S—O   d. S—Cl

12.4 Stable Electron Configurations and Charges on Ions

QUESTIONS

29. In virtually every stable compound, each of the atoms has achieved an electron configuration analogous to that of the _____ elements.

30. Metals form positive ions by losing enough electrons to achieve the electron configuration of the _____ noble gas.

31. Nonmetals form negative ions by (losing/gaining) enough electrons to achieve the electron configuration of the next noble gas.

PROBLEMS

33. Write the electron configuration for each of the following atoms and for the simple ion that the element most commonly forms. In each case, indicate which noble gas has the same electron configuration as the ion.
    a. sodium, $Z = 11$
    b. chlorine, $Z = 17$
    c. sulfur, $Z = 16$
    d. potassium, $Z = 19$
    e. iodine, $Z = 53$

35. What simple ion does each of the following elements most commonly form?
    a. lithium, $Z = 3$
    b. magnesium, $Z = 12$
    c. sulfur, $Z = 16$
    d. iodine, $Z = 53$

37. On the basis of their electron configurations, predict the formula of the simple binary ionic compound likely to form when the following pairs of elements react with each other.
    a. aluminum, Al, and oxygen, O
    b. cesium, Cs, and oxygen, O
    c. barium, Ba, and chlorine, Cl
    d. francium, Fr, and oxygen, O
    e. radium, Ra, and chlorine, Cl
    f. barium, Ba, and tellurium, Te

39. Which noble gas has the same electron configuration as each of the ions in the following compounds?
    a. barium oxide, BaO
    b. sodium iodide, NaI
    c. potassium fluoride, KF
    d. magnesium sulfide, MgS

32. Explain how the atoms in *covalent* molecules achieve configurations similar to those of the noble gases. How does this differ from the situation in ionic compounds?

34. Write the electron configuration for each of the following atoms and for the simple ion that the element most commonly forms. In each case, indicate which noble gas has the same electron configuration as the ion.
    a. rubidium, $Z = 37$
    b. oxygen, $Z = 8$
    c. strontium, $Z = 38$
    d. cesium, $Z = 55$
    e. barium, $Z = 56$

36. What simple ion does each of the following elements most commonly form?
    a. calcium, $Z = 20$
    b. nitrogen, $Z = 7$
    c. bromine, $Z = 35$
    d. scandium, $Z = 21$

38. On the basis of their electron configurations, predict the formula of each simple binary ionic compound likely to form when the following pairs of elements react with each other.
    a. calcium, Ca, and oxygen, O
    b. strontium, Sr, and fluorine, F
    c. aluminum, Al, and sulfur, S
    d. rubidium, Rb, and nitrogen, N
    e. sodium, Na, and tellurium, Te

40. Name the noble gas atom that has the same electron configuration as each of the ions in the following compounds.
    a. aluminum sulfide, Al_2S_3
    b. magnesium nitride, Mg_3N_2
    c. rubidium oxide, Rb_2O
    d. cesium iodide, CsI

12.5 Ionic Bonding and Structures of Ionic Compounds

QUESTIONS

41. Why is the formula written for an ionic compound such as NaCl the *empirical* formula, rather than a "molecular" formula?

43. How do the sizes of ions compare with the sizes of the atoms from which they form? Explain.

42. Describe in general terms the structure of ionic solids such as NaCl. How are the ions packed in the crystal?

44. Explain how a species such as SO_4^{2-} or CO_3^{2-} can exhibit both ionic and polar covalent bonding at the same time.

PROBLEMS

45. For each of the following pairs, indicate which is larger.
    a. Na or Na^+ c. Mg or Mg^{2+}
    b. F or F^- d. Al or Al^{3+}

46. For each of the following pairs, indicate which is larger.
    a. Li^+ or F^- c. Ca^{2+} or O^{2-}
    b. Na^+ or Cl^- d. Cs^+ or I^-

47. For each of the following pairs, indicate which is smaller.
    a. Fe or Fe^{3+}
    b. Cl or Cl^-
    c. Al^{3+} or Na^+
    d. O or O^{2-}

48. For each of the following pairs, indicate which is smaller.
    a. Cs^+ or I^-
    b. Na^+ or Br^-
    c. Cs or Cs^+
    d. O or O^{2-}

12.6 and 12.7 Lewis Structures

QUESTIONS

49. Why are the *valence* electrons of an atom the only electrons likely to be involved in bonding to other atoms?

50. Explain what the "duet" and "octet" rules are and how they are used to describe the arrangement of electrons in a molecule.

51. What type of structure must each atom in a compound usually exhibit for the compound to be stable?

52. When elements in the second and third periods occur in compounds, what number of electrons in the valence shell represents the most stable electron arrangement? Why?

53. A pair of valence electrons that exists on a given atom but is not used for attachment to another atom is called a _____ or nonbonding pair.

54. When a double bond forms between two atoms, the atoms share _____ pairs of electrons.

PROBLEMS

55. Write the Lewis structure for each of the following atoms.
    a. Si ($Z = 14$)
    b. As ($Z = 33$)
    c. Mg ($Z = 12$)
    d. Na ($Z = 11$)
    e. Ca ($Z = 20$)
    f. Al ($Z = 13$)

56. Write the Lewis structure for each of the following atoms.
    a. Rb ($Z = 37$)
    b. Cl ($Z = 17$)
    c. Kr ($Z = 36$)
    d. Ba ($Z = 56$)
    e. P ($Z = 15$)
    f. At ($Z = 85$)

57. What is the *total* number of valence electrons in each of the following molecules?
    a. PH_3
    b. CCl_4
    c. C_2H_6
    d. OF_2

58. Give the *total* number of valence electrons in each of the following molecules.
    a. CBr_4
    b. NO_2
    c. C_6H_6
    d. H_2O_2

59. Write a Lewis structure for each of the following simple molecules. Show all bonding valence electron pairs as a line and all nonbonding valence electron pairs as dots.
    a. PH_3
    b. SF_2
    c. HBr
    d. CCl_4

60. Write a Lewis structure for each of the following simple molecules. Show all bonding valence electron pairs as lines, and all nonbonding valence electron pairs as dots.
    a. NH_3
    b. CI_4
    c. NCl_3
    d. $SiBr_4$

61. Write a Lewis structure for each of the following simple molecules. Show all bonding valence electron pairs as a line and all nonbonding valence electron pairs as dots.
    a. CO
    b. CO_2
    c. O_2
    d. H_2Se

62. Write a Lewis structure for each of the following simple molecules. Show all bonding valence electron pairs as lines, and all nonbonding valence electron pairs as dots.
    a. H_2S
    b. SiF_4
    c. C_2H_4
    d. C_3H_8

63. Write a Lewis structure for each of the following simple molecules. Show all bonding valence electron pairs as a line and all nonbonding valence electron pairs as dots. For those molecules that exhibit resonance, draw the various possible resonance forms.
    a. Cl_2O
    b. CO_2
    c. SO_3

64. Write a Lewis structure for each of the following simple molecules. Show all bonding valence electron pairs as lines, and all nonbonding valence electrons as dots. For those molecules that exhibit resonance, draw the various possible resonance forms.
    a. NO_2
    b. H_2SO_4
    c. N_2O_4

65. Write a Lewis structure for each of the following polyatomic ions. Show all bonding valence electron pairs as a line and all nonbonding valence electron pairs as dots. For those ions that exhibit resonance, draw the various possible resonance forms.
    a. sulfate ion, SO_4^{2-}
    b. phosphate ion, PO_4^{3-}
    c. sulfite ion, SO_3^{2-}

66. Write a Lewis structure for each of the following polyatomic ions. Show all bonding valence electron pairs as lines and all nonbonding valence electrons as dots. For those ions that exhibit resonance, draw the various possible resonance forms.
    a. chlorate ion, ClO_3^-
    b. peroxide ion, O_2^{2-}
    c. acetate ion, $C_2H_3O_2^-$

67. Write a Lewis structure for each of the following polyatomic ions. Show all bonding valence electron pairs as a line and all nonbonding valence electron pairs as dots. For those ions that exhibit resonance, draw the various possible resonance forms.
    a. nitrite ion
    b. hydrogen carbonate ion
    c. hydroxide ion

68. Write a Lewis structure for each of the following polyatomic ions. Show all bonding valence electron pairs as a line and all nonbonding valence electron pairs as dots. For those ions that exhibit resonance, draw the various possible resonance forms.
    a. cyanide ion, CN^-
    b. hydrogen sulfate ion, HSO_4^-
    c. azide ion, N_3^-

12.8 Molecular Structure

QUESTIONS

69. What is the geometric structure of the water molecule, H_2O? How many pairs of valence electrons are there on the oxygen atom in the water molecule? What is the approximate H—O—H bond angle in water?

71. What is the geometric structure of the boron trifluoride molecule, BF_3? How many pairs of valence electrons are present on the boron atom in BF_3? What are the approximate F—B—F bond angles in BF_3?

70. What is the geometric structure of the ammonia molecule? How many pairs of electrons surround the nitrogen atom in NH_3? What is the approximate H—N—H bond angle in ammonia?

72. What is the geometric structure of the CH_4 molecule? How many pairs of valence electrons are present on the carbon atom of CH_4? Refer to Figure 12.11 and estimate the H—C—H bond angles in CH_4.

12.9 Molecular Structure: The VSEPR Model

QUESTIONS

73. Why is the geometric structure of a molecule important, especially for biological molecules?

75. How is the structure around a given atom related to repulsion between valence electron pairs on the atom?

77. Although the valence electron pairs in ammonia, NH_3, have a tetrahedral arrangement, the overall geometric structure of the ammonia molecule is *not* described as being tetrahedral. Explain.

74. What general principles determine the molecular structure (shape) of a molecule?

76. Why are all diatomic molecules *linear,* regardless of the number of valence electron pairs on the atoms involved?

78. Although the BF_3 and NF_3 molecules both contain the same number of atoms, the BF_3 molecule is flat, whereas the NF_3 molecule is pyramidal. Explain.

PROBLEMS

79. For the indicated atom in each of the following molecules, give the number and the arrangement of the electron pairs around that atom.
    a. N in NI_3
    b. S in H_2S
    c. Cl in HCl

81. Using the simple VSEPR theory, predict the molecular structure of each of the following molecules.
    a. GeH_4 b. ICl   c. NI_3

83. Using the simple VSEPR theory, predict the molecular structure of each of the following polyatomic ions.
    a. sulfate ion, SO_4^{2-}
    b. phosphate ion, PO_4^{3-}
    c. ammonium ion, NH_4^+

80. For the indicated atom in each of the following molecules or ions, give the number and arrangement of the electron pairs around that atom.
    a. P in PH_3
    b. Cl in ClO_4^-
    c. O in H_2O

82. Using the simple VSEPR theory, predict the molecular structure of each of the following molecules.
    a. GeH_4 b. PCl_3 c. Cl_2O

84. Using the simple VSEPR theory, predict the molecular structure of each of the following polyatomic ions.
    a. chlorate ion
    b. hydrogen phosphate ion (hydrogen is bonded to oxygen)
    c. sulfite ion

85. For each of the following molecules or ions, indicate what bond angle the VSEPR theory leads us to expect between the central atom and any two adjacent hydrogen atoms.
    a. H_2O
    b. NH_3
    c. NH_4^+
    d. CH_4

86. For each of the following molecules, indicate what bond angle the VSEPR theory leads us to expect between the central atom and any two adjacent oxygen atoms.
    a. H_2CO_3
    b. H_2SO_4 (hydrogen is bonded to oxygen)
    c. PO_4^{3-}
    d. HCO_3^- (hydrogen is bonded to oxygen)

Additional Problems

87. Explain briefly how substances with ionic bonding differ in properties from substances with covalent bonding.

88. Explain the difference between a covalent bond formed between two atoms of the same element and a covalent bond formed between atoms of two different elements.

89. What is *resonance*? Give three examples of molecules or ions that exhibit resonance, drawing Lewis structures for each of the possible resonance forms.

90. When two atoms share two pairs of electrons, a _____ bond is said to exist between them.

91. The geometric arrangement of electron pairs around a given atom is determined principally by the tendency to minimize _____ between the electron pairs.

92. In each case, which of the following pairs of bonded elements forms the more polar bond?
    a. S—F or S—Cl
    b. N—O or P—O
    c. C—H or Si—H

93. For each case, which of the following pairs of bonded elements forms the more polar bond?
    a. Br—Cl or Br—F
    b. As—S or As—O
    c. Pb—C or Pb—Si

94. The quantity of energy necessary to _____ a chemical bond is referred to as the bond energy.

95. A _____ chemical bond represents the equal sharing of a pair of electrons between two nuclei.

96. For each of the following pairs of elements, identify which element would be expected to be more electronegative. It should not be necessary to look at a table of actual electronegativity values.
    a. Be or Ba
    b. N or P
    c. F or Cl

97. On the basis of the electronegativity values given in Figure 12.3, indicate whether each of the following bonds would be expected to be ionic, covalent, or polar covalent.

    a. H—O   c. H—H
    b. O—O   d. H—Cl

98. Which of the following molecules contain polar covalent bonds?
    a. carbon monoxide, CO   d. phosphorus, P_4
    b. chlorine, Cl_2
    c. iodine monochloride, ICl

99. On the basis of the electronegativity values given in Figure 12.3, indicate which is the more polar bond in each of the following pairs.
    a. N—P or N—O   c. N—S or N—C
    b. N—C or N—O   d. N—F or N—S

100. In each of the following diatomic molecules, which end of the molecule is positive relative to the other end?
    a. hydrogen fluoride, HF
    b. iodine monochloride, ICl
    c. nitrogen(II) oxide, NO

101. For each of the following bonds, draw a figure indicating the direction of the bond dipole and indicating which end of the bond is positive and which is negative.
    a. N—Cl   c. N—S
    b. N—P   d. N—C

102. Write the electron configuration for each of the following atoms and for the simple ion that the element most commonly forms. In each case, indicate which noble gas has the same electron configuration as the ion.
    a. aluminum, $Z = 13$
    b. bromine, $Z = 35$
    c. calcium, $Z = 20$
    d. lithium, $Z = 3$
    e. fluorine, $Z = 9$

103. What simple ion does each of the following elements most commonly form?
    a. barium, $Z = 56$
    b. rubidium, $Z = 37$
    c. aluminum, $Z = 13$
    d. oxygen, $Z = 8$

104. On the basis of their electron configurations, predict the formula of each simple binary ionic compound likely to form when the following pairs of elements react with each other.
    a. sodium, Na, and selenium, Se
    b. rubidium, Rb, and fluorine, F
    c. potassium, K, and tellurium, Te
    d. barium, Ba, and selenium, Se
    e. potassium, K, and astatine, At
    f. francium, Fr, and chlorine, Cl

105. Which noble gas has the same electron configuration as each of the ions in the following compounds?
    a. calcium bromide, $CaBr_2$
    b. aluminum selenide, Al_2Se_3
    c. strontium oxide, SrO
    d. potassium sulfide, K_2S

106. For each of the following pairs, indicate which is smaller.
    a. Rb^+ or Na^+
    b. Mg^{2+} or Al^{3+}
    c. F^- or I^-
    d. Na^+ or K^+

107. Write the Lewis structure for each of the following atoms.
    a. He ($Z = 2$)   d. Ne ($Z = 10$)
    b. Br ($Z = 35$)   e. I ($Z = 53$)
    c. Sr ($Z = 38$)   f. Ra ($Z = 88$)

108. What is the *total* number of valence electrons in each of the following molecules?
    a. SiH_4 c. N_2O_4
    b. SF_6 d. AsI_3

109. Write a Lewis structure for each of the following simple molecules. Show all bonding valence electron pairs as a line and all nonbonding valence electron pairs as dots.
    a. GeH_4 c. NI_3
    b. ICl   d. PF_3

110. Write a Lewis structure for each of the following simple molecules. Show all bonding valence electron pairs as a line and all nonbonding valence electron pairs as dots.
    a. N_2H_4
    b. C_2H_6
    c. NCl_3
    d. $SiCl_4$

111. Write a Lewis structure for each of the following simple molecules. Show all bonding valence electron pairs as a line and all nonbonding valence electron pairs as dots. For those molecules that exhibit resonance, draw the various possible resonance forms.
    a. SO_2
    b. N_2O (N in center)
    c. O_3

112. Write a Lewis structure for each of the following polyatomic ions. Show all bonding valence electron pairs as a line and all nonbonding valence electron pairs as dots. For those ions that exhibit resonance, draw the various possible resonance forms.
    a. nitrate ion
    b. carbonate ion
    c. ammonium ion

113. Why is the molecular structure of H_2O nonlinear, whereas that of BeF_2 is linear, even though both molecules consist of three atoms?

114. For the indicated atom in each of the following molecules, give the number and the arrangement of the electron pairs around that atom.
    a. C in CCl_4
    b. Ge in GeH_4
    c. B in BF_3

115. Using the simple VSEPR theory, predict the molecular structure of each of the following molecules.
    a. Cl_2O
    b. OF_2
    c. $SiCl_4$

116. Using the simple VSEPR theory, predict the molecular structure of each of the following polyatomic ions.
    a. chlorate ion
    b. chlorite ion
    c. perchlorate ion

117. For each of the following molecules, indicate what bond angle the VSEPR theory leads us to expect between the central atom and any two adjacent fluorine atoms.
    a. BeF_2
    b. BF_3
    c. NF_3
    d. CF_4

118. Using the simple VSEPR theory, predict the molecular structure of each of the following molecules or ions containing multiple bonds.
    a. SO_2
    b. SO_3
    c. HCO_3^- (hydrogen is bonded to oxygen)
    d. HCN

119. Using the simple VSEPR theory, predict the molecular structure of each of the following molecules or ions containing multiple bonds.
    a. CO_3^{2-}
    b. HNO_3 (hydrogen is bonded to oxygen)
    c. NO_2^-
    d. C_2H_2

Solutions to Self-Check Exercises

SELF-CHECK EXERCISE 12.1

Using the electronegativity values given in Figure 12.3, we choose the bond in which the atoms exhibit the largest difference in electronegativity. (Electronegativity values are shown in parentheses.)

a.   H—C > H—P
   (2.1)(2.5)   (2.1)(2.1)

b.   O—I > O—F
   (3.5)(2.5)   (3.5)(4.0)

c.   S—O > N—O
   (2.5)(3.5)   (3.0)(3.5)

d.   N—H > Si—H
   (3.0)(2.1)   (1.8)(2.1)

SELF-CHECK EXERCISE 12.2

H has one electron, and Cl has seven valence electrons. This gives a total of eight valence electrons. We first draw in the bonding pair:

H—Cl, which could be drawn as H : Cl

We have six electrons yet to place. The H already has two electrons, so we place three lone pairs around the chlorine to satisfy the octet rule.

H—Cl :   or   H : Cl :

SELF-CHECK EXERCISE 12.3

STEP 1

O_3: 3(6) = 18 valence electrons

STEP 2

O—O—O

STEP 3

O=O—O :   and   : O—O=O

This molecule shows resonance (it has two valid Lewis structures).

SELF-CHECK EXERCISE 12.4

Molecule or Ion	Total Valence Electrons	Draw Single Bonds	Calculate Number of Electrons Remaining	Use Remaining Electrons to Achieve Noble Gas Configurations	Check Atom	Electrons
a. NF_3	$5 + 3(7) = 26$	F—N⟨F,F⟩	$26 - 6 = 20$	:F—N—F: (with :F:)	N / F	8 / 8
b. O_2	$2(6) = 12$	O—O	$12 - 2 = 10$	:O=O:	O	8
c. CO	$4 + 6 = 10$	C—O	$10 - 2 = 8$	:C≡O:	C / O	8 / 8
d. PH_3	$5 + 3(1) = 8$	H₂P—H	$8 - 6 = 2$	H—P—H (H below)	P / H	8 / 2
e. H_2S	$2(1) + 6 = 8$	H—S—H	$8 - 4 = 4$	H—S—H	S / H	8 / 2
f. SO_4^{2-}	$6 + 4(6) + 2 = 32$	O—S—O (O top, O bottom)	$32 - 8 = 24$	$[:O—S—O:]^{2-}$	S / O	8 / 8
g. NH_4^+	$5 + 4(1) - 1 = 8$	H—N—H (H top, H bottom)	$8 - 8 = 0$	$[H—N—H]^+$	N / H	8 / 2
h. ClO_3^-	$7 + 3(6) + 1 = 26$	O—Cl⟨O,O⟩	$26 - 6 = 20$	$[:O—Cl—O:]^-$ (:O: below)	Cl / O	8 / 8
i. SO_2	$6 + 2(6) = 18$	O—S—O	$18 - 4 = 14$	O=S—O: and :O—S=O	S / O	8 / 8

SELF-CHECK EXERCISE 12.5

a. NH_4^+
The Lewis structure is $[H—N—H$ with H top and bottom$]^+$

(See Self-Check Exercise 12.4.) There are four pairs of electrons around the nitrogen. This requires a tetrahedral arrangement of electron pairs. The NH_4^+ ion has a tetrahedral molecular structure (case 3 in Table 12.4), because all electron pairs are shared.

b. SO_4^{2-}
The Lewis structure is

$$
\left[\begin{array}{c}
\ddot{\text{:O:}} \\
| \\
\text{:O} - \text{S} - \text{O:} \\
| \\
\text{:O:}
\end{array}\right]^{2-}
$$

(See Self-Check Exercise 12.4.) The four electron pairs around the sulfur require a tetrahedral arrangement. The SO_4^{2-} has a tetrahedral molecular structure (case 3 in Table 12.4).

c. NF_3
The Lewis structure is

$$
\begin{array}{c}
\ddot{\text{N}} \\
\diagup\ |\ \diagdown \\
\text{:F}\ \text{:F:}\ \text{F:}
\end{array}
$$

(See Self-Check Exercise 12.4.) The four pairs of electrons on the nitrogen require a tetrahedral arrangement. In this case only three of the pairs are shared with the fluorine atoms, leaving one lone pair. Thus the molecular structure is a trigonal pyramid (case 4 in Table 12.4).

d. H_2S
The Lewis structure is $\text{H} - \ddot{\text{S}} - \text{H}$

(See Self-Check Exercise 12.4.) The four pairs of electrons around the sulfur require a tetrahedral arrangement. In this case two pairs are shared with hydrogen atoms, leaving two lone pairs. Thus the molecular structure is bent or V-shaped (case 5 in Table 12.4).

e. ClO_3^-
The Lewis structure is

$$
\left[\begin{array}{c}
\text{:O} - \ddot{\text{Cl}} - \text{O:} \\
| \\
\text{:O:}
\end{array}\right]^{-}
$$

(See Self-Check Exercise 12.4.) The four pairs of electrons require a tetrahedral arrangement. In this case, three pairs are shared with oxygen atoms, leaving one lone pair. Thus the molecular structure is a trigonal pyramid (case 4 in Table 12.4).

f. BeF_2
The Lewis structure is $\text{:F} - \text{Be} - \text{F:}$

The two electron pairs on beryllium require a linear arrangement. Because both pairs are shared by fluorine atoms, the molecular structure is also linear (case 1 in Table 12.4).

CHAPTER

13 Gases

CONTENTS

One goal of any chemistry course should be to clarify for the students just what the scientific method is and how it works. This material provides an excellent opportunity to help students understand laws and models and the differences between them.

The relationship between the temperature of a gas and its volume explains how these hot-air balloons can float safely through the atmosphere.

A woman undergoing a test to see how efficiently her body uses oxygen from the air.

Pollutants being emitted from industrial smokestacks.

Dry air (air from which the water vapor has been removed) is 78.1% N_2 molecules, 20.9% O_2 molecules, 0.9% Ar atoms, 0.03% CO_2 molecules, along with smaller amounts of Ne, He, CH_4, Kr, and other trace components.

If you have the resources, you can show that the mercury in a tube inverted in a dish of mercury really doesn't run out.

Matter exists in three distinct physical states: gas, liquid, and solid. As we saw in Chapter 3, these states have very different properties (see Table 3.1 and Section 3.2), many of which are already familiar to you. In this chapter we will focus on the properties of the gaseous state.

We live immersed in a gaseous solution. The earth's atmosphere is a mixture of gases that consists mainly of elemental nitrogen, N_2, and oxygen, O_2. The atmosphere both supports life and acts as a waste receptacle for the exhaust gases that accompany many industrial processes. The chemical reactions of these waste gases in the atmosphere lead to various types of pollution, including smog and acid rain.

The gases in the atmosphere also shield us from harmful radiation from the sun and keep the earth warm by reflecting heat radiation back toward the earth. In fact, there is now great concern that an increase in atmospheric carbon dioxide, a product of the combustion of fossil fuels, is causing a dangerous warming of the earth. (See Chemistry in Focus: "Atmospheric Effects," in Chapter 11.)

In this chapter we will look carefully at the properties of gases. First we will see how measurements of gas properties lead to various types of laws—statements that show how the properties are related to each other. Then we will construct a model to explain why gases behave as they do. This model will show how the behavior of the individual particles of a gas lead to the observed properties of the gas itself (a collection of many, many particles).

The study of gases provides an excellent example of the scientific method in action. It illustrates how observations lead to natural laws, which in turn can be accounted for by models.

13.1   Pressure

AIM: To demonstrate atmospheric pressure and explain how barometers work.
To define the various units of pressure.

A gas uniformly fills any container, is easily compressed, and mixes completely with any other gas (see Section 3.1). One of the most obvious properties of a gas is that it exerts pressure on its surroundings. For example, when you blow up a balloon, the air inside pushes against the elastic sides of the balloon and keeps it firm.

The gases most familiar to us form the earth's atmosphere. The pressure exerted by this gaseous mixture that we call air can be dramatically demonstrated by the experiment shown in Figure 13.1. A small volume of water is placed in a metal can and the water is boiled, which fills the can with steam. The can is then sealed and allowed to cool. Why does the can collapse as it cools? It is the atmospheric pressure that crumples the can. When the can is cooled after being sealed so that no air can flow in, the water vapor (steam) inside the can con-

You can perform the demonstration pictured in Figure 13.1 by boiling some water in an open aluminum soda can, then quickly inverting it (unsealed) in ice water. A bunsen burner or a hot plate will work as a heat source.

(a)

(b)

Figure 13.1
The pressure exerted by the gases in the atmosphere can be demonstrated by boiling water in an aluminum beverage can (a), and then turning off the heat and sealing the can. As the can cools, the water vapor condenses, lowering the gas pressure inside the can. This causes the can to crumple (b).

denses to a very small volume of liquid water. As a gas, the water vapor filled the can, but when it is condensed to a liquid, the liquid does not come close to filling the can. The H_2O molecules formerly present as a gas are now collected in a much smaller volume of liquid, and there are very few molecules of gas left to exert pressure outwards and counteract the air pressure. As a result, the pressure exerted by the gas molecules in the atmosphere smashes the can.

As a gas, water occupies 1200 times as much space as it does as a liquid at 25 °C and atmospheric pressure.

A device to measure atmospheric pressure, the **barometer,** was invented in 1643 by an Italian scientist named Evangelista Torricelli (1608–1647), who had been a student of the famous astronomer Galileo. Torricelli's barometer is constructed by filling a glass tube with liquid mercury and inverting it in a dish of mercury, as shown in Figure 13.2(a). Notice that a large quantity of mercury stays in the tube. In fact, at sea level the height of this column of mercury averages 760 mm. Why does this mercury stay in the tube, seemingly in defiance of gravity? Figure 13.2(b) illustrates how the pressure exerted by the atmospheric gases on the surface of mercury in the dish keeps the mercury in the tube.

A nineteenth century engraving of Evangelista Torricelli, the man who invented the barometer.

Figure 13.2
(a) When a glass tube is filled with mercury and inverted in a dish of mercury at sea level, the mercury flows out of the tube until a column approximately 760 mm high remains (the height varies with atmospheric conditions). (b) A diagram showing how the pressure of the atmosphere balances the weight of the column of mercury in the tube.

(a)   (b)

Soon after Torricelli died, a German physicist named Otto von Guericke invented an air pump. In a famous demonstration for the King of Prussia in 1663, Guericke placed two hemispheres together, pumped the air out of the resulting sphere through a valve, and showed that teams of horses could not pull the hemispheres apart. Then, after secretly opening the air valve, Guericke easily separated the hemispheres by hand. The King of Prussia was so impressed that he awarded Guericke a lifetime pension!

Atmospheric pressure results from the mass of the air being pulled toward the center of the earth by gravity—in other words, it results from the weight of the air. Changing weather conditions cause the atmospheric pressure to vary, so the height of the column of Hg supported by the atmosphere at sea level varies; it is not always 760 mm. The meteorologist who says a "low" is approaching means that the atmospheric pressure is going to decrease. This condition often occurs in conjunction with a storm.

Atmospheric pressure also varies with altitude. For example, when Torricelli's experiment is done in Breckenridge, Colorado (elevation 9600 feet), the atmosphere supports a column of mercury only about 520 mm high because the air is "thinner." That is, there is less air pushing down on the earth's surface at Breckenridge than at sea level.

Units of Pressure

Mercury is used to measure pressure because of its high density. By way of comparison, the column of water required to measure a given pressure would be 13.6 times as high as a mercury column used for the same purpose.

Because instruments used for measuring pressure (see Figure 13.3) often contain mercury, the most commonly used units for pressure are based on the height of the mercury column (in millimeters) that the gas pressure can support. The unit **mm Hg** (millimeters of mercury) is often called the **torr** in honor of Torricelli. The terms torr and mm Hg are used interchangeably by chemists. A related unit for pressure is the **standard atmosphere** (abbreviated atm).

Conversion problems with pressure units provide a good way to start students thinking about gases. This type of problem helps students to prepare for pressure conversions in more complex gas law problems.

$$1 \text{ standard atmosphere} = 1.000 \text{ atm} = 760.0 \text{ mm Hg} = 760.0 \text{ torr}$$

Figure 13.3
A device (called a manometer) for measuring the pressure of a gas in a container. The pressure of the gas is given by h (the difference in mercury levels) in units of torr (equivalent to mm Hg). (a) Gas pressure = atmospheric pressure − h. (b) Gas pressure = atmospheric pressure + h.

The SI unit for pressure is the **pascal** (abbreviated Pa).

$$1 \text{ standard atmosphere } = 101{,}325 \text{ Pa}$$

Thus 1 atmosphere is about 100,000 or 10^5 pascals. Because the pascal is so small and is not commonly used in the United States, we will use it sparingly in this book. A unit of pressure that is employed in the engineering sciences and that we use for measuring tire pressure is pounds per square inch, abbreviated psi.

$$1.000 \text{ atm } = 14.69 \text{ psi}$$

Sometimes we need to convert from one unit of pressure to another. We do this by using conversion factors. The process is illustrated in Example 13.1.

There are so many decisions to make when solving gas law problems that providing the relevant pressure conversions on exams can help students focus on solving the problems.

1.000 atm

760.0 mm Hg
760.0 torr
14.69 psi
101,325 Pa

EXAMPLE 13.1   Pressure Unit Conversions

The pressure of a tire is measured to be 28 psi. Represent this pressure in atmospheres, torr, and pascals.

SOLUTION

To convert from pounds per square inch to atmospheres, we need the equivalence statement

$$1.000 \text{ atm } = 14.69 \text{ psi}$$

which leads to the conversion factor

$$\frac{1.000 \text{ atm}}{14.69 \text{ psi}}$$

$$28 \text{ psi} \times \frac{1.000 \text{ atm}}{14.69 \text{ psi}} = 1.9 \text{ atm}$$

To convert from atmospheres to torr, we use the equivalence statement

$$1.000 \text{ atm } = 760.0 \text{ torr}$$

which leads to the conversion factor

$$\frac{760.0 \text{ torr}}{1.000 \text{ atm}}$$

$$1.9 \text{ atm} \times \frac{760.0 \text{ torr}}{1.000 \text{ atm}} = 1.4 \times 10^3 \text{ torr}$$

To change from torr to pascals, we need the equivalence statement

$$1.000 \text{ atm } = 101{,}325 \text{ Pa}$$

Checking the air pressure in a tire.

$$1.9 \times 760.0 = 1444$$

$$1444 \xrightarrow{\text{Round off}} 1400 = 1.4 \times 10^3$$

$1.9 \times 101{,}325 = 192{,}517.5$

$192{,}517.5$ ➤ Round off ➤ $190{,}000 = 1.9 \times 10^5$

EXAMPLE 13.1, CONTINUED

which leads to the conversion factor

$$\frac{101{,}325 \text{ Pa}}{1.000 \text{ atm}}$$

$$1.9 \text{ atm} \times \frac{101{,}325 \text{ Pa}}{1.000 \text{ atm}} = 1.9 \times 10^5 \text{ Pa}$$

NOTE: The best way to check a problem like this is to make sure the final units are the ones required.

SELF-CHECK EXERCISE 13.1

On a summer day in Breckenridge, Colorado, the atmospheric pressure is 525 mm Hg. What is this air pressure in atmospheres?

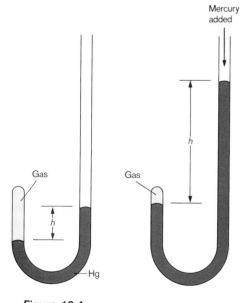

Figure 13.4
A J-tube similar to the one used by Boyle. The pressure on the trapped gas can be changed by adding or withdrawing mercury.

13.2   Pressure and Volume: Boyle's Law

AIM: To describe the law that relates the pressure and volume
of a gas.
To do calculations involving this law.

The first careful experiments on gases were performed by the Irish scientist Robert Boyle (1627–1691). Using a J-shaped tube closed at one end (Figure 13.4), which he reportedly set up in the multi-story entryway of his house, Boyle studied the relationship between the pressure of the trapped gas and its volume. Representative values from Boyle's experiments are given in Table 13.1. The units given for the volume (cubic inches) and pressure (inches of mercury) are the ones Boyle used. Keep in mind that the metric system was not in use at this time.

First let's examine Boyle's observations (Table 13.1) for general trends. Note that as the pressure increases, the volume of the trapped gas decreases. In fact, if you compare the data from experiments 1 and 4, you can see that as the pressure is doubled (from 29.1 to 58.2), the volume of the gas is halved (from 48.0 to 24.0). The same relationship can be seen in experiments 2 and 5 and in experiments 3 and 6 (approximately).

We can see the relationship between the volume of a gas and its pressure more clearly by looking at the product of the values of these two properties

In this chapter, students learn to distinguish between curves in which the data points demonstrate an inverse relationship and those in which the points demonstrate a direct relationship. It can help to show the differences in appearances between those two kinds of curves.

Table 13.1	A Sample of Boyle's Observations (moles of gas and temperature both constant)			
Experiment	Pressure (in. Hg)	Volume (in³)	Pressure × Volume (in. Hg) × (in³)	
			Actual	*Rounded**
1	29.1	48.0	1396.8	1.40×10^3
2	35.3	40.0	1412.0	1.41×10^3
3	44.2	32.0	1414.4	1.41×10^3
4	58.2	24.0	1396.8	1.40×10^3
5	70.7	20.0	1414.0	1.41×10^3
6	87.2	16.0	1395.2	1.40×10^3
7	117.5	12.0	1410.0	1.41×10^3

*Three significant figures are allowed in the product because both of the numbers that are multiplied together have three significant figures.

$(P \times V)$ using Boyle's observations. This product is shown in the last column of Table 13.1. Note that for all the experiments,

$$P \times V = 1.4 \times 10^3 \text{ (in. Hg)} \times \text{in}^3$$

with only a slight variation due to experimental error. Other similar measurements on gases show the same behavior. This means that the relationship of the pressure and volume of a gas can be expressed in words as

Pressure times volume equals a constant

or in terms of an equation as

$$PV = k$$

which is called **Boyle's law,** and where k is a constant at a specific temperature for a given amount of gas. For the data we used from Boyle's experiment, $k = 1.41 \times 10^3$(in. Hg) $\times$ in³.

It is often easier to visualize the relationships between two properties if we make a graph. Figure 13.5 uses the data given in Table 13.1 to show how pressure is related to volume. This relationship, called a plot or a graph, shows that V decreases as P increases. When this type of relationship exists, we say that volume and pressure are inversely related or *inversely proportional;* when one increases, the other decreases.

Boyle's law means that if we know the volume of a gas at a given pressure, we can predict the new volume if the pressure is changed, *provided that neither the temperature nor the amount of gas is changed.* For example, if we represent

For Boyle's law to hold, the amount of gas (moles) must not be changed. The temperature must also be constant.

The fact that the constant is sometimes 1.40×10^3 instead of 1.41×10^3 is due to experimental error (uncertainties in measuring the values of P and V).

If circumstances allow, experimentally determine Boyle's law in class using a J-tube with a movable mercury reservoir. Have the students graph the results themselves.

I like to demonstrate that $PV = k$ by providing pressure and volume data similar to those in Table 13.1, then calculating the products and constructing a small table.

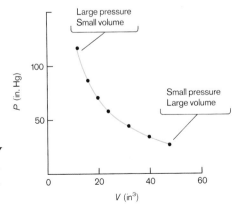

Figure 13.5
A plot of P versus V from Boyle's data in Table 13.1.

Some students may have trouble rearranging the gas law equations to isolate the asked-for variable on one side.

the original pressure and volume as P_1 and V_1 and the final values as P_2 and V_2, using Boyle's law we can write

$$P_1 V_1 = k \quad \text{and} \quad P_2 V_2 = k$$

We can also say

$$P_1 V_1 = k = P_2 V_2$$

or simply

$$P_1 V_1 = P_2 V_2$$

Students' biggest problem with the material in this chapter is working with the various gas law equations. They mix up P_1 and P_2, and V_1 and V_2. This problem occurs in part because students try to memorize too many different equations instead of deriving the one they need from the ideal gas equation.

This is really another way to write Boyle's law. We can solve for the final volume (V_2) by dividing both sides of the equation by P_2.

$$\frac{P_1 V_1}{P_2} = \frac{\cancel{P_2} V_2}{\cancel{P_2}}$$

Canceling the P_2 terms on the right gives

$$\frac{P_1}{P_2} \times V_1 = V_2$$

or

$$V_2 = V_1 \times \frac{P_1}{P_2}$$

This equation tells us that we can calculate the new gas volume (V_2) by multiplying the original volume (V_1) by the ratio of the original pressure to the final pressure (P_1/P_2), as illustrated in Example 13.2.

In summary, examination of CO_2 leads us to the conclusion that, in using the VSEPR model for molecules with double bonds, each double bond should be treated the same as a single bond. In other words, although a double bond involves four electrons, these electrons are restricted to the space between a given pair of atoms. Therefore these four electrons do not function as two independent pairs but are "tied together" to form a single effective repulsive unit.

We reach this same conclusion by considering the known structures of other molecules that contain double bonds. For example, consider the ozone molecule, which has 18 valence electrons and exhibits the resonance structures

$$:\!\overset{..}{\text{O}}\!-\!\overset{..}{\text{O}}\!=\!\overset{..}{\text{O}}\cdot \longleftrightarrow \cdot\overset{..}{\text{O}}\!=\!\overset{..}{\text{O}}\!-\!\overset{..}{\text{O}}\!:$$

The ozone molecule is known to have a bond angle close to 120°. Recall that 120° angles represent the minimum repulsion for three pairs of electrons

This indicates that the double bond in the ozone molecule is behaving as one effective repulsive unit:

EXAMPLE I3.2   **Calculating Volume Using
                      Boyle's Law**

Freon-12 (the common name for the compound CCl_2F_2) is widely used in refrigeration systems. Consider a 1.5-L sample of gaseous CCl_2F_2 at a pressure of 56 torr. Then the pressure is changed to 150 torr at a constant temperature.

a.   Will the volume of the gas increase or decrease?
b.   What will be the new volume of the gas?

SOLUTION

a.   As the first step in a gas law problem, always write down the information given, in the form of a table showing the initial and final conditions.

Initial Conditions	Final Conditions
$P_1 = 56$ torr	$P_2 = 150$ torr
$V_1 = 1.5$ L	$V_2 = ?$

Drawing a picture also is often helpful. Notice that the pressure is increased from 56 torr to 150 torr, so the volume must decrease:

We can verify this by using Boyle's law in the form

$$V_2 = V_1 \times \frac{P_1}{P_2}$$

Note that V_2 is obtained by "correcting" V_1 using the ratio P_1/P_2. Because P_1 is less than P_2, the ratio P_1/P_2 is a fraction that is less than 1. Thus V_2 must be a fraction of (smaller than) V_1; the volume decreases.

Strongly suggest that students check their work after solving a problem by asking, Does this answer make sense? Go through an example with them to show them how they can predict in advance whether volume and pressure should increase or decrease.

EXAMPLE 13.2, CONTINUED

b.   We can calculate V_2 as follows:

$$V_2 = V_1 \times \frac{P_1}{P_2} = 1.5 \text{ L} \times \frac{\overset{P_1}{\downarrow}}{\underset{\uparrow}{150 \text{ torr}}} = 0.56 \text{ L}$$

The volume of the gas decreases from 1.5 L to 0.56 L. This change is in the expected direction.

The fact that the volume decreases in Example 13.2 makes sense because the pressure was increased. To help catch errors, make it a habit to check whether an answer to a problem makes physical sense.

Neon lights in Hong Kong.

SELF-CHECK EXERCISE 13.2

A sample of neon to be used in a neon sign has a volume of 1.51 L at a pressure of 635 torr. Calculate the volume of the gas after it is pumped into the glass tubes of the sign, where it shows a pressure of 785 torr.

EXAMPLE 13.3	Calculating Pressure Using Boyle's Law

In an automobile engine the gaseous fuel–air mixture enters the cylinder and is compressed by a moving piston before it is ignited. In a certain engine the initial cylinder volume is 0.725 L. After the piston moves up, the volume is 0.075 L. The fuel–air mixture initially has a pressure of 1.00 atm. Calculate the pressure of the compressed fuel–air mixture, assuming that the temperature and the amount of gas both remain constant.

SOLUTION

We summarize the given information in the following table:

Initial Conditions	Final Conditions
$P_1 = 1.00$ atm	$P_2 = ?$
$V_1 = 0.725$ L	$V_2 = 0.075$ L

Then we solve Boyle's law in the form $P_1 V_1 = P_2 V_2$ for P_2 by dividing both sides by V_2 to give the equation

$$P_2 = P_1 \times \frac{V_1}{V_2} = 1 \text{ atm} \times \frac{0.725 \cancel{L}}{0.075 \cancel{L}} = 9.7 \text{ atm}$$

$$P_1 V_1 = P_2 V_2$$

$$\frac{P_1 V_1}{V_2} = \frac{P_2 \cancel{V_2}}{\cancel{V_2}}$$

$$P_1 \times \frac{V_1}{V_2} = P_2$$

$$\frac{0.725}{0.075} = 9.666\ldots$$

9.666 → Round off → 9.7

Note that the pressure must increase because the volume gets smaller. Pressure and volume are inversely related.

13.3 Volume and Temperature: Charles's Law

AIM: To define absolute zero.
To describe the law relating the volume and temperature of a sample of gas at constant moles and pressure, and to do calculations involving that law.

In the century following Boyle's findings, scientists continued to study the properties of gases. The French physicist Jacques Charles (1746–1823), who was the first person to fill a balloon with hydrogen gas and who made the first solo flight, showed that the volume of a given amount of gas (at constant pressure) increases with the temperature of the gas. That is, the volume increases when the temperature increases. A plot of the volume of a given sample of gas (at constant pressure) versus its temperature (in Celsius degrees) gives a straight line. This type of relationship is called *linear,* and this behavior is shown for several gases in Figure 13.6 below.

The solid lines in Figure 13.6 are based on actual measurements of temperature and volume for the gases listed. As we cool the gases they eventually liquefy, so we cannot determine any experimental points below this temperature. However, when we extend each straight line (which is called *extrapolation* and is shown here by a dashed line), something very interesting happens. *All* of the lines extrapolate to zero volume at the same temperature: -273 °C. This suggests that -273 °C is the lowest possible temperature, because a negative volume is physically impossible. In fact, experiments have shown that matter cannot be cooled to temperatures lower than -273 °C. Therefore this temperature is defined as **absolute zero** on the Kelvin scale.

The air in a balloon expands when it is heated, making the balloon buoyant.

Many students will not be familiar with the word extrapolate. Draw curves to show absolute zero on the X-axis; you can demonstrate the meaning of the word in this context.

Temperatures such as 0.0001 K have been obtained in the laboratory, but 0 K has never been reached.

It is useful to show a plot of V(L) versus T(°C) for several gases and then extrapolate each curve separately to show they all meet at the same temperature. This is a dramatic demonstration that -273 °C is not an arbitrary number and was derived from observations about real gases.

Figure 13.6
Plots of V (L) versus T (°C) for several gases.

Figure 13.7
Plots of V versus T as in Figure 13.6, except that here the Kelvin scale is used for temperature.

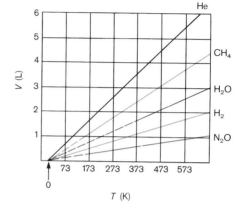

One of the most common mistakes students make when working gas problems is to forget that degrees Celsius or Fahrenheit must be converted to kelvins. Every time you solve a problem, make a point of converting units before proceeding.

When the volumes of the gases shown in Figure 13.6 are plotted versus temperature on the Kelvin scale rather than the Celsius scale, the plots shown in Figure 13.7 result. These plots show that the volume of each gas is *directly proportional to the temperature* (in kelvins) and extrapolates to zero when the temperature is 0 K. Let's illustrate this statement with an example. Suppose we have 1 L of gas at 300 K. When we double the temperature of this gas to 600 K (without changing its pressure), the volume also doubles, to 2 L. Verify this type of behavior by looking carefully at the lines for various gases shown in Figure 13.7.

The direct proportionality between volume and temperature (in kelvins) is represented by the equation known as **Charles's law:**

$$V = bT$$

where T is in kelvins and b is the proportionality constant. Charles's law holds for a given sample of gas at constant pressure. It tells us that (for a given amount of gas at a given pressure) the volume of the gas is directly proportional to the temperature on the Kelvin scale:

$$V = bT \quad \text{or} \quad \frac{V}{T} = b = \text{constant}$$

From Figure 13.7 for Helium

V (L)	T (K)	b
0.7	73	0.01
1.7	173	0.01
2.7	273	0.01
3.7	373	0.01
5.7	573	0.01

Notice that in the second form, this equation states that the *ratio* of V to T (in kelvins) must be constant. (This is shown for helium in the margin.) Thus when we triple the temperature (in kelvins) of a sample of gas, the volume of the gas triples also.

$$\frac{V}{T} = \frac{3 \times V}{3 \times T} = b = \text{constant}$$

We can also write Charles's law in terms V_1 and T_1 (the initial conditions) and V_2 and T_2 (the final conditions).

$$\frac{V_1}{T_1} = b \quad \text{and} \quad \frac{V_2}{T_2} = b$$

Thus

$$\frac{V_1}{T_1} = \frac{V_2}{T_2}$$

We will illustrate the use of this equation in Examples 13.4 and 13.5.

Charles's law in the form $V_1/T_1 = V_2/T_2$ applies only when the amount of gas (moles) and the pressure are both constant.

EXAMPLE 13.4   Calculating Volume Using Charles's Law, I

A 2.0-L sample of air is collected at 298 K and then cooled to 278 K. (The pressure is held constant at 1.0 atm.)

a. Does the volume increase or decrease?
b. Calculate the volume of air at 278 K.

It is both time-consuming to write gas law problems on the board and time-consuming to solve them. I usually put the problem itself on an overhead transparency so that I can spend more time discussing how to approach the solution of the problem.

SOLUTION

a. Because the gas is cooled, the volume of the gas must decrease:

$$\frac{V}{T} = \text{constant}$$

$\uparrow$
T is decreased, so
V must decrease to
maintain a constant
ratio.

$\dfrac{V_1}{T_1}$ | Temperature smaller, volume smaller | $\dfrac{V_2}{T_2}$

b. To calculate the new volume, V_2, we will use Charles's law in the form

$$\frac{V_1}{T_1} = \frac{V_2}{T_2}$$

We are given the following information:

Initial Conditions	Final Conditions
$T_1 = 298$ K	$T_2 = 278$ K
$V_1 = 2.0$ L	$V_2 = ?$

We want to solve the equation

$$\frac{V_1}{T_1} = \frac{V_2}{T_2}$$

for V_2. We can do this by multiplying both sides by T_2 and canceling.

EXAMPLE 13.4, CONTINUED

$$T_2 \times \frac{V_1}{T_1} = \frac{V_2}{\cancel{T_2}} \times \cancel{T_2} = V_2$$

Thus

$$V_2 = T_2 \times \frac{V_1}{T_1} = 278 \cancel{K} \times \frac{2.0 \text{ L}}{298 \cancel{K}} = 1.9 \text{ L}$$

Note that the volume gets smaller when the temperature decreases, just as we predicted.

EXAMPLE 13.5   **Calculating Volume Using Charles's Law, II**

A sample of gas at 15 °C (at 1 atm) has a volume of 2.58 L. The temperature is then raised to 38 °C (at 1 atm).
a.   Does the volume of the gas increase or decrease?
b.   Calculate the new volume.

SOLUTION

a.   In this case we have a given sample (constant amount) of gas that is heated from 15 °C to 38 °C *while the pressure is held constant.* We know from Charles's law that the volume of a given sample of gas is directly proportional to the temperature (at constant pressure). So the increase in temperature will *increase* the volume; the new volume will be greater than 2.58 L.
b.   To calculate the new volume, we use Charles's law in the form

$$\frac{V_1}{T_1} = \frac{V_2}{T_2}$$

We are given the following information:

Initial Conditions	Final Conditions
$T_1 = 15$ °C	$T_2 = 38$ °C
$V_1 = 2.58$ L	$V_2 = ?$

As is often the case, the temperatures are given in Celsius degrees. However, in order for us to use Charles's law, the temperature *must be in kelvins.* Thus we must convert by adding 273 to each temperature.

Initial Conditions	Final Conditions
$T_1 = 15$ °C $= 15 + 273 = 288$ K	$T_2 = 38$ °C $= 38 + 273 = 311$ K
$V_1 = 2.58$ L	$V_2 = ?$

EXAMPLE 13.5, CONTINUED

Solving for V_2 gives

$$V_2 = V_1 \times \frac{T_2}{T_1} = 2.58 \text{ L} \left(\frac{311 \text{ K}}{288 \text{ K}} \right) = 2.79 \text{ L}$$

The new volume (2.79 L) is greater than the initial volume (2.58 L), as we expected.

SELF-CHECK EXERCISE 13.3

A child blows a bubble that contains air at 28 °C and has a volume of 23 cm^3 at 1 atm. As the bubble rises, it encounters a pocket of cold air (temperature 18 °C). If there is no change in pressure, will the bubble get larger or smaller as the air inside cools to 18 °C? Calculate the new volume of the bubble.

Notice from Example 13.5 that we adjust the volume of a gas for a temperature change by multiplying the original volume by the ratio of the kelvin temperatures—final (T_2) over initial (T_1). Remember to check whether your answer makes sense. When the temperature increases (at constant pressure), the volume must increase, and vice versa.

EXAMPLE 13.6   **Calculating Temperature Using Charles's Law**

In former times, gas volume was used as a way to measure temperature using devices called gas thermometers. Consider a gas that has a volume of 0.675 L at 35 °C and 1 atm pressure. What is the temperature (in units of °C) of a room where this gas has a volume of 0.535 L at 1 atm pressure?

SOLUTION

The information given in the problem is

Initial Conditions	Final Conditions
$T_1 = 35 \text{ °C} = 35 + 273 = 308 \text{ K}$	$T_2 = ?$
$V_1 = 0.675 \text{ L}$	$V_2 = 0.535 \text{ L}$
$P_1 = 1 \text{ atm}$	$P_2 = 1 \text{ atm}$

The pressure remains constant, so we can use Charles's law in the form

$$\frac{V_1}{T_1} = \frac{V_2}{T_2}$$

EXAMPLE 13.6, CONTINUED

and solve for T_2. First we multiply both sides by T_2.

$$T_2 \times \frac{V_1}{T_1} = \frac{V_2}{T_2} \times T_2 = V_2$$

Next we multiply both sides by T_1.

$$T_1 \times T_2 \times \frac{V_1}{T_1} = T_1 \times V_2$$

This gives

$$T_2 \times V_1 = T_1 \times V_2$$

Now we divide both sides by V_1 (multiply by $1/V_1$):

$$\frac{1}{V_1} \times T_2 \times V_1 = \frac{1}{V_1} \times T_1 \times V_2$$

and obtain

$$T_2 = T_1 \times \frac{V_2}{V_1}$$

We have now isolated T_2 on one side of the equation, and we can do the calculation.

$$T_2 = T_1 \times \frac{V_2}{V_1} = (308 \text{ K}) \times \frac{0.535 \text{ L}}{0.675 \text{ L}} = 244 \text{ K}$$

To convert from units of K to units of °C, we subtract 273 from the Kelvin temperature.

$$T_{°C} = T_K - 273 = 244 - 273 = -29 \text{ °C}$$

The room is very cold.

13.4 Volume and Moles: Avogadro's Law

AIM: To describe the law relating the volume and the number of moles of a sample of gas at constant temperature and pressure, and to do calculations involving this law.

With the introduction of Avogadro's law, all four variables found in gas law problems have been used. From this point on, I construct a table before beginning to solve problems. I show values for all the known variables so that students have an easier time determining which variable the problem asks them to find.

What is the relationship between the volume of a gas and the number of molecules present in the gas sample? Experiments show that when the number of moles of gas is doubled (at constant temperature and pressure), the volume dou-

V

(a)

$2V$

(b)

$3V$

(c)

Figure 13.8
The relationship between volume V and number of moles n. As the number of moles is increased from 1 to 2 (a to b), the volume doubles. When the number of moles is tripled (c), the volume is also tripled. The temperature and pressure remain the same in these cases.

bles. In other words, the volume of a gas is directly proportional to the number of moles if temperature and pressure remain constant. Figure 13.8 illustrates this relationship, which can also be represented by the equation

$$V = an \quad \text{or} \quad \frac{V}{n} = a$$

where V is the volume of the gas, n is the number of moles, and a is the proportionality constant. Note that this equation means that the ratio of V to n is constant as long as the temperature and pressure remain constant. Thus when the number of moles of gas is increased by a factor of 5, the volume also increases by a factor of 5:

$$\frac{V}{n} = \frac{5 \times V}{5 \times n} = a = \text{constant}$$

and so on. In words, this equation means that *for a gas at constant temperature and pressure, the volume is directly proportional to the number of moles of gas.* This relationship is called **Avogadro's law** after the Italian scientist Amadeo Avogadro, who first postulated it in 1811.

For cases where the number of moles of gas is changed from an initial amount to another amount (at constant T and P), we can represent Avogadro's law as

$$\underbrace{\frac{V_1}{n_1}}_{\substack{\uparrow \\ \text{Initial} \\ \text{amount}}} = a = \underbrace{\frac{V_2}{n_2}}_{\substack{\uparrow \\ \text{Final} \\ \text{amount}}}$$

The use of diagrams such as those in Figure 13.8 can make explaining the gas laws much easier.

or

$$\frac{V_1}{n_1} = \frac{V_2}{n_2}$$

We will illustrate the use of this equation in Example 13.7

For Avogadro's law problems, temperature and pressure are constant. When I work example problems, I provide both initial and final values for both temperature and pressure so that I can point out that the values don't change.

EXAMPLE 13.7   Using Avogadro's Law in Calculations

Suppose we have a 12.2-L sample containing 0.50 mol of oxygen gas, O_2, at a pressure of 1 atm and a temperature of 25 °C. If all of this O_2 is converted to ozone, O_3, at the same temperature and pressure, what will be the volume of the ozone formed?

SOLUTION

To do this problem we need to compare the moles of gas originally present to the moles of gas present after the reaction. We know that 0.50 mol of O_2 is present initially. To find out how many moles of O_3 will be present after the reaction, we need to use the balanced equation for the reaction.

$$3O_2(g) \rightarrow 2O_3(g)$$

We calculate the moles of O_3 produced by using the appropriate mole ratio from the balanced equation.

$$0.50 \text{ mol } O_2 \times \frac{2 \text{ mol } O_3}{3 \text{ mol } O_2} = 0.33 \text{ mol } O_3$$

Avogadro's law states that

$$\frac{V_1}{n_1} = \frac{V_2}{n_2}$$

where V_1 is the volume of n_1 moles of O_2 gas and V_2 is the volume of n_2 moles of O_3 gas. In this case we have

Initial Conditions	Final Conditions
$n_1 = 0.50$ mol	$n_2 = 0.33$ mol
$V_1 = 12.2$ L	$V_2 = ?$

Solving Avogadro's law for V_2 gives

$$\frac{V_1}{n_1} = \frac{V_2}{n_2}$$

$$n_2 \times \frac{V_1}{n_1} = \frac{V_2}{n_2} \times n_2$$

$$V_1 \times \frac{n_2}{n_1} = V_2$$

$$V_2 = V_1 \times \frac{n_2}{n_1} = 12.2 \text{ L} \left(\frac{0.33 \text{ mol}}{0.50 \text{ mol}}\right) = 8.1 \text{ L}$$

Note that the volume decreases, as it should, because fewer molecules are present in the gas after O_2 is converted to O_3.

SELF-CHECK EXERCISE 13.4

Consider two samples of nitrogen gas (composed of N_2 molecules). Sample 1 contains 1.5 mol of N_2 and has a volume of 36.7 L at 25 °C and 1 atm. Sample 2 has a volume of 16.5 L at 25 °C and 1 atm. Calculate the number of moles of N_2 in Sample 2.

13.5   The Ideal Gas Law

AIM: To define the ideal gas law and to use it in calculations.

We have considered three laws that describe the behavior of gases as it is revealed by experimental observations.

Boyle's law: $PV = k$ or $V = \dfrac{k}{P}$ (at constant T and n)

Charles's law: $V = bT$ (at constant P and n)

Avogadro's law: $V = an$ (at constant T and P)

Constant n means a constant number of moles of gas.

These relationships, which show how the volume of a gas depends on pressure, temperature, and number of moles of gas present, can be combined as follows:

$$V = R\left(\frac{Tn}{P}\right)$$

where R is the combined proportionality constant and is called the **universal gas constant.** When the pressure is expressed in atmospheres and the volume in liters, R always has the value 0.08206 L atm/K mol. We can rearrange the above equation by multiplying both sides by P:

$$P \times V = \cancel{P} \times R\left(\frac{Tn}{\cancel{P}}\right)$$

to obtain the **ideal gas law** written in its usual form,

$$\boxed{PV = nRT}$$

Remind students that when they solve problems that use the universal gas constant, R, the units of all variables in the problem must be converted to the units in which R is expressed. When you work examples, show how three units of R cancel, leaving the answer expressed in the desired units.

$$R = 0.08206\frac{\text{L atm}}{\text{K mol}}$$

The ideal gas law involves all the important characteristics of a gas: its pressure (P), volume (V), number of moles (n), and temperature (T) in kelvins. Knowledge of any three of these properties is enough to define completely the condition of the gas, because the fourth property can be determined from the ideal gas law.

It is important to recognize that the ideal gas law is based on experimental measurements of the properties of gases. A gas that obeys this equation is said to behave *ideally.* That is, this equation defines the behavior of an **ideal gas.** Most gases obey this equation closely at pressures of approximately 1 atm or lower, when the temperature is approximately 0 °C or higher. You should assume ideal gas behavior when working problems involving gases in this text.

The ideal gas law can be used to solve a variety of problems. Example 13.8 demonstrates one type, where you are asked to find one property characterizing the condition of a gas given the other three properties.

You can tell students that the answers to problems that are calculated with the ideal gas law may or may not be close to the experimental value actually measured in a lab. The ideal gas law does not predict the behavior of all gases.

EXAMPLE 13.8   Using the Ideal Gas Law in Calculations

A sample of hydrogen gas, H_2, has a volume of 8.56 L at a temperature of 0 °C and a pressure of 1.5 atm. Calculate the number of moles of H_2 present in this gas sample. (Assume that the gas behaves ideally.)

SOLUTION

In this problem we are given the pressure, volume, and temperature of the gas: $P = 1.5$ atm, $V = 8.56$ L, and $T = 0$ °C. Remember that the temperature must be changed to the Kelvin scale.

$$T = 0 \,°C = 0 + 273 = 273 \text{ K}$$

We can calculate the number of moles of gas present by using the ideal gas law, $PV = nRT$. We solve for n by dividing both sides by RT:

$$\frac{PV}{RT} = n\frac{RT}{RT}$$

to give

$$\frac{PV}{RT} = n$$

Thus

$$n = \frac{PV}{RT} = \frac{(1.5 \text{ atm}) (8.56 \text{ L})}{\left(0.08206 \dfrac{\text{L atm}}{\text{K mol}}\right)(273 \text{ K})} = 0.57 \text{ mol}$$

Students can have a difficult time rearranging the ideal gas law to isolate the required variable on one side. By continuing to remind them about the process to follow, you can help strengthen that skill.

SELF-CHECK EXERCISE 13.5

A weather balloon contains 1.10×10^5 mol of helium and has a volume of 2.70×10^6 L at 1.00 atm pressure. Calculate the temperature of the helium in the balloon in kelvins and in Celsius degrees.

EXAMPLE 13.9   Ideal Gas Law Calculations Involving Conversion of Units

What volume is occupied by 0.250 mol of carbon dioxide gas at 25 °C and 371 torr?

SOLUTION

We can use the ideal gas law to calculate the volume, but we must first convert pressure to atmosphere and temperature to the Kelvin scale.

EXAMPLE 13.9, CONTINUED

$$P = 371 \text{ torr} = 371 \text{ torr} \times \frac{1.000 \text{ atm}}{760.0 \text{ torr}} = 0.488 \text{ atm}$$

$$T = 25 \text{ °C} = 25 + 273 = 298 \text{ K}$$

We solve for V by dividing both sides of the ideal gas law ($PV = nRT$) by P.

$$V = \frac{nRT}{P} = \frac{(0.250 \text{ mol})\left(0.08206 \frac{\text{L atm}}{\text{K mol}}\right)(298 \text{ K})}{0.488 \text{ atm}} = 12.5 \text{ L}$$

$$PV = nRT$$
$$\frac{PV}{P} = \frac{nRT}{P}$$
$$V = \frac{nRT}{P}$$

The volume of the sample of CO_2 is 12.5 L.

SELF-CHECK EXERCISE 13.6

Radon, a radioactive gas formed naturally in the soil, can cause lung cancer. It can pose a hazard to humans by seeping into houses, and there is concern about this problem in many areas of the country. A 1.5-mol sample of radon gas has a volume of 21.0 L at 33 °C. What is the pressure of the gas?

Note that R has units of L atm/K mol. Accordingly, whenever we use the ideal gas law, we must express the volume in units of liters, the temperature in kelvins, and the pressure in atmospheres. When we are given data in other units, we must first convert to the appropriate units.

The ideal gas law can also be used to calculate the changes that will occur when the conditions of the gas are changed.

EXAMPLE 13.10   **Using the Ideal Gas Law Under Changing Conditions**

Suppose we have a 0.240-mol sample of ammonia gas at 25 °C with a volume of 3.5 L at a pressure of 1.68 atm. The gas is compressed to a volume of 1.35 L at 25 °C. Use the ideal gas law to calculate the final pressure.

SOLUTION

In this case we have a sample of ammonia gas in which the conditions are changed. We are given the following information:

Students are encouraged to use the ideal gas law to solve most gas law problems, even when conditions are changing. Boyle's, Charles', and Avogadro's laws can all be derived from the ideal gas law. If students can derive the necessary formula from the ideal gas law, they will not need to memorize many different formulas.

Initial Conditions	Final Conditions
$V_1 = 3.5 \text{ L}$	$V_2 = 1.35 \text{ L}$
$P_1 = 1.68 \text{ atm}$	$P_2 = ?$
$T_1 = 25 \text{ °C} = 25 + 273 = 298 \text{ K}$	$T_2 = 25 \text{ °C} = 25 + 273 = 298 \text{ K}$
$n_1 = 0.240 \text{ mol}$	$n_2 = 0.240 \text{ mol}$

EXAMPLE 13.10, continued

Note that both n and T remain constant—only P and V change. Thus we could simply use Boyle's law ($P_1 V_1 = P_2 V_2$) to solve for P_2. However, we will use the ideal gas law to solve this problem in order to introduce the idea that one equation—the ideal gas equation—can be used to do almost any gas problem. The key idea here is that in using the ideal gas law to describe a change in conditions for a gas, we always *solve the ideal gas equation in such a way that the variables that change are on one side of the equals sign and the constant terms are on the other side.* That is, we start with the ideal gas equation in the conventional form ($PV = nRT$) and rearrange it so that all the terms that change are moved to one side and all the terms that do not change are moved to the other side. In this case the pressure and volume change, and the temperature and number of moles remain constant (as does R, by definition). So we write the ideal gas law as

$$\underbrace{PV}_{\uparrow \atop \text{Change}} = \underbrace{nRT}_{\uparrow \atop \text{Remain constant}}$$

Because nR, and T remain the same in this case, we can write $P_1 V_1 = nRT$ and $P_2 V_2 = nRT$. Combining these gives

$$P_1 V_1 = nRT = P_2 V_2 \quad \text{or} \quad P_1 V_1 = P_2 V_2$$

and

$$P_2 = P_1 \times \frac{V_1}{V_2} = (1.68 \text{ atm}) \left(\frac{3.5 \text{ L}}{1.35 \text{ L}} \right) = 4.4 \text{ atm}$$

CHECK: Does this answer make sense? The volume was decreased (at constant temperature and constant number of moles), which means that the pressure should increase, as the calculation indicates.

SELF-CHECK EXERCISE 13.7

A sample of methane gas that has a volume of 3.8 L at 5 °C is heated to 86 °C at constant pressure. Calculate its new volume.

Note that in solving Example 13.10, we actually obtained Boyle's law ($P_1 V_1 = P_2 V_2$) from the ideal gas equation. You might well ask, "Why go to all this trouble?" The idea is to learn to use the ideal gas equation to solve all types of gas law problems. This way you will never have to ask yourself, "Is this a Boyle's law problem or a Charles's law problem?"

We continue to practice using the ideal gas law in Example 13.11. Remember, the key idea is to rearrange the equation so that the quantities that change are moved to one side of the equation and those that remain constant are moved to the other.

EXAMPLE 13.11   Calculating Volume Changes Using the Ideal Gas Law

A sample of diborane gas, B_2H_6, a substance that bursts into flame when exposed to air, has a pressure of 0.454 atm at a temperature of -15 °C and a volume of 3.48 L. If conditions are changed so that the temperature is 36 °C and the pressure is 0.616 atm, what will be the new volume of the sample?

SOLUTION

We are given the following information:

Initial Conditions	Final Conditions
P_1 = 0.454 atm	P_2 = 0.616 atm
V_1 = 3.48 L	V_2 = ?
T_1 = -15 °C = $273 - 15 = 258$ K	T_2 = 36 °C = $273 + 36 = 309$ K

Note that the value of n is not given. However, we know that n is constant (that is, $n_1 = n_2$) because no diborane gas is added or taken away. Thus in this experiment, n is constant and P, V, and T change. Therefore we rearrange the ideal gas equation ($PV = nRT$) by dividing both sides by T:

$$\frac{PV}{T} = nR$$

$PV = nRT$

$\dfrac{PV}{T} = \dfrac{nRT}{T}$

$\dfrac{PV}{T} = nR$

Change   Constant

which leads to the equation

$$\frac{P_1V_1}{T_1} = nR = \frac{P_2V_2}{T_2}$$

or

$$\frac{P_1V_1}{T_1} = \frac{P_2V_2}{T_2}$$

We can now solve for V_2 by dividing both sides by P_2 and multiplying both sides by T_2.

$$\frac{1}{P_2} \times \frac{P_1V_1}{T_1} = \frac{P_2V_2}{T_2} \times \frac{1}{P_2} = \frac{V_2}{T_2}$$

$$T_2 \times \frac{P_1V_1}{P_2T_1} = \frac{V_2}{T_2} \times T_2 = V_2$$

That is,

$$\frac{T_2P_1V_1}{P_2T_1} = V_2$$

EXAMPLE 13.11, CONTINUED

It is sometimes convenient to think in terms of the ratios of the initial temperature and pressure and the final temperature and pressure. That is,

$$V_2 = \frac{T_2 P_1 V_1}{T_1 P_2} = V_1 \times \frac{T_2}{T_1} \times \frac{P_1}{P_2}$$

Substituting the information given yields

$$V_2 = \frac{309 \text{ K}}{258 \text{ K}} \times \frac{0.454 \text{ atm}}{0.616 \text{ atm}} \times 3.48 \text{ L} = 3.07 \text{ L}$$

Always convert the temperature to the Kelvin scale and the pressure to atmospheres when applying the ideal gas law.

SELF-CHECK EXERCISE 13.8

A sample of argon gas with a volume of 11.0 L at a temperature of 13 °C and a pressure of 0.747 atm is heated to 56 °C and a pressure of 1.18 atm. Calculate the final volume.

The equation obtained in Example 13.11,

$$\frac{P_1 V_1}{T_1} = \frac{P_2 V_2}{T_2}$$

is often called the **combined gas law** equation. It holds when the amount of gas (moles) is held constant. While it may be convenient to remember this equation, it is not necessary because you can always use the ideal gas equation.

13.6 Dalton's Law of Partial Pressures

Dalton's law of partial pressures is conceptually very easy. But students have a difficult time applying the law to problems.

AIM: To state the relationship between the partial and total pressures of a gas mixture, and to use this relationship in calculations.

Many important gases contain a mixture of components. One notable example is air. Scuba divers use another important mixture, helium and oxygen. Normal air is not used, because the nitrogen present dissolves in the blood in large quantities as a result of the high pressures experienced by the diver under several hundred feet of water. When the diver returns too quickly to the surface, the nitrogen bubbles out of the blood just as soda fizzes when it's opened, and the diver gets "the bends"—a very painful and potentially fatal condition. Because helium gas is only sparingly soluble in blood, it does not cause this problem.

A diver's voice sounds something like Donald Duck because of the effects of the helium (with a much smaller density than air) surrounding the vocal cords.

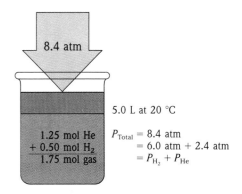

Figure 13.9
When two gases are present, the total pressure is the sum of the partial pressures of the gases.

Studies of gaseous mixtures show that each component behaves independently of the others. In other words, a given amount of oxygen exerts the same pressure in a 1.0-L vessel whether it is alone or in the presence of nitrogen (as in the air) or helium.

Among the first scientists to study mixtures of gases was John Dalton. In 1803 Dalton summarized his observations in this statement: *For a mixture of gases in a container, the total pressure exerted is the sum of the partial pressures of the gases present. The* **partial pressure** *of a gas is the pressure that the gas would exert if it were alone in the container.* This statement, known as **Dalton's law of partial pressures,** can be expressed as follows for a mixture containing three gases:

$$P_{total} = P_1 + P_2 + P_3$$

where the subscripts refer to the individual gases (gas 1, gas 2, and gas 3). The pressures P_1, P_2, and P_3 are the partial pressures; that is, each gas is responsible for only part of the total pressure (Figure 13.9).

Assuming that each gas behaves ideally, we can calculate the partial pressure of each gas from the ideal gas law:

$$P_1 = \frac{n_1 RT}{V}, \qquad P_2 = \frac{n_2 RT}{V}, \qquad P_3 = \frac{n_3 RT}{V}$$

The total pressure of the mixture, P_{total}, can be represented as

$$
\begin{aligned}
P_{total} = P_1 + P_2 + P_3 &= \frac{n_1 RT}{V} + \frac{n_2 RT}{V} + \frac{n_3 RT}{V} \\
&= n_1\left(\frac{RT}{V}\right) + n_2\left(\frac{RT}{V}\right) + n_3\left(\frac{RT}{V}\right) \\
&= (n_1 + n_2 + n_3)\left(\frac{RT}{V}\right) \\
&= n_{total}\left(\frac{RT}{V}\right)
\end{aligned}
$$

$$PV = nRT$$

$$\frac{PV}{V} = \frac{nRT}{V}$$

$$P = \frac{nRT}{V}$$

Figure 13.10
The total pressure of a mixture of gases depends on the number of moles of gas particles (atoms or molecules) present, not on the identities of the particles. Note that these three samples show the same total pressure because each contains 1.75 mol of gas. The detailed nature of the mixture is unimportant.

where n_{total} is the sum of the numbers of moles of the gases in the mixture. Thus for a mixture of ideal gases, it is the *total number of moles of particles* that is important, not the *identity* of the individual gas particles. This idea is illustrated in Figure 13.10.

The fact that the pressure exerted by an ideal gas is affected by the number of gas particles and is independent of the nature of the gas particles tells us two important things about ideal gases.

1. The volume of the individual gas particle (atom or molecule) must not be very important.
2. The forces among the particles must not be very important.

If these factors were important, the pressure of the gas would depend on the nature of the individual particles. For example, an argon atom is much larger than a helium atom. Yet 1.75 mol of argon gas in a 5.0-L container at 20 °C exerts the same pressure as 1.75 mol of helium gas in a 5.0-L container at 20 °C.

The same idea applies to the forces among the particles. Although the forces among gas particles depend on the nature of the particles, this seems to have little influence on the behavior of an ideal gas. We will see that these observations strongly influence the model that we will construct to explain ideal gas behavior.

Problems such as Example 13.12 are hard for students because two separate calculations are required before P_{total} can be determined. When you work a similar problem, explain that first the number of moles of each gas must be determined and then the partial pressures.

EXAMPLE 13.12   Using Dalton's Law of Partial Pressures, I

Mixtures of helium and oxygen are used in the "air" tanks of underwater divers for deep dives. For a particular dive, 46 L of O_2 at 25 °C and 1.0 atm and 12 L of

EXAMPLE 13.12, CONTINUED

He at 25 °C and 1.0 atm were both pumped into a 5.0-L tank. Calculate the partial pressure of each gas and the total pressure in the tank at 25 °C.

SOLUTION

Because the partial pressure of each gas depends on the moles of that gas present, we must first calculate the number of moles of each gas by using the ideal gas law in the form

$$n = \frac{PV}{RT}$$

From the above description we know that $P = 1.0$ atm, $V = 46$ L for O_2 and 12 L for He, and $T = 25 + 273 = 298$ K. Also, $R = 0.08206$ L atm/K mol (as always).

$$\text{Moles of } O_2 = n_{O_2} = \frac{(1.0 \text{ atm})(46 \text{ L})}{(0.08206 \text{ L atm/K mol})(298 \text{ K})} = 1.9 \text{ mol}$$

$$\text{Moles of He} = n_{He} = \frac{(1.0 \text{ atm})(12 \text{ L})}{(0.08206 \text{ L atm/K mol})(298 \text{ K})} = 0.49 \text{ mol}$$

The tank containing the mixture has a volume of 5.0 L, and the temperature is 25 °C (298 K). We can use these data and the ideal gas law to calculate the partial pressure of each gas.

$$P = \frac{nRT}{V}$$

$$P_{O_2} = \frac{(1.9 \text{ mol})(0.08206 \text{ L atm/K mol})(298 \text{ K})}{5.0 \text{ L}} = 9.3 \text{ atm}$$

$$P_{He} = \frac{(0.49 \text{ mol})(0.08206 \text{ L atm/K mol})(298 \text{ K})}{5.0 \text{ L}} = 2.4 \text{ atm}$$

The total pressure is the sum of the partial pressures.

$$P_{total} = P_{O_2} + P_{He} = 9.3 \text{ atm} + 2.4 \text{ atm} = 11.7 \text{ atm}$$

Divers use a mixture of oxygen and helium in their breathing tanks for dives to depths greater than 150 feet.

$$PV = nRT$$

$$\frac{PV}{RT} = \frac{nR\cancel{T}}{R\cancel{T}}$$

$$\frac{PV}{RT} = n$$

SELF-CHECK EXERCISE 13.9

A 2.0-L flask contains a mixture of nitrogen gas and oxygen gas at 25 °C. The total pressure of the gaseous mixture is 0.91 atm, and the mixture is known to contain 0.050 mol of N_2. Calculate the partial pressure of oxygen and the moles of oxygen present.

KCIO₃ (with a
small amount of MnO₂)

Oxygen plus
water vapor

Figure 13.11

The production of oxygen by thermal decomposition of $KClO_3$. MnO_2 is mixed with the $KClO_3$ to make the reaction proceed at a more rapid rate.

The calculation of the pressure of a gas collected over water provides a good test of students' understanding of Dalton's law of partial pressures.

Table 13.2	The Vapor Pressure of Water as a Function of Temperature	
T (°C)	P (torr)
0.0	4.579
10.0	9.209
20.0	17.535
25.0	23.756
30.0	31.824
40.0	55.324
60.0	149.4
70.0	233.7
90.0	525.8

A mixture of gases occurs whenever a gas is collected by displacement of water. For example, Figure 13.11 shows the collection of oxygen gas that was produced by the decomposition of solid potassium chlorate. The gas is collected by bubbling it into a bottle that is initially filled with water. Thus the gas in the bottle is really a mixture of water vapor and oxygen. (Water vapor is present because molecules of water escape from the surface of the liquid and collect as a gas in the space above the liquid.) Therefore, the total pressure exerted by this mixture is the sum of the partial pressure of the gas being collected and the partial pressure of the water vapor. The partial pressure of the water vapor is called the vapor pressure of water. Because water molecules are more likely to escape from hot water than from cold water, the vapor pressure of water increases with temperature. This is shown by the values of vapor pressure at various temperatures shown in Table 13.2.

EXAMPLE 13.13   Using Dalton's Law of Partial Pressures, II

A sample of solid potassium chlorate, $KClO_3$, was heated in a test tube (see Figure 13.11) and decomposed according to the reaction

$$2KClO_3(s) \rightarrow 2KCl(s) + 3O_2(g)$$

The oxygen produced was collected by displacement of water at 22 °C. The resulting mixture of O_2 and H_2O vapor had a total pressure of 754 torr and a

EXAMPLE 13.13, CONTINUED

volume of 0.65 L. Calculate the partial pressure of O_2 in the gas collected and the number of moles of O_2 present. The vapor pressure of water at 22 °C is 21 torr.

SOLUTION

We know the total pressure (754 torr) and the partial pressure of water (vapor pressure = 21 torr). We can find the partial pressure of O_2 from Dalton's law of partial pressures:

$$P_{total} = P_{O_2} + P_{H_2O} = P_{O_2} + 21 \text{ torr} = 754 \text{ torr}$$

or

$$P_{O_2} + 21 \text{ torr} = 754 \text{ torr}$$

We can solve for P_{O_2} by subtracting 21 torr from both sides of the equation.

$$P_{O_2} = 754 \text{ torr} - 21 \text{ torr} = 733 \text{ torr}$$

Next we solve the ideal gas law for the number of moles of O_2.

$$n_{O_2} = \frac{P_{O_2}V}{RT}$$

In this case, $P_{O_2} = 733$ torr. We change the pressure to atmospheres as follows:

$$\frac{733 \text{ torr}}{760 \text{ torr/atm}} = 0.964 \text{ atm}$$

Then,

$$V = 0.650 \text{ L}$$
$$T = 22 °C = 22 + 273 = 295 \text{ K}$$
$$R = 0.08206 \text{ L atm/K mol}$$

so

$$n_{O_2} = \frac{(0.964 \text{ atm})(0.650 \text{ L})}{(0.08206 \text{ L atm/K mol})(295 \text{ K})} = 2.59 \times 10^{-2} \text{ mol}$$

$PV = nRT$

$\frac{PV}{RT} = \frac{nRT}{RT}$

$\frac{PV}{RT} = n$

SELF-CHECK EXERCISE 13.10

Consider a sample of hydrogen gas collected over water at 25 °C where the vapor pressure of water is 24 torr. The volume occupied by the gaseous mixture is 0.500 L, and the total pressure is 0.950 atm. Calculate the partial pressure of H_2 and the number of moles of H_2 present.

Now is a good time to discuss the distinction between theories and models. The gas laws allow students to predict the behavior of gases, whereas the kinetic molecular theory attempts to explain the behavior of gases.

 Laws and Models: A Review

AIM: To discuss the relationship between laws and models (theories).

In this chapter we have considered several properties of gases and have seen how the relationships among these properties can be expressed by various laws written in the form of mathematical equations. The most useful of these is the ideal gas equation, which relates all the important gas properties. However, under certain conditions gases do not obey the ideal gas equation. For example, at high pressures and/or low temperatures, the properties of gases deviate significantly from the predictions of the ideal gas equation. On the other hand, as the pressure is lowered and/or the temperature is increased, almost all gases show close agreement with the ideal gas equation. This means that an ideal gas is really a hypothetical substance. At low pressures and/or high temperatures, real gases *approach* the behavior expected for an ideal gas.

At this point we want to build a model (a theory) to explain *why* a gas behaves as it does. We want to answer the question *What are the characteristics of the individual gas particles that cause a gas to behave as it does?* However, before we do this let's briefly review the scientific method. Recall that a law is a generalization about behavior that has been observed in many experiments. Laws are very useful; they allow us to predict the behavior of similar systems. For example, a chemist who prepares a new gaseous compound can assume that that substance will obey the ideal gas equation (at least at low P and/or high T).

However, laws do not tell us *why* nature behaves the way it does. Scientists try to answer this question by constructing theories (building models). The models in chemistry are speculations about how individual atoms or molecules (microscopic particles) cause the behavior of macroscopic systems (collections of atoms and molecules in large enough numbers so that we can observe them).

A model is considered successful if it explains known behavior and predicts correctly the results of future experiments. But a model can never be proved absolutely true. In fact, by its very nature *any model is an approximation* and is destined to be modified, at least in part. Models range from the simple (to predict approximate behavior) to the extraordinarily complex (to account precisely for observed behavior). In this text, we use relatively simple models that fit most experimental results.

 The Kinetic Molecular Theory of Gases

AIM: To present the basic postulates of the kinetic molecular theory.

A relatively simple model that attempts to explain the behavior of an ideal gas is the **kinetic molecular theory.** This model is based on speculations about the behavior of the individual particles (atoms or molecules) in a gas. The assumptions (postulates) of the kinetic molecular theory can be stated as follows:

Postulates of the Kinetic Molecular Theory of Gases

1. Gases consist of tiny particles (atoms or molecules).
2. These particles are so small, compared to the distances between them, that the volume (size) of the individual particles can be assumed to be negligible (zero).
3. The particles are in constant random motion, colliding with the walls of the container. These collisions with the walls cause the pressure exerted by the gas.
4. The particles are assumed not to attract or to repel each other.
5. The average kinetic energy of the gas particles is directly proportional to the Kelvin temperature of the gas.

Students should probably memorize the postulates of the kinetic molecular theory. They will have difficulty understanding how the KM theory explains observed behavior if they don't have the postulates in mind as they study this material.

Kinetic~ of or pertaining *relating* to the motion of material bodies + the associated forces + energy.

   The kinetic energy referred to in postulate 5 is the energy associated with the motion of a particle. Kinetic energy (KE) is given by the equation $KE = \frac{1}{2}mv^2$, where m is the mass of the particle and v is the velocity (speed) of the particle. The greater the mass or velocity of a particle, the greater its kinetic energy. Postulate 5 means that if a gas is heated to higher temperatures, the average speed of the particles increases; therefore their kinetic energy increases.
   Although real gases do not conform exactly to the five assumptions listed above, we will see in the next section that these postulates do indeed explain *ideal* gas behavior—behavior shown by real gases at high temperatures and/or low pressures.

13.9 The Implications of the Kinetic Molecular Theory

AIM: To define the term *temperature.*
      To show how the kinetic molecular theory explains the gas laws.

Because models are useful only as long as they work, you can tell students that in the event that some part of the KM theory had not explained the behavior of gases observed by the gas laws, then some or all of the theory would have been discarded or revised.

In this section we will discuss the *qualitative* relationships between the kinetic molecular (KM) theory and the properties of gases. That is, without going into the mathematical details, we will show how the kinetic molecular theory explains some of the observed properties of gases.

The Meaning of Temperature

In Chapter 2 we introduced temperature very practically as something we measure with a thermometer. We know that as the temperature of an object increases, the object feels "hotter" to the touch. But what does temperature really mean? How does matter change when it gets "hotter"? The kinetic molecular theory allows us to answer this very important question. As postulate 5 of the KM theory states, the temperature of a gas reflects how rapidly, on average, its individual gas particles are moving. At high temperatures the particles move very fast and hit the walls of the container frequently, whereas at low temperatures the particles' motions are more sluggish and they collide with the walls of the container much less often. Therefore, temperature really is a measure of the motions of the gas particles. In fact, the Kelvin temperature of a gas is directly proportional to the average kinetic energy of the gas particles.

The Relationship Between Pressure and Temperature

To see how the meaning of temperature given above helps to explain gas behavior, picture a gas in a rigid container. As the gas is heated to a higher temperature, the particles move faster, hitting the walls more often. And, of course, the impacts become more forceful as the particles move faster. If the pressure is due to collisions with the walls, the gas pressure should increase as temperature is increased.

Is this what we observe when we measure the pressure of a gas as it is heated? Yes. A given sample of gas in a rigid container (if the volume is not changed) shows an increase in pressure as its temperature is increased.

The Relationship Between Volume and Temperature

Diagrams showing gas particles in containers can be effectively used to illustrate all three parts of this section.

Now picture the gas in a container with a movable piston. As shown in Figure 13.12(a), the gas pressure P_{gas} is just balanced by an external pressure P_{ext}. What

Figure 13.12

(a) A gas confined in a cylinder with a movable piston. The gas pressure P_{gas} is just balanced by the external pressure P_{ext}. That is, $P_{gas} = P_{ext}$. (b) The temperature of the gas is increased at constant pressure P_{ext}. The increased particle motions at the higher temperature push back the piston, increasing the volume of the gas.

happens when we heat the gas to a higher temperature? As the temperature increases, the particles move faster, causing the gas pressure to increase. As soon as the gas pressure P_{gas} becomes greater than P_{ext} (the pressure holding the piston), the piston moves up until $P_{gas} = P_{ext}$. Therefore, the KM model predicts that the volume of the gas will increase as we raise its temperature at a constant pressure. This agrees with experimental observations (as summarized by Charles's law).

EXAMPLE 13.14   **Using the Kinetic Molecular Theory to Explain Gas Law Observations**

Use the KM theory to predict what will happen to the pressure of a gas when its volume is decreased (n and T constant). Does this prediction agree with the experimental observations?

SOLUTION

When we decrease the gas's volume (make the container smaller), the particles hit the walls more often because they do not have to travel so far between the walls. This would suggest an increase in pressure. This prediction on the basis of the model is in agreement with experimental observations of gas behavior (as summarized by Boyle's law).

In this section we have seen that the predictions of the kinetic molecular theory generally fit the behavior observed for gases. This makes it a useful and successful model.

13.10 Gas Stoichiometry

AIM: To define the molar volume of an ideal gas.
To define STP.
To use these concepts and the ideal gas equation.

We have seen repeatedly in this chapter just how useful the ideal gas equation is. For example, if we know the pressure, volume, and temperature for a given sample of gas, we can calculate the number of moles present: $n = PV/RT$. This fact makes it possible to do stoichiometric calculations for reactions involving gases. We will illustrate this process in Example 13.15.

A brief review of the principles of stoichiometry is usually necessary before beginning a discussion of the variations that gases introduce.

EXAMPLE 13.15   **Gas Stoichiometry: Calculating Volume**

Calculate the volume of oxygen gas produced at 1.00 atm and 25 °C by the complete decomposition of 10.5 g of potassium chlorate. The balanced equation for the reaction is

$$2KClO_3(s) \rightarrow 2KCl(s) + 3O_2(g)$$

SOLUTION

This is a stoichiometry problem very much like those we considered in Chapter 10. The only difference is that in this case, we want to calculate the volume of a gaseous product rather than the number of grams. To do so, we can use the relationship between moles and volume given by the ideal gas law.

We'll summarize the steps required to do this problem in the following schematic:

Students have been accustomed to using the molar mass to calculate the moles of a substance. They must now learn to use the ideal gas law to calculate the moles of a substance.

$$\frac{10.5}{122.6} = 0.085644$$

0.085644 ⟹ Round off ⟹ 0.0856

$0.0856 = 8.56 \times 10^{-2}$

STEP 1
To find the moles of $KClO_3$ in 10.5 g, we use the molar mass of $KClO_3$ (122.6 g).

$$10.5 \text{ g } KClO_3 \times \frac{1 \text{ mol } KClO_3}{122.6 \text{ g } KClO_3} = 8.56 \times 10^{-2} \text{ mol } KClO_3$$

STEP 2
To find the moles of O_2 produced, we use the mole ratio of O_2 to $KClO_3$ derived from the balanced equation.

$$8.56 \times 10^{-2} \text{ mol } KClO_3 \times \frac{3 \text{ mol } O_2}{2 \text{ mol } KClO_3} = 1.28 \times 10^{-1} \text{ mol } O_2$$

STEP 3
To find the volume of oxygen produced, we use the ideal gas law $PV = nRT$, where

$P = 1.00$ atm
$V = ?$
$n = 1.28 \times 10^{-1}$ mol, the moles of O_2 we calculated
$R = 0.08206$ L atm/K mol
$T = 25 °C = 25 + 273 = 298$ K

EXAMPLE 13.15, CONTINUED

Solving the ideal gas law for V gives

$$V = \frac{nRT}{P} = \frac{(1.28 \times 10^{-1} \text{ mol})\left(0.08206\frac{\text{L atm}}{\text{K mol}}\right)(298 \text{ K})}{1.00 \text{ atm}} = 3.13 \text{ L}$$

Thus 3.13 L of O_2 will be produced at 1.00 atm and 25 °C from 10.5 g of $KClO_3$.

SELF-CHECK EXERCISE 13.11

Calculate the volume of hydrogen produced at 1.50 atm and 19 °C by the reaction of 26.5 g of zinc with excess hydrochloric acid according to the balanced equation

$$Zn(s) + 2HCl(aq) \rightarrow ZnCl_2(aq) + H_2(g)$$

In dealing with the stoichiometry of reactions involving gases, it is useful to define the volume occupied by 1 mol of a gas under certain specified conditions. For 1 mol of an ideal gas at 0 °C (273 K) and 1 atm, the volume of the gas given by the ideal gas law is

$$V = \frac{nRT}{P} = \frac{(1.00 \text{ mol})(0.08206 \text{ L atm/K mol})(273 \text{ K})}{1.00 \text{ atm}} = 22.4 \text{ L}$$

This volume of 22.4 L is called the **molar volume** of an ideal gas.

The conditions 0 °C and 1 atm are called **standard temperature and pressure** (abbreviated **STP**). Properties of gases are often given under these conditions. Remember, the molar volume of an ideal gas is 22.4 L *at STP.* That is, 22.4 L contains 1 mol of an ideal gas at STP.

STP: 0 °C and 1 atm

After students have learned about molar volume at STP, 22.4 L, they will tend to use this number for all temperature and pressure conditions. I usually show how gas problems at STP can be solved with either 22.4 L or with the ideal gas law, so that if students are not sure whether molar volume at STP can be safely used, they can always use the ideal gas law and arrive at the correct answer.

EXAMPLE 13.16	**Gas Stoichiometry: Calculations Involving Gases at STP**

A sample of nitrogen gas has a volume of 1.75 L at STP. How many moles of N_2 are present?

SOLUTION

We could solve this problem by using the ideal gas equation, but we can take a shortcut by using the molar volume of an ideal gas at STP. Because 1 mol of an ideal gas at STP has a volume of 22.4 L, a 1.75 L sample of N_2 at STP contains

EXAMPLE 13.16, CONTINUED

considerably less than 1 mol. We can find how many moles by using the equivalence statement

$$1.000 \text{ mol} = 22.4 \text{ L} \quad (STP)$$

which leads to the conversion factor we need:

$$1.75 \text{ L N}_2 \times \frac{1.000 \text{ mol N}_2}{22.4 \text{ L N}_2} = 7.81 \times 10^{-2} \text{ mol N}_2$$

SELF-CHECK EXERCISE 13.12

Ammonia is commonly used as a fertilizer to provide a source of nitrogen for plants. A sample of $NH_3(g)$ occupies 5.00 L at 25 °C and 15.0 atm. What volume will this sample occupy at STP?

Standard conditions (STP) and molar volume are also useful in carrying out stoichiometric calculations on reactions involving gases, as shown in Example 13.17.

EXAMPLE 13.17	**Gas Stoichiometry: Reactions Involving Gases at STP**

Quicklime, CaO, is produced by heating calcium carbonate, $CaCO_3$. Calculate the volume of CO_2 produced at STP from the decomposition of 152 g of $CaCO_3$ according to the reaction

$$CaCO_3(s) \rightarrow CaO(s) + CO_2(g)$$

SOLUTION

The strategy for solving this problem is summarized by the following schematic:

STEP 1
    Using the molar mass of $CaCO_3$ (100.1 g), we calculate the number of moles of $CaCO_3$.

EXAMPLE 13.17, CONTINUED

$$152 \text{ g } \cancel{CaCO_3} \times \frac{1 \text{ mol CaCO}_3}{100.1 \text{ g } \cancel{CaCO_3}} = 1.52 \text{ mol CaCO}_3$$

STEP 2

Each mole of $CaCO_3$ produces 1 mol of CO_2, so 1.52 mol of CO_2 will be formed.

STEP 3

We can convert the moles of CO_2 to volume by using the molar volume of an ideal gas, because the conditions are STP.

$$1.52 \text{ mol } \cancel{CO_2} \times \frac{22.4 \text{ L CO}_2}{1 \text{ mol } \cancel{CO_2}} = 34.1 \text{ L CO}_2$$

Thus the decomposition of 152 g of $CaCO_3$ produces 34.1 L of CO_2 at STP.

Note that the final step in Example 13.17 involved calculating the volume of gas from the number of moles. Because the conditions were specified as STP, we were able to use the molar volume of a gas at STP. If the conditions of a problem are different from STP, we must use the ideal gas law to compute the volume, as we did in Section 13.5.

Remember that the molar volume of an ideal gas is 22.4 L at STP.

CHAPTER REVIEW

Key Terms

barometer (p. 421)
mm Hg (p. 422)
torr (p. 422)
standard atmosphere (p. 422)
pascal (p. 423)
Boyle's law (p. 425)
absolute zero (p. 429)
Charles's law (p. 430)
Avogadro's law (p. 435)
universal gas constant (p. 437)
ideal gas law (p. 437)
ideal gas (p. 437)
combined gas law (p. 442)

partial pressures (p. 443)
Dalton's law of partial pressures (p. 443)
kinetic molecular theory (p. 449)
molar volume (p. 453)
standard temperature and pressure (STP)
  (p. 453)

Summary

1. Atmospheric pressure is measured with a barometer. The most commonly used units of pressure are mm Hg (torr), atmospheres, and pascals (the SI unit).
2. Boyle's law states that the volume of a given amount of gas is inversely proportional to its pressure (at constant temperature): $PV = k$ or $P = k/V$. That is, as pressure increases, volume decreases.
3. Charles's law states that, for a given amount of gas at constant pressure, the volume is directly proportional to the temperature (in kelvins): $V = bT$. At -273 °C (0 K), the volume of a gas extrapolates to zero, and this temperature is called absolute zero.
4. Avogadro's law states that for a gas at constant temperature and pressure, the volume is directly proportional to the number of moles of gas: $V = an$.
5. These three laws can be combined into the ideal gas law, $PV = nRT$, where R is called the universal gas constant. This equation makes it possible to calculate any one of the properties—volume, pressure, temperature, or moles of gas present—given the other three. A gas that obeys this equation is said to behave ideally.
6. From the ideal gas equation we can derive the combined gas law,

$$\frac{P_1 V_1}{T_1} = \frac{P_2 V_2}{T_2}$$

which holds when the amount of gas (moles) remains constant.
7. The pressure of a gas mixture is described by Dalton's law of partial pressures, which states that the total pressure of the mixture of gases in a container is the sum of the partial pressures of the gases that make up the mixture.
8. The kinetic molecular theory of gases is a model that accounts for ideal gas behavior. This model assumes that a gas consists of tiny particles with negligible volumes, that there are no interactions among particles, and that the particles are in constant motion, colliding with the container walls to produce pressure.

Questions and Problems

All even-numbered exercises have answers in the back of this book and solutions in the Solutions Guide.

13.1 Pressure

QUESTIONS

1. What are the three physical states of matter? Summarize the properties of each of these states. How are the three states of matter similar, and how do they differ?

2. What is meant by "the pressure of the atmosphere"? What causes this pressure? How do we measure atmospheric pressure? Is the atmospheric pressure constant everywhere on the surface of the earth?

3. One standard atmosphere of pressure is equivalent to approximately _____ pascals.
5. A pressure of 760 mm Hg is equivalent to approximately _____ kilopascals.

4. One standard atmosphere of pressure is equivalent to _____ mm Hg.
6. During stormy weather, the atmospheric pressure is "low"; this means a barometer will read _____ than 760 mm Hg.

PROBLEMS

7. Convert the following pressures into atmospheres.
   a. 540 mm Hg
   b. 9.00×10^5 Pa
   c. 890. torr
   d. 133 kPa
9. Convert the following pressures into mm Hg.
   a. 0.403 atm
   b. 103,400 Pa
   c. 205 kPa
   d. 842 torr
11. Convert the following pressures into pascals.
   a. 744 mm Hg
   b. 132 kPa
   c. 0.994 atm
   d. 599 torr

8. Convert the following pressures to units of atmospheres.
   a. 110.2 kPa
   b. 74.2 cm Hg
   c. 441 mm Hg
   d. 921 torr
10. Convert the following pressures to units of mm Hg.
   a. 1.02 atm
   b. 121.4 kPa
   c. 792 torr
   d. 1.09×10^4 Pa
12. Convert the following pressures to units of kilopascals.
   a. 2.07×10^6 Pa
   b. 795 mm Hg
   c. 10.9 atm
   d. 659 torr

13.2 Pressure and Volume: Boyle's Law

QUESTIONS

13. When the pressure on a sample of gas is increased (at constant temperature), the volume of the sample _____.
15. The volume of a sample of ideal gas is _____ proportional to the pressure on it at constant temperature.

14. When the volume of a sample of gas is decreased, the pressure inside the sample of gas _____.
16. A mathematical expression that summarizes Boyle's law is _____.

PROBLEMS

17. For each of the following sets of pressure/volume data, calculate the missing quantity. Assume the temperature and the amount of gas remain constant.
   a. $V = 43.0$ L at 1.04 atm; $V = ?$ at 2.94 atm
   b. $V = 234$ mL at 723 mm Hg; $V = 434$ mL at ? mm Hg
   c. $V = 23.5$ L at 655 torr; $V = ?$ at 1.04 atm
19. For each of the following sets of pressure/volume data, calculate the missing quantity. Assume the temperature and the amount of gas remain constant.
   a. $V = 55$ mL at 190 torr; $V = ?$ at 1.0 atm
   b. $V = 100.$ L at 1.043 kPa; $V = ?$ at 1.0 atm
   c. $V = 245$ mL at 1.0233 atm; $V = ?$ at 385 torr
21. A 4.20-L sample of gas has its pressure decreased from 3.43 atm to 1.29 atm. What does the volume of the gas become?

23. An aerosol can contains 400. mL of compressed gas at 5.2 atm pressure. When the gas is sprayed into a large plastic bag, the bag inflates to a volume of 2.14 L. What is the pressure of gas inside the plastic bag?

18. For each of the following sets of pressure/volume data, calculate the missing quantity. Assume the temperature and the mass of gas remain constant.
   a. $V = 541$ mL at 1.00 atm; $V = ?$ at 699 torr
   b. $V = 2.32$ L at 110.2 kPa; $V = ?$ at 0.995 atm
   c. $V = 4.15$ mL at 135 atm; $V = 10.0$ mL at ? mm Hg
20. For each of the following sets of pressure/volume data, calculate the missing quantity. Assume the temperature and the mass of gas remain constant.
   a. $V = 561$ mL at 1.82 atm; $V = ?$ at 245 mm Hg
   b. $V = 561$ mL at 1.82 atm; $V = ?$ at 1.82 kPa
   c. $V = 561$ mL at 1.82 atm; $V = ?$ at 2.45×10^4 Pa
22. If the pressure exerted on the gas in a weather balloon decreases from 1.01 atm to 0.562 atm as it rises, by what factor will the volume of the gas in the balloon increase as it rises?
24. What pressure (in atmospheres) is required to compress 1.00 L of gas at 760. mm Hg pressure to a volume of 50.0 mL?

13.3 Volume and Temperature: Charles's Law

QUESTIONS

25. When the temperature of a sample of ideal gas is increased under constant pressure conditions, the _____ of the sample also increases.

26. The lowest possible temperature that can exist is referred to as _____ and is equivalent to -273 °C.

27. The volume of a sample of ideal gas is _____ proportional to its temperature (K) at constant pressure.

28. A mathematical expression that summarizes Charles's law is _____.

PROBLEMS

29. When 500. mL of helium gas at 25 °C is cooled at constant pressure to 10.0 K, what does the volume of the gas become?

30. If 525 mL of gas at 25 °C is heated to 50 °C, at constant pressure, calculate the new volume of the sample.

31. For each of the following sets of volume/temperature data, calculate the missing quantity. Assume the pressure and the amount of gas remain constant.
    a. $V = 10.$ L at 25 °C; $V = ?$ at 250. °C
    b. $V = 250.$ mL at 300. K; $V = ?$ at 0°C
    c. $V = 35$ L at 1500 °C; $V = ?$ at 1 K

32. For each of the following sets of volume/temperature data, calculate the missing quantity. Assume the pressure and the mass of gas remain constant.
    a. $V = 25.0$ L at 0 °C; $V = 50.0$ L at ? °C
    b. $V = 247$ mL at 25 °C; $V = 255$ mL at ? °C
    c. $V = 1.00$ mL at -272 °C; $V = ?$ at 25 °C

33. For each of the following sets of volume/temperature data, calculate the missing quantity. Assume the pressure and the amount of gas remain constant.
    a. $V = 25$ mL at 25 °C; $V = ?$ at 0 °C
    b. $V = 10.2$ L at 100. °C; $V = ?$ at 100 K
    c. $V = 551$ mL at 75 °C; $V = 1.00$ mL at ? °C

34. For each of the following sets of volume/temperature data, calculate the missing quantity. Assume the pressure and the mass of gas remain constant.
    a. $V = 2.01 \times 10^2$ L at 1,150 °C; $V = 5.00$ L at ? °C
    b. $V = 44.2$ mL at 298 K; $V = ?$ at 0 K
    c. $V = 44.2$ mL at 298 K; $V = ?$ at 0 °C

35. To what temperature must 500. mL of gas at 22 °C be cooled, at constant pressure, so that the volume of the gas is reduced to 1.00 mL?

36. A 113 L sample of helium at 27 °C is cooled at constant pressure to -78 °C. Calculate the new volume of the helium.

37. The label on an aerosol spray can contains a warning that the can should not be heated to over 130 °F because of the danger of explosion. Although the pressure in an aerosol can also increases if it is heated (which contributes to the danger of explosion), calculate the potential volume of the gas contained in a 500.-mL aerosol can when it is heated from 25 °C to 54 °C (approximately 130 °F).

38. As we noted in Example 13.6, gas volume was formerly used as a way to measure temperature by applying Charles's law. Suppose a sample in a gas thermometer has a volume of 135 mL at 11 °C. Indicate what temperature would correspond to each of the following volumes, assuming that the pressure remains constant: 113 mL, 142 mL, 155 mL, 127 mL.

13.4 Volume and Moles: Avogadro's Law

QUESTIONS

39. At conditions of constant temperature and pressure, the volume of a sample of ideal gas is _____ proportional to the number of moles of gas present.

40. A mathematical expression that summarizes Avogadro's law is _____.

PROBLEMS

41. If 5.00 g of O_2 gas has a volume of 7.20 L at a certain temperature and pressure, what volume does 15.0 g of O_2 have under the same conditions?

42. If 0.500 mol of nitrogen gas occupies a volume of 11.2 L at 0 °C, what volume will 2.00 mol of nitrogen occupy under the same conditions?

43. If 3.25 mol of argon gas occupies a volume of 100.L at a particular temperature and pressure, what volume does 14.15 mol of argon occupy under the same conditions?

44. If 46.2 g of oxygen gas occupies a volume of 100. L at a particular temperature and pressure, what volume will 5.00 g of oxygen gas occupy under the same conditions?

13.5 The Ideal Gas Law

QUESTIONS

45. What is an "ideal" gas? Under what conditions of pressure do real gases behave most nearly ideally?

47. Why must we always express the temperature of a gas in kelvins when using the ideal gas law?

46. What are the *units* of the universal gas constant (R) when the constant has the numerical value 0.08206?

48. Show how Boyle's, Charles's, and Avogadro's gas laws may be derived from the ideal gas law.

PROBLEMS

49. Given each of the following sets of values for three of the gas variables, calculate the unknown quantity.
    a. $P = 10.4$ atm; $V = 256$ mL; $n = 0.302$ mol; $T = ?$ °C
    b. $P = ?$ atm; $V = 22.4$ L; $n = 1.00$ mol; $T = 273$ K
    c. $P = 755$ torr; $V = ?$ L; $n = 0.341$ mol; $T = 22$ °C

51. Given each of the following sets of values for three of the gas variables, calculate the unknown quantity.
    a. $P = 7.74 \times 10^3$ Pa; $V = 12.2$ mL; $n = ?$ mol; $T = 298$ K
    b. $P = ?$ mm Hg; $V = 43.0$ mL; $n = 0.421$ mol; $T = 223$ K
    c. $P = 455$ mm Hg; $V = ?$ mL; $n = 4.4 \times 10^{-2}$ mol; $T = 331$ °C

53. What volume is occupied by 2.0 g of He at 25 °C and a pressure of 775 mm Hg?

55. What mass of hydrogen gas, H_2, is needed to fill an 80.0-L tank to a pressure of 150. atm at 27 °C?

57. At what temperature will a 1.0-g sample of neon gas exert a pressure of 500. torr in a 5.0-L container?

59. What is the pressure in a 25-L vessel containing 1.0 kg of oxygen gas at 300. K?

61. When 500. mL of O_2 gas at 25 °C and 1.045 atm is cooled to −40. °C and the pressure is increased to 2.00 atm, what is the new volume of the gas sample?

63. What is the final pressure for a sample of gas when 500. mL of the gas is cooled from 25 °C and 1.00 atm to −272 °C with no change in the volume of the sample?

50. Given each of the following sets of values for an ideal gas, calculate the unknown quantity.
    a. $P = 782$ mm Hg; $V = ?$; $n = 0.210$ mol; $T = 27$ °C
    b. $P = ?$ mm Hg; $V = 644$ mL; $n = 0.0921$ mol; $T = 303$ K
    c. $P = 745$ mm Hg; $V = 11.2$ L; $n = 0.401$ mol; $T = ?$ K

52. Given each of the following sets of values for an ideal gas, calculate the unknown quantity.
    a. $P = 1.01$ atm; $V = ?$; $n = 0.00831$ mol; $T = 25$ °C
    b. $P = ?$ atm; $V = 602$ mL; $n = 8.01 \times 10^{-3}$ mol; $T = 310$ K
    c. $P = 0.998$ atm; $V = 629$ mL; $n = ?$ mol; $T = 35$ °C

54. What volume is occupied by 5.03 g of O_2 at 28 °C and a pressure of 0.998 atm?

56. Suppose two 200.0-L tanks are to be filled separately with the gases helium and hydrogen. What mass of each gas is needed to produce a pressure of 135 atm in its respective tank at 24 °C?

58. At what temperature does 16.3 g of nitrogen gas have a pressure of 1.25 atm in a 25.0-L tank?

60. Calculate the pressure in a 212-L tank containing 51.3 lb of argon gas at 25 °C?

62. What will be the new volume if 125 mL of He gas at 100 °C and 0.981 atm is cooled to 25 °C and the pressure is increased to 1.15 atm?

64. At what temperature does 5.00 g of H_2 occupy a volume of 50.0 L at a pressure of 761 mm Hg?

13.6 Dalton's Law of Partial Pressures

QUESTIONS

65. Explain why the measured properties of a mixture of gases depend only on the total number of moles of particles, not on the identity of the individual gas particles. How is this observation summarized as a law?

66. We often collect small samples of gases in the laboratory by bubbling the gas into a bottle or flask containing water. Explain why the gas becomes saturated with water vapor and how we must take the presence of water vapor into account when calculating the properties of the gas sample.

67. If 4.0 g of $O_2(g)$ and 4.0 g of $He(g)$ are placed in a 5.0-L vessel at 65 °C, what will be the partial pressure of each gas and the total pressure in the vessel?

69. A tank contains a mixture of 3.0 mol of N_2, 2.0 mol of O_2, and 1.0 mol of CO_2 at 25 °C and a total pressure of 10.0 atm. Calculate the partial pressure (in torr) of each gas in the mixture.

71. A sample of oxygen gas is saturated with water vapor at 27 °C. The total pressure of the mixture is 772 torr, and the vapor pressure of water is 26.7 torr at 27 °C. What is the partial pressure of the oxygen gas?

73. A 500.-mL sample of O_2 gas at 24 °C was prepared by decomposing a 3% aqueous solution of hydrogen peroxide, H_2O_2, in the presence of a small amount of manganese catalyst by the reaction

$$2H_2O_2(aq) \rightarrow 2H_2O(g) + O_2(g)$$

The oxygen thus prepared was collected by displacement of water. The total pressure of gas collected was 755 mm Hg. What is the partial pressure of O_2 in the mixture? How many moles of O_2 are in the mixture? (The vapor pressure of water at 24 °C is 23 mm Hg.)

68. A gaseous mixture contains 5.23 g of N_2 and 4.41 g of O_2. What volume does this mixture occupy at 25 °C and 1.00 atm pressure?

70. What pressure exists in a 25.0-L tank containing 50.1 g of O_2 and 26.3 g of N_2 at 27 °C?

72. A sample of oxygen gas is collected by displacement of water at 25 °C and 1.02 atm total pressure. If the vapor pressure of water is 23.756 mm Hg at 25 °C, what is the partial pressure of the oxygen gas in the sample?

74. Small quantities of hydrogen gas can be prepared in the laboratory by the addition of aqueous hydrochloric acid to metallic zinc.

$$Zn(s) + 2HCl(aq) \rightarrow ZnCl_2(aq) + H_2(g)$$

Typically, the hydrogen gas is bubbled through water for collection and becomes saturated with water vapor. Suppose 240. mL of hydrogen gas is collected at 30. °C and has a total pressure of 1.032 atm by this process. What is the partial pressure of hydrogen gas in the sample? How many moles of hydrogen gas are present in the sample? How many grams of zinc must have reacted to produce this quantity of hydrogen? (The vapor pressure of water is 32 torr at 30 °C.)

13.7 Laws and Models: A Review

QUESTIONS

75. What is a scientific *law?* What is a *theory?* How do these concepts differ? Does a law explain a theory, or does a theory attempt to explain a law?

76. When is a scientific theory considered to be successful? Are all theories successful? Will a theory that has been successful in the past necessarily be successful in the future?

13.8 The Kinetic Molecular Theory of Gases

QUESTIONS

77. The observed behavior of gases indicates that the volumes of the actual molecules present are _____ compared to the bulk volume of the gas sample.

79. Temperature is a measure of the average _____ of the molecules in a sample of gas.

78. The kinetic molecular theory of gases proposes that the observed pressure of the gas arises from _____ of the gas molecules with the walls of their container.

80. The kinetic molecular theory of gases suggests that gas particles exert _____ attractive or repulsive forces on each other.

13.9 The Implications of the Kinetic Molecular Theory

QUESTIONS

81. How is the phenomenon of temperature explained on the basis of the kinetic molecular theory? What microscopic property of gas molecules is reflected in the temperature measured?

82. Explain, in terms of the kinetic molecular theory, how an increase in the temperature of a gas confined to a rigid container causes an increase in the pressure of the gas.

13.10 Gas Stoichiometry

QUESTIONS

83. What is the *molar volume* of a gas? Do all gases that behave ideally have the same molar volume?

84. What conditions are considered "standard temperature and pressure" (STP) for gases? Suggest a reason why these particular conditions might have been chosen for STP.

PROBLEMS

85. Consider the following chemical equation.

$$CaCO_3(s) \rightarrow CaO(s) + CO_2(g)$$

What volume of carbon dioxide gas is produced at 25 °C and a pressure of 1.02 atm when 10.0 g of calcium carbonate is decomposed?

86. Consider the following reaction:

$$P_4(s) + 6H_2(g) \rightarrow 4PH_3(g)$$

What volume of hydrogen gas, at 25 °C and 753 mm Hg, is required to react exactly with 2.51 g of phosphorus?

87. Consider the following *unbalanced* chemical equation.

$$C_6H_6(l) + O_2(g) \rightarrow CO_2(g) + H_2O(g)$$

What volume of oxygen gas, measured at 31 °C and 745 mm Hg, is needed to react with 5.00 g of C_6H_6? What volume of each product is produced under the same conditions?

88. Zinc metal reacts vigorously with chlorine gas to form zinc chloride:

$$Zn(s) + Cl_2(g) \rightarrow ZnCl_2(s)$$

What volume of chlorine gas at 35 °C and 1.01 atm is required to react completely with 1.13 g of zinc?

89. Consider the following *unbalanced* chemical equation.

$$Si(s) + N_2(g) \rightarrow Si_3N_4(s)$$

What volume of nitrogen gas, measured at 100. °C and 802 torr, is required to react with 115 g of Si?

90. When calcium carbonate is heated strongly, carbon dioxide gas is evolved:

$$CaCO_3(s) \rightarrow CaO(s) + CO_2(g)$$

If 4.74 g of calcium carbonate is heated, what volume of $CO_2(g)$ would be produced when collected at 26 °C and 0.997 atm?

91. Many transition metal salts are hydrates: they contain a fixed number of water molecules bound per formula unit of the salt. For example, copper(II) sulfate most commonly exists as the pentahydrate, $CuSO_4 \cdot 5H_2O$. If 5.00 g of $CuSO_4 \cdot 5H_2O$ is heated strongly so as to drive off all of the waters of hydration as water vapor, what volume will this water vapor occupy at 350. °C and a pressure of 1.04 atm?

92. If water is added to magnesium nitride, ammonia gas is produced when the mixture is heated.

$$Mg_3N_2(s) + 3H_2O(l) \rightarrow 3MgO(s) + 2NH_3(g)$$

If 10.3 g of magnesium nitride is treated with water, what volume of ammonia gas would be collected at 24 °C and 752 mm Hg?

93. What volume does 16.0 g of O_2 occupy at STP?

94. What volume does a mixture of 2.01 g He(g) and 4.14 g $N_2(g)$ occupy at 45 °C and 1.02 atm?

95. An ideal gas has a volume of 50. mL at 100. °C and a pressure of 690 torr. Calculate the volume of this sample of gas at STP.

96. A sample of hydrogen gas has a volume of 145 mL when measured at 44 °C and 1.47 atm. What volume would the hydrogen sample occupy at STP?

97. A mixture contains 5.00 g *each* of O_2, N_2, CO_2, and Ne gas. Calculate the volume of this mixture at STP. Calculate the partial pressure of each gas in the mixture at STP.

98. A gaseous mixture contains 6.25 g of He and 4.97 g of Ne. What volume does the mixture occupy at STP? Calculate the partial pressure of each gas in the mixture at STP.

99. What volume of oxygen gas at STP is required for the complete combustion of 1.00 g of methane, CH_4?

$$CH_4(g) + 2O_2(g) \rightarrow CO_2(g) + 2H_2O(g)$$

100. What volume of chlorine gas at STP is required for the complete reaction with 10.2 g of nitrogen in the following reaction?

$$N_2(g) + 3Cl_2(g) \rightarrow 2NCl_3(g)$$

101. During the making of steel, iron(II) oxide is reduced to metallic iron by treatment with carbon monoxide gas.

$$FeO(s) + CO(g) \rightarrow Fe(s) + CO_2(g)$$

What volume of carbon monoxide, measured at STP, is needed to react with 1.45×10^2 kg of iron(II) oxide? What volume of $CO_2(g)$ is produced?

102. Potassium permanganate, $KMnO_4$, is produced commercially by oxidizing aqueous potassium manganate, K_2MnO_4, with chlorine gas. The *unbalanced* chemical equation is

$$K_2MnO_4(aq) + Cl_2(g) \rightarrow KMnO_4(s) + KCl(aq)$$

What volume of $Cl_2(g)$, measured at STP, is needed to produce 10.0 g of $KMnO_4$?

Additional Problems

103. When doing any calculation involving gas samples, we must express the temperature in terms of the _____ temperature scale.

104. Two moles of ideal gas occupy a volume that is _____ the volume of 1 mol of ideal gas under the same temperature and pressure conditions.

105. Summarize the postulates of the kinetic molecular theory for gases. How does the kinetic molecular theory account for the observed properties of temperature and pressure?

106. Give a formula or equation that represents each of the following gas laws.
    a. Boyle's law
    b. Charles's law
    c. Avogadro's law
    d. the ideal gas law
    e. the combined gas law

107. For a mixture of gases in the same container, the total pressure exerted by the mixture of gases is the _____ of the pressures that those gases would exert if they were alone in the container under the same conditions.

108. A helium tank contains 25.2 L of helium at 8.40 atm pressure. Determine how many 1.50-L balloons at 755 mm Hg can be inflated with the gas in the tank, assuming that the tank will also have to contain He at 755 mm Hg after the balloons are filled (that is, it is not possible to completely empty the tank). The temperature is 25 °C in all cases.

109. As weather balloons rise from the earth's surface, the pressure of the atmosphere becomes less, tending to cause the volume of the balloon to expand. However, the temperature is much lower in the upper atmosphere than at sea level. Would this temperature effect tend to make such a balloon expand or contract? Weather balloons do, in fact, expand as they rise. What does this tell you?

110. Consider the following chemical equation.

$$2NO_2(g) \rightarrow N_2O_4(g)$$

If 25.0 mL of NO_2 gas is completely converted to N_2O_4 gas under the same conditions, what volume will the N_2O_4 occupy?

111. Carbon dioxide gas, in the dry state, may be produced by heating calcium carbonate.

$$CaCO_3(s) \rightarrow CaO(s) + CO_2(g)$$

What volume of CO_2, collected dry at 55 °C and a pressure of 774 torr, is produced by complete thermal decomposition of 10.0 g of $CaCO_3$?

112. Carbon dioxide gas, saturated with water vapor, can be produced by the addition of aqueous acid to calcium carbonate.

$$CaCO_3(s) + 2H^+(aq) \rightarrow Ca^{2+}(aq) + H_2O(l) + CO_2(g)$$

How many moles of $CO_2(g)$, collected at 60. °C and 774 torr total pressure, are produced by the complete reaction of 10.0 g of $CaCO_3$ with acid? What volume does this wet CO_2 occupy? What volume would the CO_2 occupy at 774 torr if a desiccant (a chemical drying agent) were added to remove the water? (The vapor pressure of water at 60 °C is 149.4 mm Hg.)

113. Sulfur trioxide, SO_3, is produced in enormous quantities each year for use in the synthesis of sulfuric acid.

$$S(s) + O_2(g) \rightarrow SO_2(g)$$
$$2SO_2(g) + O_2(g) \rightarrow 2SO_3(g)$$

What volume of $O_2(g)$ at 350. °C and a pressure of 5.25 atm is needed to completely convert 5.00 g of sulfur to sulfur trioxide?

114. Calculate the volume of $O_2(g)$ produced at 25 °C and 630. torr when 50.0 g of $KClO_3(s)$ is heated in the presence of a small amount of MnO_2 catalyst.

115. If 10.0 g of liquid helium at 1.7 K is completely vaporized, what volume does the helium occupy at STP?

116. Convert the following pressures into atmospheres.
    a. 665 mm Hg
    b. 124 kPa
    c. 2.540×10^6 Pa
    d. 803 torr

117. Convert the following pressures into mm Hg.
     a. 0.903 atm
     b. 2.1240×10^6 Pa
     c. 445 kPa
     d. 342 torr
118. Convert the following pressures into pascals.
     a. 645 mm Hg
     b. 221 kPa
     c. 0.876 atm
     d. 32 torr
119. For each of the following sets of pressure/volume data, calculate the missing quantity. Assume the temperature and the amount of gas remain constant.
     a. $V = 123$ L at 4.56 atm; $V = ?$ at 1002 mm Hg
     b. $V = 634$ mL at 25.2 mm Hg; $V = 166$ mL at ? atm
     c. $V = 44.3$ L at 511 torr; $V = ?$ at 1.05 kPa
120. For each of the following sets of pressure/volume data, calculate the missing quantity. Assume the temperature and the amount of gas remain constant.
     a. $V = 255$ mL at 1.00 mm Hg; $V = ?$ at 2.00 torr
     b. $V = 1.3$ L at 1.0 kPa; $V = ?$ at 1.0 atm
     c. $V = 1.3$ L at 1.0 kPa; $V = ?$ at 1.0 mm Hg
121. A particular balloon is designed by its manufacturer to be inflated to a volume of no more than 2.5 L. If the balloon is filled with 2.0 L of helium at sea level, is released, and rises to an altitude at which the atmospheric pressure is only 500. mm Hg, will the balloon burst?
122. What pressure is needed to compress the air in a 1.105-L cylinder at 755 torr to a volume of 1.00 mL?
123. An expandable vessel contains 729 mL of gas at 22 °C. What volume will the gas sample in the vessel have if it is placed in a boiling water bath (100. °C)?
124. For each of the following sets of volume/temperature data, calculate the missing quantity. Assume the pressure and the amount of gas remain constant.
     a. $V = 100.$ mL at 75 °C; $V = ?$ at −75 °C
     b. $V = 500.$ mL at 100 °C; $V = 600.$ mL at ? °C
     c. $V = 10{,}000$ L at 25 °C; $V = ?$ at 0 K
125. For each of the following sets of volume/temperature data, calculate the missing quantity. Assume the pressure and the amount of gas remain constant.
     a. $V = 22.4$ L at 0 °C; $V = 44.4$ L at ? K
     b. $V = 1.0 \times 10^{-3}$ mL at −272 °C; $V = ?$ at 25 °C
     c. $V = 32.3$ L at −40 °C; $V = 1000.$ L at ? °C
126. A 25-L sample of nitrogen gas is cooled from 25 °C to a final temperature of −100. °C. What is the new volume of the gas?
127. If 4.0 g of helium gas occupies a volume of 22.4 L at 0 °C and a pressure of 1.0 atm, what volume does 3.0 g of He occupy under the same conditions?
128. If 23.2 g of a given gas occupies a volume of 93.2 L at a particular temperature and pressure, what mass of the gas occupies a volume of 10.4 L under the same conditions?

129. Given each of the following sets of values for three of the gas variables, calculate the unknown quantity.
     a. $P = 21.2$ atm; $V = 142$ mL; $n = 0.432$ mol; $T = ?$ K
     b. $P = ?$ atm; $V = 1.23$ mL; $n = 0.000115$ mol; $T = 293$ K
     c. $P = 755$ mm Hg; $V = ?$ mL; $n = 0.473$ mol; $T = 131$ °C
130. Given each of the following sets of values for three of the gas variables, calculate the unknown quantity.
     a. $P = 1.034$ atm; $V = 21.2$ mL; $n = 0.00432$ mol; $T = ?$ K
     b. $P = ?$ atm; $V = 1.73$ mL; $n = 0.000115$ mol; $T = 182$ K
     c. $P = 1.23$ mm Hg; $V = ?$ L; $n = 0.773$ mol; $T = 152$ °C
131. What volume is occupied by 32.0 g of SO_2 at 22 °C and a pressure of 1.054 atm?
132. Suppose three 100.-L tanks are to be filled separately with the gases CH_4, N_2, and CO_2, respectively. What mass of each gas is needed to produce a pressure of 120. atm in its tank at 27 °C?
133. At what temperature does 4.00 g of helium gas have a pressure of 1.00 atm in a 22.4-L vessel?
134. What is the pressure in a 100.-mL flask containing 55 mg of oxygen gas at 26 °C?
135. A weather balloon is filled with 1.0 L of helium at 23 °C and 1.0 atm. What volume does the balloon have when it has risen to a point in the atmosphere where the pressure is 220 torr and the temperature is −31 °C?
136. At what temperature does 100. mL of N_2 at 300. K and 1.13 atm occupy a volume of 500. mL at a pressure of 1.89 atm?
137. If 1.0 mol of $N_2(g)$ is injected into a 5.0-L tank already containing 50. g of O_2 at 25 °C, what will be the total pressure in the tank?
138. A mixture contains 5.0 g *each* of O_2, N_2, CO_2, and Ne gas. Calculate the volume of this mixture at 29 °C and a pressure of 755 mm Hg.
139. A flask of hydrogen gas is collected at 1.023 atm and 35 °C by displacement of water from the flask. The vapor pressure of water at 35 °C is 42.2 mm Hg. What is the partial pressure of hydrogen gas in the flask?
140. Consider the following chemical equation.

$$N_2(g) + 3H_2(g) \rightarrow 2NH_3(g)$$

What volumes of nitrogen gas and hydrogen gas, each measured at 11 °C and 0.998 atm, are needed to produce 5.00 g of ammonia?

141. Consider the following *unbalanced* chemical equation.

$$C_6H_{12}O_6(s) + O_2(g) \rightarrow CO_2(g) + H_2O(g)$$

What volume of oxygen gas, measured at 28 °C and 0.976 atm, is needed to react with 5.00 g of $C_6H_{12}O_6$? What volume of each product is produced under the same conditions?

142. Consider the following *unbalanced* chemical equation.

$$Cu_2S(s) + O_2(g) \rightarrow Cu_2O(s) + SO_2(g)$$

What volume of oxygen gas, measured at 27.5 °C and 0.998 atm, is required to react with 25 g of copper(I) sulfide? What volume of sulfur dioxide gas is produced under the same conditions?

143. When sodium bicarbonate, $NaHCO_3(s)$, is heated, sodium carbonate is produced, with the evolution of water vapor and carbon dioxide gas.

$$2NaHCO_3(s) \rightarrow Na_2CO_3(s) + H_2O(g) + CO_2(g)$$

What total volume of gas, measured at 29 °C and 769 torr, is produced when 1.00 g of $NaHCO_3(s)$ is completely converted to $Na_2CO_3(s)$?

144. What volume does 35 moles of N_2 occupy at STP?

145. A sample of oxygen gas has a volume of 125 L at 25 °C and a pressure of 0.987 atm. Calculate the volume of this oxygen sample at STP.

146. A mixture contains 5.0 g of He, 1.0 g of Ar, and 3.5 g of Ne. Calculate the volume of this mixture at STP. Calculate the partial pressure of each gas in the mixture at STP.

147. What volume of hydrogen gas at STP is required for the complete reaction of 10.0 g of nitrogen in the following reaction?

$$N_2(g) + 3H_2(g) \rightarrow 2NH_3(g)$$

148. Concentrated hydrogen peroxide solutions are explosively decomposed by traces of transition metal ions (such as Mn or Fe):

$$2H_2O_2(aq) \rightarrow 2H_2O(l) + O_2(g)$$

What volume of pure $O_2(g)$, collected at 27 °C and 764 torr, would be generated by decomposition of 125 g of a 50.0% by mass hydrogen peroxide solution?

Solutions to Self-Check Exercises

SELF-CHECK EXERCISE 13.1

We know that 1.000 atm = 760.0 mm Hg. So

$$525 \text{ mm Hg} \times \frac{1.000 \text{ atm}}{760.0 \text{ mm Hg}} = 0.691 \text{ atm}$$

SELF-CHECK EXERCISE 13.2

Initial Conditions	Final Conditions
P_1 = 635 torr	P_2 = 785 torr
V_1 = 1.51 L	V_2 = ?

Solving Boyle's law ($P_1 V_1 = P_2 V_2$) for V_2 gives

$$V_2 = V_1 \times \frac{P_1}{P_2}$$

$$= 1.51 \text{ L} \times \frac{635 \text{ torr}}{785 \text{ torr}} = 1.22 \text{ L}$$

Note that the volume decreased, as the increase in pressure led us to expect.

SELF-CHECK EXERCISE 13.3

Because the temperature of the gas inside the bubble decreases (at constant pressure), the bubble gets smaller. The conditions are

Initial Conditions	Final Conditions
$T_1 = 28\ °C = 28 + 273 = 301\ K$	$T_2 = 18\ °C = 18 + 273 = 291\ K$
$V_1 = 23\ cm^3$	$V_2 = ?$

Solving Charles's law,

$$\frac{V_1}{T_1} = \frac{V_2}{T_2}$$

for V_2 gives

$$V_2 = V_1 \times \frac{T_2}{T_1} = 23\ cm^3 \times \frac{291\ \cancel{K}}{301\ \cancel{K}} = 22\ cm^3$$

SELF-CHECK EXERCISE 13.4

Because the temperature and pressure of the two samples are the same, we can use Avogadro's law in the form

$$\frac{V_1}{n_1} = \frac{V_2}{n_2}$$

The following information is given:

Sample 1	Sample 2
$V_1 = 36.7\ L$	$V_2 = 16.5\ L$
$n_1 = 1.5\ mol$	$n_2 = ?$

We can now solve Avogadro's law for the value of n_2 (the moles of N_2 in Sample 2):

$$n_2 = n_1 \times \frac{V_2}{V_1} = 1.5\ mol \times \frac{16.5\ \cancel{L}}{36.7\ \cancel{L}} = 0.67\ mol$$

Here n_2 is smaller than n_1, which makes sense in view of the fact that V_2 is smaller than V_1.

NOTE: We isolate n_2 from Avogadro's law as given above by multiplying both sides of the equation by n_2 and then by n_1/V_1:

$$\left(n_2 \times \frac{n_1}{V_1} \right) \frac{V_1}{n_1} = \left(n_2 \times \frac{n_1}{V_1} \right) \frac{V_2}{n_2}$$

to give $n_2 = n_1 \times V_2/V_1$.

SELF-CHECK EXERCISE I3.5

We are given the following information:

$$P = 1.00 \text{ atm}$$
$$V = 2.70 \times 10^6 \text{ L}$$
$$n = 1.10 \times 10^5 \text{ mol}$$

We solve for T by dividing both sides of the ideal gas law by nR:

$$\frac{PV}{nR} = \frac{\cancel{nR}T}{\cancel{nR}} = T$$

to give

$$T = \frac{PV}{nR} = \frac{(1.00\,\cancel{\text{atm}})(2.70 \times 10^6\,\cancel{L})}{(1.10 \times 10^5\,\cancel{\text{mol}})\left(0.08206\dfrac{\cancel{L}\,\cancel{\text{atm}}}{K\,\cancel{\text{mol}}}\right)}$$

$$= 299 \text{ K}$$

The temperature of the helium is 299 K, or $299 - 273 = 26$ °C.

SELF-CHECK EXERCISE I3.6

We are given the following information about the radon sample:

$$n = 1.5 \text{ mol}$$
$$V = 21.0 \text{ L}$$
$$T = 33 \text{ °C} = 33 + 273 = 306 \text{ K}$$
$$P = ?$$

We solve the ideal gas law ($PV = nRT$) for P by dividing both sides of the equation by V:

$$P = \frac{nRT}{V} = \frac{(1.5\,\cancel{\text{mol}})\left(0.08206\dfrac{\cancel{L}\,\text{atm}}{K\,\cancel{\text{mol}}}\right)(306\,\cancel{K})}{21.0\,\cancel{L}}$$

$$= 1.8 \text{ atm}$$

SELF-CHECK EXERCISE I3.7

To solve this problem we take the ideal gas law and separate those quantities that change from those that remain constant (on opposite sides of the equation). In this case volume and temperature change, and number of moles and pressure (and of course R) remain constant. So $PV = nRT$ becomes $V/T = nR/P$, which leads to

$$\frac{V_1}{T_1} = \frac{nR}{P} \qquad \text{and} \qquad \frac{V_2}{T_2} = \frac{nR}{P}$$

Combining these gives

$$\frac{V_1}{T_1} = \frac{nR}{P} = \frac{V_2}{T_2} \qquad \text{or} \qquad \frac{V_1}{T_1} = \frac{V_2}{T_2}$$

We are given

Initial Conditions	Final Conditions
$T_1 = 5\,°C = 5 + 273 = 278$ K	$T_2 = 86\,°C = 86 + 273 = 359$ K
$V_1 = 3.8$ L	$V_2 = ?$

Thus

$$V_2 = \frac{T_2 V_1}{T_1} = \frac{(359\ \cancel{K})(3.8\ \text{L})}{278\ \cancel{K}} = 4.9\ \text{L}$$

CHECK: Is the answer sensible? In this case the temperature was increased (at constant pressure), so the volume should increase. The answer makes sense.

Note that this problem could be described as a "Charles's law problem." The real advantage of using the ideal gas law is that you need to remember only *one* equation to do virtually any problem involving gases.

SELF-CHECK EXERCISE 13.8

We are given the following information:

Initial Conditions	Final Conditions
$P_1 = 0.747$ atm	$P_2 = 1.18$ atm
$T_1 = 13\,°C = 13 + 273 = 286$ K	$T_2 = 56\,°C = 56 + 273 = 329$ K
$V_1 = 11.0$ L	$V_2 = ?$

In this case the number of moles remains constant. Thus we can say

$$\frac{P_1 V_1}{T_1} = nR \quad \text{and} \quad \frac{P_2 V_2}{T_2} = nR$$

or

$$\frac{P_1 V_1}{T_1} = \frac{P_2 V_2}{T_2}$$

Solving for V_2 gives

$$V_2 = V_1 \times \frac{T_2}{T_1} \times \frac{P_1}{P_2} = (11.0\ \text{L})\left(\frac{329\ \cancel{K}}{286\ \cancel{K}}\right)\left(\frac{0.747\ \cancel{\text{atm}}}{1.18\ \cancel{\text{atm}}}\right) = 8.01\ \text{L}$$

SELF-CHECK EXERCISE 13.9

As usual when dealing with gases, we can use the ideal gas equation $PV = nRT$. First consider the information given:

$$P = 0.91\ \text{atm} = P_{\text{total}}$$
$$V = 2.0\ \text{L}$$
$$T = 25\,°C = 25 + 273 = 298\ \text{K}$$

Given this information, we can calculate the number of moles of gas in the mixture: $n_{total} = n_{N_2} + n_{O_2}$. Solving for n in the ideal gas equation gives

$$n_{total} = \frac{P_{total}V}{RT} = \frac{(0.91 \text{ atm})(2.0 \text{ L})}{\left(0.08206\dfrac{\text{L atm}}{\text{K mol}}\right)(298 \text{ K})} = 0.074 \text{ mol}$$

We also know that 0.050 mol of N_2 is present. Because

$$n_{total} = n_{N_2} + n_{O_2} = 0.074 \text{ mol}$$
$$\uparrow$$
$$(0.050 \text{ mol})$$

we can calculate the moles of O_2 present.

$$0.050 \text{ mol} + n_{O_2} = 0.074 \text{ mol}$$
$$n_{O_2} = 0.074 \text{ mol} - 0.050 \text{ mol} = 0.024 \text{ mol}$$

Now that we know the moles of oxygen present, we can calculate the partial pressure of oxygen from the ideal gas equation.

$$P_{O_2} = \frac{n_{O_2}RT}{V} = \frac{(0.024 \text{ mol})\left(0.08206\dfrac{\text{L atm}}{\text{K mol}}\right)(298 \text{ K})}{2.0 \text{ L}}$$

$$= 0.29 \text{ atm}$$

Although it is not requested, note that the partial pressure of the N_2 must be 0.62 atm, because

$$\underbrace{0.62 \text{ atm}}_{\substack{\uparrow \\ P_{N_2}}} + \underbrace{0.29 \text{ atm}}_{\substack{\uparrow \\ P_{O_2}}} = \underbrace{0.91 \text{ atm}}_{\substack{\uparrow \\ P_{total}}}$$

SELF-CHECK EXERCISE 13.10

The volume is 0.500 L, the temperature is 25 °C (or 25 + 273 = 298 K), and the total pressure is given as 0.950 atm. Of this total pressure, 24 torr is due to the water vapor. We can calculate the partial pressure of the H_2 because we know that

$$P_{total} = P_{H_2} + P_{H_2O} = 0.950 \text{ atm}$$
$$\uparrow$$
$$24 \text{ torr}$$

Before we carry out the calculation, however, we must convert the pressures to the same units. Converting P_{H_2O} to atmospheres gives

$$24 \text{ torr} \times \frac{1.000 \text{ atm}}{760.0 \text{ torr}} = 0.032 \text{ atm}$$

Thus

$$P_{total} = P_{H_2} + P_{H_2O} = 0.950 \text{ atm} = P_{H_2} + 0.032 \text{ atm}$$

and

$$P_{H_2} = 0.950 \text{ atm} - 0.032 \text{ atm} = 0.918 \text{ atm}$$

Now that we know the partial pressure of the hydrogen gas, we can use the ideal gas equation to calculate the moles of H_2.

$$n_{H_2} = \frac{P_{H_2}V}{RT} = \frac{(0.918 \text{ atm})(0.500 \text{ L})}{\left(0.08206\dfrac{\text{L atm}}{\text{K mol}}\right)(298 \text{ K})}$$

$$= 0.0188 \text{ mol} = 1.88 \times 10^{-2} \text{ mol}$$

The sample of gas contains 1.88×10^{-2} mol of H_2, which exerts a partial pressure of 0.918 atm.

SELF-CHECK EXERCISE 13.11

We will solve this problem by taking the following steps:

STEP 1
Using the atomic mass of zinc (65.38), we calculate the moles of zinc in 26.5 g.

$$26.5 \text{ g Zn} \times \frac{1 \text{ mol Zn}}{65.38 \text{ g Zn}} = 0.405 \text{ mol Zn}$$

STEP 2
Using the balanced equation, we next calculate the moles of H_2 produced.

$$0.405 \text{ mol Zn} \times \frac{1 \text{ mol } H_2}{1 \text{ mol Zn}} = 0.405 \text{ mol } H_2$$

STEP 3
Now that we know the moles of H_2, we can compute the volume of H_2 by using the ideal gas law, where

$$P = 1.50 \text{ atm}$$
$$V = ?$$
$$n = 0.405 \text{ mol}$$
$$R = 0.08206 \text{ L atm/K mol}$$
$$T = 19 \text{ °C} = 19 + 273 = 292 \text{ K}$$

$$V = \frac{nRT}{P} = \frac{(0.405 \text{ mol})\left(0.08206\dfrac{\text{L atm}}{\text{K mol}}\right)(292 \text{ K})}{1.50 \text{ atm}}$$

$$= 6.47 \text{ L of } H_2$$

SELF-CHECK EXERCISE 13.12

Although there are several possible ways to do this problem, the most convenient method involves using the molar volume at STP. First we use the ideal gas equation to calculate the moles of NH_3 present:

$$n = \frac{PV}{RT}$$

where $P = 15.0$ atm, $V = 5.00$ L, and $T = 25 + 273 = 298$ K.

$$n = \frac{(15.0 \text{ atm})(5.00 \text{ L})}{\left(0.08206 \dfrac{\text{L atm}}{\text{K mol}}\right)(298 \text{ K})} = 3.07 \text{ mol}$$

We know that at STP each mole of gas occupies 22.4 L. Therefore 3.07 mol have the volume

$$3.07 \text{ mol} \times \frac{22.4 \text{ L}}{1 \text{ mol}} = 68.8 \text{ L}$$

The volume of the ammonia at STP is 68.8 L.

Dry ice (solid CO_2) as it sublimates and acidifies water.

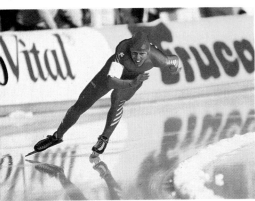

Flying, swimming, and skating are done in contact with water in its various forms.

You have only to think about water to appreciate how different the three states of matter are. Flying, swimming, and ice skating are all done in contact with water in its various states. We swim in liquid water and skate on water in its solid form (ice). Airplanes fly in an atmosphere containing water in the gaseous state (water vapor). To allow these various activities, the arrangements of the water molecules must be significantly different in their gas, liquid, and solid forms.

In Chapter 13 we saw that the particles of a gas are far apart, in rapid random motion, and have little effect on each other. Solids are obviously very different from gases. Gases have low densities, have high compressibilities, and completely fill a container. Solids have much greater densities than gases, are compressible only to a very slight extent, and are rigid; a solid maintains its shape regardless of its container. These properties indicate that the components of a solid are close together and exert large attractive forces on each other.

The properties of liquids lie somewhere between those of solids and of gases—but not midway between, as can be seen from some of the properties of the three states of water. For example, it takes about seven times more energy to change liquid water to steam (a gas) at 100 °C than to melt ice to form liquid water at 0 °C.

$$H_2O(s) \rightarrow H_2O(l) \qquad \text{energy required} \cong 6 \text{ kJ/mol}$$
$$H_2O(l) \rightarrow H_2O(g) \qquad \text{energy required} \cong 41 \text{ kJ/mol}$$

These values indicate that going from the liquid to the gaseous state involves a much greater change than going from the solid to the liquid. Therefore, we can conclude that the solid and liquid states are more similar than the liquid and gaseous states. This is also demonstrated by the densities of the three states of water (Table 14.1). Note that water in its gaseous state is about 3000 times less dense than in the solid and liquid states and that the latter two states have very similar densities.

We find in general that the liquid and solid states show many similarities and are strikingly different from the gaseous state (see Figure 14.1). The best way to picture the solid state is in terms of closely packed, highly ordered particles in contrast to the widely spaced, randomly arranged particles of a gas. The liquid

Table 14.1   Densities of the Three States of Water

State	Density (g/cm^3)
solid (0 °C, 1 atm)	0.9168
liquid (25 °C, 1 atm)	0.9971
gas (400 °C, 1 atm)	3.26×10^{-4}

Gas

Liquid

Solid

state lies in between, but its properties indicate that it much more closely resembles the solid than the gaseous state. It is useful to picture a liquid in terms of particles that are generally quite close together, but with a more disordered arrangement than for the solid state and with some empty spaces. For most substances, the solid state has a higher density than the liquid, as Figure 14.1 suggests. However, water is an exception to this rule. Ice has an unusual amount of empty space and so is less dense than liquid water, as indicated in Table 14.1.

In this chapter we will explore the important properties of liquids and solids. We will illustrate many of these properties by considering one of the earth's most important substances: water.

Water and Its Phase Changes

AIM: To discuss some of the important features of water.

In the world around us we see many solids (soil, rocks, trees, concrete, and so on), and we are immersed in the gases of the atmosphere. But the liquid we most commonly see is water; it is virtually everywhere, covering about 70% of the earth's surface. Approximately 97% of the earth's water is found in the oceans, which are actually mixtures of water and huge quantities of dissolved salts.

Water is one of the most important substances on earth. It is crucial for sustaining the reactions within our bodies that keep us alive, but it also affects our lives in many indirect ways. The oceans help moderate the earth's temperature. Water cools automobile engines and nuclear power plants. Water provides a means of transportation on the earth's surface and acts as a medium for the growth of the myriads of creatures we use as food, and much more.

Pure water is a colorless, tasteless substance that at 1 atm pressure freezes to form a solid at 0 °C and vaporizes completely to form a gas at 100 °C. This means that (at 1 atm pressure) the liquid range of water occurs between the temperatures 0 °C and 100 °C.

What happens when we heat liquid water? First the temperature of the water rises. Just as with gas molecules, the motions of the water molecules increase as it is heated. Eventually the temperature of the water reaches 100 °C; now bubbles develop in the interior of the liquid, float to the surface, and burst—the boiling point has been reached. An interesting thing happens at the boiling point: even though heating continues, the temperature stays at 100 °C until all the water has changed to vapor. Only when all of the water has changed to the gaseous state does the temperature begin to rise again. (We are now heating the vapor.) At 1 atm pressure, liquid water always changes to gaseous water at 100 °C, the **normal boiling point** for water.

The water we drink often has a taste because of the substances dissolved in it.

You might want to briefly review Section 3.6, which covers the specific heat capacity of water.

Many common phenomena that are interesting to discuss in class are based on the phase changes of water and the properties of ice, liquid water, and water vapor. The density of ice compared with that of liquid water explains why ice floats. The volume of ice compared with that of liquid water explains why pipes freeze and pot holes form in winter. Foods cook more quickly in a pressure cooker than they do boiling at atmospheric pressure.

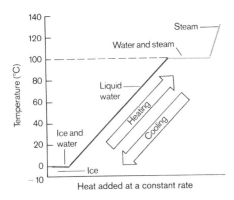

Figure 14.2
The heating/cooling curve for water heated at a constant rate. The plateau at the boiling point is longer than the plateau at the melting point, because it takes almost seven times as much energy (and thus seven times the heating time) to vaporize liquid water than to melt ice.

At 0 °C the density of liquid water is $\dfrac{1.00 \text{ g}}{1.00 \text{ mL}} = 1.00$ g/mL, and the density of ice is $\dfrac{1.00 \text{ g}}{1.09 \text{ mL}} = 0.917$ g/mL.

The experiment just described is represented in Figure 14.2, which is called the **heating/cooling curve** for water. Going from left to right on this graph means energy is being added (heating). Going from right to left on the graph means that energy is being removed (cooling).

When liquid water is cooled, the temperature decreases until it reaches 0 °C where the liquid begins to freeze (see Figure 14.2). The temperature remains at 0 °C until all the liquid water has changed to ice and then begins to drop again as cooling continues. At 1 atm pressure, water freezes (or, in the opposite process, ice melts) at 0 °C. This is called the **normal freezing point** of water. Liquid and solid water can coexist indefinitely if the temperature is held at 0 °C. However, at temperatures below 0 °C liquid water freezes, while at temperatures above 0 °C ice melts.

Interestingly, water expands when it freezes. That is, one gram of ice at 0 °C has a greater volume that one gram of liquid water at 0 °C. This has very important practical implications. For instance, water in a confined space can break its container when it freezes and expands. This accounts for the bursting of water pipes and engine blocks that are left unprotected in freezing weather.

The expansion of water when it freezes also explains why ice cubes float. Recall that density is defined as mass/volume. When one gram of liquid water freezes, its volume becomes greater (it expands). Therefore the *density* of one gram of ice is less than the density of one gram of water, because in the case of ice we divide by a slightly larger volume.

The lower density of ice also means that ice floats on the surface of lakes as they freeze, providing a layer of insulation that helps to prevent lakes and rivers from freezing solid in the winter. This means that aquatic life continues to have liquid water available through the winter.

Harbor seals on an ice floe in Alaska.

14.2 Energy Requirements for the Changes of State

AIM: To discuss interactions among water molecules.
To explain and use heat of fusion and heat of vaporization.

It is important to recognize that changes of state from solid to liquid and from liquid to gas are *physical* changes. No *chemical* bonds are broken in these processes. Ice, water, and steam all contain H_2O molecules. When water is boiled to form steam, water molecules are separated from each other (see Figure 14.3) but the individual molecules remain intact.

Intermolecular forces are introduced now to help explain the differences in energy requirements for melting and for vaporization. However, the detailed discussion of intermolecular forces occurs in Section 14.3.

14.2 Energy Requirements for the Changes of State   **475**

The bonding forces that hold the atoms of a molecule together are called **intramolecular** (within the molecule) **forces.** The forces that occur among molecules that cause them to aggregate to form a solid or a liquid are called **intermolecular** (between the molecules) **forces.** These two types of forces are illustrated in Figure 14.4.

It takes energy to melt ice and to vaporize water, because intermolecular forces between water molecules must be overcome. In ice the molecules are virtually locked in place, although they can vibrate about their positions. When energy is added, the vibrational motions increase, and the molecules eventually achieve the greater movement and disorder characteristic of liquid water. The ice has melted. As still more energy is added, the gaseous state is eventually reached, in which the individual molecules are far apart and interact relatively little. However, the gas still consists of water molecules. It would take *much* more energy to overcome the covalent bonds and decompose the water molecules into their component atoms.

The energy required to melt 1 mol of a substance is called the **molar heat of fusion.** For ice, the molar heat of fusion is 6.02 kJ/mol. The energy required to change 1 mol of liquid to its vapor is called the **molar heat of vaporization.** For water, the molar heat of vaporization is 40.6 kJ/mol at 100 °C. Notice in Figure 14.2 that the plateau that corresponds to the vaporization of water is much longer than that for the melting of ice. This occurs because it takes much more energy (almost seven times as much) to vaporize a mole of water than to melt a mole of ice. This is consistent with our models of solids, liquids, and gases (see Figure 14.1). In liquids, the particles (molecules) are relatively close together, so most of the intermolecular forces are still present. However, when the molecules go from the liquid to the gaseous state, they must be moved far apart. To separate the molecules enough to form a gas, virtually all of the intermolecular forces must be overcome, and this requires large quantities of energy.

Remember that temperature is a measure of the random motions (average kinetic energy) of the particles in a substance.

Figure 14.3
Both liquid water and gaseous water contain H_2O molecules. In liquid water the H_2O molecules are close together, whereas in the gaseous state the molecules are widely separated. The bubbles contain gaseous water.

If you have the time and the facilities, you might measure the temperature of ice and liquid water as heat is steadily applied and have students plot the temperature data versus time.

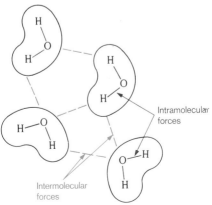

Figure 14.4
Intramolecular (bonding) forces occur between the atoms in a molecule and hold the molecule together. Intermolecular forces occur between molecules. These are the forces that cause water to condense to a liquid or form a solid at low enough temperatures. Intermolecular forces are typically much weaker than intramolecular forces.

In calculating the total amount of energy required to change temperature when a phase change is also involved, students often forget to calculate the energy required to change the phase. If you show the heating/cooling curve for water on an overhead transparency while calculating the energy required for each change, you can help remind students that a change in energy accompanies a phase change.

CHEMISTRY IN FOCUS

Whales Need Changes of State

Sperm whales are prodigious divers. They commonly dive a mile or more into the ocean, hovering at that depth in search of schools of squid or fish. To remain motionless at a given depth, the whale must have the same density as the surrounding water. Because the density of seawater increases with depth, the sperm whale has a system that automatically increases its density as it dives. This system involves the spermaceti organ found in the whale's head. Spermaceti is a waxy substance with the formula

$$CH_3-(CH_2)_{15}-O-\underset{\underset{O}{\|}}{C}-(CH_2)_{14}CH_3$$

which is a liquid above 30 °C. At the

ocean surface the spermaceti in the whale's head is a liquid, warmed by the flow of blood through the spermaceti organ. When the whale dives, this blood flow decreases and the colder water causes the spermaceti to begin freezing. Because solid spermaceti is more dense than the liquid state, the sperm whale's density increases as it dives, matching the increase in the water's density.* When the whale wants to resurface, blood flow through the spermaceti organ increases, remelting the spermaceti and making the whale more buoyant. So the sperm whale's sophisticated density-regulating mechanism is based on a simple change of state.

A sperm whale.

*For most substances, the solid state is more dense than the liquid state. Water is an important exception.

EXAMPLE 14.1	Calculating Energy Changes: Solid to Liquid

Calculate the energy required to melt 8.5 g of ice at 0 °C. The molar heat of fusion for ice is 6.02 kJ/mol.

SOLUTION

The molar heat of fusion is the energy required to melt *1 mol* of ice. In this problem we have 8.5 g of solid water. We must find out how many moles of ice

EXAMPLE 14.1, CONTINUED

this mass represents. Because the molar mass of water is 16 + 2(1) = 18, we know that 1 mol of water has a mass of 18 g, so we can convert 8.5 g of H_2O to moles of H_2O.

$$8.5 \text{ g } H_2O \times \frac{1 \text{ mol } H_2O}{18 \text{ g } H_2O} = 0.47 \text{ mol } H_2O$$

Because 6.02 kJ of energy is required to melt a mole of solid water, our sample will take about half this amount (we have approximately half a mole of ice). To calculate the exact amount of energy required, we will use the equivalence statement

6.02 kJ required for 1 mol of H_2O

which leads to the conversion factor we need:

$$0.47 \text{ mol } H_2O \times \frac{6.02 \text{ kJ}}{\text{mol } H_2O} = 2.8 \text{ kJ}$$

This can be represented symbolically as

EXAMPLE 14.2   Calculating Energy Changes: Liquid to Gas

Calculate the energy (in kJ) required to heat 25 g of liquid water from 25 °C to 100. °C and change it to steam at 100. °C. The specific heat capacity of liquid water is 4.18 J/g °C, and the molar heat of vaporization of water is 40.6 kJ/mol.

Specific heat capacity was discussed in Section 3.6.

SOLUTION

This problem can be split into two parts: (1) heating the water to its boiling point and (2) converting the liquid water to vapor at the boiling point.

STEP 1: HEATING TO BOILING
We must first supply energy to heat the liquid water from 25 °C to 100. °C. Because 4.18 J is required to heat *one* gram of water by *one* Celsius degree, we must multiply by both the mass of water (25 g) and the temperature change (100. °C − 25 °C = 75 °C):

Energy required (Q)	=	Specific heat capacity (s)	×	Mass of water (m)	×	Temperature change (ΔT)

EXAMPLE 14.2, CONTINUED

which we can represent by the equation

$$Q = s \times m \times \Delta T$$

Thus

$$Q = 4.18 \frac{J}{g \cdot °C} \times 25 \text{ g} \times 75 \text{ °C} = 7.8 \times 10^3 \text{ J}$$

          ↑   ↑   ↑   ↑

Energy required   ·Specific   Mass   Temperature
to heat 25 g of   heat   of   change
water from 25 °C capacity   water
to 100. °C

$$= 7.8 \times 10^3 \text{ J} \times \frac{1 \text{ kJ}}{1000 \text{ J}} = 7.8 \text{ kJ}$$

STEP 2: VAPORIZATION

Now we must use the molar heat of vaporization to calculate the energy required to vaporize the 25 g of water at 100. °C. The heat of vaporization is given *per mole* rather than per gram, so we must first convert the 25 g of water to moles.

$$25 \text{ g } H_2O \times \frac{1 \text{ mol } H_2O}{18 \text{ g } H_2O} = 1.4 \text{ mol } H_2O$$

We can now calculate the energy required to vaporize the water.

$$\frac{40.6 \text{ kJ}}{\text{mol } H_2O} \times 1.4 \text{ mol } H_2O = 57 \text{ kJ}$$

        ↑   ↑

Molar heat   Moles of
of   water
vaporization

The total energy is the sum of the two steps.

$$7.8 \text{ kJ} + 57 \text{ kJ} = 65 \text{ kJ}$$

    ↑   ↑

Heat from   Change to
25 °C to   vapor
100. °C

SELF-CHECK EXERCISE 14.1

Calculate the total energy required to melt 15 g of ice at 0 °C, heat the water to 100. °C, and vaporize it to steam at 100. °C.

HINT: Break the process into three steps and then take the sum.

CHEMISTRY IN FOCUS

Boiling Points

*O*ne very interesting fact about water is that its boiling point changes with elevation. Although water boils at 100 °C in Champaign, Illinois, its boiling point is only 90 °C in Breckenridge, Colorado. The reason for this difference is that Champaign is very near sea level and Breckenridge is nearly 10,000 ft above sea level. The variation of water's boiling point with elevation is illustrated in the accompanying table.

Because of this phenomenon, cooking in the mountains is very different from cooking at sea level. It takes much longer to cook an egg or a potato in boiling water at high elevations, because the cooking is taking place at a much lower temperature. For example, because liquid water cannot be heated to a temperature higher than its boiling point, potatoes boil in Breckenridge, Colorado, at 90 °C, instead of 100 °C as they would in Champaign, Illinois.

These campers in the mountains of Pakistan will find that their cooking water boils at a temperature significantly lower than 100 °C.

Boiling Point of Water at Various Locations

Location	Feet Above Sea Level	Boiling Point (°C)
top of Mt. Everest, Tibet	29,028	70
top of Mt. McKinley, Alaska	20,320	79
top of Mt. Whitney, California	14,494	85
Leadville, Colorado	10,150	89
Boulder, Colorado	5,430	94
New York City, New York	10	100
Death Valley, California	−282	100.3

14.3 Intermolecular Forces

AIM: To explain dipole–dipole attraction, hydrogen bonding, and London dispersion forces.

To describe the effect of these forces on the properties of liquids.

We have seen that covalent bonding forces within molecules arise from the sharing of electrons, but how do intermolecular forces arise? Actually several types of

The polarity of a molecule was discussed in Section 12.3.

I always review the meaning of dipole before discussing intermolecular forces. The review helps to remind students that electronegativity and arrangement of atoms are factors that cause some molecules but not others to have dipoles.

See Section 12.2 for a discussion of electronegativity.

Many students believe that dipole–dipole attractions and hydrogen bonds are two unrelated kinds of intermolecular forces, possibly because we put so much emphasis on the formation of hydrogen bonds.

intermolecular forces exist. To illustrate one type, we will consider the forces that exist among water molecules.

As we saw in Chapter 12, water is a polar molecule—it has a dipole moment. When molecules with dipole moments are put together, they orient themselves to take advantage of their charge distributions. Molecules with dipole moments can attract each other by lining up so that the positive and negative ends are close to each other, as shown in Figure 14.5(a). This is called a **dipole–dipole attraction.** In the liquid, the dipoles find the best compromise between attraction and repulsion, as shown in Figure 14.5(b).

Dipole–dipole forces are typically only about 1% as strong as covalent or ionic bonds, and they become weaker as the distance between the dipoles increases. In the gas phase, where the molecules are usually very far apart, these forces are relatively unimportant.

Particularly strong dipole–dipole forces occur between molecules in which hydrogen is bound to a highly electronegative atom, such as nitrogen, oxygen, or fluorine. Two factors account for the strengths of these interactions: the great polarity of the bond and the close approach of the dipoles, which is made possible by the very small size of the hydrogen atom. Because dipole–dipole attractions of this type are so unusually strong, they are given a special name—**hydrogen bonding.** Figure 14.6 illustrates hydrogen bonding among water molecules.

Hydrogen bonding has a very important effect on various physical properties. For example, the boiling points for the covalent compounds of hydrogen

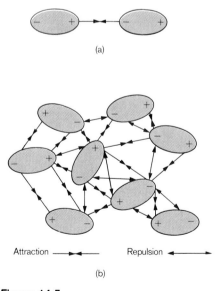

Figure 14.5
(a) The interaction of two polar molecules.
(b) The interaction of many dipoles in a liquid.

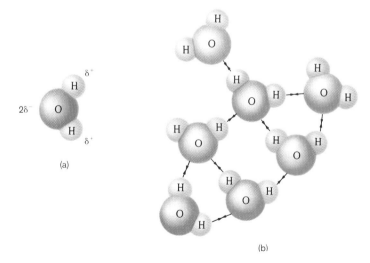

Figure 14.6
(a) The polar water molecule. (b) Hydrogen bonding among water molecules. The small size of the hydrogen atoms allows the molecules to get very close and thus to produce strong interactions.

with the elements in Group 6 are given in Figure 14.7. Note that the boiling point of water is much higher than would be expected from the trend shown by the other members of the series. Why? Because the especially large electronegativity value of the oxygen atom compared to that of the other group members causes the O—H bonds to be much more polar than the S—H, Se—H, or Te—H bonds. This leads to very strong hydrogen bonding forces among the water molecules. An unusually large quantity of energy is required to overcome these interactions and separate the molecules to produce the gaseous state. That is, water molecules tend to remain together in the liquid state even at relatively high temperatures; hence the very high boiling point of water.

However, even molecules without dipole moments must exert forces on each other. We know this because all substances—even the noble gases—exist in the liquid and solid states at very low temperatures. There must be forces to hold the atoms or molecules as close together as they are in these condensed states. The forces that exist among noble gas atoms and nonpolar molecules are called **London dispersion forces.** To understand the origin of these forces, consider a pair of noble gas atoms. Although we usually assume that the electrons of an atom are uniformly distributed about the nucleus (see Figure 14.8a), this is apparently not true at every instant. Atoms can develop a temporary dipolar arrangement of charge as the electrons move around the nucleus (see Figure 14.8b). This *instantaneous dipole* can then *induce* a similar dipole in a neighboring atom, as shown in Figure 14.8(c). The interatomic attraction thus formed is both weak and short-lived, but it can be very significant for large atoms, as we will see.

Figure 14.7
The boiling points of the covalent hydrides of elements in Group 6.

Students often think that any molecule that contains a hydrogen atom or that contains either a nitrogen, fluorine, or oxygen atom forms hydrogen bonds. You might use examples such as NH_3, CH_3OH, and HF to show that in each case the H is bonded directly to the N, O, or F. Produce overhead transparencies showing how the partial positive end of one hydrogen bonding molecule is attracted to the partial negative end of another. Show them that molecules such as $(CH_3)_3N$ do not form hydrogen bonds because there is not a hydrogen bonded to the nitrogen.

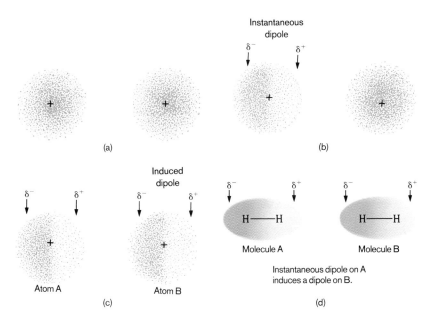

(a)

(b)

(c)

(d)

Instantaneous dipole

δ^- δ^+

Induced dipole

δ^- δ^+ δ^- δ^+

Atom A   Atom B

δ^- δ^+

H——H

Molecule A

δ^- δ^+

H——H

Molecule B

Instantaneous dipole on A induces a dipole on B.

Figure 14.8
(a) Two atoms with spherical electron probability. These atoms have no polarity. (b) The atom on the left develops an instantaneous dipole when more electrons happen to congregate on the left than on the right. (c) The dipole in the atom on the left induces a dipole in the atom on the right. (d) Nonpolar molecules such as H_2 can also develop instantaneous and induced dipoles.

On a time-line of intermolecular forces, I list the characteristics of molecules that exhibit that particular force. All molecules with N—H, O—H, or F—H bonds form hydrogen bonds; all molecules with dipoles form dipole–dipole attractions; and all molecules form London dispersion forces.

Table 14.2 The Freezing Points of the Group 8 Elements	
Element	Freezing Point (°C)
helium*	−269.7
neon	−248.6
argon	−189.4
krypton	−157.3
xenon	−111.9

*Helium will freeze if the pressure is increased above 1 atm.

The motions of the atoms must be greatly slowed down before the weak London dispersion forces can lock the atoms into place to produce a solid. This explains, for instance, why the noble gas elements have such low freezing points (see Table 14.2).

Nonpolar molecules such as H_2, N_2, and I_2, none of which has a permanent dipole moment, also attract each other by London dispersion forces (see Figure 14.8d).

14.4 Evaporation and Vapor Pressure

AIM: To explain the relationship among vaporization, condensation, and vapor pressure.

We all know that a liquid can evaporate from an open container. This is clear evidence that the molecules of a liquid can escape the liquid's surface and form a gas. This process, which is called **vaporization** or **evaporation,** requires energy to overcome the relatively strong intermolecular forces in the liquid.

The fact that vaporization requires energy has great practical significance; in fact, one of the most important roles that water plays in our world is to act as a coolant. Because of the strong hydrogen bonding among its molecules in the liquid state, water has an unusually large heat of vaporization (41 kJ/mol). A significant portion of the sun's energy is spent evaporating water from the oceans, lakes, and rivers rather than warming the earth. The vaporization of water is also crucial to our body's temperature-control system, which relies on the evaporation of perspiration.

Vapor Pressure

When we place a given amount of liquid in a container and then close it, we observe that the amount of liquid at first decreases slightly but eventually be-

Water is used to absorb heat from nuclear reactors. The water is then cooled in cooling towers before it is returned to the environment.

comes constant. The decrease occurs because there is a transfer of molecules from the liquid to the vapor phase (Figure 14.9). However, as the number of vapor molecules increases, it becomes more and more likely that some of them will return to the liquid. The process by which vapor molecules form a liquid is called **condensation.** Eventually, the same number of molecules are leaving the liquid as are returning to it: the rate of condensation equals the rate of evaporation. *At this point no further change occurs in the amounts of liquid or vapor, because the two opposite processes exactly balance each other;* the system is at *equilibrium.* Note that this system is highly *dynamic* on the molecular level— molecules are constantly escaping from and entering the liquid. However, there is no *net* change because the two opposite processes just *balance* each other. As an analogy, consider two island cities connected by a bridge. Suppose the traffic flow on the bridge is the same in both directions. There is motion—we can see the cars traveling across the bridge—but the number of cars in each city is not changing because an equal number enter and leave each one. The result is no *net* change in the number of autos in each city: an equilibrium exists.

The pressure of the vapor present at equilibrium with its liquid is called the *equilibrium vapor pressure* or, more commonly, the **vapor pressure** of the liquid. A simple barometer can be used to measure the vapor pressure of a liquid, as shown in Figure 14.10. Because mercury is so dense, any common liquid injected at the bottom of the column of mercury floats to the top where it produces a vapor, and the pressure of this vapor pushes some mercury out of the tube. When the system reaches equilibrium, the vapor pressure can be determined from the change in the height of the mercury column.

In effect, we are using the space above the mercury in the tube as a closed container for each liquid. However, in this case as the liquid vaporizes, the vapor formed creates a pressure that pushes some mercury out of the tube and lowers the mercury level. The mercury level stops changing when the excess liquid

Vapor, not gas, is the term we customarily use for the gaseous state of a substance that exists naturally as a solid or liquid at 25 °C and 1 atm.

A system at equilibrium is dynamic on the molecular level, but shows no visible changes.

(a)   (b)   (c)

Figure 14.9
Behavior of a liquid in a closed container.
(a) Net evaporation occurs at first, so the amount of liquid decreases slightly. (b) As the number of vapor molecules increases, the rate of condensation increases. (c) Finally the rate of condensation equals the rate of evaporation. The system is at equilibrium.

(a)   (b)   (c)

Figure 14.10
(a) It is easy to measure the vapor pressure of a liquid by using a simple barometer of the type shown here. (b) The water vapor pushed the mercury level down 24 mm (760 − 736), so the vapor pressure of water is 24 mm Hg at this temperature. (c) Diethyl ether is much more volatile than water and thus shows a higher vapor pressure. In this case, the mercury level has been pushed down 545 mm (760 − 215), so the vapor pressure of diethyl ether is 545 mm Hg at this temperature.

Heavy molecules move at lower speeds than light ones and so produce a lower vapor pressure.

To compare the relative vapor pressures of two molecules, students must be able to use several pieces of information. They must be able to judge the strengths of the intermolecular forces in each molecule, and compare the molar masses.

floating on the mercury comes to equilibrium with the vapor. The change in the mercury level (in millimeters) from its initial position (before the liquid was injected) to its final position is equal to the vapor pressure of the liquid.

The vapor pressures of liquids vary widely (see Figure 14.10). Liquids with high vapor pressures are said to be *volatile*—they evaporate rapidly.

The vapor pressure of a liquid at a given temperature is determined by the *intermolecular forces* that act among the molecules. Liquids in which the intermolecular forces are large have relatively low vapor pressures, because such molecules need high energies to escape to the vapor phase. For example, although water is a much smaller molecule than diethyl ether, C_2H_5—O—C_2H_5, the strong hydrogen-bonding forces in water cause its vapor pressure to be much lower than that of ether (see Figure 14.10).

EXAMPLE 14.3	**Using Knowledge of Intermolecular Forces to Predict Vapor Pressure**

Predict which substance in each of the following pairs will show the largest vapor pressure at a given temperature.

a. $H_2O(l)$, $CH_3OH(l)$ b. $CH_3OH(l)$, $CH_3CH_2CH_2CH_2OH(l)$

SOLUTION

a.   Water contains two polar O—H bonds; methanol (CH_3OH) only has one. Therefore the hydrogen bonding among H_2O molecules is expected to be much stronger than that among CH_3OH molecules. This gives water a lower vapor pressure than methanol.

b.   Each of these molecules has one polar O—H bond. However, because $CH_3CH_2CH_2CH_2OH$ is a much larger molecule than CH_3OH, it is much less likely to escape from its liquid. Thus $CH_3CH_2CH_2CH_2OH(l)$ has a lower vapor pressure than $CH_3OH(l)$.

Bring some examples of crystalline solids to class. The regular shapes of crystals of ionic salts and molecular solids such as sucrose can be seen with a magnifying glass, although not with the naked eye. If you have examples of larger crystals such as quartz or calcite, students enjoy seeing them; you can use SiO_2 and $CaCO_3$ as examples. A graphite electrode or a piece of copper wire can be used to illustrate atomic solids.

14.5   The Solid State: Types of Solids

AIM: To describe various types of crystalline solids.

Solids play a very important role in our lives. The concrete we drive on, the trees that shade us, the windows we look through, the paper that holds this print, the diamond in an engagement ring, and the plastic lenses in eyeglasses are all important solids. Most solids, such as wood, paper, and glass, contain mixtures of various components. However, some natural solids, such as diamonds and table salt, are nearly pure substances.

Many substances form **crystalline solids**—those with a regular arrangement of their components. This is illustrated by the partial structure of sodium chloride shown in Figure 14.11. The highly ordered arrangement of the components in a crystalline solid produces beautiful, regularly shaped crystals such as those shown in Figure 14.12.

There are many different types of crystalline solids. For example, both sugar and salt have beautiful crystals that we can easily see. However, although both dissolve readily in water, the properties of the resulting solutions are quite different. The salt solution readily conducts an electric current; the sugar solution does not. This behavior arises from the different natures of the components in these two solids. Common salt, NaCl, is an ionic solid that contains Na^+ and Cl^- ions. When solid sodium chloride dissolves in water, sodium ions and chloride ions are distributed throughout the resulting solution. These ions are free to move through the solution to conduct an electric current. Table sugar (sucrose), on the other hand, is composed of neutral molecules that are dispersed throughout the water when the solid dissolves. No ions are present, and the resulting solution does not conduct electricity. These examples illustrate two important types of crystalline solids: **ionic solids,** represented by sodium chloride; and **molecular solids,** represented by sucrose.

A third type of crystalline solid is represented by elements such as graphite and diamond (both pure carbon), boron, silicon, and all metals. These substances, which contain atoms of only one element covalently bonded to each other, are called **atomic solids.**

We have seen that crystalline solids can be grouped conveniently into three classes as shown in Figure 14.13 (p. 486). Notice that the names of the three classes come from the components of the solid. An ionic solid contains ions, a molecular solid contains molecules, and an atomic solid contains atoms. Examples of the three types of solids are shown in Figure 14.14 (p. 486).

The properties of a solid are determined primarily by the nature of the forces that hold the solid together. For example, although argon, copper, and diamond are all atomic solids (their components are atoms), they have strikingly different properties. Argon has a very low melting point ($-189\ °C$), whereas diamond and copper melt at high temperatures (about $3500\ °C$ and $1083\ °C$,

Cl⁻   Na⁺

Figure 14.11

The regular arrangement of sodium and chloride ions in sodium chloride, a crystalline solid.

The internal forces in a solid determine many of the properties of the solid.

(a)

(b)

(c)

Figure 14.12

Several crystalline solids: (a) quartz, SiO_2; (b) table salt, NaCl; and (c) iron pyrite, FeS_2.

Figure 14.13
The classes of crystalline solids.

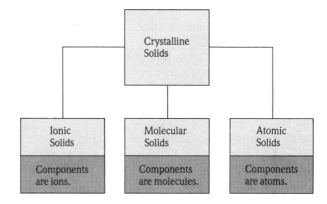

Figure 14.14
Examples of three types of crystalline solids. Only part of the structure is shown in each case. The structures continue in three dimensions with the same patterns. (a) An ionic solid. The spheres represent alternating Na⁺ and Cl⁻ ions in solid sodium chloride. (b) A molecular solid. Each unit of three spheres represents an H_2O molecule in ice. The dashed lines show the hydrogen bonding among the polar water molecules. (c) An atomic solid. Each sphere represents a carbon atom in diamond.

respectively). Copper is an excellent conductor of electricity, whereas argon and diamond are both insulators. The shape of copper can easily be changed; it is both malleable (will form thin sheets) and ductile (can be pulled into a wire). Diamond, on the other hand, is the hardest natural substance known. The marked differences in properties among these three atomic solids are due to differences in bonding. We will explore the bonding in solids in the next section.

14.6 Bonding in Solids

AIM: To discuss the intermolecular forces in crystalline solids.
         To describe how the bonding in metals determines
           metallic properties.

We have seen that crystalline solids can be divided into three classes, depending on the fundamental particle or unit of the solid. Ionic solids consist of oppositely

If you have a ball and stick model of an ionic compound you might want to show it in class, even if you previously used it when discussing ionic compounds.

Styrofoam balls of two different sizes painted two different colors can be glued together to show how the smaller cations fit in the empty spaces between larger anions.

14.6 Bonding in Solids   **487**

charged ions packed together, molecular solids contain molecules, and atomic solids have atoms as their fundamental particles. Examples of the various types of solids are given in Table 14.3.

Ionic Solids

Ionic solids are stable substances with high melting points that are held together by the strong forces that exist between oppositely charged ions. The structures of ionic solids can be visualized best by thinking of the ions as spheres packed together as efficiently as possible. For example, in NaCl the larger Cl^- ions are packed together much like one would pack balls in a box. The smaller Na^+ ions occupy the small spaces ("holes") left among the spherical Cl^- ions, as represented in Figure 14.15.

Ionic solids were also discussed in Section 12.5.

When spheres are packed together, there are many small empty spaces (holes) left among the spheres.

Molecular Solids

In a molecular solid the fundamental particle is a molecule. Examples of molecular solids include ice (contains H_2O molecules), dry ice (contains CO_2 molecules), sulfur (contains S_8 molecules), and white phosphorus (contains P_4 molecules). The latter two substances are shown in Figure 14.16 (p. 488).

It is difficult to demonstrate the three-dimensional structures of molecular solids in two dimensions on the blackboard. Using models can help.

Molecular solids tend to melt at relatively low temperatures because the intermolecular forces that exist among the molecules are relatively weak. If the molecule has a dipole moment, dipole–dipole forces hold the solid together. In solids with nonpolar molecules, London dispersion forces hold the solid together.

Part of the structure of solid phosphorus is represented in Figure 14.17 (p. 488). Note that the distances between P atoms in a given molecule are much shorter than the distances between the P_4 molecules. This is because the covalent bonds *between atoms* in the molecule are so much stronger than the London dispersion forces *between molecules.*

Table 14.3 Examples of the Various Types of Solids.

Type of Solid	Examples	Fundamental Unit(s)
ionic	sodium chloride, $NaCl(s)$	Na^+, Cl^- ions
ionic	ammonium nitrate, $NH_4NO_3(s)$	NH_4^+, NO_3^- ions
molecular	dry ice, $CO_2(s)$	CO_2 molecules
molecular	ice, $H_2O(s)$	H_2O molecules
atomic	diamond, $C(s)$	C atoms
atomic	iron, $Fe(s)$	Fe atoms
atomic	argon, $Ar(s)$	Ar atoms

Cl^- Na^+

Figure 14.15
The packing of Cl^- and Na^+ ions in solid sodium chloride.

Figure 14.16
(a) Sulfur crystals contain S_8 molecules.
(b) White phosphorus contains P_4 molecules. It is so reactive with the oxygen in air that it must be stored under water.

(a)   (b)

Figure 14.17
A representation of part of the structure of solid phosphorus, a molecular solid that contains P_4 molecules.

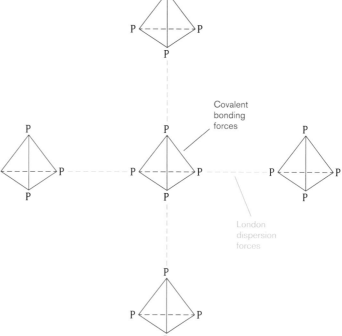

Atomic Solids

The properties of atomic solids vary greatly because of the different ways in which the fundamental particles, the atoms, can interact with each other. For example, the solids of the Group 8 elements have very low melting points (see Table 14.2), because these atoms, having filled valence orbitals, cannot form covalent bonds with each other. So the forces in these solids are the relatively weak London dispersion forces.

On the other hand, diamond, a form of solid carbon, is one of the hardest substances known and has an extremely high melting point (about 3500 °C). The incredible hardness of diamond arises from the very strong covalent carbon–carbon bonds in the crystal, which lead to a giant molecule. In fact, the entire crystal can be viewed as one huge molecule. A small part of the diamond structure is represented in Figure 14.14. In diamond each carbon atom is bound covalently to four other carbon atoms to produce a very stable solid. Several other elements also form solids where the atoms join together covalently to form giant molecules. Silicon and boron are examples.

At this point you might be asking yourself, "Why aren't solids such as a crystal of diamond, which is a 'giant molecule,' classed as molecular solids?" The answer is that, by convention, a solid is classed as a molecular solid only if (like ice, dry ice, sulfur, and phosphorus) it contains small molecules.

Bonding in Metals

Metals represent another type of atomic solid. Metals have familiar physical properties: they can be pulled into wires, can be hammered into sheets, and are efficient conductors of heat and electricity. However, although the shapes of most pure metals can be changed relatively easily, metals are also durable and have high melting points. These facts indicate that it is difficult to separate metal atoms but relatively easy to slide them past each other. In other words, the bonding in most metals is *strong* but *nondirectional.*

The simplest picture that explains these observations is the **electron sea model,** which pictures a regular array of metal atoms in a "sea" of valence electrons that are shared among the atoms in a nondirectional way and that are quite mobile in the metal crystal. The mobile electrons can conduct heat and electricity, and the cations can be moved rather easily, as, for example, when the metal is hammered into a sheet or pulled into a wire.

Because of the nature of the metallic crystal, other elements can be introduced relatively easily to produce substances called alloys. An **alloy** is best defined as *a substance that contains a mixture of elements and has metallic properties.* There are two common types of alloys.

In a **substitutional alloy** some of the host metal atoms are *replaced* by other metal atoms of similar sizes. For example, in brass approximately one-third of the atoms in the host copper metal have been replaced by zinc atoms, as shown in Figure 14.18(a). Sterling silver (93% silver and 7% copper), pewter (85% tin, 7% copper, 6% bismuth, and 2% antimony), and plumber's solder (67% lead and 33% tin) are other examples of substitutional alloys.

An **interstitial alloy** is formed when some of the interstices (holes) among the closely packed metal atoms are occupied by atoms much smaller than the host atoms, as shown in Figure 14.18(b). Steel, the best-known interstitial alloy, contains carbon atoms in the "holes" of an iron crystal. The presence of interstitial

There are many common alloys you can bring to class, such as a sterling silver spoon, a piece of brass, or a piece of steel.

Figure 14.18

Two types of alloys. (a) Brass is a substitutional alloy in which copper atoms in the host crystal are replaced by the similarly sized zinc atoms. (b) Steel is an interstitial alloy in which carbon atoms occupy interstices (holes) among the closely packed iron atoms.

atoms changes the properties of the host metal. Pure iron is relatively soft, ductile, and malleable because of the absence of strong directional bonding. The spherical metal atoms can be moved rather easily with respect to each other. However, when carbon, which forms strong directional bonds, is introduced into an iron crystal, the presence of the directional carbon–iron bonds makes the resulting alloy harder, stronger, and less ductile than pure iron. The amount of carbon directly affects the properties of steel. *Mild steels* (containing less than 0.2% carbon) are still ductile and malleable and are used for nails, cables, and chains. *Medium steels* (containing 0.2–0.6% carbon) are harder than mild steels and are used in rails and structural steel beams. *High-carbon steels* (containing 0.6–1.5% carbon) are tough and hard and are used for springs, tools, and cutlery.

Styrofoam balls of various sizes and colors can be used to demonstrate alloys of metals as well as to demonstrate structure of ionic compounds.

Many types of steel also contain elements in addition to iron and carbon. Such steels are often called *alloy steels* and can be viewed as being mixed interstitial (carbon) and substitutional (other metals) alloys. An example is stainless steel, which has cobalt and nickel atoms substituted for some of the iron atoms. The addition of these metals greatly increases the steel's resistance to corrosion.

EXAMPLE 14.4	Identifying Types of Crystalline Solids

Name the type of crystalline solid formed by each of the following substances:

a. ammonia
b. iron
c. cesium fluoride

d. argon
e. sulfur

SOLUTION

a. Solid ammonia contains NH_3 molecules, so it is a molecular solid.
b. Solid iron contains iron atoms as the fundamental particles. This is an atomic solid.
c. Solid cesium fluoride contains the Cs^+ and F^- ions. It is an ionic solid.
d. Solid argon contains argon atoms, which cannot form covalent bonds to each other. This is an atomic solid.
e. Sulfur contains S_8 molecules, so it is a molecular solid.

SELF-CHECK EXERCISE 14.2

Name the type of crystalline solid formed by each of the following:

a. sulfur trioxide
b. barium oxide
c. gold

CHAPTER REVIEW

Key Terms

normal boiling point (p. 473)
heating/cooling curve (p. 474)
normal freezing point (p. 474)
intramolecular forces (p. 475)
intermolecular forces (p. 475)
molar heat of fusion (p. 475)
molar heat of vaporization (p. 475)
dipole–dipole attraction (p. 480)
hydrogen bonding (p. 480)
London dispersion forces (p. 481)
vaporization (evaporation) (p. 482)
condensation (p. 483)

vapor pressure (p. 483)
crystalline solid (p. 485)
ionic solid (p. 485)
molecular solid (p. 485)
atomic solid (p. 485)
alloy (p. 489)

Summary

1. Liquids and solids exhibit some similarities and are very different from the gaseous state.
2. The temperature at which a liquid changes its state to a gas (at 1 atm pressure) is called the normal boiling point of that liquid. Similarly, the temperature at which a liquid freezes (at 1 atm pressure) is the normal freezing point. Changes of state are physical changes, not chemical changes.
3. To convert a substance from the solid to the liquid and then to the gaseous state requires the addition of energy. Forces among the molecules in a solid or a liquid must be overcome by the input of energy. The energy required to melt 1 mol of a substance is called the molar heat of fusion, and the energy required to change 1 mol of liquid to the gaseous state is called the molar heat of vaporization.
4. There are several types of intermolecular forces. Dipole–dipole interactions occur when molecules with dipole moments attract each other. A particularly strong dipole–dipole interaction called hydrogen bonding occurs in molecules that contain hydrogen bonded to a very electronegative element such as N, O, or F. London dispersion forces occur when instantaneous dipoles in atoms or nonpolar molecules lead to relatively weak attractions.
5. The change of a liquid to its vapor is called vaporization or evaporation. The process where vapor molecules form a liquid is called condensation. In a closed container, the pressure of the vapor over its liquid reaches a constant value called the vapor pressure of the liquid.

6. Many solids are crystalline (contain highly regular arrangements of their components). The three types of crystalline solids are ionic, molecular, and atomic solids. In ionic solids, the ions are packed together in a way that maximizes the attractions of oppositely charged ions and minimizes the repulsions among identically charged ions. Molecular solids are held together by dipole–dipole attractions if the molecules are polar and by London dispersion forces if the molecules are nonpolar. Atomic solids are held together by covalent bonding forces or London dispersion forces, depending on the atoms present.

Questions and Problems

All even-numbered exercises have answers in the back of this book and solutions in the Solutions Guide.

14.1 Water and Its Phase Changes

QUESTIONS

1. Approximately what fraction of the earth's surface is covered by water? How does water help to moderate the environment of the earth?

2. Describe some uses, both in nature and in industry, of water as a *cooling* agent.

3. What are the freezing point and the boiling point of water at 1 atm pressure?

4. Discuss some implications of the fact that, unlike most substances, water *expands* in volume when it freezes.

5. Describe, on both a microscopic and a macroscopic basis, what happens to a sample of water as it is heated from room temperature to 50 °C above its normal boiling point.

6. Figure 14.2 presents the *cooling curve* for water. Discuss the meaning of the different portions of this curve (for example, explain what each flat section and each sloping section represents).

14.2 Energy Requirements for the Changes of State

QUESTIONS

7. Are changes of state (melting, vaporization, and so on) physical or chemical changes? Explain how we know this.

8. Describe what happens on a microscopic basis when a liquid is *boiled.*

9. Explain the difference between *intra*molecular forces and *inter*molecular forces. Which forces must be overcome in order to melt a solid or to vaporize a liquid?

10. Energy must be applied to a liquid to make it boil. Where does this energy finally reside?

11. Discuss the similarities and differences between the arrangements of molecules and the forces between molecules in liquid water versus steam, and in liquid water versus ice.

12. The energy required to melt 1 mol of ice is ≈6 kJ/mol, whereas the energy required to vaporize 1 mol of water is ≈41 kJ/mol. Why is nearly seven times more energy needed to vaporize the same amount of water as to melt it?

PROBLEMS

13. The following data have been collected for substance X. Construct a heating curve for substance X. (The drawing does not need to be absolutely to scale, but it should clearly show relative differences.)

normal melting point	−15 °C
molar heat of fusion	2.5 kJ/mol
normal boiling point	134 °C
molar heat of vaporization	55.3 kJ/mol

14. Consider the data for substance X given in Problem 13. If the molar mass of substance X is 52 g/mol, what quantity of heat is required to melt 10.0 g of substance X at −15 °C? What quantity of heat is required to vaporize 25.0 g of substance X at 134 °C?

15. The molar heat of fusion of elemental iodine is 16.7 kJ/mol at its normal melting point of 114 °C. What quantity of heat is required to melt 1.0 g of iodine at 114 °C?

16. The molar heats of fusion and vaporization for water are 6.02 kJ/mol and 40.6 kJ/mol, respectively, and the specific heat capacity of liquid water is 4.18 J/g °C. What total quantity of heat energy is required to melt 5.0 g of ice at 0 °C, to heat the liquid water from 0 °C to 100 °C, and to vaporize the liquid water at 100 °C?

17. Given that the specific heat capacities of ice and steam are 2.06 J/g °C and 2.03 J/g °C, respectively, and considering the information about water given in Problem 16, calculate the total quantity of heat evolved when 10.0 g of steam at 200. °C is condensed, cooled, and frozen to ice at −50. °C.

18. The heat of fusion of aluminum is 3.95 kJ/g. What is the *molar* heat of fusion of aluminum? What quantity of energy is needed to melt 10.0 g of aluminum? What quantity of energy is required to melt 10.0 mol of aluminum?

14.3 Intermolecular Forces

QUESTIONS

19. What is a *dipole–dipole* attraction? Give three examples of liquid substances in which you would expect dipole–dipole attractions to be large.

20. How is the strength of dipole–dipole interactions related to the *distance* between polar molecules? Are dipole–dipole forces short-range or long-range forces?

21. Why are dipole–dipole attractions between polar molecules not important in the vapor phase?

22. Water has a much higher boiling point than would be expected. Explain.

23. List some physical properties of a substance that are influenced by the presence of strong hydrogen bonding within the substance.

24. Although the noble gas elements are monatomic and could not give rise to dipole–dipole forces or hydrogen bonding, these elements still can be liquefied and solidified. Explain.

PROBLEMS

25. Discuss the types of intermolecular forces acting in the liquid state of each of the following substances.
    a. P_4 c. H_2S
    b. HCl   d. HF

26. Discuss the types of intermolecular forces acting in the liquid state of each of the following substances.
    a. Kr   c. NF_3
    b. S_8 d. H_2O

27. The boiling points of the noble gas elements are listed below. Comment on the trend in the boiling points. Why do the boiling points vary in this manner?

He	−268.9 °C	Kr	−152.3 °C
Ne	−245.9 °C	Xe	−107.1 °C
Ar	−185.7 °C	Rn	−61.8 °C

28. The heats of fusion of three substances are listed below. Explain the trend this list reflects.

HI	2.87 kJ/mol
HBr	2.41 kJ/mol
HCl	1.99 kJ/mol

29. Deuterium is the hydrogen isotope with mass number 2 (a neutron is present in deuterium that is not present in normal hydrogen). How would you expect the boiling and melting points of deuterium oxide, D_2O, often called "heavy water," to differ from those of ordinary water, H_2O?

30. When 50 mL of liquid water at 25 °C is added to 50 mL of ethanol (ethyl alcohol) also at 25 °C, the combined volume of the mixture is considerably *less* than 100 mL. Give a possible explanation.

14.4 Evaporation and Vapor Pressure

QUESTIONS

31. Describe, on a microscopic basis, the processes of *evaporation* and *condensation*. Which process requires the input of energy?

32. What is *vapor pressure*? On a microscopic basis, how does a vapor pressure develop in a closed flask containing a small amount of liquid? What processes are going on in the flask?

33. What do we mean by a *dynamic equilibrium*? Describe how the development of a vapor pressure above a liquid represents such an equilibrium.

34. How is the vapor pressure of a liquid affected by the *molar mass* of the liquid's molecules? All other considerations being equal, which would tend to be more volatile, a high molar mass substance or a low molar mass substance?

PROBLEMS

35. Which substance in each pair would be expected to be more volatile at a particular temperature? Explain your reasoning.
    a. $H_2O(l)$ or $H_2S(l)$
    b. $H_2O(l)$ or $CH_3OH(l)$
    c. $CH_3OH(l)$ or $CH_3CH_2OH(l)$

37. Although water and ammonia differ in molar mass by only one unit, the boiling point of water is over 100 °C higher than that of ammonia. Explain.

36. Which substance in each pair would be expected to show the largest vapor pressure at a given temperature? Explain your reasoning.
    a. $H_2O(l)$ or $HF(l)$
    b. $CH_3OCH_3(l)$ or $CH_3CH_2OH(l)$
    c. $CH_3OH(l)$ or $CH_3SH(l)$

38. Two molecules that contain the same number of each kind of atom but that have different molecular structures are said to be *isomers* of each other. For example, ethyl alcohol and dimethyl ether (shown below) both have the formula C_2H_6O and are isomers. Based on considerations of intermolecular forces, which substance would you expect to be more volatile? Which would you expect to have the higher boiling point? Explain.

    dimethyl ether $\quad$ ethyl alcohol
    CH_3-O-CH_3 $\qquad$ CH_3-CH_2-OH

14.5 The Solid State: Types of Solids

QUESTIONS

39. What are crystalline solids? What sort of microscopic structure do such solids have? How is this microscopic structure reflected in the macroscopic appearance of such solids?

40. On the basis of the smaller units that make up the crystals, cite three types of crystalline solids. For each type of crystalline solid, give an example of a substance that forms that type of solid.

14.6 Bonding in Solids

QUESTIONS

41. What are the fundamental particles in ionic solids? Give two examples of ionic solids, and tell what individual particles make them up.

43. What are the fundamental particles in atomic solids? Give two examples of substances that exist as atomic solids, and tell what individual particles make them up.

45. What sorts of forces exist between the individual particles in an ionic solid? Are these forces relatively strong or relatively weak?

47. Explain and compare the intermolecular forces that exist in a sample of solid krypton, Kr, with those that exist in diamond, C.

49. What is an *alloy*? Explain the differences in structure between substitutional and interstitial alloys. Give an example of each type.

42. What are the fundamental particles in molecular solids? Give two examples of molecular solids, and tell what individual particles make them up.

44. Ionic solids typically have melting points hundreds of degrees higher than the melting points of molecular solids. Explain.

46. Ordinary ice (solid water) melts at 0 °C, whereas dry ice (solid carbon dioxide) melts at a much lower temperature. Explain the differences in the melting points of these two substances on the basis of the intermolecular forces involved.

48. How do we explain that a metal (such as copper) conducts electricity well in the solid state, whereas an ionic substance (such as copper sulfate) does *not* conduct electricity in the solid state even though the ionic substance consists of electrically charged particles.

50. Explain how the properties of a metal may be modified by alloying the metal with some other substance. Discuss, in particular, how the properties of iron are modified in producing the various types of steel.

Additional Problems

Column 1

51. boiling point at pressure of 1 atm
52. energy required to melt 1 mol of a substance
53. forces between atoms in a molecule
54. forces between molecules in a solid
55. instantaneous dipole forces for nonpolar molecules
56. lining up of opposite charges on adjacent polar molecules
57. maximum pressure of vapor that builds up in a closed container
58. mixture of elements having metallic properties overall
59. repeating arrangement of component species in a solid
60. solids that melt at relatively low temperatures

Column 2

a. alloy
b. specific heat
c. crystalline solid
d. dipole–dipole attraction
e. equilibrium vapor pressure
f. intermolecular
g. intramolecular
h. ionic solids
i. London dispersion forces
j. molar heat of fusion
k. molar heat of vaporization
l. molecular solids
m. normal boiling point
n. semiconductor

61. Given the densities and conditions of ice, liquid water, and steam listed in Table 14.1, calculate the volume of 1.0 g of water under each of these circumstances.
62. As you will see in Chapter 20, in carbon compounds a given group of atoms can often be arranged in more than one way. This means that more than one structure may be possible for the same atoms. For example, the molecules dimethyl ether and ethanol both have the same number of each type of atom, but they have different structures and are said to be isomers of one another.

$$\text{dimethyl ether} \quad CH_3—O—CH_3$$
$$\text{ethanol} \quad CH_3—CH_2—OH$$

Which substance would you expect to have the larger vapor pressure? Why?
63. Which of the substances in each of the following sets would be expected to have the highest boiling point? Explain why.
a. Ga, KBr, O_2 b. Hg, $NaCl$, He c. H_2, O_2, H_2O
64. Which of the substances in each of the following sets would be expected to have the lowest melting point. Explain why.
a. H_2, N_2, O_2 c. Cl_2, Br_2, I_2
b. Xe, $NaCl$, C(diamond)
65. When a person has a severe fever, one therapy to reduce the fever is an "alcohol rub." Explain how the evaporation of alcohol from the person's skin removes heat energy from the body.
66. What is steel? How do the properties of steel differ from the properties of its constituents?
67. Some properties of aluminum are summarized in the following list.

normal melting point	658 °C
heat of fusion	3.95 kJ/g
normal boiling point	2467 °C
heat of vaporization	10.52 kJ/g
specific heat of the solid	0.902 J/g °C

a. Calculate the quantity of energy required to heat 1.00 mol of aluminum from 25 °C to its normal melting point.
b. Calculate the quantity of heat required to melt 1.00 mol of aluminum at 658 °C.
c. Calculate the amount of heat required to vaporize 1.00 mol of aluminum at 2467 °C.
68. What are some important uses of water, both in nature and in industry? What is the liquid range for water?
69. Describe, on both a microscopic and a macroscopic basis, what happens to a sample of water as it is cooled from room temperature to 50 °C below its normal freezing point.
70. Cake mixes and other packaged foods that require cooking often contain special directions for use at high altitudes. Typically these directions indicate that the food should be cooked longer above 5000 ft in altitude. Explain why it takes longer to cook something at higher altitudes.
71. Why is there no change in *intra*molecular forces when a solid is melted? Are intramolecular forces stronger or weaker than intermolecular forces?
72. What do we call the energies required, respectively, to melt and to vaporize 1 mol of a substance? Which of these energies is always larger for a given substance? Why?
73. The molar heat of vaporization of carbon disulfide, CS_2, is 28.4 kJ/mol at its normal boiling point of 46 °C. How much energy (heat) is required to vaporize 1.0 g of CS_2 at 46 °C? How much heat is evolved when 50. g of CS_2 is condensed from the vapor to the liquid form at 46 °C? Which is stronger, a dipole–dipole attraction between two

molecules or a covalent bond between two atoms within the same molecule? Explain.

75. What is *hydrogen bonding* and how does it arise? Give three examples of substances in which you would expect hydrogen bonding to be important in the liquid state of the substance.

76. What are *London dispersion forces* and how do they arise in a nonpolar molecule? Are London forces typically stronger or weaker than dipole–dipole attractions between polar molecules? Are London forces stronger or weaker than covalent bonds? Explain.

77. Discuss the types of intermolecular forces acting in the liquid state of each of the following substances.

   a. N_2
   b. NH_3
   c. He
   d. CO_2 (linear, nonpolar)

78. Explain how the evaporation of water acts as a coolant for the earth.

79. What do we mean when we say a liquid is *volatile*? Do volatile liquids have large or small vapor pressures? What sorts of intermolecular forces occur in highly volatile liquids?

80. Although methane, CH_4, and ammonia, NH_3, differ in molar mass by only one unit, the boiling point of ammonia is over 100 °C higher than that of methane (a nonpolar molecule). Explain.

81. Which type of solid is likely to have the highest melting point, an ionic solid, a molecular solid, or an atomic solid? Explain.

82. What sorts of intermolecular forces exist in a crystal of ice? How do these forces differ from the sorts of intermolecular forces that exist in a crystal of solid oxygen?

83. Discuss the *electron sea* model for metals. How does this model account for the fact that metals are very good conductors of electricity?

Solutions to Self-Check Exercises

SELF-CHECK EXERCISE 14.1

Energy to melt the ice:

$$15 \text{ g } H_2O \times \frac{1 \text{ mol } H_2O}{18 \text{ g } H_2O} = 0.83 \text{ mol } H_2O$$

$$0.83 \text{ mol } H_2O \times 6.02 \frac{kJ}{\text{mol } H_2O} = 5.0 \text{ kJ}$$

Energy to heat the water from 0 °C to 100 °C:

$$4.18 \frac{J}{\text{g °C}} \times 15 \text{ g} \times 100 \text{ °C} = 6300 \text{ J}$$

$$6300 \text{ J} \times \frac{1 \text{ kJ}}{1000 \text{ J}} = 6.3 \text{ kJ}$$

Energy to vaporize the water at 100 °C:

$$0.83 \text{ mol } H_2O \times 40.6 \frac{kJ}{\text{mol } H_2O} = 34 \text{ kJ}$$

Total energy required:

$$5.0 \text{ kJ} + 6.3 \text{ kJ} + 34 \text{ kJ} = 45 \text{ kJ}$$

SELF-CHECK EXERCISE 14.2

a. Contains SO_3 molecules—a molecular solid.
b. Contains Ba^{2+} and O^{2-} ions—an ionic solid.
c. Contains Au atoms—an atomic solid.

15 Solutions

A close-up of a dye dissolving in water.

CONTENTS

Most of the important chemistry that keeps plants, animals, and humans functioning occurs in aqueous solutions. We also encounter many chemical solutions in our daily lives: air, tap water, shampoo, orange soda, coffee, gasoline, cough syrup, and many others.

A **solution** is a homogeneous mixture, a mixture in which the components are uniformly intermingled. This means that a sample from one part is the same as a sample from any other part. For example, the first sip of lowfat milk is the same as the last sip.

The atmosphere that surrounds us is a gaseous solution containing $O_2(g)$, $N_2(g)$, and other gases randomly dispersed. Solutions can also be solids. For example, the various types of steel are mixtures of iron, carbon, and other substances such as manganese, chromium, and molybdenum. Brass is a homogeneous mixture—a solution—of copper and zinc.

These examples illustrate that a solution can be a gas, a liquid, or a solid (see Table 15.1). The substance present in the largest amount is called the **solvent,** and the other substance or substances are called **solutes.** For example, when we dissolve a teaspoon of sugar in a glass of water, the sugar is the solute and the water is the solvent.

Aqueous solutions are solutions with water as the solvent. Because they are so important, we will concentrate in this chapter on the properties of aqueous solutions.

You may have to work hard to convince students that a mixture of either solids or gases can be called a solution. Initially, they believe that all solutions contain water.

15.1   Solubility

AIM: To describe the process of dissolving.
        To explain why certain components dissolve in water.

What happens when you put a teaspoon of sugar in your iced tea and stir it, or when you add salt to water for cooking vegetables? Why do the sugar and salt "disappear" into the water? What does it mean when something dissolves—that is, when a solution forms?

We saw in Chapter 7 that when sodium chloride dissolves in water, the resulting solution conducts an electric current. This convinces us that the solution contains *ions* that can move (this is how the electric current is conducted). The dissolving of solid sodium chloride in water is represented in Figure 15.1. Notice that in the solid state the ions are packed closely together. However, when the solid dissolves, the ions are separated and dispersed throughout the solution. The strong ionic forces that hold the sodium chloride crystal together are overcome by the strong attractions between the ions and the polar water molecules. This process is represented in Figure 15.2 (p. 500). Notice that each polar water

Figure 15.1
When solid sodium chloride dissolves, the ions are dispersed randomly throughout the solution.

Dissolving an ionic compound such as $CuSO_4$ in water can demonstrate the solubility of ionic compounds in water and the homogeneity of solutions.

Table 15.1 Various Types of Solutions

Example	State of Solution	Original State of Solute	State of Solvent
air, natural gas	gas	gas	gas
vodka in water, antifreeze in water	liquid	liquid	liquid
brass, steel	solid	solid	solid
carbonated water (soda)	liquid	gas	liquid
seawater, sugar solution	liquid	solid	liquid

molecule orients itself in a way to maximize its attraction with a Cl^- or Na^+ ion. The negative end of a water molecule is attracted to a Na^+ ion, while the positive end is attracted to a Cl^- ion. The strong forces holding the positive and negative ions in the solid are replaced by strong water–ion interactions, and the solid dissolves (the ions disperse).

It is important to remember that when an ionic substance (such as a salt) dissolves in water, it breaks up into *individual* cations and anions, which are dispersed in the water. For instance, when ammonium nitrate, NH_4NO_3, dissolves in water, the resulting solution contains NH_4^+ and NO_3^- ions, which move around independently. This process can be represented as

$$NH_4NO_3(s) \xrightarrow{H_2O(l)} NH_4^+(aq) + NO_3^-(aq)$$

where (aq) indicates that the ions are surrounded by water molecules.

Water also dissolves many nonionic substances. Sugar is one example of a nonionic solute that is very soluble in water. Another example is ethanol, C_2H_5OH. Wine, beer, and mixed drinks are aqueous solutions of ethanol (and other substances). Why is ethanol so soluble in water? The answer lies in the structure of the ethanol molecule (Figure 15.3a on page 501). The molecule contains a polar O—H bond like those in water, which makes it very compatible with water. Just as hydrogen bonds form among water molecules in pure water (see Figure 14.6), ethanol molecules can form hydrogen bonds with water molecules in a solution of the two. This is shown in Figure 15.3(b).

The sugar molecule (common table sugar has the chemical name sucrose) is shown in Figure 15.4 (p. 501). Notice that this molecule has many polar O—H groups, each of which can hydrogen-bond to a water molecule. Because of the attractions between sucrose and water molecules, solid sucrose is quite soluble in water.

Many substances do not dissolve in water. For example, when petroleum leaks from a damaged tanker, it does not disperse uniformly in the water (does not dissolve) but rather floats on the surface because its density is less than that of

Cations are positive ions. Anions are negative ions.

Remind students that polyatomic ions such as NO_3^- stay together when a compound is dissolved in water because they contain covalent bonds.

Oil floats on the water as it is released from a damaged oil tanker in Alaska.

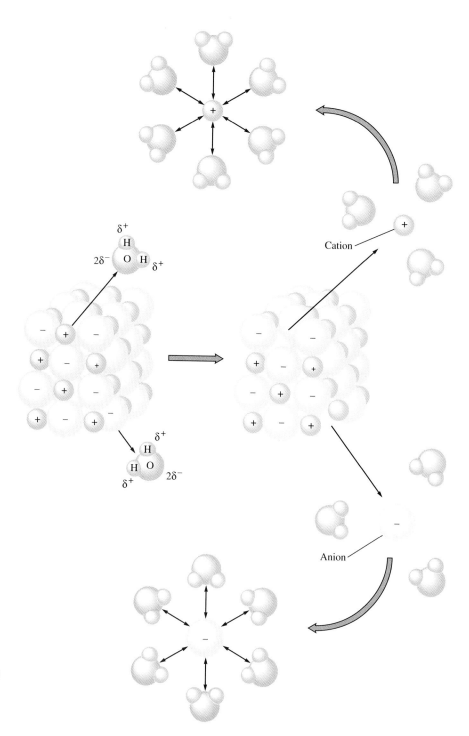

Figure 15.2
Polar water molecules interact with the positive and negative ions of a salt. These interactions replace the strong ionic forces holding the ions together in the undissolved solid, thus assisting in the dissolving process.

Figure 15.3

(a) The ethanol molecule contains a polar O—H bond similar to those in the water molecule. (b) The polar water molecule interacts strongly with the polar O—H bond in ethanol.

Figure 15.4

The structure of common table sugar (called sucrose). The large number of polar O—H groups in the molecule causes sucrose to be very soluble in water.

water. Petroleum is a mixture of molecules like the one shown in Figure 15.5. Since carbon and hydrogen have very similar electronegativities, the bonding electrons are shared almost equally and the bonds are essentially nonpolar. The resulting molecule with its nonpolar bonds cannot form attractions to the polar water molecules and this prevents it from being soluble in water. This situation is represented in Figure 15.6 (p. 502).

Notice in Figure 15.6 that the water molecules in liquid water are associated with each other by hydrogen-bonding interactions. In order for a solute to dissolve in water, a "hole" must be made in the water structure for each solute particle. This will occur only if the lost water–water interactions are replaced by similar water–solute interactions. In the case of sodium chloride, strong interactions occur between the polar water molecules and the Na^+ and Cl^- ions. This

Students readily remember that like dissolves like. It is harder for them to correctly decide which substances are really alike and which are not. Compare the structures of some small molecules such as CH_3OH with that of water. If students can see that some molecules are structurally similar to the polar water molecule, they will be prepared to determine those molecules that will be soluble in water, and those that will not.

CH₃ CH₂ CH₂ CH₂ CH₂ CH₂ CH₂ CH₂ CH₂ CH₃
   CH₂ CH₂ CH₂ CH₂ CH₂ CH₂ CH₂ CH₂

Figure 15.5

A molecule typical of those found in petroleum. The bonds are not polar.

Oil layer

Water

Figure 15.6
An oil layer floating on water. For a substance to dissolve, the water–water hydrogen bonds must be broken to make a "hole" for each solute particle. However, the water–water interactions will break only if they are replaced by similar strong interactions with the solute.

While you talk, you can demonstrate both unsaturated and saturated solutions by dissolving more and more of a compound such as sucrose in water until no more dissolves.

Concentrated and dilute solutions of compounds such as $CuSO_4$ or $NiNO_3$ can be used to demonstrate the meaning of the terms concentrated and dilute. By preparing several dilutions, you can show that one solution can be concentrated with respect to a more dilute solution, but can also be more dilute than a third solution.

You can mix together water containing food color with a nonpolar substance such as vegetable oil to demonstrate that unlike substances do not mix. This simple demonstration doesn't generate any chemical waste requiring special disposal.

allows the sodium chloride to dissolve. In the case of ethanol or sucrose, hydrogen-bonding interactions can occur between the O—H groups on these molecules and water molecules, making these substances soluble as well. But oil molecules are not soluble in water, because the many water–water interactions that would have to be broken to make "holes" for these large molecules are not replaced by favorable water–solute interactions.

These considerations account for the observed behavior *"like dissolves like."* In other words, we observe that a given solvent usually dissolves solutes that have polarities similar to itself. For example, water dissolves most polar solutes, because the solute–solvent interactions formed in the solution are similar to the water–water interactions present in the pure solvent. Likewise, nonpolar solvents dissolve nonpolar solutes. For example, dry cleaning solvents used for removing grease stains from clothes are nonpolar liquids. "Grease" is composed of nonpolar molecules, so a nonpolar solvent is needed to remove a grease stain.

15.2   Solution Composition: An Introduction

AIM: To define qualitative terms associated with the concentration of a solution.

Even for very soluble substances, there is a limit to how much solute can be dissolved in a given amount of solvent. For example, when you add sugar to a glass of water, the sugar rapidly disappears at first. However, as you continue to add more sugar, at some point the solid no longer dissolves but collects at the bottom of the glass. When a solution contains as much solute as will dissolve at that temperature, we say it is **saturated.** If a solid solute is added to a solution already saturated with that solute, the added solid does not appear to dissolve. A solution that has *not* reached the limit of solute that will dissolve in it is said to be **unsaturated.** When more solute is added to an unsaturated solution, it dissolves.

Although a chemical compound always has the same composition, a solution is a mixture and the amounts of the substances present can vary in different solutions. For example, coffee can be strong or weak. Strong coffee has more coffee dissolved in a given amount of water than weak coffee. To describe a solution completely, we must specify the amounts of solvent and solute. We sometimes use the qualitative terms *concentrated* and *dilute* to describe a solution. A relatively large amount of solute is dissolved in a **concentrated** solution (strong coffee is concentrated). A relatively small amount of solute is dissolved in a **dilute** solution (weak coffee is dilute).

Although these qualitative terms serve a useful purpose, we often need to know the exact amount of solute present in a given amount of solution. In the next several sections, we will consider various ways to describe the composition of a solution.

15.3   Solution Composition: Mass Percent

AIM: To define the concentration term *mass percent* and show how to calculate it.

Describing the composition of a solution means giving the amount of solute present in a given quantity of the solution. We typically give the amount of solute in terms of mass (number of grams) or in terms of moles. The quantity of solution is defined in terms of mass or volume.

One common way of describing a solution's composition is **mass percent** (sometimes called *weight percent*), which expresses the mass of solute present in a given mass of solution. The definition of mass percent is:

$$\text{Mass percent} = \frac{\text{mass of solute}}{\text{mass of solution}} \times 100\%$$

$$= \frac{\text{grams of solute}}{\text{grams of solute} + \text{grams of solvent}} \times 100\%$$

For example, suppose a solution is prepared by dissolving 1.0 g of sodium chloride in 48 g of water. The solution has a mass of 49 g (48 g of H_2O plus 1.0 g of NaCl), and there is 1.0 g of solute (NaCl) present. The mass percent of solute, then, is

$$\frac{1.0 \text{ g solute}}{49 \text{ g solution}} \times 100\% = 0.020 \times 100\% = 2.0\% \text{ NaCl}$$

When calculating mass percent, students often forget to calculate the total mass of solution, that is, to add the mass of the solute to the mass of the solvent. The most common mistake students make is to use only the mass of the solvent. They have this problem in part because their ideas about percentages are vague and incorrect.

The mass of the solution is the sum of the masses of the solute and the solvent.

EXAMPLE 15.1   Solution Composition: Calculating Mass Percent

A solution is prepared by mixing 1.00 g of ethanol, C_2H_5OH, with 100.0 g of water. Calculate the mass percent of ethanol in this solution.

SOLUTION

In this case we have 1.00 g of solute (ethanol) and 100.0 g of solvent (water). We now apply the definition of mass percent.

With each problem I work, I ask students to identify which substance is the solute and which is the solvent.

EXAMPLE I5.I, CONTINUED

$$\text{Mass percent } C_2H_5OH = \left(\frac{\text{grams of } C_2H_5OH}{\text{grams of solution}}\right) \times 100\%$$

$$= \left(\frac{1.00 \text{ g } C_2H_5OH}{100.0 \text{ g } H_2O + 1.00 \text{ g } C_2H_5OH}\right) \times 100\%$$

$$= \frac{1.00 \text{ g}}{101.0 \text{ g}} \times 100\%$$

$$= 0.990\% \ C_2H_5OH$$

SELF-CHECK EXERCISE I5.I

A 135-g sample of seawater is evaporated to dryness, leaving 4.73 g of solid residue (the salts formerly dissolved in the seawater). Calculate the mass percent of solute present in the original seawater.

EXAMPLE I5.2	Solution Composition: Determining Mass of Solute

Cow's milk typically contains 4.5% by mass of the sugar lactose, $C_{12}H_{22}O_{11}$. Calculate the mass of lactose present in 175 g of milk.

SOLUTION

We are given the following information:

$$\text{Mass of solution (milk)} = 175 \text{ g}$$
$$\text{Mass percent of solute (lactose)} = 4.5\%$$

We need to calculate the mass of solute (lactose) present in 175 g of milk. Using the definition of mass percent, we have

$$\text{Mass percent} = \frac{\text{grams of solute}}{\text{grams of solution}} \times 100\%$$

We now substitute the quantities we know:

$$\text{Mass percent} = \frac{\text{grams of solute}}{175 \text{ g}} \times 100\% = 4.5\%$$

EXAMPLE 15.2, CONTINUED

We now solve for grams of solute by multiplying both sides by 175 g:

$$175 \text{ g} \times \frac{\text{grams of solute}}{175 \text{ g}} \times 100\% = 4.5\% \times 175 \text{ g}$$

and then dividing both sides by 100%:

$$\text{Grams of solute} \times \frac{100\%}{100\%} = \frac{4.5\%}{100\%} \times 175 \text{ g}$$

to give

$$\text{Grams of solute} = 0.045 \times 175 \text{ g} = 7.9 \text{ g lactose}$$

SELF-CHECK EXERCISE 15.2

What mass of water must be added to 425 g of formaldehyde to prepare a 40.0% (by mass) solution of formaldehyde? This solution, called formalin, is used to preserve biological specimens.

HINT: Substitute the known quantities into the definition for mass percent, and then solve for the unknown quantity (mass of solvent).

A grasshopper preserved in a formaldehyde solution.

15.4   Solution Composition: Molarity

AIM: To define molarity.
To use molarity to calculate the number of moles of solute present.

When a solution is described in terms of mass percent, the amount of solution is given in terms of its mass. However, it is often more convenient to measure the volume of a solution than to measure its mass. Because of this, chemists often describe a solution in terms of concentration. We define the *concentration* of a solution as the amount of solute in a *given volume* of solution. The most commonly used expression of concentration is **molarity (*M*).** Molarity describes the amount of solute in moles and the volume of the solution in liters. Molarity is *the number of moles of solute per volume of solution in liters.* That is

$$M = \text{molarity} = \frac{\text{moles of solute}}{\text{liters of solution}} = \frac{\text{mol}}{\text{L}}$$

A solution that is 1.0 molar (written as 1.0 *M*) contains 1.0 mol of solute per liter of solution.

Stress that the definition of molarity, *M*, which is moles of solute per volume of liters of solution, can be used as a conversion factor. If we know any two quantities, we can calculate the third.

EXAMPLE 15.3   Solution Composition: Calculating Molarity, I

Calculate the molarity of a solution prepared by dissolving 11.5 g of solid NaOH in enough water to make 1.50 L of solution.

SOLUTION

We are given the following information:

$$\text{Mass of solute} = 11.5 \text{ g NaOH}$$
$$\text{Volume of solution} = 1.50 \text{ L}$$

Because we are asked to calculate the molarity of the solution, we start by writing the definition of molarity.

$$M = \frac{\text{moles of solute}}{\text{liters of solution}}$$

We have the mass (in grams) of solute, so we need to convert the mass of solute to moles (using the molar mass of NaOH). Then we can divide the number of moles by the volume in liters.

We compute the number of moles of solute, using the molar mass of NaOH (40.0 g).

$$11.5 \text{ g NaOH} \times \frac{1 \text{ mol NaOH}}{40.0 \text{ g NaOH}} = 0.288 \text{ mol NaOH}$$

Then we divide by the volume of the solution in liters.

$$\text{Molarity} = \frac{\text{moles of solute}}{\text{liters of solution}} = \frac{0.288 \text{ mol NaOH}}{1.50 \text{ L solution}} = 0.192 \ M \text{ NaOH}$$

EXAMPLE 15.4   Solution Composition: Calculating Molarity, II

Remind students that they need to convert volume expressed in milliliters to liters before they work molarity problems.

Calculate the molarity of a solution prepared by dissolving 1.56 g of gaseous HCl into enough water to make 26.8 mL of solution.

EXAMPLE 15.4, CONTINUED

SOLUTION

We are given

Mass of solute (HCl) = 1.56 g
Volume of solution = 26.8 mL

Molarity is defined as

$$\frac{\text{Moles of solute}}{\text{Liters of solution}}$$

so we must change 1.56 g of HCl to moles of HCl, and then we must change 26.8 mL to liters (because molarity is defined in terms of liters). First we calculate the number of moles of HCl (molar mass = 36.5 g).

$$1.56 \text{ g HCl} \times \frac{1 \text{ mol HCl}}{36.5 \text{ g HCl}} = 0.0427 \text{ mol HCl}$$

$$= 4.27 \times 10^{-2} \text{ mol HCl}$$

Next we change the volume of the solution from milliliters to liters, using the equivalence statement 1 L = 1000 mL, which gives the appropriate conversion factor.

$$26.8 \text{ mL} \times \frac{1 \text{ L}}{1000 \text{ mL}} = 0.0268 \text{ L}$$

$$= 2.68 \times 10^{-2} \text{ L}$$

Finally, we divide the moles of solute by the liters of solution.

$$\text{Molarity} = \frac{4.27 \times 10^{-2} \text{ mol HCl}}{2.68 \times 10^{-2} \text{ L solution}} = 1.59 \text{ } M \text{ HCl}$$

Some students cannot proceed when the quantity of solute is presented in grams. They have forgotten that they can calculate moles of solute from a quantity in grams. It is helpful to briefly review this concept.

SELF-CHECK EXERCISE 15.3

Calculate the molarity of a solution prepared by dissolving 1.00 g of ethanol, C_2H_5OH, in enough water to give a final volume of 101 mL.

It is important to realize that the description of a solution's composition may not accurately reflect the true chemical nature of the solute as it is present in the dissolved state. Solute concentration is always written in terms of the form of the solute *before* it dissolves. For example, describing a solution as 1.0 M NaCl means that the solution was prepared by dissolving 1.0 mol of solid NaCl in enough water to make 1.0 L of solution; it does not mean that the solution contains 1.0 mol of NaCl units. Actually the solution contains 1.0 mol of Na^+ ions and 1.0 mol of Cl^- ions. That is, it contains 1.0 M Na^+ and 1.0 M Cl^-.

A solution of $Co(NO_3)_2$ in water.

When students are asked to determine the molarity of ions, they should always write the equation for the dissociation of the ions in solution before they solve the problem. Some students will not realize that 1 mole of $FeCl_3$ forms 3 mol Cl^- unless they see the equation written.

EXAMPLE 15.5   Solution Composition: Calculating Ion Concentration from Molarity

Give the concentrations of all the ions in each of the following solutions:

a.   0.50 M Co(NO$_3$)$_2$ b.   1 M FeCl$_3$

SOLUTION

Remember, ionic compounds separate into the component ions when they dissolve in water.

a.   When solid Co(NO$_3$)$_2$ dissolves, it produces ions as follows:

$$Co(NO_3)_2(s) \xrightarrow{H_2O(l)} Co^{2+}(aq) + 2NO_3^-(aq)$$

which we can represent as

$$1 \text{ mol } Co(NO_3)_2(s) \xrightarrow{H_2O(l)} 1 \text{ mol } Co^{2+}(aq) + 2 \text{ mol } NO_3^-(aq)$$

Co(NO$_3$)$_2$

⬇

Co^{2+}

NO$_3^-$ NO$_3^-$

Therefore, a solution that is 0.50 M Co(NO$_3$)$_2$ contains 0.50 M Co^{2+} and (2 × 0.50) M NO$_3^-$, or 1.0 M NO$_3^-$.

b.   When solid FeCl$_3$ dissolves, it produces ions as follows:

$$FeCl_3(s) \xrightarrow{H_2O(l)} Fe^{3+}(aq) + 3Cl^-(aq)$$

or

FeCl$_3$

⬇

Fe^{3+}

Cl$^-$ Cl$^-$ Cl$^-$

$$1 \text{ mol } FeCl_3(s) \xrightarrow{H_2O(l)} 1 \text{ mol } Fe^{3+}(aq) + 3 \text{ mol } Cl^-(aq)$$

A solution that is 1 M FeCl$_3$ contains 1 M Fe^{3+} ions and 3 M Cl$^-$ ions.

SELF-CHECK EXERCISE 15.4

Give the concentrations of the ions in each of the following solutions:

a.   0.10 M Na$_2$CO$_3$ b.   0.010 M Al$_2$(SO$_4$)$_3$

Often we need to determine the number of moles of solute present in a given volume of a solution of known molarity. To do this, we use the definition of molarity. When we multiply the molarity of a solution by the volume (in liters), we get the moles of solute present in that sample:

$$M = \frac{\text{moles of solute}}{\text{liters of solution}}$$

Liters × M → Moles of solute

$$\text{Liters of solution} \times \text{molarity} = \cancel{\text{liters of solution}} \times \frac{\text{moles of solute}}{\cancel{\text{liters of solution}}}$$

$$= \text{moles of solute}$$

Using the definition of molarity to solve for moles or liters requires rearranging the equation. You might want to work an example to show how this is done.

EXAMPLE 15.6	Solution Composition: Calculating Number of Moles from Molarity

How many moles of Ag^+ ions are present in 25 mL of a 0.75 M AgNO$_3$ solution?

SOLUTION

In this problem we know

$$\text{Molarity of the solution} = 0.75 \ M$$
$$\text{Volume of the solution} = 25 \ \text{mL}$$

We need to calculate the moles of Ag^+ present. To solve this problem, we must first recognize that a 0.75 M AgNO$_3$ solution contains 0.75 M Ag^+ ions and 0.75 M NO$_3^-$ ions. Next we must express the volume in liters. That is, we must convert from mL to L.

$$25 \ \text{mL} \times \frac{1 \ \text{L}}{1000 \ \text{mL}} = 0.025 \ \text{L} = 2.5 \times 10^{-2} \ \text{L}$$

Now we multiply the volume times the molarity.

$$2.5 \times 10^{-2} \ \text{L solution} \times \frac{0.75 \ \text{mol} \ Ag^+}{\text{L solution}} = 1.9 \times 10^{-2} \ \text{mol} \ Ag^+$$

SELF-CHECK EXERCISE 15.5

Calculate the number of moles of Cl^- ions in 1.75 L of 1.0×10^{-3} M AlCl$_3$.

A **standard solution** is a solution *whose concentration is accurately known.* When the appropriate solute is available in pure form, a standard solution can be prepared by weighing out a sample of solute, transferring it completely to a *volumetric flask* (a flask of accurately known volume), and adding enough solvent to bring the volume up to the mark on the neck of the flask. This procedure is illustrated in Figure 15.7.

EXAMPLE 15.7	Solution Composition: Calculating Mass from Molarity

To analyze the alcohol content of a certain wine, a chemist needs 1.00 L of an aqueous 0.200 M K$_2$Cr$_2$O$_7$ (potassium dichromate) solution. How much solid K$_2$Cr$_2$O$_7$ (molar mass = 294.2 g) must be weighed out to make this solution?

Figure 15.7
Steps involved in the preparation of a standard aqueous solution. (a) Put a weighed amount of a substance (the solute) into the volumetric flask, and add a small quantity of water. (b) Dissolve the solid in the water by gently swirling the flask (*with the stopper in place*). (c) Add more water (with gentle swirling) until the level of the solution just reaches the mark etched on the neck of the flask. Then mix the solution thoroughly by inverting the flask several times.

It might be helpful to show in class how a solution of known concentration is prepared using a volumetric flask, balance, solute, and water.

EXAMPLE 15.7, CONTINUED

SOLUTION

We know the following:

$$\text{Molarity of the solution} = 0.200 \ M$$
$$\text{Volume of the solution} = 1.00 \ L$$

We need to calculate the number of grams of solute ($K_2Cr_2O_7$) present (and thus the mass needed to make the solution). First we determine the number of moles of $K_2Cr_2O_7$ present by multiplying the volume (in liters) by the molarity.

$$1.00 \ \cancel{\text{L solution}} \times \frac{0.200 \ \text{mol} \ K_2Cr_2O_7}{\cancel{\text{L solution}}} = 0.200 \ \text{mol} \ K_2Cr_2O_7$$

Then we convert the moles of $K_2Cr_2O_7$ to grams, using the molar mass of $K_2Cr_2O_7$ (294.2 g).

$$0.200 \ \cancel{\text{mol} \ K_2Cr_2O_7} \times \frac{294.2 \ \text{g} \ K_2Cr_2O_7}{\cancel{\text{mol} \ K_2Cr_2O_7}} = 58.8 \ \text{g} \ K_2Cr_2O_7$$

Therefore, to make 1.00 L of 0.200 M $K_2Cr_2O_7$, the chemist must weigh out 58.8 g of $K_2Cr_2O_7$ and dissolve it in enough water to make 1.00 L of solution. This is most easily done by using a 1.00-L volumetric flask (see Figure 15.7).

SELF-CHECK EXERCISE 15.6

Formalin is an aqueous solution of formaldehyde, HCHO, used as a preservative for biological specimens. How many grams of formaldehyde must be used to prepare 2.5 L of 12.3 M formalin?

Liters × M → Moles of solute

15.5 Dilution

AIM: To show how to calculate the concentration of a solution made by diluting a stock solution.

To save time and space in the laboratory, solutions that are routinely used are often purchased or prepared in concentrated form (called *stock solutions*). Water (or another solvent) is then added to achieve the molarity desired for a particular solution. The process of adding more solvent to a solution is called **dilution.** For example, the common laboratory acids are purchased as concentrated solutions and diluted with water as they are needed. A typical dilution calculation involves determining how much water must be added to an amount of stock solution to achieve a solution of the desired concentration. The key to doing these calcula-

The molarities of stock solutions of the common concentrated acids are:

Sulfuric (H_2SO_4)	18 M
Nitric (HNO_3)	16 M
Hydrochloric (HCl)	12 M

tions is to remember that *only water is added in the dilution.* The amount of solute in the final, more dilute, solution is the *same* as the amount of solute in the original concentrated stock solution. That is,

Moles of solute after dilution = moles of solute before dilution

The number of moles of solute stays the same but more water is added, increasing the volume, so the molarity decreases.

$$M = \frac{\text{moles of solute}}{\text{volume (L)}}$$

Remains constant ↓ (above "moles of solute")

↑ Decreases (below M) ↑ Increases (water added) (below volume)

For example, suppose we want to prepare 500. mL of 1.00 *M* acetic acid, $HC_2H_3O_2$, from a 17.5 *M* stock solution of acetic acid. What volume of the stock solution is required?

The first step is to determine the number of moles of acetic acid needed in the final solution. We do this by multiplying the volume of the solution by its molarity.

$$\text{Volume of dilute solution (liters)} \times \text{molarity of dilute solution} = \text{moles of solute present}$$

The number of moles of solute present in the more dilute solution equals the number of moles of solute that must be present in the more concentrated (stock) solution, because this is the only source of acetic acid.

Because molarity is defined in terms of liters, we must first change 500. mL to liters and then multiply the volume (in liters) by the molarity.

$$500. \text{ mL solution} \times \frac{1 \text{ L solution}}{1000 \text{ mL solution}} = 0.500 \text{ L solution}$$

↑ $V_{\text{dilute solution}}$ (in mL)   ↑ Convert mL to L   ↑ 0.500 L solution

$$0.500 \text{ L solution} \times \frac{1.00 \text{ mol } HC_2H_3O_2}{\text{L solution}} = 0.500 \text{ mol } HC_2H_3O_2$$

↑ $M_{\text{dilute solution}}$

Now we need to find the volume of 17.5 *M* acetic acid that contains 0.500 mol of $HC_2H_3O_2$. We will call this unknown volume *V*. Because volume × molarity = moles, we have

$$V \text{ (in liters)} \times \frac{17.5 \text{ mol } HC_2H_3O_2}{\text{L solution}} = 0.500 \text{ mol } HC_2H_3O_2$$

The important reminder here for students: The quantity of solute removed from the concentrated solution is the same quantity of solute that will be present in the dilute solution.

Dilution with water doesn't alter the number of moles of solute present.

If you have the time and facilities, it is useful to perform an actual dilution in class. Using a colored ionic compound as solute can enhance the effect. You can use this opportunity to demonstrate the use of a pipet and volumetric glassware.

Liters × *M* → Moles of solute

Figure 15.8
(a) 28.6 mL of 17.5 M acetic acid solution is transferred to a volumetric flask that already contains some water. (b) Water is added to the flask (with swirling) to bring the volume to the calibration mark, and the solution is mixed by inverting the flask several times. (c) The resulting solution is 1.00 M acetic acid.

(a)   (b)   (c)

Solving for V $\left(\text{by dividing both sides by } \dfrac{17.5 \text{ mol}}{\text{L solution}}\right)$ gives

$$V = \frac{0.500 \text{ mol } HC_2H_3O_2}{\dfrac{17.5 \text{ mol } HC_2H_3O_2}{\text{L solution}}} = 0.0286 \text{ L, or } 28.6 \text{ mL, of solution}$$

Therefore, to make 500. mL of a 1.00 M acetic acid solution, we take 28.6 mL of 17.5 M acetic acid and dilute it to a total volume of 500. mL. This process is illustrated in Figure 15.8. Because the moles of solute remain the same before and after dilution, we can write

The process of dilution is of practical importance to almost everyone in the sciences. Using the formula $M_1 \times V_1 = M_2 \times V_2$ makes the calculation relatively straightforward. Students commonly forget to change the volume units from liters to milliliters or from milliliters to liters. They incorrectly report the volume of concentrated solution needed. When asked for the volume of water needed to prepare a dilute solution, students often report the volume of concentrated solution needed, or report the total volume. These common mistakes can be remedied only by practice.

Initial Conditions   Final Conditions
$$M_1 \quad \times \quad V_1 \quad = \text{moles of solute} = \quad M_2 \quad \times \quad V_2$$
↑   ↑   ↑   ↑
Molarity   Volume   Molarity   Volume
before   before   after   after
dilution   dilution   dilution   dilution

We can check our calculations on acetic acid by showing that $M_1 \times V_1 = M_2 \times V_2$. In the above example, $M_1 = 17.5 \ M$, $V_1 = 0.0286 \ L$, $V_2 = 0.500 \ L$, and $M_2 = 1.00 \ M$, so

$$M_1 \times V_1 = 17.5 \ \frac{\text{mol}}{\text{L}} \times 0.0286 \ L = 0.500 \text{ mol}$$

$$M_2 \times V_2 = 1.00 \ \frac{\text{mol}}{\text{L}} \times 0.500 \ L = 0.500 \text{ mol}$$

and therefore

$$M_1 \times V_1 = M_2 \times V_2$$

This shows that the volume (V_2) we calculated is correct.

EXAMPLE 15.8   Calculating Concentrations of Diluted Solutions

What volume of 16 M sulfuric acid must be used to prepare 1.5 L of a 0.10 M H_2SO_4 solution?

SOLUTION

We can summarize what we are given as follows:

Initial Conditions (concentrated)	Final Conditions (dilute)
$M_1 = 16 \dfrac{\text{mol}}{\text{L}}$	$M_2 = 0.10 \dfrac{\text{mol}}{\text{L}}$
$V_1 = ?$	$V_2 = 1.5 \text{ L}$

We know that

$$\text{Moles of solute} = M_1 \times V_1 = M_2 \times V_2$$

and we can solve the equation

$$M_1 \times V_1 = M_2 \times V_2$$

for V_1 by dividing both sides by M_1:

$$\frac{\cancel{M_1} \times V_1}{\cancel{M_1}} = \frac{M_2 \times V_2}{M_1}$$

to give

$$V_1 = \frac{M_2 \times V_2}{M_1}$$

Now we substitute the known values of M_2, V_2, and M_1.

$$V_1 = \frac{\left(0.10 \ \dfrac{\text{mol}}{\text{L}}\right)(1.5 \text{ L})}{16 \ \dfrac{\text{mol}}{\text{L}}} = 9.4 \times 10^{-3} \text{ L}$$

$$9.4 \times 10^{-3} \text{ L} \times \frac{1000 \text{ mL}}{1 \text{ L}} = 9.4 \text{ mL}$$

Therefore $V_1 = 9.4 \times 10^{-3}$ L, or 9.4 mL. To make 1.5 L of 0.10 M H_2SO_4 using 16 M H_2SO_4, we must take 9.4 mL of the concentrated acid and dilute it with water to a final volume of 1.5 L. The correct way to do this is to add the 9.4 mL of acid to about 1 L of water and then dilute to 1.5 L by adding more water.

Approximate dilutions can be carried out using a calibrated beaker. Here concentrated sulfuric acid is being added to water to make a dilute solution.

It is always best to add concentrated acid to water, not water to the acid. That way, if any splashing occurs accidentally, it is dilute acid that splashes.

EXAMPLE 15.8, CONTINUED

SELF-CHECK EXERCISE 15.7

What volume of 12 M HCl must be taken to prepare 0.75 L of 0.25 M HCl?

15.6 Stoichiometry of Solution Reactions

AIM: To describe the strategy for solving stoichiometric problems for solution reactions.

Much of solution stoichiometry builds on material presented in previous chapters. You should probably assume that students do not remember all the steps used to solve stoichiometry problems and you should go over the basics once again.

The boxed steps for solving stoichiometry problems provide a good review of the principles of stoichiometry.

See Section 7.3 for a discussion of net ionic equations.

Step 1 asks students to write net ionic equations for reactions involving ions. You might want to review how to write net ionic equations. Remind students to balance all equations.

Because so many important reactions occur in solution, it is important to be able to do stoichiometric calculations for solution reactions. The principles needed to perform these calculations are very similar to those developed in Chapter 10. It is helpful to think in terms of the following steps:

Steps for Solving Stoichiometric Problems Involving Solutions

STEP 1
Write the balanced equation for the reaction. For reactions involving ions, it is best to write the net ionic equation.

STEP 2
Calculate the moles of reactants.

STEP 3
Determine which reactant is limiting.

STEP 4
Calculate the moles of other reactants or products, as required.

STEP 5
Convert to grams or other units, if required.

EXAMPLE I5.9   **Solution Stoichiometry:**
Calculating Mass of Reactants
and Products

Calculate the mass of solid NaCl that must be added to 1.50 L of a 0.100 M $AgNO_3$ solution to precipitate all of the Ag^+ ions in the form of AgCl. Calculate the mass of AgCl formed.

SOLUTION

STEP I   *Write the balanced equation for the reaction.*
When added to the $AgNO_3$ solution (which contains Ag^+ and NO_3^- ions), the solid NaCl dissolves to yield Na^+ and Cl^- ions. Solid AgCl forms according to the following balanced net ionic reaction:

$$Ag^+(aq) + Cl^-(aq) \rightarrow AgCl(s)$$

When you work solution stoichiometry problems in class, demonstrate the reaction. In general, students find it easier to think about the reaction when they can see the precipitate form.

This reaction was discussed in Section 7.2.

STEP 2   *Calculate the moles of reactants.*
In this case we must add enough Cl^- ions to just react with all the Ag^+ ions present, so we must calculate the moles of Ag^+ ions present in 1.50 L of a 0.100 M $AgNO_3$ solution. (Remember that a 0.100 M $AgNO_3$ solution contains 0.100 M Ag^+ ions and 0.100 M NO_3^- ions.)

$$1.50 \; \cancel{L} \times \frac{0.100 \; \text{mol} \; Ag^+}{\cancel{L}} = 0.150 \; \text{mol} \; Ag^+$$

↑
Moles of Ag^+
present in 1.5 L of
0.100 M $AgNO_3$

Liters × M → Moles of solute

STEP 3   *Determine which reactant is limiting.*
In this situation we want to add just enough Cl^- to react with the Ag^+ present. That is, we want to precipitate *all* the Ag^+ in the solution. Thus the Ag^+ present determines the amount of Cl^- needed.

STEP 4   *Calculate the moles of Cl^- required.*
We have 0.150 mol of Ag^+ ions and, because one Ag^+ ion reacts with one Cl^- ion, we need 0.150 mol of Cl^-:

$$0.150 \; \cancel{\text{mol} \; Ag^+} \times \frac{1 \; \text{mol} \; Cl^-}{1 \; \cancel{\text{mol} \; Ag^+}} = 0.150 \; \text{mol} \; Cl^-$$

so 0.150 mol of AgCl will be formed.

$$0.150 \; \text{mol} \; Ag^+ + 0.150 \; \text{mol} \; Cl^- \rightarrow 0.150 \; \text{mol} \; AgCl$$

STEP 5   *Convert to grams of NaCl required.*
To produce 0.150 mol Cl^-, we need 0.150 mol NaCl. We calculate the mass

EXAMPLE 15.9, CONTINUED

of NaCl required as follows:

$$0.150 \; \cancel{\text{mol NaCl}} \times \frac{58.4 \text{ g NaCl}}{\cancel{\text{mol NaCl}}} = 8.76 \text{ g NaCl}$$

The mass of AgCl formed is

$$0.150 \; \cancel{\text{mol AgCl}} \times \frac{143.3 \text{ g AgCl}}{\cancel{\text{mol AgCl}}} = 21.5 \text{ g AgCl}$$

Limiting-reactant problems were first introduced in Section 10.4.

Barium chromate precipitating.

See Section 7.2 for a discussion of this reaction.

EXAMPLE 15.10   **Solution Stoichiometry: Determining Limiting Reactants and Calculating Mass of Products**

When $Ba(NO_3)_2$ and K_2CrO_4 react in aqueous solution, the yellow-brown solid $BaCrO_4$ is formed. Calculate the mass of $BaCrO_4$ that forms when 3.50×10^{-3} mol of solid $Ba(NO_3)_2$ is dissolved in 265 mL of 0.0100 M K_2CrO_4 solution.

SOLUTION

STEP 1
The original K_2CrO_4 solution contains the ions K^+ and CrO_4^{2-}. When the $Ba(NO_3)_2$ is dissolved in this solution, Ba^{2+} and NO_3^- ions are added. The Ba^{2+} and CrO_4^{2-} ions react to form solid $BaCrO_4$. The balanced net ionic equation is

$$Ba^{2+}(aq) + CrO_4^{2-}(aq) \rightarrow BaCrO_4(s)$$

STEP 2
Next we determine the moles of reactants. We are told that 3.50×10^{-3} mol of $Ba(NO_3)_2$ is added to the K_2CrO_4 solution. Each formula unit of $Ba(NO_3)_2$ contains one Ba^{2+} ion, so 3.50×10^{-3} mol of $Ba(NO_3)_2$ gives 3.50×10^{-3} mol of Ba^{2+} ions in solution.

Because $V \times M$ = moles of solute, we can compute the moles of K_2CrO_4 in the solution from the volume and molarity of the original solution. First

EXAMPLE 15.10, CONTINUED

we must convert the volume of the solution (265 mL) to liters.

$$265 \text{ mL} \times \frac{1 \text{ L}}{1000 \text{ mL}} = 0.265 \text{ L}$$

Next we determine the number of moles of K_2CrO_4, using the molarity of the K_2CrO_4 solution (0.0100 M).

$$0.265 \text{ L} \times \frac{0.0100 \text{ mol } K_2CrO_4}{\text{L}} = 2.65 \times 10^{-3} \text{ mol } K_2CrO_4$$

We know that

so the solution contains 2.65×10^{-3} mol of CrO_4^{2-} ions.

STEP 3
The balanced equation tells us that one Ba^{2+} ion reacts with one CrO_4^{2-}. Because the number of moles of CrO_4^{2-} ions (2.65×10^{-3}) is smaller than the number of moles of Ba^{2+} ions (3.50×10^{-3}), the CrO_4^{2-} will run out first.

$$Ba^{2+}(aq) \ + \ CrO_4^{2-}(aq) \ \rightarrow \ BaCrO_4(s)$$

3.50×10^{-3} mol	2.65×10^{-3} mol

↑
Smaller (runs
out first)

Therefore the CrO_4^{2-} is limiting.

STEP 4
The 2.65×10^{-3} mol of CrO_4^{2-} ions will react with 2.65×10^{-3} mol of Ba^{2+} ions to form 2.65×10^{-3} mol of $BaCrO_4$.

2.65×10^{-3} mol Ba^{2+}	+	2.65×10^{-3} mol CrO_4^{2-}	⇨	2.65×10^{-3} mol $BaCrO_4(s)$

EXAMPLE 15.10, CONTINUED

STEP 5
The mass of $BaCrO_4$ formed is obtained from its molar mass (253.3 g) as follows:

$$2.65 \times 10^{-3} \text{ mol BaCrO}_4 \times \frac{253.3 \text{ g BaCrO}_4}{\text{mol BaCrO}_4} = 0.671 \text{ g BaCrO}_4$$

SELF-CHECK EXERCISE 15.8

When aqueous solutions of Na_2SO_4 and $Pb(NO_3)_2$ are mixed, $PbSO_4$ precipitates. Calculate the mass of $PbSO_4$ formed when 1.25 L of 0.0500 M $Pb(NO_3)_2$ and 2.00 L of 0.0250 M Na_2SO_4 are mixed.

HINT: Calculate the moles of Pb^{2+} and SO_4^{2-} in the mixed solution, decide which ion is limiting, and calculate the moles of $PbSO_4$ formed.

15.7   Neutralization Reactions

AIM: To show how to do calculations involved in acid–base reactions.

It can be interesting to perform a titration in class. This is a good opportunity to demonstrate to students a faint pink phenolphthalein end point.

So far we have considered the stoichiometry of reactions in solution that result in the formation of a precipitate. Another common type of solution reaction occurs between an acid and a base. We introduced these reactions in Section 7.4. Recall from that discussion that an acid is a substance that furnishes H^+ ions. A strong acid, such as hydrochloric acid, HCl, dissociates completely in water.

$$HCl(aq) \rightarrow H^+(aq) + Cl^-(aq)$$

Strong bases are water-soluble metal hydroxides, which are completely dissociated in water. An example is NaOH, which dissolves in water to give Na^+ and OH^- ions.

$$NaOH(s) \xrightarrow{H_2O(l)} Na^+(aq) + OH^-(aq)$$

When a strong acid and strong base react, the net ionic reaction is

$$H^+(aq) + OH^-(aq) \rightarrow H_2O(l)$$

An acid–base reaction is often called a **neutralization reaction.** When just enough strong base is added to react exactly with the strong acid in a solution, we say the acid has been *neutralized.* One product of this reaction is always water. The steps in dealing with the stoichiometry of any neutralization reaction are the same as those we followed in the previous section.

EXAMPLE 15.11	Solution Stoichiometry: Calculating Volume in Neutralization Reactions

What volume of a 0.100 M HCl solution is needed to neutralize 25.0 mL of a 0.350 M NaOH solution?

Most neutralization problems contain so much information that it is hard for students to know how to begin. I encourage students to use the steps presented in Section 15.6 to solve these problems. Without using a systematic method for solving problems, many students become discouraged.

SOLUTION

STEP 1   *Write the balanced equation for the reaction.*
Hydrochloric acid is a strong acid, so all the HCl molecules dissociate to produce H^+ and Cl^- ions. Also, when the strong base NaOH dissolves, the solution contains Na^+ and OH^- ions. When these two solutions are mixed, the H^+ ions from the hydrochloric acid react with the OH^- ions from the sodium hydroxide solution to form water. The balanced net ionic equation for the reaction is

$$H^+(aq) \; + \; OH^-(aq) \rightarrow H_2O(l)$$

STEP 2   *Calculate the moles of reactants.*
In this problem we are given a volume (25.0 mL) of 0.350 M NaOH, and we want to add just enough 0.100 M HCl to provide just enough H^+ ions to react with all the OH^-. Therefore we must calculate the number of moles of OH^- ions in the 25.0-mL sample of 0.350 M NaOH. To do this, we first change the volume to liters and multiply by the molarity.

$$25.0 \; \cancel{mL \; NaOH} \times \frac{1 \; L}{1000 \; \cancel{mL}} \times \frac{0.350 \; mol \; OH^-}{L \; \cancel{NaOH}} = 8.75 \times 10^{-3} \; mol \; OH^-$$

Moles of OH^- present in 25.0 mL of 0.350 M NaOH

STEP 3   *Determine which reactant is limiting.*
This problem requires the addition of just enough H^+ ions to react exactly with the OH^- ions present, so the number of moles of OH^- ions present determines the number of moles of H^+ that must be added. The OH^- ions are limiting.

STEP 4   *Calculate the moles of H^+ required.*
The balanced equation tells us that the H^+ and OH^- ions react in a $1:1$ ratio, so 8.75×10^{-3} mol of H^+ ions is required to neutralize (exactly react with) the 8.75×10^{-3} mol of OH^- ions present.

STEP 5   *Calculate the volume of 0.100 M HCl required.*
Next we must find the volume (V) of 0.100 M HCl required to furnish this amount of H^+ ions. Because the volume (in liters) times the molarity gives

EXAMPLE 15.11, CONTINUED

the number of moles, we have

$$V \times \frac{0.100 \text{ mol H}^+}{\text{L}} = 8.75 \times 10^{-3} \text{ mol H}^+$$

↑
Unknown
volume
(in liters)

↑
Moles of
H⁺ needed

Now we must solve for V by dividing both sides of the equation by 0.100.

$$V \times \frac{0.100 \text{ mol H}^+}{0.100 \text{ L}} = \frac{8.75 \times 10^{-3} \text{ mol H}^+}{0.100}$$

$$V = 8.75 \times 10^{-2} \text{ L}$$

Changing to milliliters, we have

$$V = 8.75 \times 10^{-2} \text{ L} \times \frac{1000 \text{ ml}}{\text{L}} = 87.5 \text{ mL}$$

Therefore, 87.5 mL of 0.100M HCl is required to neutralize 25.0 mL of 0.350 M NaOH.

SELF-CHECK EXERCISE 15.9

Calculate the volume of 0.10 M HNO$_3$ needed to neutralize 125 mL of 0.050 M KOH.

15.8   Solution Composition: Normality

AIM: To explain normality and equivalent weight.
To use these concepts in stoichiometric calculations.

Normality is another unit of concentration that is sometimes used, especially when dealing with acids and bases. The use of normality focuses mainly on the H⁺ and OH⁻ available in an acid–base reaction. Before we discuss normality, however, we need to define some terms. One **equivalent of an acid** is the *amount of that acid that can furnish 1 mol of H⁺ ions.* Similarly, one **equivalent of a base** is defined as the *amount of that base that can furnish 1 mol of OH⁻ ions.* The **equivalent weight** of an acid or base is the mass in grams of 1 equivalent (equiv) of that acid or base.

The common strong acids are HCl, HNO$_3$, and H$_2$SO$_4$. For HCl and HNO$_3$ each molecule of acid furnishes one H$^+$ ion, so 1 mol of HCl can furnish 1 mol of H$^+$ ions. This means that

Furnishes 1 mol of H$^+$

$\downarrow$

$$1 \text{ mol HCl} = 1 \text{ equiv HCl}$$
$$\text{Molar mass (HCl)} = \text{equivalent weight (HCl)}$$

Likewise, for HNO$_3$,

$$1 \text{ mol HNO}_3 = 1 \text{ equiv HNO}_3$$
$$\text{Molar mass (HNO}_3) = \text{equivalent weight (HNO}_3)$$

However, H$_2$SO$_4$ can furnish *two* H$^+$ ions per molecule, so 1 mol of H$_2$SO$_4$ can furnish *two* mol of H$^+$. This means that

Students have a hard time remembering whether one-half mole or two moles of a diprotic acid is equal to one equivalent of acid.

| 1 mol H$_2$SO$_4$ | furnishes | 2 mol H$^+$ |

| $\frac{1}{2}$ mol H$_2$SO$_4$ | furnishes | 1 mol H$^+$ |

| $\frac{1}{2}$ mol H$_2$SO$_4$ | = | 1 equiv H$_2$SO$_4$ | | 1 mol H$^+$ |

Because each mole of H$_2$SO$_4$ can furnish 2 mol of H$^+$, we only need to take $\frac{1}{2}$ mol of H$_2$SO$_4$ to get 1 equiv of H$_2$SO$_4$. Therefore,

$$\tfrac{1}{2} \text{ mol HSO}_4 = 1 \text{ equiv H}_2\text{SO}_4$$

If students take the time to write the equation for the dissociation of one mole of acid, they will find it easier to see the number of moles of protons produced.

and

$$\text{Equivalent weight (H}_2\text{SO}_4) = \tfrac{1}{2} \text{ molar mass (H}_2\text{SO}_4)$$
$$= \tfrac{1}{2}(98 \text{ g}) = 49 \text{ g}$$

The equivalent weight of H$_2$SO$_4$ is 49 g.

The common strong bases are NaOH and KOH. For NaOH and KOH, each formula unit furnishes one OH$^-$ ion, so we can say

$$1 \text{ mol NaOH} = 1 \text{ equiv NaOH}$$
$$\text{Molar mass (NaOH)} = \text{equivalent weight (NaOH)}$$

$$1 \text{ mol KOH} = 1 \text{ equiv KOH}$$
$$\text{Molar mass (KOH)} = \text{equiv weight (KOH)}$$

These ideas are summarized in Table 15.2 on the following page.

Table 15.2	The Molar Masses and Equivalent Weights of the Common Strong Acids and Bases	
	Molar Mass (g)	Equivalent Weight (g)
Acid		
HCl	36.5	36.5
HNO_3	63.0	63.0
H_2SO_4	98.0	$49.0 = \dfrac{98.0}{2}$
Base		
NaOH	40.0	40.0
KOH	56.1	56.1

EXAMPLE 15.12   Solution Stoichiometry: Calculating Equivalent Weight

Phosphoric acid, H_3PO_4, can furnish three H^+ ions per molecule. Calculate the equivalent weight of H_3PO_4.

SOLUTION

The key point here involves how many protons (H^+ ions) each molecule of H_3PO_4 can furnish.

Because each H_3PO_4 can furnish three H^+ ions, 1 mol of H_3PO_4 can furnish 3 mol of H^+ ions:

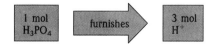

so 1 equiv of H_3PO_4 (the amount that can furnish 1 mol of H^+) is one-third of a mole.

This means the equivalent weight of H_3PO_4 is one-third its molar mass.

EXAMPLE 15.12, CONTINUED

$$
\boxed{\begin{array}{c}\text{Equivalent} \\ \text{weight}\end{array}} = \boxed{\dfrac{\text{Molar mass}}{3}}
$$

$$
\text{Equivalent weight (H}_3\text{PO}_4) = \frac{\text{molar mass (H}_3\text{PO}_4)}{3}
$$

$$
= \frac{98.0 \text{ g}}{3} = 32.7 \text{ g}
$$

Normality (N) is defined as the number of equivalents of solute per liter of solution.

$$
\text{Normality} = N = \frac{\text{number of equivalents}}{1 \text{ liter of solution}} = \frac{\text{equivalents}}{\text{liter}} = \frac{\text{equiv}}{\text{L}}
$$

This means that a $1\,N$ solution contains 1 equivalent of solute per liter of solution. Notice that when we multiply the volume of a solution in liters by the normality, we get the number of equivalents.

$$
N \times V = \frac{\text{equiv}}{\cancel{\text{L}}} \times \cancel{\text{L}} = \text{equiv}
$$

Solution stoichiometry calculations involving normality depend on first determining the number of equivalents. If students don't understand this basic concept, they will not be able to apply it to more complicated problems.

EXAMPLE 15.13   **Solution Stoichiometry: Calculating Normality**

A solution of sulfuric acid contains 86 g of H_2SO_4 per liter of solution. Calculate the normality of this solution.

SOLUTION

We want to calculate the normality of this solution, so we focus on the definition of normality, the number of equivalents per liter:

$$
N = \frac{\text{equiv}}{\text{L}}
$$

This definition leads to two questions we need to answer:

1. What is the number of equivalents?
2. What is the volume?

We know the volume; it is 1.0 L. To find the number of equivalents present, we must calculate the number of equivalents represented by 86 g of H_2SO_4. To do

Whenever you need to calculate the concentration of a solution, first write the appropriate definition. Then decide how to calculate the quantities shown in the definition.

EXAMPLE 15.13, CONTINUED

this calculation, we focus on the definition of the equivalent: it is the amount of acid that furnishes 1 mol of H^+. Because H_2SO_4 can furnish two H^+ ions per molecule, 1 equiv of H_2SO_4 is $\frac{1}{2}$ mol of H_2SO_4, so

$$\text{Equivalent weight } (H_2SO_4) = \frac{\text{molar mass } (H_2SO_4)}{2}$$

$$= \frac{98.0 \text{ g}}{2} = 49.0 \text{ g}$$

We have 86 g of H_2SO_4.

$$86 \text{ g } H_2SO_4 \times \frac{1 \text{ equiv } H_2SO_4}{49.0 \text{ g } H_2SO_4} = 1.8 \text{ equiv } H_2SO_4$$

$$N = \frac{\text{equiv}}{L} = \frac{1.8 \text{ equiv } H_2SO_4}{1.0 \text{ L}} = 1.8 \text{ N } H_2SO_4$$

We know that 86 g is more than 1 equiv of H_2SO_4 (49 g), so this answer makes sense.

SELF-CHECK EXERCISE 15.10

Calculate the normality of a solution containing 23.6 g of KOH in 755 mL of solution.

The main advantage of using equivalents is that 1 equiv of acid contains the same number of available H^+ ions as the number of OH^- ions present in 1 equiv of base. That is,

> 0.75 equiv (base) will react exactly with 0.75 equiv (acid).
> 0.23 equiv (base) will react exactly with 0.23 equiv (acid).
> And so on.

In each of these cases, the number of H^+ ions furnished by the sample of acid is the same as the *number of* OH^- ions furnished by the sample of base. The point is that *n equivalents of any acid will exactly neutralize n equivalents of any base.*

Because we know that equal equivalents of acid and base are required for neutral-

EXAMPLE 15.14, CONTINUED

ization, we can say that

$$\text{equiv (acid)} = \text{equiv (base)}$$

That is,

$$N_{\text{acid}} \times V_{\text{acid}} = \text{equiv (acid)} = \text{equiv (base)} = N_{\text{base}} \times V_{\text{base}}$$

Therefore, for any neutralization reaction, the following relationship holds:

$$N_{\text{acid}} \times V_{\text{acid}} = N_{\text{base}} \times V_{\text{base}}$$

EXAMPLE 15.14	**Solution Stoichiometry: Using Normality in Calculations**

What volume of a 0.075 N NaOH solution is required to react exactly with 0.135 L of 0.45 N H_3PO_4?

SOLUTION

We know that for neutralization, equiv (acid) = equiv (base), or

$$N_{\text{acid}} \times V_{\text{acid}} = N_{\text{base}} \times V_{\text{base}}$$

We want to calculate for the volume of base, V_{base}, so we solve for V_{base} by dividing both sides by N_{base}.

$$\frac{N_{\text{acid}} \times V_{\text{acid}}}{N_{\text{base}}} = \frac{N_{\text{base}} \times V_{\text{base}}}{N_{\text{base}}} = V_{\text{base}}$$

Now we can substitute the given values $N_{\text{acid}} = 0.45$ N, $V_{\text{acid}} = 0.135$ L, and $N_{\text{base}} = 0.075$ N into the equation.

$$V_{\text{base}} = \frac{N_{\text{acid}} \times V_{\text{acid}}}{N_{\text{base}}} = \frac{\left(0.45\ \frac{\text{equiv}}{L}\right)(0.135\ \text{L})}{0.075\ \frac{\text{equiv}}{L}} = 0.81\ \text{L}$$

This gives $V_{\text{base}} = 0.81$ L, so 0.81 L of 0.075 N KOH is required to react exactly with 0.135 L of 0.45 N H_3PO_4.

SELF-CHECK EXERCISE 15.11

What volume of 0.50 N H_2SO_4 is required to react exactly with 0.250 L of 0.80 N KOH?

CHAPTER REVIEW

Key Terms

solution (p. 498)
solvent (p. 498)
solute (p. 498)
aqueous solution (p. 498)
saturated (p. 502)
unsaturated (p. 502)
concentrated (p. 502)
dilute (p. 502)
mass percent (p. 503)
molarity (*M*) (p. 505)
standard solution (p. 509)

dilution (p. 510)
neutralization reaction (p. 518)
equivalent of an acid (p. 520)
equivalent of a base (p. 520)
equivalent weight (p. 520)
normality (p. 523)

Summary

1. A solution is a homogeneous mixture. The solubility of a solute in a given solvent depends on the interactions between the solvent and solute particles. Water dissolves many ionic compounds and compounds with polar molecules, because strong forces occur between the solute and the polar water molecules. Nonpolar solvents tend to dissolve nonpolar solutes. "Like dissolves like."

2. Solution composition can be described in many ways. Two of the most important are in terms of mass percent of solute:

$$\text{Mass percent} = \frac{\text{mass of solute}}{\text{mass of solution}} \times 100\%$$

and molarity:

$$\text{Molarity} = \frac{\text{moles of solute}}{\text{liters of solution}}$$

3. A standard solution is one whose concentration is accurately known. Solutions are often made from a stock solution by dilution. When a solution is diluted, only solvent is added, which means that

$$\text{Moles of solute after dilution} = \text{moles of solute before dilution}$$

4. Normality is defined as the number of equivalents per liter of solution. One equivalent of acid is the amount of acid that furnishes 1 mol of H^+ ions. One equivalent of base is the amount of base that furnishes 1 mol of OH^- ions.

Questions and Problems

All even-numbered exercises have answers in the back of this book and solutions in the Solutions Guide.

15.1 Solubility

QUESTIONS

1. What does it mean to say that a solution is a *homogeneous mixture*? Give two examples of homogeneous mixtures (solutions) and two examples of *non*homogeneous mixtures. How do the two types of mixtures compare in properties?

3. In a solution, the substance present in the largest amount is called the _____, whereas the other substances present are called the _____ .

5. Why are some molecular solids not soluble in water? Give two examples of molecular solids that are soluble in water and two examples of molecular solids that are not soluble in water. Account for the solubility (or insolubility) of each example.

7. A substance such as NaCl dissolves in water because the strong ionic forces that exist in solid NaCl can be overcome by, and replaced by, forces between _____ and the ions.

2. Discuss how an *ionic* solute dissolves in water. How are the strong interionic forces in the solid overcome to permit the solid to dissolve? How are the dissolved positive and negative ions shielded from one another, preventing them from recombining to form the solid?

4. A metallic alloy, such as nickel steel, is an example of a _____ solution.

6. Why are some molecular solids (such as sugar) very soluble in water? What feature in the structure of sugar molecules makes their interaction with water favor the forming of a solution?

8. When an ionic substance such as potassium bromide, KBr, dissolves, the resulting solution contains separate, hydrated ions that behave _____ of one another.

15.2 Solution Composition: An Introduction

QUESTIONS

9. A solution that contains as much solute as will dissolve at a given temperature is said to be _____ .

11. A solution is a homogeneous mixture and, unlike a compound, has _____ composition.

10. A solution that has not reached its limit of dissolved solute is said to be _____ .

12. The label "concentrated H_2SO_4" on a bottle means that there is a relatively _____ amount of H_2SO_4 present in the solution.

15.3 Solution Composition: Mass Percent

QUESTIONS

13. A solution that is 5% by mass NaCl contains 5 g of NaCl per _____ g of solution.

14. A solution has been prepared to be 9% (by mass) glucose. For every 100.0 g of solution present there are _____ g glucose.

PROBLEMS

15. Calculate the mass percent of NaCl in each of the following solutions.
   a. 1.0 g of NaCl in 48 g of water
   b. 2.0 g of NaCl in 96 g of water
   c. 1.0 kg of NaCl in 48 kg of water
   d. 5.0 g of NaCl in 240. g of water

16. Calculate the mass percent of NH_4NO_3 in each of the following solutions.
   a. 1.0 g NH_4NO_3 in 10 g water
   b. 1.0 g NH_4NO_3 in 25 g water
   c. 1.0 g NH_4NO_3 in 75 g water
   d. 1.0 g NH_4NO_3 in 100 g water

17. Calculate the mass, in grams, of $CaCl_2$ present in each of the following solutions.
    a. 155 g of 1.0% $CaCl_2$ solution
    b. 1.0 kg of 5.5% $CaCl_2$ solution
    c. 250. g of 0.10% $CaCl_2$ solution
    d. 15 mg of 10.% $CaCl_2$ solution

18. For a 15.0% (by mass) NaCl solution, calculate the indicated quantity.
    a. the mass of NaCl in 150 g of the solution
    b. the amount of solution needed to obtain 35.0 g NaCl
    c. the mass of NaCl needed to make 1,000 g of the solution
    d. the mass of NaCl contained in 1,000 g of the solution

19. In general chemistry laboratories, a sodium hydrogen carbonate solution is typically available for neutralizing spills of acid. If 50. g of $NaHCO_3$ is mixed with 950. g of water, what is the mass percent of the solution?

20. A certain alloy is made by dissolving 5.31 g of copper and 4.03 g of zinc in 145 g of iron. Calculate the percent of each component in the alloy.

21. The manufacturer's label on organic chemical reagents often indicates the residue remaining after the reagent is burned. Such a residue must be due to impurities because when a pure organic substance is burned it produces gaseous water and carbon dioxide. If 1.00 g of an organic substance is ignited, and a residue weighing 10.3 mg remains, what is the mass percent of nonflammable impurity in the substance?

22. A 151 g-portion of a sodium chloride solution is evaporated, producing 5.35 g of dry solid NaCl. Calculate the percentage of NaCl in the original solution.

23. How many grams of Na_2CO_3 are required to prepare 500. g of a 5.5% by mass Na_2CO_3 solution?

24. How many grams of KBr are contained in 125 g of a 6.25% (by mass) KBr solution?

25. How many grams of sugar are contained in 250. g of an aqueous mixture that contains 10.% sugar and 5.0% alcohol?

26. How many grams of $CuCl_2$ are required to prepare 1250. g of a 1.25% (by mass) $CuCl_2$ solution?

27. What mass of solute is contained in 75 g of 5.0% by mass HCl solution?

28. A hexane solution contains as impurities 5.2% (by mass) heptane and 2.9% (by mass) pentane. Calculate the mass of each component present in 93 g of the solution.

15.4 Solution Composition: Molarity

QUESTIONS

29. A solution that is labeled "1 M HCl" contains 1 mol of dissolved HCl per _____ of the solution.

30. A solution that is 0.50 M in $CuCl_2$ contains _____ mol of Cu^{2+} and _____ mol of Cl^- per liter.

31. What is a *standard* solution? Describe the steps involved in preparing a standard solution.

32. If you were to prepare exactly 1.00 L of a 5 M NaCl solution, you would *not* need exactly 1.00 L of water. Explain.

PROBLEMS

33. For each of the following solutions, the number of moles of solute is given, followed by the total volume of solution prepared. Calculate the molarity.
    a. 0.50 mol of NaCl; 0.200 L
    b. 0.50 mol of NaCl; 0.125 L
    c. 0.25 mol of NaCl; 100. mL
    d. 0.75 mol of NaCl; 300. mL

34. For each of the following solutions, the number of moles of solute is given, followed by the total volume of solution prepared. Calculate the molarity.
    a. 0.50 mol KBr; 250 mL
    b. 0.50 mol KBr; 500 mL
    c. 0.50 mol KBr; 750 mL
    d. 0.50 mol KBr; 1.0 L

35. For each of the following solutions, the mass of the solute is given, followed by the total volume of solution prepared. Calculate the molarity.
    a. 5.0 g of $CaCl_2$; 2.5 L
    b. 1.1 kg of KBr; 4.5 L
    c. 1.5 g of $NaNO_3$; 75 mL
    d. 4.5 g of Na_2SO_4; 125 mL

36. For each of the following solutions the mass of the solute is given, followed by the total volume of the solution prepared. Calculate the molarity.
    a. 4.25 g $CuCl_2$; 125 mL
    b. 0.101 g $NaHCO_3$; 11.3 mL
    c. 52.9 g Na_2CO_3; 1.15 L
    d. 0.14 mg KOH; 1.5 mL

37. If 155 g of sucrose, $C_{12}H_{22}O_{11}$, is dissolved in enough water to make 1.00 L of solution, calculate the molarity.

38. If 275 g of $CuSO_4$ is dissolved in enough water to make 4.25 L, what is the molarity of the solution?

39. 1.0 g of $AgNO_3$ is dissolved in enough water to make a final volume of 18 mL. Calculate the molarity of the $AgNO_3$ solution.

40. An alcoholic iodine solution ("tincture" of iodine) is prepared by dissolving 5.15 g of iodine crystals in enough alcohol to make a volume of 225 mL. Calculate the molarity of iodine in the solution.

41. 1.5 mg of NaCl is dissolved in enough water to make 1.0 mL of solution. What is the molarity of NaCl in the solution?

42. If 495 g of NaOH is dissolved to a final total volume of 20.0 L, what is the molarity of the solution?

43. How many *moles* of the indicated solute does each of the following solutions contain?
    a. 10.0 L of 0.550 M $NaHCO_3$ solution
    b. 5.0 L of 12 M HCl solution
    c. 250. mL of 19.4 M NaOH solution
    d. 125 mL of 17.0 M acetic acid, $HC_2H_3O_2$

44. What number of *moles* of the indicated solute does each of the following solutions contain?
    a. 122 mL of 0.451 M HCl solution
    b. 2.78 L of 0.101 M KOH solution
    c. 9.7 mL of 0.45 M sugar solution
    d. 425 L of 1.55 M NaCl solution

45. How many *grams* of the indicated solute does each of the following solutions contain?
    a. 2.00 L of 1.33 M NaCl solution
    b. 0.050 mL (approximately 1 drop) of 6.0 M HCl solution
    c. 125 mL of 3.05 M HNO_3 solution
    d. 1.25 L of 0.503 M NaBr solution

46. What mass in *grams* of the indicated solute does each of the following solutions contain?
    a. 145 mL of 0.0221 EDTA solution (Molar mass EDTA = 292.2 g)
    b. 2.25 L of 0.135 M $CuSO_4$ solution
    c. 1.02 mL of 0.452 M H_3PO_4 solution
    d. 24 mL of 1.0 M HCl solution

47. If 50. g of NaCl is available, what volume of 2.0 M NaCl solution can be prepared?

48. What volume of 0.25 M NaCl solution can be prepared from 5.2 g NaCl(s)?

49. Calculate the number of moles of the indicated ion present in each of the following solutions.
    a. Na^+ ion in 1.00 L of 0.251 M Na_2SO_4 solution
    b. Cl^- ion in 5.50 L of 0.10 M $FeCl_3$ solution
    c. NO_3^- ion in 100. mL of 0.55 M $Ba(NO_3)_2$ solution
    d. NH_4^+ ion in 250. mL of 0.350 M $(NH_4)_2SO_4$ solution

50. Calculate the number of moles of *each* ion present in each of the following solutions.
    a. 10.2 mL of 0.451 M $AlCl_3$ solution
    b. 5.51 L of 0.103 M Na_3PO_4 solution
    c. 1.75 mL of 1.25 M $CuCl_2$ solution
    d. 25.2 mL of 0.00157 M $Ca(OH)_2$ solution

51. Standard silver nitrate solutions are used in the titration analysis of samples containing chloride ion. How many grams of silver nitrate are needed to prepare 250. mL of standard 0.100 M $AgNO_3$ solution?

52. The substance Na_2H_2EDTA (EDTA is ethylenediaminetetraacetate, $C_{10}H_{12}N_2O_8$) is used in the titration analysis of samples containing "hard water." How many grams of Na_2H_2 EDTA are needed to prepare 250 mL of 0.0200 M "EDTA" solution?

15.5 Dilution

QUESTIONS

53. When a concentrated stock solution is diluted to prepare a less concentrated reagent, the number of _____ is the same both before and after the dilution.

54. When the volume of a given solution is doubled (by adding water), the new concentration of solute is _____ the original concentration.

PROBLEMS

55. Calculate the new molarity that results when each of the following solutions is diluted with water to a final total volume of 1.00 L.
    a. 125 mL of 0.105 M HCl
    b. 275 mL of 0.500 M $Ca(NO_3)_2$
    c. 0.500 L of 0.750 M H_3PO_4
    d. 15 mL of 18.0 M H_2SO_4

56. Calculate the new molarity that results when 250 mL of water is added to each of the following solutions.
    a. 125 mL of 0.251 M HCl
    b. 445 mL of 0.499 M H_2SO_4
    c. 5.25 L of 0.101 M HNO_3
    d. 11.2 mL of 14.5 M $HC_2H_3O_2$

57. How many mL of 3.0 M H_2SO_4 solution can be prepared by using 100. mL of 18.0 M H_2SO_4?

58. Concentrated sulfuric acid is typically 18.1 M H_2SO_4. Calculate the volume (in mL) of concentrated sulfuric acid needed to prepare 125 mL of a 0.100 M H_2SO_4 solution?

59. Sodium hydroxide is frequently sold commercially as 19 M solution. How many milliliters of this 19 M solution are needed to prepare 20. L of 0.200 M NaOH solution?

60. If 75 mL of 0.211 M NaOH is diluted to a final volume of 125 mL, what is the concentration of NaOH in the diluted solution?

61. How much *water* must be added to 500. mL of 0.200 M HCl to produce a 0.150 M solution? Assume the volumes are additive.

62. When 5.0 L of water is added to 1.0 L of 6.0 M HCl, what is the molarity of the resulting solution? (Assume that the volumes are additive.)

15.6 Stoichiometry of Solution Reactions

PROBLEMS

63. One way to determine the amount of chloride ion in a water sample is to titrate the sample with standard $AgNO_3$ solution to produce solid AgCl.

$$Ag^+(aq) + Cl^-(aq) \rightarrow AgCl(s)$$

If a 25.0-mL water sample requires 27.2 mL of 0.104 M $AgNO_3$ in such a titration, what is the concentration of Cl^- in the sample?

64. What volume (in mL) of 0.25 M Na_2SO_4 solution is needed to precipitate all the barium, as $BaSO_4(s)$, from 12.5 mL of 0.15 M $Ba(NO_3)_2$ solution?

$$Ba(NO_3)_2(aq) + Na_2SO_4(aq) \rightarrow BaSO_4(s) + 2NaNO_3(aq)$$

65. How many grams of NaCl are required to precipitate all the silver ion present in 27.2 mL of 0.104 M $AgNO_3$ solution?

$$AgNO_3(aq) + NaCl(aq) \rightarrow AgCl(s) + NaNO_3(aq)$$

66. What mass (in grams) of $AgNO_3$ is required to precipitate all of the chloride ion, as AgCl, from 135 mL of 0.101 M NaCl solution?

$$AgNO_3(aq) + NaCl(aq) \rightarrow AgCl(s) + NaNO_3(aq)$$

67. When aqueous solutions of lead(II) ion are treated with potassium chromate solution, a bright yellow precipitate of lead(II) chromate, $PbCrO_4$, forms. How many grams of lead chromate form when a 1.00-g sample of $Pb(NO_3)_2$ is added to 25.0 mL of 1.00 M K_2CrO_4 solution?

68. Aluminum ion may be precipitated from aqueous solution by addition of hydroxide ion, forming $Al(OH)_3$. A large excess of hydroxide ion must not be added, however, because the precipitate of $Al(OH)_3$ will redissolve to form a soluble aluminum/hydroxide complex. How many grams of solid NaOH should be added to 10.0 mL of 0.250 M $AlCl_3$ to just precipitate all the aluminum?

15.7 Neutralization Reactions

PROBLEMS

69. What volume of 0.200 M HCl solution is needed to exactly neutralize 25.0 mL of 0.150 M NaOH solution?

70. What volume of 0.175 M HCl solution is needed to exactly neutralize 24.9 mL of 0.451 M NaOH solution?

71. If 33.82 mL of an HNO_3 solution of unknown molarity requires 29.95 mL of 0.100 M NaOH solution to be neutralized, what is the concentration of the HNO_3 solution?

72. The total acidity in water samples can be determined by neutralization with standard sodium hydroxide solution. What is the total concentration of hydrogen ion, H^+, present in a water sample if 100. mL of the sample requires 7.2 mL of 2.5×10^{-3} M NaOH to be neutralized?

73. What volume of 1.00 M NaOH is required to neutralize each of the following solutions?
    a. 25.0 mL of 0.154 M acetic acid, $HC_2H_3O_2$
    b. 35.0 mL of 0.102 M hydrofluoric acid, HF
    c. 10.0 mL of 0.143 M phosphoric acid, H_3PO_4
    d. 35.0 mL of 0.220 M sulfuric acid, H_2SO_4

74. What volume of 0.101 M HNO_3 is required to neutralize each of the following solutions?
    a. 12.7 mL of 0.501 M NaOH
    b. 24.9 mL of 0.00491 M $Ba(OH)_2$
    c. 49.1 mL of 0.103 M NH_3
    d. 1.21 L of 0.102 M KOH

15.8 Solution Composition: Normality

QUESTIONS

75. One _____ of an acid (or base) is the amount of the acid (or base) required to provide 1 mol of hydrogen ion (or hydroxide ion).

77. Explain why the equivalent weight of H_2SO_4 is half the molar mass of this substance. How many hydrogen ions are produced when each H_2SO_4 molecule reacts with excess OH^- ions?

76. A solution that contains 1 equivalent of acid or base per liter is said to be a _____ solution.

78. How many equivalents of hydroxide ion are needed to react with 1.53 equivalents of hydrogen ion? How did you know this when no balanced chemical equation was provided for the reaction?

PROBLEMS

79. For each of the following solutions, the mass of solute taken is indicated, as well as the total volume of solution prepared. Calculate the normality of each solution.
    a. 36.5 g of HCl; 1.0 L
    b. 49 g of H_2SO_4; 1.0 L
    c. 49 g of H_3PO_4; 1.0 L

81. Calculate the normality of each of the following solutions.
    a. 0.250 M HCl
    b. 0.105 M H_2SO_4
    c. 5.3 $\times$ 10^{-2} M H_3PO_4

83. A solution of phosphoric acid, H_3PO_4, is found to contain 35.2 g of H_3PO_4 per liter of solution. Calculate the molarity and normality of the solution.

85. How many milliliters of 0.50 N NaOH are required to neutralize exactly 15.0 mL of 0.35 N H_2SO_4?

87. What volume of 0.25 N KOH is needed to neutralize 50. mL of 0.10 N $HClO_4$? What volume of 0.25 N NaOH is needed to neutralize the same amount of $HClO_4$?

80. For each of the following solutions, the mass of solute taken is indicated, along with the total volume of solution prepared. Calculate the normality of each solution.
    a. 0.113 g NaOH; 10.2 mL
    b. 12.5 mg $Ca(OH)_2$; 100 mL
    c. 12.4 g H_2SO_4; 155 mL

82. Calculate the normality of each of the following solutions.
    a. 4.01 $\times$ 10^{-3} M H_3PO_4
    b. 0.000101 M $Ba(OH)_2$
    c. 0.101 M $HC_2H_3O_2$

84. A solution of the sparingly soluble base $Ca(OH)_2$ is prepared in a volumetric flask by dissolving 5.21 mg of $Ca(OH)_2$ to a total volume of 1,000 mL. Calculate the molarity and normality of the solution.

86. How many milliliters of 0.103 M H_2SO_4 are required to neutralize exactly 4.25 L of 0.104 M NaOH?

88. Suppose that 27.34 mL of standard 0.1021 M NaOH are required to neutralize 25.00 mL of unknown H_2SO_4 solution. Calculate the molarity and the normality of the unknown solution.

Additional Problems

89. A mixture is prepared by mixing 50.0 g of ethanol, 50.0 g of water, and 5.0 g of sugar. What is the mass percent of each component in the mixture? How many grams of the mixture should one take in order to have 1.5 g of sugar? How many grams of the mixture should one take in order to have 10.0 g of alcohol?

90. Explain the difference in meaning between the following two solutions: "50. g of NaCl dissolved in 1.0 L of water" and "50. g of NaCl dissolved in enough water to make 1.0 L of solution." For which solution can the molarity be calculated directly (using the molar mass of NaCl)?

91. Suppose 50.0 mL of 0.250 M $CoCl_2$ solution is added to 25.0 mL of 0.350 M $NiCl_2$ solution. Calculate the concentration, in moles per liter, of each of the ions present after mixing. Assume that the volumes are additive.

92. How many grams of 5.0% by mass NaCl solution can be prepared by dilution of 75 g of 25% NaCl solution?

93. Calculate the mass of AgCl formed, and the concentration of silver ion remaining in solution, when 10.0 g of solid $AgNO_3$ is added to 50. mL of 1.0 $\times$ 10^{-2} M NaCl solution. Assume there is no volume change upon addition of the solid.

94. Calculate the mass of solid $BaSO_4$ formed, and the concentration of barium ion remaining in solution, when 10. mL of 0.50 M $Ba(NO_3)_2$ is added to 10. mL of 0.20 M H_2SO_4 solution. Assume the volumes are additive.

95. Many metal ions form insoluble sulfide compounds when a solution of the metal ion is treated with hydrogen sulfide gas. For example, nickel(II) precipitates nearly quantitatively as NiS when H_2S gas is bubbled through a nickel ion

solution. How many milliliters of gaseous H_2S at STP are needed to precipitate all the nickel ion present in 10. mL of 0.050 M $NiCl_2$ solution?

96. Strictly speaking, the solvent is the component of a solution that is present in the largest amount on a *mole* basis. For solutions involving water, water is almost always the solvent because there tend to be many more water molecules present than molecules of any conceivable solute. To see why this is so, calculate the number of moles of water present in 1.0 L of water. Recall that the density of water is very nearly 1.0 g/mL under most conditions.

97. Ammonia is typically sold by chemical supply houses as the saturated solution, which has a concentration of 14.5 mol/L. What volume of NH_3 at STP is required to prepare 100. mL of concentrated ammonia solution?

98. What volume of hydrogen chloride gas at STP is required to prepare 500. mL of 0.100 M HCl solution?

99. What do we mean when we say that "like dissolves like"? Do two molecules have to be identical to be able to form a solution in one another?

100. The concentration of a solution of HCl is 33.1% by mass, and its density was measured to be 1.147 g/mL. How many milliliters of the HCl solution are required to obtain 10.0 g of HCl?

101. An experiment calls for 1.00 g of silver nitrate, but all that is available in the laboratory is a 0.50% solution of $AgNO_3$. Assuming the density of the silver nitrate solution to be very nearly that of water because it is so dilute, determine how many milliliters of the solution should be used.

102. What is the molarity of NaCl in the solution that results when 10.0 g of NaCl is placed in a calibrated flask and dissolved and then water is added to the 100. mL mark?

103. A solution is 0.1% by mass calcium chloride. Therefore 100. g of the solution contains _____ g of calcium chloride.

104. Calculate the mass percent of KNO_3 in each of the following solutions.
    a. 5.0 g of KNO_3 in 75 g of water
    b. 2.5 mg of KNO_3 in 1.0 g of water
    c. 11 g of KNO_3 in 89 g of water
    d. 11 g of KNO_3 in 49 g of water

105. Calculate the total mass of 25% by mass sugar solution that one should take in order to obtain the quantity of sugar listed.
    a. 25 g of sugar
    b. 5.0 g of sugar
    c. 100. g of sugar
    d. 1.0 kg of sugar

106. A certain grade of steel is made by dissolving 5.0 g of carbon and 1.5 g of nickel per 100. g of molten iron. What is the mass percent of each component in the finished steel?

107. A sugar solution is prepared in such a way that it contains 10.% dextrose by mass. What quantity of this solution do we need to obtain 25 g of dextrose?

108. How many grams of Na_2CO_3 are contained in 500. g of a 5.5% by mass Na_2CO_3 solution?

109. How many grams of $AgNO_3$ are required to prepare 100. g of a 1% $AgNO_3$ solution?

110. A solution contains 7.5% by mass NaCl and 2.5% by mass KBr. What mass of *each* solute is contained in 125 g of the solution?

111. A solution that is 1.0 M in Na_2SO_4 contains _____ mol of sodium ion and _____ mol of sulfate ion per liter.

112. For each of the following solutions, the number of moles of solute is given, followed by the total volume of solution prepared. Calculate the molarity.
    a. 0.10 mol of $CaCl_2$; 25 mL
    b. 2.5 mol of KBr; 2.5 L
    c. 0.55 mol of $NaNO_3$; 755 mL
    d. 4.5 mol of Na_2SO_4; 1.25 L

113. For each of the following solutions, the mass of the solute is given, followed by the total volume of solution prepared. Calculate the molarity.
    a. 5.0 g of $BaCl_2$; 2.5 L
    b. 3.5 g of KBr; 75 mL
    c. 21.5 g of Na_2CO_3; 175 mL
    d. 55 g of $CaCl_2$; 1.2 L

114. If 125 g of sucrose, $C_{12}H_{22}O_{11}$, is dissolved in enough water to make 450. mL of solution, calculate the molarity.

115. Concentrated hydrochloric acid is made by pumping hydrogen chloride gas into distilled water. If concentrated HCl contains 439 g of HCl per liter, what is the molarity?

116. If 1.5 g of NaCl is dissolved in enough water to make 1.0 L of solution, what is the molarity of NaCl in the solution?

117. How many *moles* of the indicated solute does each of the following solutions contain?
    a. 1.5 L of 3.0 M H_2SO_4 solution
    b. 35 mL of 5.4 M NaCl solution
    c. 5.2 L of 18 M H_2SO_4 solution
    d. 0.050 L of 1.1 × 10^{-3} M NaF solution

118. How many *grams* of the indicated solute does each of the following solutions contain?
    a. 3.8 L of 1.5 M KCl solution
    b. 15 mL of 5.4 M NaCl solution
    c. 20. L of 12.1 M HCl solution
    d. 25 mL of 0.100 M $HClO_4$ solution

119. If 10. g of $AgNO_3$ is available, what volume of 0.25 M $AgNO_3$ solution can be prepared?
120. Calculate the number of moles of *each* ion present in each of the following solutions.
    a. 1.25 L of 0.250 M Na_3PO_4 solution
    b. 3.5 mL of 6.0 M H_2SO_4 solution
    c. 25 mL of 0.15 M $AlCl_3$ solution
    d. 1.50 L of 1.25 M $BaCl_2$ solution
121. Calcium carbonate, $CaCO_3$, can be obtained in a very pure state. Standard solutions of calcium ion are usually prepared by dissolving calcium carbonate in acid. What mass of $CaCO_3$ should be taken to prepare 500. mL of 0.0200 M calcium ion solution?
122. Calculate the new molarity when 150. mL of water is added to each of the following solutions.
    a. 125 mL of 0.200 M HBr
    b. 155 mL of 0.250 M $Ca(C_2H_3O_2)_2$
    c. 0.500 L of 0.250 M H_3PO_4
    d. 15 mL of 18.0 M H_2SO_4
123. How many mL of 12.1 M HCl are needed to prepare 100. mL of 0.100 M HCl solution?
124. When 50. mL of 5.4 M NaCl is diluted to a final volume of 300. mL, what is the concentration of NaCl in the diluted solution?
125. When 10. L of water is added to 3.0 L of 6.0 M H_2SO_4, what is the molarity of the resulting solution? Assume the volumes are additive.
126. How many milliliters of 0.10 M Na_2S solution are required to precipitate all the nickel, as NiS, from 25.0 mL of 0.20 M $NiCl_2$ solution?

$$NiCl_2(aq) + Na_2S(aq) \rightarrow NiS(s) + 2NaCl(aq)$$

127. How many grams of $Ba(NO_3)_2$ are required to precipitate all the sulfate ion present in 15.3 mL of 0.139 M H_2SO_4 solution?

$$Ba(NO_3)_2(aq) + H_2SO_4(aq) \rightarrow BaSO_4(s) + 2HNO_3(aq)$$

128. What volume of 0.150 M HNO_3 solution is needed to exactly neutralize 35.0 mL of 0.150 M NaOH solution?
129. What volume of 0.250 M HCl is required to neutralize each of the following solutions?
    a. 25.0 mL of 0.103 M sodium hydroxide, NaOH
    b. 50.0 mL of 0.00501 M calcium hydroxide, $Ca(OH)_2$
    c. 20.0 mL of 0.226 M ammonia, NH_3
    d. 15.0 mL of 0.0991 M potassium hydroxide, KOH
130. For each of the following solutions, the mass of solute taken is indicated, as well as the total volume of solution prepared. Calculate the normality of each solution.
    a. 15.0 g of HCl; 500. mL
    b. 49.0 g of H_2SO_4; 250. mL
    c. 10.0 g of H_3PO_4; 100. mL
131. Calculate the normality of each of the following solutions.
    a. 0.50 M acetic acid, $HC_2H_3O_2$
    b. 0.00250 M sulfuric acid, H_2SO_4
    c. 0.10 M potassium hydroxide, KOH
132. A sodium dihydrogen phosphate solution was prepared by dissolving 5.0 g of NaH_2PO_4 in enough water to make 500. mL of solution. What are the molarity and normality of the resulting solution?
133. How many milliliters of 0.50 M NaOH are needed to neutralize exactly 15.0 mL of 0.35 M H_2SO_4?
134. If 27.5 mL of 3.5×10^{-2} N $Ca(OH)_2$ solution are needed to neutralize 10.0 mL of nitric acid solution of unknown concentration, what is the normality of the nitric acid?

Solutions to Self-Check Exercises

SELF-CHECK EXERCISE 15.1

$$\text{Mass percent} = \frac{\text{mass of solute}}{\text{mass of solution}} \times 100$$

For this sample, the mass of solution is 135 g and the mass of the solute is 4.73 g, so

$$\text{Mass percent} = \frac{4.73 \text{ g solute}}{135 \text{ g solution}} \times 100$$

$$= 3.50\%$$

Using the definition of mass percent, we have

$$\frac{\text{Mass of solute}}{\text{Mass of solution}} = \frac{\text{grams of solute}}{\text{grams of solute } + \text{ grams of solvent}} \times 100\% = 40.0\%$$

There are 425 grams of solute (formaldehyde). Substituting, we have

$$\frac{425 \text{ g}}{425 \text{ g } + \text{ grams of solvent}} \times 100\% = 40.0\%$$

We must now solve for grams of solvent (water). This will take some patience, but we can do it if we proceed step by step. First we divide both sides by 100%.

$$\frac{425 \text{ g}}{425 \text{ g } + \text{ grams of solvent}} \times \frac{\cancel{100\%}}{\cancel{100\%}} = \frac{40.0\%}{100\%} = 0.400$$

Now we have

$$\frac{425 \text{ g}}{425 \text{ g } + \text{ grams of solvent}} = 0.400$$

Next we multiply both sides by (425 g + grams of solvent).

$$\cancel{(425 \text{ g } + \text{ grams of solvent})} \times \frac{425 \text{ g}}{\cancel{425 \text{ g } + \text{ grams of solvent}}}$$

$$= 0.400 \times (425 \text{ g } + \text{ grams of solvent})$$

This gives

$$425 \text{ g } = 0.400 \times (425 \text{ g } + \text{ grams of solvent})$$

Carrying out the multiplication gives

$$425 \text{ g } = 170. \text{ g } + 0.400 \text{ (grams of solvent)}$$

Now we subtract 170. g from both sides:

$$425 \text{ g } - 170. \text{ g } = \cancel{170. \text{ g}} - \cancel{170. \text{ g}} + 0.400 \text{ (grams of solvent)}$$

$$255 \text{ g } = 0.400 \text{ (grams of solvent)}$$

and divide both sides by 0.400.

$$\frac{255 \text{ g}}{0.400} = \frac{\cancel{0.400}}{\cancel{0.400}} \text{ (grams of solvent)}$$

We finally have the answer:

$$\frac{255 \text{ g}}{0.400} = 638 \text{ g } = \text{ grams of solvent } = \text{ mass of water needed}$$

SELF-CHECK EXERCISE 15.3

The moles of ethanol can be obtained from its molar mass (46.1).

$$1.00 \text{ g } C_2H_5OH \times \frac{1 \text{ mol } C_2H_5OH}{46.1 \text{ g } C_2H_5OH} = 2.17 \times 10^{-2} \text{ mol } C_2H_5OH$$

$$\text{Volume in liters } = 101 \text{ mL} \times \frac{1 \text{ L}}{1000 \text{ mL}} = 0.101 \text{ L}$$

$$\text{Molarity of } C_2H_5OH = \frac{\text{moles of } C_2H_5OH}{\text{liters of solution}} = \frac{2.17 \times 10^{-2} \text{ mol}}{0.101 \text{ L}}$$

$$= 0.215 \text{ } M$$

SELF-CHECK EXERCISE 15.4

When Na_2CO_3 and $Al_2(SO_4)_3$ dissolve in water, they produce ions as follows:

$$Na_2CO_3(s) \xrightarrow{H_2O(l)} 2Na^+(aq) + CO_3^{2-}(aq)$$

$$Al_2(SO_4)_3(s) \xrightarrow{H_2O(l)} 2Al^{3+}(aq) + 3SO_4^{2-}(aq)$$

Therefore, in a 0.10 M Na_2CO_3 solution, the concentration of Na^+ ions is $2 \times 0.10 \text{ } M = 0.20 \text{ } M$ and the concentration of CO_3^{2-} ions is 0.10 M. In a 0.010 M $Al_2(SO_4)_3$ solution, the concentration of Al^{3+} ions is $2 \times 0.010 \text{ } M = 0.020 \text{ } M$ and the concentration of SO_4^{2-} ions is $3 \times 0.010 \text{ } M = 0.030 \text{ } M$.

SELF-CHECK EXERCISE 15.5

When solid $AlCl_3$ dissolves, it produces ions as follows:

$$AlCl_3(s) \xrightarrow{H_2O(l)} Al^{3+}(aq) + 3Cl^-(aq)$$

so a 1.0×10^{-3} M $AlCl_3$ solution contains 1.0×10^{-3} M Al^{3+} ions and 3.0×10^{-3} M Cl^- ions.

To calculate the moles of Cl^- ions in 1.75 L of the 1.0×10^{-3} M $AlCl_3$ solution, we must multiply the volume by the molarity.

$$1.75 \text{ L solution} \times 3.0 \times 10^{-3} \text{ } M \text{ Cl}^- = 1.75 \text{ L solution} \times \frac{3.0 \times 10^{-3} \text{ mol Cl}^-}{\text{L solution}}$$

$$= 5.2 \times 10^{-3} \text{ mol Cl}^-$$

SELF-CHECK EXERCISE 15.6

We must first determine the number of moles of formaldehyde in 2.5 L of 12.3 M formalin. Remember that volume of solution (in liters) times molarity gives moles of solute. In this case, the volume of solution is 2.5 L and the molarity is 12.3 mol of HCHO per liter of solution.

$$2.5 \; \cancel{\text{L solution}} \times \frac{12.3 \; \text{mol HCHO}}{\cancel{\text{L solution}}} = 31 \; \text{mol HCHO}$$

Next, using the molar mass of HCHO (30.0), we convert 31 mol of HCHO to grams.

$$31 \; \cancel{\text{mol HCHO}} \times \frac{30.0 \; \text{g HCHO}}{1 \; \cancel{\text{mol HCHO}}} = 9.3 \times 10^2 \; \text{g HCHO}$$

Therefore 2.5 L of 12.3 M formalin contains 9.3×10^2 g of formaldehyde. We must weigh out 930 g of formaldehyde and dissolve it in enough water to make 2.5 L of solution.

SELF-CHECK EXERCISE 15.7

We are given the following information:

$$M_1 = 12 \; \frac{\text{mol}}{\text{L}} \qquad\qquad M_2 = 0.25 \; \frac{\text{mol}}{\text{L}}$$
$$V_1 = ? \; \text{(what we need to find)} \qquad V_2 = 0.75 \; \text{L}$$

Using the fact that the moles of solute do not change upon dilution, we know that

$$M_1 \times V_1 = M_2 \times V_2$$

Solving for V_1 by dividing both sides by M_1 gives

$$V_1 = \frac{M_2 \times V_2}{M_1} = \frac{0.25 \; \frac{\text{mol}}{\cancel{\text{L}}} \times 0.75 \; \text{L}}{12 \; \frac{\text{mol}}{\cancel{\text{L}}}}$$

and

$$V_1 = 0.016 \; \text{L} = 16 \; \text{mL}$$

SELF-CHECK EXERCISE 15.8

STEP 1

When the aqueous solutions of Na_2SO_4 (containing Na^+ and SO_4^{2-} ions) and $Pb(NO_3)_2$ (containing Pb^{2+} and NO_3^- ions) are mixed, solid $PbSO_4$ is formed.

$$Pb^{2+}(aq) + SO_4^{2-}(aq) \rightarrow PbSO_4(s)$$

STEP 2

We must first determine whether Pb^{2+} or SO_4^{2-} is the limiting reactant by calculating the moles of Pb^{2+} and SO_4^{2-} ions present. Because 0.0500 M $Pb(NO_3)_2$ contains 0.0500 M Pb^{2+} ions, we can calculate the moles of Pb^{2+} ions in 1.25 L of this solution as follows:

$$1.25 \; \cancel{\text{L}} \times \frac{0.0500 \; \text{mol} \; Pb^{2+}}{\cancel{\text{L}}} = 0.0625 \; \text{mol} \; Pb^{2+}$$

The 0.0250 M Na_2SO_4 solution contains 0.0250 M SO_4^{2-} ions, and the number of moles of SO_4^{2-} ions in 2.00 L of this solution is

$$2.00 \, \cancel{L} \times \frac{0.0250 \text{ mol } SO_4^{2-}}{\cancel{L}} = 0.0500 \text{ mol } SO_4^{2-}$$

STEP 3
Pb^{2+} and SO_4^{2-} react in a $1:1$ ratio, so the amount of SO_4^{2-} ions is limiting because SO_4^{2-} is present in the smaller number of moles.

STEP 4
The Pb^{2+} ions are present in excess, and only 0.0500 mol of solid $PbSO_4$ will be formed.

STEP 5
We calculate the mass of $PbSO_4$ by using the molar mass of $PbSO_4$ (303.3 g).

$$0.0500 \, \cancel{\text{mol } PbSO_4} \times \frac{303.3 \text{ g } PbSO_4}{1 \, \cancel{\text{mol } PbSO_4}} = 15.2 \text{ g } PbSO_4$$

SELF-CHECK EXERCISE 15.9

STEP 1
Because nitric acid is a strong acid, the nitric acid solution contains H^+ and NO_3^- ions. The KOH solution contains K^+ and OH^- ions. When these solutions are mixed, the H^+ and OH^- react to form water.

$$H^+(aq) + OH^-(aq) \rightarrow H_2O(l)$$

STEP 2
The number of moles of OH^- present in 125 mL of 0.050 M KOH is

$$125 \, \cancel{mL} \times \frac{1 \, \cancel{L}}{1000 \, \cancel{mL}} \times \frac{0.050 \text{ mol } OH^-}{\cancel{L}} = 6.3 \times 10^{-3} \text{ mol } OH^-$$

STEP 3
H^+ and OH^- react in a $1:1$ ratio, so we need 6.3×10^{-3} mol of H^+ from the 0.100 M HNO_3.

STEP 4
6.3×10^{-3} mol of OH^- requires 6.3×10^{-3} mol of H^+ to form 6.3×10^{-3} mol of H_2O.

Therefore,

$$V \times \frac{0.100 \text{ mol } H^+}{L} = 6.3 \times 10^{-3} \text{ mol } H^+$$

where V represents the volume in liters of 0.100 M HNO_3 required. Solving for V, we have

$$V = \frac{6.3 \times 10^{-3} \, \cancel{\text{mol } H^+}}{\dfrac{0.100 \, \cancel{\text{mol } H^+}}{L}} = 6.3 \times 10^{-2} \text{ L}$$

$$= 6.3 \times 10^{-2} \, \cancel{L} \times \frac{1000 \text{ mL}}{\cancel{L}} = 63 \text{ mL}$$

From the definition of normality, $N =$ equiv/L, we need to calculate (1) the equivalents of KOH and (2) the volume of the solution in liters. To find the number of equivalents, we use the equivalent weight of KOH, which is 56.1 g (see Table 15.2).

$$23.6 \text{ g KOH} \times \frac{1 \text{ equiv KOH}}{56.1 \text{ g KOH}} = 0.421 \text{ equiv KOH}$$

Next we convert the volume to liters.

$$755 \text{ mL} \times \frac{1 \text{ L}}{1000 \text{ mL}} = 0.755 \text{ L}$$

Finally, we substitute these values into the equation that defines normality.

$$\text{Normality} = \frac{\text{equiv}}{\text{L}} = \frac{0.421 \text{ equiv}}{0.755 \text{ L}} = 0.558 \ N$$

To solve this problem, we use the relationship

$$N_{acid} \times V_{acid} = N_{base} \times V_{base}$$

where

$$N_{acid} = 0.50 \ \frac{\text{equiv}}{\text{L}}$$

$$V_{acid} = ?$$

$$N_{base} = 0.80 \ \frac{\text{equiv}}{\text{L}}$$

$$V_{base} = 0.250 \text{ L}$$

We solve the equation

$$N_{acid} \times V_{acid} = N_{base} \times V_{base}$$

for V_{acid} by dividing both sides by N_{acid}.

$$\frac{N_{acid} \times V_{acid}}{N_{acid}} = \frac{N_{base} \times V_{base}}{N_{acid}}$$

$$V_{acid} = \frac{N_{base} \times V_{base}}{N_{acid}} = \frac{\left(0.80 \ \frac{\text{equiv}}{L}\right) \times \left(0.250 \text{ L}\right)}{0.50 \ \frac{\text{equiv}}{L}}$$

$$V_{acid} = 0.40 \text{ L}$$

Therefore, 0.40 L of 0.50 N H_2SO_4 is required to neutralize 0.250 L of 0.80 N KOH.

CHAPTER

16 Equilibrium

A superlight aerogel "floats" on beaten egg whites.

CONTENTS

Refrigeration preserves food by slowing the rates of detrimental reactions.

Chemistry is mostly about reactions—processes in which groups of atoms are reorganized. So far we have learned to describe chemical reactions by using balanced equations and to calculate amounts of reactants and products. However, there are many important characteristics of reactions that we have not yet considered.

For example, why do refrigerators prevent food from spoiling? That is, why do the chemical reactions that cause food to decompose occur more slowly at lower temperatures? On the other hand, how can a chemical manufacturer speed up a chemical reaction that runs too slowly to be economical?

Another question that arises is why chemical reactions carried out in a closed vessel appear to stop at a certain point. For example, when the reaction of brown nitrogen dioxide to form colorless dinitrogen tetroxide,

$$2NO_2(g) \rightarrow N_2O_4(g)$$
Brown   Colorless

is carried out in a closed container, the brown color at first fades but stops changing after a time and then stays the same color indefinitely if left undisturbed (see Figure 16.1). We will account for all of these important observations about reactions in this chapter.

16.1   How Chemical Reactions Occur

AIM: To describe the collision model of how chemical reactions occur.

Figure 16.1

(a) A sample containing a large quantity of brown NO_2 gas. (b) As the reaction to form colorless N_2O_4 occurs, the sample becomes lighter brown. (c) After equilibrium is reached [$2NO_2(g) \rightleftharpoons N_2O_4(g)$], the color remains the same.

In writing the equation for a chemical reaction, we put the reactants on the left and the products on the right with an arrow between them. But how do the atoms in the reactants reorganize to form the products?

Chemists believe that molecules react by colliding with each other. Some collisions are violent enough to break bonds, allowing the reactants to rearrange

(a)

(b)

(c)

The collision model can be effectively communicated by drawings. You can show molecules colliding and then how atoms in molecules can recombine to form new molecules.

Figure 16.2
Visualizing the reaction $2BrNO(g) \rightarrow 2NO(g) + Br_2(g)$. (a) Two BrNO molecules approach each other at high speeds. (b) The collision occurs. (c) The energy of the collision causes the Br—N bonds to break and allows the Br—Br bond to form. (d) The products: one Br_2 and two NO molecules.

to form the products. For example, consider the reaction

$$2BrNO(g) \rightleftharpoons 2NO(g) + Br_2(g)$$

which we picture to occur as shown in Figure 16.2. Notice that the Br—N bonds in two BrNO molecules must be broken and a new Br—Br bond must be formed during a collision in order for the reactants to become products.

The idea that reactions occur during molecular collisions, which is called the **collision model,** explains many characteristics of chemical reactions. For example, it explains why a reaction proceeds faster if the concentrations of the reacting molecules are increased (higher concentrations lead to more collisions and therefore to more reaction events). The collision model also explains why reactions go faster at higher temperatures, as we will see in the next section.

16.2 Conditions That Affect Reaction Rates

AIM: To describe activation energy.
To explain how a catalyst speeds up a reaction.

It is easy to see why reactions speed up when the *concentrations* of reacting molecules are increased: higher concentrations (more molecules per unit volume) lead to more collisions and so to more reaction events. But reactions also speed up when the *temperature* is increased. Why? The answer lies in the fact that not all collisions possess enough energy to break bonds. A minimum energy called the **activation energy (E_a)** is needed for a reaction to occur (see Figure 16.3). If a given collision possesses an energy greater than E_a, that collision can result in a reaction. If a collision has an energy less than E_a, the molecules will bounce apart unchanged.

There are many concepts presented in this section. Two reaction curves are presented in Figures 16.3 and 16.4. In one, the energy of the products is less than the energy of the reactants. In the other, the energy of the products is greater than the energy of the reactants. You should explain why the two curves look different.

Figure 16.3
When molecules collide, a certain minimum energy called the activation energy (E_a) is needed for a reaction to occur. If the energy contained in a collision of two BrNO molecules is greater than E_a, the reaction can go "over the hump" to form products. If the collision energy is less than E_a, the colliding molecules bounce apart unchanged.

Sprinkling flour, lycopodium powder, or granular iron into a burner flame provides a nice demonstration of the dependence of combustion rate on particle size. This is a good time to discuss coal-dust and grain-dust explosions.

Recall from Section 13.8 that the average kinetic energy of a collection of molecules is directly proportional to the temperature (K).

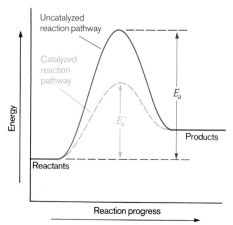

Figure 16.4

Comparison of the activation energies for an uncatalyzed reaction (E_a) and for the same reaction with a catalyst present (E_a'). Note that a catalyst works by lowering the activation energy for a reaction.

Although O atoms are too reactive to exist near the earth's surface, they do exist in the upper atmosphere.

The purpose and use of catalysts can be illustrated by discussing catalytic converters in automobiles. Metals such as platinum and palladium help convert uncombusted hydrocarbons to carbon dioxide and water. In the process, the metals are not used up, but are used over and over again.

The reason that a reaction occurs faster as the temperature is increased is that the speeds of the molecules increase with temperature. So at higher temperatures, the average collision is more energetic. This makes it more likely that a given collision will possess enough energy to break bonds and to produce the molecular rearrangements needed for a reaction to occur:

Is it possible to speed up a reaction without changing the temperature or the reactant concentrations? Yes, by using something called a **catalyst,** *a substance that speeds up a reaction without being consumed.* This may sound too good to be true, but it is a very common occurrence. In fact, you would not be alive now if your body did not contain thousands of catalysts called **enzymes.** Enzymes allow our bodies to speed up complicated reactions that would be too slow to sustain life at normal body temperatures.

Although we cannot consider the details here, a catalyst works because it provides a new pathway for the reaction—a pathway that has a lower activation energy than the original pathway, as illustrated in Figure 16.4. Because of the lower activation energy, more collisions will have enough energy to allow a reaction. This in turn leads to a faster reaction.

A very important example of a reaction involving a catalyst occurs in our atmosphere; it is the breakdown of ozone, O_3, catalyzed by chlorine atoms. Ozone is one constituent of the earth's upper atmosphere that is especially crucial, because it absorbs harmful high-energy radiation from the sun. There are natural processes that result in both the formation and the destruction of ozone in the upper atmosphere. The natural balance of all these opposing processes has resulted in an amount of ozone that has been relatively constant over the years. However, the ozone level now seems to be decreasing, especially over Antarctica (Figure 16.5), apparently because chlorine atoms act as catalysts for the decomposition of ozone to oxygen by the following pair of reactions:

$$Cl + O_3 \rightarrow ClO + O_2$$
$$O + ClO \rightarrow Cl + O_2$$

Sum: $Cl + O_3 + O + \cancel{ClO} \rightarrow \cancel{ClO} + O_2 + \cancel{Cl} + O_2$

When species that appear on both sides of the equation are canceled, the end result is the reaction

$$O + O_3 \rightarrow 2O_2$$

Notice that a chlorine atom is used up in the first reaction but a chlorine atom is formed again by the second reaction. Therefore, the amount of chlorine does not change as the overall process occurs. This means that the chlorine atom is a true catalyst: it participates in the process but is not consumed. Estimates show that

Figure 16.5
A photo showing the ozone "hole" over Antarctica.

Figure 16.6
The establishment of the equilibrium vapor pressure over a liquid in a closed container. (a) At first there is a net transfer of molecules from the liquid state to the vapor state. (b) After a while, the amount of the substance in the vapor state becomes constant—both the pressure of the vapor and the level of the liquid stay the same. This is the equilibrium state. (c) The equilibrium state is very dynamic. The vapor pressure and liquid level remain constant because exactly the same number of molecules escape the liquid as return to it.

one chlorine atom can catalyze the destruction of about one million ozone molecules per second.

The chlorine atoms that promote this damage to the ozone layer are present because of pollution. Specifically, they come from the decomposition of compounds called freons, such as CF_2Cl_2, which are widely used in refrigerators and air conditioners. The freons leak into the atmosphere, where they are decomposed by light to produce chlorine atoms and other substances.

16.3 The Equilibrium Condition

AIM: To discuss how equilibrium is established.

Equilibrium is a word that implies balance or steadiness. When we say that someone is maintaining his or her equilibrium, we are describing a state of balance among various opposing forces. The term is used in a similar but more specific way in chemistry. Chemists define **equilibrium** as the *exact balancing of two processes, one of which is the opposite of the other.*

We first encountered the concept of equilibrium in Section 14.4, when we described the way vapor pressure develops over a liquid in a closed container (see Figure 14.9). This equilibrium process is summarized in Figure 16.6. The *equilib-*

Diagrams such as Figure 16.6 clearly show equilibrium between two states of matter. But it is difficult to show with diagrams the changes that occur as molecules evaporate and condense at equilibrium.

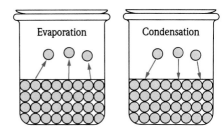

rium state occurs when the rate of evaporation exactly equals the rate of condensation.

So far in this book we have assumed that reactions proceed to completion—that is, until one of the reactants "runs out." Indeed, many reactions *do* proceed essentially to completion. For such reactions we can assume that the reactants are converted to products until the limiting reactant is completely consumed. On the other hand, there are many chemical reactions that "stop" far short of completion when they are allowed to take place in a closed container. An example is the reaction of nitrogen dioxide to form dinitrogen tetroxide.

$$NO_2(g) \; + \; NO_2(g) \rightarrow N_2O_4(g)$$
$$\text{Brown} \qquad\qquad \text{Colorless}$$

The reactant NO_2 is a dark brown gas, and the product N_2O_4 is a colorless gas. Imagine an experiment where pure NO_2 is placed in an empty, sealed glass vessel at 25 °C. The initial dark brown color will decrease in intensity as the NO_2 is converted to colorless N_2O_4 (see Figure 16.1). However, even over a long period of time, the contents of the reaction vessel do not become colorless. Instead, the intensity of the brown color eventually becomes constant, which means that the concentration of NO_2 is no longer changing. This simple observation is a clear indication that the reaction has "stopped" short of completion. In fact, the reaction has not stopped. Rather, the system has reached **chemical equilibrium,** *a dynamic state where the concentrations of all reactants remain constant.*

This situation is similar to the one where a liquid in a closed container develops a constant vapor pressure, except that in this case two opposite chemical reactions are involved. When pure NO_2 is first placed in the closed flask, there is no N_2O_4 present. As collisions between NO_2 molecules occur, N_2O_4 is formed and the concentration of N_2O_4 in the container increases. However, the reverse reaction can also occur. A given N_2O_4 molecule can decompose to two NO_2 molecules.

$$N_2O_4(g) \rightarrow NO_2(g) \; + \; NO_2(g)$$

That is, chemical reactions are *reversible;* they can occur in either direction. We usually indicate this fact by using double arrows.

$$2NO_2(g) \underset{\text{Reverse}}{\overset{\text{Forward}}{\rightleftharpoons}} N_2O_4(g)$$

In this case the double arrows mean that either two NO_2 molecules can combine to form an N_2O_4 molecule (the *forward* reaction) or an N_2O_4 molecule can decompose to give two NO_2 molecules (the *reverse* reaction).

Equilibrium is reached whether pure NO_2, pure N_2O_4, or a mixture of NO_2 and N_2O_4 is initially placed in a closed container. In any of these cases, conditions will eventually be reached in the container such that N_2O_4 is being formed and is decomposing at exactly the same rate. This leads to chemical equilibrium, a dynamic situation where the concentrations of reactants and products remain the same indefinitely, as long as the conditions are not changed.

When you discuss chemical equilibrium, point out that equilibrium is reached only in *closed* systems and that industrial processes are designed to maximize the amount of product formed and to avoid the limitation of equilibrium.

Students often believe that equilibrium is a static condition and that at equilibrium change has ceased to occur. You cannot stress enough that no net change occurs, but chemical reactions continue to take place.

16.4 Chemical Equilibrium: A
Dynamic Condition

AIM: To discuss the characteristics of chemical equilibrium.

Because no changes occur in the concentrations of reactants or products in a reaction system at equilibrium, it may appear that everything has stopped. However, this is not the case. On the molecular level there is frantic activity. Equilibrium is not static but is a highly *dynamic* situation. Consider again the analogy between chemical equilibrium and two island cities connected by a single bridge. Suppose the traffic flow on the bridge is the same in both directions. It is obvious that there is motion (we can see the cars traveling across the bridge), but the number of cars in each city does not change because there is an equal flow of cars entering and leaving. The result is no *net* change in the number of cars in each of the two cities.

Equilibrium is a dynamic situation.

To see how this concept applies to chemical reactions, let's consider the reaction between steam and carbon monoxide in a closed vessel at a high temperature where the reaction takes place rapidly.

$$H_2O(g) + CO(g) \rightleftharpoons H_2(g) + CO_2(g)$$

Assume that the same number of moles of gaseous CO and gaseous H_2O are placed in a closed vessel and allowed to react (Figure 16.7a).

H_2O	CO	H_2O	CO
CO	H_2O	H_2O	CO
CO	H_2O	CO	H_2O
H_2O	CO	H_2O	CO
CO	H_2O	CO	H_2O

(a)

H_2	CO_2	H_2O	CO
CO	H_2O	CO_2	H_2
CO	H_2O	CO	H_2O
H_2	CO_2	H_2O	CO
CO	H_2O	CO	H_2O

(b)

H_2	CO_2	CO_2	H_2
CO	H_2O	H_2	CO_2
CO_2	H_2	CO	H_2O
H_2	CO_2	H_2	CO_2
CO	H_2O	CO_2	H_2

(c)

H_2O	CO	CO_2	H_2
CO_2	H_2	H_2O	CO
H_2	CO_2	CO	H_2O
CO_2	H_2	H_2	CO_2
H_2	CO_2	CO_2	H_2

(d)

Figure 16.7
The reaction of H_2O and CO to form CO_2 and H_2 as time passes. (a) Equal numbers of moles of H_2O and CO are mixed in a closed container. (b) The reaction begins to occur, and some products (H_2 and CO_2) are formed. (c) The reaction continues as time passes and more reactants are changed to products. (d) Although time continues to pass, the numbers of reactant and product molecules are the same as in (c). No further changes are seen as time continues to pass. The system has reached equilibrium.

When CO and H_2O, the reactants, are mixed, they immediately begin to react to form the products, H_2 and CO_2. This leads to a decrease in the concentrations of the reactants, but the concentrations of the products, which were initially at zero, are increasing (Figure 16.7b). After a certain period of time, the concentrations of reactants and products no longer change at all—equilibrium has been reached (Figure 16.7c and d). Unless the system is somehow disturbed, no further changes in concentrations will occur.

Why does equilibrium occur? We saw earlier in this chapter that molecules react by colliding with one another, and that the more collisions, the faster the reaction. This is why the speed of a reaction depends on concentrations. In this case the concentrations of H_2O and CO are lowered as the forward reaction occurs—that is, as products are formed.

$$H_2O \ + \ CO \rightarrow H_2 \ + \ CO_2$$

As the concentrations of the reactants decrease, the forward reaction slows down (Figure 16.8). But as in the traffic-on-the-bridge analogy, there is also movement in the reverse direction.

$$H_2 \ + \ CO_2 \rightarrow H_2O \ + \ CO$$

Initially in this experiment, no H_2 and CO_2 are present, so this reverse reaction cannot occur. However, as the forward reaction proceeds, the concentrations of H_2 and CO_2 build up and the speed (or rate) of the reverse reaction increases (Figure 16.8) as the forward reaction slows down. Eventually the concentrations reach levels at which *the rate of the forward reaction equals the rate of the reverse reaction.* The system has reached equilibrium.

A double arrow ($\rightleftharpoons$) is used to show that a reaction is occurring in both directions.

Diagrams that show reaction rates versus time can reinforce some students views that change stops when equilibrium is reached. Point out that the y-axis is labeled reaction rate and that the flat curve means that the speed of the reaction no longer changes, not that no reaction occurs.

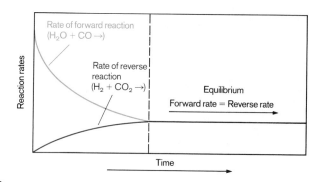

Figure 16.8
The changes with time in the rates of the forward and reverse reactions for $H_2O(g)$ and $CO(g) \rightleftharpoons H_2(g) \ + \ CO_2(g)$ when equal numbers of moles of $H_2O(g)$ and $CO(g)$ are mixed. At first, the rate of the forward reaction decreases and the rate of the reverse reaction increases. Equilibrium is reached when the forward rate and the reverse rate become the same.

The Equilibrium Constant: An Introduction

AIM: To introduce the law of chemical equilibrium and to show how to calculate values for the equilibrium constant.

Science is based on the results of experiments. The development of the equilibrium concept is typical. On the basis of their observations of many chemical reactions, two Norwegian chemists, Cato Maximilian Guldberg and Peter Waage, proposed in 1864 the **law of chemical equilibrium** (originally called the *law of mass action*) as a general description of the equilibrium condition. Guldberg and Waage postulated that for a reaction of the type

$$aA + bB \rightleftharpoons cC + dD$$

where A, B, C, and D represent chemical species and *a, b, c,* and *d* are their coefficients in the balanced equation, the law of mass action is represented by the following **equilibrium expression:**

$$K = \frac{[C]^c\,[D]^d}{[A]^a\,[B]^b}$$

The square brackets indicate the concentrations of the chemical species *at equilibrium* (in units of mol/L), and *K* is a constant called the **equilibrium constant.** Note that the equilibrium expression is a special ratio of the concentrations of the products to the concentrations of the reactants. Each concentration is raised to a power corresponding to its coefficient in the balanced equation.

The law of chemical equilibrium as proposed by Guldberg and Waage is based on experimental observations. Experiments on many reactions showed that the equilibrium condition could always be described by this special ratio, called the equilibrium expression.

To see how to construct an equilibrium expression, consider the reaction where ozone changes to oxygen:

$$2O_3(g) \rightleftharpoons 3O_2(g)$$

Reactant Coefficient   Product Coefficient

To obtain the equilibrium expression, we place the concentration of the product in the numerator and the concentration of the reactant in the denominator.

$$\frac{[O_2]}{[O_3]} \begin{matrix} \leftarrow \text{Product} \\ \leftarrow \text{Reactant} \end{matrix}$$

It is important to begin this section using balanced equations, with symbols to indicate the physical states of reactants and products. In Section 16.6 the topic of heterogeneous equilibria is introduced and students will need to distinguish among solids, liquids, and gases.

With practice, students can write the correct equilibrium expression for a reaction and can calculate *K* from equilibrium concentrations. They have difficulty understanding what *K* represents, though. Sometimes students cannot distinguish between initial and equilibrium concentrations and will use either set to calculate a value for *K*.

Square brackets, [ ], indicate concentration units of mol/L.

Although most students can write a correct equilibrium expression using letter symbols, it is harder for them to use an actual balanced equation to generate an equilibrium expression.

Then we use the coefficients as powers.

$$K = \frac{[O_2]^3}{[O_3]^2}$$ Coefficients become powers.

EXAMPLE 16.1   Writing Equilibrium Expressions

Write the equilibrium expression for the following reactions.

a. $H_2(g) + F_2(g) \rightleftharpoons 2HF(g)$ b. $N_2(g) + 3H_2(g) \rightleftharpoons 2NH_3(g)$

SOLUTION

Applying the law of chemical equilibrium, we place products over reactants (using square brackets to denote concentrations in units of moles per liter) and raise each concentrations to the power that corresponds to the coefficient in the balanced chemical equation.

a. $K = \dfrac{[HF]^2}{[H_2][F_2]}$ ← Product (coefficient of 2 becomes power of 2)
                 ← Reactants (coefficients of 1 become powers of 1)

Note that when a coefficient (power) of 1 occurs, it is not written but is understood.

b. $K = \dfrac{[NH_3]^2}{[N_2][H_2]^3}$

SELF-CHECK EXERCISE 16.1

Write the equilibrium expression for the following reaction.

$$4NH_3(g) + 7O_2(g) \rightleftharpoons 4NO_2(g) + 6H_2O(g)$$

What does the equilibrium expression mean? It means that, for a given reaction at a given temperature, the special ratio of the concentrations of the products to reactants defined by the equilibrium expression will always be equal to the same number—namely, the equilibrium constant K. For example, consider a series of experiments on the ammonia synthesis reaction

$$N_2(g) + 3H_2(g) \rightleftharpoons 2NH_3(g)$$

carried out at 500 °C to measure the concentrations of N_2, H_2, and NH_3 present at equilibrium. The results of these experiments are shown in Table 16.1. In this table, subscript zeros next to square brackets are used to indicate *initial concentrations:* the concentrations of reactants and products originally mixed together before any reaction has occurred.

Table 16.1 Results of Three Experiments for the Reaction $N_2(g) + 3H_2(g) \rightleftharpoons 2NH_3(g)$ at 500 °C

Experi-ment	Initial Concentrations			Equilibrium Concentrations			$\dfrac{[NH_3]^2}{[N_2][H_2]^3} = K^*$
	$[N_2]_0$	$[H_2]_0$	$[NH_3]_0$	$[N_2]$	$[H_2]$	$[NH_3]$	
I	1.000 M	1.000 M	0	0.921 M	0.763 M	0.157 M	$\dfrac{(0.157)^2}{(0.921)(0.763)^3} = 0.0602$
II	0	0	1.000 M	0.399 M	1.197 M	0.203 M	$\dfrac{(0.203)^2}{(0.399)(1.197)^3} = 0.0602$
III	2.00 M	1.00 M	3.00 M	2.59 M	2.77 M	1.82 M	$\dfrac{(1.82)^2}{(2.59)(2.77)^3} = 0.0602$

*The units for K are customarily omitted.

Consider the results of experiment I. One mole each of N_2 and H_2 were sealed into a 1-L vessel at 500 °C and allowed to reach chemical equilibrium. At equilibrium the concentrations in the flask were found to be $[N_2] = 0.921\ M$, $[H_2] = 0.763\ M$, and $[NH_3] = 0.157\ M$. The equilibrium expression for the reaction

$$N_2(g) + 3H_2(g) \rightleftharpoons 2NH_3(g)$$

is

$$K = \frac{[NH_3]^2}{[N_2][H_2]^3} = \frac{(0.157)^2}{(0.921)(0.763)^3}$$
$$= 0.0602 = 6.02 \times 10^{-2}$$

Similarly, as shown on Table 16.1, we can calculate for experiments II or III that K, the equilibrium constant, has the value 6.02×10^{-2}. In fact, whenever N_2, H_2, and NH_3 are mixed together at this temperature, the system *always* comes to an equilibrium position such that

$$K = 6.02 \times 10^{-2}$$

regardless of the amounts of the reactants and products that are mixed together initially.

It is important to see from Table 16.1 that the *equilibrium concentrations are not always the same.* However, even though the individual sets of equilibrium

When you calculate K from concentrations of reactants and products, it is important to tabulate both initial and equilibrium concentrations so that students can distinguish between the two sets of concentration data. When both initial and equilibrium concentrations are presented in a problem, students often confuse the two sets of data.

concentrations are quite different for the different situations, the *equilibrium constant, which depends on the ratio of the concentrations, remains the same.*

Each *set of equilibrium concentrations* is called an **equilibrium position.** It is essential to distinguish between the equilibrium constant and the equilibrium positions for a given reaction system. There is only *one* equilibrium constant for a particular system at a particular temperature, but there is an *infinite* number of equilibrium positions. The specific equilibrium position adopted by a system depends on the initial concentrations; the equilibrium constant does not.

Note that in the above discussion, the equilibrium constant was given without units. In certain cases the units are included when the values of equilibrium constants are given, and in other cases they are omitted. We will not discuss the reasons for this. We will omit the units in this text.

EXAMPLE 16.2   Calculating Equilibrium Constants

The following results were collected for two experiments involving the reaction at 600 °C between gaseous sulfur dioxide and oxygen to form gaseous sulfur trioxide:

$$2SO_2(g) + O_2(g) \rightleftharpoons 2SO_3(g)$$

	Experiment I		Experiment II	
	Initial	*Equilibrium*	*Initial*	*Equilibrium*
	$[SO_2]_0 = 2.00\ M$	$[SO_2] = 1.50\ M$	$[SO_2]_0 = 0.500\ M$	$[SO_2] = 0.590\ M$
	$[O_2]_0 = 1.50\ M$	$[O_2] = 1.25\ M$	$[O_2]_0 = 0$	$[O_2] = 0.045\ M$
	$[SO_3]_0 = 3.00\ M$	$[SO_3] = 3.50\ M$	$[SO_3]_0 = 0.350\ M$	$[SO_3] = 0.260\ M$

The law of chemical equilibrium predicts that the value of *K* should be the same for both experiments. Verify this by calculating the equilibrium constant observed for each experiment.

SOLUTION

The balanced equation for the reaction is

$$2SO_2(g) + O_2(g) \rightleftharpoons 2SO_3(g)$$

From the law of chemical equilibrium, we can write the equilibrium expression

$$K = \frac{[SO_3]^2}{[SO_2]^2[O_2]}$$

EXAMPLE 16.2, CONTINUED

For experiment I we calculate the value of K by substituting the observed *equilibrium* concentrations,

$$[SO_3] = 3.50 \ M$$
$$[SO_2] = 1.50 \ M$$
$$[O_2] = 1.25 \ M$$

into the equilibrium expression.

$$K_I = \frac{(3.50)^2}{(1.50)^2(1.25)} = 4.36$$

For experiment II at equilibrium,

$$[SO_3] = 0.260 \ M$$
$$[SO_2] = 0.590 \ M$$
$$[O_2] = 0.045 \ M$$

and

$$K_{II} = \frac{(0.260)^2}{(0.590)^2(0.045)} = 4.32$$

Notice that K_I and K_{II} are almost exactly the same, as we expected. That is, the value of K is constant, within differences due to rounding off and due to experimental error. These experiments show *two different equilibrium positions* for this system, but K, the equilibrium constant, is, indeed, constant.

16.6 Heterogeneous Equilibria

AIM: To show the role that liquids and solids play in constructing the equilibrium expression.

So far we have discussed equilibria only for systems in the gaseous state, where all reactants and products are gases. These are examples of **homogeneous equilibria** in which all substances are in the same state. However, many equilibria involve more than one state and are called **heterogeneous equilibria.** For example, the thermal decomposition of calcium carbonate in the commercial preparation of lime occurs by a reaction involving solids and gases.

$$CaCO_3(s) \rightleftharpoons CaO(s) + CO_2(g)$$
$$\uparrow$$
Lime

Students often forget to check the physical states of products and reactants before writing equilibrium expressions; they include pure liquids and solids by mistake. You will need to remind them at every opportunity to check before they write.

In terms of amount produced, lime is among the top ten chemicals manufactured in the United States.

Straightforward application of the law of equilibrium leads to the equilibrium expression

$$K = \frac{[CO_2][CaO]}{[CaCO_3]}$$

Students have a difficult time understanding why solids and liquids are not included in equilibrium expressions. The amounts of solids and liquids change, but equilibrium expressions are defined as concentrations.

The concentrations of pure liquids and solids are constant.

However, experimental results show that the *position of a heterogeneous equilibrium does not depend on the amounts of pure solids or liquids present.* The fundamental reason for this behavior is that the concentrations of pure solids and liquids cannot change. In other words, we might say that the concentrations of pure solids and liquids are constants. Therefore, we can write the equilibrium expression for the decomposition of solid calcium carbonate as

$$K' = \frac{[CO_2]C_1}{C_2}$$

where C_1 and C_2 are constants representing the concentrations of the solids CaO and $CaCO_3$, respectively. This expression can be rearranged to give

$$\frac{C_2 K'}{C_1} = K = [CO_2]$$

where the constants C_2, K', and C_1 are combined into a single constant K. This leads us to the following general statement: the concentrations of pure solids or pure liquids involved in a chemical reaction *are not included in the equilibrium expression* for the reaction. This applies *only* to pure solids or liquids. It does not apply to solutions or gases, because their concentrations can vary.

For example, consider the decomposition of liquid water to gaseous hydrogen and oxygen:

$$2H_2O(l) \rightleftharpoons 2H_2(g) + O_2(g)$$

where

$$K = [H_2]^2[O_2]$$

Some substances such as water can appear in different balanced equations as either a liquid or as a gas. It is not safe for students to assume that water will always appear in the equilibrium expression as a gas.

Water is not included in the equilibrium expression because it is a pure liquid. However, when the reaction is carried out under conditions where the water is a gas rather than a liquid:

$$2H_2O(g) \rightleftharpoons 2H_2(g) + O_2(g)$$

we have

$$K = \frac{[H_2]^2[O_2]}{[H_2O]^2}$$

because the concentration of water vapor can change.

EXAMPLE 16.3   **Writing Equilibrium Expressions for Heterogeneous Equilibria**

Write the expressions for K for the following processes.

a. The decomposition of solid phosphorus pentachloride to liquid phosphorus trichloride and chlorine gas.
b. Deep-blue solid copper(II) sulfate pentahydrate is heated to drive off water vapor and form white solid copper(II) sulfate.

SOLUTION

a. The reaction is

$$PCl_5(s) \rightleftharpoons PCl_3(l) + Cl_2(g)$$

In this case, neither the pure solid PCl_5 nor the pure liquid PCl_3 is included in the equilibrium expression. The equilibrium expression is

$$K = [Cl_2]$$

b. The reaction is

$$CuSO_4 \cdot 5H_2O(s) \rightleftharpoons CuSO_4(s) + 5H_2O(g)$$

The two solids are not included. The equilibrium expression is

$$K = [H_2O]^5$$

(a)

(b)

As solid copper (II) sulfate pentahydrate, $CuSO_4 \cdot 5H_2O$, is heated, it loses H_2O eventually forming white $CuSO_4$.

EXAMPLE 16.3, CONTINUED

SELF-CHECK EXERCISE 16.2

Write the equilibrium expression for each of the following reactions.

a. $2KClO_3(s) \rightleftharpoons 2KCl(s) + 3O_2(g)$
    This reaction is often used to produce oxygen gas in the laboratory.

b. $NH_4NO_3(s) \rightleftharpoons N_2O(g) + 2H_2O(g)$

c. $CO_2(g) + MgO(s) \rightleftharpoons MgCO_3(s)$

d. $SO_3(g) + H_2O(l) \rightleftharpoons H_2SO_4(l)$

16.7 Le Châtelier's Principle

AIM: To show how to predict the changes that occur when a system at equilibrium is disturbed.

It is important to understand the factors that control the *position* of a chemical equilibrium. For example, when a chemical is manufactured, the chemists and chemical engineers in charge of production want to choose conditions that favor the desired product as much as possible. That is, they want the equilibrium to lie far to the right (toward products). When the process for the synthesis of ammonia was being developed, extensive studies were carried out to determine how the equilibrium concentration of ammonia depended on the conditions of temperature and pressure.

The best way for students to understand Le Châtelier's principle is for you to work through an example. The definition by itself is not easily related to a chemical reaction.

In this section we will explore how various changes in conditions affect the equilibrium position of a reaction system. We can predict the effects of changes in concentration, pressure, and temperature on a system at equilibrium by using **Le Châtelier's principle,** which states that when *a change is imposed on a system at equilibrium, the position of the equilibrium shifts in a direction that tends to reduce the effect of that change.*

The Effect of a Change in Concentration

Let us consider the ammonia synthesis reaction. Suppose there is an equilibrium position described by these concentrations:

$$[N_2] = 0.399 \ M \qquad [H_2] = 1.197 \ M \qquad [NH_3] = 0.202 \ M$$

What will happen if 1.000 mol/L of N_2 is suddenly injected into the system? We can begin to answer this question by remembering that for the system at equilibrium, the rates of the forward and reverse reactions exactly balance:

$$N_2(g) + 3H_2(g) \rightleftharpoons 2NH_3(g)$$

as indicated here by arrows of the same length. When the N_2 is added, there are suddenly more collisions between N_2 and H_2 molecules. This increases the rate of the forward reaction (shown here by the greater length of the arrow pointing in that direction),

$$N_2(g) + 3H_2(g) \rightleftharpoons 2NH_3(g)$$

and the reaction produces more NH_3. As the concentration of NH_3 increases, the reverse reaction also speeds up (as more collisions between NH_3 molecules occur) and the system again comes to equilibrium. However, the new equilibrium position has more NH_3 than was present in the original position. We say that the equilibrium shifts to the *right*—toward the products. The original and new equilibrium positions are shown below.

Equilibrium Position I		Equilibrium Position II
$[N_2] = 0.399\ M$	1.000 mol/L of N_2 added	$[N_2] = 1.348\ M$
$[H_2] = 1.197\ M$		$[H_2] = 1.044\ M$
$[NH_3] = 0.203\ M$		$[NH_3] = 0.304\ M$

Students find it very hard to understand the new equilibrium position established after injection of 1.000 mol/L NH_3. After the reaction has shifted to the right and equilibrium has been reestablished, students often believe that the concentration of NH_3 should return to a value equal to the concentration before the equilibrium was disturbed.

Note that the equilibrium does in fact shift to the right; the concentration of H_2 decreases (from 1.197 M to 1.044 M), the concentration of NH_3 increases (from 0.203 M to 0.304 M), and, of course, because nitrogen was added, the concentration of N_2 shows an increase relative to the original amount present.

It is important to note at this point that, although the equilibrium shifted to a new position, the *value of K did not change.* We can demonstrate this by inserting the equilibrium concentrations from position I and II into the equilibrium expression.

Position I:

$$K = \frac{[NH_3]^2}{[N_2][H_2]^3} = \frac{(0.203)^2}{(0.399)(1.197)^3} = 0.0602$$

Position II:

$$K = \frac{[NH_3]^2}{[N_2][H_2]^3} = \frac{(0.304)^2}{(1.348)(1.044)^3} = 0.0602$$

These values of K are the same. Therefore, although the equilibrium *position* shifted when we added more N_2, the *equilibrium constant K* remained the same.

Could we have predicted this shift by using Le Châtelier's principle? Because the change in this case was to add nitrogen, Le Châtelier's principle predicts that the system will shift in a direction that *consumes* nitrogen. This tends

Figure 16.9
(a) The initial equilibrium mixture of N_2, H_2, and NH_3. (b) Addition of N_2. (c) The new equilibrium position for the system containing more N_2 (because of the addition of N_2), less H_2, and more NH_3 than in (a).

(a)   (b)   (c)

to offset the original change—the addition of N_2. Therefore Le Châtelier's principle correctly predicts that adding nitrogen will cause the equilibrium to shift to the right (Figure 16.9) as some of the added nitrogen is consumed.

If ammonia had been added instead of nitrogen, the system would have shifted to the left, consuming ammonia. Another way of stating Le Châtelier's principle, then, is to say that *when a reactant or product is added to a system at equilibrium, the system shifts away from the added component.* On the other hand, if *a reactant or product is removed, the system shifts toward the removed component.* For example, if we had removed nitrogen, the system would have shifted to the left and the amount of ammonia present would have been reduced.

A system at equilibrium shifts in the direction that compensates for any imposed change.

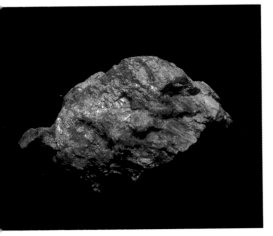

An ore sample containing arsenic.

EXAMPLE 16.4   **Using Le Châtelier's Principle: Changes in Concentration**

Arsenic, As_4, is obtained from nature by first reacting its ore with oxygen (called *roasting*) to form solid As_4O_6. (As_4O_6, a toxic compound fatal in doses of 0.1 g or more, is the "arsenic" made famous in detective stories.) The As_4O_6 is then reduced using carbon:

$$As_4O_6(s) + 6C(s) \rightleftharpoons As_4(g) + 6CO(g)$$

Predict the direction of the shift in the equilibrium position for this reaction that occurs in response to each of the following changes in conditions.

a.   Addition of carbon monoxide
b.   Addition or removal of $C(s)$ or $As_4O_6(s)$
c.   Removal of $As_4(g)$

You can demonstrate the equilibrium reaction of $CoCl_2$ with water in the atmosphere by using a novelty "barometer" purchased from a gift shop. The commercial dessicant Drierite with moisture indicator also contains $CoCl_2$. Its use in laboratories is another interesting application of this equilibrium reaction.

EXAMPLE 16.4, CONTINUED

SOLUTION

a. Le Châtelier's principle predicts a shift away from the substance whose concentration is increased. The equilibrium position will shift to the left when carbon monoxide is added.
b. Because the amount of a pure solid has no effect on the equilibrium position, changing the amount of carbon or tetraarsenic hexoxide will have no effect.
c. When gaseous arsenic is removed, the equilibrium position will shift to the right to form more products. In industrial processes, the desired product is often continuously removed from the reaction system to increase the yield.

When blue anhydrous $CoCl_2$ reacts with water, pink $CoCl_2 \cdot 6H_2O$ is formed.

SELF-CHECK EXERCISE 16.3

Novelty devices for predicting rain contain cobalt(II) chloride and are based on the following equilibrium:

$$CoCl_2(s) + 6H_2O(g) \rightleftharpoons CoCl_2 \cdot 6H_2O(s)$$
Blue   Pink

What color will this indicator be when rain is imminent?

(a)

The Effect of a Change in Volume

When the volume of a gas is decreased (when a gas is compressed), the pressure increases. This occurs because the molecules present are now contained in a smaller space and they hit the walls of their container more often, giving a greater pressure. Therefore, when the volume of a gaseous reaction system at equilibrium is suddenly reduced, leading to a sudden increase in pressure, by Le Châtelier's principle the system will shift in the direction that reduces the pressure.
For example, consider the reaction

$$CaCO_3(s) \rightleftharpoons CaO(s) + CO_2(g)$$

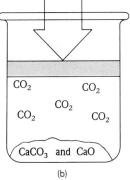

(b)

in a container with a movable piston (Figure 16.10). If the volume is suddenly decreased by pushing in the piston, the pressure of the CO_2 gas initially increases. How can the system offset this pressure increase? By shifting to the left—the direction that reduces the amount of gas present. That is, a shift to the left will use up CO_2 molecules, thus lowering the pressure. (There will then be fewer molecules present to hit the walls, because more of the CO_2 molecules have combined with CaO and thus have become part of the solid $CaCO_3$.)

Figure 16.10
The reaction system $CaCO_3(s) \rightleftharpoons CaO(s) + CO_2(g)$. (a) The system is initially at equilibrium. (b) Then the piston is pushed in, decreasing the volume and increasing the pressure. The system shifts in the direction that consumes CO_2 molecules, thus lowering the pressure again.

Shifts in equilibrium brought about by volume changes also produce changes in pressure.

(a)

(b)

(c)

Figure 16.11
(a) A mixture of $NH_3(g)$, $N_2(g)$, and $H_2(g)$ at equilibrium. (b) The volume is suddenly decreased. (c) The new equilibrium position for the system containing more NH_3 and less N_2 and H_2. The reaction $N_2(g) + 3H_2(g) \rightleftharpoons 2NH_3(g)$ shifts to the right (toward the side with fewer molecules) when the container volume is decreased.

Therefore, when the volume of a gaseous reaction system at equilibrium is decreased (thus increasing the pressure), *the system shifts in the direction that gives the smaller number of gas molecules.* So a decrease in the system volume leads to a shift that decreases the total number of gaseous molecules in the system.

Suppose we are running the reaction

$$N_2(g) + 3H_2(g) \rightleftharpoons 2NH_3(g)$$

and we have a mixture of the gases nitrogen, hydrogen, and ammonia at equilibrium (Figure 16.11(a)). If we suddenly reduce the volume, what will happen to the equilibrium position? Because the decrease in volume initially increases the pressure, the system moves in the direction that lowers its pressure. The reaction system can reduce its pressure by reducing the number of gas molecules present. This means that the reaction

$$N_2(g) + 3H_2(g) \rightleftharpoons 2NH_3(g)$$

4 gaseous molecules $\rightleftharpoons$ 2 gaseous molecules

shifts to the right, because in this direction four molecules (one of nitrogen and three of hydrogen) react to produce two molecules (of ammonia), thus *reducing the total number of gaseous molecules present.* The equilibrium position shifts to the right—toward the side of the reaction that involves the smaller number of gaseous molecules in the balanced equation.

The opposite is also true. When the container volume is increased (which lowers the pressure of the system), the system shifts so as to increase its pressure. An increase in volume in the ammonia synthesis system produces a shift to the left to increase the total number of gaseous molecules present (to increase the pressure).

EXAMPLE 16.5   Using Le Châtelier's Principle: Changes in Volume

Predict the shift in equilibrium position that will occur for each of the following processes when the volume is reduced.

a. The preparation of liquid phosphorus trichloride by the reaction

$$P_4(s) + 6Cl_2(g) \rightleftharpoons 4PCl_3(l)$$

6 gaseous molecules $\rightleftharpoons$ 0 gaseous molecules

SOLUTION

P_4 and PCl_3 are a pure solid and a pure liquid, respectively, so we need to consider only the effect on Cl_2. If the volume is decreased, the Cl_2 pressure will initially increase, so the position of the equilibrium will shift to the right, consuming gaseous Cl_2 and lowering the pressure (to counteract the original change).

EXAMPLE 16.5, CONTINUED

b.   The preparation of gaseous phosphorus pentachloride according to the
     equation

$$PCl_3(g) + Cl_2(g) \rightleftharpoons PCl_5(g)$$

2 gaseous molecules ⇌ 1 gaseous molecule

SOLUTION

Decreasing the volume (increasing the pressure) will shift this equilibrium to the
right, because the product side contains only one gaseous molecule while the
reactant side has two. That is, the system will respond to the decreased volume
(increased pressure) by lowering the number of molecules present.

c.   The reaction of phosphorus trichloride with ammonia:

$$PCl_3(g) + 3NH_3(g) \rightleftharpoons P(NH_2)_3(g) + 3HCl(g)$$

SOLUTION

Both sides of the balanced reaction equation have four gaseous molecules. A
change in volume will have no effect on the equilibrium position. There is no
shift in this case, because the system cannot change the number of molecules
present by shifting in either direction.

SELF-CHECK EXERCISE 16.4

For each of the following reactions, predict the direction the equilibrium will
shift when the volume of the container is increased.

a. $H_2(g) + F_2(g) \rightleftharpoons 2HF(g)$
b. $CO(g) + 2H_2(g) \rightleftharpoons CH_3OH(g)$
c. $2SO_3(g) \rightleftharpoons 2SO_2(g) + O_2(g)$

_____

The Effect of a Change in Temperature

It is important to remember that although the changes we have just discussed
may alter the equilibrium *position,* they do not alter the equilibrium *constant.*
For example, the addition of a reactant shifts the equilibrium position to the right
but has no effect on the value of the equilibrium constant; the new equilibrium
concentrations satisfy the original equilibrium constant. This was demonstrated
earlier in this section for the addition of N_2 to the ammonia synthesis reaction.
     The effect of temperature on equilibrium is different, however, because *the
value of K changes with temperature.* We can use Le Châtelier's principle to
predict the direction of the change in *K.*

For reactions involving only gas particles,
students can't predict the effects on the
equilibrium of changing the volume of the
system without a balanced equation.

Students often ask how they will know whether a reaction is exothermic or endothermic. You should tell them that this piece of information will be provided to them in the problem.

To do this we need to classify reactions according to whether they produce heat or absorb heat. A reaction that produces heat (heat is a "product") is said to be *exothermic.* A reaction that absorbs heat is called *endothermic.* Because heat is needed for an endothermic reaction, energy (heat) can be regarded as a "reactant" in this case.

In an exothermic reaction, heat is treated as a product. For example, the synthesis of ammonia from nitrogen and hydrogen is exothermic (produces heat). We can represent this by treating energy as a product:

$$N_2(g) + 3H_2(g) \rightleftharpoons 2NH_3(g) + 92 \text{ kJ}$$

Energy released

Le Châtelier's principle predicts that, when we add energy to this system at equilibrium by heating it, the shift will be in the direction that consumes energy—that is, to the left.

On the other hand, for an endothermic reaction (one that absorbs energy), such as the decomposition of calcium carbonate,

$$CaCO_3(s) + 556 \text{ kJ} \rightleftharpoons CaO(s) + CO_2(g)$$

Energy needed

energy is treated as a reactant. In this case an increase in temperature causes the equilibrium to shift to the right.

In summary, to use Le Châtelier's principle to describe the effect of a temperature change on a system at equilibrium, *simply treat energy as a reactant (in an endothermic process) or as a product (in an exothermic process),* and *predict the direction of the shift* in the same way you would if an actual reactant or product were being added or removed.

EXAMPLE 16.6   Using Le Châtelier's Principle: Changes in Temperature

For each of the following reactions, predict how the equilibrium will shift as the temperature is increased.

a. $N_2(g) + O_2(g) \rightleftharpoons 2NO(g)$ (endothermic)

SOLUTION

This is an endothermic reaction, so energy can be viewed as a reactant.

$$N_2(g) + O_2(g) + \text{energy} \rightleftharpoons 2NO$$

EXAMPLE 16.6, CONTINUED

Thus the equilibrium will shift to the right as the temperature is increased (energy added).

b. $2SO_2(g) + O_2(g) \rightleftharpoons 2SO_3(g)$ (exothermic)

SOLUTION

This is an exothermic reaction, so energy can be regarded as a product.

$$2SO_2(g) + O_2(g) \rightleftharpoons 2SO_3(g) + \text{energy}$$

As the temperature is increased, the equilibrium will shift to the left.

SELF-CHECK EXERCISE 16.5

For the exothermic reaction

$$2SO_2(g) + O_2(g) \rightleftharpoons 2SO_3(g)$$

predict the equilibrium shift caused by each of the following changes.

a. SO_2 is added.
b. SO_3 is removed.
c.   The volume is decreased.
d.   The temperature is decreased.

(a)

We have seen how Le Châtelier's principle can be used to predict the effect of several types of changes on a system at equilibrium. To summarize these ideas, Table 16.2 shows how various changes affect the equilibrium position of the endothermic reaction $N_2O_4(g) \rightleftharpoons 2NO_2(g)$. The effect of a temperature change on this system is depicted in Figure 16.12.

Table 16.2 Shifts in the Equilibrium Position for the Reaction
Energy + $N_2O_4(g) \rightleftharpoons 2NO_2(g)$

Change	Shift
addition of $N_2O_4(g)$	right
addition of $NO_2(g)$	left
removal of $N_2O_4(g)$	left
removal of $NO_2(g)$	right
decrease in container volume	left
increase in container volume	right
increase in temperature	right
decrease in temperature	left

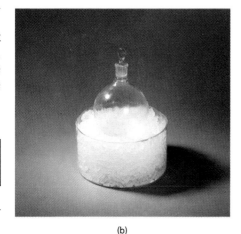

(b)

Figure 16.12
Shifting the $N_2O_4(g) \rightleftharpoons 2NO_2(g)$ equilibrium by changing the temperature. (a) At 100 °C the flask is quite brown due to a large amount of NO_2 present. (b) At 0 °C the equilibrium is shifted toward colorless $N_2O_4(g)$.

16.8   Applications Involving the Equilibrium Constant

AIM: To show how to calculate equilibrium concentrations from equilibrium constants.

Knowing the value of the equilibrium constant for a reaction allows us to do many things. For example, the size of K tells us the inherent tendency of the reaction to occur. A value of K much larger than 1 means that, at equilibrium, the reaction system will consist of mostly products—the equilibrium lies to the right. For example, consider a general reaction of the type

$$A(g) \rightarrow B(g)$$

where

$$K = \frac{[B]}{[A]}$$

If K for this reaction is 10,000 (10^4), then at equilibrium,

It is helpful to show final ratios for the products over reactants when calculating K. The numerical value of K can indicate whether the reaction goes mainly to the right or mainly to the left. Students often view the value of K as a meaningless number.

$$\frac{[B]}{[A]} = 10,000 \qquad or \qquad \frac{[B]}{[A]} = \frac{10,000}{1}$$

That is, at equilibrium [B] is 10,000 times greater than [A]. This means that the reaction strongly favors the product B. Another way of saying this is that the reaction goes essentially to completion. That is, virtually all of A becomes B.

On the other hand, a small value of K means that the system at equilibrium consists largely of reactants—the equilibrium position is far to the left. The given reaction does not occur to any significant extent.

Another way we use the equilibrium constant is to calculate the equilibrium concentrations of reactants and products. For example, if we know the value of K and the concentrations of all the reactants and products except one, we can calculate the missing concentration. This is illustrated in Example 16.7

Some students will have difficulty rearranging the equilibrium expression to isolate the desired variable.

EXAMPLE 16.7   Calculating Equilibrium Concentration Using Equilibrium Expressions

Gaseous phosphorus pentachloride decomposes to chlorine gas and gaseous phosphorus trichloride. In a certain experiment, at a temperature where $K = 8.96 \times 10^{-2}$, the equilibrium concentrations of PCl_5 and PCl_3 were found to be

EXAMPLE 16.4, CONTINUED

6.70×10^{-3} *M* and 0.300 *M*, respectively. Calculate the concentration of Cl_2 present at equilibrium.

SOLUTION

For this reaction, the balanced equation is

$$PCl_5(g) \rightleftharpoons PCl_3(g) + Cl_2(g)$$

and the equilibrium expression is

$$K = \frac{[PCl_3][Cl_2]}{[PCl_5]} = 8.96 \times 10^{-2}$$

We know that

$$[PCl_5] = 6.70 \times 10^{-3} \ M$$
$$[PCl_3] = 0.300 \ M$$

We want to calculate $[Cl_2]$. We will rearrange the equilibrium expression to solve for the concentration of Cl_2. First we divide both sides of the expression

$$K = \frac{[PCl_3][Cl_2]}{[PCl_5]}$$

by $[PCl_3]$ to give

$$\frac{K}{[PCl_3]} = \frac{[\cancel{PCl_3}][Cl_2]}{[\cancel{PCl_3}][PCl_5]} = \frac{[Cl_2]}{[PCl_5]}$$

Next we multiply both sides by $[PCl_5]$.

$$\frac{K[PCl_5]}{[PCl_3]} = \frac{[Cl_2][\cancel{PCl_5}]}{[\cancel{PCl_5}]} = [Cl_2]$$

Then we can calculate $[Cl_2]$ by substituting the known information.

$$[Cl_2] = K \times \frac{[PCl_5]}{[PCl_3]} = (8.96 \times 10^{-2})\frac{(6.70 \times 10^{-3})}{(0.300)}$$
$$[Cl_2] = 2.00 \times 10^{-3}$$

The equilibrium concentration of Cl_2 is 2.00×10^{-3} *M*.

Once again, students won't be able to solve these problems correctly unless they have a balanced equation.

Toothpaste containing sodium fluoride, an additive that helps prevent tooth decay.

16.9   Solubility Equilibria

AIM: To calculate the solubility product of a salt, given its solubility; and vice versa.

Solubility is a very important phenomenon. Consider the following examples.

- Because sugar and table salt dissolve readily in water, we can flavor foods easily.
- Because calcium sulfate is less soluble in hot water than in cold water, it coats tubes in boilers, reducing thermal efficiency.
- When food lodges between teeth, acids form that dissolve tooth enamel, which contains the mineral hydroxyapatite, $Ca_5(PO_4)_3OH$. Tooth decay can be reduced by adding fluoride to toothpaste. Fluoride replaces the hydroxide in hydroxyapatite to produce the corresponding fluorapatite, $Ca_5(PO_4)_3F$, and calcium fluoride, CaF_2, both of which are less soluble in acids than the original enamel.
- The use of a suspension of barium sulfate improves the clarity of X rays of the digestive tract. Barium sulfate contains the toxic ion Ba^{2+}, but its very low solubility makes ingestion of solid $BaSO_4$ safe.

In this section we will consider the equilibria associated with dissolving solids in water to form aqueous solutions. When a typical ionic solid dissolves in water, it dissociates completely into separate cations and anions. For example, calcium fluoride dissolves in water as follows:

$$CaF_2(s) \xrightarrow{H_2O(l)} Ca^{2+}(aq) + 2F^-(aq)$$

When the solid salt is first added to the water, no Ca^{2+} and F^- ions are present. However, as dissolving occurs, the concentrations of Ca^{2+} and F^- increase, and it becomes more and more likely that these ions will collide and re-form the solid. Thus two opposite (competing) processes are occurring: the dissolving reaction shown above and the reverse reaction to re-form the solid:

$$Ca^{2+}(aq) + 2F^-(aq) \rightarrow CaF_2(s)$$

Ultimately, equilibrium is reached.

$$CaF_2(s) \rightleftharpoons Ca^{2+} + 2F^-(aq)$$

No more solid dissolves and the solution is said to be saturated.

We can write an equilibrium expression for this process according to the law of chemical equilibrium.

$$K_{sp} = [Ca^{2+}][F^-]^2$$

where $[Ca^{2+}]$ and $[F^-]$ are expressed in mol/L. The constant K_{sp} is called the **solubility product constant,** or simply the **solubility product.**

This X ray of the large intestine has been enhanced by the patient's consumption of barium sulfate.

Because CaF_2 is a pure solid, it is not included in the equilibrium expression. It may seem strange at first that the amount of excess solid present does not affect the position of the solubility equilibrium. Surely more solid means more surface area exposed to the solvent, which would seem to result in greater solubility. This is not the case, however, because dissolving and re-forming of the solid both occur at the surface of the excess solid. When a solid dissolves, it is the ions at the surface that go into solution. And when the ions in solution re-form the solid, they do so on the surface of the solid. So doubling the surface area of the solid doubles not only the rate of dissolving but also doubles the rate of re-formation of the solid. The amount of excess solid present therefore has no effect on the equilibrium position. Similarly, although either increasing the surface area by grinding up the solid or stirring the solution speeds up the attainment of equilibrium, neither procedure changes the *amount* of solid dissolved at equilibrium.

Pure liquids and pure solids are never included in an equilibrium expression.

EXAMPLE 16.8   Writing Solubility Product Expressions

Write the balanced equation describing the reaction for dissolving each of the following solids in water. Also write the K_{sp} expression for each solid.

a. $PbCl_2(s)$ c. $Bi_2S_3(s)$
b. $Ag_2CrO_4(s)$

Remind students about the polyatomic ions from Chapter 5 when they try to write the solubility reaction for salts. Initially, students forget that $Na_2Cr_2O_4$ forms two Na^+ ions and a $Cr_2O_7^{2-}$ ion and instead they tend to write $2Na^+$, Cr^{2-}, and O_7^2.

SOLUTION

a. $PbCl_2(s) \rightleftharpoons Pb^{2+}(aq) + 2Cl^-(aq);$ $K_{sp} = [Pb^{2+}][Cl^-]^2$
b. $Ag_2CrO_4(s) \rightleftharpoons 2Ag^+(aq) + CrO_4^{2-}(aq);$ $K_{sp} = [Ag^+]^2[CrO_4^{2-}]$
c. $Bi_2S_3(s) \rightleftharpoons 2Bi^{3+}(aq) + 3S^{2-}(aq);$ $K_{sp} = [Bi^{3+}]^2[S^{2-}]^3$

SELF-CHECK EXERCISE 16.6

Write the balanced equation for the reaction describing the dissolving of each of the following solids in water. Also write the K_{sp} expression for each solid.

a. $BaSO_4(s)$ c. $Ag_3PO_4(s)$
b. $Fe(OH)_3(s)$

EXAMPLE 16.9   Calculating Solubility Products

Copper(I) bromide, CuBr, has a measured solubility of 2.0×10^{-4} mol/L at 25 °C. That is, when excess CuBr(s) is placed in 1.0 L of water, we can determine that 2.0×10^{-4} mol of the solid dissolves to produce a saturated solution. Calculate the solid's K_{sp} value.

When you choose examples of K_{sp} problems to work in class, it is probably a good idea to use examples in which the ratio of anions to cations is 1:1. Using salts with other ratios causes students at this level a good deal of difficulty.

EXAMPLE 16.9, CONTINUED

SOLUTION

At first it is not obvious how to use the given information to solve the problem, but think about what happens when the solid dissolves. When the solid CuBr is placed in contact with water, it dissolves to form the separated Cu^+ and Br^- ions:

$$CuBr(s) \rightleftharpoons Cu^+(aq) + Br^-(aq)$$

where

$$K_{sp} = [Cu^+][Br^-]$$

Therefore we can calculate the value of K_{sp} if we know $[Cu^+]$ and $[Br^-]$, the equilibrium concentrations of the ions. We know that the measured solubility of CuBr is 2.0×10^{-4} mol/L. This means that 2.0×10^{-4} mol of solid CuBr dissolves per 1.0 L of solution to come to equilibrium. The reaction is

$$CuBr(s) \rightarrow Cu^+(aq) + Br^-(aq)$$

so

2.0×10^{-4} mol/L CuBr(s)
$$\rightarrow 2.0 \times 10^{-4} \text{mol/L } Cu^+(aq) + 2.0 \times 10^{-4} \text{mol/L } Br^-(aq)$$

We can now write the equilibrium concentrations:

$$[Cu^+] = 2.0 \times 10^{-4} \text{ mol/L}$$

and

$$[Br^-] = 2.0 \times 10^{-4} \text{ mol/L}$$

These equilibrium concentrations allow us to calculate the value of K_{sp} for CuBr.

$$K_{sp} = [Cu^+][Br^-] = (2.0 \times 10^{-4})(2.0 \times 10^{-4})$$
$$= 4.0 \times 10^{-8}$$

The units for K_{sp} values are omitted.

SELF-CHECK EXERCISE 16.7

Calculate the K_{sp} value for barium sulfate, $BaSO_4$, which has a solubility of 3.9×10^{-5} mol/L at 25 °C.

Solubilities must be expressed in mol/L in K_{sp} calculations.

We have seen that the known solubility of an ionic solid can be used to calculate its K_{sp} value. The reverse is also possible: the solubility of an ionic solid can be calculated if its K_{sp} value is known.

EXAMPLE 16.10   Calculating Solubility from K_{sp} Values

The K_{sp} value for the solid $AgI(s)$ is 1.5×10^{-16} at 25 °C. Calculate the solubility of $AgI(s)$ in water at 25 °C.

SOLUTION

The solid AgI dissolves according to the equation

$$AgI(s) \rightleftharpoons Ag^+(aq) + I^-(aq)$$

and the corresponding equilibrium expression is

$$K_{sp} = 1.5 \times 10^{-16} = [Ag^+][I^-]$$

Because we do not know the solubility of this solid, we will assume that x moles per liter dissolves to reach equilibrium. Therefore

$$x\frac{mol}{L} AgI(s) \rightarrow x\frac{mol}{L} Ag^+(aq) + x\frac{mol}{L} I^-(aq)$$

and at equilibrium,

$$[Ag^+] = x\frac{mol}{L}$$

$$[I^-] = x\frac{mol}{L}$$

Substituting these concentrations into the equilibrium expression gives

$$K_{sp} = 1.5 \times 10^{-16} = [Ag^+][I^-] = (x)(x) = x^2$$

Thus

$$x^2 = 1.5 \times 10^{-16}$$

$$x = \sqrt{1.5 \times 10^{-16}} = 1.2 \times 10^{-8} \text{ mol/L}$$

The solubility of $AgI(s)$ is 1.2×10^{-8} mol/L.

SELF-CHECK EXERCISE 16.8

The K_{sp} value for lead chromate, $PbCrO_4$, is 2.0×10^{-16} at 25 °C. Calculate its solubility at 25 °C.

CHAPTER REVIEW

Key Terms

collision model (p. 541)
activation energy (E_a) (p. 541)
catalyst (p. 542)
enzyme (p. 542)
equilibrium (p. 543)
chemical equilibrium (p. 544)
law of chemical equilibrium (p. 547)

equilibrium expression (p. 547)
equilibrium constant (p. 547)
equilibrium position (p. 550)
homogeneous equilibria (p. 551)
heterogeneous equilibria (p. 551)
Le Châtelier's principle (p. 554)
solubility product (p. 564)

Summary

1. Chemical reactions can be described by the collision model, which assumes that molecules must collide in order to react. In terms of this model, a certain threshold energy, called the activation energy (E_a), must be overcome for a collision to form products.

2. A catalyst is a substance that speeds up a reaction without being consumed. A catalyst operates by providing a lower energy pathway for the reaction in question. Enzymes are biological catalysts.

3. When a chemical reaction is carried out in a closed vessel, the system achieves chemical equilibrium, the state where the concentrations of both reactants and products remain constant over time. Equilibrium is a highly dynamic state; reactants are converted continually into products, and vice versa, as molecules collide with each other. At equilibrium, the rates of the forward and reverse reactions are equal.

4. The law of chemical equilibrium is a general description of the equilibrium condition. It states that for a reaction of the type

$$aA + bB \rightleftharpoons cC + dD$$

the equilibrium expression is given by

$$K = \frac{[C]^c [D]^d}{[A]^a [B]^b}$$

where K is the equilibrium constant.

5. For each reaction system at a given temperature, there is only one value for the equilibrium constant, but there is an infinite number of possible equilibrium positions. An equilibrium position is defined as a particular set of equilibrium concentrations that satisfy the equilibrium expression. A specific equilibrium position depends on the initial concentrations. The amount of a pure liquid or a pure solid is never included in the equilibrium expression.

6. Le Châtelier's principle allows us to predict the effects of changes in concentration, pressure, and temperature on a system at equilibrium. This principle states that when a change is imposed on a system at equilibrium, the equilibrium position will shift in a direction that tends to compensate for the imposed change.

7. The principle of equilibrium can also be applied when an excess of a solid is added to water to form a saturated solution. The solubility product (K_{sp}) is an equilibrium constant defined by the law of chemical equilibrium. Solubility is an equilibrium position, and the K_{sp} value of a solid can be determined by measuring its solubility. Conversely, the solubility of a solid can be determined if its K_{sp} value is known.

Questions and Problems

All even-numbered exercises have answers in the back of this book and solutions in the Solutions Guide.

16.1 How Chemical Reactions Occur

QUESTIONS

1. What is meant by the *collision model* for chemical reactions? Do all collisions between molecules result in the formation of products?

2. For the reaction $N_2(g) + 3H_2(g) = 2NH_3$, list the types of bonds that must be broken and the type of bonds that must form in order for the chemical reaction to take place.

16.2 Conditions That Affect Reaction Rates

QUESTIONS

3. Not all collisions between reactant molecules are successful in forming products. What criterion must be met for a collision to be effective? What happens when this criterion is not met? Explain.

4. What does the *activation energy* for a reaction represent? How is the activation energy related to whether or not a collision between molecules is successful?

5. What is a catalyst? What are the catalysts found in living cells called? How does a catalyst speed up a reaction? Draw an energy diagram (see Figures 16.3 and 16.4) showing the effect of a catalyst on a reaction.

6. How do chlorine atoms catalyze the decomposition of ozone, O_3, in the upper atmosphere? Where does most of the chlorine involved in this process come from? What are the implications of the destruction of the ozone layer for society? What steps can be taken to prevent this effect?

16.3 The Equilibrium Condition

QUESTIONS

7. How does *equilibrium* represent the balancing of opposing processes? Give an example of an "equilibrium" encountered in everyday life, showing how the processes involved oppose each other.

8. What do chemists mean by a state of *equilibrium?* Give an example of a *physical* equilibrium and of a *chemical* equilibrium.

9. What does it mean to say that chemical reactions are *reversible?* Are all chemical reactions, in principle, reversible? Are some reactions more likely to occur in one direction than in the other?

10. What is *chemical equilibrium?* How do we recognize a system that has reached a state of chemical equilibrium? When writing a chemical equation, how do we indicate reactions that come to equilibrium?

16.4 Chemical Equilibrium: A Dynamic Condition

QUESTIONS

11. When a reaction system has reached chemical equilibrium, the concentrations of the reactants and products no longer change with time. Why does the amount of product no longer increase, even though large concentrations of the reactants may still be present?

12. What does it mean to say that the condition of chemical equilibrium is a *dynamic* situation? Although the reaction overall may appear to have stopped, what is still going on in the system?

16.5 The Equilibrium Constant: An Introduction

QUESTIONS

13. In general terms, what does the equilibrium constant for a reaction represent? What is the algebraic form of the equilibrium constant for a typical reaction? What do square brackets indicate when we write an equilibrium constant?

14. There is only one value of the equilibrium constant for a particular system at a particular temperature, but there is an infinite number of equilibrium positions. Explain.

PROBLEMS

15. Write the equilibrium expression for each of the following reactions.
    a. $2H_2(g) + O_2(g) \rightleftharpoons 2H_2O(g)$
    b. $Cl_2(g) + F_2(g) \rightleftharpoons 2ClF(g)$
    c. $2SO_2(g) + O_2(g) \rightleftharpoons 2SO_3(g)$

16. Write the equilibrium expression for each of the following reactions.
    a. $N_2(g) + 3Cl_2(g) \rightleftharpoons 2NCl_3(g)$
    b. $H_2(g) + I_2(g) \rightleftharpoons 2HI(g)$
    c. $N_2(g) + 2H_2(g) \rightleftharpoons N_2H_4(g)$

17. Write the equilibrium expression for each of the following reactions.
    a. $2NO(g) + O_2(g) \rightleftharpoons 2NO_2(g)$
    b. $2C_8H_{18}(g) + 25O_2(g) \rightleftharpoons 16CO_2(g) + 18H_2O(g)$
    c. $N_2H_4(g) + O_2(g) \rightleftharpoons N_2(g) + 2H_2O(g)$

18. Write the equilibrium expression for each of the following reactions.
    a. $P_4(g) + 6Cl_2(g) \rightleftharpoons 4PCl_3(g)$
    b. $N_2(g) + O_2(g) \rightleftharpoons 2NO(g)$
    c. $C_3H_8(g) + 5O_2(g) \rightleftharpoons 3CO_2(g) + 4H_2O(g)$

19. Suppose that, for the hypothetical reaction

$$A_2(g) + 2B(g) \rightleftharpoons 2AB(g)$$

it is determined that, at a certain temperature, the equilibrium concentrations are $[A_2] = 0.00340$ M, $[B] = 0.103$ M, and $[AB] = 1.212$ M. Calculate the value of K for the reaction.

20. Ammonia, a very important industrial chemical, is produced by the direct combination of the following elements under carefully controlled conditions:

$$N_2(g) + 3H_2(g) \rightleftharpoons 2NH_3(g)$$

Suppose, in an experiment, that the reaction mixture is analyzed after equilibrium is reached and it is found, at the particular temperature, that $[NH_3(g)] = 0.34$ M, $[H_2(g)] = 2.1 \times 10^{-3}$ M, and $[N_2(g)] = 4.9 \times 10^{-4}$ M. Calculate the value of K at this temperature.

21. At high temperatures, elemental nitrogen and oxygen react with each other to form nitrogen monoxide.

$$N_2(g) + O_2(g) \rightleftharpoons 2NO(g)$$

Suppose the system is analyzed at a particular temperature, and the equilibrium concentrations are found to be $[N_2] = 0.041$ M, $[O_2] = 0.0078$ M, and $[NO] = 4.7 \times 10^{-4}$ M. Calculate the value of K for the reaction.

22. Sulfur trioxide is generated in massive quantities by the chemical industry each year for the production of sulfuric acid. Sulfur trioxide is produced by the oxidation of sulfur dioxide:

$$SO_2(g) + O_2(g) \rightleftharpoons SO_3(g)$$

Suppose, in an experiment, that the reaction mixture is analyzed after equilibrium is reached and it is found that $[SO_3(g)] = 4.2 \times 10^{-1}$ M, $[SO_2(g)] = 2.3 \times 10^{-3}$ M, and $[O_2(g)] = 3.7 \times 10^{-5}$ M. Calculate K for the conditions of this experiment.

16.6 Heterogeneous Equilibria

QUESTIONS

23. What is a *homogeneous* equilibrium system? Give an example of a homogeneous equilibrium reaction. What is a *heterogeneous* equilibrium system? Write two chemical equations that represent heterogeneous equilibria.

24. Explain why the position of a heterogeneous equilibrium does not depend on the amounts of pure solid or pure liquid reactants or products present.

PROBLEMS

25. Write the equilibrium expression for each of the following heterogeneous equilibria.
a. $2C(s) + 2H_2(g) \rightleftharpoons C_2H_4(g)$
b. $B_2O_3(s) + 3H_2O(l) \rightleftharpoons B_2H_6(g) + 3O_2(g)$
c. $BaCO_3(s) \rightleftharpoons BaO(s) + CO_2(g)$

27. Write the equilibrium expression for each of the following heterogeneous equilibria.
a. $2AlCl_3(s) \rightleftharpoons 2Al(s) + 3Cl_2(g)$
b. $Ca(OH)_2(s) + CO_2(g) \rightleftharpoons CaCO_3(s) + H_2O(g)$
c. $UO_2(s) + 4HF(g) \rightleftharpoons UF_4(g) + 2H_2O(g)$

26. Write the equilibrium expression for each of the following heterogenous equlibria.
a. $S(s) + O_2(g) \rightleftharpoons SO_2(g)$
b. $C_5H_{12}(l) + 8O_2(g) \rightleftharpoons 5CO_2(g) + 6H_2O(l)$
c. $N_2O_4(l) \rightleftharpoons 2NO_2(g)$

28. Write the equilibrium expression for each of the following heterogeneous equilibria.
a. $N_2(g) + 3Br_2(l) \rightleftharpoons 2NBr_3(s)$
b. $SOCl_2(l) + H_2O(l) \rightleftharpoons SO_2(g) + 2HCl(g)$
c. $2HI(g) \rightleftharpoons H_2(g) + I_2(s)$

16.7 Le Châtelier's Principle

QUESTIONS

29. In your own words, describe what Le Châtelier's principle tells us about how we can change the position of a reaction system at equilibrium.

31. For an equilibrium involving gaseous substances, what effect, in general terms, is realized when the volume of the system is decreased?

30. Discuss in general terms the effect on the net amount of product when an additional amount of one of the reactants is added to an equilibrium system. Does the value of the equilibrium constant change in this situation?

32. What is the effect on the equilibrium position if an endothermic reaction is performed at a higher temperature? Does the net amount of product increase or decrease? Does the value of the equilibrium constant change if the temperature is increased?

PROBLEMS

33. For the reaction system

$$2H_2(g) + O_2(g) \rightleftharpoons 2H_2O(g)$$

which has already reached equilibrium, predict the effect that each of the following changes will have on the position of the equilibrium. Tell whether the equilibrium will shift to the right, will shift to the left, or will not be affected.
a. The pressure of hydrogen is increased by injecting 1.0 mol of hydrogen into the reaction vessel.
b. The amount of oxygen is decreased by addition of an "oxygen scrubber" (a chemical that absorbs oxygen).
c. The water vapor is liquefied and removed.

34. Suppose the reaction system

$$SO_2(g) + O_2(g) \rightleftharpoons SO_3(g)$$

has already reached equilibrium. Predict the effect of each of the following changes on the position of the equilibrium. Tell whether the equilibrium will shift to the right, will shift to the left, or will not be affected.
a. Additional $SO_2(g)$ is added to the system.
b. The $SO_3(g)$ present is liquefied and removed from the system.
c. A very efficient catalyst is used.

35. Suppose the reaction system

$$CH_4(g) + 2O_2(g) \rightleftharpoons CO_2(g) + H_2O(l)$$

has already reached equilibrium. Predict the effect of each of the following changes on the position of the equilibrium. Tell whether the equilibrium will shift to the right, will shift to the left, or will not be affected.
   a. Any liquid water present is removed from the system.
   b. CO_2 is added to the system by dropping a chunk of dry ice into the reaction vessel.
   c. The reaction is performed in a metal cylinder fitted with a piston, and the piston is compressed in order to decrease the total volume of the system.
   d. Additional $O_2(g)$ is added to the system from a cylinder of pure O_2.

37. The reaction that represents the burning of natural gas

$$CH_4(g) + 2O_2(g) \rightleftharpoons CO_2(g) + H_2O(g)$$

is, of course, exothermic: we make use of the heat evolved to heat our homes and cook our food. What would be the effect on the position of this equilibrium if the reaction were performed at a lower temperature? at a higher temperature?

39. The reaction

$$2B_5H_9(l) + 12O_2(g) \rightleftharpoons 9H_2O(g) + 5B_2O_3(s)$$

is very strongly exothermic. Will an increase in temperature shift the equilibrium position toward products or toward reactants?

41. Plants synthesize the sugar dextrose according to the following reaction by absorbing radiant energy from the sun (photosynthesis).

$$6CO_2(g) + 6H_2O(g) \rightleftharpoons C_6H_{12}O_6(s) + 6O_2(g)$$

Will an increase in temperature tend to favor or discourage the production of $C_6H_{12}O_6(s)$?

36. Suppose the reaction system

$$UO_2(s) + 4HF(g) \rightleftharpoons UF_4(g) + 2H_2O(g)$$

has already reached equilibrium. Predict the effect of each of the following changes on the position of the equilibrium. Tell whether the equilibrium will shift to the right, will shift to the left, or will not be affected.
   a. Additional $UO_2(s)$ is added to the system.
   b. 5.0 mol of $Xe(g)$ is added to the system.
   c. The reaction is performed in a glass reaction vessel; $HF(g)$ attacks and reacts with glass.
   d. Water vapor is removed.
   e. The size of the reaction vessel is increased.

38. The reaction

$$NH_3(g) + HCl(g) \rightleftharpoons NH_4Cl(s)$$

is highly exothermic. Will a decrease in temperature of the system tend to increase or decrease the amount of product formed?

40. The reaction

$$4NO(g) + 6H_2O(g) \rightleftharpoons 4NH_3(g) + 5O_2(g)$$

is strongly endothermic. Will an increase in temperature shift the equilibrium position toward products or toward reactants?

42. Chemists have expended a great deal of effort to maximize the production of ammonia from the elements

$$N_2(g) + 3H_2(g) \rightleftharpoons 2NH_3(g) + 92 \text{ kJ}$$

Should this process be performed at high or low temperatures in order to enhance the production of ammonia?

16.8 Applications Involving the Equilibrium Constant

QUESTIONS

43. Suppose a reaction has the equilibrium constant $K = 1.3 \times 10^8$. What does the magnitude of this constant tell you about the relative concentrations of products and reactants that will be present once equilibrium is reached? Is this reaction likely to be a good source of the products?

44. Suppose a reaction has the equilibrium constant $K = 4.5 \times 10^{-6}$ at a particular temperature. If an experiment is set up with this reaction, will there be large relative concentrations of products present at equilibrium? Is this reaction useful as a means of producing the products? How might the reaction be made more useful?

PROBLEMS

45. For the reaction

$$2SO_2(g) + O_2(g) \rightleftharpoons 2SO_3(g)$$

it is found that, at equilibrium, $[SO_2] = 0.59\ M$, $[O_2] = 4.5 \times 10^{-2}\ M$, and $[SO_3] = 0.26\ M$ at a particular temperature. Calculate the equilibrium constant K for the reaction at this temperature.

47. For the reaction

$$3O_2(g) \rightleftharpoons 2O_3(g)$$

$K = 1.8 \times 10^{-7}$ at a particular temperature. If the equilibrium system is analyzed and it is found that $[O_2] = 0.0012\ M$, what is the concentration of O_3 in the system?

49. The equilibrium constant for the reaction

$$H_2(g) + F_2(g) \rightleftharpoons 2HF(g)$$

has the value 2.1×10^3 at a particular temperature. When the system is analyzed at equilibrium at this temperature, the concentrations of $H_2(g)$ and $F_2(g)$ are both found to be $0.0021\ M$. What is the concentration of $HF(g)$ in the equilibrium system under these conditions?

51. For the reaction

$$N_2(g) + O_2(g) \rightleftharpoons 2NO(g)$$

the equilibrium constant K has the value 1.71×10^{-3} at a particular temperature. If the concentrations of $N_2(g)$ and $O_2(g)$ are both $0.0342\ M$ in an equilibrium mixture at this temperature, what is the concentration of $NO(g)$ under these conditions?

46. For the reaction

$$N_2(g) + 3Cl_2(g) \rightleftharpoons 2NCl_3(g)$$

an analysis of an equilibrium mixture is performed. It is found that $[NCl_3(g)] = 1.9 \times 10^{-1}\ M$, $[N_2(g)] = 1.4 \times 10^{-3}\ M$, and $[Cl_2(g)] = 4.3 \times 10^{-4}\ M$. Calculate K for the reaction.

48. For the reaction

$$CaCO_3(s) \rightleftharpoons CaO(s) + CO_2(g)$$

it is found that at equilibrium $[CO_2] = 2.1 \times 10^{-3}\ M$ at a particular temperature. Calculate K for the reaction at this temperature.

50. For the reaction

$$2H_2O(g) \rightleftharpoons 2H_2(g) + O_2(g)$$

$K = 2.4 \times 10^{-3}$ at a given temperature. At equilibrium it is found that $[H_2O(g)] = 1.1 \times 10^{-1}\ M$ and $[H_2] = 1.9 \times 10^{-2}\ M$. What is the concentration of $O_2(g)$ under these conditions?

52. For the reaction

$$2NO_2(g) \rightleftharpoons N_2O_4(g)$$

the equilibrium constant K has the value 1.2×10^4 at a particular temperature. If the concentration of $N_2O_4(g)$ is found to be $0.45\ M$ in the equilibrium system, what is the concentration of $NO_2(g)$ under these conditions?

16.9 Solubility Equilibria

QUESTIONS

53. Explain how the dissolving of an ionic solute in water represents an equilibrium process.

55. Why does the amount of excess solid solute present in a solution not affect the amount of solute that ultimately dissolves in a given amount of solvent?

54. What is the special name given to the equilibrium constant for the dissolving of an ionic solute in water?

56. Why does neither stirring nor grinding a solute affect the amount of solute that ultimately dissolves in a given amount of solvent?

PROBLEMS

57. Write the balanced chemical equations describing the dissolving of the following solids in water. Write the expression for K_{sp} for each process.
  a. $PbI_2(s)$ c. $Ag_2S(s)$
  b. $Ca_3(PO_4)_2(s)$ d. $BaSO_4(s)$

58. Write balanced chemical equations that describe the dissolving of the following solids in water. Write the K_{sp} expression for each process.
  a. $PbCl_2$ c. $Ba(OH)_2$
  b. NiS d. Fe_2S_3

59. Silver bromide, AgBr(s), dissolves in water to the extent of approximately 5.8×10^{-7} M at 25 °C. Calculate K_{sp} for AgBr at this temperature.

60. Zinc carbonate, $ZnCO_3(s)$, dissolves in water to give a solution that is 1.7×10^{-5} M at 22 °C. Calculate K_{sp} for $ZnCO_3(s)$ at this temperature.

61. Approximately 3.3×10^{-4} g of copper(II) sulfide, CuS(s), dissolves per liter of water at 18 °C. Calculate K_{sp} for CuS(s) at this temperature.

62. Copper(II) chromate, $CuCrO_4(s)$, dissolves in water to give a solution containing 1.1×10^{-5} g per liter at 23 °C. Calculate K_{sp} for $CuCrO_4(s)$

63. Silver acetate, $AgC_2H_3O_2$, dissolves in water to the extent of approximately 10.6 g/L at 25 °C. Calculate the solubility of silver acetate in moles per liter, and calculate K_{sp} for $AgC_2H_3O_2$ at 25 °C.

64. K_{sp} for strontium carbonate, $SrCO_3(s)$, is 1.1×10^{-10} at 25 °C. Calculate the solubility of $SrCO_3(s)$ in moles per liter at 25 °C.

65. Approximately 1.5×10^{-3} g of iron(II) hydroxide, $Fe(OH)_2(s)$, dissolves per liter of water at 18 °C. Calculate K_{sp} for $Fe(OH)_2(s)$ at this temperature.

66. Chromate ion is used as a qualitative test for lead(II) ion, forming a bright yellow precipitate of lead chromate, $PbCrO_4(s)$. K_{sp} for $PbCrO_4(s)$ is 2.8×10^{-13} at 25 °C. Calculate the solubility of $PbCrO_4$ in grams per liter at 25 °C.

67. Magnesium fluoride dissolves in water to the extent of approximately 8.0×10^{-2} g/L at 25 °C. Calculate the solubility of $MgF_2(s)$ in moles per liter, and calculate K_{sp} for MgF_2 at 25 °C.

68. Lead(II) chloride, $PbCl_2(s)$, dissolves in water to the extent of approximately 3.6×10^{-2} M at 20 °C. Calculate K_{sp} for $PbCl_2(s)$, and calculate its solubility in grams per liter.

69. The addition of chromate ion, $CrO_4^{2-}(aq)$, to a solution containing barium ion results in the formation of a bright yellow precipitate of barium chromate, $BaCrO_4(s)$, and is used as a confirmatory test in qualitative inorganic analysis. K_{sp} for $BaCrO_4(s)$ is 8.5×10^{-11} at 25 °C. Calculate the concentration of barium ion in a saturated solution of $BaCrO_4$.

70. The solubility product of iron(III) hydroxide is very small: $K_{sp} = 4 \times 10^{-38}$ at 25 °C. A classical method of analysis for unknown samples containing iron is to add NaOH or NH_3. This precipitates $Fe(OH)_3$, which can then be filtered and weighed. To demonstrate that the concentration of iron remaining in solution in such a sample is very small, calculate the solubility of $Fe(OH)_3$ in moles per liter and in grams per liter.

Additional Problems

71. In order for two molecules to react with each other, the molecules must _____ with each other.

72. As the _____ of a reaction mixture is increased, the reaction speeds up because more of the molecules present have enough energy to react.

73. The minimum energy required for molecules to react with each other is called the _____ energy.

74. A _____ speeds up a reaction without being consumed.

75. Equilibrium may be defined as the _____ of two processes, one of which is the opposite of the other.

76. When a chemical system has reached equilibrium, the concentrations of all reactants and products remain _____ with time.

77. Chemical reactions are said to be _____, which means they can occur in either direction.

78. Chemical equilibrium is said to be a _____ process, which means that even though the system may macroscop-

ically appear to have stopped, the forward and reverse reactions are still occurring.

79. At the point of chemical equilibrium, the rate of the forward reaction _____ the rate of the reverse reaction.

80. Equilibria involving reactants or products in more than one state are said to be _____.

81. According to Le Châtelier's principle, when a large excess of a gaseous reactant is added to a reaction system at equilibrium, the amounts of products _____.

82. Addition of an inert substance (one that does not participate in the reaction) does not change the _____ of an equilibrium.

83. When the volume of a vessel containing a gaseous equilibrium system is decreased, the _____ of the gaseous substances present is initially increased.

84. Increasing the temperature for an _____ process favors conversion of products to reactants.

85. The equilibrium expression that describes the solubility equilibrium for a sparingly soluble salt is referred to as the _____.

86. For a given reaction at a given temperature, the special ratio of products to reactants defined by the equilibrium constant is always equal to the same number. Explain why this is true, no matter what initial concentrations of reactants (or products) may have been taken in setting up an experiment.

87. Many sugars undergo a process called mutarotation, in which the sugar molecules interconvert between two isomeric forms, finally reaching an equilibrium between them. This is true for the simple sugar glucose, $C_6H_{12}O_6$, which exists in solution in isomeric forms called alpha-glucose and beta-glucose. If a solution of glucose at a certain temperature is analyzed, and it is found that the concentration of alpha-glucose is twice the concentration of beta-glucose, what is the value of K for the interconversion reaction?

88. Suppose $K = 4.5 \times 10^{-3}$ at a certain temperature for the reaction

$$PCl_5(g) \rightleftharpoons PCl_3(g) + Cl_2(g)$$

If it is found that the concentration of PCl_5 is twice the concentration of PCl_3, what must be the concentration of Cl_2 under these conditions?

89. For the reaction

$$CaCO_3(s) \rightleftharpoons CaO(s) + CO_2(g),$$

the equilibrium constant K has the form $K = [CO_2]$. Using a handbook to find density information about $CaCO_3(s)$ and $CaO(s)$, show that the *concentrations* of the two solids (the number of moles contained in 1 L of volume) are constant.

90. If the K_{sp} for silver chromate is 9.1×10^{-12} at 25 °C, what is the solubility of Ag_2CrO_4 in grams per liter?

91. Teeth and bones are composed, to a first approximation, of calcium phosphate, $Ca_3(PO_4)_2(s)$. The K_{sp} for this salt is 1.3×10^{-32} at 25 °C. Calculate the concentration of calcium ion in a saturated solution of $Ca_3(PO_4)_2$.

92. Under what circumstances can we compare the solubilities of two salts by directly comparing the values of their solubility products?

93. How does the collision model account for the fact that a reaction proceeds faster when the concentrations of the reactants are increased?

94. How does an increase in temperature result in an increase in the number of successful collisions between reactant molecules? What does an increase in temperature mean on a molecular basis?

95. Explain why the development of a vapor pressure above a liquid in a closed container represents an equilibrium. What are the opposing processes? How do we recognize when the system has reached a state of equilibrium?

96. Write the equilibrium expression for each of the following reactions.
    a. $H_2(g) + Br_2(g) \rightleftharpoons 2HBr(g)$
    b. $2H_2(g) + S_2(g) \rightleftharpoons 2H_2S(g)$
    c. $H_2(g) + C_2N_2(g) \rightleftharpoons 2HCN(g)$

97. Write the equilibrium expression for each of the following reactions.
    a. $2O_3(g) \rightleftharpoons 3O_2(g)$
    b. $CH_4(g) + 2O_2(g) \rightleftharpoons CO_2(g) + 2H_2O(g)$
    c. $C_2H_4(g) + Cl_2(g) \rightleftharpoons C_2H_4Cl_2(g)$

98. At high temperatures, elemental bromine, Br_2, dissociates into individual bromine atoms.

$$Br_2(g) \rightleftharpoons 2Br(g)$$

Suppose that in an experiment at 2000 °C, it is found that $[Br_2] = 0.97\ M$ and $[Br] = 0.034\ M$ at equilibrium. Calculate the value of K.

99. Gaseous phosphorus pentachloride decomposes according to the reaction

$$PCl_5(g) \rightleftharpoons PCl_3(g) + Cl_2(g)$$

The equilibrium system was analyzed at a particular temperature, and the concentrations of the substances present were determined to be $[PCl_5] = 1.1 \times 10^{-2} M$, $[PCl_3] = 0.325\ M$, and $[Cl_2] = 3.9 \times 10^{-3}\ M$. Calculate the value of K for the reaction.

100. Write the equilibrium expression for each of the following heterogeneous equilibria.
     a. $4Al(s) + 3O_2(g) \rightleftharpoons 2Al_2O_3(s)$
     b. $NH_3(g) + HCl(g) \rightleftharpoons NH_4Cl(s)$
     c. $2Mg(s) + O_2(g) \rightleftharpoons 2MgO(s)$

101. Write the equilibrium expression for each of the following heterogeneous equilibria.
     a. $P_4(s) + 5O_2(g) \rightleftharpoons P_4O_{10}(s)$
     b. $CO_2(g) + 2NaOH(s) \rightleftharpoons Na_2CO_3(s) + H_2O(g)$
     c. $NH_4NO_3(s) \rightleftharpoons N_2O(g) + 2H_2O(g)$

102. What is the effect on the position of a reaction system at equilibrium when an exothermic reaction is performed at a higher temperature? Does the value of the equilibrium constant change in this situation?

103. Suppose the reaction system

$$2NO(g) + O_2(g) \rightleftharpoons 2NO_2(g)$$

has already reached equilibrium. Predict the effect of each of the following changes on the position of the equilibrium. Tell whether the equilibrium will shift to the right, will shift to the left, or will not be affected.
   a. Additional oxygen is injected into the system.
   b. NO_2 is removed from the reaction vessel.
   c. 1.0 mol of helium is injected into the system.

104. The reaction

$$PCl_3(l) + Cl_2(g) \rightleftharpoons PCl_5(s)$$

liberates 124 kJ of energy per mol of PCl_3 reacted. Will an increase in temperature shift the equilibrium position toward products or toward reactants?

105. For the process

$$CO(g) + H_2O(g) \rightleftharpoons CO_2(g) + H_2(g)$$

it is found that the equilibrium concentrations at a particular temperature are $[H_2] = 1.4\ M$, $[CO_2] = 1.3\ M$, $[CO] = 0.71\ M$, and $[H_2O] = 0.66\ M$. Calculate the equilibrium constant K for the reaction under these conditions.

106. For the reaction

$$N_2(g) + 3H_2(g) \rightleftharpoons 2NH_3(g)$$

$K = 1.3 \times 10^{-2}$ at a given temperature. If the system at equilibrium is analyzed and the concentrations of N_2 and H_2 are both found to be 0.10 M, what is the concentration of NH_3 in the system?

107. The equilibrium constant for the reaction

$$2NOCl(g) \rightleftharpoons 2NO(g) + Cl_2(g)$$

has the value 9.2×10^{-6} at a particular temperature. The system is analyzed at equilibrium, and it is found that the concentrations of $NOCl(g)$ and $NO(g)$ are 0.44 M and $1.5 \times 10^{-3}\ M$, respectively. What is the concentration of $Cl_2(g)$ in the equilibrium system under these conditions?

108. Write balanced chemical equations describing the dissolving of the following solids in water. Write the expression for K_{sp} for each process.
   a. $MgCO_3(s)$ c. $Ca(OH)_2(s)$
   b. $Al_2(SO_4)_3(s)$ d. $Ag_2SO_4(s)$

109. Nickel(II) sulfide, $NiS(s)$, dissolves in water to the extent of approximately $4.0 \times 10^{-5}\ M$ at 18 °C. Calculate K_{sp} for NiS at this temperature.

110. Approximately 9.0×10^{-4} g of silver chloride, $AgCl(s)$, dissolves per liter of water at 10 °C. Calculate K_{sp} for $AgCl(s)$ at this temperature.

111. Mercuric sulfide, HgS, is one of the least soluble salts known, with $K_{sp} = 1.6 \times 10^{-54}$ at 25 °C. Calculate the solubility of HgS in moles per liter and in grams per liter.

112. Approximately 0.14 g of nickel(II) hydroxide, $Ni(OH)_2(s)$, dissolves per liter of water at 20 °C. Calculate K_{sp} for $Ni(OH)_2(s)$ at this temperature.

Solutions to Self-Check Exercises

SELF-CHECK EXERCISE 16.1

Applying the law of chemical equilibrium gives

$$K = \frac{[NO_2]^4 [H_2O]^6}{[NH_3]^4 [O_2]^7}$$

Coefficient of NO_2 ↘   ↙ Coefficient of H_2O

↖ Coefficient of O_2

↑ Coefficient of NH_3

SELF-CHECK EXERCISE 16.2

a. $K = [O_2]^3$ The solids are not included.

b. $K = [N_2O][H_2O]^2$ The solid is not included. Water is gaseous in this reaction, so it is included.

c. $K = \dfrac{1}{[CO_2]}$ The solids are not included.

d. $K = \dfrac{1}{[SO_3]}$ Water and H_2SO_4 are pure liquids and so are not included.

SELF-CHECK EXERCISE 16.3

When rain is imminent, the concentration of water vapor in the air increases. This shifts the equilibrium to the right, forming $CoCl_2 \cdot 6H_2O(s)$, which is pink.

SELF-CHECK EXERCISE 16.4

a.   No change. Both sides of the equation contain the same number of gaseous components. The system cannot change its pressure by shifting its equilibrium position.

b.   Shifts to the left. The system can increase the number of gaseous components present, and so increase the pressure, by shifting to the left.

c.   Shifts to the right to increase the number of gaseous components and thus its pressure.

SELF-CHECK EXERCISE 16.5

a.   Shifts to the right away from added SO_2.

b.   Shifts to the right to replace removed SO_3.

c.   Shifts to the right to decrease its pressure.

d.   Shifts to the right. Energy is a product in this case, so a decrease in temperature favors the forward reaction (which produces energy).

Gargoyles damaged by acid rain.

CONTENTS

Students have had an introduction to the reactions of acids and bases in Section 7.4 and to naming acids in Section 5.9.

The discussion of acid and base chemistry provides an opportunity to discuss and to bring to class many examples from everyday life. Vinegar, antacids, and lemon juice are common substances whose properties and reactions are important.

Lemon juice is sour because of its acid content.

This is an important section. Many times students cannot solve acid–base problems because they cannot identify either the acid–conjugate base pair or the base–conjugate acid pair.

Don't taste chemical reagents!

The label on a bottle of concentrated hydrochloric acid.

Acids are very important substances. They cause lemons to be sour, digest food in the stomach (and sometimes cause heartburn), dissolve rock to make fertilizer, dissolve your tooth enamel to form cavities, and clean the deposits out of your coffee maker. Acids are essential industrial chemicals. In fact, the chemical in first place in terms of the amount manufactured in the United States is sulfuric acid, H_2SO_4. Eighty *billion* pounds of this material are used every year in the manufacture of fertilizers, detergents, plastics, pharmaceuticals, storage batteries, and metals.

In this chapter we will consider the most important properties of acids and of their opposites, the bases.

17.1   Acids and Bases

AIM: To discuss two models of acids and bases and to show the relationship of conjugate acid–base pairs.

Acids were first recognized as substances that taste sour. Vinegar tastes sour because it is a dilute solution of acetic acid; citric acid is responsible for the sour taste of a lemon. Bases, sometimes called *alkalis,* are characterized by their bitter taste and slippery feel. Most hand soaps and commercial preparations for unclogging drains are highly basic.

The first person to recognize the essential nature of acids and bases was Svante Arrhenius. On the basis of his experiments with electrolytes, Arrhenius postulated that **acids** *produce hydrogen ions in aqueous solution,* whereas **bases** *produce hydroxide ions* (review Section 7.4).

For example, when hydrogen chloride gas is dissolved in water it dissociates completely.

$$HCl(g) \xrightarrow{H_2O} H^+(aq) + Cl^-(aq)$$

This solution is the strong acid known as hydrochloric acid. On the other hand, when solid sodium hydroxide is dissolved in water, it produces a solution containing Na^+ and OH^- ions.

$$NaOH(s) \xrightarrow{H_2O} Na^+(aq) + OH^-(aq)$$

This solution is called a strong base.

Although the **Arrhenius concept of acids and bases** was a major step forward in understanding acid–base chemistry, this concept is limited because it allows for only one kind of base—the hydroxide ion. A more general definition of acids and bases was suggested by the Danish chemist Johannes Brønsted and the English chemist Thomas Lowry. In the **Brønsted–Lowry model,** *an acid is a*

Students have virtually no problem identifying Arrhenius acids and bases. They do have problems recognizing Brønsted–Lowry bases such as $C_2H_3O_2^-$ and NH_3 because the atom donating electrons is not always the same. There is no easily recognizable OH^-. Identifying conjugate acids of amine bases is also difficult. Students have trouble thinking of ions such as $CH_3NH_3^+$ as acids.

proton (H^+) donor, and a base is a proton acceptor. According to the Brønsted–Lowry model, the general reaction that occurs when an acid is dissolved in water can best be represented as an acid (HA) donating a proton to a water molecule to form a new acid (the **conjugate acid**) and a new base (the **conjugate base**).

$$HA(aq) + H_2O(l) \rightleftharpoons H_3O^+(aq) + A^-(aq)$$

| Acid | Base | Conjugate acid | Conjugate base |

This model emphasizes the significant role of the polar water molecule in pulling the proton from the acid. Note that the conjugate base is everything that remains of the acid molecule after a proton is lost. The conjugate acid is formed when the proton is transferred to the base. A **conjugate acid–base pair** consists of two substances related to each other by the donating and accepting of a single proton. In the above equation there are two conjugate acid–base pairs: HA (acid) and A^- (base), and H_2O (base) and H_3O^+ (acid). For example, when hydrogen chloride is dissolved in water it behaves as an acid.

Acid–conjugate base pair

$$HCl(aq) + H_2O(l) \rightleftharpoons H_3O^+(aq) + Cl^-(aq)$$

Base–conjugate acid pair

In this case HCl is the acid that loses an H^+ ion to form Cl^-, its conjugate base. On the other hand, H_2O (behaving as a base) gains an H^+ ion to form H_3O^+ (the conjugate acid).

How can water act as a base? Remember that the oxygen of the water molecule has two unshared electron pairs, either of which can form a covalent bond with an H^+ ion. When gaseous HCl dissolves in water, the following reaction occurs.

Note that an H^+ ion is transferred from the HCl molecule to the water molecule to form H_3O^+, which is called the **hydronium ion.**

Recall that (aq) means the substance is hydrated—it has water molecules clustered around it.

In previous discussions of the dissociation of strong acids, no mention was made of the formation of H_3O^+. This is students' first introduction to the hydronium ion.

EXAMPLE 17.1   **Identifying Conjugate Acid–Base Pairs**

Which of the following represent conjugate acid–base pairs?

a. HF, F^-
b. NH_4^+, NH_3

c. HCl, H_2O

EXAMPLE 17.1, CONTINUED

SOLUTION

a. and b.   HF, F^- and NH_4^+, NH_3 are conjugate acid–base pairs because the
two species differ by one H^+.

$$HF \rightarrow H^+ + F^-$$
$$NH_4^+ \rightarrow H^+ + NH_3$$

c.   HCl and H_2O are not a conjugate acid–base pair because they are not
related by the removal or addition of one H^+. The conjugate base of HCl
is Cl^-. The conjugate acid of H_2O is H_3O^+.

EXAMPLE 17.2   **Writing Conjugate Bases**

Write the conjugate base for each of the following:

a. $HClO_4$
b. H_3PO_4
c. $CH_3NH_3^+$

SOLUTION

To get the conjugate base for an acid, we must remove an H^+ ion.

a. $HClO_4 \rightarrow H^+ + ClO_4^-$
     Acid   Conjugate base

b. $H_3PO_4 \rightarrow H^+ + H_2PO_4^-$
     Acid   Conjugate base

c. $CH_3NH_3^+ \rightarrow H^+ + CH_3NH_2$
     Acid   Conjugate base

SELF-CHECK EXERCISE 17.1

Which of the following represent conjugate acid–base pairs?

a. H_2O, H_3O^+
b. OH^-, HNO_3
c. H_2SO_4, SO_4^{2-}
d. $HC_2H_3O_2$, $C_2H_3O_2^-$

17.2   Acid Strength

AIM: To relate acid strength to the position of the dissociation equilibrium.
To relate acid strength to the strength of the conjugate base.

When an acid is placed in water, the reaction

$$HA(aq) + H_2O(l) \rightleftharpoons H_3O^+(aq) + A^-(aq)$$

can occur in both directions.

1. In the forward reaction the acid is donating a proton to a water molecule, which behaves as a base.
2. In the reverse reaction the acid is the hydronium ion, H_3O^+, and the base is A^-, the conjugate base of HA.

Therefore, these two reactions represent *a competition for the proton between the two bases H_2O and A^-*. If H_2O is a much stronger base than A^- (that is, if H_2O has a much greater affinity for H^+ than does A^-), the equilibrium position will be far to the right. Virtually all of the acid dissolved will be in the ionized form. Conversely, if A^- is a much stronger base than H_2O, the equilibrium position will lie far to the left. In this case most of the acid dissolved will be present at equilibrium as HA (that is, molecules of HA). We can represent acid strength by using the following schematic.

Strong acid

$$HA + H_2O \quad\Longrightarrow\quad H_3O^+ + A^-$$

In this case A^- has a much lower attraction for H^+ than does H_2O. That is, A^- is a much weaker base than H_2O.

Weak acid

$$HA + H_2O \quad\Longleftarrow\quad H_3O^+ + A^-$$

In this case A^- has a much larger attraction for H^+ than does H_2O, and the HA molecules remain intact.

We can determine what is actually going on in a solution by measuring its ability to conduct an electric current. Recall from Chapter 7 that a solution can

The concept of weak and strong acids can be effectively demonstrated several ways. You can use a conductivity set up with a light bulb and various solutions of weak and strong acids. This demonstration was discussed in Chapter 5 for comparing solutions of ionic and nonionic compounds. By the brightness of the light in the solutions, you can show that strong acids produce many ions, whereas weak acids produce few. You can also use universal indicator in solutions of acids and bases to show how much H^+ or OH^- is present. If you can bring a pH meter to class, students will enjoy reading the pH from the instrument. Students are often amazed when the pH of water is not seven. If you use deionized water for your demonstration, you may notice that the pH is less than seven.

A strong acid is completely dissociated in water. No HA molecules remain. Only H^+ and A^- are present.

A weak acid is mostly undissociated in water.

A solution of hydrochloric acid readily conducts an electric current, showing that it is a strong electrolyte.

A strong acid has a weak conjugate base.

conduct a current in proportion to the number of ions that are present (see Figure 7.2). When 1 mol of solid sodium chloride is dissolved in 1 L of water, the resulting solution is an excellent conductor of an electric current because the Na^+ and Cl^- ions separate completely. We call NaCl a strong electrolyte. Similarly, when 1 mol of hydrogen chloride is dissolved in 1 L of water, the resulting solution is an excellent conductor. Therefore hydrogen chloride is also a strong electrolyte, which means that each HCl molecule must dissociate to produce H^+ and Cl^- ions. This tells us that the equilibrium

$$HCl(aq) + H_2O(l) \rightleftharpoons H_3O^+(aq) + Cl^-(aq)$$

lies far to the right. (Accordingly, the arrow pointing right is longer than the arrow pointing left.) In solution there are virtually no HCl molecules, only H^+ and Cl^- ions. This shows that Cl^- is a very poor base compared to the H_2O molecule; it has virtually no ability to attract H^+ ions in water. This aqueous solution of hydrogen chloride (called *hydrochloric acid*) is a strong acid.

In general, the strength of an acid is defined by the equilibrium position of its dissociation reaction:

$$HA(aq) + H_2O(l) \rightleftharpoons H_3O^+(aq) + A^-(aq)$$

A **strong acid** is one for which *this equilibrium lies far to the right.* This means that almost all the original HA is dissociated at equilibrium (see Figure 17.1a). There is an important connection between the strength of an acid and that of its conjugate base. *A strong acid contains a relatively weak conjugate base*—one

Figure 17.1
Graphical representation of the behavior of acids of different strengths in aqueous solution. (a) A strong acid is completely dissociated. (b) In contrast, only a small fraction of the molecules of a weak acid are dissociated at equilibrium.

that has a low attraction for protons. A strong acid can be described as an acid whose conjugate base is a much weaker base than water (Figure 17.2). In this case the water molecules win the competition for the H^+ ions.

In contrast to hydrochloric acid, when acetic acid, $HC_2H_3O_2$, is dissolved in water, the resulting solution conducts an electric current only weakly. That is, acetic acid is a weak electrolyte, which means that only a few ions are present. In other words, the acid dissociation equilibrium

$$HC_2H_3O_2(aq) + H_2O(l) \rightleftharpoons H_3O^+(aq) + C_2H_3O_2^-(aq)$$

lies far to the left (thus the arrow pointing left is longer). In fact, measurements show that only about one in one hundred (1%) of the $HC_2H_3O_2$ molecules is dissociated in a 0.1 *M* solution of acetic acid. Thus acetic acid is a weak acid. When acetic acid molecules are placed in water, almost all of the molecules remain undissociated. This tells us that the acetate ion, $C_2H_3O_2^-$, is an effective base—it very successfully attracts H^+ ions in water. This means that acetic acid remains largely in the form of $HC_2H_3O_2$ molecules in solution. A **weak acid** is one for which the acid dissociation equilibrium

$$HA(aq) + H_2O(l) \rightleftharpoons H_3O^+(aq) + A^-(aq)$$

lies far to the left. Most of the acid originally placed in the solution is still present as HA at equilibrium. That is, a weak acid dissociates only to a very small extent in aqueous solution (see Figure 17.1b). In contrast to a strong acid, a weak acid has a conjugate base that is a much stronger base than water. In this case a water molecule is not very successful in pulling an H^+ ion away from the conjugate base. *A weak acid contains a relatively strong conjugate base* (Figure 17.2).

The various ways of describing the strength of an acid are summarized in Table 17.1.

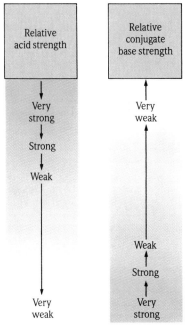

Figure 17.2
The relationship of acid strength and conjugate base strength for the dissociation reaction
$HA(aq) + H_2O(l) \rightleftharpoons H_3O^+(aq) + A^-(aq)$
   Acid   Conjugate
                                         base

Table 17.1	Ways to Describe Acid Strength	
Property	Strong Acid	Weak Acid
position of the dissociation equilibrium	far to the right	far to the left
strength of the conjugate base compared to that of water	A^- a much weaker base than H_2O	A^- a much stronger base than H_2O

An acetic acid solution does not allow much electric current to flow, showing that it contains only a few ions.

Perchloric acid can explode when handled improperly.

H
|
O
|
P
H—O/ \O—H
O
Phosphoric acid

H   O
\ //
H—C—C
/   \
H   O—H
Acetic acid

N
/ \
O   O—H
Nitrous acid

Cl—O
\
H
Hypochlorous acid

For some of the strong and weak acids that students are not likely to have encountered previously, it helps to show the equation for their dissociation in water.

The common strong acids are sulfuric acid, $H_2SO_4(aq)$; hydrochloric acid, $HCl(aq)$; nitric acid, $HNO_3(aq)$; and perchloric acid, $HClO_4(aq)$. Sulfuric acid is actually a **diprotic acid,** an acid that can furnish two protons. The acid H_2SO_4 is a strong acid that is virtually 100% dissociated in water:

$$H_2SO_4(aq) \rightarrow H^+(aq) + HSO_4^-(aq)$$

The HSO_4^- ion is also an acid but it is a weak acid:

$$HSO_4^-(aq) \rightleftharpoons H^+(aq) + SO_4^{2-}(aq)$$

Most of the HSO_4^- ions remain undissociated.

Most acids are **oxyacids,** in which the acidic hydrogen is attached to an oxygen atom (several oxyacids are shown in the margin). The strong acids we have mentioned, except hydrochloric acid, are typical examples. **Organic acids,** those with a carbon-atom backbone, commonly contain the **carboxyl group:**

O
//
—C
\
O—H

Acids of this type are usually weak. An example is acetic acid, CH_3COOH, which is often written as $HC_2H_3O_2$.

There are some important acids in which the acidic proton is attached to an atom other than oxygen. The most significant of these are the hydrohalic acids HX, where X represents a halogen atom. Examples are $HCl(aq)$, a strong acid, and $HF(aq)$, a weak acid.

17.3   Water as an Acid and a Base

AIM: To discuss the ionization of water.

The idea that water can act both as an acid and as a base is an important one for students to understand.

A substance is said to be *amphoteric* if it can behave either as an acid or as a base. Water is the most common **amphoteric substance.** We can see this clearly in the **ionization of water,** which involves the transfer of a proton from one water molecule to another to produce a hydroxide ion and a hydronium ion.

$$H_2O(l) + H_2O(l) \rightleftharpoons H_3O^+(aq) + OH^-(aq)$$

In this reaction one water molecule acts as an acid by furnishing a proton, and the other acts as a base by accepting the proton. The autoionization reaction for water leads to the equilibrium expression

$$K_w = [H_3O^+][OH^-]$$

The ionization of water is often written in shorthand form as

$$H_2O(l) \rightleftharpoons H^+(aq) + OH^-(aq)$$

which leads to the K_w expression

$$K_w = [H^+][OH^-]$$

The constant K_w, called the **ion-product constant** (or the *dissociation constant*), always refers to the ionization of water. The concentration of water does not appear in the K_w expression, because the concentration of water remains essentially constant. Recall that pure liquids are never included in an equilibrium expression (see Section 16.8).

The amount of H_2O that ionizes is so slight that the $[H_2O]$ is virtually unchanged.

Experiment shows that at 25 °C,

$$[H^+] = [OH^-] = 1.0 \times 10^{-7}\ M$$

which means that at 25 °C,

$$K_w = [H^+][OH^-] = (1.0 \times 10^{-7})(1.0 \times 10^{-7}) = 1.0 \times 10^{-14}$$

The number of H_2O molecules that ionize is very small: only about two molecules in a billion.

$$\boxed{K_w = 1.0 \times 10^{-14}}$$

The units are customarily omitted when the value of the constant is given and used.

It is important to recognize the meaning of K_w. In any aqueous solution at 25 °C, *no matter what it contains,* the product of $[H^+]$ and $[OH^-]$ must always equal 1.0×10^{-14}. This means that if the $[H^+]$ goes up, the $[OH^-]$ must go down so that the product of the two is still 1.0×10^{-14}. For example, if HCl gas is dissolved in water, increasing the $[H^+]$, the $[OH^-]$ must decrease. There are three possible situations we might encounter in an aqueous solution:

$$K_w = [H^+][OH^-]$$
$$= 1.0 \times 10^{-14}$$

1. A **neutral solution**, where $[H^+] = [OH^-]$
2. An **acidic solution**, where $[H^+] > [OH^-]$
3. A **basic solution**, where $[OH^-] > [H^+]$

In each case, however, $K_w = [H^+][OH^-] = 1.0 \times 10^{-14}$.

EXAMPLE 17.3   Calculating Ion Concentrations in Water

Calculate $[H^+]$ or $[OH^-]$ as required for each of the following solutions at 25 °C, and state whether the solution is neutral, acidic, or basic.

a. $1.0 \times 10^{-5}\ M\ OH^-$ c. $10.0\ M\ H^+$
b. $1.0 \times 10^{-7}\ M\ OH^-$

SOLUTION

a.   We know that $K_w = [H^+][OH^-] = 1.0 \times 10^{-14}$. We need to calculate the $[H^+]$. However, the $[OH^-]$ is given—it is 1.0×10^{-5} M—so we will solve for $[H^+]$ by dividing both sides by $[OH^-]$.

$$K_w = [H^+][OH^-]$$
$$\frac{K_w}{[OH^-]} = [H^+]$$

EXAMPLE 17.3, CONTINUED

$$[H^+] = \frac{1.0 \times 10^{-14}}{[OH^-]} = \frac{1.0 \times 10^{-14}}{1.0 \times 10^{-5}} = 1.0 \times 10^{-9} \ M$$

Because $[OH^-] = 1.0 \times 10^{-5} \ M$ is greater than $[H^+] = 1.0 \times 10^{-9}$ M, the solution is basic. (Remember: the more negative the exponent, the smaller the number.)

$K_w = [H^+][OH^-]$

$\frac{K_w}{[H^+]} = [OH^-]$

b. Again the $[OH^-]$ is given, so we solve the K_w expression for $[H^+]$.

$$[H^+] = \frac{1.0 \times 10^{-14}}{[OH^-]} = \frac{1.0 \times 10^{-14}}{1.0 \times 10^{-7}} = 1.0 \times 10^{-7} \ M$$

Here $[H^+] = [OH^-] = 1.0 \times 10^{-7} \ M$, so the solution is neutral.

c. In this case the $[H^+]$ is given, so we solve for $[OH^-]$.

$$[OH^-] = \frac{1.0 \times 10^{-14}}{[H^+]} = \frac{1.0 \times 10^{-14}}{10.0} = 1.0 \times 10^{-15} \ M$$

Now we compare $[H^+] = 10.0 \ M$ to $[OH^-] = 1.0 \times 10^{-15} \ M$. Because $[H^+]$ is greater than $[OH^-]$, the solution is acidic.

SELF-CHECK EXERCISE 17.2

Calculate $[H^+]$ in a solution in which $[OH^-] = 2.0 \times 10^{-2} \ M$. Is this solution acidic, neutral, or basic?

EXAMPLE 17.4	**Using the Ion-Product Constant in Calculations**

Is it possible for an aqueous solution at 25 °C to have $[H^+] = 0.010 \ M$ and $[OH^-] = 0.010 \ M$?

SOLUTION

The concentration 0.010 M can also be expressed as $1.0 \times 10^{-2} \ M$. Thus if $[H^+] = [OH^-] = 1.0 \times 10^{-2} M$, the product

$$[H^+][OH^-] = (1.0 \times 10^{-2})(1.0 \times 10^{-2}) = 1.0 \times 10^{-4}$$

This is not possible. The product of $[H^+]$ and $[OH^-]$ must always be 1.0×10^{-14} in water at 25 °C, so a solution could not have $[H^+] = [OH^-] = 0.010$ M. If H^+ and OH^- are added to water in these amounts, they will react with each other to form H_2O:

$$H^+ + OH^- \rightarrow H_2O$$

until the product $[H^+][OH^-] = 1.0 \times 10^{-14}$.

EXAMPLE 17.4, CONTINUED

This is a general result. When H^+ and OH^- are added to water in amounts such that the product of their concentrations is greater than 1.0×10^{-14}, they will react to form water until enough H^+ and OH^- are consumed so that $[H^+][OH^-] = 1.0 \times 10^{-14}$.

17.4   The pH Scale

AIM: To define pH and pOH.
To show how to find pOH and pH for various solutions.
To show how to use a calculator in these calculations.

To express small numbers conveniently, chemists often use the "p scale," which is based on common logarithms (base 10 logs). In this system if N represents some number, then

$$pN = -\log N = (-1) \times \log N$$

That is, the p means to take the log of the number that follows and multiply the result by -1. For example, to express the number 1.0×10^{-7} on the p scale, we first take the log of 1.0×10^{-7}. On most calculators this means entering the number and then pressing the log key.

1. Enter 1.0×10^{-7}.

2. Press the $\boxed{\log}$ key.

Now the calculator shows the log of the number 1.0×10^{-7}, which is

$$-7.00$$

Next we must multiply by -1. On most calculators this is done by using the $+/-$ key (the $+/-$ key just reverses the sign, which is what multiplying by -1 really does). In this case we get the result

$$7.00$$

So

$$p(1.0 \times 10^{-7}) = -\log (1.0 \times 10^{-7}) = 7.00$$

Because the $[H^+]$ in an aqueous solution is typically quite small, using the p scale in the form of the **pH scale** provides a convenient way to represent solution acidity. The pH is defined as

$$\boxed{pH = -\log[H^+]}$$

The pH scale provides a compact way to represent solution acidity.

To obtain the pH value of a solution, we must compute the negative log of the $[H^+]$. On a typical calculator, this involves the following steps:

Steps for Calculating pH on a Calculator

STEP 1: Enter the $[H^+]$.

STEP 2: Press the $\boxed{\text{log}}$ key.

STEP 3: Press the $\boxed{+/-}$ (change-of-sign) key.

In the case where $[H^+] = 1.0 \times 10^{-5}\ M$, following the above steps gives a pH value of 5.00.

To represent pH to the appropriate number of significant figures, you need to know the following rule for logarithms: *the number of decimal places for a log must be equal to the number of significant figures in the original number.* Thus

The rules for determining the correct number of significant figures to use when calculating pH and $[H^+]$ are confusing, in part because students have a poor understanding of logarithms.

$$[H^+] = \overbrace{1.0}^{\substack{2 \\ \text{significant} \\ \text{figures}}} \times 10^{-5}\ M$$

and

$$pH = 5.\underbrace{00}_{\substack{2 \\ \text{decimal} \\ \text{places}}}$$

EXAMPLE 17.5   Calculating pH

Calculate the pH value for each of the following solutions at 25 °C.

a. A solution in which $[H^+] = 1.0 \times 10^{-9}\ M$
b. A solution in which $[OH^-] = 1.0 \times 10^{-6}\ M$

SOLUTION

a. For this solution $[H^+] = 1.0 \times 10^{-9}$.

STEP 1: Enter the number.

EXAMPLE 17.5, CONTINUED

STEP 2: Push the $\boxed{\log}$ key to give -9.00.

STEP 3: Push the $\boxed{+/-}$ key to give 9.00.

$$pH = 9.00$$

b. In this case we are given the $[OH^-]$. Thus we must first calculate $[H^+]$ from the K_w expression. We solve

$$K_w = [H^+][OH^-] = 1.0 \times 10^{-14}$$

for $[H^+]$ by dividing both sides by $[OH^-]$.

$$[H^+] = \frac{1.0 \times 10^{-14}}{[OH^-]} = \frac{1.0 \times 10^{-14}}{1.0 \times 10^{-6}} = 1.0 \times 10^{-8}$$

Now that we know the $[H^+]$, we can calculate the pH by taking the three steps listed above. Doing so yields pH = 8.00.

SELF-CHECK EXERCISE 17.3

Calculate the pH value for each of the following solutions at 25 °C.

a. A solution in which $[H^+] = 1.0 \times 10^{-3}\ M$
b. A solution in which $[OH^-] = 5.0 \times 10^{-5}\ M$

Table 17.2	The Relationship of the H$^+$ Concentration of a Solution to Its pH
[H$^+$]	**pH**
1.0×10^{-1}	1.00
1.0×10^{-2}	2.00
1.0×10^{-3}	3.00
1.0×10^{-4}	4.00
1.0×10^{-5}	5.00
1.0×10^{-6}	6.00
1.0×10^{-7}	7.00

Because the pH scale is a log scale based on 10, *the pH changes by 1 for every power-of-10 change in the [H$^+$]*. For example, a solution of pH 3 has an H$^+$ concentration of $10^{-3}\ M$, which is 10 times that of a solution of pH 4 ($[H^+] = 10^{-4}\ M$) and 100 times that of a solution of pH 5. This is illustrated in Table 17.2. Also note from Table 17.2 that *the pH decreases as the [H$^+$] increases.* That is, a lower pH means a more acidic solution. The pH scale and the pH values for several common substances are shown in Figure 17.3 (p. 592).

We often measure the pH of a solution by using a pH meter, an electronic device with a probe that can be inserted into a solution of unknown pH. A pH meter is shown in Figure 17.4 (p. 592). Colored indicator paper is also commonly used to measure the pH of a solution when less accuracy is needed. A drop of the solution to be tested is placed on this special paper, which promptly turns to a color characteristic of a given pH (see Figure 17.5, p. 593).

Log scales similar to the pH scale are used for representing other quantities. For example,

$$\boxed{pOH = -\log[OH^-]}$$

The pH decreases as [H$^+$] increases, and vice versa.

Because most students still do not have a good feel for the pH scale, their first calculations of pH will generate numbers that may not be very meaningful to them. You might want to point out that each one unit change in pH results in a tenfold change in hydrogen ion concentration.

The symbol p means $-\log$.

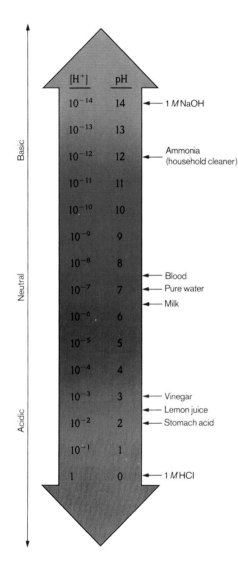

Basic

Neutral

Acidic

$[H^+]$	pH	
10^{-14}	14	← 1 M NaOH
10^{-13}	13	
10^{-12}	12	← Ammonia (household cleaner)
10^{-11}	11	
10^{-10}	10	
10^{-9}	9	
10^{-8}	8	← Blood
10^{-7}	7	← Pure water
10^{-6}	6	← Milk
10^{-5}	5	
10^{-4}	4	
10^{-3}	3	← Vinegar
10^{-2}	2	← Lemon juice ← Stomach acid
10^{-1}	1	
1	0	← 1 M HCl

Figure 17.3
The pH scale and pH values of some common substances.

Therefore, in a solution in which

$$[OH^-] = 1.0 \times 10^{-12} \; M$$

the pOH is

$$-\log[OH^-] = -\log(1.0 \times 10^{-12}) = 12.00$$

EXAMPLE 17.6   Calculating pH and pOH

Calculate the pH and pOH for each of the following solutions at 25 °C.

a. $1.0 \times 10^{-3} \; M \; OH^-$ b. $1.0 \; M \; H^+$

SOLUTION

a.   We are given the $[OH^-]$, so we can calculate the pOH value by taking $-\log[OH^-]$.

$$pOH = -\log[OH^-] = -\log(1.0 \times 10^{-3}) = 3.00$$

To calculate the pH, we must first solve the K_w expression for $[H^+]$.

$$[H^+] = \frac{K_w}{[OH^-]} = \frac{1.0 \times 10^{-14}}{1.0 \times 10^{-3}} = 1.0 \times 10^{-11} \; M$$

Now we compute the pH.

$$pH = -\log[H^+] = -\log(1.0 \times 10^{-11}) = 11.00$$

b.   In this case we are given the $[H^+]$ and we can compute the pH.

$$pH = -\log[H^+] = -\log(1.0) = 0$$

Figure 17.4
A pH meter. The electrodes on the left are placed in the solution with unknown pH. The difference between the $[H^+]$ in the solution sealed into one of the electrodes and the $[H^+]$ in the solution being analyzed is translated into an electrical potential and registered on the meter as a pH reading.

EXAMPLE 17.6, CONTINUED

We next solve the K_w expression for $[OH^-]$.

$$[OH^-] = \frac{K_w}{[H^+]} = \frac{1.0 \times 10^{-14}}{1.0} = 1.0 \times 10^{-14}\ M$$

Now we compute the pOH.

$$pOH = -\log[OH^-] = -\log(1.0 \times 10^{-14}) = 14.00$$

We can obtain a convenient relationship between pH and pOH by starting with the K_w expression $[H^+][OH^-] = 1.0 \times 10^{-14}$ and taking $-\log$ of both sides.

$$-\log([H^+][OH^-]) = -\log(1.0 \times 10^{-14})$$

Because the log of a product equals the sum of the logs of the terms—that is, $\log(A \times B) = \log A + \log B$—we have

$$\underbrace{-\log[H^+]}_{pH}\ \underbrace{-\log[OH^-]}_{pOH} = -\log(1.0 \times 10^{-14}) = 14.00$$

which gives the equation

$$\boxed{pH + pOH = 14.00}$$

This means that once we know either the pH or the pOH for a solution, we can calculate the other. For example, if a solution has a pH of 6.00, the pOH is calculated as follows:

$$pH + pOH = 14.00$$
$$pOH = 14.00 - pH$$
$$pOH = 14.00 - 6.00 = 8.00$$

EXAMPLE 17.7   **Calculating pOH from pH**

The pH of blood is about 7.4. What is the pOH of blood?

SOLUTION

$$pH + pOH = 14.00$$
$$pOH = 14.00 - pH$$
$$= 14.00 - 7.4$$
$$= 6.6$$

The pOH of blood is 6.6.

Figure 17.5
Indicator paper being used to measure the pH of a solution. The pH is determined by comparing the color that the solution turns the paper to the color chart.

Some students are never able to perform calculations with pOH and $[OH^-]$ but can calculate pH and $[H^+]$.

Red blood cells can exist only over a narrow range of pH.

Measuring the pH of the water in a pond.

This operation may involve a 10^X key on some calculators.

Students who have calculators with both a log and an inverse log key will find calculating $[H^+]$ easier than those students whose calculators have a log key but use the 10^X key to calculate the inverse log. You may need to explain to those students how they can use their calculators to find the inverse log.

EXAMPLE 17.7, CONTINUED

SELF-CHECK EXERCISE 17.4

A sample of rain in an area with severe air pollution has a pH of 3.5. What is the pOH of this rainwater?

It is also possible to find the $[H^+]$ or $[OH^-]$ from the pH or pOH. To find the $[H^+]$ from the pH, we must go back to the definition of pH:

$$pH = -\log[H^+]$$

or

$$-pH = \log[H^+]$$

To arrive at $[H^+]$ on the right-hand side of this equation we must "undo" the log operation. This is called taking the *antilog* or the *inverse* log.

> Inverse log $(-pH)$ = inverse log $(\log[H^+])$
> Inverse log $(-pH)$ = $[H^+]$

There are different methods for carrying out the inverse log operation on various calculators. One common method is the two-key $\boxed{\text{inv}}$ $\boxed{\text{log}}$ sequence. (Consult the user's manual for your calculator to find out how to do the antilog or inverse log operation.) The steps in going from pH to $[H^+]$ are as follows:

Steps for Calculating $[H^+]$ from pH

STEP 1: Enter the pH.

STEP 2: Change the sign of the pH by using the $\boxed{+/-}$ key.

STEP 3: Take the inverse log (antilog) of $-pH$ to give $[H^+]$ by using the $\boxed{\text{inv}}$ $\boxed{\text{log}}$ keys in that order. (Your calculator may require different keys for this operation.)

For practice, we will convert pH $= 7.0$ to $[H^+]$.

STEP 1: pH $= 7.0$

STEP 2: $-pH = -7.0$

STEP 3: $\boxed{\text{inv}}$ $\boxed{\text{log}}$ -7.0 gives 1.0×10^{-7}

$$[H^+] = 1.0 \times 10^{-7} \ M.$$

This process is illustrated further in Example 17.8.

EXAMPLE 17.8   Calculating [H⁺] from pH

The pH of a human blood sample was measured to be 7.41. What is the $[H^+]$ in this blood?

SOLUTION

STEP 1: pH $= 7.41$

STEP 2: $-$pH $= -7.41$

STEP 3: $[H^+] =$ inverse log of -7.41

$\boxed{\text{inv}}$ $\boxed{\text{log}}$ $-7.41 = 3.9 \times 10^{-8}$

$$[H^+] = 3.9 \times 10^{-8} \ M.$$

Notice that because the pH has two decimal places, we need two significant figures for $[H^+]$.

SELF-CHECK EXERCISE 17.5

The pH of rainwater in a polluted area was found to be 3.50. What is the $[H^+]$ for this rainwater?

A similar procedure is used to change from pOH to $[OH^-]$, as shown in Example 17.9.

EXAMPLE 17.9   Calculating [OH⁻] from pOH

The pOH of the water in a fish tank is found to be 6.59. What is the $[OH^-]$ for this water?

SOLUTION

We use the same steps as for converting pH to $[H^+]$, except that we use the pOH.

STEP 1: pOH $= 6.59$

STEP 2: $-$pOH $= -6.59$

EXAMPLE 17.9, CONTINUED

STEP 3: [inv] [log] $-6.59 = 2.6 \times 10^{-7}$

$$[OH^-] = 2.6 \times 10^{-7} \ M$$

Note that two significant figures are required.

SELF-CHECK EXERCISE 17.6

The pOH of a liquid drain cleaner was found to be 10.50. What is the $[OH^-]$ for this cleaner?

17.5 Calculating the pH of Strong Acid Solutions

To calculate the pH of strong acids, students must know that strong acids dissociate completely to form hydrogen ions and anions. They must remember this fact to determine the hydrogen ion concentration.

Students should memorize the short list of strong acids on page 212.

AIM: To show how to calculate the pH of solutions of strong acids.

In this section we will learn to calculate the pH for a solution containing a strong acid of known concentration. For example, if we know a solution contains 1.0 M HCl, how can we find the pH of the solution? To answer this question we must know that when HCl dissolves in water, each molecule dissociates into H^+ and Cl^- ions. That is, we must know that HCl is a strong acid. Thus, although the label on the bottle says 1.0 M HCl, the solution contains virtually no HCl molecules. A 1.0 M HCl solution contains H^+ and Cl^- ions rather than HCl molecules. Typically, container labels indicate the substance(s) used to make up the solution but do not necessarily describe the solution components after dissolution. In this case,

$$1.0 \ M \ HCl \rightarrow 1.0 \ M \ H^+ \text{ and } 1.0 \ M \ Cl^-$$

Therefore, the $[H^+]$ in the solution is 1.0 M. The pH is then

$$pH = -\log[H^+] = -\log(1.0) = 0$$

EXAMPLE 17.10	Calculating the pH of Strong Acid Solutions

Calculate the pH of 0.10 M HNO_3.

EXAMPLE 17.10, CONTINUED

SOLUTION

HNO_3 is a strong acid, so the ions in solution are H^+ and NO_3^-. In this case,

$$0.10 \ M \ HNO_3 \rightarrow 0.10 \ M \ H^+ \text{ and } 0.10 \ M \ NO_3^-$$

Thus

$$[H^+] = 0.10 \ M \quad \text{and} \quad pH = -\log(0.10) = 1.00$$

SELF-CHECK EXERCISE 17.7

Calculate the pH of a solution of $5.0 \times 10^{-3} \ M$ HCl.

<hr>

17.6 Buffered Solutions

AIM: To explain the general characteristics of buffered solutions.

A **buffered solution** is one that resists a change in its pH even when a strong acid or base is added to it. For example, when 0.01 mol of HCl is added to 1 L of pure water, the pH changes from its initial value of 7 to 2, a change of 5 pH units. However, when 0.01 mol of HCl is added to a solution containing both 0.1 M acetic acid ($HC_2H_3O_2$) and 0.1 M sodium acetate ($NaC_2H_3O_2$), the pH changes from an initial value of 4.74 to 4.66, a change of only 0.08 pH units. The latter solution is buffered—it undergoes only a very slight change in pH when a strong acid or base is added to it.

Buffered solutions are vitally important to living organisms whose cells can survive only in a very narrow pH range. Many goldfish have died because their owners did not realize the importance of buffering the aquarium water at an appropriate pH. For humans to survive, the pH of the blood must be maintained between 7.35 and 7.45. This narrow range is maintained by several different buffering systems.

A solution is **buffered** by the *presence of a weak acid and its conjugate base.* An example of a buffered solution is an aqueous solution that contains acetic acid and sodium acetate. The sodium acetate is a salt that furnishes acetate ions (the conjugate base of acetic acid) when it dissolves. To see how this system acts as a buffer, we must recognize that the species present in this solution are

$$HC_2H_3O_2, \quad \underbrace{Na^+, \quad C_2H_3O_2^-}$$

When $NaC_2H_3O_2$ is dissolved, it produces the separated ions.

Buffering power can be easily demonstrated by comparing the behavior of water to the behavior of $HC_2H_3O_2/NaC_2H_3O_2$ solution when H^+ or OH^- is added. Add H^+ or OH^- dropwise to water and measure the pH with pH meter or universal indicator. Add the same amount of H^+ or OH^- to a buffered solution and measure the pH in the same way.

Water: pH = 7
0.01 M HCl: pH = 2

For goldfish to survive, the pH of the water must be carefully controlled.

Students can obtain a good understanding of buffer action and composition from the qualitative discussion in the text. To reinforce understanding of buffer composition, students will examine pairs of substances and predict which will form buffers. After they have identified pairs of substances that form buffers, students can write equations describing reaction of the buffers with added H^+ and OH^-.

What happens in this solution when a strong acid such as HCl is added? In pure water, the H^+ ions from the HCl would accumulate, thus lowering the pH.

$$HCl \xrightarrow{100\%} H^+ + Cl^-$$

However, this buffered solution contains $C_2H_3O_2^-$ ions, which are basic. That is, $C_2H_3O_2^-$ has a strong affinity for H^+, as evidenced by the fact that $HC_2H_3O_2$ is a weak acid. This means that the $C_2H_3O_2^-$ and H^+ ions do not exist together in large numbers. Because the $C_2H_3O_2^-$ ion has a high affinity for H^+, these two combine to form $HC_2H_3O_2$ molecules. Thus the H^+ from the added HCl does not accumulate in solution but reacts with the $C_2H_3O_2^-$ as follows:

$$H^+(aq) + C_2H_3O_2^-(aq) \rightarrow HC_2H_3O_2(aq)$$

Next consider what happens when a strong base such as sodium hydroxide is added to the buffered solution. If this base were added to pure water, the OH^- ions from the solid would accumulate and greatly change (raise) the pH.

$$NaOH \xrightarrow{100\%} Na^+ + OH^-$$

However, in the buffered solution the OH^- ion, which has a *very strong* affinity for H^+, reacts with $HC_2H_3O_2$ molecules as follows:

$$HC_2H_3O_2(aq) + OH^-(aq) \rightarrow H_2O(l) + C_2H_3O_2^-(aq)$$

This happens because, although $C_2H_3O_2^-$ has a strong affinity for H^+, OH^- has a much stronger affinity for H^+ and thus can remove H^+ ions from acetic acid molecules.

Note that the buffering materials dissolved in the solution prevent added H^+ or OH^- from building up in the solution. Any added H^+ is trapped by $C_2H_3O_2^-$ to form $HC_2H_3O_2$. Any added OH^- reacts with $HC_2H_3O_2$ to form H_2O and $C_2H_3O_2^-$.

The general properties of a buffered solution are summarized in Table 17.3.

Table 17.3 The Characteristics of a Buffer

1. The solution contains a weak acid HA and its conjugate base A^-.
2. The buffer resists changes in pH by reacting with any added H^+ or OH^- so that these ions do not accumulate.
3. Any added H^+ reacts with the base A^-.
$$H^+(aq) + A^-(aq) \rightarrow HA(aq)$$
4. Any added OH^- reacts with the weak acid HA.
$$OH^-(aq) + HA(aq) \rightarrow H_2O(l) + A^-(aq)$$

CHAPTER REVIEW

Key Terms

acid (p. 580)
base (p. 580)
Arrhenius concept of acids and
  bases (p. 580)
Brønsted–Lowry model (p. 580)
conjugate acid (p. 581)
conjugate base (p. 581)
conjugate acid–base pair (p. 581)
hydronium ion (p. 581)
strong acid (p. 584)
weak acid (p. 585)
diprotic acid (p. 586)

oxyacid (p. 586)
organic acid (p. 586)
carboxyl group (p. 586)
amphoteric substance (p. 586)
ion-product constant, K_w (p. 587)
pH scale (p. 589)
buffered solution (p. 597)

Summary

1. Acids and bases in water are commonly described by two different models. Arrhenius postulated that acids produce H^+ ions in solutions and that bases produce OH^- ions. The Brønsted–Lowry model is more general: an acid is a proton donor, and a base is a proton acceptor. Water acts as a Brønsted–Lowry base when it accepts a proton from an acid to form a hydronium ion:

$$HA(aq) + H_2O(l) \rightleftharpoons H_3O^+(aq) + A^-(aq)$$

Acid   Base   Conjugate   Conjugate
                  acid   base

A conjugate base is everything that remains of the acid molecule after the proton is lost. A conjugate acid is formed when a proton is transferred to the base. Two substances related in this way are called a conjugate acid–base pair.

2. A strong acid or base is one for which the dissociation equilibrium lies far to the right. A weak acid is one for which the dissociation equilibrium lies far to the left. Strong acids have weak conjugate bases. Weak acids have relatively strong conjugate bases.

3. Water is an amphoteric substance—it can behave either as an acid or as a base. The ionization of water reveals this property; one water molecule transfers a proton to another water molecule to produce a hydronium ion and a hydroxide ion.

$$H_2O(l) + H_2O(l) \rightleftharpoons H_3O^+(aq) + OH^-(aq)$$

This leads to the equilibrium expression

$$K_w = [H_3O^+][OH^-] = [H^+][OH^-]$$

where K_w is called the ion-product constant. It has been shown experimentally that at 25 °C,

$$[H^+] = [OH^-] = 1.0 \times 10^{-7} \, M$$

so $K_w = 1.0 \times 10^{-14}$.

4. In an acidic solution, $[H^+]$ is greater than $[OH^-]$. In a basic solution, $[OH^-]$ is greater than $[H^+]$. In a neutral solution, $[H^+] = [OH^-]$.

5. To describe $[H^+]$ in aqueous solutions, we use the pH scale.

$$pH = -\log[H^+]$$

Note that the pH decreases as $[H^+]$ (acidity) increases.

6. The pH of strong acid solutions can be calculated directly from the concentration of the acid, because 100% dissociation occurs in aqueous solution.

7. A buffered solution is one that resists a change in its pH even when a strong acid or base is added to it. A buffered solution contains a weak acid and its conjugate base.

Questions and Problems

All even-numbered exercises have answers in the back of this book and solutions in the Solutions Guide.

17.1 Acids and Bases

QUESTIONS

1. What are some physical properties that historically led chemists to classify various substances as acids and bases?

2. In the Arrhenius definition, what characterizes an acid? What characterizes a base? Why are the Arrhenius definitions too restrictive?

3. What specific ion is the basis for the Brønsted–Lowry model for acid–base reactions? What is an acid in the Brønsted–Lowry model? What is a base?

4. How do the substances in a conjugate acid–base pair differ from one another?

5. How does a Brønsted–Lowry acid form its conjugate base?

6. When an acid is dissolved in water, what ion does the water form? What is the relationship of this ion to water itself?

PROBLEMS

7. Which of the following represent conjugate acid–base pairs? For those pairs that themselves are not conjugates, write the correct conjugate acid or base for each species in the pair.
   a. HCl, Cl^-
   b. Cl^-, ClO^-
   c. NH_3, NH_2^-
   d. HCN, CN^-

8. Which of the following represent conjugate acid–base pairs? For those pairs that themselves are not conjugates, write the correct conjugate acid or base for each species in the pair.
   a. HSO_4^-, SO_4^{2-}
   b. HBr, BrO^-
   c. $H_2PO_4^-$, PO_4^{3-}
   d. HNO_3, NO_2^-

9. In each of the following chemical equations, identify the conjugate acid–base pairs.
   a. $HNO_3 + H_2O \rightleftharpoons H_3O^+ + NO_3^-$
   b. $H_2SO_4 + H_2O \rightleftharpoons HSO_4^- + H_3O^+$
   c. $NH_4^+ + H_2O \rightleftharpoons H_3O^+ + NH_3$

11. Write the conjugate *acid* for each of the following:
   a. Cl^-
   b. HSO_4^-
   c. NH_2^-
   d. CN^-

13. Write the conjugate *base* for each of the following:
   a. HSO_4^-
   b. NH_3
   c. $H_2PO_4^-$
   d. $HC_2O_4^{2-}$

15. Write chemical equations showing the ionization in water for each of the following acids.
   a. HCl
   b. $HClO_4$
   c. HNO_3
   d. HCN

10. In each of the following chemical equations, identify the conjugate acid–base pairs.
   a. $HSO_4^- + H_2O \rightleftharpoons SO_4^{2-} + H_3O^+$
   b. $CH_3NH_2 + HCl \rightleftharpoons CH_3NH_3^+ + Cl^-$
   c. $CH_3COO^- + H_2O \rightleftharpoons CH_3COOH + OH^-$

12. Write the conjugate *acid* for each of the following:
   a. SO_4^{2-}
   b. PO_4^{3-}
   c. CH_3NH_2
   d. F^-

14. Write the conjugate *base* for each of the following:
   a. HNO_3
   b. HPO_4^{2-}
   c. H_3O^+
   d. H_2CO_3

16. Write chemical equations showing the ionization in water of each of the following acids.
   a. $HBrO_4$
   b. HF
   c. $CH_3NH_3^+$
   d. $H_2PO_4^-$

17.2 Acid Strength

QUESTIONS

17. What does it mean to say that an acid is *strong* in aqueous solution? What does this reveal about the ability of the acid's anion to attract protons?

19. How is the strength of an acid related to the fact that a competition for protons exists in aqueous solution between water molecules and the anion of the acid?

21. For a weak acid dissolved in water, where does the dissociation equilibrium lie?

23. What group of atoms characterizes organic acids? Are most organic acids strong or weak?

25. Which of the following acids have relatively *weak* conjugate bases?
   a. HF
   b. HCl
   c. H_2SO_4
   d. CH_3COOH

18. What does it mean to say that an acid is *weak* in aqueous solution? What does this reveal about the ability of the acid's anion to attract protons?

20. A strong acid has a weak conjugate base, whereas a weak acid has a relatively strong conjugate base. Explain.

22. Name four strong acids. For each of these, write the equation showing the acid dissociating in water.

24. Most acids are oxyacids, in which the acidic proton is bonded to an oxygen atom. Write the formulas of three acids that are oxyacids. Write the formulas of three acids that are *not* oxyacids.

26. Which of the following bases have relatively *strong* conjugate acids?
   a. SO_4^{2-}
   b. Br^-
   c. CN^-
   d. CH_3COO^-

17.3 Water as an Acid and a Base

QUESTIONS

27. What does it mean to say that a substance is *amphoteric*? Write an equation to show that water is an amphoteric substance.

29. What happens to the hydroxide ion concentration when we increase the hydrogen ion concentration of water by adding an acid? Explain.

28. Write the expression for the ion product constant, K_w, for water. What does the value of K_w tell us about the extent of dissociation of pure water?

30. How do we characterize solutions as acidic, basic, or neutral in terms of the relative concentrations of hydrogen ion and hydroxide ion?

31. Calculate $[H^+]$ in each of the following solutions, and indicate whether the solution is acidic, basic, or neutral.
    a. $[OH^-] = 1.41 \times 10^{-6} M$
    b. $[OH^-] = 9.35 \times 10^{-9} M$
    c. $[OH^-] = 0.0221 M$
    d. $[OH^-] = 4.33 \times 10^{-3} M$

32. Calculate $[H^+]$ in each of the following solutions, and indicate whether the solution is acidic, basic, or neutral.
    a. $[OH^-] = 3.99 \times 10^{-5} M$
    b. $[OH^-] = 2.91 \times 10^{-9} M$
    c. $[OH^-] = 7.23 \times 10^{-2} M$
    d. $[OH^-] = 9.11 \times 10^{-7} M$

33. Calculate $[OH^-]$ in each of the following solutions, and indicate whether the solution is acidic, basic, or neutral.
    a. $[H^+] = 6.34 \times 10^{-8} M$
    b. $[H^+] = 7.0 \times 10^{-1} M$
    c. $[H^+] = 9.21 \times 10^{-3} M$
    d. $[H^+] = 0.0010 M$

34. Calculate $[OH^-]$ in each of the following solutions, and indicate whether the solution is acidic, basic, or neutral.
    a. $[H^+] = 1.27 \times 10^{-3} M$
    b. $[H^+] = 4.49 \times 10^{-8} M$
    c. $[H^+] = 3.87 \times 10^{-5} M$
    d. $[H^+] = 8.16 \times 10^{-11} M$

35. For each pair of concentrations, tell which represents the more acidic solution.
    a. $[H^+] = 1.2 \times 10^{-3} M$ or $[H^+] = 4.5 \times 10^{-4} M$
    b. $[H^+] = 2.6 \times 10^{-6} M$ or $[H^+] = 4.3 \times 10^{-8} M$
    c. $[H^+] = 0.000010 M$ or $[H^+] = 0.0000010 M$

36. For each pair of concentrations, tell which represents the more basic solution.
    a. $[H^+] = 1.21 \times 10^{-8} M$ or $[OH^-] = 0.10 M$
    b. $[H^+] = 0.00100 M$ or $[OH^-] = 1.78 \times 10^{-13} M$
    c. $[H^+] = 4.23 \times 10^{-6} M$ or $[H^+] = 1.15 \times 10^{-7} M$

17.4 The pH Scale

37. Why do scientists tend to express the acidity of a solution in terms of its pH, rather than in terms of the molarity of hydrogen ion present? How is pH defined mathematically?

38. Using Figure 17.3, list the approximate pH of five "everyday" solutions. How do the familiar properties (such as sour taste for acids) of these solutions correspond to their pH?

39. For a hydrogen ion concentration of $2.33 \times 10^{-6} M$, how many *decimal places* should we give when expressing the pH of the solution?

40. As the hydrogen ion concentration of a solution *increases*, does the pH of the solution increase or decrease? Explain.

41. Calculate the pH corresponding to each of the hydrogen ion concentrations given. Tell whether each solution is acidic, basic, or neutral.
    a. $[H^+] = 6.34 \times 10^{-8} M$
    b. $[H^+] = 7.0 \times 10^{-1} M$
    c. $[H^+] = 9.21 \times 10^{-3} M$
    d. $[H^+] = 1.00 \times 10^{-7} M$

42. Calculate the pH corresponding to each of the hydrogen ion concentrations given. Tell whether each solution is acidic, basic, or neutral.
    a. $[H^+] = 0.00100 M$
    b. $[H^+] = 2.19 \times 10^{-4} M$
    c. $[H^+] = 9.18 \times 10^{-11} M$
    d. $[H^+] = 4.71 \times 10^{-7} M$

43. Calculate the pH corresponding to each of the hydroxide ion concentrations given. Tell whether each solution is acidic, basic, or neutral.
    a. $[OH^-] = 4.22 \times 10^{-3} M$
    b. $[OH^-] = 1.01 \times 10^{-13} M$
    c. $[OH^-] = 1.0 \times 10^{-7} M$
    d. $[OH^-] = 3.98 \times 10^{-10} M$

44. Calculate the pH corresponding to each of the hydroxide ion concentrations given. Tell whether each solution is acidic, basic, or neutral.
    a. $[OH^-] = 4.22 \times 10^{-3} M$
    b. $[OH^-] = 1.99 \times 10^{-7} M$
    c. $[OH^-] = 0.000100 M$
    d. $[OH^-] = 2.47 \times 10^{-11} M$

45. Calculate the pH corresponding to each of the pOH values listed, and indicate whether each solution is acidic, basic, or neutral.
    a. pOH = 4.32   c. pOH = 1.81
    b. pOH = 8.90   d. pOH = 13.1

46. Calculate the pOH corresponding to each of the pH values listed, and indicate whether each solution is acidic, basic, or neutral.
    a. pH = 6.49   c. pH = 11.21
    b. pH = 1.93   d. pH = 7.00

47. Given each hydrogen or hydroxide ion concentration listed, calculate the concentration of the complementary ion and the pH and pOH of the solution.
    a. $[H^+] = 3.22 \times 10^{-6}\ M$
    b. $[OH^-] = 4.90 \times 10^{-8}\ M$
    c. $[H^+] = 0.00015\ M$
    d. $[OH^-] = 1.04 \times 10^{-12}\ M$

48. Given each hydrogen or hydroxide ion concentration listed, calculate the concentration of the complementary ion and the pH and pOH of the solution.
    a. $[H^+] = 1.00 \times 10^{-7}\ M$
    b. $[OH^-] = 4.39 \times 10^{-5}\ M$
    c. $[H^+] = 4.29 \times 10^{-11}\ M$
    d. $[OH^-] = 7.36 \times 10^{-2}\ M$

49. Calculate the hydrogen ion concentration, in moles per liter, for solutions with each of the following pH values.
    a. pH = 4.32   c. pH = 1.02
    b. pH = 5.93   d. pH = 13.1

50. Calculate the hydrogen ion concentration, in moles per liter, for solutions with each of the following pH values.
    a. pH = 4.11   c. pH = 7.00
    b. pH = 9.54   d. pH = 13.21

51. Calculate the hydrogen ion concentration, in moles per liter, for solutions with each of the following pOH values.
    a. pOH = 4.95   c. pOH = 12.94
    b. pOH = 7.00   d. pOH = 1.02

52. Calculate the hydrogen ion concentration, in moles per liter, for solutions with each of the following pOH values.
    a. pOH = 3.91   c. pOH = 1.15
    b. pOH = 12.56   d. pOH = 8.77

53. Calculate the hydrogen ion concentration, in moles per liter, for solutions with each of the following pH or pOH values.
    a. pH = 3.90   c. pH = 9.21
    b. pOH = 3.90   d. pOH = 12.9

54. Calculate the hydrogen ion concentration, in moles per liter, for solutions with each of the following pH or pOH values.
    a. pH = 4.38   c. pH = 7.21
    b. pOH = 4.38   d. pOH = 5.39

17.5 Calculating the pH of Strong Acid Solutions

QUESTIONS

55. When 1 mol of gaseous hydrogen chloride is dissolved in enough water to make 1 L of solution, approximately how many HCl molecules remain in the solution? Explain.

56. A bottle of acid solution is labeled "3 M HNO$_3$." What are the substances that are actually present in the solution? Are any HNO$_3$ molecules present? Why or why not?

PROBLEMS

57. Calculate the hydrogen ion concentration and the pH of each of the following solutions of strong acids.
    a. 0.010 M HCl
    b. 0.050 M HNO$_3$
    c. 0.00025 M HClO$_4$
    d. $3.2 \times 10^{-4}\ M$ HCl

58. Calculate the hydrogen ion concentration and the pH of each of the following solutions of strong acids.
    a. 0.00010 M HCl
    b. 0.0050 M HNO$_3$
    c. $4.21 \times 10^{-5}\ M$ HClO$_4$
    d. $6.33 \times 10^{-3}\ M$ HNO$_3$

17.6 Buffered Solutions

QUESTIONS

59. What characteristic properties do buffered solutions possess?

60. What two components make up a buffered solution? Give an example of a combination that would serve as a buffered solution.

61. Which component of a buffered solution is capable of combining with an added strong acid? Using your example from Question 60, show how this component would react with added HCl.

PROBLEMS

63. Which of the following combinations would act as buffered solutions?
    a. $CH_3COOH(HC_2H_3O_2)$ and NaCl
    b. $CH_3COOH(NaC_2H_3O_2)$ and $NaCH_3COO$
    c. $HClO_4$ and $KClO_3$
    d. HNO_2 and $NaNO_2$

Additional Problems

65. The concepts of acid–base equilibria were developed in this chapter for aqueous solutions (in aqueous solutions, water is the solvent and is intimately involved in the equilibria). However, the Brønsted–Lowry acid–base theory can be extended easily to other solvents. One such solvent that has been investigated in depth is liquid ammonia, NH_3.
    a. Write a chemical equation indicating how HCl behaves as an acid in liquid ammonia.
    b. Write a chemical equation indicating how OH^- behaves as a base in liquid ammonia.

66. *Strong bases* are bases that completely ionize in water to produce hydroxide ion, OH^-. The strong bases include the hydroxides of the Group 1 elements. For example, if 1.0 mol of NaOH is dissolved per liter, the concentration of OH^- ion is 1.0 *M*. Calculate $[OH^-]$, pOH, and pH for each of the following strong base solutions.
    a. 0.10 *M* NaOH
    b. 2.0×10^{-4} *M* KOH
    c. 6.2×10^{-3} *M* CsOH
    d. 0.0001 *M* NaOH

67. Which of the following conditions indicate an *acidic* solution?
    a. pH < 7.0   c. $[H^+] > [OH^-]$
    b. pOH < 7.0   d. $[H^+] > 1.0 \times 10^{-7}$ *M*

68. Which of the following conditions indicate a *basic* solution?
    a. pOH < 7.0   c. $[OH^-] < [H^+]$
    b. pH > 7.0   d. $[H^+] < 1.0 \times 10^{-7}$ *M*

69. Buffered solutions are mixtures of a weak acid and its conjugate base. Explain why a mixture of a *strong* acid and its conjugate base (such as HCl and Cl^-) is not buffered.

70. Which of the following acids are classified as *strong* acids?
    a. HNO_3 c. HCl   e. $HClO_4$
    b. CH_3COOH d. HF

62. Which component of a buffered solution consumes added strong base? Using your example from Question 60, show how this component would react with added NaOH.

64. For those combinations in Question 63 that behave as buffered solutions, write equations showing how the components of the buffer consume added strong acid (HCl) or strong base (NaOH).

71. Is it possible for a solution to have $[H^+] = 0.002$ *M* and $[OH^-] = 5.2 \times 10^{-6}$ *M* at 25 °C? Explain.

72. In spite of HCl being a strong acid, the pH of a 1.0×10^{-7} *M* HCl solution is *not* exactly 7.00. On the basis of the equilibria involved, suggest a reason why.

73. According to Arrhenius, bases are species that produce _____ ion in aqueous solution.

74. According to the Brønsted–Lowry theory, a base is a species that _____ protons.

75. A conjugate acid–base pair consists of two substances related by the donating and accepting of a _____.

76. Acetate ion, $C_2H_3O_2^-$, has a stronger affinity for protons than does water. Therefore, when dissolved in water, acetate ion behaves as a _____.

77. An acid such as HCl that strongly conducts an electric current when dissolved in water is said to be a _____ acid.

78. Organic acids, which behave as typical weak acids, contain the _____ group.

79. Because of _____, even pure water contains measurable quantities of H^+ and OH^-.

80. The ion-product constant for water, K_w, has the value _____ at 25 °C.

81. The number of _____ in the logarithm of a number is equal to the number of significant figures in the number.

82. A solution with pH = 2 has a _____ hydrogen ion concentration than a solution with pH = 4.

83. A 0.20 *M* HCl solution contains _____ *M* hydrogen ion and _____ *M* chloride ion concentrations.

84. A buffered solution is one that resists a change in _____ when either a strong acid or a strong base is added to it.

85. A solution is buffered when a weak acid and its _____ are present in comparable amounts.

86. When sodium hydroxide, NaOH, is added dropwise to a buffered solution, the _____ component of the buffer consumes the added hydroxide ion.

87. When hydrochloric acid, HCl, is added dropwise to a buffered solution, the _____ component of the buffer consumes the added hydrogen ion.

88. Which of the following represent conjugate acid–base pairs? For those pairs that themselves are not conjugates, write the correct conjugate acid or base for each species in the pair.
    a. H_2O, OH^-
    b. H_2SO_4, SO_4^{2-}
    c. H_3PO_4, $H_2PO_4^-$
    d. $HC_2H_3O_2$, $C_2H_3O_2^-$

89. In each of the following chemical equations, identify the conjugate acid–base pairs.
    a. $CH_3NH_2 + H_2O \rightleftharpoons CH_3NH_3^+ + OH^-$
    b. $CH_3COOH + NH_3 \rightleftharpoons CH_3COO^- + NH_4^+$
    c. $HF + NH_3 \rightleftharpoons F^- + NH_4^+$

90. Write the conjugate *acid* for each of the following:
    a. NH_3
    b. CH_3COO^-
    c. $H_2PO_4^-$
    d. $HC_2O_4^-$

91. Write the conjugate *base* for each of the following:
    a. H_3PO_4
    b. HCO_3^-
    c. HF
    d. H_2SO_4

92. Write chemical equations showing the ionization in water for each of the following acids.
    a. CH_3CH_2COOH (Only the last H is acidic.)
    b. NH_4^+
    c. H_2SO_4
    d. H_3PO_4

93. Which of the following bases have relatively *strong* conjugate acids?
    a. F^-
    b. Cl^-
    c. HSO_4^-
    d. NO_3^-

94. Calculate $[H^+]$ in each of the following solutions, and indicate whether the solution is acidic, basic, or neutral.
    a. $[OH^-] = 4.22 \times 10^{-3}$ M
    b. $[OH^-] = 1.01 \times 10^{-13}$ M
    c. $[OH^-] = 3.05 \times 10^{-7}$ M
    d. $[OH^-] = 6.02 \times 10^{-6}$ M

95. Calculate $[OH^-]$ in each of the following solutions, and indicate whether the solution is acidic, basic, or neutral.
    a. $[H^+] = 4.21 \times 10^{-7}$ M
    b. $[H^+] = 0.00035$ M
    c. $[H^+] = 0.00000010$ M
    d. $[H^+] = 9.9 \times 10^{-6}$ M

96. For each pair of concentrations, tell which represents the more basic solution.
    a. $[H^+] = 0.000013$ M or $[OH^-] = 0.0000032$ M
    b. $[H^+] = 1.03 \times 10^{-6}$ M or $[OH^-] = 1.54 \times 10^{-8}$ M
    c. $[OH^-] = 4.02 \times 10^{-7}$ M or $[OH^-] = 0.0000001$ M

97. Calculate the pH corresponding to each of the hydrogen ion concentrations given. Tell whether each solution is acidic, basic, or neutral.

a. $[H^+] = 4.21 \times 10^{-7}$ M
b. $[H^+] = 0.00035$ M
c. $[H^+] = 0.00000010$ M
d. $[H^+] = 1.21 \times 10^{-3}$ M

98. Calculate the pH corresponding to each of the hydroxide ion concentrations given. Tell whether each solution is acidic, basic, or neutral.
    a. $[OH^-] = 1.4 \times 10^{-6}$ M
    b. $[OH^-] = 9.35 \times 10^{-9}$ M
    c. $[OH^-] = 2.21 \times 10^{-1}$ M
    d. $[OH^-] = 7.98 \times 10^{-12}$ M

99. Calculate the pOH corresponding to each of the pH values listed, and indicate whether each solution is acidic, basic, or neutral.
    a. pH = 1.02
    b. pH = 13.4
    c. pH = 9.03
    d. pH = 7.20

100. Given each hydrogen or hydroxide ion concentration listed, calculate the concentration of the complementary ion and the pH and pOH of the solution.
    a. $[H^+] = 5.72 \times 10^{-4}$ M
    b. $[OH^-] = 8.91 \times 10^{-5}$ M
    c. $[H^+] = 2.87 \times 10^{-12}$ M
    d. $[OH^-] = 7.22 \times 10^{-8}$ M

101. Calculate the hydrogen ion concentration, in moles per liter, for solutions with each of the following pH values.
    a. pH = 8.34
    b. pH = 5.90
    c. pH = 2.65
    d. pH = 12.6

102. Calculate the hydrogen ion concentration, in moles per liter, for solutions with each of the following pOH values.
    a. pOH = 7.95
    b. pOH = 14.00
    c. pOH = 5.00
    d. pOH = 3.95

103. Calculate the hydrogen ion concentration, in moles per liter, for solutions with each of the following pH or pOH values.
    a. pOH = 0.90
    b. pH = 0.90
    c. pOH = 10.3
    d. pH = 5.33

104. Calculate the hydrogen ion concentration and the pH of each of the following solutions of strong acids.
    a. 1.4×10^{-3} M $HClO_4$
    b. 3.0×10^{-5} M HCl
    c. 5.0×10^{-2} M HNO_3
    d. 0.0010 M HCl

105. Which of the following combinations would act as buffered solutions?
    a. HF and NaF
    b. HCN and KCN
    c. HCl and NaCl
    d. HNO_3 and KNO_3

Solutions to Self-Check Exercises

SELF-CHECK EXERCISE 17.1

The conjugate acid–base pairs are

$$H_2O, \qquad H_3O^+$$

Base   Conjugate acid

and

$$HC_2H_3O_2, \quad C_2H_3O_2^-$$

Acid   Conjugate base

The members of both pairs differ by one H^+.

SELF-CHECK EXERCISE 17.2

Because $[H^+][OH^-] = 1.0 \times 10^{-14}$, we can solve for $[H^+]$.

$$[H^+] = \frac{1.0 \times 10^{-14}}{[OH^-]} = \frac{1.0 \times 10^{-14}}{2.0 \times 10^{-2}} = 5.0 \times 10^{-13} \ M$$

This solution is basic: $[OH^-] = 2.0 \times 10^{-2}$ is greater than $[H^+] = 5.0 \times 10^{-13} \ M$.

SELF-CHECK EXERCISE 17.3

a. Because $[H^+] = 1.0 \times 10^{-3} \ M$, we get pH $= 3.00$ by using the regular steps.

b. Because $[OH^-] = 5.0 \times 10^{-5} \ M$, we can find $[H^+]$ from the K_w expression.

$$[H^+] = \frac{K_w}{[OH^-]} = \frac{1.0 \times 10^{-14}}{5.0 \times 10^{-5}} = 2.0 \times 10^{-10} \ M$$

Then we follow the regular steps to get pH from $[H^+]$.

1. Enter 2.0×10^{-10}.

2. Push $\boxed{\text{log}}$.

3. Push $\boxed{+/-}$.

$$\text{pH} = 9.70$$

SELF-CHECK EXERCISE 17.4

$$\text{pOH} + \text{pH} = 14.00$$
$$\text{pOH} = 14.00 - \text{pH} = 14.00 - 3.5$$
$$\text{pOH} = 10.5$$

SELF-CHECK EXERCISE 17.5

STEP I: $pH = 3.50$

STEP 2: $-pH = -3.50$

STEP 3:   [inv] [log] $-3.50 = 3.2 \times 10^{-4}$

$$[H^+] = 3.2 \times 10^{-4}$$

SELF-CHECK EXERCISE 17.6

STEP I: $pOH = 10.50$

STEP 2: $-pOH = -10.50$

STEP 3:   [inv] [log] $-10.50 = 3.2 \times 10^{-11}$

$$[OH^-] = 3.2 \times 10^{-11} \, M$$

SELF-CHECK EXERCISE 17.7

Because HCl is a strong acid, it is completely dissociated:

$$5.0 \times 10^{-3} \, M \, HCl \rightarrow 5.0 \times 10^{-3} \, M \, H^+ \text{ and } 5.0 \times 10^{-3} \, M \, Cl^-$$

so $[H^+] = 5.0 \times 10^{-3} \, M.$

$$pH = -\log(5.0 \times 10^{-3}) = 2.30$$

Oxidation–Reduction Reactions and Electrochemistry

CONTENTS

This chapter presents a more detailed discussion of oxidation–reduction chemistry, which was introduced in Chapter 8. The students will understand this material better if they review Section 8.1. You might want to spend some time in class reviewing loss and gain of electrons and how to determine the number of electrons a neutral atom gains or loses when it becomes an ion.

An oxidation–reduction reaction at work.

By discussing the economics of oxidation–reduction reactions, you can get the attention of students, many of whom are interested in keeping their automobiles from rusting or in the development of lightweight batteries that are easy to recharge.

What do a forest fire, rusting steel, combustion in an automobile engine, and the metabolism of food in a human body have in common? All of these important processes involve oxidation–reduction reactions. In fact, virtually all of the processes that provide energy to heat buildings, power vehicles, and allow people to work and play depend on oxidation–reduction reactions. And every time you start your car, turn on your calculator, look at your digital watch, or listen to a radio at the beach, you are depending on an oxidation–reduction reaction to power the battery in each of these devices.

In this chapter we will explore the properties of oxidation–reduction reactions, and we will see how these reactions are used to power batteries.

18.1   Oxidation–Reduction Reactions

AIM: To characterize metal–nonmetal oxidation–reduction reactions.

In Section 8.1 we discussed the chemical reactions between metals and nonmetals. For example, sodium chloride is formed by the reaction of elemental sodium and chlorine.

$$2Na(s) + Cl_2(g) \rightarrow 2NaCl(s)$$

Because elemental sodium and chlorine contain uncharged atoms and because sodium chloride is known to contain Na^+ and Cl^- ions, this reaction must involve a transfer of electrons from sodium atoms to chlorine atoms.

A raging forest fire and the power generated by an alkaline AA battery and a mercury battery result from examples of oxidation–reduction reactions.

Reactions like this one, in which one or more electrons are transferred, are called **oxidation–reduction reactions,** *or* **redox reactions. Oxidation** is defined as a *loss of electrons.* **Reduction** is defined as a *gain of electrons.* In the reaction of elemental sodium and chlorine, each sodium atom loses one electron, forming a 1+ ion. Therefore sodium is oxidized. Each chlorine atom gains one electron, forming a negative chloride ion, and is thus reduced. Whenever a metal reacts with a nonmetal to form an ionic compound, electrons are transferred from the metal to the nonmetal. So these reactions are always oxidation–reduction reactions where the metal is oxidized (loses electrons) and the nonmetal is reduced (gains electrons).

Some students use the mnemonic OIL RIG:

Oxidation Is Loss;
Reduction Is Gain.

You should remind students that oxidation does not occur unless it is coupled with reduction.

An introduction to the terms oxidation and reduction is presented here. Make use of one of the mnemonic devices "LEO (Loss Equals Oxidation) the lion says GER (Gain Equals Reduction)" or "OIL RIG." These aids may sound silly, but they are very useful.

EXAMPLE 18.1   Identifying Oxidation and Reduction in a Reaction

In the following reactions, identify which element is oxidized and which element is reduced.

a. $2Mg(s) + O_2(g) \rightarrow 2MgO(s)$ b. $2Al(s) + 3I_2(s) \rightarrow 2AlI_3(s)$

SOLUTION

a.   We have learned that Group 2 metals form 2+ cations and that Group 6 nonmetals form 2− anions, so we can predict that magnesium oxide contains Mg^{2+} and O^{2-} ions. This means that in the reaction given, each Mg loses two electrons to form Mg^{2+} and so is oxidized. Also each O gains two electrons to form O^{2-} and so is reduced.

b.   Aluminum iodide contains the Al^{3+} and I^- ions. Thus aluminum atoms lose electrons (are oxidized). Iodine atoms gain electrons (are reduced).

SELF-CHECK EXERCISE 18.1

For the following reactions, identify the element oxidized and the element reduced.

a. $2Cu(s) + O_2(g) \rightarrow 2CuO(s)$ b. $2Cs(s) + F_2(g) \rightarrow 2CsF(s)$

Magnesium burns in oxygen to produce a bright, white flame.

Although we can identify reactions between metals and nonmetals as redox reactions, it is more difficult to decide whether a given reaction between nonmetals is a redox reaction. In fact, many of the most significant redox reactions involve only nonmetals. For example, combustion reactions such as methane burning in oxygen:

$$CH_4(g) + 2O_2(g) \rightarrow CO_2(g) + 2H_2O(g) + energy$$

are oxidation–reduction reactions. Even though none of the reactants or products in this reaction is ionic, the reaction does involve a transfer of electrons from carbon to oxygen. To explain this, we must introduce the concept of oxidation states.

18.2   Oxidation States

AIM: To describe how to assign oxidation states.

The concept of **oxidation states** (sometimes called *oxidation numbers*) lets us keep track of electrons in oxidation–reduction reactions by assigning charges to

When you introduce oxidation states, it helps to discuss the oxidation states of ions before discussing those of covalently bonded molecules. The transfer of electrons between ions establishes the concept of assigning electrons in a molecule to a particular atom.

the various atoms in a compound. Sometimes these charges are quite apparent. For example, in a binary ionic compound the ions have easily identified charges: in sodium chloride, sodium is $+1$ and chlorine is -1; in magnesium oxide, magnesium is $+2$ and oxygen is -2; and so on. In such binary ionic compounds the oxidation states are simply the charges of the ions.

Ion	Oxidation State
Na^+	$+1$
Cl^-	-1
Mg^{2+}	$+2$
O^{2-}	-2

In a pure element, all of the atoms are uncharged (neutral). For example, sodium metal contains neutral sodium atoms, and chlorine gas is made up of Cl_2 molecules each of which contains two neutral chlorine atoms. Therefore an atom in a pure element has no charge and is assigned an oxidation state of zero.

In a covalent compound such as water, although no ions are actually present, chemists find it useful to assign imaginary charges to the elements in the compound. The oxidation states of the elements in these compounds are equal to the imaginary charges we determine by assuming that the most electronegative atom (see Section 12.2) in a bond controls or possesses *both* of the shared electrons. For example, in the O—H bonds in water, it is assumed for purposes of assigning oxidation states that the much more electronegative oxygen atom controls both of the shared electrons in each bond. This gives the oxygen eight valence electrons.

In effect we say that each hydrogen has lost its single electron to the oxygen. This gives each hydrogen an oxidation state of $+1$ and the oxygen an oxidation state of -2 (the oxygen atom has formally gained two electrons). In virtually all covalent compounds, oxygen is assigned an oxidation state of -2 and hydrogen is assigned an oxidation state of $+1$.

Because fluorine is so electronegative, it is always assumed to control any shared electrons. So fluorine is always assumed to have a complete octet of electrons and is assigned an oxidation state of -1. That is, for purposes of assigning oxidation states, fluorine is always imagined to be F^- in its covalent compounds.

The most electronegative elements are F, O, N, and Cl. In general, we give each of these elements an oxidation state equal to its charge as an anion (fluorine is -1, chlorine is -1, oxygen is -2, and nitrogen is -3). When two of these elements are found in the same compound, we assign them in order of electronegativity, starting with the element that has the largest electronegativity.

$$F \; > \; O \; > \; N \; > \; Cl$$

Greatest   Least
electronegativity   electronegativity

For example, in the compound NO_2, because oxygen has a greater electronegativity than nitrogen, we assign each oxygen an oxidation state of -2. This gives a total "charge" of -4 (2×-2) on the two oxygen atoms. Because the NO_2 molecule has zero overall charge, the N must be $+4$ to exactly balance the -4 on the oxygens. In NO_2, then, the oxidation state of *each* oxygen is -2 and the oxidation state of the nitrogen is $+4$.

The rules for assigning oxidation states are given below and are illustrated in Table 18.1 (p. 614). Application of these rules allows us to assign oxidation states in most compounds. The principles are illustrated by Example 18.2.

Rules for Assigning Oxidation States

1. The oxidation state of an atom in a pure element is 0.

2. The oxidation state of a monatomic ion is the same as its charge.

3. Oxygen is assigned an oxidation state of -2 in most of its covalent compounds. Important exception: peroxides (compounds containing the $O_2{}^{2-}$ group), in which each oxygen is assigned an oxidation state of -1.

4. In its covalent compounds with nonmetals, hydrogen is assigned an oxidation state of $+1$.

5. In binary compounds, the element with the greater electronegativity is assigned a negative oxidation state equal to its charge as an anion in its ionic compounds.

6. For an electrically neutral compound, the sum of the oxidation states must be zero.

7. For an ionic species, the sum of the oxidation states must equal the overall charge.

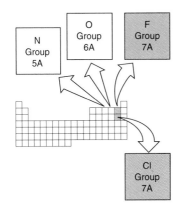

As you show how to determine the oxidation states of elements in ions and compounds, put the rules for determining the oxidation states on an overhead transparency. Students can see which rules you used and how you applied them. They will need to memorize the rules on this page.

Students are always concerned with exceptions to the rules. Show examples of the exceptions for oxygen and hydrogen, such as H_2O_2 and NaH, so that students can see what they look like.

Students should spend time studying Table 18.1 to make sure they understand how oxidation states for each compound were assigned.

Table 18.1 Examples of Oxidation States		
Substance	Oxidation States	Comments
sodium metal, Na	Na 0	rule 1
phosphorus, P	P 0	rule 1
sodium fluoride, NaF	Na +1 F −1	rule 2 rule 2
magnesium sulfide, MgS	Mg +2 S −2	rule 2 rule 2
carbon monoxide, CO	C +2 O −2	 rule 3
sulfur dioxide, SO_2	S +4 O −2	 rule 3
hydrogen peroxide, H_2O_2	H +1 O −1	 rule 3 (exception)
ammonia, NH_3	H +1 N −3	rule 4 rule 5
hydrogen sulfide, H_2S	H +1 S −2	rule 4 rule 5
hydrogen iodide, HI	H +1 I −1	rule 4 rule 5
sodium carbonate, Na_2CO_3	Na +1 O −2 C +4	rule 2 rule 3 For CO_3^{2-}, the sum of the oxidation states is $+4 + 3(-2) = -2$. rule 7
ammonium chloride, NH_4Cl	N −3 H +1 Cl −1	rule 5 rule 4 For NH_4^+, the sum of the oxidation states is $-3 + 4(+1) = +1$. rule 7 rule 2

EXAMPLE 18.2   **Assigning Oxidation States**

Assign oxidation states to all atoms in the following molecules or ions.

a. CO_2 b. SF_6 c. NO_3^-

SOLUTION

a.   Rule 3 takes precedence here: oxygen is assigned an oxidation state of
     -2. We determine the oxidation state for carbon by recognizing that
     because CO_2 has no charge, the sum of the oxidation states for oxygen
     and carbon must be 0 (rule 6). Each oxygen is -2 and there are two
     oxygen atoms, so the carbon atom must be assigned an oxidation state
     of $+4$.

CHECK: $+4 + 2(-2) = 0$

b.   Because fluorine has the greater electronegativity, we assign its
     oxidation state first. Its charge as an anion is always -1, so we assign
     -1 as the oxidation state of each fluorine atom (rule 5). The sulfur
     must then be assigned an oxidation state of $+6$ to balance the total of
     -6 from the six fluorine atoms (rule 7).

CHECK: $+6 + 6(-1) = 0$

c.   Oxygen has a greater electronegativity than nitrogen, so we assign its
     oxidation state of -2 first (rule 5). Because the overall charge on NO_3^-
     is -1 and because the sum of the oxidation states of the three oxygens
     is -6, the nitrogen must have an oxidation state of $+5$.

$$NO_3^-$$

$+5$ -2 for *each* oxygen gives -6 total

CHECK: $+5 + 3(-2) = -1$

This is correct; NO_3^- has a -1 charge.

SELF-CHECK EXERCISE 18.2

Assign oxidation states to all atoms in the following molecules or ions.

a. SO_3 c. N_2O_5 e. C_2H_6
b. SO_4^{2-} d. PF_3

18.3 Oxidation–Reduction Reactions Between Nonmetals

AIM: To define *oxidation* and *reduction* in terms of oxidation states.

To identify oxidizing and reducing agents.

We have seen that oxidation–reduction reactions are characterized by a transfer of electrons. In some cases, the transfer literally occurs to form ions, such as in the reaction

$$2Na(s) + Cl_2(g) \rightarrow 2NaCl(s)$$

We can use oxidation states to verify that electron transfer has occurred.

$$2Na(s) + Cl_2(g) \rightarrow 2NaCl(s)$$

Oxidation state: 0   0   +1   −1
(element) (element) (Na⁺)(Cl⁻)

Thus in this reaction, we represent the electron transfer as follows:

Students can now predict whether or not reactions are oxidation–reduction reactions by using changes in oxidation state. Previously, students relied on the presence of molecular oxygen to help identify combustion reactions.

In other cases the electron transfer occurs in a different sense, such as in the combustion of methane (the oxidation state for each atom is given below each reactant and product).

$$CH_4(g) + 2O_2(g) \rightarrow CO_2(g) + 2H_2O(g)$$

Oxidation state: −4 +1 (each H)   0   +4 −2 (each O) +1 (each H) −2

Note that the oxidation state of oxygen in O_2 is 0 because the oxygen is in elemental form. In this reaction there are no ionic compounds, but we can still describe the process in terms of the transfer of electrons. Note that carbon undergoes a change in oxidation state from −4 in CH_4 to +4 in CO_2. Such a change can be accounted for by a loss of eight electrons:

C (in CH_4)   [Loss of 8e⁻ ➡]   C (in CO_2)

−4   +4

or, in equation form,

$$CH_4 \rightarrow CO_2 + 8e^-$$

−4   +4

Oxidation
Loss of electrons

Reduction
Gain of electrons

Oxidizing Agent

The reactant that:
   accepts electrons
   contains the element reduced

Reducing Agent

The reactant that:
   furnishes electrons
   contains the element oxidized

In a redox reaction, an oxidizing agent is reduced (gains electrons) and a reducing agent is oxidized (loses electrons).

Note that when the oxidizing or reducing agent is named, the *whole compound* is specified, not just the element that undergoes the change in oxidation state.

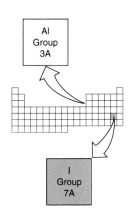

EXAMPLE 18.3   **Identifying Oxidizing and Reducing Agents, I**

When powdered aluminum metal is mixed with pulverized iodine crystals and a drop of water is added, the resulting reaction produces a great deal of energy. The mixture bursts into flames, and a purple smoke of I_2 vapor is produced from the excess iodine. The equation for the reaction is

$$Al(s) + 3I_2(s) \rightarrow 2AlI_3(s)$$

For this reaction, identify the atoms that are oxidized and those that are reduced, and specify the oxidizing and reducing agents.

SOLUTION

The first step is to assign oxidation states.

$$Al(s) + 3I_2(s) \rightarrow 2AlI_3(s)$$

$$\begin{array}{cccc} 0 & 0 & +3 & -1 \text{ (each I)} \end{array}$$

Free elements $AlI_3(s)$ is a salt that contains Al^{3+} and I^- ions.

Because each aluminum atom changes its oxidation state from 0 to +3 (an increase in oxidation state), aluminum is *oxidized* (loses electrons). On the other hand, the oxidation state of each iodine atom decreases from 0 to −1, and iodine is *reduced* (gains electrons). Because Al furnishes electrons for the reduction of iodine, it is the *reducing agent.* I_2 is the *oxidizing agent* (the reactant that accepts the electrons).

EXAMPLE 18.4   Identifying Oxidizing and Reducing Agents, II

Metallurgy, the process of producing a metal from its ore, always involves oxidation–reduction reactions. In the metallurgy of galena (PbS), the principal lead-containing ore, the first step is the conversion of lead sulfide to its oxide (a process called *roasting*).

$$2PbS(s) + 3O_2(g) \rightarrow 2PbO(s) + 2SO_2(g)$$

The oxide is then treated with carbon monoxide to produce the free metal.

$$PbO(s) + CO(g) \rightarrow Pb(s) + CO_2(g)$$

For each reaction, identify the atoms that are oxidized and those that are reduced, and specify the oxidizing and reducing agents.

SOLUTION

For the first reaction, we can assign the following oxidation states:

$$PbS(s) + 3O_2(g) \rightarrow 2PbO(s) + 2SO_2(g)$$

$$\begin{array}{cccccc} +2 & -2 & 0 & +2 & -2 & +4 \quad -2 \text{ (each O)} \end{array}$$

The oxidation state for the sulfur atom increases from -2 to $+4$, so sulfur is oxidized (loses electrons). The oxidation state for each oxygen atom decreases from 0 to -2. Oxygen is reduced (gains electrons). The oxidizing agent (accepts electrons) is O_2, and the reducing agent (furnishes electrons) is PbS.

For the second reaction, we have

$$PbO(s) + CO(g) \rightarrow Pb(s) + CO_2(g)$$

$$\begin{array}{cccccc} +2 & -2 & +2 & -2 & 0 & +4 \quad -2 \text{ (each O)} \end{array}$$

Lead is reduced (gains electrons; its oxidation state decreases from $+2$ to 0), and carbon is oxidized (loses electrons; its oxidation state increases from $+2$ to $+4$). PbO is the oxidizing agent (accepts electrons), and CO is the reducing agent (furnishes electrons).

SELF-CHECK EXERCISE 18.3

Ammonia, NH_3, which is widely used as a fertilizer, is prepared by reaction of the elements:

$$N_2(g) + 3H_2(g) \rightarrow 2NH_3(g)$$

Is this an oxidation–reduction reaction? If so, specify the oxidizing agent and the reducing agent.

18.4   Balancing Oxidation–Reduction Reactions by the Half-Reaction Method

AIM: To show how to balance oxidation–reduction equations by using half-reactions.

Many oxidation–reduction reactions can be balanced readily by trial and error. That is, we use the procedure described in Chapter 6 to find a set of coefficients that give the same number of each type of atom on both sides of the equation.

However, the oxidation–reduction reactions that occur in aqueous solution are often so complicated that it becomes very tedious to balance them by trial and error. In this section we will develop a systematic approach for balancing the equations for these reactions.

To balance the equations for oxidation–reduction reactions that occur in aqueous solution, we separate the reaction into two half-reactions. **Half-reactions** are equations that have electrons as reactants or products. One half-reaction represents a reduction process and the other half-reaction represents an oxidation process. In a reduction half-reaction, electrons are shown on the reactant side (electrons are gained by a reactant in the equation). In an oxidation half-reaction, the electrons are shown on the product side (electrons are lost by a reactant in the equation).

For example, consider the unbalanced equation for the oxidation–reduction reaction between the cerium(IV) ion and the tin(II) ion.

$$Ce^{4+}(aq) + Sn^{2+}(aq) \rightarrow Ce^{3+}(aq) + Sn^{4+}(aq)$$

This reduction can be separated into a half-reaction involving the substance being *reduced*:

$$e^- + Ce^{4+}(aq) \rightarrow Ce^{3+}(aq) \quad \text{(reduction half-reaction)}$$

and a half-reaction involving the substance being *oxidized*:

$$Sn^{2+}(aq) \rightarrow Sn^{4+}(aq) + 2e^- \quad \text{(oxidation half-reaction)}$$

Notice that Ce^{4+} must gain one electron to become Ce^{3+}, so one electron is shown as a reactant along with Ce^{4+} in this half-reaction. On the other hand, for Sn^{2+} to become Sn^{4+} it must lose two electrons. This means that two electrons must be shown as products in this half-reaction.

The key principle in balancing oxidation–reduction reactions is that the number of electrons lost (from the reactant that is oxidized) must equal the number of electrons gained (from the reactant that is reduced).

The first source of error is incorrectly identifying the half-reactions. Sometimes the half-reactions are obvious, but often, the identification of the half-reactions depends on correctly assigning oxidation states to the atoms.

Ce^{4+} gains 1e⁻ to form Ce^{3+} and is thus reduced.

Sn^{2+} loses 2e⁻ to form Sn^{4+} and is thus oxidized.

In the half-reaction shown above, one electron is gained by each Ce^{4+} while two electrons are lost by each Sn^{2+}. We must equalize the number of electrons gained and lost. To do this, we first multiply the reduction half-reaction by 2.

$$2e^- + 2Ce^{4+} \rightarrow 2Ce^{3+}$$

Then we add this half-reaction to the oxidation half-reaction.

$$2e^- + 2Ce^{4+} \rightarrow 2Ce^{3+}$$
$$\underline{Sn^{2+} \rightarrow Sn^{4+} + 2e^-}$$
$$2e^- + 2Ce^{4+} + Sn^{2+} \rightarrow 2Ce^{3+} + Sn^{4+} + 2e^-$$

Finally, we cancel the $2e^-$ on each side to give the overall balanced equation

$$2e^- + 2Ce^{4+} + Sn^{2+} \rightarrow 2Ce^{3+} + Sn^{4+} + 2e^-$$
$$2Ce^{4+} + Sn^{2+} \rightarrow 2Ce^{3+} + Sn^{4+}$$

We can now summarize what we have said about the method for balancing oxidation–reduction reactions in aqueous solution:

- Separate the reaction into an oxidation half-reaction and a reduction half-reaction.
- Balance the half-reactions separately.
- Equalize the number of electrons gained and lost.
- Add the half-reactions together and cancel electrons to give the overall balanced equation.

It turns out that most oxidation–reduction reactions occur in solutions that are distinctly basic or distinctly acidic. We will cover only the acidic case in this text, because it is the most common. The detailed procedure for balancing the equations for oxidation–reduction reactions that occur in acidic solution are given below, and Example 18.5 illustrates the use of these steps.

Students often balance the atoms correctly but not the charge, because they add electrons to the wrong side of the half-reaction. Sometimes students forget to equalize the number of electrons gained and lost between the two half reactions.

After both half-reactions are added together, remind students to cancel identical species appearing on both sides of the balanced equation. By checking both the numbers of atoms and the charges on each side, students can know whether or not the equation is really balanced. The check is worth the time it takes.

The Half-Reaction Method for Balancing Equations for Oxidation–Reduction Reactions Occurring in Acidic Solution

Students become discouraged because this method for balancing equations takes so long and their mistakes are so numerous. Reassure them that they will become faster with practice. It will help if they follow the same steps each time, even for relatively simple reactions. Following steps provides them with a method that works when they encounter difficult reactions.

STEP 1
Identify and write the equations for the oxidation and reduction half-reactions.

STEP 2
For each half-reaction:

a. Balance all of the elements except hydrogen and oxygen.
b. Balance oxygen using H_2O.

c. Balance hydrogen using H^+.
d. Balance the charge using electrons.

STEP 3
If necessary, multiply one or both balanced half-reactions by an integer to equalize the number of electrons transferred in the two half-reactions.

STEP 4
Add the half-reactions, and cancel identical species that appear on both sides.

STEP 5
Check to be sure the elements and charges balance.

(a)   (b)

(a) A solution containing MnO_4^- ions and (b) a solution containing Fe^{2+} ions.

H_2O and H^+ will be added to this equation as we balance it. We do not have to worry about them now.

As you work examples, you can ask students to identify the oxidation and reduction half-reactions, the species oxidized and reduced, and the oxidizing and reducing agents.

Note that the left side contains oxygen but the right side does not. This will be taken care of later when we add water.

| EXAMPLE 18.5 | **Balancing Oxidation–Reduction Reactions Using the Half-Reaction Method, I** |

Balance the equation for the reaction between permanganate and iron(II) ions in acidic solution. The net ionic equation for this reaction is

$$MnO_4^-(aq) + Fe^{2+}(aq) \xrightarrow{\text{Acid}} Fe^{3+}(aq) + Mn^{2+}(aq)$$

This reaction is used to analyze iron ore for its iron content.

SOLUTION

STEP I
Identify and write equations for the half-reactions. The oxidation states for the half-reaction involving the permanganate ion show that manganese is reduced.

$$MnO_4^- \rightarrow Mn^{2+}$$
$$\quad +7 \quad -2 \qquad +2$$
$$\text{(each O)}$$

Because manganese changes from an oxidation state of $+7$ to $+2$, it is reduced. So this is the *reduction half-reaction.* It will have electrons as reactants, although we will not write them yet. The other half-reaction involves the oxidation of iron(II) to the iron(III) ion and is the *oxidation half-reaction.*

$$Fe^{2+} \rightarrow Fe^{3+}$$
$$\quad +2 \qquad +3$$

EXAMPLE 18.5, CONTINUED

This reaction will have electrons as products, although we will not write them yet.

STEP 2
Balance each half-reaction. For the reduction reaction, we have

$$MnO_4^- \rightarrow Mn^{2+}$$

a. The manganese is already balanced.
b. We balance oxygen by adding $4H_2O$ to the right side of the equation.

$$MnO_4^- \rightarrow Mn^{2+} + 4H_2O$$

c. Next we balance hydrogen by adding $8H^+$ to the left side.

$$8H^+ + MnO_4^- \rightarrow Mn^{2+} + 4H_2O$$

d. All of the elements have been balanced, but we need to balance the charge using electrons. At this point we have the following charges for reactants and products in the reduction half-reaction.

$$8H^+ + MnO_4^- \rightarrow Mn^{2+} + 4H_2O$$
$$\underbrace{8+ \quad + \quad 1-}_{7+} \quad \underbrace{2+ \quad + \quad 0}_{2+}$$

We can equalize the charges by adding five electrons to the left side.

$$\underbrace{5e^- + 8H^+ + MnO_4^-}_{2+} \rightarrow \underbrace{Mn^{2+} + 4H_2O}_{2+}$$

Both the *elements* and the *charges* are now balanced, so this represents the balanced reduction half-reaction. The fact that five electrons appear on the reactant side of the equation makes sense, because five electrons are required to reduce MnO_4^- (in which Mn has an oxidation state of +7) to Mn^{2+} (in which Mn has an oxidation state of +2).

For the oxidation reaction,

$$Fe^{2+} \rightarrow Fe^{3+}$$

the elements are balanced, so all we have to do is balance the charge.

$$\underbrace{Fe^{2+}}_{2+} \rightarrow \underbrace{Fe^{3+}}_{3+}$$

One electron is needed on the right side to give a net 2+ charge on both sides.

$$\underbrace{Fe^{2+}}_{2+} \rightarrow \underbrace{Fe^{3+} + e^-}_{2+}$$

Because each of these equations takes so long to balance in class, it is impossible to go over very many examples. Students must work problems on their own to gain speed and skill; they will probably need extra help with homework.

The H^+ comes from the acidic solution in which the reaction is taking place.

Always add electrons to the side of the half-reaction with excess positive charge.

EXAMPLE 18.5, CONTINUED

STEP 3

Equalize the number of electrons transferred in the two half-reactions. Because the reduction half-reaction involves a transfer of five electrons and the oxidation half-reaction involves a transfer of only one electron, the oxidation half-reaction must be multiplied by 5.

$$5Fe^{2+} \rightarrow 5Fe^{3+} + 5e^-$$

The number of electrons gained in the reduction half-reaction must equal the number of electrons lost in the oxidation half-reaction.

STEP 4

Add the half-reactions and cancel identical species.

$$5e^- + 8H^+ + MnO_4^- \rightarrow Mn^{2+} + 4H_2O$$
$$5Fe^{2+} \rightarrow 5Fe^{3+} + 5e^-$$
$$\overline{5e^- + 8H^+ + MnO_4^- + 5Fe^{2+} \rightarrow Mn^{2+} + 5Fe^{3+} + 4H_2O + 5e^-}$$

Note that the electrons cancel out (as they must) to give the final balanced equation

$$5Fe^{2+}(aq) + MnO_4^-(aq) + 8H^+(aq)$$
$$\rightarrow 5Fe^{3+}(aq) + Mn^{2+}(aq) + 4H_2O(l)$$

Note that we show the physical states of the reactants and products—(*aq*) and (*l*) in this case—only in the final balanced equation.

STEP 5

Check to be sure that elements and charges balance.

Elements   5Fe, 1Mn, 4O, 8H → 5Fe, 1Mn, 4O, 8H
Charges   17+ → 17+

The equation is balanced.

EXAMPLE 18.6 **Balancing Oxidation–Reduction Reactions Using the Half-Reaction Method, II**

When an automobile engine is started, it uses the energy supplied by a lead storage battery. This battery uses an oxidation–reduction reaction between elemental lead (lead metal) and lead(IV) oxide to provide the power to start the engine. The unbalanced equation for a simplified version of the reaction is

$$Pb(s) + PbO_2(s) + H^+(aq) \rightarrow Pb^{2+}(aq) + H_2O(l)$$

Balance this equation using the half-reaction method.

EXAMPLE 18.6, CONTINUED

SOLUTION

STEP 1
First we identify and write the two half-reactions. One half-reaction must be

$$Pb \rightarrow Pb^{2+}$$

and the other is

$$PbO_2 \rightarrow Pb^{2+}$$

The first reaction involves the oxidation of Pb to Pb^{2+}. The second reaction involves the reduction of Pb^{4+} in (PbO_2) to Pb^{2+}.

Because Pb^{2+} is the only lead-containing product, it must be the product in both half-reactions.

STEP 2
Now we will balance each half-reaction separately.

The oxidation half-reaction

$$Pb \rightarrow Pb^{2+}$$

a.–c. All the elements are balanced.
d. The charge on the left is zero and that on the right is +2, so we must add $2e^-$ to the right to give zero overall charge.

$$Pb \rightarrow Pb^{2+} + 2e^-$$

This half-reaction is balanced.

The reduction half-reaction

$$PbO_2 \rightarrow Pb^{2+}$$

a. All elements are balanced except O.
b. The left side has two oxygen atoms and the right side has none, so we add $2H_2O$ to the right side.

$$PbO_2 \rightarrow Pb^{2+} + 2H_2O$$

c. Now we balance hydrogen by adding $4H^+$ to the left.

$$4H^+ + PbO_2 \rightarrow Pb^{2+} + 2H_2O$$

d. Because the left side has a +4 overall charge and the right side has a +2 charge, we must add $2e^-$ to the left side.

$$2e^- + 4H^+ + PbO_2 \rightarrow Pb^{2+} + 2H_2O$$

The half-reaction is balanced.

STEP 3
Because each half-reaction involves $2e^-$, we can simply add the half-reactions as they are.

Copper metal reacting with nitric acid. The solution is colored by the presence of Cu^{2+} ions. The brown gas is NO_2, which is formed when NO reacts with O_2 in the air.

EXAMPLE 18.6, CONTINUED

STEP 4

$$\begin{array}{rcl} Pb & \to & Pb^{2+} + 2e^- \\ 2e^- + 4H^+ + PbO_2 & \to & Pb^{2+} + 2H_2O \\ \hline 2e^- + 4H^+ + Pb + PbO_2 & \to & 2Pb^{2+} + 2H_2O + 2e^- \end{array}$$

Canceling electrons gives the balanced overall equation

$$Pb(s) + PbO_2(s) + 4H^+(aq) \to 2Pb^{2+}(aq) + 2H_2O(l)$$

where the appropriate states are also indicated.

STEP 5
The elements and the charges both balance.

$$\begin{array}{rl} \text{Elements} & 2Pb, 2O, 4H \to 2Pb, 2O, 4H \\ \text{Charges} & 4+ \to 4+ \end{array}$$

The equation is correctly balanced.

SELF-CHECK EXERCISE 18.4

Copper metal reacts with nitric acid, $HNO_3(aq)$, to give aqueous copper(II) nitrate, water, and nitrogen monoxide gas as products. Write and balance the equation for this reaction.

18.5 Electrochemistry: An Introduction

AIM: To define the term *electrochemistry*.
To identify the components of an electrochemical (galvanic) cell.

Our lives would be very different without batteries. We would have to crank the engines in our cars by hand, wind our watches, and buy very long extension cords if we wanted to listen to a radio on a picnic. Indeed, our society seems to run increasingly on batteries. In this section and the next, we will find out how these important devices produce electrical energy.

A battery uses the energy from an oxidation–reduction reaction to produce an electric current. This is an important illustration of **electrochemistry,** *the study of the interchange of chemical and electrical energy.*

Electrochemistry involves two types of processes:

1. The production of an electric current from a chemical (oxidation–reduction) reaction
2. The use of an electric current to produce a chemical change

To understand how a redox reaction can be used to generate a current, let's reconsider the aqueous reaction between MnO_4^- and Fe^{2+} that we worked with in Example 18.6. We can break this redox reaction into the following half-reactions:

$$8H^+ + MnO_4^- + 5e^- \rightarrow Mn^{2+} + 4H_2O \quad \text{(reduction)}$$
$$Fe^{2+} \rightarrow Fe^{3+} + e^- \quad \text{(oxidation)}$$

When the reaction between MnO_4^- and Fe^{2+} occurs in solution, electrons are transferred directly as the reactants collide. No useful work is obtained from the chemical energy involved in the reaction. How can we harness this energy? The key is to *separate the oxidizing agent (electron acceptor) from the reducing agent (electron donor),* thus requiring the electron transfer to occur through a wire. That is, to get from the reducing agent to the oxidizing agent, the electrons must travel through a wire. The current produced in the wire by this electron flow can be directed through a device, such as an electric motor, to do useful work.

> The energy involved in a chemical reaction is customarily not shown in the balanced equation. In the reaction of MnO_4^- with Fe^{2+}, energy is released that can be used to do useful work.

For example, consider the system illustrated in Figure 18.1. If our reasoning has been correct, electrons should flow through the wire from Fe^{2+} to MnO_4^-. However, when we construct the apparatus as shown, no flow of electrons occurs. Why? The problem is that if the electrons flowed from the right to the left compartment, the left compartment would become negatively charged and the right compartment would experience a build-up of positive charge (Figure 18.2).

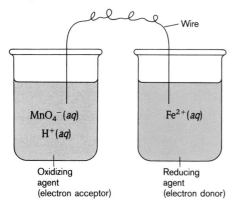

Figure 18.1
Schematic of a method for separating the oxidizing and reducing agents in a redox reaction. (The solutions also contain other ions to balance the charge.) This cell is incomplete at this point.

Figure 18.2
Electron flow under these conditions would lead to a build-up of negative charge on the left and positive charge on the right, which is not feasible without a huge input of energy.

Problem

Solution

(a) (b)

Figure 18.3
Here the ion flow between the two solutions keeps the charge neutral as electrons are transferred. This can be accomplished by having negative ions (anions) flow in the opposite direction to the electrons or by having positive ions (cations) flow in the same direction as the electrons. Both actually occur in a working battery.

Reusable salt bridges can be made from glass tubing bent in the shape of a U and filled with 0.5% agar in 1 M KNO_3. The agar must be heated to almost 100 °C until it dissolves. If the agar-filled U-tubes are not allowed to dry out, they can be used for several weeks.

The name *galvanic cell* honors Luigi Galvani (1737–1798), an Italian scientist generally credited with the discovery of electricity. These cells are sometimes called *voltaic cells* after Alessandro Volta (1745–1827), another Italian, who first constructed cells of this type around 1800.

Anode: The electrode where oxidation occurs.
Cathode: The electrode where reduction occurs.

Figure 18.4
A salt bridge or a porous-disk connection allows ions to flow, completing the electrical circuit. (a) The salt bridge contains a strong electrolyte either as a gel or as a solution; both ends are covered with membrane that allows only ions to pass. (b) The porous disk allows ion flow but does not permit overall mixing of the solutions in the two compartments.

Creating a charge separation of this type would require large amounts of energy. Therefore electron flow does not occur under these conditions.

We can, however, solve this problem very simply. The solutions must be connected (without allowing them to mix extensively) so that *ions* can also flow to keep the net charge in each compartment zero (Figure 18.3). This can be accomplished by using a salt bridge (a U-shaped tube filled with a strong electrolyte) or a porous disk in a tube connecting the two solutions (see Figure 18.4). Either of these devices allows ion flow but prevents extensive mixing of the solutions. When we make a provision for ion flow, the circuit is complete. Electrons flow through the wire from reducing agent to oxidizing agent, and ions in the two aqueous solutions flow from one compartment to the other to keep the net charge zero.

Thus an **electrochemical battery,** also called a **galvanic cell,** is a device powered by an oxidation–reduction reaction where the oxidizing agent is separated from the reducing agent so that the electrons must travel through a wire from the reducing agent to the oxidizing agent (Figure 18.5).

Notice that in a battery, the reducing agent loses electrons (which flow through the wire toward the oxidizing agent) and so is oxidized. The electrode where oxidation occurs is called the **anode.** At the other electrode, the oxidizing agent gains electrons and is thus reduced. The electrode where reduction occurs is called the **cathode.**

We have seen that an oxidation–reduction reaction can be used to generate an electric current. In fact, this type of reaction is used to produce electric currents in many space vehicles. An oxidation–reduction reaction that can be used

Students should know that electrochemical cells differ from ordinary oxidation–reduction reactions. In electrochemical cells the oxidation and reduction half-reactions are separated from each other.

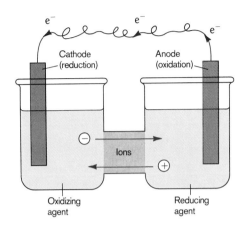

Cathode (reduction)

Anode (oxidation)

Ions

Oxidizing agent

Reducing agent

Figure 18.5
Schematic of a battery (galvanic cell).

Alessandro Volta.

for this purpose is hydrogen and oxygen reacting to form water.

$$2H_2(g) + O_2(g) \rightarrow 2H_2O(l)$$

Oxidation states: 0   0   +1   −2
(each H)

Notice from the changes in oxidation states that in this reaction, hydrogen is oxidized and oxygen reduced. The opposite process can also occur. We can *force* a current through water to produce hydrogen and oxygen gas.

$$2H_2O(l) \xrightarrow[\text{energy}]{\text{Electrical}} 2H_2(g) + O_2(g)$$

This process, where *electrical energy is used to produce a chemical change,* is called **electrolysis.**

In the remainder of this chapter, we will discuss both types of electrochemical processes. In the next section we will concern ourselves with the practical galvanic cells we know as batteries.

The concept of an electrochemical cell is best presented by demonstrations. Set up two beakers, one with 1 M $CuSO_4$ and a strip of Cu metal and one with $ZnSO_4$ and a strip of Zn. Attach a voltmeter or a cheap multimeter. A salt bridge can be made from a strip of filter paper, soaked in 1 M KNO_3, with one end in the $CuSO_4$ solution and the other end in the $ZnSO_4$ solution. You can determine the voltage with the salt bridge and then remove the filter paper to show that ion transfer is necessary for current flow.

18.6   Batteries

AIM: To discuss the composition and operation of commonly used batteries.

In the last section we saw that a galvanic cell is a device that uses an oxidation–reduction reaction to generate an electric current by separating the oxidizing agent from the reducing agent. In this section we will consider several specific galvanic cells and their applications.

Sections 18.6–18.8 explain important practical applications of oxidation–reduction reactions.

If you can obtain one, a used lead storage battery that has been sawed in half is very interesting to students and provides a good way to explain the parts and functions of this important battery.

Remember: The oxidizing agent accepts electrons and the reducing agent furnishes electrons.

e^- flow

H_2SO_4 electrolyte solution

Pb metal grid (anode)

PbO_2 coated onto a lead grid (cathode)

Figure 18.6
In a lead storage battery each cell consists of several lead grids that are connected by a metal bar. These lead grids furnish electrons (the lead atoms lose electrons to form Pb^{2+} ions, which combine with SO_4^{2-} ions to give solid $PbSO_4$). Because the lead is oxidized, it functions as the anode of the cell. The substance that gains electrons is PbO_2; it is coated onto lead grids, several of which are hooked together by a metal bar. The PbO_2 formally contains Pb^{4+}, which is reduced to Pb^{2+}, which in turn combines with SO_4^{2-} to form solid $PbSO_4$. The PbO_2 accepts electrons, so it functions as the cathode.

Lead Storage Battery

Since about 1915 when self-starters were first used in automobiles, the **lead storage battery** has been a major factor in making the automobile a practical means of transportation. This type of battery can function for several years under temperature extremes from -30 °F to 100 °F and under incessant punishment from rough roads. The fact that this same type of battery has been in use for so many years in the face of all of the changes in science and technology over that span of time attests to how well it does its job.

In the lead storage battery, the reducing agent is lead metal, Pb, and the oxidizing agent is lead(IV) oxide, PbO_2. We have already considered a simplified version of this reaction in Example 18.6. In an actual lead storage battery, sulfuric acid, H_2SO_4, furnishes the H^+ needed in the reaction; it also furnishes SO_4^{2-} ions that react with the Pb^{2+} ions to form solid $PbSO_4$. A schematic of one cell of the lead storage battery is shown in Figure 18.6.

In this cell the anode is constructed of lead metal, which is oxidized. In the cell reaction, lead atoms lose two electrons each to form Pb^{2+} ions, which combine with SO_4^{2-} ions present in the solution to give solid $PbSO_4$.

The cathode of this battery has lead(IV) oxide coated onto lead grids. Lead atoms in the $+4$ oxidation state in PbO_2 accept two electrons each (are reduced) to give Pb^{2+} ions that also form solid $PbSO_4$.

In the cell the anode and cathode are separated (so that the electrons must flow through an external wire) and bathed in sulfuric acid. The half-reactions that occur at the two electrodes and the overall cell reaction are shown below.

Anode reaction: $Pb + H_2SO_4 \rightarrow PbSO_4 + 2H^+ + 2e^-$ (oxidation)

Cathode reaction:

$PbO_2 + H_2SO_4 + 2e^- + 2H^+ \rightarrow PbSO_4 + 2H_2O$ (reduction)

Overall reaction:

$Pb(s) + PbO_2(s) + 2H_2SO_4(aq) \rightarrow 2PbSO_4(s) + 2H_2O(l)$

The tendency for electrons to flow from the anode to the cathode in a battery depends on the ability of the reducing agent to release electrons and on the ability of the oxidizing agent to capture electrons. If a battery consists of a reducing agent that releases electrons readily and an oxidizing agent with a high affinity for electrons, the electrons are driven through the connecting wire with great force and can provide much electrical energy. It is useful to think of the analogy of water flowing through a pipe. The greater the pressure on the water, the more vigorously the water flows. The "pressure" on electrons to flow from one electrode to the other in a battery is called the **potential** of the battery and is measured in volts. For example, each cell in a lead storage battery produces about 2 volts of potential. In an actual automobile battery, six of these cells are connected to produce about 12 volts of potential.

The Lead Storage Battery

*T*he lead storage battery used in automobiles has several interesting characteristics. First, the overall battery reaction shows that sulfuric acid is consumed as the battery is used to produce electrical energy. This lowers the density of the electrolyte solution from its fully charged value of about 1.28 g/cm^3. As a result, the condition of the battery can be monitored by measuring the density of the sulfuric acid solution. Because the solid lead sulfate formed in the cell reaction during discharge adheres to the grid surfaces of the electrodes, the battery can be recharged by forcing current in the opposite direction to reverse the cell reaction. A car's battery is continuously charged by an alternator driven by the automobile engine. (The charging of a battery is discussed in Section 18.8.)

It is possible to "jump-start" an automobile with a dead battery by connecting its battery to the battery in a running automobile. This process can be dangerous, however, because the current causes electrolysis of water in the dead battery, producing hydrogen and oxygen gases (see Section 18.8 for details). Disconnecting the jumper cables after the disabled car starts causes an arc that can ignite the gaseous mixture. If this happens, the battery may explode, ejecting corrosive sulfuric acid. This problem can be avoided by connecting the ground jumper cable to a part of the engine remote from the battery. Any arc produced when this cable is disconnected is harmless.

Traditional types of storage batteries require periodic "topping off," because the water in the electrolyte solution is depleted by the electrolysis that accompanies the charging process. More recently designed batteries have electrodes made of an alloy of calcium and lead that inhibits the electrolysis of water. These batteries can be sealed; they require no addition of water.

It is rather amazing that in the 75 years in which lead storage batteries have been used, no better system has been found. Although a lead storage battery provides excellent service, it has a useful lifetime of 3 to 5 years in an automobile. While it might seem that the battery could undergo an indefinite number of discharge/charge cycles, physical damage from road shock and chemical side reactions eventually cause it to fail.

Jump-starting a car.

Dry Cell Batteries

It can be interesting to bring to class a variety of small dry cell batteries for students to examine. Although students are accustomed to small batteries, this could be the first time they have thought about why they work.

The calculators, electronic watches, portable radios, and tape players that are so familiar to us are all powered by small, efficient **dry cell batteries.** They are called dry cells because they do not contain a liquid electrolyte. The common dry cell battery was invented more than 100 years ago by George Leclanché (1839–1882), a French chemist. In its *acid version,* the dry cell battery contains a zinc inner case that acts as the anode and a carbon (graphite) rod in contact with a moist paste of solid MnO_2, solid NH_4Cl, and carbon that acts as the cathode (Figure 18.7). The half-cell reactions are complex but can be approximated as follows:

Anode reaction: $\qquad\qquad\qquad\qquad Zn \rightarrow Zn^{2+} + 2e^-$ $\qquad\qquad$ (oxidation)
Cathode reaction:
$$2NH_4^+ + 2MnO_2 + 2e^- \rightarrow Mn_2O_3 + 2NH_3 + H_2O \quad \text{(reduction)}$$

This cell produces a potential of about 1.5 volts.

In the *alkaline version* of the dry cell battery, the NH_4Cl is replaced with KOH or NaOH. In this case the half-reactions can be approximated as follows:

Anode reaction: $\qquad\quad Zn + 2OH^- \rightarrow ZnO(s) + H_2O + 2e^-$ $\quad$ (oxidation)
Cathode reaction:
$$2MnO_2 + H_2O + 2e^- \rightarrow Mn_2O_3 + 2OH^- \qquad \text{(reduction)}$$

The alkaline dry cell lasts longer, mainly because the zinc anode corrodes less rapidly under basic conditions than under acidic conditions.

Other types of dry cell batteries include the *silver cell,* which has a Zn anode and a cathode that employs Ag_2O as the oxidizing agent in a basic environment. *Mercury cells,* often used in calculators, have a Zn anode and a cathode involving HgO as the oxidizing agent in a basic medium (see Figure 18.8).

Figure 18.7
A common dry cell battery.

Figure 18.8
A mercury battery of the type used in small calculators.

An especially important type of dry cell is the *nickel–cadmium battery,* in which the electrode reactions are

Anode reaction: $Cd + 2OH^- \rightarrow Cd(OH)_2 + 2e^-$ (oxidation)
Cathode reaction: $NiO_2 + 2H_2O + 2e^- \rightarrow Ni(OH)_2 + 2OH^-$ (reduction)

In this cell, as in the lead storage battery, the products adhere to the electrodes. Therefore, a nickel–cadmium battery can be recharged an indefinite number of times, because the products can be turned back into reactants by the use of an external source of current.

Some dry cell batteries are rechargable and some are not. A discussion about why this is so can focus students' attentions on the differences between the types of dry cell batteries.

18.7   Corrosion

AIM: To explain the electrochemical nature of corrosion and to describe some ways of preventing it.

Most metals are found in nature in compounds with nonmetals such as oxygen and sulfur. For example, iron exists as iron ore (which contains Fe_2O_3 and other oxides of iron).

Corrosion can be viewed as the process of returning metals to their natural state—the ores from which they were originally obtained. Corrosion involves oxidation of the metal. Because corroded metal often loses its strength and attractiveness, this process causes great economic loss. For example, approximately one-fifth of the iron and steel produced annually is used to replace rusted metal.

Because most metals react with O_2, we might expect them to corrode so fast in air that they wouldn't be useful. It is surprising, therefore, that the problem of corrosion does not virtually prevent the use of metals in air. Part of the explanation is that most metals develop a thin oxide coating, which tends to protect their internal atoms against further oxidation. The best example of this is aluminum. Aluminum readily loses electrons, so it should be very easily oxidized by O_2. Given this fact, why is aluminum so useful for building airplanes, bicycle frames, and so on? Aluminum is such a valuable structural material because it forms a thin adherent layer of aluminum oxide, Al_2O_3, which greatly inhibits further corrosion. Thus aluminum protects itself with this tough oxide coat. Many other metals, such as chromium, nickel, and tin, do the same thing.

Iron can also form a protective oxide coating. However, this oxide is not a very effective shield against corrosion, because it scales off easily, exposing new metal surfaces to oxidation. Under normal atmospheric conditions, copper forms an external layer of greenish copper sulfate or carbonate called *patina. Silver tarnish* is silver sulfide, Ag_2S, which in thin layers gives the silver surface a richer appearance. Gold shows no appreciable corrosion in air.

Preventing corrosion is an important way of conserving our natural supplies of metals and energy. The primary means of protection is the application of a

For an easily prepared demonstration of corrosion, drop a steel nail and a galvanized steel nail into two separate containers of dilute HCl a day or so before class. Students can compare the amount of rust formed on the two nails.

Some metals, such as copper, gold, silver, and platinum, are relatively difficult to oxidize. They are often called noble metals.

Polished silver and tarnished silver.

One of the most interesting aspects of corrosion is its prevention. It is possible to cover quite a bit of corrosion chemistry while discussing various ways to prevent it.

An alloy is a mixture of elements with metallic properties.

Figure 18.9
Cathodic protection of an underground pipe.

The lead storage battery is a practical example of electrolysis. Most students know that batteries are "recharged" when the automobile engine runs. This makes a good starting point for your discussion of electrolysis.

The electrolysis of water is an effective demonstration. Students know that water does not spontaneously decompose to molecular hydrogen and oxygen.

An electrolytic cell uses electrical energy to produce a chemical change that would not otherwise occur.

coating—most often paint or metal plating—to protect the metal from oxygen and moisture. Chromium and tin are often used to plate steel because they oxidize to form a durable, effective oxide coating.

Alloying is also used to prevent corrosion. *Stainless steel* contains chromium and nickel, both of which form oxide coatings that protect the steel.

Cathodic protection is the method most often employed to protect steel in buried fuel tanks and pipelines. A metal that furnishes electrons more easily than iron, such as magnesium, is connected by a wire to the pipeline or tank that is to be protected (Figure 18.9). Because the magnesium is a better reducing agent than iron, electrons flow through the wire from the magnesium to the iron pipe. Thus the electrons are furnished by the magnesium rather than by the iron, keeping the iron from being oxidized. As oxidation of the magnesium occurs, the magnesium dissolves, so it must be replaced periodically.

18.8 Electrolysis

AIM: To describe the process of electrolysis and the commercial preparation of aluminum.

Unless it is recharged, a battery "runs down" because the substances in it that furnish and accept electrons (to produce the electron flow) are consumed. For example, in the lead storage battery (see Section 18.6), PbO_2 and Pb are consumed to form $PbSO_4$ as the battery runs.

$$PbO_2(s) + Pb(s) + 2H_2SO_4(aq) \rightarrow 2PbSO_4(s) + 2H_2O(l)$$

However, one of the most useful characteristics of the lead storage battery is that it can be recharged. *Forcing* current through the battery in the direction opposite to the normal direction, reverses the oxidation–reduction reaction. That is, $PbSO_4$ is consumed and PbO_2 and Pb are formed in the charging process. This recharging is done continuously by the automobile's alternator, which is powered by the engine.

The process of **electrolysis** involves *forcing a current through a cell to produce a chemical change that would not otherwise occur.*

One important example of this type of process is the electrolysis of water. Water is a very stable substance that can be broken down into its elements by using an electric current (Figure 18.10).

$$2H_2O(l) \xrightarrow[\text{current}]{\substack{\text{Forced} \\ \text{electric}}} 2H_2(g) + O_2(g)$$

The electrolysis of water to produce hydrogen and oxygen occurs whenever a current is forced through an aqueous solution. Thus when the lead storage battery is charged, or "jumped," potentially explosive mixtures of H_2 and O_2 are produced by the current flow through the solution in the battery. This is why it is very important not to produce a spark near the battery during these operations.

Another important use of electrolysis is in the production of metals from their ores. The metal produced in the greatest quantities by electrolysis is aluminum.

Aluminum is one of the most abundant elements on earth, ranking third behind oxygen and silicon. Because aluminum is a very reactive metal, it is found in nature as its oxide in an ore called *bauxite* (named after Les Baux, France, where it was discovered in 1821). Production of aluminum metal from its ore proved to be more difficult than the production of most other metals. In 1782 Lavoisier, the pioneering French chemist, recognized aluminum as a metal "whose affinity for oxygen is so strong that it cannot be overcome by any known reducing agent." As a result, pure aluminum metal remained unknown. Finally, in 1854, a process was found for producing metallic aluminum by using sodium, but aluminum remained a very expensive rarity. In fact, it is said that Napoleon III served his most honored guests with aluminum forks and spoons, while the others had to settle for gold and silver utensils!

The breakthrough came in 1886 when two men, Charles M. Hall in the United States (Figure 18.11) and Paul Heroult in France, almost simultaneously discovered a practical electrolytic process for producing aluminum, which greatly increased the availability of aluminum for many purposes. Table 18.2 (p. 636) shows how dramatically the price of aluminum dropped after this discovery. The

Figure 18.10
The electrolysis of water produces hydrogen gas at the cathode (on the left) and oxygen gas at the anode (on the right). A nonreacting strong electrolyte such as Na_2SO_4 is needed to furnish ions to allow the flow of current.

This is a good time to remind students that recycling aluminum cans prevents us from expending unnecessary energy in the electrolysis of aluminum ore.

Figure 18.11
Charles Martin Hall (1863–1914) was a student at Oberlin College in Ohio when he first became interested in aluminum. One of his professors commented that anyone who could find a way to manufacture aluminum cheaply would make a fortune, and Hall decided to give it a try. The 21-year-old worked in a wooden shed near his house with an iron frying pan as a container, a blacksmith's forge as a heat source, and galvanic cells constructed from fruit jars. Using these crude galvanic cells, Hall found that he could produce aluminum by passing a current through a molten mixture of Al_2O_3 and Na_3AlF_6. By a strange coincidence, Paul Heroult, a French chemist who was born and died in the same years as Hall, made the same discovery at about the same time.

Table 18.2	The Price of Aluminum Over the Last Century
Date	Price of Aluminum ($/lb)*
1855	100,000
1885	100
1890	2
1895	0.50
1970	0.30
1980	0.80

*Note the precipitous drop in price after the discovery of the Hall–Heroult process in 1886.

effect of the electrolysis process is to reduce Al^{3+} ions to neutral Al atoms that form aluminum metal. The aluminum produced in this electrolytic process is 99.5% pure. To be useful as a structural material, aluminum is alloyed with metals such as zinc (for trailer and aircraft construction) and manganese (for cooking utensils, storage tanks, and highway signs). The production of aluminum consumes about 4.5% of all electricity used in the United States.

CHAPTER REVIEW

Key Terms

oxidation–reduction (redox) reactions (p. 610)
oxidation (p. 610)
reduction (p. 610)
oxidation states (p. 611)
oxidizing agent (electron acceptor) (p. 617)
reducing agent (electron donor) (p. 617)
half-reactions (p. 620)
electrochemistry (p. 626)
electrochemical battery (galvanic cell) (p. 628)

anode (p. 628)
cathode (p. 628)
electrolysis (p. 629)
lead storage battery (p. 630)
dry cell battery (p. 632)
corrosion (p. 633)
cathodic protection (p. 634)

Summary

1. Oxidation–reduction reactions involve a transfer of electrons. Oxidation states provide a way to keep track of electrons in these reactions. A set of rules is used to assign oxidation states.

2. Oxidation is an increase in oxidation state (a loss of electrons); reduction is a decrease in oxidation state (a gain of electrons). An oxidizing agent accepts electrons, and a reducing agent donates electrons. Oxidation and reduction always occur together.

3. Oxidation–reduction equations can be balanced by inspection or by the half-reaction method. This method involves splitting a reaction into two parts (the oxidation half-reaction and the reduction half-reaction).

4. Electrochemistry is the study of the interchange of chemical and electrical energy that occurs through oxidation–reduction reactions.

5. When an oxidation–reduction reaction occurs with the reactants in the same solution, the electrons are transferred directly, and no useful work

can be obtained. However, when the oxidizing agent is separated from the reducing agent, so that the electrons must flow through a wire from one to the other, chemical energy is transformed into electrical energy. The opposite process, in which electrical energy is used to produce chemical change, is called electrolysis.

6. A galvanic (electrochemical) cell is a device in which chemical energy is transformed into useful electrical energy. Oxidation occurs at the anode of a cell; reduction occurs at the cathode.

7. A battery is a galvanic cell, or group of cells, that serves as a source of electric current. The lead storage battery has a lead anode and a cathode of lead coated with PbO_2, both immersed in a solution of sulfuric acid. Dry cell batteries do not have liquid electrolytes but contain a moist paste instead.

8. Corrosion involves the oxidation of metals to form mainly oxides and sulfides. Some metals, such as aluminum, form a thin protective oxide coating that inhibits their further corrosion. Corrosion of iron can be prevented by a coating (such as paint), by alloying, and by cathodic protection.

Questions and Problems

All even-numbered exercises have answers in the back of this book and solutions in the Solutions Guide.

18.1 Oxidation–Reduction Reactions

QUESTIONS

1. What name is given to reactions in which electrons are transferred? Give two examples of such reactions.

2. How are *oxidation* and *reduction* defined? Write an equation showing an atom being oxidized and one showing an atom being reduced.

3. For each of the following oxidation–reduction reactions of metals with nonmetals, identify which element is oxidized and which is reduced.
   a. $2Mg(s) + O_2(g) \rightarrow 2MgO(s)$
   b. $Mg(s) + Cu(NO_3)_2(aq) \rightarrow Mg(NO_3)_2(aq) + Cu(s)$
   c. $2Fe(s) + 3Br_2(l) \rightarrow 2FeBr_3(s)$
   d. $2Na(s) + Cl_2(g) \rightarrow 2NaCl(s)$

4. For each of the following oxidation–reduction reactions of metals with nonmetals, identify which element is oxidized and which is reduced.
   a. $6Na(s) + N_2(g) \rightarrow 2Na_3N(s)$
   b. $Mg(s) + Cl_2(g) \rightarrow MgCl_2(s)$
   c. $2Al(s) + 3Br_2(l) \rightarrow 2AlBr_3(s)$
   d. $4Fe(s) + 3O_2(g) \rightarrow 2Fe_2O_3(s)$

5. For each of the following oxidation–reduction reactions of metals with nonmetals, identify which element is oxidized and which is reduced.
   a. $2Na(s) + I_2(s) \rightarrow 2NaI(s)$
   b. $2Co(s) + 3Cl_2(g) \rightarrow 2CoCl_3(s)$
   c. $2Ba(s) + O_2(g) \rightarrow 2BaO(s)$
   d. $2Fe(s) + 3S(s) \rightarrow Fe_2S_3(s)$

6. For each of the following oxidation–reduction reactions of metals with nonmetals, identify which element is oxidized and which is reduced.
   a. $Ca(s) + S(s) \rightarrow CaS(s)$
   b. $Cu(s) + F_2(g) \rightarrow CuF_2(s)$
   c. $Ca(s) + C(s) \rightarrow CaC_2(s)$
   d. $2Zn(s) + O_2(g) \rightarrow 2ZnO(s)$

18.2 Oxidation States

QUESTIONS

7. What is an oxidation state? Why do we define such a concept?

8. What is the oxidation state of the atoms in a pure element? Does this depend on whether the element occurs as a diatomic molecule (O_2, N_2) or as a larger molecule (P_4, S_8)?

9. Explain why, although it is not an ionic compound, we still assign oxygen an oxidation state of -2 in water, H_2O. Give an example of a compound in which oxygen is *not* in the -2 oxidation state.

10. Why is fluorine always assigned an oxidation state of -1? What oxidation number is *usually* assigned to the other halogen elements when they occur in compounds? In an *interhalogen* compound involving fluorine (such as ClF), which atom has a negative oxidation state?

11. In ammonia, NH_3, nitrogen is assigned a -3 oxidation state. In nitrogen trifluoride, NF_3, nitrogen is assigned a $+3$ oxidation state. Why does nitrogen show such a difference?

12. Why must the sum of all the oxidation states of the atoms in a neutral *molecule* be zero? What must be the sum of the oxidation states in an *ion*?

PROBLEMS

13. Assign oxidation states to all of the atoms in each of the following:
    a. Cl_2
    b. CH_4
    c. BaO
    d. HCl

14. Assign oxidation states to all of the atoms in each of the following:
    a. NCl_3
    b. SF_6
    c. PCl_5
    d. SiH_4

15. Assign oxidation states to all of the atoms in each of the following:
    a. O_2
    b. O_3
    c. $FeCl_3$
    d. $FeCl_2$

16. Assign oxidation states to all of the atoms in each of the following:
    a. Cl_2
    b. Cl_2O
    c. HCl
    d. ClF

17. Assign oxidation states to all of the atoms in each of the following:
    a. NO_3^-
    b. NH_4^+
    c. NO_2
    d. N_2O_4

18. Assign oxidation states to all of the atoms in each of the following:
    a. HNO_3
    b. HPO_4^{2-}
    c. HSO_4^-
    d. O_2^{2-}

19. Assign oxidation states to all of the atoms in each of the following:
    a. MnO_4^-
    b. $MnCl_2$
    c. MnO_2
    d. $Ca(MnO_4)_2$

20. Assign oxidation states to all of the atoms in each of the following:
    a. $CuCl_2$
    b. $CrCl_3$
    c. $HCrO_4^-$
    d. Cr_2O_3

21. Assign oxidation states to all of the atoms in each of the following:
    a. SO_4^{2-}
    b. CO_3^{2-}
    c. SO_3^{2-}
    d. HCO_3^-

22. Assign oxidation states to all of the atoms in each of the following:
    a. ClO_4^-
    b. ClO_3^-
    c. ClO_2^-
    d. ClO^-

18.3 Oxidation–Reduction Reactions Between Nonmetals

QUESTIONS

23. Oxidation can be defined as a loss of electrons or as an increase in oxidation state. Explain why the two definitions mean the same thing, and give an example to support your explanation.

24. Reduction can be defined as a gain of electrons or as a decrease in oxidation state. Explain why the two definitions mean the same thing, and give an example to support your explanation.

25. What is an oxidizing agent? What is a reducing agent?

26. Is an oxidizing agent itself oxidized or reduced? Is a reducing agent itself oxidized or reduced?

27. Does an oxidizing agent donate or accept electrons? Does a reducing agent donate or accept electrons?

28. Does an oxidizing agent increase or decrease its own oxidation state when it acts on another atom? Does a reducing agent increase or decrease its own oxidation state when it acts on another atom?

PROBLEMS

29. In each of the following reactions, identify which element is oxidized and which is reduced by assigning oxidation states.
    a. $N_2(g) + 3H_2(g) \rightarrow 2NH_3(g)$
    b. $2K(s) + 2H_2O(l) \rightarrow 2KOH(aq) + H_2(g)$
    c. $N_2(g) + 2O_2(g) \rightarrow 2NO_2(g)$
    d. $2P(s) + 3Cl_2(g) \rightarrow 2PCl_3(l)$

30. In each of the following reactions, identify which element is oxidized and which is reduced by assigning oxidation states.
    a. $2C(s) + O_2(g) \rightarrow 2CO(g)$
    b. $C_2H_2(g) + 2Br_2(l) \rightarrow C_2H_2Br_4(l)$
    c. $CH_4(g) + I_2(s) \rightarrow CH_3I(s) + HI(g)$
    d. $P_4(s) + 6Cl_2(g) \rightarrow 4PCl_3(g)$

31. In each of the following reactions, identify which element is oxidized and which is reduced by assigning oxidation states.
    a. $4NH_3(g) + 7O_2(g) \rightarrow 4NO_2(g) + 6H_2O(l)$
    b. $2B(s) + 3H_2(g) \rightarrow B_2H_6(g)$
    c. $Xe(g) + 2F_2(g) \rightarrow XeF_4(s)$
    d. $Cl_2(g) + F_2(g) \rightarrow 2ClF(g)$

32. In each of the following reactions, identify which element is oxidized and which is reduced by assigning oxidation states.
    a. $Cu(s) + 2AgNO_3(aq) \rightarrow 2Ag(s) + Cu(NO_3)_2(aq)$
    b. $N_2(g) + 3F_2(g) \rightarrow 2NF_3(g)$
    c. $2Fe_2O_3(s) + 3S(s) \rightarrow 4Fe(s) + 3SO_2(g)$
    d. $2H_2O_2(l) \rightarrow 2H_2O(l) + O_2(g)$

33. In ordinary photography, the light-sensitive portion of the film is a thin coating of a silver halide (usually silver bromide). When the film is exposed to light, the reaction

$$2AgBr(s) \rightarrow 2Ag(s) + Br_2(g)$$

occurs. Identify which element is reduced and which is oxidized.

34. Iron ores, usually oxides of iron, are converted to the pure metal by reaction in a blast furnace with carbon (coke). The carbon is first reacted with air to form carbon monoxide, which in turn reacts with the iron oxides as follows:

$$Fe_2O_3(s) + 3CO(g) \rightarrow 2Fe(l) + 3CO_2(g)$$

Identify the atoms that are oxidized and reduced, and specify the oxidizing and reducing agents.

35. When copper is recovered from its ores, it is sometimes initially converted to copper(I) sulfide, Cu_2S. When the copper(I) sulfide is heated in air, the sulfur reacts with oxygen to form gaseous sulfur dioxide, leaving relatively pure metallic copper as a residue.

$$Cu_2S(s) + O_2(g) \rightarrow 2Cu(l) + SO_2(g)$$

Identify the atoms that are oxidized and reduced, and specify the oxidizing and reducing agents.

36. Elemental bromine can be prepared by treatment of seawater with chlorine gas.

$$Cl_2(g) + 2NaBr(aq) \rightarrow Br_2(aq) + 2NaCl(aq)$$

Identify the atoms that are oxidized and reduced, and specify the oxidizing and reducing agents.

18.4 Balancing Oxidation–Reduction Reactions by the Half-Reaction Method

QUESTIONS

37. In what *two* respects must oxidation–reduction reactions be balanced?

38. Why is a systematic method for balancing oxidation–reduction reactions necessary? Why can't these equations be balanced readily by inspection?

39. What is a half-reaction? What does each of the two half-reactions that make up an overall process represent?

40. Why must the number of electrons lost in the oxidation equal the number of electrons gained in the reduction? Is it possible to have "left-over" electrons in a reaction?

PROBLEMS

41. Balance each of the following half-reactions.
    a. $Ag^+(aq) \rightarrow Ag(s)$ c. $Zn(s) \rightarrow Zn^{2+}(aq)$
    b. $Cu^{2+}(aq) \rightarrow Cu(s)$ d. $Au(s) \rightarrow Au^{3+}(aq)$

42. Balance each of the following half-reactions.
    a. $N_2(g) \rightarrow N^{3-}(s)$ c. $Zn(s) \rightarrow Zn^{2+}(aq)$
    b. $O_2^{2-}(aq) \rightarrow O_2(g)$ d. $F_2(g) \rightarrow F^-(aq)$

43. Balance each of the following half-reactions, which take place in acidic solution.
    a. $MnO_4^-(aq) \rightarrow Mn^{2+}(aq)$
    b. $CrO_4^{2-}(aq) \rightarrow Cr^{3+}(aq)$
    c. $Cr_2O_7^{2-}(aq) \rightarrow Cr^{3+}(aq)$
    d. $OCl^-(aq) \rightarrow Cl_2(g)$

44. Balance each of the following half-reactions, which take place in acidic solution.
    a. $O_2(g) \rightarrow H_2O(l)$
    b. $IO_3^-(aq) \rightarrow I_2(s)$
    c. $VO^{2+}(aq) \rightarrow V^{3+}(aq)$
    d. $BiO^+(aq) \rightarrow Bi(s)$

45. Balance each of the following oxidation–reduction reactions, which take place in acidic solution, by using the "half-reaction" method.
    a. $S_2O_3^{2-}(aq) + I_2(aq) \rightarrow S_4O_6^{2-}(aq) + I^-(aq)$
    b. $VO_2^+(aq) + Zn(s) \rightarrow VO^{2+}(aq) + Zn^{2+}(aq)$
    c. $Cu(s) + NO_3^-(aq) \rightarrow Cu^{2+}(aq) + NO(g)$
    d. $Ag^+(aq) + Zn(s) \rightarrow Ag(s) + Zn^{2+}(aq)$

46. Balance each of the following oxidation–reduction reactions, which take place in acidic solution, by using the "half-reaction" method.
    a. $BrO_3^-(aq) + HBr^-(aq) \rightarrow Br_2(l)$
    b. $I^-(aq) + O_2(g) \rightarrow HI_2(s)$
    c. $Cr_2O_7^{2-}(aq) + I^-(aq) \rightarrow I_2(s) + Cr^{3+}(aq)$
    d. $Mn^{3+}(aq) \rightarrow MnO_2(s) + Mn^{2+}(aq)$

47. Iodide ion, I^-, is one of the most easily oxidized species. Balance each of the following oxidation–reduction reactions, which take place in acidic solution, by using the "half-reaction" method.
    a. $IO_3^-(aq) + I^-(aq) \rightarrow I_2(aq)$
    b. $Cr_2O_7^{2-}(aq) + I^-(aq) \rightarrow Cr^{3+}(aq) + I_2(aq)$
    c. $Cu^{2+}(aq) + I^-(aq) \rightarrow CuI(s) + I_2(aq)$

48. Cerium(IV) ion is one of the most important of the strong oxidizing agents. Balance each of the following oxidation–reduction reactions of the Ce^{4+} ion, all of which take place in acidic solution.
    a. $Ce^{4+}(aq) + H_3AsO_3(aq) \rightarrow Ce^{3+}(aq) + H_3AsO_4(aq)$
    b. $Ce^{4+}(aq) + Fe^{2+}(aq) \rightarrow Ce^{3+}(aq) + Fe^{3+}(aq)$
    c. $Ce^{4+}(aq) + I^-(aq) \rightarrow Ce^{3+}(aq) + I_2(s)$

18.5 Electrochemistry: An Introduction

QUESTIONS

49. How is an oxidation–reduction reaction set up as a galvanic cell (battery)? How is the transfer of electrons between reducing agent and oxidizing agent made useful?

50. What is a salt bridge? Why is a salt bridge necessary in a galvanic cell? Can some other method be used in place of the salt bridge?

51. In which direction do electrons flow in a galvanic cell, anode to cathode or vice versa?

52. Which electrode in a galvanic cell is the anode? Which is the cathode?

PROBLEMS

53. Consider the oxidation–reduction reaction

$$Zn(s) + 2Ag^+(aq) \rightarrow Zn^{2+}(aq) + 2Ag(s)$$

Sketch a galvanic cell that uses this reaction. Which metal ion is reduced? Which metal is oxidized? What half-reaction takes place at the anode in the cell? What half-reaction takes place at the cathode?

54. Consider the oxidation–reduction reaction

$$Zn(s) + Pb^{2+}(aq) \rightarrow Zn^{2+}(aq) + Pb(s)$$

Sketch a galvanic cell that uses this reaction. Which metal ion is reduced? Which metal is oxidized? What half-reaction takes place at the anode in the cell? What half-reaction takes place at the cathode?

18.6 Batteries

QUESTIONS

55. Write the chemical equation for the overall cell reaction that occurs in a lead storage automobile battery. What species is oxidized in such a battery? What species is reduced? Why can such a battery be "recharged"?

56. Why does an alkaline dry cell battery typically last longer than a normal dry cell? Write the chemical equation for the overall cell reaction in such an alkaline dry cell.

18.7 Corrosion

QUESTIONS

57. What process is represented by the *corrosion* of a metal? Why is corrosion undesirable?

58. Explain how some metals, notably aluminum, naturally resist complete oxidation by the atmosphere.

59. Pure iron ordinarily rusts quickly, but steel does not corrode nearly so fast. How does steel resist corrosion?

60. What is *cathodic protection,* and how is it applied to prevent oxidation of steel tanks and pipes?

18.8 Electrolysis

QUESTIONS

61. What is *electrolysis*? What sorts of reactions is electrolysis capable of causing?

62. What reactions go on during the recharging of an automobile battery?

63. How is aluminum metal produced by electrolysis? Why cannot simpler chemical reduction methods be used?

64. Jewelry is often manufactured by plating an expensive metal such as gold over a cheaper metal. How might such a process be set up as an electrolysis reaction?

Additional Problems

65. Reactions in which one or more _____ are transferred between species are called oxidation–reduction reactions.

66. Oxidation may be described as a _____ of electrons or as an increase in _____.

67. Reduction may be described as a _____ of electrons or as a decrease in _____.

68. In assigning oxidation states for a covalently bonded molecule, we assume that the more _____ element controls both electrons of the covalent bond.

69. The oxidation state of an atom in a(n) _____ is zero.

70. The sum of the oxidation states for an electrically neutral compound is _____.

71. An oxidizing agent causes the (oxidation/reduction) of another species, and the oxidizing agent itself is (oxidized/reduced).

72. In order to function as a good reducing agent, a species must easily _____ electrons.

73. When we balance an oxidation–reduction equation, the number of electrons lost by the reducing agent must _____ the number of electrons gained by the oxidizing agent.

74. In order to obtain useful electrical energy from an oxidation–reduction process, we must set up the reaction in such a way that the oxidation half-reaction and the reduction half-reaction are physically _____ one another.

75. An electrochemical cell that produces a current from an oxidation–reduction reaction is more properly called a _____ cell.

76. Oxidation occurs at the _____ of a galvanic cell.

77. Reduction occurs at the _____ of a galvanic cell.

78. When an electric current from an external source is used to force an otherwise nonspontaneous reaction to occur, the process is called _____.

79. The "pressure" on electrons to flow from one electrode to the other in a battery is called the _____ of the battery.

80. "Jump-starting" a dead automobile battery can be dangerous if precautions are not taken, because of the production of an explosive mixture of _____ and _____ gases in the battery.

81. The common acid dry cell battery typically contains an inner casing made of _____ metal, which functions as the anode.

82. Corrosion of a metal represents its _____ by species present in the atmosphere.

83. Although aluminum is a reactive metal, pure aluminum ordinarily does not corrode much in air because a protective layer of _____ builds up on the metal's surface.

84. For each of the following unbalanced oxidation–reduction chemical equations, balance the equation, and identify which species is undergoing oxidation and which is undergoing reduction. Balance each equation by inspection.

a. $Fe(s) + O_2(g) \rightarrow Fe_2O_3(s)$
b. $Al(s) + Cl_2(g) \rightarrow AlCl_3(l)$
c. $Mg(s) + P_4(s) \rightarrow Mg_3P_2(s)$

85. In each of the following reactions, identify which element is oxidized and which is reduced.

a. $Zn(s) + 2HCl(aq) \rightarrow ZnCl_2(aq) + H_2(g)$
b. $2CuI(s) \rightarrow CuI_2(s) + Cu(s)$
c. $6Fe^{2+}(aq) + Cr_2O_7^{2-}(aq) + 14H^+(aq) \rightarrow$
$6Fe^{3+}(aq) + 2Cr^{3+}(aq) + 7H_2O(l)$

86. In each of the following reactions, identify which element is oxidized and which is reduced.
    a. $2Al^{3+}(aq) + 3Mg(s) \rightarrow 2Al(s) + 3Mg^{2+}(aq)$
    b. $2AlCl_3(s) + 3Mg(s) \rightarrow 2Al(s) + 3MgCl_2(s)$
    c. $2KClO_3(s) \rightarrow 2KCl(s) + 3O_2(g)$

87. Carbon compounds containing double bonds (such compounds are called alkenes) react readily with many other reagents. In each of the following reactions, identify which atoms are oxidized and which are reduced, and specify the oxidizing and reducing agents.

    $CH_2{=}CH_2(g) + Cl_2(g) \rightarrow ClCH_2{-}CH_2Cl(l)$
    $CH_2{=}CH_2(g) + Br_2(g) \rightarrow BrCH_2{-}CH_2Br(l)$
    $CH_2{=}CH_2(g) + HBr(g) \rightarrow CH_3{-}CH_2Br(l)$
    $CH_2{=}CH_2(g) + H_2(g) \rightarrow CH_3{-}CH_3(g)$

88. Balance each of the following oxidation–reduction equations by inspection.
    a. $C_3H_8(g) + O_2(g) \rightarrow CO_2(g) + H_2O(g)$
    b. $CO(g) + H_2(g) \rightarrow CH_3OH(l)$
    c. $SnO_2(s) + C(s) \rightarrow Sn(s) + CO(g)$
    d. $C_2H_5OH(l) + O_2(g) \rightarrow CO_2(g) + H_2O(g)$

89. Balance each of the following oxidation–reduction reactions, which take place in acidic solution.
    a. $MnO_4^-(aq) + H_2O_2(aq) \rightarrow Mn^{2+}(aq) + O_2(g)$
    b. $BrO_3^-(aq) + Cu^+(aq) \rightarrow Br^-(aq) + Cu^{2+}(aq)$
    c. $HNO_2(aq) + I^-(aq) \rightarrow NO(g) + I_2(aq)$

90. For each of the following oxidation–reduction reactions of metals with nonmetals, identify which element is oxidized and which is reduced.
    a. $4Na(s) + O_2(g) \rightarrow 2Na_2O(s)$
    b. $Fe(s) + H_2SO_4(aq) \rightarrow FeSO_4(aq) + H_2(g)$
    c. $2Al_2O_3(s) \rightarrow 4Al(s) + 3O_2(g)$
    d. $3Mg(s) + N_2(g) \rightarrow Mg_3N_2(s)$

91. For each of the following oxidation–reduction reactions of metals with nonmetals, identify which element is oxidized and which is reduced.
    a. $3Zn(s) + N_2(g) \rightarrow Zn_3N_2(s)$
    b. $Co(s) + S(s) \rightarrow CoS(s)$
    c. $4K(s) + O_2(g) \rightarrow 2K_2O(s)$
    d. $4Ag(s) + O_2(g) \rightarrow 2Ag_2O(s)$

92. Assign oxidation states to all of the atoms in each of the following:
    a. NH_3 c. CO_2
    b. CO d. NF_3

93. Assign oxidation states to all of the atoms in each of the following:
    a. C_2H_4 c. C_2H_6
    b. C_2H_2 d. C_2H_6O

94. Assign oxidation states to all of the atoms in each of the following:
    a. SO_2 c. H_2S
    b. SO_3 d. H_2SO_4

95. Assign oxidation states to all of the atoms in each of the following:
    a. $CrCl_3$ c. $K_2Cr_2O_7$
    b. K_2CrO_4 d. $Cr(C_2H_3O_2)_2$

96. Assign oxidation states to all of the atoms in each of the following:
    a. BiO^+ c. NO_2^-
    b. PO_4^{3-} d. Hg_2^{2+}

97. In each of the following reactions, identify which element is oxidized and which is reduced by assigning oxidation states.
    a. $C(s) + O_2(g) \rightarrow CO_2(g)$
    b. $2CO(g) + O_2(g) \rightarrow 2CO_2(g)$
    c. $CH_4(g) + 2O_2(g) \rightarrow CO_2(g) + 2H_2O(g)$
    d. $C_2H_2(g) + 2H_2(g) \rightarrow C_2H_6(g)$

98. In each of the following reactions, identify which element is oxidized and which is reduced by assigning oxidation states.
    a. $2B_2O_3(s) + 6Cl_2(g) \rightarrow 4BCl_3(l) + 3O_2(g)$
    b. $GeH_4(g) + O_2(g) \rightarrow Ge(s) + 2H_2O(g)$
    c. $C_2H_4(g) + Cl_2(g) \rightarrow C_2H_4Cl_2(l)$
    d. $O_2(g) + 2F_2(g) \rightarrow 2OF_2(g)$

99. Balance each of the following half-reactions.
    a. $I^-(aq) \rightarrow I_2(s)$ c. $P_4(s) \rightarrow P^{3-}(s)$
    b. $O_2(g) \rightarrow O^{2-}(s)$ d. $Cl_2(g) \rightarrow Cl^-(aq)$

100. Balance each of the following half-reactions, which take place in acidic solution.
    a. $SiO_2(s) \rightarrow Si(s)$
    b. $S(s) \rightarrow H_2S(aq)$
    c. $NO_3^-(aq) \rightarrow HNO_2(aq)$
    d. $NO_3^-(aq) \rightarrow NO(g)$

101. Balance each of the following oxidation–reduction reactions, which take place in acidic solution, by using the "half-reaction" method.
    a. $I^-(aq) + MnO_4^-(aq) \rightarrow I_2(aq) + Mn^{2+}(aq)$
    b. $S_2O_8^{2-}(aq) + Cr^{3+}(aq) \rightarrow SO_4^{2-}(aq) + Cr_2O_7^{2-}(aq)$
    c. $BiO_3^-(aq) + Mn^{2+}(aq) \rightarrow Bi^{3+}(aq) + MnO_4^-(aq)$

102. Potassium permanganate, $KMnO_4$, is one of the most widely used oxidizing agents. Balance each of the following oxidation–reduction reactions of the permanganate ion in acid solution by using the "half-reaction" method.
    a. $MnO_4^-(aq) + C_2O_4^{2-}(aq) \rightarrow Mn^{2+}(aq) + CO_2(g)$
    b. $MnO_4^-(aq) + Fe^{2+}(aq) \rightarrow Mn^{2+}(aq) + Fe^{3+}(aq)$
    c. $MnO_4^-(aq) + Cl^-(aq) \rightarrow Mn^{2+}(aq) + Cl_2(g)$

103. Consider the oxidation–reduction reaction

$$Mg(s) + Cu^{2+}(aq) \rightarrow Mg^{2+}(aq) + Cu(s)$$

Sketch a galvanic cell that uses this reaction. Which metal ion is reduced? Which metal is oxidized? What half-reaction takes place at the anode in the cell? What half-reaction takes place at the cathode?

Solutions to Self-Check Exercises

SELF-CHECK EXERCISE 18.1

a. CuO contains Cu^{2+} and O^{2-} ions, so copper is oxidized $(Cu \rightarrow Cu^{2+} + 2e^-)$ and oxygen is reduced $(O + 2e^- \rightarrow O^{2-})$.

b. CsF contains Cs^+ and F^- ions. Thus cesium is oxidized $(Cs \rightarrow Cs^+ + e^-)$ and fluorine is reduced $(F + e^- \rightarrow F^-)$.

SELF-CHECK EXERCISE 18.2

a. SO_3
We assign oxygen first. Each O is assigned an oxidation state of -2, giving a total of -6 (3×-2) for the three oxygen atoms. Because the molecule has zero charge overall, the sulfur must have an oxidation state of $+6$.
CHECK: $+6 + 3(-2) = 0$

b. SO_4^{2-}
As in part a., each oxygen is assigned an oxidation state of -2, giving a total of -8 (4×-2) on the four oxygen atoms. The anion has a net charge of -2, so the sulfur must have an oxidation state of $+6$.
CHECK: $+6 + 4(-2) = -2$
SO_4^{2-} has a charge of -2, so this is correct.

c. N_2O_5
We assign oxygen before nitrogen because oxygen is the more electronegative. Thus each O is assigned an oxidation state of -2, giving a total of -10 (5×-2) on the five oxygen atoms. Therefore the oxidation states of the *two* nitrogen atoms must total $+10$, because N_2O_5 has no overall charge. Each N is assigned an oxidation state of $+5$.
CHECK: $2(+5) + 5(-2) = 0$

d. PF_3
First we assign the fluorine an oxidation state of -1, giving a total of -3 (3×-1) on the three fluorine atoms. Thus P must have an oxidation state of $+3$.
CHECK: $+3 + 3(-1) = 0$

e. C_2H_6

In this case it is best to recognize that hydrogen is always $+1$ in compounds with nonmetals. Thus each H is assigned an oxidation state of $+1$, which means that the six H atoms account for a total of $+6$ ($6 \times +1$). Therefore the *two* carbon atoms must account for -6, and each carbon is assigned an oxidation state of -3.

CHECK: $2(-3) + 6(+1) = 0$

SELF-CHECK EXERCISE 18.3

We can tell whether this is an oxidation–reduction reaction by comparing the oxidation states of the elements in the reactants and products:

$$N_2 + 3H_2 \rightarrow 2NH_3$$

Oxidation states: 0   0 -3 $+1$ (each H)

Nitrogen goes from 0 to -3. Thus it gains three electrons and is reduced. Each hydrogen atom goes from 0 to $+1$ and is thus oxidized, so this is an oxidation–reduction reaction. The oxidizing agent is N_2 (it takes electrons from H_2). The reducing agent is H_2 (it gives electrons to N_2).

$$N_2 \ + \ 3H_2 \ \rightarrow \ 2NH_3$$

$6e^-$

SELF-CHECK EXERCISE 18.4

The unbalanced equation for this reaction is

$$Cu(s) + HNO_3(aq) \rightarrow Cu(NO_3)_2(aq) + H_2O(l) + NO(g)$$

Copper   Nitric acid   Aqueous   Water   Nitrogen
metal   copper(II)   monoxide
                              nitrate
                          (contains Cu^{2+})

STEP I

The oxidation half-reaction is

$$Cu + HNO_3 \ \rightarrow \ Cu(NO_3)_2$$

Oxidation states: 0 $+1$ $+5$ -2 $+2$ $+5$ -2
                          (each 0)   (each 0)

The copper goes from 0 to $+2$ and thus is oxidized. This reduction reaction is

$$HNO_3 \ \rightarrow \ NO$$

Oxidation states: $+1$ $+5$ -2 $+2$ -2
                          (each 0)

In this case nitrogen goes from $+5$ in HNO_3 to $+2$ in NO and so is reduced. Notice two things about these reactions:

1. The HNO_3 must be included in the oxidation half-reaction to supply NO_3^- in the product $Cu(NO_3)_2$.
2. Although water is a product in the overall reaction, it does not need to be included in either half-reaction at the beginning. It will appear later as we balance the equation.

STEP 2

Balance the oxidation half-reaction.

$$Cu + HNO_3 \rightarrow Cu(NO_3)_2$$

a. Balance nitrogen first.

$$Cu + 2HNO_3 \rightarrow Cu(NO_3)_2$$

b. Balancing nitrogen also caused oxygen to balance.

c. Balance hydrogen using H^+.

$$Cu + 2HNO_3 \rightarrow Cu(NO_3)_2 + 2H^+$$

d. Balance the charge using e^-.

$$Cu + 2HNO_3 \rightarrow Cu(NO_3)_2 + 2H^+ + 2e^-$$

This is the balanced oxidation half-reaction.

Balance the reduction half-reaction.

$$HNO_3 \rightarrow NO$$

a. All elements are balanced except hydrogen and oxygen.

b. Balance oxygen using H_2O.

$$HNO_3 \rightarrow NO + 2H_2O$$

c. Balance hydrogen using H^+.

$$3H^+ + HNO_3 \rightarrow NO + 2H_2O$$

d. Balance the charge using e^-.

$$3e^- + 3H^+ + HNO_3 \rightarrow NO + 2H_2O$$

This is the balanced reduction half-reaction.

STEP 3

We equalize electrons by multiplying the oxidation half-reaction by 3:

$$3 \times (Cu + 2HNO_3 \rightarrow Cu(NO_3)_2 + 2H^+ + 2e^-)$$

gives

$$3Cu + 6HNO_3 \rightarrow 3Cu(NO_3)_2 + 6H^+ + 6e^-$$

and the reduction half-reaction by 2:

$$2 \times (3e^- + 3H^+ + HNO_3 \rightarrow NO + 2H_2O)$$

gives

$$6e^- + 6H^+ + 2HNO_3 \rightarrow 2NO + 4H_2O$$

STEP 4

We can now add the balanced half-reactions, which both involve a six-electron change.

$$3Cu + 6HNO_3 \rightarrow 3Cu(NO_3)_2 + 6H^+ + 6e^-$$
$$\underline{6e^- + 6H^+ + 2HNO_3 \rightarrow 2NO + 4H_2O}$$
$$\cancel{6e^-} + \cancel{6H^+} + 3Cu + 8HNO_3 \rightarrow 3Cu(NO_3)_2 + 2NO + 4H_2O + \cancel{6H^+} + \cancel{6e^-}$$

Canceling species common to both sides gives the balanced overall equation

$$3Cu(s) + 8HNO_3(aq) \rightarrow 3Cu(NO_3)_2(aq) + 2NO(g) + 4H_2O(l)$$

STEP 5

Check the elements and charges.

Elements $3Cu, 8H, 8N, 24O \rightarrow 3Cu, 8H, 8N, 24O$

Charges $0 \rightarrow 0$

19 Radioactivity and Nuclear Energy

CONTENTS

This colorful image of the Sun's corona, or hot outer atmosphere,
is from a photo taken by the Skylab 3 space station.

648   19 Radioactivity and Nuclear Energy

A PET scanner measures brain activity by monitoring the positron emissions of radioactively labeled glucose that has been administered to the patient.

Isotopes are introduced in Section 4.7.

The atomic number (Z) is the number of protons in a nucleus; the mass number (A) is the sum of the numbers of protons and neutrons in a nucleus.

Because the chemistry of an atom is determined by the number and arrangement of its electrons, the properties of the nucleus do not strongly affect the chemical behavior of an atom. Therefore, you might be wondering why there is a chapter on the nucleus in a chemistry textbook. The reason for this chapter is that the nucleus is very important to all of us—a quick reading of any daily newspaper will testify to that. This chapter considers those aspects of the nucleus that you should know about.

Several facts about the nucleus are immediately impressive: its very small size, its very large density, and the energy that holds it together. The radius of a typical nucleus is about 10^{-13} cm, only a hundred-thousandth the radius of a typical atom. In fact, if the nucleus of the hydrogen atom were the size of a ping-pong ball, the electron in the $1s$ orbital would be, on the average, 0.5 km (0.3 mile) away. The density of the nucleus is equally impressive; it is approximately 1.6×10^{14} g/cm^3. A sphere of nuclear material the size of a ping-pong ball would have a mass of 2.5 *billion tons!* Finally, the energies involved in nuclear processes are typically millions of times larger than those associated with normal chemical reactions, a fact that makes nuclear processes potentially attractive for generating energy.

The nucleus is believed to be made of particles called **nucleons (neutrons** and **protons**). Recall from Chapter 4 that the number of protons in a nucleus is called the **atomic number** (Z) and that the sum of the numbers of neutrons and protons is the **mass number** (A). Atoms that have identical atomic numbers but different mass numbers are called **isotopes.** The general term **nuclide** is applied to each unique atom, and we represent it as follows:

Mass number
↓
$_{Z}^{A}\text{X}$ ← Element symbol
↑
Atomic number

where X represents the symbol for a particular element. For example, the following nuclides constitute the common isotopes of carbon: carbon-12, $_{6}^{12}$C; carbon-13, $_{6}^{13}$C; and carbon-14, $_{6}^{14}$C. Notice that all the carbon nuclides have six protons ($Z = 6$) and that they have six, seven, and eight neutrons, respectively.

19.1   Radioactive Decay

AIM: To classify types of radioactive decay.
To show how to write nuclear equations that describe radioactive decay.

Students often have misconceptions about what radioactivity is and why some substances are radioactive and others are not.

Many nuclei are **radioactive;** that is, they spontaneously decompose, forming a different nucleus and producing one or more particles. An example is carbon-14, which decays as shown in the equation

$$^{14}_{6}\text{C} \rightarrow \,^{14}_{7}\text{N} + \,^{0}_{-1}\text{e}$$

where $^{0}_{-1}\text{e}$ represents an electron, which in nuclear terminology is called a **beta particle,** or **β particle.** This **nuclear equation,** which is typical of those representing radioactive decay, is quite different from the chemical equations we have written before. Recall that in a balanced chemical equation the atoms must be conserved. In a nuclear equation *both the atomic number (Z) and the mass number (A) must be conserved.* That is, the sums of the Z values on both sides of the arrow must be equal, and the same restriction applies to the A values. For example, in the above equation, the sum of the Z values is 6 on both sides of the arrow (6 and 7 − 1), and the sum of the A values is 14 on both sides of the arrow (14 and 14 + 0). Notice that the mass number for the β particle is zero; the mass of the electron is so small that it can be neglected here. Of the approximately 2000 known nuclides, only 279 do not undergo radioactive decay. Tin has the largest number of nonradioactive isotopes—ten.

Types of Radioactive Decay

There are several different types of radioactive decay. One frequently observed decay process involves production of an **alpha (α) particle,** which is a helium nucleus ($^{4}_{2}\text{He}$). **Alpha-particle production** is a very common mode of decay for heavy radioactive nuclides. For example, $^{222}_{88}\text{Ra}$, radium-222, decays by α-particle production to give radon-218.

$$^{222}_{88}\text{Ra} \rightarrow \,^{4}_{2}\text{He} + \,^{218}_{86}\text{Rn}$$

Notice in this equation that the mass number is conserved (222 = 4 + 218) and the atomic number is conserved (88 = 2 + 86). Another α-particle producer is $^{230}_{90}\text{Th}$.

$$^{230}_{90}\text{Th} \rightarrow \,^{4}_{2}\text{He} + \,^{226}_{88}\text{Ra}$$

Notice that the loss of an α particle results in a loss of 4 in mass number (A) and a loss of 2 in atomic number (Z).

    β-particle production is another common decay process. For example, the thorium-234 nuclide produces a β particle as it changes to protactinium-234.

$$^{234}_{90}\text{Th} \rightarrow \,^{234}_{91}\text{Pa} + \,^{0}_{-1}\text{e}$$

Iodine-131 is also a β-particle producer.

$$^{131}_{53}\text{I} \rightarrow \,^{0}_{-1}\text{e} + \,^{131}_{54}\text{Xe}$$

Recall that the β particle is assigned mass number 0, because its mass is tiny compared to that of a proton or neutron. The value of Z is −1 for the β particle, so the atomic number for the new nuclide is greater by 1 than the atomic number for the original nuclide. Therefore *the net effect of β-particle production is to change a neutron to a proton.*

Students usually can balance nuclear equations without too much difficulty. They should memorize the names, symbols, and values of A and Z for the common nuclear particles, as given in Table 19.1.

Over 85% of all known nuclides are radioactive.

When discussing the formation of ions in chemical reactions we stressed that the number of electrons may change but that the number of protons always remains the same. In some nuclear reactions, however, the number of protons is effectively increased by one. Students can become confused by this difference.

Notice that Z and A both balance in each of these nuclear equations.

Loss of a β particle results in no change in mass number (A), and an increase of 1 in atomic number (Z).

A **gamma ray,** or γ **ray,** is a high-energy photon of light. A nuclide in an excited nuclear energy state can release excess energy by producing a gamma ray, and γ-ray production often accompanies nuclear decays of various types. For example, in the α-particle decay of $^{238}_{92}U$:

$$^{238}_{92}U \rightarrow {}^4_2He + {}^{234}_{90}Th + 2{}^0_0\gamma$$

A gamma ray is a high energy photon.

The $^0_0\gamma$ notation indicates $Z = 0$ and $A = 0$ for a γ ray. A gamma ray is often simply indicated by γ.

two γ rays of different energies are produced in addition to the α particle (^4_2He). Gamma rays are photons of light and so have zero charge and zero mass number.

Loss of a γ ray results in no change in mass number (A), and no change in atomic number (Z).

The **positron** is a particle with the same mass as the electron but opposite charge. An example of a nuclide that decays by positron production is sodium-22.

$$^{22}_{11}Na \rightarrow {}^0_1e + {}^{22}_{10}Ne$$

Note that the production of a positron appears to change a proton to a neutron.

Loss of a positron results in no change in mass number (A), and a decrease of 1 in atomic number (Z).

Electron capture is a process where one of the inner-orbital electrons is captured by the nucleus, as illustrated by the process

$$^{201}_{80}Hg + {}^0_{-1}e \rightarrow {}^{201}_{79}Au + {}^0_0\gamma$$

↑
Inner-orbital electron

This reaction would have been of great interest to the alchemists, but unfortunately it does not occur often enough to make it a practical means of changing mercury to gold. Gamma rays are always produced along with electron capture.

Table 19.1 lists the common types of radioactive decay, with examples.

It can be interesting to discuss alchemy here. Alchemists attempted to convert base metals to precious metals by chemical means, but we now know that only by nuclear reactions can we convert one element to another. In a reaction presented on page 650, Hg is converted to Au. Although it is not a practical process, it is fun to discuss.

Table 19.1 Various Types of Radioactive Processes

Process	Example
β-particle (electron) production	$^{227}_{89}Ac \rightarrow {}^{227}_{90}Th + {}^0_{-1}e$
positron production	$^{13}_7N \rightarrow {}^{13}_6C + {}^0_1e$
electron capture	$^{73}_{33}As + {}^0_{-1}e \rightarrow {}^{73}_{32}Ge$
α-particle production	$^{210}_{84}Po \rightarrow {}^{206}_{82}Pb + {}^4_2He$
γ-ray production	excited nucleus $\rightarrow$ ground-state nucleus $+ {}^0_0\gamma$ (excess energy) (lower energy)

Figure 19.1
The decay series from $^{238}_{92}U$ to $^{206}_{82}Pb$. Each nuclide in the series except $^{206}_{82}Pb$ is radioactive, and the successive transformations (shown by the arrows) continue until $^{206}_{82}Pb$ is finally formed. The horizontal blue arrows indicate β-particle production (Z increases by 1 and A is unchanged). The diagonal red arrows signify α-particle production (both A and Z decrease).

Often a radioactive nucleus cannot reach a stable (nonradioactive) state through a single decay process. In such a case, a **decay series** occurs until a stable nuclide is formed. A well-known example is the decay series that starts with $^{238}_{92}U$ and ends with $^{206}_{82}Pb$, as shown in Figure 19.1. Similar series exist for $^{235}_{92}U$:

$$^{235}_{92}U \xrightarrow[\text{decays}]{\text{Series of}} {}^{207}_{82}Pb$$

and for $^{232}_{90}Th$:

$$^{232}_{90}Th \xrightarrow[\text{decays}]{\text{Series of}} {}^{208}_{82}Pb$$

EXAMPLE 19.1 Writing Nuclear Equations, I

Write balanced nuclear equations for each of the following processes.

a. $^{11}_{6}C$ produces a positron.
b. $^{214}_{83}Bi$ produces a β particle.
c. $^{237}_{93}Np$ produces an α particle.

EXAMPLE 19.1, CONTINUED

SOLUTION

a. We must find the product nuclide represented by $^A_Z X$ in the following equation:

$$^{11}_{6}C \rightarrow {}^{0}_{1}e + {}^A_Z X$$

↑
Positron

The key to solving this problem is to recognize that both A and Z must be conserved. That is, we can find the identity of $^A_Z X$ by recognizing that the sums of the Z and the A values must be the same on both sides of the equation. Thus for X, Z must be 5 because $Z + 1 = 6$. A must be 11 because $11 + 0 = 11$. Therefore $^A_Z X$ is $^{11}_{5}B$. (The fact that Z is 5 tells us that the nuclide is boron. See the periodic table on the inside front cover of the book.) So the balanced equation is $^{11}_{6}C \rightarrow {}^{0}_{1}e + {}^{11}_{5}B$

Students should check their balanced nuclear equations. The algebraic sum of A and of Z must be the same on each side.

CHECK:
Left Side		*Right Side*
$Z = 6$	$\rightarrow$	$Z = 5 + 1 = 6$
$A = 11$		$A = 11 + 0 = 11$

b. Knowing that a β particle is represented by $_{-1}^{0}e$, we can write

$$^{214}_{83}Bi \rightarrow {}^{0}_{-1}e + {}^A_Z X$$

where $Z - 1 = 83$ and $A + 0 = 214$. This means $Z = 84$ and $A = 214$. We can now write

$$^{214}_{83}Bi \rightarrow {}^{0}_{-1}e + {}^{214}_{84}X$$

Using the periodic table, we find that $Z = 84$ for the element polonium, so $^{214}_{84}X$ must be $^{214}_{84}Po$.

CHECK:
Left Side		*Right Side*
$Z = 83$	$\rightarrow$	$Z = 84 - 1 = 83$
$A = 214$		$A = 214 + 0 = 214$

c. Because an α particle is represented by ^4_2He, we can write

$$^{237}_{93}Np \rightarrow {}^4_2He + {}^A_Z X$$

where $A + 4 = 237$ or $A = 237 - 4 = 233$, and $Z + 2 = 93$ or $Z = 93 - 2 = 91$. Thus $A = 233$, $Z = 91$, and the balanced equation must be

$$^{237}_{93}Np \rightarrow {}^4_2He + {}^{233}_{91}Pa$$

CHECK:
Left Side		*Right Side*
$Z = 93$	$\rightarrow$	$Z = 91 + 2 = 93$
$A = 237$		$A = 233 + 4 = 237$

EXAMPLE 19.1, CONTINUED

SELF-CHECK EXERCISE 19.1

The decay series for $^{238}_{92}U$ is represented in Figure 19.1. Write the balanced nuclear equation for each of the following radioactive decays.

a.   Alpha-particle production by $^{226}_{88}Ra$
b.   Beta-particle production by $^{214}_{82}Pb$

EXAMPLE 19.2	Writing Nuclear Equations, II

Students have more difficulty supplying the missing particle when two isotopes are given than they do supplying missing isotopes when an isotope and a particle are given.

In each of the following nuclear reactions, supply the missing particle.

a. $^{195}_{79}Au + ? \rightarrow {}^{195}_{78}Pt$
b. $^{38}_{19}K \rightarrow {}^{38}_{18}Ar + ?$

SOLUTION

a. A does not change and Z for Pt is 1 lower than Z for Au, so the missing particle must be an electron.

$$^{195}_{79}Au + {}^{0}_{-1}e \rightarrow {}^{195}_{78}Pt$$

CHECK: _Left Side_   _Right Side_

$$\begin{array}{cc} Z = 79 - 1 = 78 & Z = 78 \\ A = 195 + 0 = 195 & A = 195 \end{array}$$

This is an example of electron capture.

b.   For Z and A to be conserved, the missing particle must be a positron.

$$^{38}_{19}K \rightarrow {}^{38}_{18}Ar + {}^{0}_{1}e$$

CHECK: _Left Side_   _Right Side_

$$\begin{array}{cc} Z = 19 & Z = 18 + 1 = 19 \\ A = 38 & A = 38 + 0 = 38 \end{array}$$

Potassium-38 decays by positron production.

SELF-CHECK EXERCISE 19.2

Supply the missing species in each of the following nuclear equations.

a. $^{222}_{86}Rn \rightarrow {}^{218}_{84}Po + ?$
b. $^{15}_{8}O \rightarrow ? + {}^{0}_{1}e$

Nuclear transformations are balanced just as nuclear decay reactions are, so they should present no additional difficulties for students.

19.2 Nuclear Transformations

AIM: To show how one element may be changed into another by particle bombardment.

In 1919 Lord Rutherford observed the first **nuclear transformation,** *the change of one element into another.* He found that bombarding $^{14}_{7}N$ with α particles produced the nuclide $^{17}_{8}O$:

$$^{14}_{7}N + {}^{4}_{2}He \rightarrow {}^{17}_{8}O + {}^{1}_{1}H$$

with a proton ($^{1}_{1}H$) released as another product. Fourteen years later, Irene Curie and her husband Frederick Joliot observed a similar transformation from aluminum to phosphorus:

$$^{27}_{13}Al + {}^{4}_{2}He \rightarrow {}^{30}_{15}P + {}^{1}_{0}n$$

where $^{1}_{0}n$ represents a neutron that is released in the process.

Notice that in both of these cases the bombarding particle is a helium nucleus (an α particle). Other small nuclei, such as $^{12}_{6}C$ and $^{15}_{7}N$, can also be used to bombard heavier nuclei and cause transformations. However, because these positive bombarding ions are repelled by the positive charge of the target nucleus, the bombarding particle must be moving at a very high speed to penetrate the target. These high speeds are achieved in various types of *particle accelerators.*

Neutrons are also employed as bombarding particles to effect nuclear transformations. However, because neutrons are uncharged (and thus not repelled by

Irene Curie and Frederick Joliot.

Stanford University's linear particle accelerator.

Table 19.2 Syntheses of Some of the Transuranium Elements

Element	Neutron Bombardment
neptunium ($Z = 93$)	$^{238}_{92}U + ^{1}_{0}n \rightarrow ^{239}_{92}U \rightarrow ^{239}_{93}Np + ^{0}_{-1}e$
americium ($Z = 95$)	$^{239}_{94}Pu + 2\,^{1}_{0}n \rightarrow ^{241}_{94}Pu \rightarrow ^{241}_{95}Am + ^{0}_{-1}e$

Element	Positive-Ion Bombardment
curium ($Z = 96$)	$^{239}_{94}Pu + ^{4}_{2}He \rightarrow ^{242}_{96}Cm + ^{1}_{0}n$
californium ($Z = 98$)	$^{242}_{96}Cm + ^{4}_{2}He \rightarrow ^{245}_{98}Cf + ^{1}_{0}n$ or
	$^{238}_{92}U + ^{12}_{6}C \rightarrow ^{246}_{98}Cf + 4\,^{1}_{0}n$
element 104	$^{249}_{98}Cf + ^{12}_{6}C \rightarrow ^{257}_{104}X + 4\,^{1}_{0}n$
element 105	$^{249}_{98}Cf + ^{15}_{7}N \rightarrow ^{260}_{105}X + 4\,^{1}_{0}n$
element 106	$^{249}_{98}Cf + ^{18}_{8}O \rightarrow ^{263}_{106}X + 4\,^{1}_{0}n$

a target nucleus), they are readily absorbed by many nuclei, producing new nuclides. The most common source of neutrons for this purpose is a fission reactor (see Section 19.8).

By using neutron and positive-ion bombardment, scientists have been able to extend the periodic table—that is, to produce chemical elements that are not present naturally. Prior to 1940, the heaviest known element was uranium ($Z = 92$), but in 1940, neptunium ($Z = 93$) was produced by neutron bombardment of $^{238}_{92}U$. The process initially gives $^{239}_{92}U$, which decays to $^{239}_{93}Np$ by β-particle production.

$$^{238}_{92}U + ^{1}_{0}n \rightarrow ^{239}_{92}U \rightarrow ^{239}_{93}Np + ^{0}_{-1}e$$

In the years since 1940, the elements with atomic numbers 93 through 109, called the **transuranium elements,** have been synthesized. Table 19.2 gives some examples of these processes.

Element 108 has not yet been reported.

19.3 Detection of Radioactivity and the Concept of Half-Life

AIM: To discuss radiation detection instruments.
To explain half-life.

The most familiar instrument for measuring radioactivity levels is the **Geiger–Müller counter,** or **Geiger counter** (Figure 19.2, p. 656). High-energy particles from radioactive decay produce ions when they travel through matter. The probe of the Geiger counter contains argon gas. The argon atoms have no charge, but

If you have one available, bring a Geiger counter to class for demonstration. A mantel from a gas camping light or a piece of orange Fiestaware pottery can provide the source of radioactive material.

Geiger counters are commonly called survey meters.

Figure 19.2
A schematic representation of a Geiger–Müller counter. The high-speed particle knocks electrons off argon atoms to form ions:

$$Ar \xrightarrow{\text{Particle}} Ar^+ + e^-$$

and a pulse of current flows.

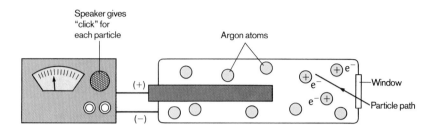

they can be ionized by a rapidly moving particle.

$$Ar(g) \xrightarrow[\text{particle}]{\text{High-energy}} Ar^+(g) + e^-$$

That is, the fast-moving particle "knocks" electrons off some of the argon atoms. Although a sample of uncharged argon atoms does not conduct a current, the ions and electrons formed by the high-energy particle allow a current to flow momentarily, so a "pulse" of current flows every time a particle enters the probe. The Geiger counter detects each pulse of current, and these events are counted.

A **scintillation counter** is another instrument often employed to detect radioactivity. This device uses a substance, such as sodium iodide, that gives off light when it is struck by a high-energy particle. A detector senses the flashes of light and thus counts the decay events.

One important characteristic of a given type of radioactive nuclide is its half-life. The **half-life** is the *time required for half of the original sample of nuclei to decay.* For example, if a certain radioactive sample contains 1000 nuclei at a given time and 500 nuclei (half of the original number) 7.5 days later, this radioactive nuclide has a half-life of 7.5 days.

A given type of radioactive nuclide always has the same half-life. However, the various radioactive nuclides have half-lives that cover a tremendous range. For example, $^{234}_{91}Pa$, protactinium-234 has a half-life of 1.2 minutes, and $^{238}_{92}U$, uranium-238, has a half-life of 4.5×10^9 (4.5 *billion*) years. This means that a sample containing 100 million $^{234}_{91}Pa$ nuclei will have only 50 million $^{234}_{91}Pa$ nuclei in it (half of 100 million) after 1.2 minutes have passed. In another 1.2 minutes, the number of nuclei will decrease to half of 50 million, or 25 million nuclei.

This means that a sample of $^{234}_{91}Pa$ with 100 million nuclei will show 50 million decay events (50 million $^{234}_{91}Pa$ nuclei will decay) over a time of 1.2 minutes. By contrast, a sample containing 100 million $^{238}_{92}U$ nuclei will undergo 50 million decay events over 4.5 billion years. Therefore $^{234}_{91}Pa$ shows much greater activity than $^{238}_{92}U$. We say that $^{234}_{91}Pa$ is much "hotter" than $^{238}_{92}U$.

Thus, a radioactive nucleus with a short half-life is much more likely than one with a long half-life to decay at any particular time.

EXAMPLE 19.3   Understanding Half-Life

Table 19.3 lists various radioactive nuclides of radium.

a. Order these nuclides in terms of activity (from most decays per day to least).

b. How long will it take for a sample containing 1.00 mol of $^{223}_{88}$Ra to reach a point where it contains only 0.25 mol of $^{223}_{88}$Ra?

Table 19.3	The Half-lives for the Radioactive Nuclides of Radium
Nuclide	**Half-life**
$^{223}_{88}$Ra	12 days
$^{224}_{88}$Ra	3.6 days
$^{225}_{88}$Ra	15 days
$^{226}_{88}$Ra	1600 years
$^{228}_{88}$Ra	6.7 years

SOLUTION

a. The shortest half-life indicates the greatest activity (the most decays over a given period of time). Therefore the order is

Most activity (shortest half-life)   Least activity (longest half-life)

$^{224}_{88}$Ra > $^{223}_{88}$Ra > $^{225}_{88}$Ra > $^{228}_{88}$Ra > $^{226}_{88}$Ra

(3.6 days)   (12 days)   (15 days)   (6.7 years)   (1600 years)

b. In one half-life (12 days), the sample will decay from 1.00 mol of $^{223}_{88}$Ra to 0.50 mol of $^{223}_{88}$Ra. In the next half-life (another 12 days), it will decay from 0.50 mol of $^{223}_{88}$Ra to 0.25 mol of $^{223}_{88}$Ra.

1.00 mol $^{223}_{88}$Ra ⟶ 12 days ⟶ 0.50 mol $^{223}_{88}$Ra ⟶ 12 days ⟶ 0.25 mol $^{223}_{88}$Ra

Therefore it will take 24 days (two half-lives) for the sample to change from 1.00 mol of $^{223}_{88}$Ra to 0.25 mol of $^{223}_{88}$Ra.

SELF-CHECK EXERCISE 19.3

Watches with numerals that "glow in the dark" were formerly made by including radioactive radium in the paint used to letter the watch faces. Assume that to make the numeral 3 on a given watch, a sample of paint containing 8.0×10^{-7} mol of $^{228}_{88}$Ra was used. This watch was then put in a drawer and forgotten. Many years later someone finds the watch and wishes to know when it was made. Analyzing the paint, this person finds 1.0×10^{-7} mol of $^{228}_{88}$Ra in the numeral 3. How much time elapsed between the making of the watch and the finding of the watch?

HINT: Use the half-life of $^{228}_{88}$Ra from Table 19.3.

A watch dial with radium paint.

19.4   Dating by Radioactivity

AIM: To show how objects can be dated by radioactivity.

Radiocarbon dating is interesting to students and provides a good illustration of the application of half-life.

Archaeologists, geologists, and others involved in reconstructing the ancient history of the earth rely heavily on the half-lives of radioactive nuclei to provide accurate dates for artifacts and rocks. A method for dating ancient articles made from wood or cloth is **radiocarbon dating,** or **carbon-14 dating,** a technique originated in the 1940s by Willard Libby, an American chemist who received the Nobel Prize for his efforts.

Radiocarbon dating is based on the radioactivity of $^{14}_{6}C$, which decays by β-particle production.

$$^{14}_{6}C \rightarrow {}^{0}_{-1}e + {}^{14}_{7}N$$

Carbon-14 is continuously produced in the atmosphere when high-energy neutrons from space collide with nitrogen-14.

$$^{14}_{7}N + {}^{1}_{0}n \rightarrow {}^{14}_{6}C + {}^{1}_{1}H$$

Carbon-14 is produced naturally in the environment; when it is incorporated into carbon dioxide, it can be fixed as carbohydrate in plants. Carbon dioxide containing carbon-14 is an example of natural sources of radioactivity in the environment with which we come in contact daily.

Just as carbon-14 is continuously produced by this process, it continuously decomposes through β-particle production. Over the years, these two opposing processes have come into balance, causing the amount of $^{14}_{6}C$ present in the atmosphere to remain approximately constant.

Carbon-14 can be used to date wood and cloth artifacts because the $^{14}_{6}C$, along with the other carbon isotopes in the atmosphere, reacts with oxygen to form carbon dioxide. A living plant consumes this carbon dioxide in the photosynthesis process and incorporates the carbon, including $^{14}_{6}C$, into its molecules. As long as the plant lives, the $^{14}_{6}C$ content in its molecules remains the same as in the atmosphere because of the plant's continuous uptake of carbon. However, as soon as a tree is cut to make a wooden bowl or a flax plant is harvested to make linen, it stops taking in carbon. There is no longer a source of $^{14}_{6}C$ to replace that lost to radioactive decay, so the material's $^{14}_{6}C$ content begins to decrease.

Because the half-life of $^{14}_{6}C$ is known to be 5730 years, a wooden bowl found in an archaeological dig that shows a $^{14}_{6}C$ content of half that found in currently living trees is approximately 5730 years old. That is, because half of the $^{14}_{6}C$ present when the tree was cut has disappeared, the tree must have been cut one half-life of $^{14}_{6}C$ ago.

Carbon-14 radioactivity is used to date human skeletons found at archaeological sites.

19.5 Medical Applications of Radioactivity

AIM: To discuss the use of radiotracers in medicine.

Although we owe the rapid advances of the medical sciences in recent decades to many causes, one of the most important has been the discovery and use of **radiotracers**—radioactive nuclides that can be introduced into organisms in food or drugs and subsequently *traced* by monitoring their radioactivity. For example, the incorporation of nuclides such as $^{14}_{6}C$ and $^{32}_{15}P$ into nutrients has yielded important information about how these nutrients are used to provide energy for the body.

Iodine-131 has proved very useful in the diagnosis and treatment of illnesses of the thyroid gland. Patients drink a solution containing a small amount of NaI that includes ^{131}I, and the uptake of the iodine by the thyroid gland is monitored with a scanner (Figure 19.3).

Thallium-201 can be used to assess the damage to the heart muscle in a person who has suffered a heart attack, because thallium becomes concentrated in healthy muscle tissue. Technetium-99, which is also taken up by normal heart tissue, is used for damage assessment in a similar way.

Radiotracers provide sensitive and nonsurgical methods for learning about biological systems, for detecting disease, and for monitoring the action and effectiveness of drugs. Some useful radiotracers are listed in Table 19.4.

Your students may find it interesting to compare the half-lives of radiotracers with that of carbon-14. The useful properties of each depend in part on large differences in half-lives.

Nuclides used as radiotracers have short half-lives so that they disappear rapidly from the body.

Table 19.4 Some Radioactive Nuclides, Their Half-Lives, and Their Medical Applications as Radiotracers*

Nuclide	Half-life	Area of the Body Studied
^{131}I	8.1 days	thyroid
^{59}Fe	45.1 days	red blood cells
^{99}Mo	67 hours	metabolism
^{32}P	14.3 days	eyes, liver, tumors
^{51}Cr	27.8 days	red blood cells
^{87}Sr	2.8 hours	bones
^{99}Tc	6.0 hours	heart, bones, liver, lungs
^{133}Xe	5.3 days	lungs
^{24}Na	14.8 hours	circulatory system

*Z is sometimes not written when listing nuclides.

(a)

(b)

Figure 19.3
After consumption of Na^{131}I, the patient's thyroid is scanned for radioactivity levels to determine the efficiency of iodine absorption. (a) Scan of radioactive iodine in a normal thyroid. (b) Scan of an enlarged thyroid.

Both fission and fusion processes demonstrate the great energy involved in nuclear reactions. During the course we have discussed the energy absorbed or given off by chemical reactions. The fact that the amount of energy involved with nuclear processes is much greater than the amount involved with chemical reactions is of interest to students.

19.6 Nuclear Energy

AIM: To introduce fusion and fission as producers of nuclear energy.

The protons and the neutrons in atomic nuclei are bound together with forces that are much greater than the forces that bind atoms together to form molecules. In fact, the energies associated with nuclear processes are more than a million times those associated with chemical reactions. This potentially makes the nucleus a very attractive source of energy.

Because medium-sized nuclei contain the strongest binding forces ($^{56}_{26}Fe$ has the strongest binding forces of all), there are two types of nuclear processes that produce energy:

1. Combining two light nuclei to form a heavier nucleus. This process is called **fusion.**
2. Splitting a heavy nucleus into two nuclei with smaller mass numbers. This process is called **fission.**

As we will see in the next several sections, these two processes can supply amazing quantities of energy with relatively small masses of materials consumed.

19.7 Nuclear Fission

AIM: To describe nuclear fission.

Nuclear fission was discovered in the late 1930s when $^{235}_{92}U$ nuclides bombarded with neutrons were observed to split into two lighter elements.

$$^{1}_{0}n + ^{235}_{92}U \rightarrow ^{142}_{56}Ba + ^{91}_{36}Kr + 3\,^{1}_{0}n$$

This process, shown schematically in Figure 19.4, releases 2.1×10^{13} joules per mole of $^{235}_{92}U$. Compared to what we get from typical fuels, this is a huge amount of energy. For example, the fission of 1 mol of $^{235}_{92}U$ produces about *26 million times* as much energy as the combustion of 1 mol of methane.

The process shown above is only one of the many fission reactions that $^{235}_{92}U$ can undergo. In fact, over 200 different isotopes of 35 different elements have been observed among the fission products of $^{235}_{92}U$.

In addition to the product nuclides, neutrons are also produced in the fission reactions of $^{235}_{92}U$. As these neutrons fly through the solid sample of uranium, they may collide with other $^{235}_{92}U$ nuclei, producing additional fission events. Each of these fission events produces more neutrons that can, in turn, produce the fission

Major General Leslie R. Groves (right), chief of the Manhattan Project in which the first atomic bomb was developed, and Dr. J. R. Oppenheimer, director of the Los Alamos laboratory and physicist at the California Institute of Technology, view the base of the steel tower on which the first atomic bomb hung when tested near Alamogordo, New Mexico in July, 1945.

The fission reaction is similar to other nuclear reactions in this chapter. Students should be able to understand the chemistry of the reaction itself. If you have time, you can easily generate a discussion among the students about the pros and cons of using nuclear energy.

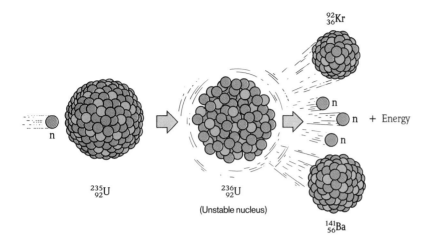

Figure 19.4
Upon capturing a neutron, the $^{235}_{92}U$ nucleus undergoes fission to produce two lighter nuclides, more neutrons (typically three), and a large amount of energy.

of more $^{235}_{92}U$ nuclei. Because each fission event produces neutrons, the process can be self-sustaining. We call it a **chain reaction** (Figure 19.5). In order for the fission process to be self-sustaining, at least one neutron from each fission event must go on to split another nucleus. If, on the average, *less than one* neutron causes another fission event, the process dies out. If *exactly one* neutron from each fission event causes another fission event, the process sustains itself at the same level and is said to be *critical.* If *more than one* neutron from each fission event causes another fission event, the process rapidly escalates and the heat build-up causes a violent explosion.

To achieve the critical state, a certain mass of fissionable material, called the **critical mass,** is needed. If the sample is too small, too many neutrons escape before they have a chance to cause a fission event, and the process stops.

During World War II, the United States carried out an intense research effort called the Manhattan Project to build a bomb based on the principles of nuclear fission. This program produced the fission bomb, which was used with devastating effect on the cities of Hiroshima and Nagasaki in 1945. Basically, a fission bomb operates by suddenly combining two subcritical masses, which results in rapidly escalating fission that produces an explosion of incredible intensity.

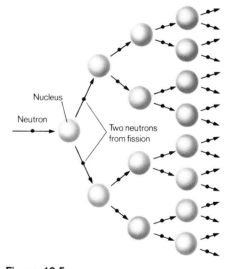

Figure 19.5
Representation of a fission process in which each event produces two neutrons that can go on to split other nuclei, leading to a self-sustaining chain reaction.

19.8   Nuclear Reactors

AIM: To explain how a nuclear reactor works.

Because of the tremendous energies involved, fission has been developed as an energy source to produce electricity in reactors where controlled fission can occur. The resulting energy is used to heat water to produce steam that runs

Students are always curious about how nuclear reactors work. This is a good place to point out the tradeoffs inherent in that technology.

Figure 19.6

A schematic diagram of a nuclear power plant. The energy from the fission process is used to boil water, producing steam for use in a turbine-driven generator. Cooling water from a lake or river is used to condense the steam after it leaves the turbine.

turbine generators, in much the same way that a coal-burning power plant generates energy by heating water to produce steam. A schematic diagram of a nuclear power plant is shown in Figure 19.6.

In the reactor core (Figure 19.7) uranium that has been enriched to approximately 3% $^{235}_{92}U$ (natural uranium contains only 0.7% $^{235}_{92}U$) is housed in metal cylinders. A *moderator* surrounding the cylinders slows the neutrons down so that the uranium fuel can capture them more efficiently. *Control rods,* composed of substances (such as cadmium) that absorb neutrons, are used to regulate the power level of the reactor. The reactor is designed so that if a malfunction occurs, the control rods are automatically inserted into the core to absorb neutrons and stop the reaction. A liquid (usually water) is circulated through the core to extract the heat generated by the energy of fission. This heat energy is then used to change water to steam, which runs turbines that in turn run electrical generators.

Although the concentration of $^{235}_{92}U$ in the fuel elements is not great enough to allow an explosion such as that which occurs in a fission bomb, a failure of the cooling system can lead to temperatures high enough to melt the reactor core. This means that the building housing the core must be designed to contain the core even in the event of such a "melt-down." A great deal of controversy now exists about the efficiency of the safety systems in nuclear power plants. Accidents such as the one at the Three Mile Island facility in Pennsylvania in 1979 and the one at Chernobyl in the Soviet Union in 1986 have led many people to question the wisdom of continuing to build fission-based power plants.

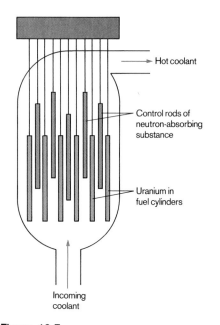

Figure 19.7

A schematic of the reactor core.

The products of fission and breeder reactors are completely different. It can be interesting to compare and contrast the two.

19.9 Nuclear Fusion   **663**

Breeder Reactors

One potential problem facing the nuclear power industry is the limited supply of $^{235}_{92}U$. Some scientists believe we have nearly depleted the uranium deposits that are rich enough in $^{235}_{92}U$ to make the production of fissionable fuel economically feasible. Because of this possibility, reactors have been developed in which fissionable fuel is actually produced while the reactor runs. In these **breeder reactors,** the major component of natural uranium, nonfissionable $^{238}_{92}U$, is changed to fissionable $^{239}_{94}Pu$. The reaction involves absorption of a neutron, followed by production of two β particles.

$$^{1}_{0}n + {}^{238}_{92}U \rightarrow {}^{239}_{92}U$$

$$^{239}_{92}U \rightarrow {}^{239}_{93}Np + {}^{0}_{-1}e$$

$$^{239}_{93}Np \rightarrow {}^{239}_{94}Pu + {}^{0}_{-1}e$$

As the reactor runs and $^{235}_{92}U$ is split, some of the excess neutrons are absorbed by $^{238}_{92}U$ to produce $^{239}_{94}Pu$. The $^{239}_{94}Pu$ is then separated out and used to fuel another reactor. Such a reactor thus "breeds" nuclear fuel as it operates.

Although breeder reactors are now used in Europe, the United States is proceeding slowly with their development because much controversy surrounds their use. One problem involves the hazards that arise in handling plutonium, which is very toxic and flames on contact with air.

Natural uranium consists mostly of $^{238}_{92}U$.

The core of a nuclear power plant.

19.9   Nuclear Fusion

AIM: To describe nuclear fusion.

The process of combining two light nuclei—called **nuclear fusion**—produces even more energy per mole than does nuclear fission. In fact, stars produce their energy through nuclear fusion. Our sun, which presently consists of 73% hydrogen, 26% helium, and 1% other elements, gives off vast quantities of energy from the fusion of protons to form helium. One possible scheme for this process is

$$^{1}_{1}H + {}^{1}_{1}H \rightarrow {}^{2}_{1}H + {}^{0}_{1}e + \text{energy}$$

$$^{1}_{1}H + {}^{2}_{1}H \rightarrow {}^{3}_{2}He + \text{energy}$$

$$^{3}_{2}He + {}^{3}_{2}He \rightarrow {}^{4}_{2}He + 2\,{}^{1}_{1}H + \text{energy}$$

$$^{3}_{2}He + {}^{1}_{1}H \rightarrow {}^{4}_{2}He + {}^{0}_{1}e + \text{energy}$$

Intense efforts are under way to develop a feasible fusion process because of the ready availability of many light nuclides (deuterium, $^{2}_{1}H$, in seawater, for example) that can serve as fuel in fusion reactors. However, initiating the fusion

Some students are familiar with the term *fusion,* although they don't know what it means. Because the reactions involved in fusion occur within the sun, they have potential as a source of clean energy on earth.

CHEMISTRY IN FOCUS

Stellar Nucleosynthesis

*H*ow did all the matter around us originate? One scientific answer to this question is a theory called *stellar nucleosynthesis*—the formation of nuclei in stars.

Many scientists believe that our universe originated as a cloud of neutrons that became unstable and produced an immense explosion, giving this model its name: *the big bang theory.* The model postulates that after the initial explosion, neutrons decomposed into protons and electrons:

$$_0^1 n \rightarrow {}_1^1 H + {}_{-1}^0 e$$

that eventually combined to form clouds of hydrogen. Over the eons, gravitational forces caused many of these hydrogen clouds to contract and heat up sufficiently to reach temperatures at which proton fusion began to occur, releasing large quantities of energy. When the tendency to expand in response to the heat from fusion and the tendency to contract in response to the forces of gravity are balanced, a stable young star such as our sun can form.

Eventually, when the star's supply of hydrogen is exhausted, the core of the star again contracts, with fur-

A painting of a supernova explosion.

ther heating, until it reaches temperatures at which fusion of helium nuclei can occur, leading to the formation of $_6^{12}C$ and $_8^{16}O$ nuclei. In turn, when the supply of helium nuclei runs out, further contraction and heating occur, until the fusion of heavier nuclei takes place. This process occurs repeatedly, forming heavier and heavier nuclei, until iron nuclei are

formed. Because the iron nucleus has the greatest binding forces of all, iron nuclei do not release excess energy when they fuse to form heavier nuclei. This means that there is no further fusion energy to sustain the star, and it cools to a small, dense *white dwarf.*

The evolution just described is characteristic of small and medium-sized stars. Much larger stars, however, become unstable at some time during their evolution and undergo a *supernova explosion.* In this explosion, nuclei with medium mass numbers are fused to produce heavy elements. Also in the explosion, light nuclei capture neutrons. These neutron-rich nuclei then produce β particles, increasing their atomic number with each β decay. This eventually leads to nuclei with large atomic numbers. In fact, almost all nuclei beyond iron are thought to originate in supernova explosions. The debris of such an explosion contains a wide variety of elements, and it is thought that this debris might eventually form a solar system such as our own.

process is much more difficult than initiating fission. The forces that bind nucleons together to form a nucleus become effective only at *very small* distances (approximately 10^{-13} cm), so for two protons to bind together and thereby release energy, they must get very close together. But protons, because they are identically charged, repel each other. This suggests that to get two protons (or two deuterons) close enough to bind together (the strong nuclear binding force is *not* related to charge), they must be "shot" at each other at speeds high enough to overcome their repulsion from each other. The repulsive forces between two $_1^2H$ nuclei are so great that temperatures of about 40 million K are thought to be necessary. Only at these temperatures are the nuclei moving fast enough to overcome the repulsions.

Currently, scientists are studying two types of systems to produce the extremely high temperatures required: high-powered lasers and heating by electric currents. At present many technical problems remain to be solved, and it is not clear whether either method will prove useful.

Recent developments have raised hope that fusion may be possible at room temperatures. In 1989 several chemists performed experiments in which an electric current was passed through liquid D_2O ("heavy water") in the presence of a palladium electrode. These experiments seemed to suggest that deuterium atoms migrate into the interior of the solid palladium, where they may undergo spontaneous fusion. Although these results are quite controversial (and possibly wrong), the amounts of energy produced by this system appear to be greater than what would be expected from known chemical reactions. Scientists are continuing to examine the possibility of "cold fusion." At present there is widespread doubt in the scientific community about the feasibility of cold fusion.

19.10   Effects of Radiation

AIM: To show how radiation damages human tissue.

Everyone knows that being hit by a train is a catastrophic event. The energy transferred in such a collision is very large. In fact, any source of energy is potentially harmful to organisms. Energy transferred to cells can break chemical bonds and cause malfunctioning of the cell systems. This fact is behind our present concern about maintaining the ozone layer in the earth's upper atmosphere, which screens out high-energy ultraviolet radiation arriving from the sun. Radioactive elements, which are sources of high-energy particles, are also potentially hazardous. However, the effects are usually quite subtle, because even though high-energy particles are involved, the quantity of energy actually deposited in tissues *per decay event* is quite small. The resulting damage is no less real, but the effects may not be apparent for years.

Radiation damage to organisms can be classified as somatic or genetic damage. *Somatic damage* is damage to the organism itself, resulting in sickness or death. The effects may appear almost immediately if a massive dose of radiation is received; for smaller doses, damage may appear years later, usually in the form of cancer. *Genetic damage* is damage to the genetic machinery of reproductive cells, creating problems which often afflict the offspring of the organism.

$_1^2H$ particles are called deuterons.

A solar flare erupts from the surface of the sun.

A presentation of the cold fusion controversy can provide the opportunity to discuss many aspects of the scientific method and of scientific ethics that students don't normally hear, but probably should.

CHEMISTRY IN FOCUS

Nuclear Waste Disposal

*O*ur society does not have a very impressive record for safe disposal of industrial wastes. We have polluted our water and air, and some land areas have become virtually uninhabitable because of the improper burial of chemical wastes. As a result, many people are wary about the radioactive wastes from nuclear reactors. The potential threats of cancer and genetic mutations make these materials especially frightening.

Because of its controversial nature, most of the nuclear waste generated over the last 40 years has been placed in temporary storage. However, in 1982 the U.S. Congress passed the Nuclear Waste Policy Act, which established a timetable for choosing and preparing sites for the deep underground disposal of radioactive materials.

The tentative disposal plan calls for incorporation of the spent nuclear fuel into blocks of glass that will be packed in corrosion-resistant metal containers and then buried in a deep, stable rock formation indicated by the rock layers in Figure 19.8.

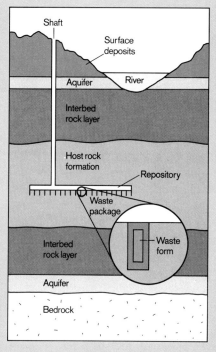

Figure 19.8

A schematic diagram for the tentative plan for deep underground isolation of nuclear waste. The disposal system would consist of a waste package buried in an underground repository.

Adapted with permission from *Chemical & Engineering News,* July 18, 1983, pp. 20–38. Copyright 1983 American Chemical Society.

There are indications that this method will isolate the waste until the radioactivity decays to safe levels. Some reassuring evidence comes from a natural fission "reactor" that has been discovered at Oklo in Gabon, Africa. Spawned about 2 billion years ago when uranium in ore deposits there formed a critical mass, this "reactor" produced fission and fusion products for several thousand years. Although some of these products have migrated away from the site in the intervening 2 billion years, most have stayed in place.

Radon test kits for home use are readily available. If you can, bring one to class and discuss the sources of radon that can be found in homes.

The biological effects of a particular source of radiation depend on several factors:

1. *The energy of the radiation.* The higher the energy content of the radiation, the more damage it can cause.

2. *The penetrating ability of the radiation.* The particles and rays produced in radioactive processes vary in their ability to penetrate human tissue: γ rays are highly penetrating, β particles can penetrate approximately 1 cm, and α particles are stopped by the skin (Figure 19.9).

3. *The ionizing ability of the radiation.* Because ions behave quite differently from neutral molecules, radiation that removes electrons from molecules in living tissues seriously disturbs their functions. The ionizing ability of radiation varies dramatically. For example, γ rays penetrate very deeply but cause only occasional ionization. On the other hand, α particles, although they are not very penetrating, are very effective at causing ionization and produce serious damage. Therefore the ingestion of a producer of α particles, such as plutonium, is particularly damaging.

4. *The chemical properties of the radiation source.* When a radioactive nuclide is ingested, its capacity to cause damage depends on how long it remains in the body. For example, $^{85}_{36}\text{Kr}$ and $^{90}_{38}\text{Sr}$ are both β-particle producers. Because krypton, being a noble gas, is chemically inert, it passes through the body quickly and does not have much time to do damage. Strontium, on the other hand, is chemically similar to calcium. It can collect in bones, where it may cause leukemia and bone cancer.

Figure 19.9
Radioactive particles and rays vary greatly in penetrating power. Gamma rays are by far the most penetrating.

Because of the differences in the behavior of the particles and rays produced by radioactive decay, we have invented a unit called the **rem** which indicates the danger the radiation poses to humans.

Table 19.5 shows the physical effects of short-term exposure to various doses of radiation, and Table 19.6 (p. 668) gives the sources and amounts of the radiation a typical person in the United States is exposed to each year. Note that natural sources contribute about twice as much as human activities to the total exposure. However, although the nuclear industry contributes only a small percentage of the total exposure, controversy surrounds nuclear power plants because of their *potential* for creating radiation hazards. These hazards arise mainly from two sources: accidents allowing the release of radioactive materials, and improper disposal of the radioactive products in spent fuel elements.

Table 19.5 Effects of Short-Term Exposures to Radiation

Dose (rem)	Clinical Effect
0–25	nondetectable
25–50	temporary decrease in white blood cell counts
100–200	strong decrease in white blood cell counts
500	death of half the exposed population within 30 days after exposure

Table 19.6 lists sources and amounts of radiation exposure humans can expect to encounter. You can discuss the amount of radiation we can normally expect to absorb from natural sources.

Generally, students are interested in knowing how X rays are used to make images and whether or not this radiation is a safe diagnostic tool.

Table 19.6   Typical Radiation Exposures for a Person Living in the United States (1 millirem = 10^{-3} rem)

	Exposure (millirems/year)
cosmic	50
from the earth	47
from building materials	3
in human tissues	21
inhalation of air	5
Total from natural sources	126
X-ray diagnosis	50
radiotherapy X rays, radioisotopes	10
internal diagnosis and therapy	1
nuclear power industry	0.2
luminous watch dials, TV tubes, industrial wastes	2
radioactive fallout	4
Total from human activities	67
Total	193 = 0.193 rems

CHAPTER REVIEW

Key Terms

nucleons (neutrons and protons) (p. 648)
atomic number (Z) (p. 648)
mass number (A) (p. 648)
isotopes (p. 648)
nuclide (p. 648)
radioactive nuclide (p. 648)
beta (β) particle (p. 649)
nuclear equation (p. 649)
alpha (α) particle (p. 649)
gamma (γ) ray (p. 650)
positron (p. 650)
electron capture (p. 650)
decay series (p. 651)
nuclear transformation (p. 654)

transuranium elements (p. 655)
Geiger–Müller counter (Geiger counter) (p. 655)
half-life (p. 656)
radiocarbon dating (carbon-14 dating) (p. 658)
radiotracers (p. 659)
fusion (p. 660)
fission (p. 660)
chain reaction (p. 661)
critical mass (p. 661)

Summary

1. Radioactivity is the spontaneous decomposition of a nucleus to form another nucleus and produce one or more particles. We can write a nuclear equation to represent radioactive decay, in which both A (mass number) and Z (atomic number) must be conserved.

2. There are several types of radioactive decay: alpha-particle production, in which an alpha particle (helium nucleus) is produced; beta-particle (or electron) production; the production of gamma rays (high-energy photons of light); and electron capture, in which one of the inner-orbital electrons is captured by the nucleus. Often a series of decays occurs before a radioactive nucleus attains a stable state.

3. The production of new elements by nuclear transformation (the change of one element into another) is carried out by bombarding various nuclei with particles in accelerators. The transuranium elements have been synthesized in this way.

4. The half-life of a radioactive nuclide is the time required for one-half of the original sample to decay. Radiocarbon dating is based on the radioactivity of carbon-14.

5. Radiotracers—radioactive nuclides that can be introduced into organisms in food or drugs and whose pathways can be traced by monitoring their radioactivity—are used diagnostically in medicine.

6. Nuclear fusion is the process of combining two light nuclei to form a heavier, more stable nucleus. Nuclear fission involves the splitting of a heavy nucleus into two (more stable) lighter nuclei. Current nuclear reactors employ controlled fission.

7. Radiation can cause either direct damage to living tissues or damage to reproductive cells that manifests itself in the organism's offspring. The biological effects of radiation depend on the energy of the radiation, the radiation's penetrating ability and ionizing ability, and the chemical properties of the source of the radiation.

Questions and Problems

All even-numbered exercises have answers in the back of this book and solutions in the Solutions Guide.

19.1 Radioactive Decay

QUESTIONS

1. How large is a typical atomic nucleus? How does the size of the nucleus compare with the size of the atom overall?

2. What particles are present in the nucleus that account for its great density? What are the mass numbers of these particles, and what are their charges?

3. What does the *atomic number* of a nucleus represent? What does the *mass number* of a nucleus represent?

4. What is an *isotope*? How do two isotopes of an element differ, and how are they similar?

5. Using Z to represent atomic number and A to represent mass number, give the general symbol for a nuclide of element X.

6. What does it mean to say that a nucleus is *radioactive*?

7. Write the charge, mass number, and symbol for a *beta particle*.

8. Write the charge, mass number, and symbol for an *alpha particle*.

9. When an unstable nucleus produces an alpha particle, by how many units does the atomic number of the nucleus change? Does the atomic number increase or decrease?

10. When an unstable nucleus produces a beta particle, by how many units does the atomic number of the nucleus change? Does the atomic number increase or decrease?

11. By how many units does the mass number of a nucleus decrease when the nucleus produces an alpha particle? By how many units does the mass number change when a beta particle is produced?

12. What does a *gamma ray* represent? Is a gamma ray a particle? Is there a change in mass or atomic number when a nucleus produces only a gamma ray?

13. What is a *positron*? What are the mass number and charge of a positron? How do the mass number and atomic number of a nucleus change when the nucleus produces a positron?

14. What do we mean when we say a nucleus has undergone an *electron capture* process? What type of electron is captured by the nucleus in this process?

PROBLEMS

15. Boron consists primarily of two isotopes, of mass numbers 10 and 12, respectively. Write the nuclear symbol for each of these isotopes.

16. Uranium consists primarily of two isotopes, with mass numbers 235 and 238, respectively. Write nuclear symbols for each of these isotopes.

17. Oxygen exists in several isotopic forms, including $^{16}_{8}O$, $^{15}_{8}O$, and $^{17}_{8}O$. Indicate the number of protons and the number of neutrons in each of these isotopes.

18. Phosphorus exists in several isotopic forms, including those with mass numbers 30, 31, and 32. Indicate the number of protons and the number of neutrons in each of these isotopes.

19. Give the nuclear symbol for each of the following:
    a. a beta particle   c. a neutron
    b. an alpha particle   d. a proton

20. Give the nuclear symbol for each of the following:
    a. an electron   c. a gamma ray
    b. a positron

21. Complete each of the following nuclear equations by supplying the missing particle (or particles).
    a. $^{230}_{90}Th \rightarrow ? + ^{226}_{88}Ra$ c. $^{14}_{7}N + ? \rightarrow ^{17}_{8}O + ^{1}_{1}H$
    b. $^{234}_{91}Pa \rightarrow ? + ^{234}_{92}U$

22. Complete each of the following nuclear equations by supplying the missing particle (or particles).
    a. $^{22}_{11}Na \rightarrow ? + ^{22}_{10}Ne$ c. $^{238}_{92}U \rightarrow ? + ^{4}_{2}He$
    b. $^{230}_{90}Th \rightarrow ^{4}_{2}He + ?$

23. Complete each of the following nuclear equations by supplying the missing particle (or particles).
    a. $? \rightarrow ^{0}_{-1}e + ^{32}_{16}S$ c. $? \rightarrow ^{60}_{28}Ni + ^{0}_{-1}e$
    b. $^{235}_{92}U + ^{1}_{0}n \rightarrow ^{139}_{56}Ba + ^{94}_{36}Kr + ?$

24. Complete each of the following nuclear equations by supplying the missing particle (or particles).
    a. $^{201}_{80}Hg + ? \rightarrow ^{201}_{79}Au$ c. $? \rightarrow ^{210}_{84}Po + ^{0}_{-1}e$
    b. $^{210}_{82}Pb \rightarrow ^{210}_{83}Bi + ?$

25. Write a balanced nuclear equation for each of the following nuclei producing a beta particle.
    a. $^{3}_{1}H$ c. $^{99}_{42}Mo$
    b. $^{60}_{27}Co$ d. $^{32}_{15}P$

26. Write a balanced nuclear equation for each of the following nuclei producing a beta particle.
    a. $^{59}_{26}Fe$ c. $^{198}_{79}Au$
    b. $^{24}_{11}Na$ d. $^{47}_{20}Ca$

27. Write a balanced nuclear equation for each of the following nuclei producing an alpha particle.
    a. $^{210}_{84}Po$ c. $^{230}_{90}Th$
    b. $^{237}_{93}Np$ d. $^{238}_{92}U$

28. Write a balanced nuclear equation for each of the following nuclei producing an alpha particle.
    a. $^{226}_{88}Ra$ c. $^{239}_{94}Pu$
    b. $^{222}_{86}Rn$ d. $^{8}_{4}Be$

19.2 Nuclear Transformations

QUESTIONS

29. What does a *nuclear transformation* represent? How is a nuclear transformation performed?

30. Why are particle accelerators needed for bombardment processes?

31. What are the elements with atomic numbers greater than 92 called? How have these elements been prepared?

32. Write a balanced nuclear equation for the bombardment of $^{27}_{13}Al$ with alpha particles to produce $^{30}_{15}P$ and a neutron.

19.3 Detection of Radioactivity and the Concept of Half-Life

QUESTIONS

33. Describe the operation of a Geiger counter. How does a Geiger counter detect radioactive particles? How does a scintillation counter differ from a Geiger counter?

35. What do we mean when we say that one radioactive nucleus is "hotter" than another? Which element would have more decay events over a given period of time?

34. What is the *half-life* of a radioactive nucleus? Does a given type of nucleus always have the same half-life? Do nuclei of different elements have the same half-life?

36. Consider the isotopes of radium listed in Table 19.3. Which isotope is most stable against decay? Which isotope is "hottest"?

PROBLEMS

37. The following isotopes (listed with their half-lives) have been used in the medical and biological sciences. Arrange these isotopes in order of their relative decay activities: ^3H (12.2 years), ^{24}Na (15 hours), ^{131}I (8 days), ^{60}Co (5.3 years), ^{14}C (5730 years).

39. Gold-198 is a beta emitter with a half-life of 2.7 days that has been used as an implant for cancer therapy. For an implant containing 50 μg of Au-198, approximately how much Au-198 remains after 1 week?

41. The nuclide $^{24}_{11}Na$ has a half-life of 15 hours. If a sample that originally contained 100. mg of $^{24}_{11}Na$ now only contains 3.1 mg, how many hours have gone by?

38. A list of several important radioactive nuclides is given in Table 19.4. Arrange these nuclides in order of their relative decay activities.

40. For the first three isotopes of radium listed in Table 19.3, given a starting amount of 1,000 mg, *estimate* the approximate amount of each isotope remaining after one month.

42. Technetium-99 has been used as a radiographic agent in bone scans ($^{99}_{43}Tc$ is absorbed by bones). If $^{99}_{43}Tc$ has a half-life of 6.0 hours, what fraction of an administered dose of $^{99}_{43}Tc$ remains in a patient's body after 2.0 days?

19.4 Dating by Radioactivity

QUESTIONS

43. What nuclide is commonly used in the dating of artifacts?

45. Why is it assumed that the amount of $^{14}_{6}C$ in the atmosphere remains constant?

44. How is $^{14}_{6}C$ produced in the atmosphere? Write a balanced equation for this process.

46. Why does an ancient wood or cloth artifact contain less $^{14}_{6}C$ than contemporary or more recently fabricated articles made of similar materials?

19.5 Medical Applications of Radioactivity

QUESTIONS

47. What is a *radiotracer*? What tracers have been used to study the conversion of nutrients into energy in living cells?

48. List four radioisotopes used in medical diagnosis or treatment, giving their nuclear symbols. Discuss why the isotopes you choose are particularly well suited to their uses.

19.6 Nuclear Energy

QUESTIONS

49. How do the forces that hold an atomic nucleus together compare in strength with the forces between atoms in a molecule?

50. Define the terms *nuclear fission* and *nuclear fusion*. Which process results in the production of a heavier nucleus? Which results in the production of several smaller nuclei?

19.7 Nuclear Fission

QUESTIONS

51. How do the energies released by nuclear processes compare in magnitude with the energies of ordinary chemical processes?

53. How is it possible for the fission of $^{235}_{92}U$, once started, to lead to a chain reaction?

52. Write an equation for the fission of $^{235}_{92}U$ by bombardment with neutrons.

54. What does it mean to say that fissionable material possesses a *critical mass*? Can a chain reaction occur when a sample has less than the critical mass?

19.8 Nuclear Reactors

QUESTIONS

55. Describe the purpose of each of the major components of a nuclear reactor (moderator, control rods, containment, cooling liquid, and so on).

57. What is a *melt-down* and how can it occur? Most nuclear reactors use water as the cooling liquid. Is there any danger of a steam explosion if the reactor core becomes overheated?

56. Can a nuclear explosion take place in a reactor? Is the concentration of fissionable material used in reactors large enough for this?

58. In a *breeder* nuclear reactor, nonfissionable $^{235}_{92}U$ is converted to fissionable $^{239}_{94}Pu$. Write nuclear equations showing this transformation.

19.9 Nuclear Fusion

QUESTIONS

59. What is the nuclear *fusion* of small nuclei? How does the energy released by fusion compare in magnitude with that released by fission?

61. Why is the development of nuclear fusion processes desirable?

60. What are some reasons why no practical fusion reactor has yet been developed?

62. What is meant by "cold" fusion? Why is cold fusion preferable to such "hot" fusion as might be induced by a high-temperature laser? From reports by the media, discuss the controversy surrounding the cold fusion experiments.

19.10 Effects of Radiation

QUESTIONS

63. Although the energy transferred per event when a living creature is exposed to radiation is small, why is such exposure dangerous?

65. Describe the relative penetrating powers of alpha, beta, and gamma radiations.

67. How do the *chemical properties* of radioactive nuclei (as opposed to the nuclear decay they undergo) influence the degree of damage they do to an organism?

64. Explain the difference between *somatic* damage from radiation and *genetic* damage. Which type causes immediate damage to the exposed individual?

66. Explain why, although gamma rays are far more penetrating than alpha particles, the latter are actually more likely to cause damage to an organism. Which radiation is more effective at causing ionization of biomolecules?

68. Although radiation has been measured in many terms, the *rem* is most commonly used in discussions of the exposure of human beings to radiation. Using Table 19.6, discuss the sources of radiation to which humans are exposed naturally, and indicate how the total exposure compares with the "acceptable" (nondetectable clinical effect) level of exposure given in Table 19.5.

Additional Problems

69. The number of protons contained in a given nucleus is called the _____.

70. A nucleus that spontaneously decomposes is said to be _____.

71. A(n) _____, when it is produced by a nucleus at high speed, is more commonly called a beta particle.

72. In a nuclear equation, both the atomic number and the _____ number must be conserved.

73. Production of a helium nucleus from a heavy atom is referred to as _____ decay.

74. The net effect of the production of a beta particle is to convert a _____ to a _____.

75. In addition to particles, many radioactive nuclei also produce high-energy _____ rays when they decay.

76. When a nuclide decomposes through a series of steps before reaching stability, the nuclide is said to have gone through a _____ series.

77. When a nuclide produces a beta particle, the atomic number of the resulting new nuclide is one unit _____ than that of the original nuclide.

78. When a nucleus undergoes alpha decay, the _____ of the nucleus decreases by four units.

79. Machines that increase the speed of species used for nuclear bombardment processes are called _____.

80. The elements with atomic numbers of 93 or greater are referred to as the _____ elements.

81. A _____ counter contains argon gas, which is ionized by radiation, making possible the conduction and measurement of electric current in the argon gas.

82. The time required for half of an original sample of a radioactive nuclide to decay is referred to as the _____ of the nuclide.

83. The radioactive nuclide that has been used in determining the age of historical artifacts is _____.

84. _____ are radioactive substances that physicians introduce into the body to enable them to study the absorption and metabolism of the substance or to analyze the functioning of an organ or gland that can make use of the substance.

85. Combining two small nuclei to form a larger nucleus is referred to as the process of nuclear _____.

86. A self-sustaining nuclear process, in which the bombarding particles needed to produce the fission of further material are themselves produced as the product of the initial fission, is called a _____ reaction.

87. The most common type of nuclear reactor uses the nuclide _____ as its fissionable material.

88. A nuclear reactor that generates additional fissionable fuel (in addition to producing heat for generating electricity) is referred to as a _____ reactor.

89. The decay series from uranium-238 to lead-206 is indicated in Figure 19.1. For each *step* of the process indicated in the figure, specify what sort of particle is emitted by the particular nucleus involved at that point in the series.

90. The U.S. Department of Energy sells an isotope of the transuranium element californium for approximately $10 per microgram to qualified researchers. What would a pound of the Cf isotope cost?

91. Each of the following isotopes has been used medically for the purpose indicated. Suggest reasons why the particular element might have been chosen for this purpose.

    a. cobalt-57   (study of the body's use of vitamin B12)
    b. calcium-47   (study of bone metabolism)
    c. iron-59   (study of red blood cell function)
    d. mercury-197   (brain scans before CAT scan became available)

92. The fission of $^{235}_{92}U$ releases 2.1×10^{13} joule per mole of $^{235}_{92}U$. Calculate the energy released per atom and per gram of $^{235}_{92}U$.

93. During the research that led to production of the two atomic bombs used against Japan in World War II, different mechanisms for obtaining a supercritical mass of fissionable material were investigated. In one type of bomb, what is essentially a gun was used to shoot one piece of fissionable material into a cavity containing another piece of fissionable material. In the second type of bomb, the fissionable material was surrounded with a high explosive that, when detonated, compressed the fissionable material into a smaller volume. Discuss what is meant by critical mass, and explain why the ability to achieve a critical mass is essential to sustaining a nuclear reaction.

94. Discuss some of the problems associated with the storage and disposal of the waste products of the nuclear industry.

95. Chlorine consists primarily of two isotopes, of mass numbers 35 and 37, respectively. Write the nuclear symbol for each of these isotopes.

96. Aluminum exists in several isotopic forms, including $^{27}_{13}Al$, $^{28}_{13}Al$, and $^{29}_{13}Al$. Indicate the number of protons and the number of neutrons in each of these isotopes.

97. Complete each of the following nuclear equations by supplying the missing particle (or particles).

    a. $^{226}_{88}Ra \rightarrow ^{222}_{86}Rn + ?$ c. $^{2}_{1}H + ^{3}_{1}H \rightarrow ^{4}_{2}He + ?$
    b. $^{222}_{86}Rn \rightarrow ^{218}_{84}Po + ?$

98. Complete each of the following nuclear equations by supplying the missing particle (or particles).

    a. $^9_4\text{Be} + ^4_2\text{He} \rightarrow ^{12}_6\text{C} + ?$ c. $^{131}_{53}\text{I} \rightarrow ^{131}_{54}\text{Xe} + ?$

    b. $^{201}_{80}\text{Hg} + ^0_{-1}\text{e} \rightarrow ?$

99. Write a balanced nuclear equation for the bombardment of $^{14}_7\text{N}$ with alpha particles to produce $^{17}_8\text{O}$ and a proton.

100. The only major use of iodine in the human body is in the production of certain hormones by the thyroid gland, and iodine from the diet concentrates in this area of the body. Iodine-131 is used in the diagnosis and treatment of thy-roid disease and has a half-life of 8 days. If a patient with thyroid disease consumes a sample of Na^{131}I containing 10 μg of ^{131}I, how long will it take for the amount of ^{131}I to decrease to approximately 1/1000 of the original amount?

101. How have $^{131}_{53}\text{I}$ and $^{201}_{81}\text{Tl}$ been used in medical diagnosis? Why are these particular nuclides especially well suited for these purposes?

102. What is a *breeder* nuclear reactor? What difficulties with such reactors have led to their not yet being used in the United States for the generation of electricity?

Solutions to Self-Check Exercises

SELF-CHECK EXERCISE 19.1

a. An alpha particle is a helium nucleus, ^4_2He. We can initially represent the production of an α particle by $^{226}_{88}\text{Ra}$ as follows:

$$^{226}_{88}\text{Ra} \rightarrow ^4_2\text{He} + ^A_Z\text{X}$$

Because we know that both A and Z are conserved, we can write

$$A + 4 = 226 \quad \text{and} \quad Z + 2 = 88$$

Solving for A gives 222 and for Z gives 86, so ^A_ZX is $^{222}_Z\text{X}$. Because Rn has $Z = 86$, ^A_ZX is $^{222}_{86}\text{Rn}$. The overall balanced equation is

$$^{226}_{88}\text{Ra} \rightarrow ^4_2\text{He} + ^{222}_{86}\text{Rn}$$

CHECK: $Z = 88 \atop A = 226$ $\rightarrow$ $Z = 86 + 2 = 88 \atop A = 222 + 4 = 226$

b. Using a similar strategy, we have

$$^{214}_{82}\text{Pb} \rightarrow ^0_{-1}\text{e} + ^A_Z\text{X}$$

Because $Z - 1 = 82$, $Z = 83$, and because $A + 0 = 214$, $A = 214$. Therefore $^A_Z\text{X} = ^{214}_{83}\text{Bi}$. The balanced equation is

$$^{214}_{82}\text{Pb} \rightarrow ^0_{-1}\text{e} + ^{214}_{83}\text{Bi}$$

CHECK: $Z = 82 \atop A = 214$ $\rightarrow$ $Z = 83 - 1 = 82 \atop A = 214 + 0 = 214$

SELF-CHECK EXERCISE 19.2

a. The missing particle must be ^4_2He (an α particle), because

$$^{222}_{86}\text{Rn} \rightarrow ^{218}_{84}\text{Po} + ^4_2\text{He}$$

is a balanced equation.

CHECK: $Z = 86$ $Z = 84 + 2 = 86$
 $A = 222 \xrightarrow{} A = 218 + 4 = 222$

b.   The missing species must be $^{15}_{7}X$ or $^{15}_{7}N$, because the balanced equation is

$$^{15}_{8}O \rightarrow {}^{15}_{7}N + {}^{0}_{1}e$$

CHECK: $Z = 8$ $Z = 7 + 1 = 8$
 $A = 15 \xrightarrow{} A = 15 + 0 = 15$

SELF-CHECK EXERCISE 19.3

Let's do this problem by thinking about the number of half-lives required to go from 8.0×10^{-7} mol to 1.0×10^{-7} mol of $^{228}_{88}Ra$.

8.0×10^{-7} mol → First half-life → 4.0×10^{-7} mol → Second half-life → 2.0×10^{-7} mol → Third half-life → 1.0×10^{-7} mol

It takes three half-lives, then, for the sample to go from 8.0×10^{-7} mol of $^{228}_{88}Ra$ to 1.0×10^{-7} mol of $^{228}_{88}Ra$. From Table 19.3, we know that the half-life of $^{228}_{88}Ra$ is 6.7 years. Therefore the elapsed time is $3(6.7 \text{ years}) = 20.1$ years, or 2.0×10^{1} years when we use the correct number of significant figures.

CHAPTER

20 Organic Chemistry

CONTENTS

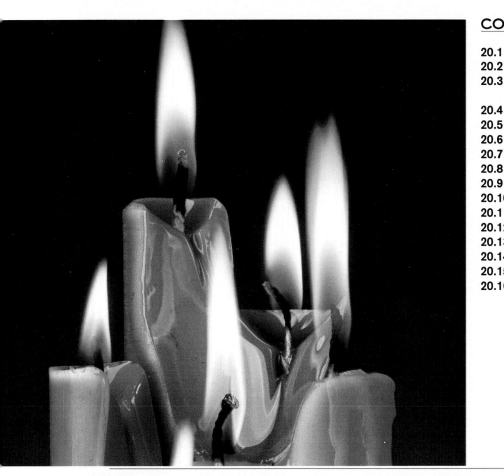

Candles are made from paraffin, a mixture of solid hydrocarbons
obtained from petroleum.

Students enjoy discussing the vital force theory, which Wöhler helped to disprove. They know that organic molecules can be synthesized in the laboratory without the intervention of a vital force. Not only will this discussion provide some historical insight, it will allow students to see how theories that seem correct can be quickly discarded when more information becomes available.

A photomicrograph showing nylon fibers. Nylon is a very important synthetic material made from organic compounds.

Two Group 4 elements, carbon and silicon, form the basis of most natural substances. Silicon, with its great affinity for oxygen, forms chains and rings containing Si—O—Si bonds to produce the silica and silicates that form the basic structure for most rocks, sands, and soils. Therefore, silicon compounds are the fundamental inorganic materials of the earth. What silicon is to the geological world, carbon is to the organic or biological world. Carbon has the unusual ability of bonding strongly to itself, forming long chains or rings of carbon atoms. In addition, carbon forms strong bonds to other nonmetals such as hydrogen, nitrogen, oxygen, sulfur, and the halogens. Because of these bonding properties, an extraordinary number of carbon compounds exists; several million are now known, and the number continues to grow rapidly. Among these many compounds are the **biomolecules,** those molecules that make possible the maintenance and reproduction of life.

The study of carbon-containing compounds and their properties is called **organic chemistry.** Although a few compounds of carbon, such as the oxides of carbon and carbonates, are considered to be inorganic substances, the vast majority of carbon compounds are designated as organic compounds—compounds that typically contain chains or rings of carbon atoms.

Originally, the distinction between inorganic and organic substances was based on whether they were produced by living systems. For example, until the early nineteenth century, it was believed that organic compounds had some sort of "life force" and could be synthesized only by living organisms. This misconception was dispelled in 1828 when the German chemist Friedrich Wöhler (1800–1882) prepared urea from the inorganic salt ammonium cyanate by simple heating.

$$NH_4OCN \xrightarrow{\text{Heat}} \underset{\displaystyle\overset{\displaystyle \|}{O}}{H_2N-C-NH_2}$$

Ammonium cyanate   Urea

Urea is a component of urine, so it is clearly an organic material formed by living things, yet here was clear evidence that it could be produced in the laboratory as well.

Organic chemistry plays a vital role in our quest to understand living systems. Beyond that, the synthetic fibers, plastics, artificial sweeteners, and medicines that we take for granted are products of industrial organic chemistry. And the energy on which we rely so heavily to power our civilization is based mostly on the combustion of organic materials found in coal and petroleum.

Because organic chemistry is such a vast subject, we can only introduce it briefly in this book. We will begin with the simplest class of organic compounds, the hydrocarbons, and then show how most other organic compounds can be considered to be derived from hydrocarbons.

A close-up photo of Velcro, a synthetic organic material used for fasteners.

This chapter introduces students to a vast area of chemistry. Students confronting organic chemistry for the first time can appreciate Wöhler's comment made several years after his synthesis of urea that "he had wandered into a deep forest from which there is seemingly no return."

20.1 Carbon Bonding

AIM: To give an overview of the types of bonds formed by the carbon atom.

So many carbon-containing compounds exist, because carbon forms strong bonds to itself and to many other elements. One carbon atom can form bonds to a maximum of four other atoms; these can be either carbon atoms or atoms of other elements. One of the hardest, toughest materials known is diamond, a form of pure carbon in which each carbon atom is bound to four other carbon atoms (see Figure 5.6).

One of the most familiar compounds of carbon is methane, CH_4, the main component of natural gas. The methane molecule consists of a carbon atom with four hydrogen atoms bound to it in a tetrahedral fashion. That is, as predicted by the VSEPR model (see Section 12.9), the four pairs of bonding electrons around the carbon have minimum repulsions when they are located at the corners of a tetrahedron.

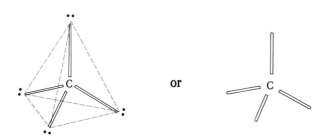

This leads to the structure for CH_4 shown in Figure 20.1. *When carbon has four atoms bound to it, these atoms will always have a tetrahedral arrangement about the carbon.*

Carbon can bond to fewer than four elements by forming one or more multiple bonds. Recall that a multiple bond involves the sharing of more than one pair of electrons. For example, a *double bond* involves sharing two pairs of electrons, as in carbon dioxide:

$$\overset{..}{\underset{..}{O}}=C=\overset{..}{\underset{..}{O}}$$

and a *triple bond* involves sharing three pairs of electrons, as in carbon monoxide:

$$:C≡O:$$

Note that carbon is bound to two other atoms in CO_2 and to only one other atom in CO.

Three-dimensional models are almost essential for giving students a feel for what organic molecules look like. Two-dimensional drawings by themselves are not adequate. Students have a difficult time mentally converting a two-dimensional drawing to a three-dimensional structure.

Figure 20.1
Methane is a tetrahedral molecule.

CO_2 and CO are classed as inorganic substances.

Use models to show the shapes of the organic molecules discussed in this section. The tetrahedral shape of methane with the carbon atom in the middle is hard to visualize without models.

Model kits with small atoms are hard for students to see from the back of a room. Painted styrofoam balls held together with applicator sticks make inexpensive models large enough to be seen by everyone.

The number of atoms surrounding a carbon when it has a double or triple bond and the overall shape of the molecule around that carbon atom can be presented without discussing bond angles.

Multiple bonding also occurs in organic molecules. Ethylene, C_2H_4, has a double bond.

$$\underset{H}{\overset{H}{\diagdown}}C=C\underset{H}{\overset{H}{\diagup}}$$

In this case each carbon is bound to three other atoms (one C atom and two H atoms). A molecule with a triple bond is acetylene, C_2H_2.

$$H—C\equiv C—H$$

Here each carbon is bound to two other atoms (one carbon atom and one hydrogen atom).

More than any other element, carbon has the ability to form chains of atoms, as illustrated by the structures of propane and butane shown in Figure 20.3. In these compounds each carbon atom is bound to four atoms in a tetrahedral fashion. We will discuss these molecules in detail in the next section.

20.2   Alkanes

AIM: To introduce the alkanes—compounds that contain saturated carbon atoms.

Hydrocarbons, as the name indicates, are compounds composed of carbon and hydrogen. Those whose carbon–carbon bonds are all single bonds are said to be **saturated,** because each carbon is bound to four atoms, the maximum number. Hydrocarbons containing carbon–carbon multiple bonds are described as being **unsaturated,** because the carbon atoms involved in a multiple bond can bond to one or more additional atoms. This is shown by the *addition* of hydrogen to ethylene.

Note that each carbon in ethylene is bonded to three atoms (one carbon and two hydrogens) but can bond to one additional atom after one bond of the carbon–carbon double bond is broken. This leads to ethane, a saturated hydrocarbon (each carbon atom is bonded to four atoms).

Saturated hydrocarbons are called **alkanes.** The simplest alkane is *methane,* CH_4, which has a tetrahedral structure (Figure 20.1). The next alkane, the one

If you show the model of an alkane with at least four carbons, students can see that the carbon atoms are not lined up in one plane as they appear to be when we write their structures on paper or on the blackboard.

It is useful to point out now that two-dimensional structures don't ecessarily tell much about the three-dimensional shape of the molecule.

containing two carbon atoms, is *ethane*, C_2H_6, shown in Figure 20.2. Note that each carbon atom in ethane is bonded to four atoms.

The next two members of the series are *propane*, with three carbon atoms and the formula C_3H_8, and *butane*, with four carbon atoms and the formula C_4H_{10}. These molecules are shown in Figure 20.3. Again, these are saturated hydrocarbons (alkanes); each carbon is bonded to four atoms.

Alkanes in which the carbon atoms form long "strings" or chains are called **normal, straight-chain,** or **unbranched hydrocarbons.** As Figure 20.3 illustrates, the chains in normal alkanes are not really straight but zig-zag, because the tetrahedral C—C—C angle is 109.5°. The normal alkanes can be represented by the structure

where *m* is an integer. Note that each member is obtained from the previous one by insertion of a *methylene*, CH_2, group. We can condense the structural formulas by omitting some of the C—H bonds. For example, the general formula for normal alkanes shown above can be condensed to

$$CH_3 - (CH_2)_{\overline{m}} CH_3$$

Figure 20.2
(a) The Lewis structure of ethane, C_2H_6.
(b) The molecular structure of ethane represented by a space-filling model and a ball-and-stick model.

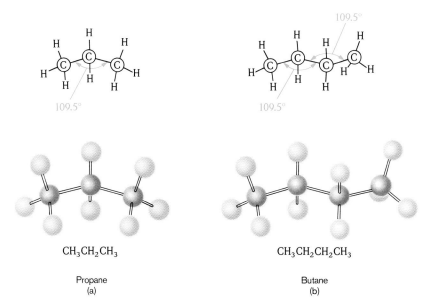

Propane
(a)

Butane
(b)

Figure 20.3
The structures of (a) propane and (b) butane.

This photo shows an actual ball-and-stick model of ethane.

EXAMPLE 20.1 Writing Formulas for Alkanes

Give the formulas for the normal (or straight-chain) alkanes with six and eight carbon atoms.

SOLUTION

The alkane with six carbon atoms can be written as

$$CH_3CH_2CH_2CH_2CH_2CH_3$$

which can be condensed to

$$CH_3-(CH_2)_4CH_3$$

Notice that the molecule contains fourteen hydrogen atoms in addition to the six carbon atoms. Therefore the formula is C_6H_{14}.

The alkane with eight carbons is

$$CH_3CH_2CH_2CH_2CH_2CH_2CH_2CH_3$$
$$1 \quad 2 \quad 3 \quad 4 \quad 5 \quad 6 \quad 7 \quad 8$$

which can be written in condensed form as

$$CH_3-(CH_2)_6CH_3$$

This molecule has eighteen hydrogens. The formula is C_8H_{18}.

SELF-CHECK EXERCISE 20.1

Give the formulas for the alkanes with ten and fifteen carbon atoms.

Students should be able to convert the condensed structure of an alkane to the expanded form with no difficulties. Writing the condensed form on the blackboard will save time in class, but only if students know how the atoms are actually joined together.

The first ten straight-chain alkanes are shown in Table 20.1. Note that all alkanes can be represented by the general formula C_nH_{2n+2}, where n represents the number of carbon atoms. For example, nonane, which has nine carbon atoms, is represented by $C_9H_{(2 \times 9)+2}$, or C_9H_{20}. The formula C_nH_{2n+2} reflects the fact that each carbon in the chain has two hydrogen atoms except the two end carbons, which have three each. Thus the number of hydrogen atoms present is twice the number of carbon atoms plus two (for the extra two hydrogen atoms on the ends).

The general formula for the alkanes is useful if students need to know the number of hydrogen atoms in an alkane with a particular number of carbons, but don't want to write the structural formula for the alkane and count the number of hydrogens.

EXAMPLE 20.2 Using the General Formula for Alkanes

Show that the alkane with fifteen carbon atoms can be represented in terms of the general formula C_nH_{2n+2}.

EXAMPLE 20.2, CONTINUED

SOLUTION

In this case $n = 15$. The formula is $C_{15}H_{2(15)+2}$, or $C_{15}H_{32}$, as was found in Self-Check Exercise 20.1.

Table 20.1 Formulas of the First Ten Straight-Chain Alkanes		
Name	Condensed Formula (C_nH_{2n+2})	Extended Formula
methane	CH_4	CH_4
ethane	C_2H_6	CH_3CH_3
propane	C_3H_8	$CH_3CH_2CH_3$
n-butane	C_4H_{10}	$CH_3CH_2CH_2CH_3$
n-pentane	C_5H_{12}	$CH_3CH_2CH_2CH_2CH_3$
n-hexane	C_6H_{14}	$CH_3CH_2CH_2CH_2CH_2CH_3$
n-heptane	C_7H_{16}	$CH_3CH_2CH_2CH_2CH_2CH_2CH_3$
n-octane	C_8H_{18}	$CH_3CH_2CH_2CH_2CH_2CH_2CH_2CH_3$
n-nonane	C_9H_{20}	$CH_3CH_2CH_2CH_2CH_2CH_2CH_2CH_2CH_3$
n-decane	$C_{10}H_{22}$	$CH_3CH_2CH_2CH_2CH_2CH_2CH_2CH_2CH_2CH_3$

To avoid having difficulty for the remainder of the semester, students should memorize the names and structures of the first 10 alkanes as given in Table 20.1.

20.3 Structural Formulas and Isomerism

AIM: To introduce structural isomers and explain how to write structural formulas.

Butane and all succeeding alkanes exhibit structural isomerism. **Structural isomerism** occurs when two molecules have the same atoms but different bonds. That is, the molecules have the same formulas but different arrangements of the atoms. For example, butane can exist as a straight-chain molecule (normal butane, or n-butane) or with a branched-chain structure (called isobutane), as shown in Figure 20.4 (p. 684). Because of their different structures, these structural isomers have different properties.

You will need to work examples in class showing that structural isomers actually have a different order of atoms and that rotating carbon–carbon single bonds does not produce a new structural isomer. It is hard to convince students that twisting down the end carbon of a normal hydrocarbon chain does not produce a new structural isomer with a methyl group on the end carbon.

The most effective way to show that *n*-butane and 2-methylpropane are different molecules is to make models. Rotate the carbons of *n*-butane to show that no matter how much you twist the carbons to new locations, you cannot create a molecule with three carbons and a branch. Then, break apart the molecule and reassemble the atoms to produce 2-methylpropane.

Figure 20.4
The structural isomers of C_4H_{10}. Each molecule is represented in three ways: a ball-and-stick structure, a space-filling structure, and a structure that shows the shared electrons as lines (a Lewis structure). (a) Normal butane (abbreviated *n*-butane). (b) The branched isomer of butane (called isobutane).

(a) (b)

EXAMPLE 20.3   **Drawing Structural Isomers of Alkanes**

Draw the structural isomers of pentane, C_5H_{12}.

SOLUTION

To find the isomeric structures for pentane, C_5H_{12}, we first write the straight carbon chain and then add the hydrogen atoms.

1. The straight-chain structure has the five carbon atoms in a row.

$$C—C—C—C—C$$

We can now add the H atoms.

CHEMISTRY IN FOCUS

Methane from Termites

*A*long with carbon dioxide, water vapor, and others, methane is one of the components of the atmosphere that helps control the earth's temperature by reflecting heat energy back toward the earth, minimizing its escape into space. This is called the greenhouse effect (see *Atmospheric Effects* in Chapter 11). Scientists are therefore very much interested in the fact that the methane content of the atmosphere seems to be increasing by about 2% annually.

One source of this increase appears to be termites! When termites consume wood, the products methane, carbon dioxide, and other compounds are given off by the microorganisms that inhabit the termites'

digestive systems. In fact, it has been estimated that termites produce 165 million tons of methane and 55 million tons of carbon dioxide every year.

Clearly, the amounts of methane and carbon dioxide produced by termites are significant and may be increasing as a result of changes in the environment that favor larger termite populations. The potential effects of these changes on the earth's atmosphere are being monitored closely by researchers. Too much methane in the atmosphere could lead to significant warming of the earth.

Termite workers.

EXAMPLE 20.3, CONTINUED

This can be written in shorthand form as

$$CH_3—CH_2—CH_2—CH_2—CH_3 \text{ or } CH_3{-}(CH_2{)}_3CH_3$$

and is called *n*-pentane.

2. Next we remove one C atom from the main chain and bond it to the second carbon in the chain.

n-Pentane

EXAMPLE 20.3, CONTINUED

Then we put on the H atoms, so that each carbon has four bonds.

This structure can be represented as

$$CH_3-\overset{\overset{\textstyle CH_3}{|}}{CH}-CH_2-CH_3$$

and is called isopentane.

Isopentane

3. Finally, we take two carbons out of the chain to give the arrangement

$$C-\overset{\overset{\textstyle C}{|}}{\underset{\underset{\textstyle C}{|}}{C}}-C$$

Adding the H atoms gives

which can be written in shorthand form as

$$CH_3-\overset{\overset{\textstyle CH_3}{|}}{\underset{\underset{\textstyle CH_3}{|}}{C}}-CH_3$$

This molecule is called neopentane.

Neopentane

EXAMPLE 20.3, CONTINUED

The space-filling and ball and stick models for these molecules are shown in the margin.

Note that all of these molecules have the formula C_5H_{12}, as required. Also note that the structures

$$CH_3-CH_2-\overset{\overset{\displaystyle CH_3}{|}}{CH}-CH_3 \qquad CH_3-\underset{\underset{\displaystyle CH_3}{|}}{CH}-CH_2-CH_3 \qquad CH_3-CH_2-\underset{\underset{\displaystyle CH_3}{|}}{CH}-CH_3$$

which might at first appear to be additional isomers, are actually identical to structure 2. All three of these structures have exactly the same skeleton of carbons as the structure shown in part 2. All of these structures have four carbons in the chain with one carbon on the side:

$$C-\underset{\underset{\displaystyle C}{|}}{C}-C-C$$

20.4 Naming Alkanes

AIM: To introduce the system for naming alkanes and substituted alkanes.

Because there are literally millions of organic compounds, it would be impossible to remember common names for all of them. Just as we did in Chapter 5 for inorganic compounds, we must learn a systematic method for naming organic compounds. We will first consider the principles applied in naming alkanes and then summarize them as a set of rules.

1. The first four members of the alkane series are called methane, ethane, propane, and butane. The names of the alkanes beyond butane are obtained by adding the suffix -*ane* to the Greek root for the number of carbon atoms.

Number	Greek Root
5	*pent*
6	*hex*
7	*hept*
8	*oct*
9	*non*
10	*dec*

Organic nomenclature is difficult; the variations are seemingly unlimited. It is not possible to memorize names and formulas for the many organic molecules. Only the basic root and substituent names can be memorized. Naming most molecules requires problem-solving skills, so lots of practice will be necessary.

The Greek roots used to describe organic molecules with from five to ten carbons will be familiar to some students already.

Therefore the alkane

$$CH_3CH_2CH_2CH_2CH_2CH_2CH_2CH_3$$

which has eight carbons in the chain, is called octane.

(oct   —   ane)

Tells us
there are
eight carbons

Tells us
it is an
alkane

2. For a branched hydrocarbon, the longest continuous chain of carbon atoms gives the root name for the hydrocarbon. For example, in the alkane

$$CH_3$$
$$|$$
$$CH_2$$ Six carbons
$$|$$
$$CH_2$$
$$|$$
$$CH_3—CH_2—CH—CH_2—CH_3$$

Five carbons

the longest continuous chain contains six carbon atoms. The specific name of this compound is not important at this point, but it will be named as a hexane (indicating a six-carbon chain).

3. Alkanes lacking one hydrogen atom can be attached to a hydrocarbon chain in place of one hydrogen atom. For example, the molecule

$$\begin{array}{ccccc} H & H & H & H & H \\ | & | & | & | & | \\ H—C—C—C—C——C—H \\ | & | & | & | & | \\ H & H & H & (CH_3) & H \end{array}$$

CH_3 substituted for H

can be viewed as a pentane (five-carbon chain) where one hydrogen atom has been replaced by a —CH_3 group, which is a methane, CH_4, molecule with a hydrogen removed. When a group is substituted for a hydrogen on an alkane chain, we call this group a *substituent*. To name the —CH_3 substituent, we start with the name of its parent alkane, drop the *-ane,* and add *-yl.* Therefore —CH_3 is called *methyl.* Likewise, when we remove one hydrogen from ethane CH_3CH_3, we get CH_3CH_2—. Dropping the *-ane* ending and adding *-yl* gives this group the name *ethyl.* Removal of a hydrogen from the end carbon of propane, $CH_3CH_2CH_3$, yields $CH_3CH_2CH_2$—, which is called the *propyl* group.

There are two ways in which the propyl group can be attached as a substituent. A hydrogen can be removed from an end carbon to give the propyl group or from the middle carbon to give the isopropyl group.

$$CH_2-CH_2-CH_3 \qquad \text{or} \qquad H_3C-\underset{H}{\overset{|}{C}}-CH_3$$

Name: Propyl   Isopropyl

When a hydrogen is removed from butane $CH_3CH_2CH_2CH_3$, we get a butyl substituent. In the case of the butyl group, there are four ways in which the atoms can be arranged. These are shown, with their respective names, in Table 20.2.

The general name for an alkane when it functions as a substituent is *alkyl.* All of the common alkyl groups are shown in Table 20.2.

4. We specify the positions of substituent groups by numbering sequentially the carbons in the longest chain of carbon atoms, starting at the end closest to the branching (the place where the first substituent occurs). For example, the compound

Methyl substituent

$$CH_3-CH_2-CH-CH_2-CH_3-CH_3$$

| Correct numbering | 1 | 2 | 3 | 4 | 5 | 6 |
| Incorrect numbering | 6 | 5 | 4 | 3 | 2 | 1 |

is called 3-methylhexane. Note that the top set of numbers is correct; the left end of the molecule is closest to the branching, and this gives the smallest number for the position of the substituent. Also note that a hyphen is written between the number and the name of the substituent.

5. When a given type of substituent occurs more than once, we indicate this by using a prefix. The prefix *di-* indicates two identical substituents, and *tri-* indicates three. For example, the compound

$$\overset{1}{CH_3}-\overset{2}{CH}-\overset{3}{CH}-\overset{4}{CH_2}-\overset{5}{CH_3}$$
$$\qquad\quad | \qquad |$$
$$\qquad\quad CH_3 \quad CH_3$$

has the root name pentane (five carbons in the longest chain). We use *di-* to indicate the two methyl substituents and use numbers to locate them on the chain. The name is 2,3-dimethylpentane.

Table 20.2 The Most Common Alkyl Substituents and Their Names

Structure	Name
$-CH_3$	methyl
$-CH_2CH_3$	ethyl
$-CH_2CH_2CH_3$	propyl
$CH_3\overset{|}{CH}CH_3$	isopropyl
$-CH_2CH_2CH_2CH_3$	butyl
$CH_3\overset{|}{CH}CH_2CH_3$	*sec*-butyl
$-CH_2-\underset{CH_3}{\overset{H}{\overset{|}{C}}}-CH_3$	isobutyl
$\underset{CH_3}{\overset{CH_3}{-\overset{|}{\underset{|}{C}}}-CH_3}$	*tert*-butyl

*The bond with one end open shows the point of attachment of the substituent to the carbon chain.

Table 20.2 contains the common organic substituent names and should be memorized. You might help students to recognize these groups when they are part of a larger molecule by circling the substituents with colored chalk or colored ink on an overhead transparency.

The names and structures of substituents derived from butane are hard for students to remember. The terms *sec* and *tert* have no meaning for beginning students in organic chemistry, so those helpful clues to structure are of no use. At this point, students must memorize the names and structures.

The following rules summarize the principles we have just developed.

You might want to put the rules for naming alkanes on an overhead transparency so that students can refer to them as you work examples.

Rules for Naming Alkanes

1. Find the longest continuous chain of carbon atoms. This chain (called the parent chain) determines the base alkane name.
2. Number the carbons in the parent chain, starting at the end closest to any branching (the first alkyl substituent). When a substituent occurs the same number of carbons from each end, use the next substituent (if any) to determine from which end to start numbering.
3. Using the appropriate name for each alkyl group, specify its position on the parent chain with a number.
4. When a given type of alkyl group occurs more than once, attach the appropriate prefix (*di-* for two, *tri-* for three, and so on) to the alkyl name.
5. The alkyl groups are listed in alphabetical order, *disregarding any prefix.*

You should work a problem similar to Example 20.4 in class. Students find drawing the structures of all the structural isomers of a molecule difficult, so a systematic approach is needed to solve this kind of problem. Students should remove one methyl at a time from the chain and try to place it on as many new carbons as possible without producing duplicates. This is a good opportunity to show that two molecules that don't look the same can actually be the same. The best way to tell if the structure shows a different isomer and not a duplicate is to name the molecule. If the molecules have the same name, then they are the same, even if the carbons are bent so that the chains appear to be different.

EXAMPLE 20.4   Naming Isomers of Alkanes

Draw the structural isomers for the alkane C_6H_{14}, and give the systematic name for each one.

SOLUTION

We proceed systematically, starting with the longest chain and then rearranging the carbons to form the shorter, branched chains.

1. $CH_3CH_2CH_2CH_2CH_2CH_3$
   1   2   3   4   5   6

   This alkane has six carbons all in the same continuous chain, so we call it hexane or, more properly, *n*-hexane, indicating that all of the carbon atoms are in the same chain.
2. We now take one carbon out of the main chain and make it a methyl substituent. This gives the molecule

$$CH_3\underset{\underset{\displaystyle CH_3}{|}}{C}HCH_2CH_2CH_3$$

The carbon skeleton is as follows:

$$\underset{\underset{\displaystyle C}{|}}{\overset{\displaystyle 1\quad 2\quad 3\quad 4\quad 5}{C-C-C-C-C}}$$

EXAMPLE 20.4, CONTINUED

Because the longest chain has five carbons, the base name is pentane. We have numbered the chain from the left, starting closest to the substituent, a methyl group. We indicate the position of the methyl group on the chain by the number 2, the number of the carbon to which it is attached. So the name is 2-methylpentane. Note that if we numbered the chain from the right end, the methyl group would be on carbon 4. We want the smallest possible number, so the numbering shown is correct.

Note that placing the —CH_3 group on carbon 4 gives the same molecule as placing it on carbon 2.

3. The methyl substituent can also be on the number-3 carbon:

$$
\begin{array}{ccccc}
1 & 2 & 3 & 4 & 5 \\
\end{array}
$$
$$CH_3CH_2CHCH_2CH_3$$
$$\quad\quad\quad | $$
$$\quad\quad\quad CH_3$$

The name is 3-methylpentane. We have now exhausted all possibilities for placing a single methyl group on pentane.

4. Next we take two carbons out of the original six-member chain.

$$CH_3CH{-}CHCH_3$$
$$\quad\quad | \quad\quad |$$
$$\quad\quad CH_3 \;\; CH_3$$

The carbon skeleton is

$$
\begin{array}{cccc}
1 & 2 & 3 & 4 \\
\end{array}
$$
$$C{-}C{-}C{-}C$$
$$\quad\; | \;\;\; |$$
$$\quad\; C \;\; C$$

The longest chain of this molecule has four carbons, so the root name is butane. Because there are two methyl groups (on carbons 2 and 3), we use the prefix *di-*. The name of the molecule is 2,3-dimethylbutane. Note that when two or more numbers are used, they are separated by a comma.

5. Two methyl groups can also be attached to the same carbon atom in the four-carbon chain to give the following molecule:

$$\quad\quad\quad CH_3$$
$$\quad\quad\quad\; |$$
$$CH_3{-}C{-}CH_2CH_3$$
$$\quad\quad\quad\; |$$
$$\quad\quad\quad CH_3$$

The carbon skeleton is

$$
\begin{array}{ccccc}
& & C & & \\
1 & 2 & | & 3 & 4 \\
\end{array}
$$
$$C{-}C{-}C{-}C$$
$$\quad\;\; |$$
$$\quad\;\; C$$

EXAMPLE 20.4, CONTINUED

The root name is butane, and there are two methyl groups on the number-2 carbon. The name is 2,2-dimethylbutane.

6. As we search for more isomers, we might try to place an ethyl substituent on the four-carbon chain to give the molecule

$$CH_3-\underset{\underset{\underset{CH_3}{|}}{\overset{|}{CH_2}}}{CH}CH_2CH_3$$

The carbon skeleton is

$$C-\underset{\underset{C}{\overset{|}{C}}}{C}-C-C$$

We might be tempted to name this molecule 2-ethylbutane, but this is incorrect. Notice that there are five carbon atoms in the longest chain.

$$C-\underset{\underset{C\,\,1}{\overset{|}{C}\,\underset{2}{^{3}}\,{}^{4}\,{}^{5}}}{C}-C-C$$

We can rearrange this carbon skeleton to give

$$\underset{1\quad2\quad3\quad4\quad5}{C-C-\overset{\overset{\displaystyle C}{|}}{C}-C-C}$$

This molecule is in fact a pentane (3-methylpentane), because the longest chain has five carbon atoms, so it is not a new isomer.

In searching for more isomers, we might try a structure such as

$$CH_3-\underset{\underset{\underset{CH_3}{|}}{\overset{\overset{\displaystyle CH_3}{|}}{C}}-CH_3}{\overset{|}{\underset{CH_2}{|}}}$$

As we have drawn it, this molecule might appear to be a propane. However, the molecule has a longest chain of four atoms (look vertically), so the correct name is 2,2-dimethylbutane.

EXAMPLE 20.4, CONTINUED

Thus there are five distinct structural isomers of C_6H_{14}: *n*-hexane, 2-methylpentane, 3-methylpentane, 2,3-dimethylbutane, and 2,2-dimethylbutane.

SELF-CHECK EXERCISE 20.2

Name the following molecules.

a. $CH_3-CH_2-CH-CH_2-CH-CH_2-CH_2-CH_3$
                                         $|$ $|$
                                        CH_3 CH_2
                                                             $|$
                                                          CH_3

b. $CH_3-CH_2-CH-CH_2-CH-CH_3$
                                      $|$ $|$
                                   CH_2 CH_2
                                      $|$ $|$
                                   CH_3 CH_3

So far we have learned how to name a compound by examining its structural formula. We must also be able to do the reverse: to write the structural formula from the name.

EXAMPLE 20.5	**Writing Structural Isomers from Names**

Write the structural formula for each of the following compounds.

a.   4-ethyl-3,5-dimethylnonane   b.   4-*tert*-butylheptane

SOLUTION

a.   The root name nonane signifies a nine-carbon chain. Therefore we have the following main chain of carbons:

$$\begin{array}{ccccccccc} 1 & 2 & 3 & 4 & 5 & 6 & 7 & 8 & 9 \\ C{-}C{-}C{-}C{-}C{-}C{-}C{-}C{-}C \end{array}$$

The name indicates an ethyl group attached to carbon 4 and two methyl groups, one on carbon 3 and one on carbon 5. This gives the following carbon skeleton:

$$\begin{array}{ccccccccc} 1 & 2 & 3 & 4 & 5 & 6 & 7 & 8 & 9 \\ C{-}C{-}C{-}C{-}C{-}C{-}C{-}C{-}C \\ & & | & | & | & & & & \\ & & C & C & C & & & & \\ & & & | & & & & & \\ & & & C & & & & & \end{array}$$

EXAMPLE 20.5, CONTINUED

When we add the hydrogen atoms, we get the final structure

$$\underset{1}{CH_3}\underset{2}{CH_2}\underset{3}{CH}-\underset{4}{CH}-\underset{5}{CH}\underset{6}{CH_2}\underset{7}{CH_2}\underset{8}{CH_2}\underset{9}{CH_3}$$

Methyl →   (CH₃) (CH₂) (CH₃)   ← Methyl
                     CH₃
                              ↖ Ethyl

b.   Heptane signifies a seven-carbon chain, and the *tert*-butyl group (see Table 20.2) is

$$H_3C-\overset{|}{\underset{|}{C}}-CH_3$$
$$CH_3$$

Thus we have the molecule

$$\underset{1}{CH_3}\underset{2}{CH_2}\underset{3}{CH_2}\underset{4}{CH}\underset{5}{CH_2}\underset{6}{CH_2}\underset{7}{CH_3}$$
$$H_3C-\overset{|}{\underset{|}{C}}-CH_3$$
$$CH_3$$

SELF-CHECK EXERCISE 20.3

Write the structural formula for 5-isopropyl-4-methyldecane.

Motor oil is an important product obtained from petroleum.

20.5   Petroleum

AIM: To describe the composition and uses of petroleum.

Woody plants, coal, petroleum, and natural gas provide a vast resource of energy that originally came from the sun. By the process of photosynthesis, plants store energy that we can claim by burning the plants themselves or, more commonly, burning the decay products that have been converted to fossil fuels. Although the United States currently depends heavily on petroleum for energy, this dependency is a relatively recent phenomenon (Figure 20.5).

Petroleum and natural gas deposits probably formed from the remains of marine organisms that lived approximately 500 million years ago. **Petroleum** is a

Figure 20.5
Energy sources used in the United States.

thick, dark liquid composed largely of hydrocarbons containing from 5 to more than 25 carbon atoms. **Natural gas,** which is usually associated with petroleum deposits, consists mostly of methane but also contains significant amounts of ethane, propane, and butane.

To be used efficiently, petroleum must be separated by boiling into portions called fractions. The smaller hydrocarbons can be boiled off at relatively low temperatures; the larger molecules require successively higher temperatures. The major uses of various petroleum fractions are shown in Table 20.3.

The petroleum era began when the demand for lamp oil during the Industrial Revolution outstripped the traditional sources: animal fats and whale oil. In response to this increased demand, Edwin Drake drilled the first oil well in 1859 at Titusville, Pennsylvania. The petroleum from this well was refined to produce *kerosene* (fraction C_{10}—C_{18}), which served as an excellent lamp oil. *Gasoline*

Distillation (separation by boiling) was discussed in Section 3.5.

If you can obtain a small sample of crude oil, bring some to class so that students can note its color and odor.

Table 20.3 Uses of the Various Petroleum Fractions

Petroleum Fraction*	Major Uses
C_5—C_{12}	gasoline
C_{10}—C_{18}	kerosene jet fuel
C_{15}—C_{25}	diesel fuel heating oil lubricating oil
>C_{25}	asphalt

*Shows the chain lengths present in each fraction.

In conjunction with your discussion of petroleum and its products, you might want to discuss octane rating in gasoline. Some gasoline additives, such as ethanol, which enhance octane rating are themselves organic molecules. Students are surprised to discover that octane itself is a very poor fuel and has a low octane rating.

(fraction C_5—C_{10}) was of limited use and was often discarded. However, the importance of these fractions was reversed when the development of the electric light reduced the need for kerosene, and the advent of the "horseless carriage" signaled the birth of the gasoline age.

As gasoline became more important, new ways were sought to increase the yield of gasoline obtained from each barrel of petroleum. William Burton invented a process called *pyrolytic cracking* at Standard Oil of Indiana. In this process the heavier molecules of the kerosene fraction are heated to about 700 °C, which causes them to break (crack) into the smaller molecules characteristic of the gasoline fraction. As cars became larger, more efficient internal combustion engines were designed. Because of the uneven burning of the gasoline then available, these engines "knocked," producing unwanted noise and even engine damage. Intensive research to find additives that would promote smoother burning yielded tetraethyl lead, $(C_2H_5)_4Pb$, a very effective "anti-knock" agent.

Adding tetraethyl lead to gasoline became a common practice, and by 1960 gasoline contained as much as 3 grams of lead per gallon. As we have discovered so often in recent years, technological advances can produce environmental problems. Lead in gasoline causes two important problems. First, it "poisons" the catalytic converters that have been added to exhaust systems to help prevent air pollution. Second, the use of leaded gasoline has greatly increased the amount of lead in the environment, where it can be ingested by animals and humans. For these reasons the use of lead in gasoline is being phased out. This has required extensive (and expensive) modifications of engines and of the gasoline-refining process.

20.6   Reactions of Alkanes

Students tend to have a hard time classifying organic reactions. As I work examples of new types of reactions, or as I review old ones, I ask students to tell me what kind of reaction is occurring so that they become accustomed to identifying the various kinds of reactions.

AIM: To describe various types of chemical reactions that alkanes undergo.

At lower temperatures, however, alkanes are not very reactive because the C—C and C—H bonds in alkanes are relatively strong. For example, at 25 °C alkanes do not react with acids, bases, or strong oxidizing agents. This chemical inertness makes alkanes valuable as lubricating materials and as the backbone for structural materials such as plastics.

At sufficiently high temperatures, however, alkanes *react vigorously with oxygen.* These **combustion reactions** are the basis for the alkanes' widespread use as fuels. For example, the combustion reaction of butane with oxygen is

$$2C_4H_{10}(g) \ + \ 13O_2(g) \rightarrow 8CO_2(g) \ + \ 10H_2O(g)$$

The alkanes can also undergo **substitution reactions**—reactions in which *one or more hydrogen atoms of the alkane are replaced* (substituted) *by different atoms.* We can represent the substitution reaction of an alkane with a halogen molecule as follows:

$$R—H + X_2 \rightarrow R—X + HX$$

where R represents an alkyl group and X represents a halogen atom. For example, methane can react successively with chlorine as follows:

$$CH_4 + Cl_2 \xrightarrow{hv} CH_3Cl + HCl$$
Chloromethane

$$CH_3Cl + Cl_2 \xrightarrow{hv} CH_2Cl_2 + HCl$$
Dichloromethane

$$CH_2Cl_2 + Cl_2 \xrightarrow{hv} CHCl_3 + HCl$$
Trichloromethane
(chloroform)

$$CHCl_3 + Cl_2 \xrightarrow{hv} CCl_4 + HCl$$
Tetrachloromethane
(carbon tetrachloride)

The *hv* above each arrow signifies that ultraviolet light is needed to furnish the energy to break the Cl—Cl bond to produce chlorine atoms:

$$Cl_2 \rightarrow Cl\cdot + Cl\cdot$$

A chlorine atom has an unpaired electron, indicated by the dot, which makes it very reactive and able to attack the C—H bond.

Notice that each step in the process involves replacement of a C—H bond by a C—Cl bond. That is, a chlorine atom *substitutes* for a hydrogen atom. The names of the products of these reactions use the term *chloro* for the chlorine substituents with a prefix that gives the number of chlorine atoms present: *di-* for two, *tri-* for three, and *tetra-* for four. No number is used to describe the chlorine positions in this case, because methane has only one carbon atom. Note that the products of the last two reactions have two names, the systematic name and the common name in parentheses.

Besides substitution reactions, alkanes can also undergo **dehydrogenation reactions** in which *hydrogen atoms are removed* and the product is an unsaturated hydrocarbon. For example, in the presence of chromium(III) oxide at high temperatures, ethane can be dehydrogenated, yielding ethylene, C_2H_4.

$$CH_3CH_3 \xrightarrow[500\ °C]{Cr_2O_3} CH_2{=}CH_2 + H_2$$
Ethylene

Students often have trouble understanding that any one of the six hydrogens on ethane can be substituted with a chlorine and the product is still chloroethane. They initially believe that six different monosubstituted molecules can be produced. A model of chloroethane can make this clearer.

The symbol *hv* signifies ultraviolet light used to furnish energy for the reaction.

Name the products of reactions between alkanes and chlorine as you write new examples. This will help students to learn to use the *chloro* substituent name and to reinforce the naming rules for alkanes.

For the dehydrogenation of alkanes, point out that a hydrogen is removed from each of two adjacent carbons and that these two carbons form the double bond.

Acetylene gas burning. The acetylene is formed by the reaction of calcium carbide, CaC_2, with water in the flask.

Figure 20.6
The ball-and-stick model of ethylene (ethene).

In writing shorthand formulas, the hydrogen atoms are often written just after the carbon to which they are attached. For example, the formula for H—C≡C—H is often written as CH≡CH.

20.7 Alkenes and Alkynes

AIM: To learn to name hydrocarbons with double bonds (alkenes) and triple bonds (alkynes).
To describe addition reactions.

We have seen that alkanes are saturated hydrocarbons—each of the carbon atoms is bound to four atoms by single bonds. Hydrocarbons that contain carbon–carbon *double bonds*

$$\diagup C = C \diagdown$$

are called **alkenes.** Hydrocarbons with carbon–carbon *triple bonds* are called **alkynes.** Alkenes and alkynes are unsaturated hydrocarbons.

Multiple carbon–carbon bonds result when hydrogen atoms are removed from alkanes. Alkenes that contain a carbon–carbon double bond have the general formula C_nH_{2n}. The simplest alkene, C_2H_4, commonly known as *ethylene,* has the Lewis structure

$$\begin{matrix} H & & H \\ \diagdown & & \diagup \\ & C = C & \\ \diagup & & \diagdown \\ H & & H \end{matrix}$$

The ball-and-stick model of ethylene is shown in Figure 20.6.

The system for naming alkenes and alkynes is similar to the one we have used for alkanes. The following rules are useful.

Rules for Naming Alkenes and Alkynes

1. Select the longest continuous chain of carbon atoms that contains the double or triple bond.
2. For an alkene, the root name of the carbon chain is the same as for the alkane, except that the *-ane* ending is replaced by *-ene.* For an alkyne, the *-ane* is replaced by *-yne.* For example, for a two-carbon chain we have

$$CH_3CH_3 \qquad CH_2{=}CH_2 \qquad CH{\equiv}CH$$
Ethane   Ethene   Ethyne

3. Number the parent chain, starting at the end closest to the double or triple bond. The location of the multiple bond is given by the lowest-numbered carbon involved in the bond. For example,

$$CH_2{=}CHCH_2CH_3$$
$$1 \quad\ 2\ 3\quad 4$$

is called 1-butene and

$$CH_3CH{=}CHCH_3$$
$$1 \quad 2 \quad\ 3\ 4$$

is called 2-butene.

4. Substituents on the parent chain are treated the same way as in naming alkanes. For example, the molecule $C1CH{=}CHCH_2CH_3$ is called 1-chloro-2-butene.

You can contrast the general formula for alkenes with that of the alkanes. The formula for the alkenes contains two fewer hydrogens than does that for the alkanes. Two carbon–hydrogen bonds in alkanes have been replaced by an additional carbon–carbon bond in alkenes.

EXAMPLE 20.6   Naming Alkenes and Alkynes

Name each of the following molecules.

a. $CH_3CH_2CHCH{=}CHCH_3$
            |
           CH_3

b. $CH_3CH_2C{\equiv}CCHCH_2CH_3$
                  |
                 CH_2
                  |
                 CH_3

Naming alkenes and alkynes is straightforward provided that students have memorized the root names for the first 10 alkanes.

SOLUTION

a. The longest chain contains six carbon atoms, and we number the carbons starting from the end closest to the double bond.

$$\overset{6}{C}H_3\overset{5}{C}H_2\overset{4}{C}H\overset{3}{C}H{=}\overset{2}{C}H\overset{1}{C}H_3$$
                   |
                  CH_3

Thus the root name for the hydrocarbon is 2-hexene. Remember to use the lowest number of the two carbon atoms involved in the double bond. There is a methyl group attached to the number-4 carbon. Therefore the name of the compound is 4-methyl-2-hexene.

b. The longest chain of carbon atoms is seven carbons long, and the chain is numbered as shown (starting from the end closest to the triple bond).

$$\overset{1}{C}H_3\overset{2}{C}H_2\overset{3}{C}{\equiv}\overset{4}{C}\overset{5}{C}H\overset{6}{C}H_2\overset{7}{C}H_3$$
                  |
                 CH_2
                  |
                 CH_3

Molecules with the same number of carbons and hydrogens but that have double or triple bonds between different carbons are sometimes isomers of each other. Students can tell which molecules are the same and which are structural isomers because the latter have different locator numbers for the double or triple bond in their names.

EXAMPLE 20.6, CONTINUED

The hydrocarbon is a 3-heptyne (we use the lowest-numbered carbon in the triple bond). Because there is an ethyl group on carbon number 5, the full name is 5-ethyl-3-heptyne.

SELF-CHECK EXERCISE 20.4

Name the following molecules.

a. $CH_3CH_2CH_2CH_2CH{=}CHCHCH_3$
$\qquad\qquad\qquad\qquad\qquad\quad\ |$
$\qquad\qquad\qquad\qquad\qquad\quad CH_3$

b. $CH_3CH_2CH_2C{\equiv}CH$

You might want to demonstrate the addition reactions of alkenes by adding a few milliliters of bromine dissolved in trichloromethane or other inert solvent to some cyclohexene. (The red color of the bromine disappears as the reaction occurs.) You can then compare the reaction that occurs after adding some of the bromine solution to a liquid alkane. (No reaction occurs and the solution stays reddish.)

Reactions of Alkenes

Because alkenes and alkynes are unsaturated, their most important reactions are **addition reactions,** in which *new atoms form single bonds to the carbons formerly involved in the double or triple bonds.* An addition reaction for an alkene changes the carbon–carbon double bond to a single bond, giving a saturated hydrocarbon (each carbon bonded to four atoms). For example, **hydrogenation reactions,** which use H_2 as a reactant, lead to the addition of a hydrogen atom to each carbon that used to be involved in the double bond.

$$CH_2{=}CHCH_3 + H_2 \xrightarrow{\text{Catalyst}} CH_3CH_2CH_3$$
$$\text{1-Propene} \qquad\qquad\qquad\qquad \text{Propane}$$

Hydrogenation of alkenes is an important industrial process, particularly in the manufacture of solid shortenings. Unsaturated fats (fats containing double bonds) are generally liquids at room temperatures, whereas saturated fats (those containing C—C single bonds) are solids. The liquid unsaturated fats are converted to solid saturated fats by hydrogenation.

Halogenation of unsaturated hydrocarbons involves the addition of halogen atoms. Here is an example:

$$CH_2{=}CHCH_2CH_2CH_3 + Br_2 \rightarrow CH_2BrCHBrCH_2CH_2CH_3$$
$$\text{1-Pentene} \qquad\qquad\qquad\qquad \text{1,2-Dibromopentane}$$

Another important reaction of certain unsaturated hydrocarbons is **polymerization,** a process in which many small molecules are joined together to form a large molecule. Polymerization will be discussed in Section 20.16.

A model that shows the six carbons of benzene in a flat ring would be useful for this section.

20.8 Aromatic Hydrocarbons   7O1

20.8   Aromatic Hydrocarbons

AIM: To introduce the aromatic hydrocarbons.

When mixtures of hydrocarbons from natural sources, such as petroleum or coal, are separated, certain of the compounds that emerge have pleasant odors and are thus known as **aromatic hydrocarbons.** When these compounds, which include wintergreen, cinnamon, and vanilla, are examined, they are all found to contain a common feature: a six-membered ring of carbon atoms called the benzene ring. **Benzene** has the formula C_6H_6 and a planar (flat) structure in which all of the bond angles are 120° (Figure 20.7).

When we examine the bonding in the benzene ring, we find that more than one Lewis structure can be drawn. That is, the double bonds can be located in different positions, as shown in Figure 20.8. Because the actual bonding is a combination of the structures represented in Figure 20.8, the benzene ring is usually shown with a circle (Figure 20.9).

Vanilla beans.

(a)

(b)

Figure 20.7
Benzene, C_6H_6, consists of six carbon atoms bonded together to form a ring. Each carbon has one hydrogen atom bonded to it. All atoms in benzene lie in the same plane. This representation does not show all the bonds between the carbon atoms in the ring.

Figure 20.8
(a) Two Lewis structures for the benzene ring. (b) As a shorthand notation, the rings are usually represented without labeling the carbon and hydrogen atoms.

Figure 20.9
To show that the bonding in the benzene ring is a combination of different Lewis structures, the ring is drawn with a circle inside.

20.9 Naming Aromatic Compounds

AIM: To explain the system for naming aromatic compounds.

Students should memorize the short list of monosubstituted benzenes with common names, especially toluene and phenol, which are found in many common household products. For example, butylated hydroxy toluene (BHT) is commonly used an an antioxidant in foods.

Substituted benzene molecules are formed by replacing one or more of the H atoms on the benzene ring with other atoms or groups of atoms. We will consider benzene rings with one substituent (called monosubstituted benzenes) first.

Monosubstituted Benzenes

The systematic method for naming monosubstituted benzenes uses the substituent name as a prefix of benzene. For example, the molecule

is called chlorobenzene, and the molecule

is called ethylbenzene.

Sometimes monosubstituted benzene compounds have special names. For example, the molecule

has the systematic name methylbenzene. However, for convenience it is given the name toluene. Likewise, the molecule which might be called hydroxybenzene, has the special name phenol. Several examples of monosubstituted benzenes are shown in Figure 20.10.

Sometimes it is more convenient to name compounds if we view the benzene ring itself as a substituent. For example, the compound

$$\overset{4}{C}H_3\overset{3}{C}H\overset{2}{C}H=\overset{1}{C}H_2$$

is most easily named as a 1-butene with a benzene ring as a substituent on the number-3 carbon. When the benzene ring is used as a substituent, it is called the **phenyl** (pronounced fen′-ill) **group.** So the name of the above compound is 3-phenyl-1-butene. As another example, the compound

$$\overset{1}{C}H_3\overset{2}{C}H\overset{3}{C}H_2\overset{4}{C}H\overset{5}{C}H_2\overset{6}{C}H_3$$
$$\qquad\qquad\qquad Cl$$

is named 4-chloro-2-phenylhexane. Remember, we start to number the chain from the end closest to the first substituent and name the substituents in alphabetical order (chloro before phenyl).

Chlorobenzene   Toluene

Bromobenzene   Phenol

Nitrobenzene   Styrene

Figure 20.10
Names of some common monosubstituted benzenes.

Many students forget that benzene when used as a substituent is called the *phenyl group* and they use the term *benzyl* instead.

Disubstituted Benzenes

When there is more than one substituent on the benzene ring, numbers are used to indicate substituent position. For example, the compound

is named 1,2-dichlorobenzene. Another naming system uses the prefix *ortho- (o-)* for two adjacent substituents, *meta- (m-)* for two substituents with one carbon between them, and *para- (p-)* for two substituents opposite each other. This means that 1,2-dichlorobenzene can also be called ortho-dichlorobenzene or

When naming disubstituted benzenes, give the names in both naming systems, using numbers and prefixes (*ortho, meta,* or *para*), so that students become accustomed to both systems.

Ortho- (*o-*) means two adjacent substituents.

You can enliven your lecture on aromatic nomenclature with some plays on words that most students find funny. Some examples are *ortho*dox, *para*dox, mercedes *benz*(ene). There are many others.

o-dichlorobenzene. Likewise, the compound

Cl

Cl

can be called 1,3-dichlorobenzene or *m*-dichlorobenzene. The compound

Meta- (*m-*) means two substituents with one carbon between them.

Cl

Cl

Para- (*p-*) means two substituents directly across the ring from each other.

is named 1,4-dichlorobenzene or *p*-dichlorobenzene.

Benzenes that have two methyl substituents have the special name xylene. So the compound

CH₃

CH₃

which might be called 1,3-dimethylbenzene, is instead called *m*-xylene (meta-xylene).

When two different substituents are present on the benzene ring, one is always assumed to be at carbon number 1, and this number is not specified in the name. For example, the compound

Cl

Br

is named 2-bromochlorobenzene, not 2-bromo-1-chlorobenzene. Various examples of disubstituted benzenes are shown in Figure 20.11.

Benzene is the simplest aromatic molecule. More complex aromatic systems can be viewed as consisting of a number of "fused" benzene rings. Some examples are given in Table 20.4.

Mothballs contain napthalene, composed of "fused" benzene rings.

Br
Br

(1,2-Dibromobenzene)
(*o*-dibromobenzene)

Br

Br

1,3-Dibromobenzene
(*m*-dibromobenzene)

Br

Br

1,4-Dibromobenzene
(*p*-dibromobenzene)

CH₃

CH₃

1,4-Dimethylbenzene
(*p*-xylene)

CH₃
CH₃

1,2-Dimethylbenzene
(*o*-xylene)

CH₃

CH₃

1,3-Dimethylbenzene
(*m*-xylene)

CH₃
NO₂

2-Nitrotoluene
(*o*-nitrotoluene)

NO₂

Br

3-Bromonitrobenzene
(*m*-bromonitrobenzene)

CH₃

Cl

3-Chlorotoluene
(*m*-chlorotoluene)

Figure 20.11
Some selected substituted benzenes and their names. Common names are given in parentheses.

Naming disubstituted benzenes can be confusing for students. When both substituents are the same, two numbers are used as locators, but when the substituents are different, only one number is used. We assume that the first substituent is located on carbon number 1. You will need to point out this distinction to your students.

Table 20.4 More Complex Aromatic Molecules

Structural Formula	Name	Use
	naphthalene	mothballs
	anthracene	dyes
	phenanthrene	dyes, explosives, and synthesis of drugs

Dyed nylon rope.

EXAMPLE 20.7	**Naming Aromatic Compounds**

Name the following compounds.

a. CH₂CH₃

CH₂CH₃

c. CH₃—CH—C≡CH

b. CH₃

Br

d. CH₃

O₂N   NO₂

NO₂

SOLUTION

a. There are ethyl groups in the 1 and 3 (or meta-) positions, so the name is 1,3-diethylbenzene or *m*-diethylbenzene.

b. The

group is called toluene. The bromine is in the 4 (or para-) position. The name is 4-bromotoluene or *p*-bromotoluene.

c. In this case we name the compound as a butyne with a phenyl substituent. The name is 3-phenyl-1-butyne.

d. We name this compound as a substituted toluene (with the —CH₃ group assumed to be on carbon number 1). So the name is 2,4,6-trinitrotoluene. This compound is more commonly known as TNT, a component of dynamite.

SELF-CHECK EXERCISE 20.5

Name the following compounds.

a. NO₂

Cl

b. CH₃CH₂CH—CH=CHCH₃

20.10 Functional Groups

AIM: To describe the common functional groups in organic molecules.

Bring to class samples of compounds containing functional groups you intend to discuss. Organic molecules have interesting odors and other observable properties.

The vast majority of organic molecules contain elements in addition to carbon and hydrogen. However, most of these substances can be viewed as **hydrocarbon derivatives,** molecules that are fundamentally hydrocarbons but that have additional atoms or groups of atoms called **functional groups.** The common functional groups are listed in Table 20.5; one example of a compound that contains that functional group is given for each. We will briefly describe some of these functional groups in the next few sections and learn to name the compounds that contain them.

The different organic functional groups look similar and the structure of each group can be abbreviated in several ways. It can be difficult for students to translate RCOOH into RC—OH.

$$RC\overset{O}{\overset{\|}{-}}OH$$

Table 20.5 The Common Functional Groups

Class	Functional Group	General Formula*	Example
halohydrocarbons	—X (F,Cl,Br,I)	R—X	CH_3I
alcohols	—OH	R—OH	CH_3OH
ethers	—O—	R—O—R′	CH_3—O—CH_3
aldehydes	$-\overset{O}{\overset{\|}{C}}-H$	$R-\overset{O}{\overset{\|}{C}}-H$	$H-\overset{O}{\overset{\|}{C}}-H$
ketones	$-\overset{O}{\overset{\|}{C}}-$	$R-\overset{O}{\overset{\|}{C}}-R'$	$CH_3-\overset{O}{\overset{\|}{C}}-CH_3$
carboxylic acids	$-\overset{O}{\overset{\|}{C}}-OH$	$R-\overset{O}{\overset{\|}{C}}-OH$	$CH_3-\overset{O}{\overset{\|}{C}}-OH$
esters	$-\overset{O}{\overset{\|}{C}}-O-$	$R-\overset{O}{\overset{\|}{C}}-O-R'$	$CH_3-\overset{O}{\overset{\|}{C}}-OCH_2CH_3$
amines	$-NH_2$	$R-NH_2$	CH_3NH_2

Students should memorize the structures of the functional groups given in Table 20.5 so that they can recognize the different types of organic molecules.

*R and R′ represent hydrocarbon fragments.

20.11 Alcohols

AIM: To describe simple alcohols and explain how to name them.

Alcohols are characterized by the presence of the —OH group. Some common alcohols are listed in Table 20.6. The systematic name for an alcohol is obtained by replacing the final -e of the parent hydrocarbon name with -ol. The position of the —OH group is specified by a number (where necessary) chosen such that it is the smallest of the substituent numbers.

The rules for naming alcohols follow.

Rules for Naming Alcohols

1. Select the longest chain of carbon atoms containing the —OH group.
2. Number the chain such that the carbon with the —OH group gets the lowest possible number.
3. Obtain the root name from the name of the parent hydrocarbon chain by replacing the final -e with -ol.
4. Name any other substituents as usual.

For example, the compound

$$CH_3CHCH_2CH_2CH_3$$
$$\qquad |$$
$$\qquad OH$$

is called 2-pentanol because the parent carbon chain is pentane. The compound

$$CH_3CH_2CHCH_2CH_2CH_3$$
$$\qquad\qquad |$$
$$\qquad\qquad OH$$

is called 3-hexanol.

Alcohols are classified according to the number of hydrocarbon fragments (alkyl groups) bonded to the carbon where the —OH group is attached. Thus we have

$$R—CH_2OH \qquad \overset{R}{\underset{R'}{\diagdown}}CHOH \qquad R'—\overset{R}{\underset{\underset{R''}{|}}{\overset{|}{C}}}OH$$

Primary alcohol   *Secondary* alcohol   *Tertiary* alcohol
(one R group)   (two R groups)   (three R groups)

where R, R', and R'' (which may be the same or different) represent hydrocarbon fragments (alkyl groups).

Table 20.6 Some Common Alcohols

Formula	Systematic Name	Common Name
CH_3OH	methanol	methyl alcohol
CH_3CH_2OH	ethanol	ethyl alcohol
$CH_3CH_2CH_2OH$	1-propanol	n-propyl alcohol

CH_3CHCH_3
   |
   OH
	2-propanol	isopropyl alcohol

EXAMPLE 20.8   Naming Alcohols

Give the systematic name for each of the following alcohols, and specify whether
the alcohol is primary, secondary, or tertiary.

a. $CH_3CHCH_2CH_3$
          |
          OH

c. CH_3
          |
   $CH_3CCH_2CH_2CH_2CH_2Br$
        |
        OH

b. $ClCH_2CH_2CH_2OH$

SOLUTION

a.   The chain is numbered as follows:

       1 2 3 4
     $CH_3CHCH_2CH_3$
        |
        OH

The compound is called 2-butanol, because the —OH group is
located at the number-2 position of a four-carbon chain. Note that the
carbon to which the —OH is attached also has two R groups (—CH_3
and —CH_2CH_3) attached. Therefore this is a *secondary* alcohol.

            H
           |
 CH_3—C—CH_2CH_3
   R | R'
       OH

b.   The chain is numbered as follows:

       3   2   1
  Cl—CH_2—CH_2—CH_2—OH

Remember that in naming an alcohol, we give the carbon with the
—OH attached the lowest possible number. The name is 3-chloro-1-
propanol. This is a *primary* alcohol.

If you write the expanded structure for the carbon that has the OH, it will be easier for students to tell whether the alcohol is primary, secondary, or tertiary.

     H
     |
  R—C—R'
     |
    OH

EXAMPLE 20.8, CONTINUED

$$R-\underset{\underset{H}{|}}{\overset{\overset{H}{|}}{C}}-OH$$

Cl—CH$_2$CH$_2$—C—OH (with H above and H below the C)

One R group attached to the carbon with the —OH group

c.   The chain is numbered as follows:

$$R-\underset{\underset{OH}{|}}{\overset{\overset{R'}{|}}{\underset{}{}}}-R''$$

(CH$_3$)

$\underset{1}{(CH_3)}-\underset{\underset{OH}{|}}{\overset{2}{C}}-\underset{3}{(CH_2}-\underset{4}{CH_2}-\underset{5}{CH_2}-\underset{6}{CH_2Br)}$

The name is 6-bromo-2-methyl-2-hexanol. This is a tertiary alcohol, because the carbon where the —OH is attached also has three R groups attached.

SELF-CHECK EXERCISE 20.6

Name each of the following alcohols, and specify whether it is primary, secondary, or tertiary.

a.   CH$_3$CH$_2$CH$_2$CH$_2$CH$_2$OH

b. $CH_3-\underset{\underset{OH}{|}}{\overset{\overset{CH_3}{|}}{C}}-CH_3$

c. $CH_3-\underset{\underset{OH}{|}}{CH}-CH_2CH_2\overset{\overset{Br}{|}}{CH}CH_3$

20.12   Properties and Uses of Alcohols

AIM: To describe how some alcohols are made and used.

Although there are many important alcohols, the simplest ones, methanol and ethanol, have the greatest commercial value. Methanol, also known as *wood alcohol* because it was formerly obtained by heating wood in the absence of air, is prepared industrially (over 20 million tons per year in the United States) by the hydrogenation of carbon monoxide (catalyzed by a ZnO/Cr$_2$O$_3$ mixture).

$$CO \ + \ 2H_2 \xrightarrow[\text{ZnO/Cr}_2\text{O}_3]{400 \ ^\circ\text{C}} CH_3OH$$

Methanol is used as a starting material for the synthesis of acetic acid and many types of adhesives, fibers, and plastics. It also can be used as a motor fuel. In fact, pure methanol has been used for many years in the engines of the cars that are driven in the Indianapolis 500 and similar races. Methanol is especially useful in racing engines because of its resistance to knock. Methanol is advantageous for regular cars because it produces less carbon monoxide (a toxic gas) in the exhaust than does gasoline. Methanol is highly toxic to humans, and swallowing it can lead to blindness and death.

Ethanol is the alcohol found in beverages such as beer, wine, and whiskey; it is produced by the fermentation of the sugar glucose in corn, barley, grapes, and so on.

Students may find it interesting to discuss some common substances associated with ethyl alcohol, such as denatured alcohol and absolute alcohol, and to compare the toxicity of methanol, ethanol, and ethylene glycol.

$$\underset{\text{Glucose}}{C_6H_{12}O_6} \xrightarrow{\text{Yeast}} \underset{\text{Ethanol}}{2CH_3CH_2OH} + 2CO_2$$

This reaction is catalyzed by the enzymes (biological catalysts) found in yeast, and it can proceed only until the alcohol content reaches approximately 13% (that found in most wines), at which point the yeast can no longer survive. Beverages with higher alcohol content are made by distilling the fermentation mixture.

Discussing the production of alcoholic beverages can interest students in the properties of alcohols.

Ethanol, like methanol, can be burned in the internal combustion engines of automobiles and is now commonly added to gasoline to form gasohol. It is also used in industry as a solvent and for the preparation of acetic acid. The commercial production of ethanol (half a million tons per year in the United States) is carried out by reaction of water with ethylene.

$$CH_2{=}CH_2 \ + \ H_2O \xrightarrow[\text{Catalyst}]{\text{Acid}} CH_3CH_2OH$$

Many alcohols are known that contain more than one —OH group. The one that is the most important commercially is ethylene glycol,

$$\begin{array}{c} H_2C\text{—OH} \\ | \\ H_2C\text{—OH} \end{array}$$

a toxic substance that is the major constituent of most automobile antifreezes.

The simplest aromatic alcohol is

Addition reactions between alkenes and various substances were discussed in Section 20.7. The reaction of alkenes with H_2O is also an addition reaction. The OH group adds to one carbon of the carbon–carbon double bond, and the H adds to the other carbon. The carbon–hydrogen and carbon–OH bonds can form because the carbon–carbon double bond breaks.

OH

commonly called **phenol.** Most of the 1 million tons of phenol produced annually in the United States is used to produce polymers for adhesives and plastics.

Some interesting aldehydes and ketones to bring to class include cinamaldehyde, citral (odor of lemons), and vanillin. These compounds have a strong odor; only a few grams or milliliters are needed.

The abbreviations for the structures of aldehydes and ketones and the many possible ways to draw the carbonyl group cause students problems. I often start with a single aldehyde and ketone, then write their structures in as many different ways as possible to show that the molecules don't change even though they may look somewhat different.

20.13 Aldehydes and Ketones

AIM: To describe the general formulas for aldehydes and ketones and some of their uses.

Aldehydes and ketones contain the **carbonyl group**

$$-\underset{\underset{O}{\|}}{C}-$$

In **ketones** this group is bonded to two carbon atoms; an example is acetone:

$$CH_3-\underset{\underset{O}{\|}}{C}-CH_3$$

The general formula for a ketone is

$$R-\underset{\underset{O}{\|}}{C}-R'$$

where R and R′ are alkyl groups that may or may not be the same. In a ketone the carbonyl group is never at the end of the hydrocarbon chain. (If it were, it would be an aldehyde.)

In **aldehydes** the carbonyl group always appears at the end of the hydrocarbon chain. There is always at least one hydrogen bonded to the carbonyl carbon atom. An example of an aldehyde is acetaldehyde:

$$CH_3-\underset{\underset{O}{\|}}{C}-H$$

The general formula for an aldehyde is

$$R-\underset{\underset{O}{\|}}{C}-H$$

We often use compact formulas for aldehydes and ketones. For example, formaldehyde (where R = H) and acetaldehyde (where R = CH_3) are usually represented as HCHO and CH_3CHO, respectively. Acetone is often written as CH_3COCH_3 or $(CH_3)_2CO$.

Many ketones have useful solvent properties (acetone is often found in nail polish remover, for example) and are frequently used in industry for this purpose. Aldehydes typically have strong odors. Vanillin is reponsible for the pleasant odor of vanilla beans; cinnamaldehyde produces the characteristic odor of cinnamon. On the other hand, the unpleasant odor of rancid butter arises from the presence of butyraldehyde. (See Figure 20.12 for the structures of these compounds.)

Vanillin

Cinnamaldehyde

Butyraldehyde

Figure 20.12
Some common aldehydes.

Primary and secondary alcohols, when oxidized, produce products with different functional groups. You can point out that this is the reason we distinguish among primary, secondary, and tertiary alcohols.

Aldehydes and ketones are most often produced commercially by the oxidation of alcohols. Oxidation of a *primary* alcohol gives the corresponding aldehyde, for example,

$$CH_3CH_2OH \xrightarrow{\text{Oxidation}} CH_3C\overset{\displaystyle O}{\underset{\displaystyle H}{\diagup}}$$

$$R{-}OH \longrightarrow R{-}C\overset{\displaystyle O}{\underset{\displaystyle H}{\diagup}}$$

And oxidation of a *secondary* alcohol results in a ketone:

$$CH_3\underset{\underset{\displaystyle OH}{|}}{C}HCH_3 \xrightarrow{\text{Oxidation}} CH_3\underset{\underset{\displaystyle O}{\|}}{C}CH_3$$

$$R\underset{\underset{\displaystyle OH}{|}}{\overset{\overset{\displaystyle H}{|}}{C}}R' \longrightarrow R\underset{\underset{\displaystyle O}{\|}}{C}R'$$

20.14 Naming Aldehydes and Ketones

AIM: To introduce the systems for naming aldehydes and ketones.

We obtain the systematic name for an aldehyde from the parent alkane by removing the final -e and adding -al. For ketones the final -e is replaced by -one, and a number indicates the position of the carbonyl group where necessary. The carbon chain in ketones is numbered such that the

$$-\underset{\underset{\displaystyle O}{\|}}{C}-$$

carbon gets the lowest possible number. In aldehydes the

$$-\underset{\underset{\displaystyle O}{\|}}{C}-H$$

is always at the end of the chain and is always assumed to be carbon number 1. The positions of other substituents are specified by numbers, as usual. The following examples illustrate these principles.

The functional group endings for alcohols, -ol, and aldehydes, -al, sound very much the same when spoken. Students sometimes have difficulty knowing whether I am discussing alcohols or aldehydes unless I write the structure of the molecule in question on the board.

In naming aldehydes and ketones, students have trouble distinguishing between aldehyde and ketone functional groups. Remind them that in ketones, the carbonyl group is located between two other carbons, whereas in aldehydes it is always on an end.

$$H{-}\underset{\underset{\displaystyle }{\|}}{\overset{\overset{\displaystyle O}{\|}}{C}}{-}H \qquad CH_3{-}\overset{\overset{\displaystyle O}{\|}}{C}{-}H \qquad CH_3{-}\overset{\overset{\displaystyle O}{\|}}{C}{-}CH_3 \qquad CH_3{-}\overset{\overset{\displaystyle O}{\|}}{C}{-}CH_2CH_2CH_3 \qquad CH_3\underset{\underset{\displaystyle Cl}{|}}{C}HCH_2\overset{\overset{\displaystyle O}{\|}}{C}{-}H$$

| Methanal | Ethanal | Propanone | 2-Pentanone | 3-Chlorobutanal |
| (formaldehyde) | (acetaldehyde) | (acetone) | | |

The names in parentheses are common names that are used much more often than the systematic names.

Another common aldehyde is benzaldehyde (an aromatic hydrocarbon), which has the structure

Benzaldehyde

An alternative system for naming ketones specifies the substituents attached to the C=O group. For example, the compound

$$CH_3\underset{\underset{O}{\|}}{C}CH_2CH_3$$

is called 2-butanone when we use the system just described. However, this molecule can also be named methyl ethyl ketone and is commonly referred to in industry as MEK (*methyl ethyl ketone*):

Another example is the use of the name ketone for the compound

which is commonly called methyl phenyl ketone.

EXAMPLE 20.9   Naming Aldehydes and Ketones

Name the following molecules.

a.
$$CH_3C{-}CHCH_3$$
with CH_3 on the $CHCH_3$ carbon and O double-bonded to the first C (Give two names.)

b.
H, O on C; ring with NO_2

c.
CH_3CH_2, O on C; attached to benzene ring

d. $CH_3CHCH_2CH_2CHO$
with Cl on the second carbon

SOLUTION

a.   We can name this molecule as a 2-butanone because the longest chain has four carbon atoms (butane root) with the

$$\diagdown C{=}O$$
$$\diagup$$

group in the number-2 position (the lowest possible number). Because the methyl group is in the number-3 position, the name is 3-methyl-2-butanone. We can also name this compound methyl isopropyl ketone.

Methyl ← CH_3 ; C with O double bond; CH with CH_3 and CH_3 → Isopropyl group

b.   We name this molecule as a substituted benzaldehyde (the nitro group is in the number-3 position): 3-nitrobenzaldehyde. It might also be named *m*-nitrobenzaldehyde.

c.   We name this molecule as a ketone: ethyl phenyl ketone.

d.   The name is 4-chloropentanal. Note that an aldehyde group is always at the end of the chain and is automatically assigned as the number-1 carbon.

EXAMPLE 20.9, CONTINUED

SELF-CHECK EXERCISE 20.7

Name the following molecules.

a.

$$CH_3CH_2\overset{\overset{\displaystyle O}{\|}}{C}\underset{\underset{\displaystyle CH_2CH_3}{|}}{CH}CH_2CH_3$$

b. $CH_3CH_2CH_2\underset{\underset{\displaystyle CH_3\overset{|}{C}HCH_3}{|}}{C}HCH_2CH_2CH_2CH_2CH_2CHO$

It is interesting to demonstrate the acidic properties of carboxylic acids using household vinegar instead of laboratory-grade acetic acid.

20.15 Carboxylic Acids and Esters

AIM: To describe the structures and names of the common carboxylic acids.

Carboxylic acids are characterized by the presence of the **carboxyl group,** —COOH, which has the structure

The general formula of a carboxylic acid is RCOOH, where R represents the hydrocarbon fragment. These molecules typically are weak acids in aqueous solution. That is, the dissociation equilibrium

$$RCOOH(aq) + H_2O(l) \rightleftharpoons H_3O^+(aq) + RCOO^-(aq)$$

lies far to the left—only a small percent of the RCOOH molecules are ionized.

We name carboxylic acids by dropping the final -e from the parent alkane (the longest chain containing the —COOH group) and adding -oic. Carboxylic acids are frequently known by their common names. For example, CH_3COOH, commonly called acetic acid, has the systematic name ethanoic acid because the parent alkane is ethane. Several carboxylic acids, their systematic names, and their common names are given in Table 20.7. Other examples of carboxylic acids are shown in Figure 20.13. Note that the —COOH group is always assigned the number-1 position in the chain.

Carboxylic acids can be produced by oxidizing primary alcohols with a strong oxidizing agent. For example, we can oxidize ethanol to acetic acid by using potassium permanganate.

$$CH_3CH_2OH \xrightarrow{\text{KMnO}_4(aq)} CH_3COOH$$

Figure 20.13
Some carboxylic acids.

Table 20.7 Several Carboxylic Acids, Their Systematic Names, and Their Common Names

Formula	Systematic Name	Common Name
HCOOH	methanoic acid	formic acid
CH_3COOH	ethanoic acid	acetic acid
CH_3CH_2COOH	propanoic acid	propionic acid
$CH_3CH_2CH_2COOH$	butanoic acid	butyric acid
$CH_3CH_2CH_2CH_2COOH$	pentanoic acid	valeric acid

The odors of lower molecular weight carboxylic acids are very noticeable. A few milliliters of dilute propanoic and butanoic acids are enough for students to appreciate the characteristic odors of cheese and rancid fats.

A carboxylic acid reacts with an alcohol to form an ester and a water molecule. For example, the reaction of acetic acid and ethanol produces the ester ethyl acetate and water.

React to form water

This reaction can be represented in general as follows:

$$RCOOH + R'OH \rightarrow RCOOR' + H_2O$$

Acid Alcohol Ester Water

An **ester** has the following general formula:

From the acid

From the alcohol

When students look at the formula for an ester it is often hard for them to tell which part of the molecule is derived from the carboxylic acid and which part from the alcohol. Writing the formula so that the alcohol part of the molecule is below the plane of the line can help students make this distinction.

Esters often have a sweet, fruity odor that contrasts markedly with the often-pungent odors of the parent carboxylic acids. For example, the odor of bananas derives from *n*-amyl acetate:

and that of oranges from *n*-octyl acetate:

Amyl is a common name for $CH_3CH_2CH_2CH_2CH_2$—.

If you have some available, bring to class some examples of fragrant esters. Or, they can be easily synthesized from the appropriate alcohol and carboxylic acid by heating a few milliliters of each with a few drops of concentrated sulfuric acid. Isobutyl formate (raspberry), octyl acetate (orange), or ethyl butyrate (pineapple) have strong fragrances.

The molecule acetylsalicylic acid is the most complex organic molecule students have seen so far. You can point out that there are two functional groups on the benzene ring, an ester and a separate carboxylic acid.

Like carboxylic acids, esters are often referred to by their common names. The name consists of the alkyl name from the alcohol followed by the acid name, where the *-ic* ending is replaced by *-ate.* For example, the ester

is made from acetic acid, CH_3COOH, and isopropyl alcohol,

and is called isopropyl acetate. The systematic name for this ester is isopropylethanoate (from ethanoic acid, the systematic name for acetic acid).

A very important ester is formed from the reaction of salicylic acid and acetic acid.

Salicylic acid   Acetic acid   Acetylsalicylic acid

The product is acetylsalicylic acid, commonly known as *aspirin,* which is manufactured in huge quantities and is widely used as an analgesic (painkiller).

20.16 Polymers

In writing the structures of polymers formed by elimination reactions, students often do not have enough experience to judge which functional groups will react.

AIM: To describe some common polymers.

Polymers are large, usually chain-like molecules that are built from small molecules called *monomers.* Polymers form the basis for synthetic fibers, rubbers, and plastics and have played a leading role in the revolution brought about in our lives by chemistry during the past 50 years. (Many important biomolecules are also polymers.)

The simplest and one of the best-known synthetic polymers is *polyethylene,* which is constructed from ethylene monomers. Its structure is

$$n\mathrm{CH_2}{=}\mathrm{CH_2} \xrightarrow{\text{Catalyst}} \left(\begin{array}{cc} \mathrm{H} & \mathrm{H} \\ | & | \\ \mathrm{C} & \mathrm{C} \\ | & | \\ \mathrm{H} & \mathrm{H} \end{array}\right)_n$$

Ethylene   Polyethylene

where n represents a large number (usually several thousand). Polyethylene is a tough, flexible plastic used for piping, bottles, electrical insulation, film for packaging, garbage bags, and many other purposes. Its properties can be varied by using substituted ethylene monomers. For example, when tetrafluorethylene is the monomer, the polymer Teflon is obtained.

$$n\left(\begin{array}{c}\mathrm{F}\quad\quad\mathrm{F}\\ \diagdown\;\diagup\\ \mathrm{C}{=}\mathrm{C}\\ \diagup\;\diagdown\\ \mathrm{F}\quad\quad\mathrm{F}\end{array}\right) \longrightarrow \left(\begin{array}{cc}\mathrm{F}&\mathrm{F}\\|&|\\\mathrm{C}&\mathrm{C}\\|&|\\\mathrm{F}&\mathrm{F}\end{array}\right)_n$$

Tetrafluorethylene   Teflon

Because of the resistance of the strong C—F bonds to chemical attack, Teflon is an inert, tough, and nonflammable material that is widely used for electrical insulation, nonstick coatings for cooking utensils, and bearings for low-temperature applications.

Other similar polyethylene-type polymers are made from monomers containing chloro, methyl, cyano, and phenyl substituents (Table 20.8, page 720). In each case, the carbon−carbon double bond in the substituted ethylene monomer becomes a single bond in the polymer. The different substituents lead to a wide variety of properties.

The polyethylene polymers illustrate one of the major types of polymerization reactions, called **addition polymerization,** in which the monomers simply "add together" to form the polymer and there are no other products.

Another common type of polymerization is **condensation polymerization,** in which a small molecule, such as water, is formed for each extension of the polymer chain. The most familiar polymer produced by condensation is *nylon.* Nylon is a **copolymer,** because two different types of monomers combine to form the chain (a **homopolymer,** by contrast, results from the polymerizing of a single type of monomer). One common form of nylon is produced when hexamethylenediamine and adipic acid react by splitting out a water molecule to form a C—N bond.

It is interesting to bring to class common household items made from the polymers you will discuss: polyethylene bags and bottles, polystyrene toys, PVC pipe, and teflon coated pans.

If you have the facilities, you might want to synthesize polymers in class. With little preparation you can make long strands of nylon-66 and polyurethane foam.

Forming polymers from monomers or writing the structure of a monomer from a polymer requires making and breaking carbon−carbon bonds. Students often have difficulty correctly deciding where to form the new bonds.

Table 20.8 Some Common Synthetic Polymers, Their Monomers, and Applications

Monomer *Name and Formula*	Polymer *Name and Formula*	Applications
ethylene $H_2C=CH_2$	polyethylene $-(CH_2-CH_2)_n-$	plastic piping, bottles, electrical insulation, toys
propylene $H_2C=C\overset{\displaystyle H}{\underset{\displaystyle CH_3}{}}$	polypropylene $-(CH-CH_2CH-CH_2)_n-$ $\quad CH_3 \qquad CH_3$	film for packaging, carpets, lab wares, toys
vinyl chloride $H_2C=C\overset{\displaystyle H}{\underset{\displaystyle Cl}{}}$	polyvinyl chloride (PVC) $-(CH_2-CH)_n-$ $\qquad\quad Cl$	piping, siding, floor tile, clothing, toys
acrylonitrile $H_2C=C\overset{\displaystyle H}{\underset{\displaystyle CN}{}}$	polyacrylnitrile (PAN) $-(CH_2-CH)_n-$ $\qquad\quad CN$	carpets, fabrics
tetrafluoroethylene $F_2C=CF_2$	Teflon $-(CF_2-CF_2)_n-$	coating for cooking utensils, electrical insulation, bearings
styrene $H_2C=C$ (phenyl)	polystyrene $-(CH_2CH)_n-$ (phenyl)	containers, thermal insulation, toys
butadiene $H_2C=C-C=CH_2$ (with H, H)	polybutadiene $-(CH_2CH=CHCH_2)_n-$	tire tread, coating resin
butadiene and styrene (see above)	styrene–butadiene rubber $-(CH-CH_2-CH_2-CH=CH-CH_2)_n-$ (phenyl)	synthetic rubber

Hexamethylenediamine   Adipic acid

+ H₂O

The molecule formed, which is called a **dimer** (two monomers joined), can undergo further condensation reactions because it has an amino group at one end and a carboxyl group at the other. Thus both ends are free to react with another monomer. Repetition of this process leads to a long chain of the type

which is the general structure of nylon. The reaction to form nylon occurs quite readily and is often used as a lecture demonstration (Figure 20.14). The properties of nylon can be varied by changing the number of carbon atoms in the chain of the acid or amine monomers.

More than 1 million tons of nylon are produced annually in the United States for use in clothing, carpets, rope, and so on. Many other types of condensation polymers are also produced. For example, Dacron is a copolymer formed from the condensation reaction of ethylene glycol (a dialcohol) and *p*-terephthalic acid (a dicarboxylic acid).

HOCH₂CH₂O⟮H   HO⟯   *p*-Terephthalic
Ethylene   ↓   acid
glycol   H₂O

The repeating unit of Dacron is

Figure 20.14

The reaction to form nylon can be carried out at the interface of two immiscible liquid layers in a beaker. The bottom layer contains adipyl chloride,

dissolved in CCl₄, and the top layer contains hexamethylenediamine,

H₂N⟮CH₂⟯₆NH₂

dissolved in water. A molecule of HCl is formed as each C—N bond forms. This is a variation of the reaction to form nylon discussed in the text.

Note that this polymerization involves a carboxylic acid and an alcohol to form an ester group,

$$
\begin{array}{c}
\quad\quad\ \ \overset{\displaystyle O}{\overset{\displaystyle \|}{}} \\
-O-C-
\end{array}
$$

Thus Dacron is called a **polyester.** By itself or blended with cotton, Dacron is widely used in fibers for the manufacture of clothing.

CHAPTER REVIEW

Key Terms

organic chemistry (p. 678)
hydrocarbon (p. 680)
alkane (p. 680)
normal (straight-chain
  or unbranched) hydrocarbons
  (p. 681)
structural isomerism (p. 683)
combustion reaction (p. 696)
substitution reaction (p. 697)
dehydrogenation reaction (p. 697)
alkene (p. 698)
alkyne (p. 698)
addition reaction (p. 700)
hydrogenation reaction (p. 700)
halogenation (p. 700)

polymerization (p. 700)
aromatic hydrocarbon (p. 701)
phenyl group (p. 703)
hydrocarbon derivative (p. 707)
functional group (p. 707)
alcohol (p. 708)
phenol (p. 711)
ketone (p. 712)
aldehyde (p. 712)
carboxylic acid (p. 716)
ester (p. 717)
polymer (p. 718)
addition polymerization (p. 719)
condensation polymerization
  (p. 719)

Summary

1. The study of carbon-containing compounds and their properties is called organic chemistry. Most organic compounds contain chains or rings of carbon atoms. The organic molecules responsible for maintaining and reproducing life are called biomolecules.
2. Hydrocarbons are organic compounds composed of carbon and hydrogen. Those that contain only C—C single bonds are saturated and are called alkanes, and those with carbon–carbon multiple bonds are unsaturated. Unsaturated hydrocarbons can become saturated by the addition of hydrogen, halogens, and or other substituents.

3. All alkanes can be represented by the general formula C_nH_{2n+2}. Methane, CH_4, is the simplest alkane, and the next three in the series are ethane, C_2H_6; propane, C_3H_8; and butane, C_4H_{10}. In a saturated hydrocarbon, each carbon atom is bonded to four other atoms. Alkanes containing long chains of carbon atoms are called normal, or straight-chain, hydrocarbons.

4. Structural isomerism in alkanes involves the formation of branched structures. Specific rules for systematically naming alkanes indicate the point of attachment of any substituent group, the length of the root chain, and so on.

5. Alkanes can undergo combustion reactions to form carbon dioxide and water or substitution reactions in which hydrogen atoms are replaced by other atoms. Alkanes can also undergo dehydrogenation to form unsaturated hydrocarbons.

6. Hydrocarbons with carbon–carbon double bonds are called alkenes. The simplest alkene is ethylene, C_2H_4. Alkynes are unsaturated hydrocarbons with a carbon–carbon triple bond. The simplest in the series is acetylene, C_2H_2.

7. Unsaturated hydrocarbons undergo addition reactions such as hydrogenation (addition of hydrogen atoms) and halogenation (addition of halogen atoms). Ethylene and substituted ethylene molecules can undergo polymerization, a process by which many molecules (monomers) are joined together to form a large chain-like molecule.

8. Organic molecules that contain elements in addition to carbon and hydrogen can be viewed as hydrocarbon derivatives: hydrocarbons with functional groups. Each functional group exhibits characteristic chemical properties.

9. Alcohols contain the —OH functional group. Aldehydes and ketones contain the carbonyl functional group,

$$\diagdown C {=} O$$
$$\diagup$$

In aldehydes this group is bonded to at least one hydrogen atom. Carboxylic acids are characterized by the carboxyl functional group,

$$-C\diagup\!\!\!{=}O$$
$$\diagdown OH$$

They can react with alcohols to form esters.

10. Polymers can be formed by addition polymerization, in which monomers add together, or by condensation polymerization, which involves the splitting out of small molecules (such as water) as the monomers react.

Questions and Problems

All even-numbered exercises have answers in the back of this book and solutions in the Solutions Guide.

20.1 Carbon Bonding

QUESTIONS

1. What makes carbon able to form so many different compounds?

2. What is the maximum number of other atoms to which a given carbon atom can be attached?

3. What does a double bond represent? Give two examples of molecules in which double bonds occur, and draw the Lewis structures of those molecules.

4. What does a triple bond represent? How many pairs of electrons are shared between the atoms in a triple bond? Draw the Lewis structure of a molecule with a triple bond.

5. When a carbon atom is bonded to four other atoms, what geometric arrangement occurs around the carbon atom? Why?

6. The structures of the organic molecules propane and butane are given in Figure 20.3. How many other atoms is each carbon bonded to in these molecules? What is the geometry around each carbon atom in these molecules?

20.2 Alkanes

QUESTIONS

7. What is a *hydrocarbon?* What are the only elements present in hydrocarbons?

8. What does it mean to say that a hydrocarbon is *saturated?* How many other atoms is each carbon atom in a saturated hydrocarbon bonded to? What name is given to the family of saturated hydrocarbons?

9. What is an *unsaturated* hydrocarbon? In what respect is such a hydrocarbon "unsaturated"? What structural feature that is not present in saturated hydrocarbons is present in an unsaturated hydrocarbon?

10. Although alkanes in which the carbon atoms form long, unbranched chains are often called straight-chain alkanes, why is such a chain not really *straight?* What is the bond angle between carbon atoms in such a chain?

PROBLEMS

11. Write structural formulas for the normal (straight-chain) alkanes with:
    a. six carbon atoms   c. twenty carbon atoms
    b. nine carbon atoms

12. Write structural formulas for the normal (straight-chain) alkanes with:
    a. eight carbon atoms   c. twelve carbon atoms
    b. ten carbon atoms

13. For each of the following straight-chain alkanes, draw the structural formula.
    a. heptane   c. propane
    b. nonane   d. decane

14. Draw the structural formula for each of the following straight-chain alkanes.
    a. butane   c. pentane
    b. octane   d. hexane

20.3 Structural Formulas and Isomerism

QUESTIONS

15. What are structural *isomers?* Which is the smallest alkane that has a structural isomer? Draw structures to illustrate the isomers.

16. What is a small chain of carbon atoms called when it is attached to a longer chain of carbon atoms?

PROBLEMS

17. Draw structural formulas for all isomeric alkanes having the general formula C_6H_{14}.

18. Draw the structural formulas for at least *five* isomeric alkanes having the formula C_8H_{18}.

20.4 Naming Alkanes

QUESTIONS

19. Give the *root names* for the alkanes with five through ten carbon atoms.
20. To what does the root name for a *branched* hydrocarbon correspond?
21. What is an *alkyl* group? How is a given alkyl group related to its parent alkane?
22. How is the *position* of substituents along the longest chain of a hydrocarbon indicated?
23. How is the *number* of multiple substituents of the same type indicated in the systematic name of a hydrocarbon?
24. In giving the name of a hydrocarbon with several substituents, in what order do we list the substituents?

PROBLEMS

25. Give the systematic name for each of the following alkanes.
   a. CH_3—CH_2—CH—CH_2—CH—CH_2—CH_3
       with CH_3 and CH_3 branches
   b. CH_3—C—CH_3
       with CH_3 above and CH_3 below
   c. CH_3—CH_2 CH_2—CH_2—CH_3
       with CH_2—CH—CH_3
   d. CH_2—CH_3
       CH_3—C—CH_2—CH_3
       CH_2—CH_3

26. Give the systematic name for each of the following alkanes.
   a. CH_3—CH_2—CH—CH_2—CH_3
       with CH_2—CH_3 branch
   b. CH_3—C—CH
       with CH_3 above and CH_2—CH_3 below
   c. CH_3 CH_3
       C
       CH_3 CH_3
   d. CH_3—CH—CH—CH—CH_3
       with CH_3 CH_3 CH_3 branches

27. Draw a structural formula for each of the following compounds.
   a. 2-methylpentane
   b. 3-methylhexane
   c. 2,3-dimethyloctane
   d. 4-propyloctane
   e. 4-isopropyloctane

28. Draw the structural formula for each of the following compounds.
   a. 3-methylpentane
   b. 4-*tert*butylheptane
   c. 5-propylnonane
   d. 2,4,6,9-tetramethyldecane

20.5 Petroleum

QUESTIONS

29. What are the major constituents of crude petroleum? What is the major constituent of natural gas? How were these mixtures formed?
30. List several petroleum fractions and tell how each fraction is primarily used.
31. What is *pyrolytic cracking,* and why is the process applied to the kerosene fraction of petroleum?
32. Why was tetraethyllead, $(C_2H_5)_4Pb$, added to gasoline in the past? Why is the use of this substance being phased out?

20.6 Reactions of Alkanes

QUESTIONS

33. Explain why alkanes are relatively unreactive.
34. What is a *combustion* reaction? How have we made use of the combustion reactions of alkanes?
35. What is a *substitution* reaction? Give an example of a reactive molecule that is able to replace the hydrogen atoms of an alkane.
36. What is a *dehydrogenation* reaction? What results when an alkane is dehydrogenated?

PROBLEMS

37. Complete and balance each of the following chemical equations that represent combustion reactions.
    a. $C_3H_8(g) + O_2(g) \rightarrow$
    b. $C_5H_{12}(l) + O_2(g) \rightarrow$
    c. $C_7H_{16}(l) + O_2(g) \rightarrow$

38. Complete and balance each of the following chemical equations.
    a. $C_2H_6(g) + O_2(g) \rightarrow ? + ?$
    b. $CH_2Cl_2(l) + Cl_2(g) \rightarrow HCl(g) + ?$
    c. $CH_2{=}CH{-}CH_2{-}CH_3(l) + Br_2(l) \rightarrow ?$

20.7 Alkenes and Alkynes

QUESTIONS

39. What is an alkene? What structural feature characterizes alkenes? Give the general formula for alkenes.

40. What is an alkyne? What structural feature characterizes alkynes? Give the general formula for alkynes.

41. How is the root name of an alkane modified to indicate that a given hydrocarbon contains a double or triple bond in its longest continuous chain?

42. How is the *location* of a double or triple bond in the longest continuous chain of an unsaturated hydrocarbon indicated?

43. Describe some addition reactions that unsaturated hydrocarbons undergo. Give the specific name of the type of addition process, and indicate what atoms are added to the double or triple bond.

44. How is a liquid fat, such as corn oil, converted into a solid form for use as a table spread or shortening?

PROBLEMS

45. Give the systematic name for each of the following unsaturated hydrocarbons.
    a. $CH_3CH{=}CH_2$
    b. $CH_2{=}CH{-}CH_2{-}CH_3$
    c. $CH{\equiv}CH$

46. Give the systematic name for each of the following unsaturated hydrocarbons.
    a. $CH{\equiv}C{-}CH_3$
    b. $CH_3{-}CH_2{-}C{\equiv}C{-}CH_2{-}CH_3$
    c. $CH_3{-}CH_2{-}CH{=}CH{-}CH_2{-}CH_3$

47. Draw structural formulas showing all the isomers of the straight-chain alkene with seven carbon atoms. Name each isomer.

48. Draw structural formulas showing all possible molecules containing six carbon atoms and having one triple bond.

20.8 Aromatic Hydrocarbons

QUESTIONS

49. What structure do all aromatic hydrocarbons have in common?

50. Benzene exhibits resonance. Explain this statement in terms of the different Lewis structures that can be drawn for benzene.

20.9 Naming Aromatic Compounds

QUESTIONS

51. How is a monosubstituted benzene named? Give the structures and names of two examples. Also give two examples of monosubstituted benzenes that have special names.

52. How is the benzene ring named if it is considered a substituent in another molecule? Give the structures and names of two examples.

53. If a benzene ring contains several substituents, how are the relative locations of the substituents numbered in the systematic name given to the molecule?

54. What do the prefixes *ortho-, meta-,* and *para-* refer to in terms of the relative location of substituents in a disubstituted benzene?

PROBLEMS

55. Draw a structural formula for each of the following aromatic or substituted aromatic compounds.
    a. 1,3-dibromobenzene
    b. 2-bromochlorobenzene
    c. 4-chloro-1,3-dimethylbenzene
    d. 1,4-di(*tert*-butyl)benzene
    e. 1,3-diphenylpropane

56. Name each of the following aromatic or substituted aromatic compounds.

20.10 Functional Groups

PROBLEMS

57. On the basis of the functional groups listed in Table 20.5, identify the family of organic compounds to which each of the following belongs.
    a. CH_3—CH—CH—CH_3
           | |
           OH CH_3
    b. CH_3—CH_2—C=O
                    |
                    CH_2—CH_3
    c. CH_3—CH_2—CH_2—CH_2—CH_2—C=O
                                          |
                                          H
    d. CH_3NH_2

58. From the functional groups listed in Table 20.5, identify the family of organic compounds to which each of the following belongs.
    a. CH_3CH_2—O—$CH_2CH_2CH_3$
    b. $CH_3CH_2CH_2CH_2$—OH
    c. $CH_3CH_2CHCH_3$—OH
    d. $CH_3CH_2CH_2CH_2COOH$

20.11 Alcohols

QUESTIONS

59. What functional group characterizes an alcohol? What ending is added to the name of the parent hydrocarbon to show that a molecule is an alcohol?

60. Distinguish among primary, secondary, and tertiary alcohols. Give a structural formula for an example of each type.

PROBLEMS

61. Give the systematic name for each of the following alcohols. Indicate whether the alcohol is primary, secondary, or tertiary.
    a. $CH_3CH_2CH_2CH_2CH_2$—OH
    b. CH_3
           |
        $CH_3CCH_2CH_3$
           |
           OH
    c. $CH_3CH_2CHCH_2CH_3$
                  |
                  OH
    d. $CH_3CH_2CH_2$—OH

62. Draw structural formulas for each of the following alcohols. Indicate whether the alcohol is primary, secondary, or tertiary.
    a. 2-butanol
    b. 1-pentanol
    c. 2-methyl-1-butanol
    d. 2-methyl-2-butanol

20.12 Properties and Uses of Alcohols

QUESTIONS

63. Why is methanol sometimes called wood alcohol? Describe the modern synthesis of methanol. What are some uses of methanol?

64. Write the equation for the fermentation of glucose to form ethanol. Why cannot ethanol solutions of greater than about 13% concentration be made directly by fermentation? How can the ethanol content be increased beyond this level in beverages?

65. Write the equation for the synthesis of ethanol from ethylene. What are some commercial uses of ethanol made by this process?

66. Give the names and structural formulas of two other commercially important alcohols. Cite the major use of each.

20.13 Aldehydes and Ketones

QUESTIONS

67. What functional group is *common* to both aldehydes and ketones?

68. What structural feature *distinguishes* aldehydes from ketones?

69. Mention some commercial uses of specific aldehydes and ketones.

70. How are aldehydes and ketones most often produced commercially? Write an equation as an example of each process.

20.14 Naming Aldehydes and Ketones

QUESTIONS

71. What ending is added to the root name of the parent hydrocarbon to indicate that the molecule contains the aldehyde functional group? What systematic ending is used to show that a molecule is a ketone?

72. Describe the alternative system for naming ketones. Give two examples of ketones, along with their names in both systems.

73. Give the systematic name for each of the following:
    a. $CH_3-CH_2-CH_2-C=O$
       $\qquad\qquad\qquad\quad |$
       $\qquad\qquad\qquad CH_2-CH_3$

    b. $CH_3-CH-CH-C=O$
       $\qquad\quad\ |\quad\ |\quad\ |$
       $\qquad\quad Cl\quad Cl\quad H$

    c. $\qquad\qquad\quad CH_3$
       $\qquad\qquad\qquad |$
       $CH_3-CH-CH-CH_2-C=O$
       $\qquad\ |\qquad\qquad\qquad |$
       $\qquad CH_3\qquad\qquad\quad H$

    d. $\qquad CH_3$
       $\qquad\quad |$
       $CH_3-CH-C=O$
       $\qquad\qquad\ |$
       $\qquad\qquad H$

    e. $CH_3-CH_2-C=O$

74. Draw a structural formula for each of the following:
    a. methyl propyl ketone
    b. 2-pentanone
    c. di(*tert*butyl)ketone
    d. 4-chlorohexanal
    e. 5-methyl-3-octanone

20.15 Carboxylic Acids and Esters

QUESTIONS

75. Draw the structure of the group that characterizes organic (carboxylic) acids. Give the general condensed formula for an organic acid.

77. What systematic ending is used to show that a molecule is a carboxylic acid? Give an example.

79. Draw the general structural formula for an ester. From what two families of organic compounds are esters synthesized?

PROBLEMS

81. Give the systematic name for each of the following:

a. $CH_3-CH-CH_2-COOH$
            |
           CH_3

b. —COOH

c. $CH_3-CH-COOH$
            |
           OH

d.
                    CH_3
                     |
$CH_3-CH_2-CH-CH-CH_2-COOH$
                 |
               CH_3

76. Are carboxylic acids typically strong acids or weak acids? Write an equation showing the acid CH_3CH_2COOH ionizing in water.

78. Complete the following reactions giving the formula of the principal organic product.

a. $CH_3CH_2OH \xrightarrow{\text{KMnO}_4(aq)}$

b. $CH_3CHCH_3 \xrightarrow{\text{oxidation}}$
        |
       OH

c. $CH_3COOH + CH_3CH_2CH_2CH_2CH_2OH \xrightarrow{\text{acid}}$

80. Draw the structure of acetylsalicylic acid, and circle the portion of the molecule that shows that it is an ester. From what acid and alcohol is acetylsalicylic acid synthesized?

82. Draw a structural formula for each of the following:
a. methyl ethanoate (methyl acetate)
b. 4-chloro-2-methylbenzoic acid
c. pentanoic acid
d. isopropyl octanoate
e. 2,5-dichlorohexanoic acid

20.16 Polymers

QUESTIONS

83. What, in general terms, is a polymer? What is a monomer?

85. What is *condensation polymerization,* and how does it differ from addition polymerization? Give an example of a common condensation polymer.

84. What is *addition polymerization?* Give an example of a common addition polymer.

86. What is a *polyester?* What structural feature characterizes polyesters? Give an example of a common polyester.

PROBLEMS

87. Draw the structures of acrylonitrile and butadiene. Also draw the structures of the basic repeating units when each of these substances polymerize. Give several uses of each polymeric substance.

88. For the polymeric substances nylon and Dacron, sketch representations of the repeating unit in each.

Additional Problems

89. The first "organic" compound to be synthesized in the laboratory, rather than being isolated from nature, was _____, which was prepared from _____.

90. An organic compound whose carbon–carbon bonds are all single bonds is said to be _____.

91. The general orientation of the four pairs of electrons around the carbon atoms in alkanes is _____.

92. Alkanes in which the carbon atoms form a single unbranched chain are said to be _____ alkanes.

93. Structural isomerism occurs when two molecules have the same number of each type of atom but exhibit different arrangements of the _____ between those atoms.

94. The systematic names of all saturated hydrocarbons have the ending _____ added to a root that indicates the number of carbon atoms in the molecule.

95. For a branched hydrocarbon, the root name for the hydrocarbon comes from the number of carbon atoms in the _____ continuous chain in the molecule.

96. The positions of substituents along the hydrocarbon framework of a molecule are indicated by the _____ of the carbon atom to which the substituents are attached.

97. Kerosene may be converted to gasoline by _____, which means that the larger, heavier kerosene components are broken down by heat into smaller, lighter gasoline fragments.

98. Tetraethyl lead, $(C_2H_5)_4Pb$, was added to gasolines in the past as an _____ agent.

99. The major use of alkanes has been in _____ reactions, as a source of heat and light.

100. With very reactive agents, such as the halogen elements, alkanes undergo _____ reactions, whereby a new atom replaces one or more hydrogen atoms of the alkane.

101. Alkenes and alkynes are characterized by their ability to undergo rapid, complete _____ reactions, by which other atoms attach themselves to the carbon atoms of the double or triple bond.

102. Unsaturated fats may be converted to saturated fats by the process of _____.

103. Benzene is the parent member of the group of hydrocarbons called _____ hydrocarbons.

104. An atom or group of atoms that imparts new and characteristic properties to an organic molecule is called a _____ group.

105. A _____ alcohol is one in which there is only one hydrocarbon group attached to the carbon atom holding the hydroxyl group.

106. The simplest alcohol, methanol, is prepared industrially by the hydrogenation of _____.

107. Ethanol is commonly prepared by the _____ of certain sugars by yeast.

108. Aldehydes and ketones both contain the _____ group, but they differ in where this group occurs along the hydrocarbon chain.

109. Aldehydes and ketones can be prepared by _____ of the corresponding alcohol.

110. Organic acids, which contain the _____ group, are typically weak acids.

111. The typically sweet-smelling compounds called _____ result from the condensation reaction of an organic acid with an _____.

112. In _____ polymerization, the monomer units simply combine over and over again to form the polymer's long chain.

113. Write the *general* formula for a normal alkane. How does a given alkane differ from the previous or the following member of the series?

114. Draw structural formulas for all isomeric alkanes having the general formula C_5H_{12}.

115. As the size of an organic molecule increases, the number of structural isomers that are possible increases dramatically. For the general alkane formula $C_{20}H_{42}$, draw structures showing the carbon framework of at least ten isomers. (You do not have to show all the hydrogen atoms.)

116. Give the systematic name of each of the following substituted alkanes.

    a. CH_3—CH—CH_2—CH_3
           |
           Cl

    b. CH_2—CH_2
       | |
       Br Br

    c. CHI_3

    d. CH_3—CH—CH—CH—CH_3
              | | |
              Cl Cl Cl

    e. Cl
              |
       CH_3—C—CH_2—CH—CH_2—CH_2—CH_3
              | |
              Cl CH_3—CH—CH_3

117. Give some examples of molecules containing "fused" benzene rings.

118. Draw a structural formula for each of the following compounds.
  a. 3-chloro-3-hexene
  b. 1,2-dibromoethane
  c. 1,1,2,2-tetrachloroethane
  d. 3-bromo-1-hexyne
  e. 1,2-dichloro-2,3,3,4-tetramethyloctane

119. Draw the structural formula(s) for, and give the name(s) of, the organic product(s) of each of the following reactions. If a mixture of several, similar products is expected, indicate the type of product expected.

  a. $CH_3{-}CH_3 + Cl_2 \xrightarrow{\text{Light}}$

  b. $CH_3{-}CH{=}CH{-}CH_3 + H_2 \xrightarrow{\text{Pt}}$

  c.
$$\begin{array}{c} CH_3{-}CH_2 \\ \qquad\qquad\searrow \\ \qquad\qquad\quad C{=}CH{-}CH_3 + Br_2 \longrightarrow \\ \qquad\qquad\nearrow \\ CH_3{-}CH_2 \end{array}$$

120. For the general formula $C_6H_{14}O$, draw the structures of three isomeric alcohols that illustrate *primary, secondary,* and *tertiary* alcohol structures.

121. The alcohol glycerol (glycerine), which is produced in the human body by the digestion of fats, has the following structure. Give the systematic name of glycerol.

$$\begin{array}{c} CH_2{-}CH{-}CH_2 \\ \ \ | \qquad | \qquad | \\ OH \quad OH \quad OH \end{array}$$

122. The sugar glucose could conceivably be given the name 2,3,4,5,6-pentahydroxyhexanal (though this name is never used, because the actual sugar glucose is only one possible isomer having this name). Draw the structure implied by this name.

123. Draw a structure corresponding to each of the following names.
  a. 2-methylpentanal
  b. 3-hydroxybutanoic acid
  c. 2-aminopropanal
  d. 2,4-hexanedione
  e. 3-methylbenzaldehyde

124. Draw a structure corresponding to each of the following names.
  a. diphenylketone
  b. 2-methyl-2-propanol
  c. methylisopropylketone
  d. 1,2-ethanediol
  e. methyl propanoate

125. Salicylic acid is an interesting molecule; it is both an acid and an alcohol, having a hydroxyl group attached to the aromatic ring (phenol). For this reason, salicylic acid can undergo two different esterification reactions, depending on which functional group of salicylic acid reacts. For ex-

ample, as discussed in this chapter, when treated with acetic acid—a stronger acid—salicylic acid behaves as an alcohol, and the ester produced is acetylsalicylic acid (aspirin). On the other hand, when reacted with methyl alcohol, salicylic acid behaves as an acid, and the ester methyl salicylate (oil of wintergreen) is produced. Methyl salicylate is also an analgesic and is part of the formulation of many liniments for sore muscles. Write an equation for the production of each of these esters of salicylic acid.

126. Alpha amino acids are organic acid molecules that also happen to contain an amino group ($-NH_2$) on the second carbon atom of the acid's chain. Proteins are condensation polymers of such alpha amino acids. The reaction by which the long chain of the protein forms is very similar to the reaction by which nylon forms, resulting in the formation of the linkage

$$\begin{array}{c} {-}N{-}C{=}O \\ \ \ | \qquad | \\ \ \ H \end{array}$$

which is called an "amide" (or "peptide") linkage. Show how the following two amino acids could react with each other to produce an amide linkage, resulting in the formation of a dimer (a "dipeptide").

$$\begin{array}{c} HO{-}C{=}O \\ \qquad\quad | \\ CH_3{-}CH{-}N{-}H + HO{-}C \\ \qquad\quad | \qquad\qquad\qquad | \\ \qquad\quad H \qquad\qquad\quad CH_2{-}NH_2 \end{array}$$

How could this dipeptide then go on to react with additional amino acids to form a polypeptide?

127. Write structural formulas for the normal (straight-chain) alkanes with:
  a. seven carbon atoms
  b. eleven carbon atoms
  c. sixteen carbon atoms

128. For each of the following numbers of carbon atoms, give the structure of the straight-chain alkane with that number of carbon atoms, and give its name.
  a. eight carbon atoms
  b. six carbon atoms
  c. four carbon atoms
  d. five carbon atoms

129. Draw structural formulas for all isomeric alkanes having the general formula C_7H_{16}.

130. Give the systematic name for each of the following alkanes.
  a.
$$\begin{array}{c} \qquad\qquad CH_3 \\ \qquad\qquad\ | \\ CH_3{-}CH{-}CH{-}CH_3 \\ \qquad\qquad\qquad | \\ \qquad\qquad\quad CH_3 \end{array}$$

b. CH_3—CH_2 CH_2—CH_3
$\diagdown$ $\diagup$
C
$\diagup$ $\diagdown$
CH_3—CH_2 CH_2—CH_3

c. CH_3 CH_3
$|$ $|$
CH—C—CH_2—CH_2—CH_3
$|$ $|$
CH_3 CH_3

d. CH_3—CH—CH—CH—CH—CH—CH_3
$|$ $|$ $|$ $|$ $|$
CH_3 CH_3 CH_3 CH_3 CH_3

131. Draw a structural formula for each of the following compounds.
    a. 2,2-dimethylhexane
    b. 2,3-dimethylhexane
    c. 3,3-dimethylhexane
    d. 3,4-dimethylhexane
    e. 2,4-dimethylhexane

132. Write the formula for the missing reactant or product in each of the following chemical equations.
    a. $CH_4(g) + Cl_2(g) \rightarrow$ _____ $+ HCl(g)$
    b. $CH_3CH_2CH_3(g) \rightarrow CH_3CH{=}CH_2(g) +$ _____
    c. $CHCl_3(l) + Cl_2(g) \rightarrow CCl_4(l) +$ _____

133. Give the systematic name for each of the following unsaturated hydrocarbons.
    a. $CH_3CH{=}CHCH_2CH_2CH_2CH_2CH_2CH_3$
    b. $CH_3CH{=}CHCH_2CH_2CH_2CH_3$
    c. $CH_3CH_2C{\equiv}CCH_3$

134. Draw structural formulas showing all the isomers of the straight-chain alkyne with eight carbon atoms. Name each isomer.

135. Name each of the following aromatic or substituted aromatic compounds.

a.

b.

c.

d.

136. On the basis of the functional groups listed in Table 20.5, identify the family of organic compounds to which each of the following belongs.

a.

b. CH_3—CH_2—CH_2—$C{=}O$

c. CH_3—CH_2—$C{=}O$
$|$
O—CH_2—CH_2—

d.

137. Draw a structural formula for each of the following alcohols. Indicate whether the alcohol is primary, secondary, or tertiary.
    a. 2-propanol
    b. 2-methyl-2-propanol
    c. 4-isopropyl-2-heptanol
    d. 2,3-dichloro-1-pentanol

138. Draw a structural formula for each of the following:
    a. 3-methylpentanal
    b. 3-methyl-2-pentanone
    c. methyl phenyl ketone
    d. 2-hydroxybutanal
    e. propanal

139. How are carboxylic acids synthesized? Give an example of an organic acid and the molecule from which it is synthesized. What sort of reaction is this?

140. Draw a structural formula for each of the following:
    a. 3-methylbutanoic acid
    b. 2-chlorobenzoic acid
    c. hexanoic acid
    d. acetic acid

141. Draw the structure of the monomer and the basic repeating unit for each of the following polymeric substances.
    a. polyethylene
    b. polyvinyl chloride
    c. Teflon
    d. polypropylene
    e. polystyrene

Solutions to Self-Check Exercises

SELF-CHECK EXERCISE 20.1

The alkane with ten carbon atoms can be represented as $CH_3\text{--}(CH_2)_8\text{--}CH_3$

and its formula is $C_{10}H_{22}$. The alkane with fifteen carbons,

$$CH_3\text{--}(CH_2)_{13}\text{--}CH_3$$

has the formula $C_{15}H_{32}$.

SELF-CHECK EXERCISE 20.2

a.

This molecule is 5-ethyl-3-methyloctane.

b.

This molecule is 5-ethyl-3-methylheptane. Note that this chain could be numbered from the opposite direction to give the name 3-ethyl-5-methylheptane. These two names are equally correct.

SELF-CHECK EXERCISE 20.3

The root name decane indicates a ten-carbon chain. There is a methyl group at the number-4 position and an isopropyl group at the number-5 position. The structural formula is

SELF-CHECK EXERCISE 20.4

a. The longest chain has eight carbon atoms with a double bond, so the root name is octene. The double bond exists between carbons 3 and 4, so the name is 3-octene. There is a methyl group on the number-2 carbon. The name is 2-methyl-3-octene.

b. The carbon chain has five carbons with a triple bond between carbons 1 and 2. The name is 1-pentyne.

SELF-CHECK EXERCISE 20.5

a. 2-chloronitrobenzene or *o*-chloronitrobenzene
b. 4-phenyl-2-hexene

SELF-CHECK EXERCISE 20.6

a. 1-pentanol; primary alcohol
b. 2-methyl-2-propanol (but this alcohol is usually called tertiary butyl alcohol); tertiary alcohol
c. 5-bromo-2-hexanol; secondary alcohol

SELF-CHECK EXERCISE 20.7

a. 4-ethyl-3-hexanone
   Because the compound is named as a hexanone, the carbonyl group is assigned the lowest possible number.
b. 7-isopropyldecanal

21 Biochemistry

CONTENTS

There is a tremendous amount of material in this chapter. You might want to pick and choose the topics that best meet the needs of your students. Make it clear just how much material you expect them to memorize. To beginning students, the complexity of this chapter can appear overwhelming. However, students tend to find much of the material interesting, because the topics are of concern to everyone.

This colorful peacock feather is composed primarily of fibrous proteins.

Biochemistry, the study of the chemistry of living systems, is a vast and exciting field in which important discoveries about how life is maintained and how diseases occur are being made every day. In particular, there has been rapid growth in the understanding of how living cells manufacture and use the molecules necessary for life.

We cannot hope to cover all the important aspects of this field in this chapter; we will concentrate here on the major types of biomolecules that support living organisms. First, however, we will survey the elements found in living systems and briefly describe the constitution of a cell.

At present, 30 elements are definitely known to be essential to human life. These **essential elements** are shown in color in Figure 21.1. The most abundant elements are hydrogen, carbon, nitrogen, and oxygen; but sodium, magnesium, potassium, calcium, phosphorus, sulfur, and chlorine are also present in relatively large amounts. Although present only in trace amounts, the first-row transition metals are essential for the action of many enzymes (biological catalysts). For example, zinc, one of the **trace elements,** is found in nearly 200 biologically important molecules. The functions of the essential elements are summarized in Table 21.1. In time, other elements will probably be found to be essential.

Life is organized around the functions of the **cell,** the smallest unit in living things that exhibits the properties normally associated with life, such as reproduction, metabolism, mutation, and sensitivity to external stimuli.

Figure 21.1

The chemical elements essential for life. Those most abundant in living systems are shown as purple. Nineteen elements, called the trace elements, are shown as green.

Table 21.1 Some Essential Elements and Their Major Functions

Element	Percent by Mass in the Human Body	Function
oxygen	65	in water and many organic compounds
carbon	18	in all organic compounds
hydrogen	10	in water and many inorganic and organic compounds
nitrogen	3	in both inorganic and organic compounds
calcium	1.5	in bone; essential to some enzymes and to muscle action
phosphorus	1.2	essential in cell membranes and to energy transfer in cells
potassium	0.2	cation in cell fluid
chlorine	0.2	anion inside and outside the cells
sulfur	0.2	in proteins
sodium	0.1	cation in cell fluid
magnesium	0.05	essential to some enzymes
iron	<0.05	in molecules that transport and store oxygen
zinc	<0.05	essential to many enzymes
cobalt	<0.05	found in vitamin B_{12}
iodine	<0.05	essential to thyroid hormones
fluorine	<0.01	in teeth and bones

Table 21.1 gives the mass percents of some elements and also their functions. It is interesting for students to see that some of the elements mentioned throughout the course actually play a role in the human body.

As the fundamental building blocks of all living systems, aggregates of cells form tissues, which in turn are assembled into the organs that make up complex living systems. Therefore, to understand how life is maintained and reproduced, we must learn how cells operate on the molecular level. This is the main thrust of biochemistry.

21.1 Proteins

AIM: To introduce proteins.

In Chapter 20 we saw that many useful synthetic materials are polymers. A great many natural materials are also polymers: starch, hair, silk and cotton fibers, and the cellulose in woody plants, to name only a few.

In this section we introduce a class of natural polymers, the **proteins,** which make up about 15% of our bodies and have molar masses that range from

Silk is a naturally occurring polymer.

Most students are aware of the functions of proteins such as those found in hair and muscle. But there are many more proteins that play vital roles and whose jobs usually go unnoticed unless there is a malfunction.

approximately 6000 to over 1,000,000 grams. Proteins have many functions in the human body. **Fibrous proteins** provide structural integrity and strength for many types of tissue and are the main components of muscle, hair, and cartilage. Other proteins, usually called **globular proteins** because of their roughly spherical shape, are the "worker" molecules of the body. These proteins transport and store oxygen and nutrients, act as catalysts for the thousands of reactions that make life possible, fight invasion of the body by foreign objects, participate in the body's many regulatory systems, and transport electrons in the complex process of metabolizing nutrients.

21.2   Primary Structure of Proteins

AIM: To describe the primary structure of proteins.

The building blocks of all proteins are the **α-amino acids:**

The R in this structure may represent H, CH_3, or more complex substituents. These molecules are called α-amino acids because the amino group ($-NH_2$) is always attached to the α-carbon, the one next to the carboxyl group ($-COOH$). The 20 amino acids most commonly found in proteins are shown in Figure 21.2.

At the pH in biological fluids, the amino acids shown in Figure 21.2 exist in a different form, with the proton of the —COOH group transferred to the —NH₂ group. For example, glycine would be in the form $H_3{}^+NCH_2COO^-$.

You might want to point out to students the atoms responsible for the hydrophilic nature of R groups of some amino acids. Students may have forgotten that oxygen and nitrogen are both electronegative atoms and can contribute to the polarity of molecules.

Students often form a bond between the carboxyl group of one amino acid and the R group of the second. Work several examples. Mark parts of each amino acid with colored chalk or colored ink on an overhead transparency, so that students can identify the parts of the molecules that react and can keep them separate from the R groups, which do not react.

Note from Figure 21.2 that the amino acids are grouped into polar and nonpolar classes on the basis of the composition of the R groups, also called the **side chains.** Nonpolar side chains contain mostly carbon and hydrogen atoms; polar side chains contain nitrogen and oxygen atoms. This difference is important because polar side chains are *hydrophilic* (water-loving), but nonpolar side chains are *hydrophobic* (water-fearing). This greatly affects the three-dimensional structure of the resulting proteins, because they exist in aqueous media in living things.

The protein polymer is built by reactions between amino acids. For example, two amino acids can react as follows, forming a C—N bond with the elimination of water.

Nonpolar R Groups (hydrophobic)

Glycine (gly)

Alanine (ala)

Proline (pro)

Phenylalanine (phe)

Isoleucine (ile)

Tryptophan (trp)

Methionine (met)

Leucine (leu)

Valine (val)

Polar R Groups (hydrophilic)

Serine (ser)

Glutamine (gln)

Tyrosine (tyr)

Histidine (his)

Asparagine (asn)

Threonine (thr)

Aspartic acid (asp)

Cysteine (cys)

Glutamic acid (glu)

Lysine (lys)

Arginine (arg)

Figure 21.2
The 20 α-amino acids found in most proteins. The R group is shown in color.

The product shown above is called a **dipeptide.** The term peptide comes from the structure

$$\overset{\overset{\displaystyle O}{\displaystyle \|}}{-C}-\overset{\overset{\displaystyle H}{\displaystyle |}}{N}-$$

which chemists call a **peptide linkage** or a peptide bond. The prefix *di-* indicates that two amino acids have been joined. Additional reactions lengthen the chain to produce a **polypeptide** and eventually a protein.

The 20 amino acids can be assembled in any order, which makes possible an enormous number of different proteins. This variety allows for the many types of proteins needed for the functions that organisms carry out.

The *order* or *sequence* of amino acids in the protein chain is called the **primary structure.** We indicate the primary structure by using three-letter codes for the amino acids (see Figure 21.2), where it is understood that the terminal carboxyl group is on the right and the terminal amino group is on the left. For example, one possible sequence for a tripeptide containing the amino acids lysine, alanine, and leucine is

which is represented in the shorthand notation (three-letter codes) as

lys-ala-leu

EXAMPLE 21.1   **Understanding Primary Structure**

Write the sequences of all possible tripeptides composed of the amino acids tyrosine (tyr), histidine (his), and cysteine (cys).

SOLUTION

There are six possible sequences:

tyr-his-cys	his-tyr-cys	cys-tyr-his
tyr-cys-his	his-cys-tyr	cys-his-tyr

You may want to ask students to memorize the names and structures of the 20 common amino acids in Figure 21.2, but it might be sufficient for them to be able to identify the side chains as polar or nonpolar.

Arranging amino acids to form the primary structures of proteins requires the ability to think of molecules in the abstract. Drawing the actual structures of dipeptides and tripeptides requires students to arrange the molecules as groups of atoms. It may take time for students to become accustomed to thinking of molecules both as groups of atoms and as abstract three-letter codes.

A striking example of the importance of the primary structure of polypeptides can be seen in the differences between *oxytocin* and *vasopressin.* Oxytocin is a hormone that triggers contraction of the uterus and milk secretion. Vasopressin raises blood pressure and regulates kidney function. Both of these molecules are nine-unit polypeptides, and they differ by only two amino acids (Figure 21.3); yet they have completely different functions in the human body.

cys-tyr-ile-gln-asn-cys-pro-leu-gly

(a)

cys-tyr-phe-gln-asn-cys-pro-arg-gly

(b)

Figure 21.3
The amino acid sequences in (a) oxytocin and (b) vasopressin. The differing amino acids are enclosed in boxes.

21.3   Secondary Structure of Proteins

AIM: To describe the secondary structure of proteins.

So far we have considered the primary structure of proteins—the order of amino acids in the chain (Figure 21.4a). A second level of structure in proteins is the arrangement in space of the chain of the long molecule. This is called the **secondary structure** of the protein (Figure 21.4b).

One common type of secondary structure resembles a spiral staircase. This spiral structure is called an **α-helix** (Figure 21.5). A spiral-like secondary structure gives the protein elasticity (springiness) and is found in the fibrous proteins in wool, hair, and tendons. Another type of secondary structure involves joining several different protein chains in an arrangement called a **pleated sheet,** as shown in Figure 21.6. Silk has this arrangement of proteins, making its fibers

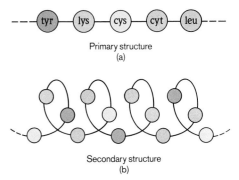

Primary structure
(a)

Secondary structure
(b)

Figure 21.4
(a) The primary structure of a protein describes the order of the amino acids in the chain. One example is shown. (b) The secondary structure of a protein describes the arrangement of the chain in space. A spiral (helical) arrangement of the protein chain is shown.

Figure 21.5
One type of secondary protein structure is like a spiral staircase and is called the α-helix. The circles represent individual amino acids.

   Some pop beads of different colors twisted to form a helix can show how the primary structure of a protein is related to the secondary structure. The connected beads represent the primary structure; the twisted spiral chain represents secondary structure.

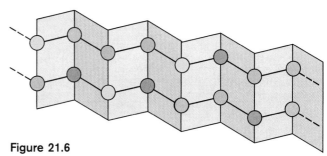

Figure 21.6
Three protein chains bound together in the pleated-sheet secondary structure. Each sphere represents an amino acid.

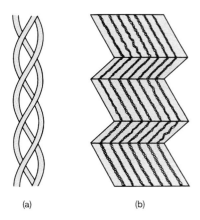

(a)   (b)

Figure 21.7
(a) Collagen, a protein found in tendons, consists of three protein chains (each with an α-helical structure) twisted together to form a super-helix. The result is a long, relatively narrow protein. (b) Many proteins are bound together in the pleated-sheet arrangement to form the elongated protein found in silk fibers.

Models, Slinkys, or three-dimensional drawings can really help students see how a protein molecule can exhibit both secondary and tertiary structure. The Slinky helix can be bent and twisted to illustrate tertiary structure, as well as secondary structure.

flexible but very strong and resistant to stretching. The pleated sheet is also found in muscle fibers.

As you might imagine, a molecule as large as a protein has a great deal of flexibility and can assume a variety of overall shapes. The function the protein is to serve influences the specific shape that it will have. For long, thin structures such as hair, wool and silk fibers, and tendons, an elongated shape is required. This may involve an α-helical secondary structure, as found in the protein α-keratin in hair and wool or in the collagen that occurs in tendons (Figure 21.7a), or it may involve a pleated-sheet secondary structure, as found in silk (Figure 21.7b). Many of the proteins in the body that have nonstructural functions (such as serving as enzymes) are globular. One is myoglobin (Figure 21.8), which absorbs an O_2 molecule and stores it for use by the cells as it is needed. Note that the secondary structure of myoglobin is basically α-helical. However, in the areas where the chain bends to give the protein its compact globular structure, the α-helix breaks down so that protein can "turn the corner."

21.4 Tertiary Structure of Proteins

AIM: To describe the tertiary structure of proteins.

The overall shape of the protein, long and narrow or globular, is called its **tertiary structure.** To make sure the difference between secondary and tertiary structure is clear, examine Figure 21.8 again. The secondary structure of the myoglobin

Figure 21.8
The protein myoglobin. Note that the secondary structure is basically α-helical except at the bends. The tertiary structure is globular.

Adapted with permission from R. E. Dickerson and I. Geis, *The Structure and Action of Proteins,* Benjamin/Cummings Publisher, Menlo Park, CA. Illustration copyright 1969 by Dickerson and Geis.

Random coil

α-Helix

protein is helical except in the regions where it bends back on itself to give the overall compact (globular) tertiary structure. If the bends did not occur, the tertiary structure would be like a long tube (tubular). In both tertiary arrangements (globular and tubular), the secondary structure of the protein is basically helical.

The amino acid *cysteine* (cys) plays a special role in stabilizing the tertiary structure of many proteins, because the —SH groups on two cysteines can react to form an S—S bond called a **disulfide linkage.**

Note that the formation of a disulfide linkage can fasten together two parts of a protein chain to form and hold a bend in the chain, for example. A practical application of the chemistry of disulfide bonds is the permanent waving of hair (Figure 21.9). The S—S linkages in the protein of hair are broken by treatment with a reducing agent. Next the hair is set in curlers to change the tertiary protein structure to the desired shape. Then treatment with an oxidizing agent causes new S—S bonds to form, which make the hair protein retain the new structure.

21.5 Functions of Proteins

AIM: To describe various functions served by proteins.

The three-dimensional structure of a protein is crucial to its function. The process of breaking down this structure is called **denaturation** (Figure 21.10, p. 744). For example, the denaturation of egg proteins occurs when an egg is cooked. Any source of energy can cause denaturation of proteins and is thus potentially dangerous to living organisms. For example, ultraviolet radiation, X-ray radiation, or nuclear radioactivity can disrupt protein structure, which may lead to cancer or genetic damage. The metals lead and mercury, which have a very high affinity for sulfur, cause protein denaturation by disrupting disulfide bonds.

The tremendous variability in the several levels of protein structure allows proteins to be tailored specifically to serve a wide range of functions, some of which are given in Table 21.2 (p. 744).

Figure 21.9
The permanent waving of hair.

Figure 21.9 describes how a permanent wave curls hair, a topic that is very interesting to students. This figure can lead to a discussion of denaturation and disruption of protein function. The formation of disulfide links demonstrates that the R groups of amino acids play an active role in protein structure, although not in primary structure.

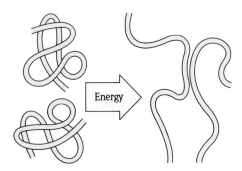

Figure 21.10
A schematic representation of the thermal denaturation of a protein.

The tertiary structure of this woman's hair has been altered by her having had a perm.

You can easily demonstrate protein denaturation by heating egg whites or by adding dilute acetic acid to nonfat dry milk dissolved in water. Most students will be familiar with the results, but will not have thought about them as chemistry.

Table 21.2 presents functions of proteins, some of which students will be familiar with. You can discuss two less familiar functions—protection and control—and reemphasize the important roles that proteins play.

Table 21.2   Common Functions of Proteins

Function	Comment/Example
1. Structure	Proteins provide the strength of tendons, bones, and skin. Cartilage, hair, wool, fingernails, and claws are mainly protein. Viruses have an outer layer of protein.
2. Movement	Proteins are the major components of muscles and are directly responsible for the ability of muscles to contract. The swimming of sperm results from the contraction of protein filaments in their tails.
3. Catalysis	Nearly all chemical reactions in living organisms are catalyzed by enzymes, which are almost always proteins.
4. Transport	Oxygen is carried from the lungs to tissues by the protein hemoglobin in red blood cells.
5. Storage	The protein ferritin stores iron in the liver, spleen, and bone marrow.
6. Energy transformation	Cytochromes are proteins found in all cells. They extract energy from food molecules by transferring electrons in a series of oxidation–reduction reactions.
7. Protection	*Antibodies* are special proteins that are synthesized in response to foreign substances and cells, such as bacterial cells. They then bind to those substances or cells and provide us with immunity to various diseases. Interferon, a small protein made and released by cells when they are exposed to a virus, protects other cells against viral infection. Blood-clotting proteins protect against bleeding (hemorrhage).
8. Control	Many *hormones* are proteins produced in the body that have specific effects on the activity of certain organs.
9. Buffering	Because proteins contain both acidic and basic groups on their side chains, they can neutralize both acids and bases and thus provide buffering for blood and tissues.

21.6   Enzymes

AIM: To describe how enzymes work.

Enzymes are proteins that catalyze specific biological reactions. Without the several hundred enzymes now known, life would be impossible. Almost all of the critical biochemical reactions would occur far too slowly at the temperatures at which life exists. Enzymes are impressive for their tremendous efficiency (they are typically 1 to 10 million times as efficient as inorganic catalysts) and their incredible selectivity (an enzyme "ignores" the thousands of molecules in body fluids that are not involved in the reaction it catalyzes).

Although the mechanisms of enzyme activity are complex and not fully understood in most cases, a simple theory called the **lock-and-key model** (Figure 21.11) seems to fit many enzymes. This model postulates that the shapes of the reacting molecule (the **substrate**) and the enzyme are such that they fit together much as a key fits a specific lock. The substrate and enzyme attach to each other in such a way that the part of the substrate where the reaction is to occur occupies the **active site** of the enzyme. After the reaction occurs, the products are liberated and the enzyme is ready for a new substrate. We can represent enzyme catalysis by the following steps:

STEP 1
The enzyme E and the substrate S come together.

$$E + S \rightleftharpoons E \cdot S$$

STEP 2
The reaction occurs to give the product P, which is released from the enzyme.

$$E \cdot S \rightarrow E + P$$

After the product is released, the enzyme is free to engage another substrate.

Because this process occurs so rapidly, only a tiny amount of enzyme is required.

The first paragraph in Section 21.6 presents those properties of enzymes that make them efficient catalysts. If you present each of these points separately, students can get a real appreciation for the role of enzymes.

An enzyme often changes shape slightly as the substrate is bound. This "clamps" the substrate into place.

Use blackboard drawings or an overhead transparency to illustrate the lock-and-key model. Show that potential substrates will not react if they do not fit the active site. You can also show that many substrate molecules react with one enzyme molecule.

Figure 21.11
Schematic diagram of the lock-and-key model.

An interesting class of enzyme inhibitors to discuss are the nerve gases that inhibit the enzyme cholinesterase. Inhibition of this enzyme causes the build-up of a chemical that leads to convulsions and death.

In some cases a substance other than the substrate can bind to the enzyme's active site. When this occurs, the enzyme is said to be *inhibited.* If the inhibition is permanent, the enzyme is said to be inactivated. Some of the most powerful toxins act by inhibiting, or inactivating, key enzymes.

Because enzymes are crucial to life and because we hope to learn how to mimic their efficiency in our industrial catalysts, the study of enzymes occupies a prominent place in chemical research.

Bread, a carbohydrate food source.

21.7   Carbohydrates

AIM: To describe the fundamental properties of carbohydrates.

The **carbohydrates** are another class of biologically important molecules. Carbohydrates serve as food sources for most organisms and as structural materials for plants. Because many carbohydrates have the empirical formula CH_2O, it was originally believed that these substances were hydrates of carbon ($C \cdot H_2O$), which accounts for their name.

Like proteins, carbohydrates occur in almost bewildering varieties. Many of the most important carbohydrates are polymers—large molecules constructed by hooking together many smaller molecules. We have seen that proteins are polymers constructed from amino acids. The polymeric carbohydrates are constructed from molecules called **simple sugars** or, more precisely, **monosaccharides.** Monosaccharides are aldehydes or ketones that contain several hydroxyl (—OH) substituents. An example of a monosaccharide is fructose, with the structure

Fructose

Fruit, a dietary source of fructose.

Fructose is a sugar found in honey and fruits. Monosaccharides can have various numbers of carbon atoms, and we name them according to the number of carbon atoms they contain by adding prefixes to the root *-ose.* The general names of monosaccharides are shown in Table 21.3. Notice that fructose is a ketone with six carbon atoms and five —OH substituents. Fructose is a member of the hexose family, where the prefix *hex-* means six.

Table 21.3 The General Names of Monosaccharides

Number of Carbon Atoms	Prefix	General Name of Sugar
3	*tri-*	triose
4	*tetr-*	tetrose
5	*pent-*	pentose
6	*hex-*	hexose
7	*hept-*	heptose
8	*oct-*	octose
9	*non-*	nonose

The most important monosaccharides found in living organisms are pentoses and hexoses. The most common pentoses and hexoses are shown in Table 21.4 below.

There are many structures presented in this section. You may want to select only certain ones for students to memorize.

Table 21.4 Some Important Monosaccharides

Pentoses (five carbon atoms)

Ribose

CHO
H—C—OH
H—C—OH
H—C—OH
CH_2OH

Arabinose

CHO
HO—C—H
H—C—OH
H—C—OH
CH_2OH

Ribulose

CH_2OH
C=O
H—C—OH
H—C—OH
CH_2OH

Hexoses (six carbon atoms)

Glucose

CHO
H—C—OH
HO—C—H
H—C—OH
H—C—OH
CH_2OH

Mannose

CHO
HO—C—H
HO—C—H
H—C—OH
H—C—OH
CH_2OH

Galactose

CHO
H—C—OH
HO—C—H
HO—C—H
H—C—OH
CH_2OH

Fructose

CH_2OH
C=O
HO—C—H
H—C—OH
H—C—OH
CH_2OH

You can point out features of the common monosaccharides by comparing the structures of fructose and glucose. Both have six carbons, but glucose has an aldehyde functional group on carbon 1, whereas fructose has a ketone on carbon 2. Both monosaccharides have OH groups on the remaining carbons.

Although we have so far represented the monosaccharides as straight-chain molecules, they usually form ring structures in aqueous solution. Figure 21.12 shows how the ring forms for fructose. Note that a new bond is formed between the oxygen of a hydroxyl group and the carbon of the ketone group. In the cyclic form, fructose is a five-membered ring containing a C—O—C bond. Glucose, a hexose that is an aldehyde, forms a six-membered ring structure. The six-membered glucose ring is shown in Figure 21.13(a). Note that the ring is nonplanar.

More complex carbohydrates are formed by the combining of monosaccharides. For example, two monosaccharides can combine to form a **disaccharide. Sucrose,** common table sugar, is a disaccharide formed from glucose and fructose by elimination of water to form a C—O—C bond between the rings that is called a **glycoside linkage** (Figure 21.13c). When sucrose is consumed in food, this reaction is reversed (the glycoside bond is broken). An enzyme in saliva catalyzes the breakdown of sucrose into its two monosaccharides.

Large polymers containing many monosaccharide units are called **polysaccharides** and can form when each ring forms two glycoside linkages, as shown in Figure 21.14. Three of the most important of these polymers are starch, cellulose, and glycogen. All of these substances are polymers of glucose; they differ in the way the glucose rings are linked together.

Figure 21.12
The formation of a ring structure for fructose.

Students have a difficult time understanding how a linear monosaccharide can cyclize. It helps to number the carbons in the linear form, then show the numbers both as the ring begins to close and after the ring has formed.

In fructose, the carbon with the ketone functional group reacts to form the ring structure. You might want to point out that when glucose reacts to form a ring, the carbon with the aldehyde group reacts.

Figure 21.13
Sucrose (c) is a disaccharide formed from glucose (a) and fructose (b).

Cyclized monosaccharides contain so many OH groups pointing in various directions that it is hard for students to focus on the site of glycosidic bond formation. You might want to use colored chalk or markers to show the locations of the specific OH groups that will form the new bond.

Figure 21.14
Starch is made by hooking together many glucose rings. Only a small part of the chain is shown here.

Starch (Figure 21.14) is the carbohydrate reservoir in plants. It is the form in which glucose is stored by the plant for later use as cellular fuel—both by the plants themselves and by organisms that eat plants.

Cellulose, the major structural component of woody plants and natural fibers such as cotton, is also a polymer of glucose. However, in cellulose the way in which the glucose rings are linked is different from the way they are linked in starch. This difference in linkage has very important consequences. The human digestive system contains enzymes than can catalyze breakage of the glycoside bonds between the glucose molecules in starch. These enzymes are *not* effective on the glycoside bonds of cellulose, however, presumably because the different structure causes a poor fit between the enzyme's active site and the carbohydrate. Interestingly, the enzymes necessary to cleave the glycoside linkages in cellulose are found in bacteria that exist in the digestive tracts of termites, cows, deer, and many other animals. Therefore, unlike humans, these animals can derive nutrition from the cellulose in wood, hay, and so on.

Glycogen is the main carbohydrate reservoir in animals. For example, it is found in muscles where it can be broken down into glucose units when energy is required for physical activity.

When starch, cellulose and glycogen form, two OH groups on each glucose molecule react. You might want to number the carbons in each ring for a small segment of polysaccharide chain to show students that the OHs on carbons 1 and 4 always react to form new bonds between glucose molecules. Just how the glucose rings have joined together can be hard to determine because of the complexity of the molecules.

The exact differences between the glucose linkages in starch, glycogen, and cellulose are not discussed. But students should know how the three polysaccharides are used and also why it is that cellulose cannot be broken down by humans, whereas starch and glycogen can.

21.8   Nucleic Acids

AIM: To describe the fundamental nucleic acid structures.

Life is possible only because each cell, when it divides, can transmit the vital information about what it is to the next generation of cells. It has been known for a long time that this process involves the chromosomes in the nucleus of the cell.

Nucleic acids are complex so how you present them will depend on how much detail you want your students to know. You can discuss both structure and function of nucleic acids without asking students to memorize the structures of individual sugars and organic bases.

Figure 21.15

The structure of the pentoses (a) deoxyribose and (b) ribose. Deoxyribose is the sugar molecule present in DNA; ribose is found in RNA. The difference in these sugars is the substitution, in ribose, of an OH for one H in deoxyribose. This OH is shown in color.

Only since 1953, however, have scientists understood the molecular basis of this intriguing cellular "talent."

The substance that stores and transmits the genetic information is a polymer called **deoxyribonucleic acid (DNA),** a huge molecule with a molar mass as high as several billion grams. Together with other similar nucleic acids called the **ribonucleic acids (RNA),** DNA carries the information needed for the synthesis of the various proteins the cell requires to carry out its life functions. The RNA molecules, which are found in the cytoplasm outside the cell nucleus, are much smaller than DNA polymers, with molar mass values of only 20,000 to 40,000 grams.

Both DNA and RNA are polymers—they are constructed by the hooking together of many smaller units. The fundamental unit in these polymers is called a **nucleotide.** Each nucleotide in turn has three parts:

1. A *nitrogen-containing organic base*
2. A *five-carbon sugar*
3. A *phosphate group*

A nucleotide can be represented as follows:

In the DNA polymer, the 5-carbon sugar is deoxyribose (Figure 21.15a); in the RNA polymer, it is ribose (Figure 21.15b). This difference in the sugar molecules

Figure 21.16

The organic bases found in DNA and RNA. Note that uracil is found only in RNA and that thymine is found only in DNA.

Uracil (U)
RNA

Cytosine (C)
DNA
RNA

Thymine (T)
DNA

Adenine (A)
DNA
RNA

Guanine (G)
DNA
RNA

(a)

(b)

Figure 21.17
(a) Adenosine is formed by the reaction of adenine and ribose. (b) The reaction of phosphoric acid with adenosine to form the ester adenosine 5-phosphoric acid, a nucleotide. (At biological pH, the phosphoric acid would not be fully protonated as is shown here.)

present in the polymers is responsible for the names DNA (deoxyribonucleic acid) and RNA (ribonucleic acid).

The organic base molecules in the nucleotides of DNA and RNA are shown in Figure 21.16. Notice that some of these bases are found in only DNA, some are found only in RNA, and some are found in both. The formation of a specific nucleotide containing the base adenine and the sugar ribose is shown in Figure 21.17. To form the DNA and RNA polymers, the nucleotides are hooked together. DNA polymers can contain up to a *billion* nucleotide units. Figure 21.18 (p. 752) shows a small portion of a DNA chain.

The key to DNA's functioning is its *double-helical structure with complementary bases on the two strands.* Complementary bases form hydrogen bonds to each other, as shown in Figure 21.19 (p. 752). Note that the structures of cytosine and guanine make them perfect partners (complementary) for hydrogen bonding and that they are *always* found as pairs on the two strands of DNA. Thymine and adenine form similar hydrogen-bonding pairs.

There is much evidence to suggest that the two strands of DNA unwind during cell division and that new complementary strands are constructed on the

From the information provided in this section, students should be able to determine that there are both differences and similarities between DNA and RNA. You can help students assimilate the properties of each by presenting them in the form of a table.

Figure 21.18
A portion of a typical nucleic acid chain, half of the DNA double helix shown in 21.19.

Repeating unit along DNA chain

(a)

Thymine   Adenine

(b)

Cytosine   Guanine

(c)

A computer graphic representing the double-helical structure of DNA.

To help students understand the structure of double-stranded DNA, you might want to use the analogy of a ladder. The sides of the ladder represent alternating deoxyribose sugars and phosphates, and the rungs represent base pairs which are hydrogen bonded to each other.

Figure 21.19
(a) The DNA double helix contains two sugar–phosphate backbones with the bases from the two strands hydrogen bonded to each other. The complementarity of the (b) thymine–adenine and (c) cytosine–guanine pairs. The hydrogen-bonding interactions are shown by dotted lines.

This might be a good time, as you are discussing how DNA unwinds and how new nucleotides pair with the exposed bases, to talk about genetic disorders and what happens if the wrong bases are inserted in the newly synthesized DNA strand.

21.9 Lipids   **753**

unraveled strands (Figure 21.20). Because the bases on the strands always pair in the same way—cytosine with guanine and thymine with adenine—each unraveled strand serves as a template for attaching each complementary base (along with the rest of its nucleotide). This process results in two new double-helix DNA structures that are identical to the original one. Each new double strand contains one strand from the original DNA double helix and one newly synthesized strand. This replication of DNA makes possible the transmission of genetic information when cells divide.

DNA and Protein Synthesis

Besides replication, the other major function of DNA is **protein synthesis.** The proteins consumed by an organism in its food are typically not the specific proteins that organism needs to maintain its existence. The nutrient proteins are broken down into their constituent amino acids, which are then used to construct those proteins that the organism needs. The information for constructing each protein needed by a particular organism is stored in that organism's DNA. A given segment of the DNA, called a **gene,** contains the code for a specific protein. This code for the primary structure of the protein (the sequence of amino acids) can be transmitted to the construction "machinery" of the cell.

DNA stores the genetic information, and RNA molecules are responsible for transmitting this information to cell components called ribosomes, where protein synthesis actually occurs. This process involves, first, the construction of a special RNA molecule called **messenger RNA (mRNA).** The mRNA is built in the cell nucleus, where a specific section of DNA (a gene) is used as the pattern. The mRNA then migrates from the nucleus into the cytoplasm of the cell, where, with the assistance of the ribosomes, the protein is synthesized.

Small RNA fragments, called **transfer RNA (tRNA)** molecules, attach themselves to specific amino acids and bring them to the growing protein chain as dictated by the pattern built into the mRNA. This process is summarized in Figure 21.21 (p. 754).

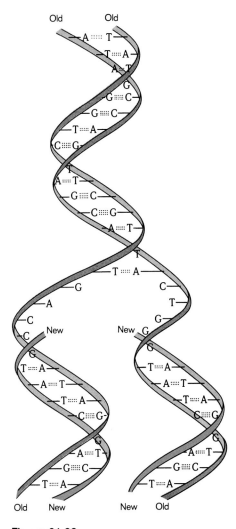

Figure 21.20
During cell division, the original DNA double helix unwinds and new complementary strands are constructed on each original strand. In this way, the two cells resulting from division have exact copies of the DNA of the original cell.

21.9 Lipids

AIM: To describe the four classes of lipids.

The **lipids** are a group of substances defined in terms of their solubility characteristics. They are water-insoluble substances that can be extracted from cells by organic solvents such as benzene. The lipids found in the human body can be divided into four classes according to their molecular structure: fats, phospholipids, waxes, and steroids.

There are many interesting things to say about lipids. You can talk about soaps and how they work, the fat content of foods, cholesterol, sex hormones, and other topics. Students like to discuss lipids, so use this opportunity to cover as much information as you need.

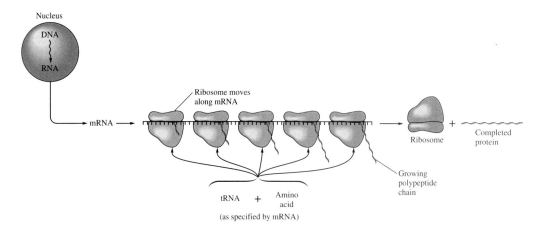

Figure 21.21
The mRNA molecule, constructed from a specific gene on the DNA, is used as the pattern for construction of a given protein with the assistance of ribosomes. The tRNA molecules attach to specific amino acids and put them in place, as dictated by the patterns on the mRNA. This sequence (left to right) shows the protein chain growing.

In Section 20.15, students learned about the formation of esters from carboxylic acids and alcohols. The formation of fats from three fatty acids and glycerol is the same reaction, but glycerol has three OH groups, each of which can react with a carboxylic (fatty) acid.

The most common **fats** are esters composed of the trihydroxy alcohol known as glycerol and long-chain carboxylic acids called **fatty acids** (Table 21.5). *Tristearin,* the most common animal fat, is typical of these substances.

$$CH_2-O\boxed{H + H-O}-\overset{\displaystyle O}{\overset{\|}{C}}-(CH_2)_{16}-CH_3 \qquad CH_2-O-\overset{\displaystyle O}{\overset{\|}{C}}-(CH_2)_{16}-CH_3$$

$$CH-O\boxed{H + H-O}-\overset{\displaystyle O}{\overset{\|}{C}}-(CH_2)_{16}-CH_3 \rightarrow CH-O-\overset{\displaystyle O}{\overset{\|}{C}}-(CH_2)_{16}-CH_3 + 3H_2O$$

$$CH_2-O\boxed{H + H-O}-\overset{\displaystyle O}{\overset{\|}{C}}-(CH_2)_{16}-CH_3 \qquad CH_2-O-\overset{\displaystyle O}{\overset{\|}{C}}-(CH_2)_{16}-CH_3$$

Glycerol   Three stearic acid   Tristearin
                  molecules   (fat)

Fats that are esters of glycerol are called **triglycerides** and have the general structure

You can point out that fats containing saturated fatty acids are found in animals and in the "tropical" vegetable oils and tend to be solids at room temperature. Vegetable oils contain more unsaturated fatty acids and are usually liquids at room temperature.

Table 21.5 Some Common Fatty Acids and Their Major Sources

Name	Formula	Major Source
Saturated		
arachidic acid	$CH_3(CH_2)_{18}-COOH$	peanut oil
butyric acid	$CH_3(CH_2)_2-COOH$	butter
caproic acid	$CH_3(CH_2)_4-COOH$	butter
lauric acid	$CH_3(CH_2)_{10}-COOH$	coconut oil
stearic acid	$CH_3(CH_2)_{16}-COOH$	animal and vegetable fats
Unsaturated		
oleic acid	$CH_3(CH_2)_7CH{=}CH(CH_2)_7-COOH$	corn oil
linoleic acid	$CH_3(CH_2)_4CH{=}CH-CH_2-CH{=}CH(CH_2)_7-COOH$	linseed oil
linolenic acid	$CH_3CH_2CH{=}CH-CH_2CH{=}CH-CH_2-CH{=}CH-(CH_2)_7COOH$	linseed oil

---

where the three R groups may be the same or different and may be saturated or unsaturated. Vegetable fats tend to be unsaturated and usually occur as oily liquids; most animal fats are saturated (contain only C—C single bonds) and occur as solids at normal temperatures.

Unsaturated fats contain one or more C=C bonds.

Triglycerides can be decomposed by treatment with aqueous sodium hydroxide. The products are glycerol and the fatty acid salts; the latter are known as soaps. This process is called **saponification.**

$$CH_2-O-\overset{\overset{\displaystyle O}{\|}}{C}-R$$
$$CH-O-\overset{\overset{\displaystyle O}{\|}}{C}-R' + 3NaOH \rightarrow$$
$$CH_2-O-\overset{\overset{\displaystyle O}{\|}}{C}-R''$$

$$CH_2OH + RCOONa$$
$$CHOH + R'COONa$$
$$CH_2OH + R''COONa$$

Triglyceride   Sodium hydroxide   Glycerol   Soaps

Lipids differ from most other organic and biomolecules in that they are not distinguished by a particular functional group but by their solubility in nonpolar solvents.

Much of what we call greasy dirt is nonpolar. Grease, for example, consists mostly of long-chain hydrocarbons. However, water, the solvent most commonly available to us, is very polar and does not dissolve "greasy dirt." We need to add something to the water that is somehow compatible with both the polar water and the nonpolar grease. Fatty-acid anions are perfect for this role, because they have a long nonpolar tail and a polar head. For example, the stearate anion can be

Like dissolves like.

represented as

$$CH_3-CH_2-CH_2-CH_2-CH_2-CH_2-CH_2-CH_2-CH_2-CH_2-CH_2-CH_2-CH_2-CH_2-CH_2-CH_2-CH_2-C\overset{O}{\underset{O^-}{\parallel}}$$

Nonpolar
tail

Polar
head

How soap works is interesting to most students; learning the composition of "bathtub ring" scum is sometimes a revelation. This material provides a good opportunity to discuss some important concepts and also talk about topics that are of universal interest.

Such ions can be dispersed in water because they form **micelles** (Figure 21.22). These aggregates of fatty acid anions have the water-incompatible tails in the interior; the anionic parts (the polar heads) point outward and interact with the polar water molecules. A soap solution does not contain *individual* fatty-acid anions dispersed in the water but rather groups of ions (micelles).

Soap dissolves grease by taking the grease molecules into the nonpolar interior of the micelle (Figure 21.23), so they can be carried away by the water. Soap thus acts to suspend the normally incompatible grease in the water. Because of this ability to assist water in suspending nonpolar materials, soap is also called a *wetting agent,* or **surfactant.**

A major disadvantage is that soap anions form precipitates in hard water (water that contains large concentrations of ions such as Ca^{2+} and Mg^{2+}). These precipitates occur because the Ca^{2+} and Mg^{2+} ions form insoluble solids with soap anions. These precipitates ("soap scum") dull clothes and drastically reduce soap's cleaning efficiency. To help alleviate this problem, a huge industry has developed to produce artificial soaps, called detergents. Detergents are similar to natural soaps in that they have a long nonpolar tail and an ionic head. However, detergent anions have the advantage of not forming insoluble solids with Ca^{2+} and Mg^{2+} ions.

Phospholipids are similar in structure to fats in that they are esters of glycerol. Unlike fats, however, they contain only two fatty acids. The third group bound to glycerol is a phosphate group, which gives phospholipids two distinct parts: the long nonpolar "tail" and the polar substituted-phosphate "head" (Figure 21.24).

Waxes are another class of lipids. Like fats and phospholipids, waxes are esters, but unlike these other lipids, they involve monohydroxy alcohols instead of glycerol. For example, *beeswax*, a substance secreted by the wax glands of bees, is mainly myricyl palmitate,

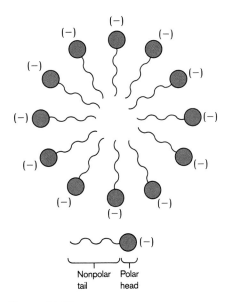

Figure 21.22
A two-dimensional "slice" of the structure of a micelle of fatty-acid anions. Cations surround the negatively charged micelle but, for the sake of simplicity, are not shown here.

$$CH_3(CH_2)_{14}-\overset{O}{\underset{}{\overset{\parallel}{C}}}-O-(CH_2)_{29}-CH_3$$

formed from palmitic acid,

$$CH_3(CH_2)_{14}-C\overset{O}{\underset{OH}{\diagdown}}$$

The structures of waxes and phospholipids are similar to those of fats, so students can usually understand their composition without much difficulty.

If you don't ask your students to memorize the structures of the steroids in Figure 21.25, you might want to ask them to memorize the structure of the steroid nucleus, which is common to all steroids and their derivatives.

21.9 Lipids   **757**

and carnaubyl alcohol

$$CH_3(CH_2)_{29}\text{—}OH$$

Waxes are solids that furnish waterproof coatings on leaves and fruit and on the skins and feathers of animals. Waxes are also important commercially. For example, whale oil is largely composed of the wax cetyl palmitate. It has been used in so many products, including cosmetics and candles, that the blue whale has been hunted almost to extinction.

Steroids are a class of lipids that have a characteristic carbon ring structure of the type

Steroids comprise four groups: cholesterol, adrenocorticoid hormones, sex hormones, and bile acids.

Cholesterol (Figure 21.25a, p. 758) is found in virtually all organisms and is the starting material for the formation of the many other steroid-based molecules, such as vitamin D (Figure 21.25b). Although cholesterol is essential for human life, it has been implicated in the formation of plaque on the walls of arteries (a process called atherosclerosis, or hardening of the arteries), which can lead eventually to clogging. This effect seems especially important in the arteries that supply blood to the heart. Blockage of these arteries leads to heart damage that often results in death from a heart attack.

The **adrenocorticoid hormones,** such as cortisol (Figure 21.25c), are synthesized in the adrenal glands (glands that lie next to each kidney) and are involved in various regulatory functions.

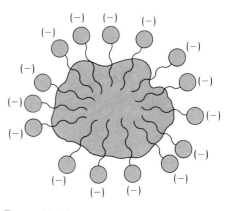

Figure 21.23
Soap micelles absorb grease molecules into their interiors so that the molecules are suspended in the water and can be washed away.

Olives have a waxy coating.

Nonpolar tails

Polar head

Figure 21.24
Lecithin, a phospholipid, with its long nonpolar tails and polar substituted-phosphate head.

Figure 21.25
Several common steroids and steroid derivatives: (a) cholesterol, (b) vitamin D₃,
(c) cortisol, (d) testosterone, (e) progesterone, (f) estradiol, (g) ethynodiol diacetate,
and (h) cholic acid.

Of the **sex hormones,** the most important male hormone is *testosterone* (Figure 21.25d), a hormone that controls the growth of the reproductive organs and hair and the development of the muscle structure and deep voice that are characteristic of males. There are two types of female sex hormones of particular significance: *progesterone* (Figure 21.25e), and a group of estrogens, one of which is *estradiol* (Figure 21.25f). Changes in the concentrations of these hormones cause the periodic changes in the ovaries and uterus that are responsible for the menstrual cycle. During pregnancy, a high level of progesterone is maintained, which prevents ovulation. This effect has led to the use of progesterone-type compounds as birth-control drugs. One of the most common of these is ethynodiol diacetate (Figure 21.25g).

The **bile acids** are produced from cholesterol in the liver and stored in the gallbladder. The primary human bile acid is *cholic acid* (Figure 21.25h), a substance that aids in the digestion of fats by emulsifying them in the intestine. Bile acids can also dissolve cholesterol ingested in food and are therefore important in limiting cholesterol in the body.

CHAPTER REVIEW

Key Terms

biochemistry (p. 736
essential elements (p. 736)
trace elements (p. 736)
protein (p. 737)
α-amino acid (p. 738)
side chain (p. 738)
dipeptide (p. 740)
peptide linkage (p. 740)
polypeptide (p. 740)
primary structure (p. 740)
secondary structure (p. 741)
α-helix (p. 741)
pleated sheet (p. 741)
tertiary structure (p. 742)
disulfide linkage (p. 743)
denaturation (p. 743)
enzyme (p. 745)
lock-and-key model (p. 745)
carbohydrate (p. 746)
monosaccharide (simple sugar)
  (p. 746)

disaccharide (p. 748)
sucrose (p. 748)
starch (p. 749)
cellulose (p. 749)
glycogen (p. 749)
deoxyribonucleic acid (DNA)
  (p. 750)
ribonucleic acid (RNA) (p. 750)
nucleotide (p. 750)
protein synthesis (p. 753)
gene (p. 753)
lipid (p. 753)
fat (p. 754)
fatty acid (p. 754)
micelle (p. 756)
surfactant (p. 756)
phospholipid (p. 756)
wax (p. 756)
steroid (p. 757)

Summary

1. Thirty elements are currently known to be essential for life. The most abundant elements in the human body are hydrogen, carbon, nitrogen, and oxygen; but calcium, phosphorus, sodium, magnesium, potassium, sulfur, and chlorine are also present in large amounts. Other elements found only in trace amounts, such as zinc, are essential for the action of many enzymes.

2. Proteins are a class of natural polymers with molar masses ranging from 6000 to 1,000,000 grams. Fibrous proteins are employed in the human body for structural purposes in muscle, hair, and cartilage. Globular proteins are molecules that transport and store oxygen and nutrients, act as catalysts, help regulate the body's systems, fight foreign objects, and so on.

3. The building blocks of proteins are the α-amino acids, which are classified as polar or nonpolar, depending on whether the side chain (R group) attached to the α-carbon is hydrophilic or hydrophobic. A protein polymer is built by successive condensation reactions that produce peptide linkages:

4. The order or sequence of amino acids in the protein chain is called the protein's primary structure. Differences in primary structure are what tailor proteins for different and very specific functions.

5. The secondary structure is the arrangement of the protein chain in space. The two most common secondary structures are the α-helix and the pleated sheet.

6. The overall shape of the protein is called its tertiary structure. Energy sources and a variety of chemicals can cause the breakdown of tertiary protein structure, which is called denaturation.

7. Enzymes are proteins that act as catalysts in biological reactions.

8. Carbohydrates serve as food sources for most organisms and as structural materials for plants. Simple carbohydrates, called monosaccharides, are most commonly five-carbon and six-carbon polyhydroxy ketones and aldehydes.

9. Monosaccharides combine to form more complex carbohydrates. For example, sucrose is a disaccharide, and starch and cellulose are polymers of glucose (polysaccharides).

10. When a cell divides, the genetic information is transmitted via deoxyribonucleic acid (DNA), which has a double-helical structure. During cell division, the double helix unravels, and a new polymer forms along each strand of the original DNA, creating two double-helical DNA molecules. The DNA contains segments called genes, which store the primary structures of specific proteins. Various types of ribonucleic acid (RNA) molecules assist in protein synthesis.

11. Lipids are water-insoluble substances found in cells. They can be divided into four classes: fats, phospholipids, waxes, and steroids.

Questions and Problems

All even-numbered exercises have answers in the back of this book and solutions in the Solutions Guide.

21.1 Proteins

QUESTIONS

1. Approximately what percentage of the human body consists of protein?

3. What uses do fibrous proteins have in the body?

2. Describe the range of molar masses shown by proteins in the body. Are such molar masses consistent with proteins being polymers?

4. How are globular proteins used in the body? What is the general shape of a globular protein?

21.2 Primary Structure of Proteins

QUESTIONS

5. Sketch the general formula for an alpha amino acid, indicating the alpha carbon atom. Circle the portion of the structure that all alpha amino acids have in common.

7. Why is it important to distinguish between hydrophilic and hydrophobic side chains? What solvent forms the natural environment for proteins in the body?

9. Why are the simple dipeptides *lys-gly* and *gly-lys* **not** isomers of one another? *Hint:* draw the structures of the dipeptides and examine the terminal groups present.

6. Write the formulas of two amino acids with polar side chains and of two amino acids with nonpolar side chains. Explain *why* the side chains you have chosen have their respective polarities.

8. Draw a general structure representing a *peptide* linkage.

10. Sketch the possible amino acid sequences possible for a small polypeptide containing *gly, ile, leu, pro,* and *phe,* with each amino acid being used only once in each case.

PROBLEMS

11. Given the formulas of the common amino acids in Figure 21.2, sketch the structures of the following simple dipeptides. Circle the peptide bonds in your structures. Label the terminal amino and carboxyl groups clearly.
    a. ala-gly   c. phe-ser
    b. gly-ala   d. ser-phe

12. Given the formulas of the common amino acids in Figure 21.2, sketch the structures of the six different tripeptides possible containing phenylalanine, alanine, and glycine (each taken only once). Circle the peptide bonds in your structures. Label the terminal amino and carboxyl groups clearly.

21.3 Secondary Structure of Proteins

QUESTIONS

13. In general terms, what does the secondary structure of a protein represent?

14. Describe the secondary protein structure known as the alpha helix. What types of proteins have this secondary structure?

15. Describe the secondary protein structure known as the pleated sheet. Give two examples of materials containing proteins with this structure.

16. How is the secondary structure of a protein related to its function in the body? Give examples.

21.4 Tertiary Structure of Proteins

QUESTIONS

17. In general terms, what does the tertiary structure of a protein describe? Clearly distinguish between the secondary and tertiary structures.

18. Describe the function of the amino acid cysteine in influencing the tertiary structure of a protein.

21.5 Functions of Proteins

QUESTIONS

19. What is meant by *denaturation* of a protein? Give three examples of situations in which proteins are denatured.

20. Describe the structure of the protein *collagen.* What function does collagen have in the body?

21. What name is given to proteins that catalyze biochemical reactions in the cell?

22. What are antibodies? Give an example of an important antibody.

23. Give several examples of proteins that serve a protective function in the body.

24. How does a "permanent wave" change the structure of the protein of hair? What amino acid undergoes reaction during a permanent wave? Which level of structure for the hair protein is affected?

21.6 Enzymes

QUESTIONS

25. How does the efficiency of an enzyme compare to that of inorganic catalysts?

26. What general name is given to the molecule acted on by a given enzyme?

27. What name is given to the specific portion of the enzyme molecule where catalysis actually occurs?

28. Describe the *lock-and-key* model for catalysis by enzymes.

21.7 Carbohydrates

QUESTIONS

29. What functional groups are present in the simple sugars (monosaccharides)?

30. How many carbon atoms are present in an *aldohexose?* In a *ketopentose?* In an *aldotriose?* In a *ketotetrose?*

31. Rather than the form of a straight chain, what structures do the monosaccharides usually assume in aqueous solution?

32. Sketch the straight-chain structures of the following monosaccharides:
    a. glucose   c. ribulose
    b. ribose   d. galactose

33. What is a disaccharide? What monosaccharide units make up the disaccharide sucrose? What is the bond called that forms between the monosaccharide units?

34. What is a polysaccharide? What monomer unit makes up the common polysaccharides starch and cellulose?

35. Sketch both the straight-chain and cyclic structures of the monosaccharide fructose. What atoms make up the ring portion of the cyclic structure?

36. In addition to using *numerical* prefixes in the general names of sugars to indicate how many carbon atoms are present, we often use the prefixes *keto-* and *aldo-* to indicate whether the sugar is a ketone or an aldehyde. For example, the monosaccharide fructose is frequently called a ketohexose to emphasize that it contains the ketone functional group. For each of the monosaccharides shown in Table 21.4, classify the sugars as aldohexoses, aldopentoses, ketohexoses, or ketopentoses.

21.8 Nucleic Acids

QUESTIONS

37. What molecule stores and transmits genetic information in the cell? How large is this molecule? Where in the cell is this molecule found?

38. Describe the structure of a typical nucleotide. What components are present in the nucleotide?

39. Sketch the structures of the sugars ribose and deoxyribose. Which molecule, DNA or RNA, contains each sugar?

40. Name the five nitrogen bases found in DNA and RNA. Which base is found commonly in RNA but not in DNA? Which base is found commonly in DNA but not in RNA?

41. Describe the double-helical structure of DNA. What type of bonding occurs *within* the chain of each strand of the double helix? What type of bonding exists *between* strands to link them together?

42. Describe the *complementary base pairing* between the two individual strands of DNA that forms the overall double-helical structure. How is complementary base pairing involved in the replication of the DNA molecule during cell division?

43. What name is applied to the specific section of the DNA molecule that contains the code for construction of a particular protein?

44. Explain the distinction between, and the functions of, messenger RNA and transfer RNA.

21.9 Lipids

QUESTIONS

45. Under what circumstances is a biomolecule classified as a *lipid?* Is the classification based on a particular characteristic structure or on some physical property?

46. Sketch the general structure of a triglyceride. What are the components that go into making up a typical triglyceride?

47. Referring to Table 21.5, list several saturated and several unsaturated fatty acids. What is the general source of most unsaturated fatty acids?

48. What is *saponification?* Write a general equation for the saponification of a triglyceride. What are the ionic products of a saponification reaction more commonly called?

49. Describe the mechanism by which fatty acid salts are able to exert a cleaning action.

50. What is a *micelle?* How do the micelles formed by soap molecules suspend greasy dirt in a solution?

51. Sketch the basic ring structure common to all steroids.

52. What steroid serves as the starting material in the body for the synthesis of other steroids? What dangers are involved in having too large a concentration of this substance in the body?

53. Give the names of several steroid sex hormones, and indicate their function in the body.

54. What is *phospholipid?* How does the structure of a phospholipid differ from that of a triglyceride. What common phospholipid behaves as an emulsifying agent and is used commonly in the food industry?

Additional Problems

Column 1	Column 2

55. polymer consisting of many amino acids
56. linkage that forms between two cysteine species
57. peptide hormone that triggers milk secretion
58. proteins with roughly spherical shape
59. sequence of amino acids in a protein
60. silk protein secondary structure
61. water-repelling amino acid side chain
62. amino acid responsible for permanent wave in hair
63. biological catalysts
64. breakdown of a protein's tertiary and/or secondary structure
65. molecule acted on by an enzyme
66. occurs when an enzyme's active site is blocked by a foreign molecule
67. special protein synthesized in response to foreign substance
68. substance that has a specific effect on a particular target organ
69. animal polymer of glucose
70. —C—O—C— bond between rings in disaccharide sugars
71. empirical formula leading to the name carbo*hydrate*
72. enzyme catalyzing the breakdown of glycoside link
73. six-carbon ketone sugars
74. structural component of plants, polymer of glucose
75. sugars consisting of two monomer units
76. six-carbon aldehyde sugars

a. aldo hexose
b. amylase
c. antibody
d. cellulose
e. CH_2O
f. cysteine
g. denaturation
h. disaccharides
i. disulfide
j. DNA
k. enzymes
l. fibrous
m. globular
n. glycogen
o. glycoside linkage
p. hormone
q. hydrophobic
r. inhibition
s. ketohexoses
t. oxytocin
u. pleated sheet
v. polypeptide
w. polysaccharides
x. primary structure
y. substrate
z. sucrose

77. The substance in the nucleus of the cell that stores and transmits genetic information is DNA, which stands for _____.
78. The basic repeating monomer units of DNA and RNA are called _____.
79. The sugar _____ is found in DNA units, whereas the sugar ribose is found in RNA.
80. The basic linkage in DNA or RNA between the sugar molecule and phosphoric acid is a phosphate _____ linkage.
81. The bases on opposite strands of DNA are said to be _____ to each other, which means the bases fit together specifically by hydrogen bonding to one another.
82. In a strand of normal DNA, the base thymine is always found paired with the base _____, whereas guanine is always found paired with _____.
83. A given segment of the DNA molecule, which contains the molecular coding for a specific protein to be synthesized, is referred to as a _____.
84. During protein synthesis, _____ RNA molecules attach to and transport specific amino acids to the appropriate position on the pattern provided by _____ RNA molecules.
85. The codes specified by DNA are responsible for assembling the correct _____ structure of proteins.
86. Substances found in the cell are classified as lipids if they are _____ in water but are _____ in nonpolar organic solvents.
87. Triglycerides are _____ of the polyhydroxyalcohol glycerol with long-chain carboxylic acids.
88. Vegetable oils tend to contain _____ fatty acids, whereas animal fats tend to be _____.
89. The process of _____ involves treating a fat with sodium hydroxide so that the fatty acids present are converted to their sodium salts.

90. Fatty acid anions make good soaps, because they have both an _____ portion that is miscible with water and a _____ portion that is miscible with grease.
91. The aggregations of fatty acid anions that form when a soap is added to water are called _____ .
92. _____ are esters of fatty acids with monohydroxy alcohols.
93. The starting material in the body for the synthesis of other steroids is _____ .
94. During pregnancy, ovulation is prevented by secretion of the hormone _____ .
95. Cholic acid and other bile acids act as _____ agents during digestion, helping to break fats up into smaller droplets that can then be acted on by enzymes.
96. What is meant by the *primary structure* of a protein?
97. How many possible primary structures exist for a small polypeptide containing four individual amino acids?
98. List three structural uses of proteins in the body.
99. What specific protein carries oxygen from the lungs to other body tissues?
100. How are proteins able to provide a buffering action?
101. What is meant by *inhibition* of an enzyme? What happens when an enzyme is irreversibly inhibited?
102. What general name is given to sugars containing five carbon atoms? six carbon atoms? three carbon atoms?
103. Although both starch and cellulose are polymers of glucose, starch is digestible by humans and cellulose is not. How do the structures of these polysaccharides differ, resulting in their different digestibility?
104. In the formation of a polynucleotide (a short portion of the DNA molecule), which components (sugar, base, or phosphate) on adjacent nucleotides bond to each other?
105. Describe the structure of a *wax*. Where do waxes occur naturally in living creatures, and what function do they serve?
106. What is the primary human bile acid, and what is its function in the body?

*U*sing Your Calculator

In this section we will review how to use your calculator to perform common mathematical operations. This discussion assumes that your calculator uses the algebraic operating system, the system used by most brands.

One very important principle to keep in mind as you use your calculator is that it is not a substitute for your brain. Keep thinking as you do the calculations. Keep asking yourself, "Does the answer make sense?"

Addition, Subtraction, Multiplication, and Division

Performing these operations on a pair of numbers always involves the following steps:

1. Enter the first number, using the numbered keys and the decimal (.) key if needed.
2. Enter the operation to be performed.
3. Enter the second number.
4. Press the "equals" key to display the answer.

For example, the operation

$$15.1 + 0.32$$

is carried out as follows:

Press	Display
15.1	15.1
+	15.1
.32	0.32
=	15.42

The answer given by the display is 15.42. If this is the final result of a calculation, you should round it off to the correct number of significant figures (15.4), as discussed in Section 2.5. If this number is to be used in further calculations, use it exactly as it appears on the display. Round off only the final answer in the calculation.

Do the following operations for practice. The detailed procedures are given below.

a. $1.5 + 32.86$
b. $23.5 - 0.41$

c. 0.33×153
d. $\dfrac{9.3}{0.56}$ or $9.3 \div 0.56$

PROCEDURES

a.
Press	Display
1.5	1.5
+	1.5
32.86	32.86
=	34.36
Rounded:	34.4

b.
Press	Display
23.5	23.5
−	23.5
.41	0.41
=	23.09
Rounded:	23.1

c.
Press	Display
.33	0.33
×	0.33
153	153
=	50.49
Rounded:	50

d.
Press	Display
9.3	9.3
÷	9.3
.56	0.56
=	16.607143
Rounded:	17

Squares, Square Roots, Reciprocals, and Logs

Now we will consider four additional operations that we often need to solve chemistry problems.

The *squaring* of a number is done with a key labeled X^2. The *square root* key is usually labeled $\sqrt{X}$. To take the *reciprocal* of a number, you need the 1/X key. The *logarithm* of a number is determined by using a key labeled log or logX.

To perform these operations, take the following steps:

1. Enter the number.
2. Press the appropriate function key.
3. The answer is displayed automatically.

For example, let's calculate the square root of 235.

Press	Display
235	235
$\sqrt{X}$	15.32971
Rounded:	15.3

We can obtain the log of 23 as follows:

Press	Display
23	23
log	1.3617278
Rounded:	1.36

Often a key on a calculator serves two functions. In this case, the first function is listed on the key and the second is shown on the calculator just above the key. For example, on some calculators the top row of keys appears as follows:

$$1/X \quad X^2$$

| 2nd | R/S | $\sqrt{X}$ | off | on/C |

To make the calculator square a number, we must use 2nd and then $\sqrt{X}$; pressing 2nd tells the calculator we want the function that is listed *above* the key. Thus we can obtain the square of 11.56 on this calculator as follows:

Press	Display
11.56	11.56
2nd then $\sqrt{X}$	133.6336
Rounded:	133.6

We obtain the reciprocal of 384 (1/384) on this calculator as follows:

Press	Display
384	384
2nd then R/S	0.0026042
Rounded:	0.00260

Your calculator may be different! See the user's manual if you are having trouble with these operations.

Chain Calculations

In solving problems you often have to perform a series of calculations—a calculation chain. This is generally quite easy if you key in the chain as you read the numbers and operations in order. For example, to perform the calculation

$$\frac{14.68 + 1.58 - 0.87}{0.0850}$$

you should use the appropriate keys as you read it to yourself:

14.68 plus 1.58 equals; minus .87 equals; divided by 0.085 equals

The details follow.

Press	Display
14.68	14.68
+	14.68
1.58	1.58
=	16.26
−	16.26
.87	0.87
=	15.39
÷	15.39
.0850	0.0850
=	181.05882
Rounded:	181

Note that you must press | = | after every operation to keep the calculation "up to date."

For more practice, consider the calculation

$$(0.360)(298) + \frac{(14.8)(16.0)}{1.50}$$

Here you are adding two numbers, but each must be obtained by the indicated calculations. One procedure is to calculate each number first and then add them. The first term is

$$(0.360)(298) = 107.28$$

The second term,

$$\frac{(14.8)(16.0)}{1.50}$$

can be computed easily by reading it to yourself. It "reads"

14.8 times 16.0 equals; divided by 1.50 equals;

and is summarized as follows:

Press	Display
14.8	14.8
×	14.8
16.0	16.0
=	236.8
÷	236.8
1.50	1.50
=	157.86667

Now we can keep this last number on the calculator and add it to 107.28 from the first calculation.

Press	Display
	157.86667
+	107.28
=	265.14667
Rounded:	265

To summarize,

$$(0.36)(298) + \frac{(14.8)(16.0)}{1.50}$$

becomes

$$107.28 + 157.86667$$

and the sum is 265.14667 or, rounded to the correct number of significant figures, 265. There are other ways to do this calculation, but this is the safest way (assuming you are careful).

A common type of chain calculation involves a number of terms multiplied together in the numerator and the denominator, as in

$$\frac{(323)(.0821)(1.46)}{(4.05)(76)}$$

There are many possible sequences by which this calculation can be carried out, but the following seems the most natural.

323 times .0821 equals; times 1.46 equals; divided by 4.05 equals; divided by 76 equals

This sequence is summarized as follows:

Press	Display
323	323
×	323
.0821	0.0821
=	26.5183
×	26.5183

Press	Display
1.46	1.46
=	38.716718
÷	38.716718
4.05	4.05
=	9.5596835
÷	9.5596835
76	76
=	0.1257853

The answer is 0.1257853, which, when rounded to the correct number of significant figures, is 0.13. Note that when two or more numbers are multiplied in the denominator, you must divide by *each* one.

Here are some additional chain calculations (with solutions) to give you more practice.

a. $15 - (0.750)(243)$

b. $\dfrac{(13.1)(43.5)}{(1.8)(63)}$

c. $\dfrac{(85.8)(0.142)}{(16.46)(18.0)} + \dfrac{(131)(0.0156)}{10.17}$

d. $(18.1)(0.051) - \dfrac{(325)(1.87)}{(14.0)(3.80)} + \dfrac{1.56 - 0.43}{1.33}$

SOLUTIONS

a. $15 - 182 = -167$
b. 5.0
c. $0.0411 + 0.201 = 0.242$
d. $0.92 - 11.4 + 0.850 = -9.6$

In performing chain calculations, take the following steps in the order listed.

1. Perform any additions and subtractions that appear inside parentheses.
2. Complete the multiplications and divisions of individual terms.
3. Add and subtract individual terms as required.

Basic Algebra

In solving chemistry problems you will use, over and over again, relatively few mathematical procedures. In this section we review the few algebraic manipulations that you will need.

Solving an Equation

In the course of solving a chemistry problem, we often construct an algebraic equation that includes the unknown quantity (the thing we want to calculate). An example is

$$(1.5)V = (0.23)(0.08206)(298)$$

We need to "solve this equation for V." That is, we need to isolate V on one side of the equals sign with all the numbers on the other side. How can we do this? The key idea in solving an algebraic equation is that *doing the same thing on both sides of the equals sign* does not change the equality. That is, it is always "legal" to do the same thing to both sides of the equation. Here we want to solve for V, so we must get the number 1.5 on the other side of the equals sign. We can do this by dividing *both sides* by 1.5.

$$\frac{(1.5)V}{1.5} = \frac{(0.23)(0.08206)(298)}{1.5}$$

Now the 1.5 in the denominator on the left cancels the 1.5 in the numerator:

$$\frac{(\cancel{1.5})V}{\cancel{1.5}} = \frac{(0.23)(0.08206)(298)}{1.5}$$

to give

$$V = \frac{(0.23)(0.08206)(298)}{1.5}$$

Using the procedures discussed in Appendix I for chain calculations, we can now obtain the value for V with a calculator.

$$V = 3.7$$

Sometimes it is necessary to solve an equation that consists of symbols. For example, consider the equation

$$\frac{P_1 V_1}{T_1} = \frac{P_2 V_2}{T_2}$$

Let's assume we want to solve for T_2. That is, we want to isolate T_2 on one side of the equation. There are several possible ways to proceed, keeping in mind that we always do the same thing on both sides of the equals sign. First we multiply both sides by T_2.

$$T_2 \times \frac{P_1 V_1}{T_1} = \frac{P_2 V_2}{\cancel{T_2}} \times \cancel{T_2}$$

This cancels T_2 on the right. Next we multiply both sides by T_1.

$$T_2 \times \frac{P_1 V_1}{\cancel{T_1}} \times \cancel{T_1} = P_2 V_2 T_1$$

This cancels T_1 on the left. Now we divide both sides by $P_1 V_1$.

$$T_2 \times \frac{\cancel{P_1 V_1}}{\cancel{P_1 V_1}} = \frac{P_2 V_2 T_1}{P_1 V_1}$$

This yields the desired equation,

$$T_2 = \frac{P_2 V_2 T_1}{P_1 V_1}$$

For practice, solve each of the following equations for the variable indicated.

a. $PV = k$; solve for P

b. $1.5x + 6 = 3$; solve for x

c. $PV = nRT$; solve for n

d. $\dfrac{P_1 V_1}{T_1} = \dfrac{P_2 V_2}{T_2}$; solve for V_2

e. $\dfrac{°F - 32}{°C} = \dfrac{9}{5}$; solve for $°C$

f. $\dfrac{°F - 32}{°C} = \dfrac{9}{5}$; solve for $°F$

SOLUTIONS

a. $\dfrac{P\cancel{V}}{\cancel{V}} = \dfrac{k}{V}$

$P = \dfrac{k}{V}$

b. $1.5x + 6 - 6 = 3 - 6$

$$1.5x = -3$$

$$\frac{\cancel{1.5}x}{\cancel{1.5}} = \frac{-3}{1.5}$$

$$x = -\frac{3}{1.5} = -2$$

c. $\dfrac{PV}{RT} = \dfrac{n\cancel{RT}}{\cancel{RT}}$

$$\frac{PV}{RT} = n$$

d. $\dfrac{P_1 V_1}{T_1} \times T_2 = \dfrac{P_2 V_2}{\cancel{T_2}} \times \cancel{T_2}$

$$\frac{P_1 V_1 T_2}{T_1 P_2} = \frac{\cancel{P_2} V_2}{\cancel{P_2}}$$

$$\frac{P_1 V_1 T_2}{T_1 P_2} = V_2$$

e. $\dfrac{°F - 32}{\cancel{°C}} \times \cancel{°C} = \dfrac{9}{5} \, °C$

$$\frac{5}{9}(°F - 32) = \frac{\cancel{5}}{\cancel{9}} \times \frac{\cancel{9}}{\cancel{5}} \, °C$$

$$\frac{5}{9}(°F - 32) = °C$$

f. $\dfrac{°F - 32}{\cancel{°C}} \times \cancel{°C} = \dfrac{9}{5} \times °C$

$$°F - \cancel{32} + \cancel{32} = \frac{9}{5} \, °C + 32$$

$$°F = \frac{9}{5} \, °C + 32$$

Scientific (Exponential) Notation

The numbers we must work with in scientific measurements are often very large or very small; thus it is convenient to express them using powers of 10. For example, the number 1,300,000 can be expressed as 1.3×10^6, which means multiply 1.3 by 10 six times, or

$$1.3 \times 10^6 = 1.3 \times \underbrace{10 \times 10 \times 10 \times 10 \times 10 \times 10}_{10^6 = 1 \text{ million}}$$

A number written in scientific notation always has the form:

A number (between 1 and 10) times
the appropriate power of ten

To represent a large number such as 20,500 in scientific notation, we must move the decimal point in such a way as to achieve a number between 1 and 10 and then multiply the result by a power of 10 to compensate for moving the decimal point. In this case, we must move the decimal point four places to the left.

$$\underset{4 \quad 3 \quad 2 \quad 1}{2 \; 0 \; 5 \; 0 \; 0}$$

to give a number between 1 and 10:

2.05

where we retain only the significant figures (the number 20,500 has three significant figures). To compensate for moving the decimal point four places to the left, we must multiply by 10^4. Thus

$$20,500 = 2.05 \times 10^4$$

As another example, the number 1985 can be expressed as 1.985×10^3. To end up with the number 1.985, which is between 1 and 10, we had to move the decimal point three places to the left. To compensate for that, we must multiply by 10^3. Some other examples are given in the accompanying table.

Number	Exponential Notation
5.6	5.6×10^0 or 5.6×1
39	3.9×10^1
943	9.43×10^2
1126	1.126×10^3

So far, we have considered numbers greater than 1. How do we represent a number such as 0.0034 in exponential notation? First, to achieve a number between 1 and 10, we start with 0.0034 and move the decimal point three places to the right.

$$\underset{1 \quad 2 \quad 3}{0 \, . \, 0 \; 0 \; 3 \; 4}$$

This yields 3.4. Then, to compensate for moving the decimal point to the right, we must multiply by a power of 10 with a negative exponent—in this case, 10^{-3}. Thus

$$0.0034 = 3.4 \times 10^{-3}$$

In a similar way, the number 0.00000014 can be written as 1.4×10^{-7}, because going from 0.00000014 to 1.4 requires that we move the decimal point seven places to the right.

Mathematical Operations with Exponentials

We next consider how various mathematical operations are performed using exponential numbers. First we cover the various rules for these operations; then we consider how to perform them on your calculator.

Multiplication and Division

When two numbers expressed in exponential notation are multiplied, the initial numbers are multiplied and the exponents of 10 are *added.*

$$(M \times 10^m)(N \times 10^n) = (MN) \times 10^{m+n}$$

For example (to two significant figures, as required),

$$(3.2 \times 10^4)(2.8 \times 10^3) = 9.0 \times 10^7$$

When the numbers are multiplied, if a result greater than 10 is obtained for the initial number, the decimal point is moved one place to the left and the exponent of 10 is increased by 1.

$$
\begin{aligned}
(5.8 \times 10^2)(4.3 \times 10^8) &= 24.9 \times 10^{10} \\
&= 2.49 \times 10^{11} \\
&= 2.5 \times 10^{11} \quad \text{(two significant figures)}
\end{aligned}
$$

Division of two numbers expressed in exponential notation involves normal division of the initial numbers and *subtraction* of the exponent of the divisor from that of the dividend. For example,

$$\underbrace{\frac{4.8 \times 10^8}{2.1 \times 10^3}}_{\substack{\uparrow \\ \text{Divisor}}} = \frac{4.8}{2.1} \times 10^{(8-3)} = 2.3 \times 10^5$$

If the initial number resulting from the division is less than 1, the decimal point is moved one place to the right and the exponent of 10 is decreased by 1. For example,

$$
\begin{aligned}
\frac{6.4 \times 10^3}{8.3 \times 10^5} = \frac{6.4}{8.3} \times 10^{(3-5)} &= 0.77 \times 10^{-2} \\
&= 7.7 \times 10^{-3}
\end{aligned}
$$

Addition and Subtraction

In order for us to add or subtract numbers expressed in exponential notation, *the exponents of the numbers must be the same.* For example, to add 1.31×10^5 and 4.2×10^4, we must rewrite one number so that the exponents of both are the same. The number 1.31×10^5 can be written 13.1×10^4: decreasing the exponent by 1 compensates for moving the decimal point one place to the right. Now we can add the numbers.

$$
\begin{array}{r}
13.1 \times 10^4 \\
+\ 4.2 \times 10^4 \\
\hline
17.3 \times 10^4
\end{array}
$$

In correct exponential notation, the result is expressed as 1.73×10^5.

To perform addition or subtraction with numbers expressed in exponential notation, we add or subtract only the initial numbers. The exponent of the result is the same as the exponents of the numbers being added or subtracted. To subtract 1.8×10^2 from 8.99×10^3, we first convert 1.8×10^2 to 0.18×10^3 so that both numbers have the same exponent. Then we subtract.

$$
\begin{array}{r}
8.99 \times 10^3 \\
-0.18 \times 10^3 \\
\hline
8.81 \times 10^3
\end{array}
$$

Powers and Roots

When a number expressed in exponential notation is taken to some power, the initial number is taken to the appropriate power and the exponent of 10 is *multiplied* by that power.

$$(N \times 10^n)^m = N^m \times 10^{m \times n}$$

For example,

$$
\begin{aligned}
(7.5 \times 10^2)^2 = (7.5)^2 \times 10^{2 \times 2} &= 56 \times 10^4 \\
&= 5.6 \times 10^5
\end{aligned}
$$

When a root is taken of a number expressed in exponential notation, the root of the initial number is taken and the exponent of 10 is divided by the number representing the root. For example, we take the square root of a number as follows:

$$\sqrt{N \times 10^n} = (N \times 10^n)^{1/2} = \sqrt{N} \times 10^{n/2}$$

For example,

$$
\begin{aligned}
(2.9 \times 10^6)^{1/2} = \sqrt{2.9} \times 10^{6/2} &= 1.7 \times 10^3
\end{aligned}
$$

Using a Calculator to Perform Mathematical Operations on Exponentials

In dealing with exponential numbers, you must first learn to enter them into your calculator. First the number is keyed in and then the exponent. There is a special key that must be pressed just before the exponent is entered. This key is often labeled [EE] or [exp]. For example, the number 1.56×10^6 is entered as follows:

Press	Display
1.56	1.56
EE or exp	1.56 00
6	1.56 06

To enter a number with a negative exponent, use the change-of-sign key   +/−   after entering the exponent number. For example, the number 7.54×10^{-3} is entered as follows:

Press	Display
7.54	7.54
EE or exp	7.54 00
3	7.54 03
+/−	7.54 −03

Once a number with an exponent is entered into your calculator, the mathematical operations are performed exactly the same as with a "regular" number. For example, the numbers 1.0×10^3 and 1.0×10^2 are multiplied as follows:

Press	Display
1.0	1.0
EE or exp	1.0 00
3	1.0 03
×	1 03
1.0	1.0
EE or exp	1.0 00
2	1.0 02
=	1 05

The answer is correctly represented as 1.0×10^5.

The numbers 1.50×10^5 and 1.1×10^4 are added as follows:

Press	Display
1.5	1.50
EE or exp	1.50 00
5	1.50 05
+	1.5 05
1.1	1.1
EE or exp	1.1 00
4	1.1 04
=	1.61 05

The answer is correctly represented as 1.61×10^5. Note that when exponential numbers are added, the calculator automatically takes into account any difference in exponents.

To take the power, root, or reciprocal of an exponential number, enter the number first, then press the appropriate key or keys. For example, the square root of 5.6×10^3 is obtained as follows:

Press	Display
5.6	5.6
EE or exp	5.6 00
3	5.6 03
$\sqrt{X}$	7.4833148 01

The answer is correctly represented as 7.5×10^1.

Practice by performing the following operations that involve exponential numbers. The answers follow the exercises.

a. $7.9 \times 10^2 \times 4.3 \times 10^4$

b. $\dfrac{5.4 \times 10^3}{4.6 \times 10^5}$

c. $1.7 \times 10^2 + 1.63 \times 10^3$

d. $4.3 \times 10^{-3} + 1 \times 10^{-4}$

e. $(8.6 \times 10^{-6})^2$

f. $\dfrac{1}{8.3 \times 10^2}$

g. $\log(1.0 \times 10^{-7})$

h. $-\log(1.3 \times 10^{-5})$

i. $\sqrt{6.7 \times 10^9}$

SOLUTIONS

a. 3.4×10^7

b. 1.2×10^{-2}

c. 1.80×10^3

d. 4.4×10^{-3}

e. 7.4×10^{-11}

f. 1.2×10^{-3}

g. -7.0

h. 4.9

i. 8.2×10^4

Graphing Functions

In interpreting the results of a scientific experiment, it is often useful to make a graph. If possible, the function to be graphed should be in a form that gives a straight line. The equation for a straight line (a *linear equation*) can be represented in the general form

$$y = mx + b$$

where y is the *dependent variable*, x is the *independent variable*, m is the *slope*, and b is the *intercept* with the y axis.

To illustrate the characteristics of a linear equation, the function $y = 3x + 4$ is plotted in Figure A.l. For this equation $m = 3$ and $b = 4$. Note that the y intercept occurs when $x = 0$. In this case the y intercept is 4, as can be seen from the equation ($b = 4$).

The slope of a straight line is defined as the ratio of the rate of change in y to that in x:

$$m = \text{slope} = \frac{\Delta y}{\Delta x}$$

For the equation $y = 3x + 4$, y changes three times as fast as x (because x has a coefficient of 3). Thus the slope in this case is 3. This can be verified from the graph. For the triangle shown in Figure A.1,

$$\Delta y = 34 - 10 = 24 \qquad \text{and} \qquad \Delta x = 10 - 2 = 8$$

Thus

$$\text{Slope} = \frac{\Delta y}{\Delta x} = \frac{24}{8} = 3$$

The above example illustrates a general method for obtaining the slope of a line from the graph of that line. Simply draw a triangle with one side parallel to the y axis and the other side parallel to the x axis, as shown in Figure A.l. Then determine the lengths of the sides to get Δy and Δx, respectively, and compute the ratio $\Delta y/\Delta x$.

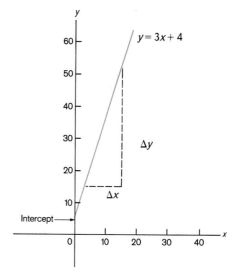

Figure A.1
Graph of the linear equation $y = 3x + 4$.

SI Units and Conversion Factors*

Length

SI unit: meter (m)

1 meter   = 1.0936 yards
1 centimeter = 0.39370 inch
1 inch   = 2.54 centimeters
             (exactly)
1 kilometer = 0.62137 mile
1 mile   = 5280 feet
             = 1.6093 kilometers

Mass

SI unit: kilogram (kg)

1 kilogram   = 1000 grams
              = 2.2046 pounds
1 pound   = 453.59 grams
              = 0.45359 kilogram
              = 16 ounces
1 atomic
mass unit   = 1.66056×10^{-27} kilograms

Volume

SI unit: cubic meter (m^3)

1 liter   = 10^{-3} m^3
          = 1 dm^3
          = 1.0567 quarts
1 gallon   = 4 quarts
          = 8 pints
          = 3.7854 liters
1 quart   = 32 fluid ounces
          = 0.94633 liter

Pressure

SI unit: pascal (Pa)

1 atmosphere = 101.325 kilopascals
               = 760 torr (mm Hg)
               = 14.70 pounds per
                 square inch

Energy

SI unit: joule (J)

1 joule   = 0.23901 calorie
1 calorie   = 4.184 joules

*Note: These conversion factors are given with more significant figures than those typically used in the body of the text.

PHOTO CREDITS

Chapter 1 p. 1, Larry Lefever / Grant Heilman Photography; p. 2 (top), Stephen Frisch / Stock Boston; p. 2 (bottom), Dr. E. R. Degginger; p. 3, Larry Lefever / Grant Heilman Photography; p. 4, Ken O'Donoghue; p. 9, Robert Harding Associate; p. 11, Claude Charlier/ Photo Researchers, Inc. **Chapter 2** p. 15, Runk / Schoenberger / Grant Heilman Photography; p. 16 (top), NASA; p. 16 (bottom), Ray Simons / Photo Researchers, Inc.; p. 20, Dr. E. R. Degginger; p. 23, Diane Schiumo / Fundamental Photographs; p. 24, Bob Daemmrich / The Image Works; p. 46, Dr. E. R. Degginger; p. 50, Dan McCoy / Rainbow; p. 51, Thomas Pantages. **Chapter 3** p. 63, Pedrick / The Image Works; p. 64 (top), Paul Silverman / Fundamental Photographs; p. 64 (middle), Brian Parker / Tom Stack & Associates; p. 64 (bottom), Runk / Schoenberger / Grant Heilman Photography, p. 65 (top), Gary Milburn / Tom Stack & Associates; p. 65 (bottom), Richard Megna/Fundamental Photographs; p. 68, Jim Pickerell / TSW / Click Chicago; p. 70, Grant Heilman / Grant Heilman Photography; p. 71 (top), Tim McCabe / Taurus Photos, Inc.; p. 71 (bottom), John Deeks / Photo Researchers, Inc.; p. 73 (both), Richard Megna / Fundamental Photographs; p. 75, Jim Richardson / West Light; p. 82, Jack Fields / Photo Researchers, Inc. **Chapter 4** p. 91, Dan McCoy / Rainbow; p. 92, The Granger Collection; p. 93, Mark Heifner / Science Stock America; p. 95, Walter Urie / West Light; p. 96, The Granger Collection; p. 100, Dr. E. R. Degginger; p. 101, The Bettmann Archive; p. 107, Dr. E. R. Degginger; p. 109, Tom Flynn / Taurus Photos, Inc.; p. 110, Beech Aircraft Corporation / Raytheon Company; p. 111, Dr. E. R. Degginger. **Chapter 5** p. 119, Victor Englebert; p. 120, Paul Silverman / Fundamental Photographs; p. 122 (top), Dr. E. R. Degginger; p. 122 (bottom), Sygma; p. 122 (middle a,b,c), Dr. E. R. Degginger; p. 123 (top), Dr. E. R. Degginger; p. 123, (bottom), Paul Silverman / Fundamental Photographs; p. 124, Robert Landau / West Light; p. 130, Dr. E. R. Degginger; p. 134, Museo Nationale Naples / Art Resource; p. 139, Thomas Pantages; p. 145, Dr. E. R. Degginger. **Chapter 6** p. 169, Richard Megna / Fundamental Photographs; p. 170 (left), Dr. E. R. Degginger; p. 170 (middle), Stephen Derr / The Image Bank; p. 170 (right), Dr. E. R. Degginger; p. 171 (left), Adam J. Stoltman / Duomo; p. 171 (center), Ken O'Donoghue; p. 171 (right), Runk / Schoenberger / Grant Heilman Photography; p. 172 (a), Richard Megna / Fundamental Photographs; p. 172 (b), Dr. E. R. Degginger; p. 172 (c), Dr. E. R. Degginger; p. 172 (d), Fundamental Photographs; p. 175 (a), Dr. E. R. Degginger; p. 175 (b) Dr. E. R. Degginger; p. 175 (c), Dr. E. R. Degginger; p. 175 (bottom), The Image Works; p. 183, Richard Megna / Fundamental Photographs. **Chapter 7** p. 193, Paul Silverman / Fundamental Photographs; p. 194, Dr. E. R. Degginger; p. 202, Ken O'Donoghue; p. 205, Courtesy, Dr. Featherstone; p. 208, Barry L. Runk / Grant Heilman Photography; p. 210, Dr. E. R. Degginger. **Chapter 8** p. 221, Richard Megna / Fundamental Pho-

tographs; p. 223, Bob Krueger / Photo Researchers, Inc.; p. 224, Dr. E. R. Degginger; p. 225, L. S. Stepanowicz / Science Stock America; p. 229, Courtesy, Morton Thiokol, Inc.; p. 233, Gary Gladstone / The Image Bank. **Chapter 9** p. 241, Richard Megna / Fundamental Photographs; p. 242 (top), Dr. E. R. Degginger; p. 242 (bottom), Dr. E. R. Degginger; p. 249 (all), Ken O'Donoghue; p. 250 (top), Thomas Pantages; p. 250 (bottom), Ken O'Donoghue; p. 253, David Parker / Science Photo Library / Photo Researchers, Inc.; p. 257, James Carmichael / Bruce Coleman, Inc.; p. 258, Dr. E. R. Degginger; p. 260, John Elk III / After Image, Inc. **Chapter 10** p. 291, Matt Meadows / Peter Arnold, Inc.; p. 292, Pedrick / The Image Works; p. 294, Tom Stack / Tom Stack & Associates; p. 303, NASA; p. 306, George Olson / The Photo File; p. 308, Grant Heilman / Grant Heilman Photography; p. 314, Ken O'Donoghue. **Chapter 11** p. 337, Michael Dalton / Fundamental Photographs; p. 338 (top), Richard Pasley / Stock Boston; p. 338 (bottom), Alfred Paseika / Taurus Photos, Inc.; p. 339 (top), Rod Allin / Tom Stack & Associates; p. 339 (bottom), S. Casenave / Vandystadt / Photo Researchers, Inc.; p. 340, Goddard Space Flight Center / NASA; p. 345, The Granger Collection; p. 346, The Granger Collection; p. 359 (top), Farrell Grehan / Photo Researchers, Inc.; p. 359 (bottom), Dan McCoy / Rainbow; p. 361, Tom Tracy / The Photo File; p. 365, Jim Zuckerman / West Light. **Chapter 12** p. 375, J&L Weber / Peter Arnold, Inc.; p. 376, Thomas Pantages; p. 388, The Bancroft Library; p. 397 (both), Donald Clegg; p. 398 (both), Digital Art / West Light. **Chapter 13** p. 419, Galen Rowell / Peter Arnold, Inc.; p. 420 (top), Elisa Leonelli / After Image, Inc.; p. 420 (bottom), Larry Lee / West Light; p. 421 (top a,b), Ken O'Donoghue. p. 421 (bottom), The Granger Collection; p. 423, Ken O'Donoghue; p. 428, Brent Bear / West Light; p. 429, Duomo; p. 445, Tom Stack / Tom Stack & Associates. **Chapter 14** p. 471, Richard Megna / Fundamental Photographs; p. 472 (top), Dian Duchin / Bruce Coleman, Inc.; p. 472 (middle), David Madison / Duomo; p. 472 (bottom), Jean-Marc Loubat / Vandystadt / Photo Researchers, Inc.; p. 474, Walt Anderson / Tom Stack & Associates; p. 476, Flip Nicklin; p. 479, Galen Rowell / After Image, Inc.; p. 482, Bartruff / The Photo File; p. 485 (left), Runk / Schoenberger / Grant Heilman Photography; p. 485 (middle), Dr. E. R. Degginger; p. 485 (right), M. Claye / Jacana / Photo Researchers, Inc.; p. 488 (both), Dr. E. R. Degginger. **Chapter 15** p. 497, Paul Silverman / Fundamental Photographs; p. 499, Michelle Barnes / The Photo File; p. 505, Richard Megna / Fundamental Photographs; p. 507, Thomas Pantages; p. 513, Thomas Pantages; p. 516, Ken O'Donoghue. **Chapter 16** p. 539, Courtesy, Lawrence Livermore National Laboratory; p. 540 (top left), Dennis Barnes / Grant Heilman Photography; p. 540 (bottom a,b,c), Ken O'Donoghue; p. 543, Goddard Space Flight Center / NASA; p. 553 (both), Richard Megna / Fundamental Photographs; p. 556, Paul Sil-

GLOSSARY

Acid a substance that produces hydrogen ions in solution; a proton donor.

Acid–base indicator a substance that marks the end point of an acid–base titration by changing color.

Acid rain rainwater with an acidic pH, a result of air pollution by sulfur dioxide.

Acidic oxide a covalent oxide that dissolves in water to give an acidic solution.

Actinide series a group of fourteen elements following actinium in the periodic table, in which the $5f$ orbitals are being filled.

Activation energy the threshold energy that must be overcome to produce a chemical reaction.

Air pollution contamination of the atmosphere, mainly by the gaseous products of transportation and the production of electricity.

Alcohol an organic compound in which the hydroxyl group is a substituent on a hydrocarbon.

Aldehyde an organic compound containing the carbonyl group bonded to at least one hydrogen atom.

Alkali metal a Group 1 metal.

Alkaline earth metal a Group 2 metal.

Alkane a saturated hydrocarbon with the general formula C_nH_{2n+2}.

Alkene an unsaturated hydrocarbon containing a carbon–carbon double bond. The general formula is C_nH_{2n}.

Alkyne an unsaturated hydrocarbon containing a carbon–carbon triple bond. The general formula is C_nH_{2n-2}.

Alloy a substance that contains a mixture of elements and has metallic properties.

Alloy steel a form of steel containing carbon plus metals such as chromium, cobalt, manganese, and molybdenum.

Alpha (α) particle a helium nucleus.

Alpha-particle production a common mode of decay for radioactive nuclides in which the mass number changes.

Amine an organic base derived from ammonia in which one or more of the hydrogen atoms are replaced by organic groups.

α-Amino acid an organic acid in which an amino group and an R group are attached to the carbon atom next to the carboxyl group.

Ampere the unit of measurement for electric current; 1 ampere is equal to 1 coulomb of charge per second.

Amphoteric substance a substance that can behave either as an acid or as a base.

Anion a negative ion.

Anode in a galvanic cell, the electrode at which oxidation occurs.

Aqueous solution a solution in which water is the dissolving medium, or solvent.

Aromatic hydrocarbon one of a special class of cyclic unsaturated hydrocarbons, the simplest of which is benzene.

Arrhenius concept a concept postulating that acids produce hydrogen ions in aqueous solution, whereas bases produce hydroxide ions.

Atmosphere the mixture of gases that surrounds the earth's surface.

Atom the fundamental unit of which elements are composed.

Atomic number the number of protons in the nucleus of an atom; each element has a unique atomic number.

Atomic radius half the distance between the atomic nuclei in a molecule consisting of identical atoms.

Atomic solid a solid that contains atoms at the lattice points.

Atomic weight the weighted average mass of the atoms in a naturally occurring element.

Aufbau principle a principle stating that as protons are added one by one to the nucleus to build up the elements, electrons are similarly added to hydrogen-like orbitals.

Autoionization the transfer of a proton from one molecule to another of the same substance.

Avogadro's law equal volumes of gases at the same temperature and pressure contain the same number of particles (atoms or molecules).

Avogadro's number the number of atoms in exactly 12 grams of pure ^{12}C, equal to 6.022×10^{23}.

Azimuthal quantum number (l) the quantum number relating to the shape of an atomic orbital; it can assume any integral value from 0 to $n - 1$ for each value of n.

Ball-and-stick model a molecular model that distorts the sizes of atoms but shows bond relationships clearly.

Barometer a device for measuring atmospheric pressure.

Base a substance that produces hydroxide ions in aqueous solution; a proton acceptor.

Basic oxide an ionic oxide that dissolves in water to produce a basic solution.

Battery a group of galvanic cells connected in series.

Beta (β) particle   an electron produced in radioactive decay.

Beta-particle production   a decay process for radioactive nuclides in which the mass number remains constant and the atomic number increases. The net effect is to change a neutron to a proton.

Binary compound   a two-element compound.

Binding energy (nuclear)   the energy required to decompose a nucleus into its component nucleons.

Biochemistry   the study of the chemistry of living systems.

Biomolecule   a molecule that functions in maintaining and/or reproducing life.

Bond (chemical bond)   the force that holds two atoms together in a compound.

Bond energy   the energy required to break a given chemical bond.

Bond length   the distance between the nuclei of the two atoms that are connected by a bond.

Bonding pair   an electron pair found in the space between two atoms.

Boyle's law   the volume of a given sample of gas at constant temperature varies inversely with the pressure.

Breeder reactor   a nuclear reactor in which fissionable fuel is produced while the reactor runs.

Brönsted–Lowry model   a model proposing that an acid is a proton donor and that a base is a proton acceptor.

Buffer capacity   the ability of a buffered solution to absorb protons or hydroxide ions without a significant change in pH.

Buffered solution   a solution that resists a change in its pH when either hydroxide ions or protons are added.

Calorie   a unit of measurement for energy; 1 calorie is the quantity of energy required to heat 1 gram of water by one Celsius degree.

Calorimetry   the science of measuring heat flow.

Carbohydrate   a polyhydroxyl ketone or polyhydroxyl aldehyde or a polymer composed of these.

Carbon steel   an alloy of iron containing up to about 1.5% carbon.

Carboxyl group   the —COOH group in an organic acid.

Carboxylic acid   an organic compound containing the carboxyl group.

Catalyst   a substance that speeds up a reaction without being consumed.

Cathode   in a galvanic cell, the electrode at which reduction occurs.

Cathode rays   the ''rays'' emanating from the negative electrode (cathode) in a partially evacuated tube; a stream of electrons.

Cathodic protection   the connection of an active metal, such as magnesium, to steel in order to protect the steel from corrosion.

Cation   a positive ion.

Cell potential (electromotive force)   the driving force in a galvanic cell that pulls electrons from the reducing agent in one compartment to the oxidizing agent in the other.

Chain reaction (nuclear)   a self-sustaining fission process caused by the production of neutrons that proceed to split other nuclei.

Charles's law   the volume of a given sample of gas at constant pressure is directly proportional to the temperature in kelvins.

Chemical change   the change of substances into other substances through a reorganization of the atoms; a chemical reaction.

Chemical equation   a representation of a chemical reaction showing the relative numbers of reactant and product molecules.

Chemical equilibrium   a dynamic reaction system in which the concentrations of all reactants and products remain constant as a function of time.

Chemical formula   a representation of a molecule in which the symbols for the elements are used to indicate the types of atoms present and subscripts are used to show the relative numbers of atoms.

Chemical kinetics   the area of chemistry that concerns reaction rates.

Chemical property   the ability of a substance to change to a different substance.

Chemical stoichiometry   the quantities of materials consumed and produced in a chemical reaction.

Collision model   a model based on the idea that molecules must collide in order to react; used to account for the observed characteristics of reaction rates.

Combustion reaction   the vigorous and exothermic oxidation–reduction reaction that takes place between certain substances (particularly organic compounds) and oxygen.

Complete ionic equation   an equation that shows as ions all substances that are strong electrolytes.

Compound   a substance with constant composition that can be broken down into elements by chemical processes.

Condensation   the process by which vapor molecules re-form a liquid.

Condensed states of matter   liquids and solids.

Conjugate acid   the species formed when a proton is added to a base.

Conjugate acid–base pair   two species related to each other by the donating and accepting of a single proton.

Conjugate base   what remains of an acid molecule after a proton is lost.

Continuous spectrum   a spectrum that exhibits all the wavelengths of visible light.

Control rods   in a nuclear reactor, rods composed of substances that absorb neutrons. These rods regulate the power level of the reactor.

Core electron   an inner electron in an atom; one that is not in the outermost (valence) principal quantum level.

Corrosion   the process by which metals are oxidized in the atmosphere.

Covalent bonding   a type of bonding in which atoms share electrons.

Critical mass   the mass of fissionable material required to produce a self-sustaining chain reaction.

Critical reaction (nuclear)   a reaction in which exactly one neutron from each fission event causes another fission event, thus sustaining the chain reaction.

Crystalline solid   a solid characterized by the regular arrangement of its components.

Dalton's law of partial pressures   for a mixture of gases in a container, the total pressure exerted is the sum of the pressures that each gas would exert if it were alone.

Denaturation   the breaking down of the three-dimensional structure of a protein, resulting in the loss of its function.

Density   a property of matter representing the mass per unit volume.

Deoxyribonucleic acid (DNA)   a huge nucleotide polymer having a double-helical structure with complementary bases on the two strands. Its major functions are protein synthesis and the storage and transport of genetic information.

Diatomic molecule   a molecule composed of two atoms.

Dilution   the process of adding solvent to lower the concentration of solute in a solution.

Dipole–dipole attraction   the attractive force resulting when polar molecules line up such that the positive and negative ends are close to each other.

Dipole moment   a property of a molecule where the charge distribution can be represented by a center of positive charge and a center of negative charge.

Disaccharide   a sugar formed from two monosaccharides joined by a glycoside linkage.

Distillation   a method for separating the components of a liquid mixture that depends on differences in the ease of vaporization of the components.

Double bond   a bond in which two atoms share two pairs of electrons.

Dry cell battery   a common battery used in calculators, watches, radios, and tape players.

Electrical conductivity   the ability to conduct an electric current.

Electrochemistry   the study of the interchange of chemical and electrical energy.

Electrolysis   a process that involves forcing a current through a cell to cause a nonspontaneous chemical reaction to occur.

Electrolyte   a material that dissolves in water to give a solution that conducts an electric current.

Electrolytic cell   a cell that uses electrical energy to produce a chemical change that would not otherwise occur.

Electromagnetic radiation   radiant energy that exhibits wave-like behavior and travels through space at the speed of light in a vacuum.

Electron   a negatively charged particle that occupies the space around the nucleus of an atom.

Electronegativity   the tendency of an atom in a molecule to attract shared electrons to itself.

Element   a substance that cannot be decomposed into simpler substances by chemical or physical means. It consists of atoms that all have the same atomic number.

Empirical formula   the simplest whole-number ratio of atoms in a compound.

End point   the point in a titration at which the indicator changes color.

Endothermic   refers to a reaction where energy (as heat) flows into the system.

Energy   the capacity to do work or to cause the flow of heat.

Enthalpy   at constant pressure, the change in enthalpy equals the energy flow as heat.

Enzyme   a large molecule, usually a protein, that catalyzes biological reactions.

Equilibrium constant   the value obtained when equilibrium concentrations of the chemical species are substituted into the equilibrium expression.

Equilibrium expression   the expression (from the law of mass action) is equal to the product of the product concentrations divided by the product of the reactant concentrations, each concentration having first been raised to a power represented by the coefficient in the balanced equation.

Equilibrium position   a particular set of equilibrium concentrations.

Equivalence point (stoichiometric point)   the point in a titration when enough titrant has been added to react exactly with the substance in solution that is being titrated.

Essential elements   the elements known to be essential to human life.

Ester   an organic compound produced by the reaction between a carboxylic acid and an alcohol.

Exothermic   refers to a reaction where energy (as heat) flows out of the system.

Exponential notation   expresses a number in the form $N \times 10^M$; a convenient method for representing a very large or very small number and for easily indicating the number of significant figures.

Fat (glyceride)   an ester composed of glycerol and fatty acids.

Fatty acid   a long-chain carboxylic acid.

Filtration   a method for separating the components of a mixture containing a solid and a liquid.

Fission   the process of using a neutron to split a heavy nucleus into two nuclei with smaller mass numbers.

Fossil fuel   a fuel that consists of carbon-based molecules derived from decomposition of once-living organisms; coal, petroleum, or natural gas.

Frequency   the number of waves (cycles) per second that pass a given point in space.

Fuel cell   a galvanic cell for which the reactants are continuously supplied.

Functional group   an atom or group of atoms in hydrocarbon derivatives that contains elements in addition to carbon and hydrogen.

Fusion   the process of combining two light nuclei to form a heavier, more stable nucleus.

Galvanic cell   a device in which chemical energy from a spontaneous redox reaction is changed to electrical energy that can be used to do work.

Galvanizing   a process in which steel is coated with zinc to prevent corrosion.

Gamma (γ) ray   a high-energy photon.

Gas   one of the three states of matter; has neither fixed shape nor fixed volume.

Geiger–Müller counter (Geiger counter)   an instrument that measures the rate of radioactive decay by registering the ions and electrons produced as a radioactive particle passes through a gas-filled chamber.

Gene   a given segment of the DNA molecule that contains the code for a specific protein.

Greenhouse effect   a warming effect exerted by certain molecules in the earth's atmosphere (particularly carbon dioxide and water).

Ground state   the lowest possible energy state of an atom or molecule.

Group (of the periodic table)   a vertical column of elements having the same valance electron configuration and similar chemical properties.

Haber process   the manufacture of ammonia from nitrogen and hydrogen, carried out at high pressure and high temperature with the aid of a catalyst.

Half-life (of a radioactive sample)   the time required for the number of nuclides in a radioactive sample to reach half the original number of nuclides.

Half-reactions   the two parts of an oxidation–reduction reaction, one representing oxidation, the other reduction.

Halogen   a Group 7 element.

Hard water   water from natural sources that contains relatively large concentrations of calcium and magnesium ions.

Heat   energy transferred between two objects because of a temperature difference between them.

Heating curve   a plot of temperature versus time for a substance, where energy is added at a constant rate.

Heisenberg uncertainty principle   a principle stating that there is a fundamental limitation to how precisely we can know both the position and the momentum of a particle at a given time.

Herbicide   a pesticide applied to kill weeds.

Heterogeneous equilibrium   an equilibrium involving reactants and/or products in more than one state.

Heterogeneous mixture   a mixture that has different properties in different regions of the mixture.

Homogeneous equilibrium   an equilibrium system where all reactants and products are in the same state.

Homogeneous mixture   a mixture that is the same throughout; a solution.

Hydration   the interaction between solute particles and water molecules.

Hydride   a binary compound containing hydrogen.

Hydrocarbon   a compound composed of carbon and hydrogen.

Hydrocarbon derivative   an organic molecule that contains one or more elements in addition to carbon and hydrogen.

Hydrogen bonding   unusually strong dipole–dipole attractions that occur among molecules in which hydrogen is bonded to a highly electronegative atom.

Hydrohalic acid   an aqueous solution of a hydrogen halide.

Hydronium ion   the H_3O^+ ion; a hydrated proton.

Hypothesis   one or more assumptions put forth to explain observed phenomena.

Ideal gas   a hypothetical gas that exactly obeys the ideal gas law. A real gas approaches ideal behavior at high temperature and/or low pressure.

Ideal gas law   an equation relating the properties of an ideal gas, expressed as $PV = nRT$, where P = pressure, V = volume, n = moles of the gas, R = the universal gas constant, and T = temperature on the Kelvin scale. This equation expresses behavior closely approached by real gases at high temperature and/or low pressure.

Indicator   a chemical that changes color and is used to mark the end point of a titration.

Intermolecular forces   relatively weak interactions that occur between molecules.

Intramolecular forces   interactions that occur within a given molecule.

Ion   an atom or a group of atoms that has a net positive or negative charge.

Ionic bonding   the attraction between oppositely charged ions.

Ionic compound   a compound that results when a metal reacts with a nonmetal to form cations and anions.

Ionic solid   a solid containing cations and anions that dissolves in water to give a solution containing the separated ions, which are mobile and thus free to conduct an electric current.

Ionization energy   the quantity of energy required to remove an electron from a gaseous atom or ion.

Ion-product constant (K_w)   the equilibrium constant for the autoionization of water; $K_w = [H^+][OH^-]$. At 25 °C, K_w equals 1.0×10^{-14}.

Isomers   species that have the same chemical formula but different properties.

Isotopes   atoms of the same element (the same number of protons) but different numbers of neutrons. They have identical atomic numbers but different mass numbers.

Joule   a unit of measurement for energy; 1 calorie = 4.184 joules.

Ketone   an organic compound containing the carbonyl group bonded to two carbon atoms.

Kinetic energy $\left(\frac{1}{2}mv^2\right)$ energy due to the motion of an object; dependent on the mass of the object and the square of its velocity.

Kinetic molecular theory   a model that assumes that an ideal gas is composed of tiny particles (molecules) in constant motion.

Lanthanide series   a group of fourteen elements following lanthanum in the periodic table, in which the $4f$ orbitals are being filled.

Lattice   a three-dimensional system of points designating the positions of the centers of the components of a solid (atoms, ions, or molecules).

Law of chemical equilibrium   a general description of the equilibrium condition; it defines the equilibrium expression.

Law of conservation of energy   energy can be converted from one form to another but can be neither created nor destroyed.

Law of conservation of mass   mass is neither created nor destroyed.

Law of constant composition   a given compound always contains elements in exactly the same proportion by mass.

Law of mass action   (also called the law of chemical equilibrium) a general description of the equilibrium condition; it defines the equilibrium expression.

Law of multiple proportions   a law stating that when two elements form a series of compounds, the ratios of the masses of the second element that combine with one gram of the first element can always be reduced to small whole numbers.

Lead storage battery   a battery (used in cars) in which the anode is lead, the cathode is lead coated with lead dioxide, and the electrolyte is a sulfuric acid solution.

Le Châtelier's principle   if a change is imposed on a system at equilibrium, the position of the equilibrium will shift in a direction that tends to reduce the effect of that change.

Lewis structure   a diagram of a molecule showing how the valence electrons are arranged among the atoms in the molecule.

Limiting reactant (limiting reagent)   the reactant that is completely consumed when a reaction is run to completion.

Line spectrum   a spectrum showing only certain discrete wavelengths.

Linear accelerator   a type of particle accelerator in which a changing electrical field is used to accelerate a positive ion along a linear path.

Lipids water-insoluble substances that can be extracted from cells by nonpolar organic solvents.

Liquid one of the three states of matter; has a fixed volume but takes the shape of its container.

London dispersion forces the relatively weak forces, which exist among noble gas atoms and nonpolar molecules, that involve an accidental dipole that induces a momentary dipole in a neighbor.

Lone pair an electron pair that is localized on a given atom; an electron pair not involved in bonding.

Magnetic quantum number m_l, the quantum number reflecting the orientation of an orbital in space relative to the other orbitals with the same l quantum number. It can have integral values between l and $-l$, including zero.

Main-group (representative) elements elements in the groups labeled 1, 2, 3, 4, 5, 6, 7, and 8 in the periodic table. The group number gives the sum of the valence s and p electrons.

Mass the quantity of matter in an object.

Mass number the total number of protons and neutrons in the atomic nucleus of an atom.

Mass percent the percent by mass of a component of a mixture or of a given element in a compound.

Matter the material of the universe.

Metal an element that gives up electrons relatively easily and is typically lustrous, malleable, and a good conductor of heat and electricity.

Metalloid an element that has both metallic and nonmetallic properties.

Metallurgy the process of separating a metal from its ore and preparing it for use.

Millimeters of mercury (mm Hg) a unit of measurement for pressure, also called a torr; 760 mm Hg = 760 torr = 101,325 Pa = 1 standard atmosphere.

Mixture a material of variable composition that contains two or more substances.

Model (theory) a set of assumptions put forth to explain the observed behavior of matter. The models of chemistry usually involve assumptions about the behavior of individual atoms or molecules.

Moderator a substance used in a nuclear reactor to slow down the neutrons.

Molar heat of fusion the energy required to melt 1 mol of a solid.

Molar heat of vaporization the energy required to vaporize 1 mol of a liquid.

Molar mass the mass in grams of one mole of a compound.

Molar volume the volume of one mole of an ideal gas; equal to 22.42 liters at standard temperature and pressure.

Molarity moles of solute per volume of solution in liters.

Mole (mol) the number equal to the number of carbon atoms in exactly 12 grams of pure ^{12}C: Avogadro's number. One mole represents 6.022×10^{23} units.

Mole ratio (stoichiometry) the ratio of moles of one substance to moles of another substance in a balanced chemical equation.

Molecular equation an equation representing a reaction in solution and showing the reactants and products in undissociated form, whether they are strong or weak electrolytes.

Molecular formula the exact formula of a molecule, giving the types of atoms and the number of each type.

Molecular solid a solid composed of neutral molecules.

Molecular structure the three-dimensional arrangement of atoms in a molecule.

Molecular weight (molar mass) the mass in grams of one mole of a substance.

Molecule a bonded collection of two or more atoms of the same element or different elements.

Monoprotic acid an acid with one acidic proton.

Natural law a statement that expresses generally observed behavior.

Net ionic equation an equation for a reaction in solution, representing strong electrolytes as ions and showing only those components that are directly involved in the chemical change.

Network solid an atomic solid containing strong directional covalent bonds.

Neutralization reaction an acid–base reaction.

Neutron a particle in the atomic nucleus with a mass approximately equal to that of the proton but with no charge.

Noble gas a Group 8 element.

Nonelectrolyte a substance that, when dissolved in water, gives a nonconducting solution.

Nonmetal an element that does not exhibit metallic characteristics. Chemically, a typical nonmetal accepts electrons from a metal.

Normal boiling point the temperature at which the vapor pressure of a liquid is exactly one atmosphere; the boiling temperature under one atmosphere of pressure.

Normal melting/freezing point the melting/freezing point of a solid at a total pressure of one atmosphere.

Normality the number of equivalents of a substance dissolved in a liter of solution.

Nuclear atom the modern concept of the atom as having a dense center of positive charge (the nucleus) and electrons moving around the outside.

Nuclear transformation the change of one element into another.

Nucleon a particle in an atomic nucleus, either a neutron or a proton.

Nucleus the small, dense center of positive charge in an atom.

Nuclide the general term applied to each unique atom; represented by $_Z^A X$, where X is the symbol for a particular element.

Octet rule the observation that atoms of nonmetals form the most stable molecules when they are surrounded by eight electrons (to fill their valence orbitals).

Orbital a representation of the space occupied by an electron in an atom; the probability distribution for the electron.

Organic acid an acid with a carbon-atom backbone and a carboxyl group.

Organic chemistry the study of carbon-containing compounds (typically containing chains of carbon atoms) and their properties.

Oxidation an increase in oxidation state (a loss of electrons).

Oxidation–reduction (redox) reaction a reaction in which one or more electrons are transferred.

Oxidation states a concept that provides a way to keep track of electrons in oxidation–reduction reactions according to certain rules.

Oxidizing agent (electron acceptor) a reactant that accepts electrons from another reactant.

Oxyacid an acid in which the acidic proton is attached to an oxygen atom.

Ozone O_3, a form of elemental oxygen much less common than O_2 in the atmosphere near the earth.

Partial pressures the independent pressures exerted by different gases in a mixture.

Particle accelerator a device used to accelerate nuclear particles to very high speeds.

Pascal the SI unit of measurement for pressure; equal to newtons per square meter.

Percent yield the actual yield of a product as a percentage of the theoretical yield.

Periodic table a chart showing all the elements arranged in columns in such a way that all the elements in a given column exhibit similar chemical properties.

pH scale a log scale based on 10 and equal to $-\log[H^+]$; a convenient way to represent solution acidity.

Phenyl group the benzene molecule minus one hydrogen atom.

Photochemical smog air pollution produced by the action of light on oxygen, nitrogen oxides, and unburned fuel from auto exhaust to form ozone and other pollutants.

Photon a ''particle'' of electromagnetic radiation.

Physical change a change in the form of a substance, but not in its chemical nature; chemical bonds are not broken in a physical change.

Physical property a characteristic of a substance that can change without the substance becoming a different substance.

Polar covalent bond a covalent bond in which the electrons are not shared equally because one atom attracts them more strongly than the other.

Polar molecule a molecule that has a permanent dipole moment.

Polyatomic ion an ion containing a number of atoms.

Polyelectronic atom an atom with more than one electron.

Polymer a large, usually chain-like molecule built from many small molecules (monomers).

Polymerization a process in which many small molecules (monomers) are joined together to form a large molecule.

Polyprotic acid an acid with more than one acidic proton. It dissociates in a stepwise manner, one proton at a time.

Porous disk a disk in a tube connecting two different solutions in a galvanic cell; it allows ion flow without extensive mixing of the solutions.

Positron production a mode of nuclear decay in which a particle is formed that has the same mass as an electron but opposite charge. The net effect is to change a proton to a neutron.

Potential energy energy due to position or composition.

Precipitation reaction a reaction in which an insoluble substance forms and separates from the solution as a solid.

Precision the degree of agreement among several measurements of the same quantity; the reproducibility of a measurement.

Primary structure (of a protein) the order (sequence) of amino acids in the protein chain.

Principal quantum number the quantum number reflecting the size and energy of an orbital; it can have any positive integer value.

Probability distribution (orbital)   a representation indicating the probabilities of finding an electron at various points in space.

Product   a substance resulting from a chemical reaction. It is shown to the right of the arrow in a chemical equation.

Protein   a natural polymer formed by condensation reactions between amino acids.

Proton   a positively charged particle in an atomic nucleus.

Pure substance   a substance with constant composition.

Radioactive decay (radioactivity)   the spontaneous decomposition of a nucleus to form a different nucleus.

Radiocarbon dating (carbon-14 dating)   a method for dating ancient wood or cloth on the basis of the radioactive decay of the nuclide $^{14}_{6}C$.

Radiotracer   a radioactive nuclide, introduced into an organism for diagnostic purposes, whose pathway can be traced by monitoring its radioactivity.

Random error   an error that has an equal probability of being high or low.

Rate of decay   the change per unit time in the number of radioactive nuclides in a sample.

Reactant   a starting substance in a chemical reaction. It appears to the left of the arrow in a chemical equation.

Reactor core   the part of a nuclear reactor where the fission reaction takes place.

Reducing agent (electron donor)   a reactant that donates electrons to another substance, reducing the oxidation state of one of its atoms.

Reduction   a decrease in oxidation state (a gain of electrons).

Rem   a unit of radiation dosage that accounts for both the energy of the dose and its effectiveness in causing biological damage (from *r*oentgen *e*quivalent for *m*an).

Resonance   a condition occurring when more than one valid Lewis structure can be written for a particular molecule. The actual electronic structure is represented not by any one of the Lewis structures but by the average of all of them.

Salt   an ionic compound.

Salt bridge   a U-tube containing an electrolyte that connects the two compartments of a galvanic cell, allowing ion flow without extensive mixing of the different solutions.

Saturated solution   a solution that contains as much solute as can be dissolved in that solution.

Scientific method   a process of studying natural phenomena that involves making observations, forming laws and theories, and testing theories by experimentation.

Scientific notation   see *Exponential notation.*

Scintillation counter   an instrument that measures radioactive decay by sensing the flashes of light that the radiation produces in a substance.

Secondary structure (of a protein)   the three-dimensional structure of the protein chain (for example, α-helix, random coil, or pleated sheet).

SI units   International System of units based on the metric system and on units derived from the metric system.

Sigma (σ) bond   a covalent bond in which the electron pair is shared in an area centered on a line running between the atoms.

Significant figures   the certain digits and the first uncertain digit of a measurement.

Silica   the fundamental silicon–oxygen compound, which has the empirical formula SiO_2 and forms the basis of quartz and certain types of sand.

Silicates   salts that contain metal cations and polyatomic silicon–oxygen anions that are usually polymeric.

Single bond   a bond in which two atoms share one pair of electrons.

Solid   one of the three states of matter; has a fixed shape and volume.

Solubility   the amount of a substance that dissolves in a given volume of solvent or solution at a given temperature.

Solubility product   the constant for the equilibrium expression representing the dissolving of an ionic solid in water.

Solute   a substance dissolved in a solvent to form a solution.

Solution   a homogeneous mixture.

Solvent   the dissolving medium in a solution.

Somatic damage   radioactive damage to an organism resulting in its sickness or death.

Specific heat   another name for specific heat capacity.

Specific heat capacity   the amount of energy required to raise the temperature of one gram of a substance by one Celsius degree.

Spectator ions   ions present in solution that do not participate directly in a reaction.

Standard atmosphere   a unit of measurement for pressure equal to 760 mm Hg.

Standard solution   a solution the concentration of which is accurately known.

Standard temperature and pressure (STP)   the condition 0 °C and 1 atmosphere of pressure.

States of matter   the three different forms in which matter can exist: solid, liquid, and gas.

Stoichiometric quantities   quantities of reactants mixed in exactly the amounts that result in their all being used up at the same time.

Stoichiometry of a reaction   the relative quantities of reactants and products involved in the reaction.

Strong acid   an acid that completely dissociates to produce an H^+ ion and the conjugate base.

Strong base   a metal hydroxide salt that completely dissociates into its ions in water.

Strong electrolyte   a material that, when dissolved in water, gives a solution that conducts an electric current very efficiently.

Structural formula   the representation of a molecule in which the relative positions of the atoms are shown and the bonds are indicated by lines.

Subcritical reaction (nuclear)   a reaction in which fewer than one of the neutrons from each fission event causes another fission event and the process dies out.

Sublimation   the process by which a substance goes directly from the solid state to the gaseous state without passing through the liquid state.

Substitution reaction (hydrocarbons)   a reaction in which an atom, usually a halogen, replaces a hydrogen atom in a hydrocarbon.

Supercooling   the process of cooling a liquid below its freezing point without its changing to a solid.

Supercritical reaction (nuclear)   a reaction in which more than one of the neutrons from each fission event causes another fission event. The process rapidly escalates to a violent explosion.

Superheating   the process of heating a liquid above its boiling point without its boiling.

Systematic error   an error that always occurs in the same direction.

Tertiary structure (of a protein)   the overall shape of a protein, long and narrow or globular, maintained by different types of intramolecular interactions.

Theoretical yield   the maximum amount of a given product that can be formed when the limiting reactant is completely consumed.

Theory (model)   a set of assumptions put forth to explain some aspect of the observed behavior of matter.

Titration   a technique in which one solution is used to analyze another.

Torr   another name for millimeter of mercury (mm Hg).

Trace elements   metals present only in trace amounts in the human body.

Transition metals   several series of elements in which inner orbitals (d or f orbitals) are being filled.

Transuranium elements   the elements beyond uranium that are made artificially by particle bombardment.

Triple bond   a bond in which two atoms share three pairs of electrons.

Uncertainty (in measurement)   the characteristic reflecting the fact that any measurement involves estimates and cannot be exactly reproduced.

Unit factor   an equivalance statement between units that is used for converting from one set of units to another.

Universal gas constant   the combined proportionality constant in the ideal gas law; 0.08206 L atm/K mol, or 8.314 J/K mol.

Unsaturated solution   a solution in which more solute can be dissolved than is dissolved already.

Valence electrons   the electrons in the outermost occupied principal quantum level of an atom.

Valence shell electron pair repulsion (VSEPR) model   a model the main postulate of which is that the structure around a given atom in a molecule is determined principally by the tendency to minimize electron-pair repulsions.

Vapor pressure   the pressure of the vapor over a liquid at equilibrium in a closed container.

Vaporization (evaporation)   the change in state that occurs when a liquid evaporates to form a gas.

Viscosity   the resistance of a liquid to flow.

Volt   the unit of measurement for electrical potential; it is defined as one joule of work per coulomb of charge transferred.

Wavelength   the distance between two consecutive peaks or troughs in a wave.

Weak acid   an acid that dissociates only to a slight extent in aqueous solution.

Weak base   a base that reacts with water to produce hydroxide ions to only a slight extent in aqueous solution.

Weak electrolyte   a material that, when dissolved in water, gives a solution that conducts only a small electric current.

Weight   the force exerted on an object by gravity.

CHAPTER 1

2. physician: understanding biochemical processes in the cell; pharmacist: understanding drug interactions; farmer: understanding use of fertilizers and pesticides 4, 6, 8. responses depend on student experience 10. (a) quantitative (b) qualitative (c) quantitative (d) qualitative (e) quantitative (f) qualitative (g) quantitative 12. A natural law is a summary of observed, measurable behavior that occurs repeatedly and consistently. A theory is our attempt to explain such behavior. 14. Most applications of chemistry involve the interpretation of observations and the solving of problems. Although memorization of some facts may aid in these endeavors, it is the ability to combine, relate, and synthesize information that is most important in the study of chemistry. 16. In real-life situations, the problems and applications likely to be encountered are not simple textbook examples. One must be able to observe an event, hypothesize a cause, and then test this hypothesis. One must be able to carry what has been learned in class forward to new, different situations.

CHAPTER 2

2. three 4. negative; positive 6. (a) -5 (b) 6 (c) -4 (d) 4 8. (a) 4.731×10^{-2} (b) 4.284×10^{3} (c) 4.201×10^{0} (d) 1.41×10^{-10} (e) 5.23×10^{1} (f) 4.909×10^{-2} (g) 5.4331×10^{7} (h) 9.81×10^{-1} 10. (a) 483 (b) 0.0007221 (c) 6.1 (d) 0.00000000911 (e) 4,221,000 (f) 0.00122 (g) 9,999 (h) 0.00001016 (i) 101,600 (j) 0.411 (k) 97,100 (l) 0.000971 12. (a) 1.312×10^{-1} (b) 1.472×10^{3} (c) 1.201×10^{-3} (d) 4.43×10^{5} (e) 7.21×10^{0} (f) 9.14×10^{-6} (g) 1.29×10^{1} (h) 1.901×10^{-11} 14. (a) 3.1×10^{3} (b) 1×10^{6} (c) 1 or 1×10^{0} (d) 1.8×10^{-5} (e) 1×10^{7} (f) 1×10^{6} (g) 1×10^{-7} (h) 1×10^{1} 16. grams 18. (a) mega (b) milli (c) nano (d) mega (e) centi (f) micro 20. 100 mi (1 km is less than 1 mi) 22. quart 24. kilogram 26. the man is slightly taller (5 ft. 9 in.) 28. d 30. d 32. 40 quarters 34. uncertainty 36. The scale of the ruler shown is only marked to the nearest tenth of a centimeter. Writing 2.850 would imply that the scale was marked to the nearest hundredth of a centimeter (and that the zero in the thousandth's place had been

estimated). 38. (a) infinite [a definition] (b) one (c) infinite [a definition] (d) one (e) infinite [a definition] 40. final 42. (a) 3.13×10^{2} (b) 3.13×10^{-4} (c) 3.13×10^{7} (d) 3.13×10^{-1} (e) 3.13×10^{-2} 44. (a) 3.42×10^{-4} (b) 1.034×10^{4} (c) 1.7992×10^{1} (d) 3.37×10^{5} 46. decimal 48. 3 50. none 52. (a) 641.0 (b) 1.327 (c) 77.34 (d) 3215 54. (a) 124 (b) 1.993×10^{-23} (c) 1.14×10^{-2} (d) 5.3×10^{-4} 56. (a) 2.045 (b) 3.8×10^{3} (c) 5.19×10^{-5} (d) 3.8418×10^{-7} 58. an infinite number (a definition) 60. 1 ft./12 in.; 12 in./1 ft. 62. 1 lb/\$0.79 64. (a) 2.44 yd. (b) 42.2 m (c) 115 in. (d) 2238 cm (e) 648.1 mi. (f) 716.9 km (g) 0.0362 km (h) 5.01×10^{4} cm 66. (a) 0.02543 kg (b) 2.74×10^{3} g (c) 6.04 lb. (d) 96.7 oz. (e) 1.177 lb. (f) 794 g (g) 2.5×10^{2} g (h) 1.62 oz. 68. 3.1×10^{2} km; 3.1×10^{5} m; 1.0×10^{6} ft. 70. 1×10^{-8} cm; 4×10^{-9} in.; 0.1 nm 72. Celsius 74. 273 76. Fahrenheit (F) 78. (a) 2 °C (b) 172 °C (c) -273 °C (d) -196 °C (e) 9727 °C (f) -271 °C 80. (a) 173 °F (b) 104 °F (c) -459 °F (d) 90 °F 82. (a) -121 °C (b) approximately 0 °F (c) 673 K (d) 255 K 84. g/cm^3 (g/mL) 86. volume 88. same 90. copper 92. (a) 28 g/cm^3 (b) 0.034 g/cm^3 (c) 0.962 g/cm^3 (d) 2.1×10^{-5} g/cm^3 94. 0.823 g/mL (g/cm^3) 96. float 98. 11.7 mL 100. (a) 1.13×10^{4} g; 1.13×10^{10} g (b) 2.16×10^{3} g; 2.16×10^{9} g (c) 8.8×10^{2} g; 8.8×10^{8} g (d) 7.87×10^{3} g; 7.87×10^{9} g 102. (a) 301,100,000,000,000,000,000,000 (b) 5,091,000,000 (c) 720 (d) 123,400 (e) 0.000432002 (f) 0.03001 (g) 0.00000029901 (h) 0.42 104. (a) centimeters (b) kilometers (c) micrometers (d) millimeters 106. (a) 5.07×10^{4} kryll (b) 0.12 blim (c) 3.7×10^{-5} blim^2 108. 20 in. 110. \$1/lb. 112. °X $= 1.26$ °C $+ 14$ 114. 3.50 g/L (3.50×10^{-3} g/cm^3) 116. 959 cm^3 118. (a) negative (b) zero (c) positive (d) negative 120. (a) 2; positive (b) 11; negative (c) 3; positive (d) 5; negative (e) 5; positive (f) zero; zero (g) 1; negative (h) 7; negative 122. (a) 2; positive (b) 3; negative (c) 1; positive (d) 8; negative (e) 4; positive (f) 1; positive (g) 6; negative (h) 1; negative 124. (a) 0.0000298 (b) 4,358,000,000 (c) 0.0000019928 (d) 602,000,000,000,000,000,000,000 (e) 0.101 (f) 0.00787 (g) 98,700,000 (h) 378.99 (i) 0.1093 (j) 2.9004 (k) 0.00039 (l) 0.00000001904 126. (a) 1×10^{-2}

(b) 1×10^2 (c) 5.5×10^{-2} (d) 3.1×10^9 (e) 1×10^3
(f) 1×10^8 (g) 2.9×10^2 (h) 3.453×10^4 **128.** Kelvin, K
130. centimeter   **132.** 0.105 m   **134.** 1 kg   **136.** 10
138. 2.8 (the hundredths place is estimated)
140. (a) 0.0000324   (b) 7,210,000   (c) 2.10×10^{-7}
(d) 550,000 (better as 5.50×10^5 to show the first zero is significant)   (e) 200. (the decimal point shows the zeros are significant)   **142.** (a) 2149.6   (b) 5.37×10^3 (c) 3.83×10^{-2}
(d) -8.64×10^5 **144.** (a) 7.6166×10^6 (b) 7.25×10^3
(c) 1.92×10^{-5} (d) 2.4482×10^{-3} **146.** 1 yr/12 mo.;
12 mo./1 yr.   **148.** (a) 25.7 kg   (b) 3.39 gal.   (c) 0.133 qt.
(d) 1.09×10^4 mL   (e) 2.03×10^3 g   (f) 0.58 qt.   **150.** for
exactly 6 gross, 864 pencils   **152.** (a) 352 K   (b) -18 °C
(c) -43 °C   (d) 257 °F   **154.** 45 g   **156.** 0.59 g/cm^3
158. (a) 23 °F   (b) 32 °F   (c) -321 °F   (d) -459 °F
(e) 187 °F   (f) -459 °F

CHAPTER **3**

2. states   **4.** container   **6.** gaseous   **8.** stronger   **10.** A gas
consists mostly of empty space. When a gas is compressed, at least
initially, it is this empty space that is being compressed.
12. chemical   **14.** chemical   **16.** When an electrical current is
transferred through water, the current causes the water molecules
to break down into their constituent elements (hydrogen gas and
oxygen gas). Anytime one substance is converted into other substances will be a chemical change.   **18.** Parts g, h, and j represent physical changes. All the other sections represent chemical
changes.   **20.** element   **22.** elements   **24.** different from
26. a variable   **28.** heterogeneous   **30.** (a) although the cotton has undoubtedly been treated with various other substances,
cotton is basically pure cellulose   (b) pure substance   (c) mixture
(d) pure substance   **32.** (a) homogeneous   (b) heterogeneous
(c) heterogeneous   (d) homogeneous   (e) the paper itself is basically homogeneous   **34.** filtration   **36.** The solution is heated
to vaporize (boil) the water. The water vapor is then cooled so
that it condenses back to the liquid state, and the liquid is collected. After all the water is vaporized from the original sample,
pure sodium chloride will remain. The process consists of physical
changes.   **38.** the calorie   **40.** As molecules of liquid water are
heated, the heat energy is converted to kinetic energy: the molecules begin to move more quickly and more randomly. Eventually,
as heating continues, the kinetic energy of the molecules will be
so large that the attractive forces that hold the molecules together
in the volume of the liquid will be overcome. The liquid will then
be said to be "boiling," as individual molecules with high kinetic
energies separate from the condensed liquid and enter the vapor.
42. temperature   **44.** 1.1×10^3 J   **46.** (a) 239.0 cal
(b) 131.5 cal   (c) 5.86×10^5 cal   (d) 0.239 cal
48. (a) 12,300 cal   (b) 290,400 cal   (c) 940,000,000 cal
(d) 4,201,000 cal   **50.** (a) 189,900,000 J   (b) 24,480,000 J
(c) 2390 J   (d) 19,750 J   **52.** 11 kJ   **54.** 1.6×10^3 g silver

56. 3.0×10^3 °C (the specific heat capacity is only known to
two significant figures)   **58.** 0.971 J/g °C   **60.** gold, 65 J; mercury, 70. J; carbon, 360 J   **62.** 1.42 J/g °C   **64.** compound
66. physical   **68.** 380,000 J (3.8×10^5 J)   **70.** Aluminum
will lose more heat because it has the higher specific heat capacity.   **72.** 24.2 °C   **74.** 22 °C   **76.** far apart   **78.** chemical
80. chemical   **82.** electrolysis   **84.** all except (b) are mixtures
86. 9.0 J   **88.** (a) 627.6 J   (b) 1.867×10^4 J   (c) 41.8 J
(d) 17.51 J   **90.** (a) 12.5 kJ   (b) 2.98 kcal   (c) 2.143 kJ
(d) 3.39 kcal   **92.** 4.4×10^3 J   **94.** $H_2O(l)$, 7.03×10^3 J;
$H_2O(s)$, 3.41×10^3 J; $H_2O(g)$, 3.4×10^3 J; Al, 1.5×10^3 J;
Fe, 7.6×10^2 J; Hg, 2.4×10^2 J; C, 1.2×10^3 J; Ag, 4.0×10^2 J; Au, 2.2×10^2 J   **96.** 0.13 J/g °C

CHAPTER **4**

2. The alchemists discovered several previously unknown elements (mercury, sulfur, antimony) and were the first to prepare
several common acids.   **4.** 108 elements are known; 88 occur
naturally, 20 are manmade. Table 4.1 lists the most common elements on the earth.   **6.** The four most abundant elements in living creatures are, respectively, oxygen, carbon, hydrogen, and nitrogen (see Table 4.2). In the nonliving world, the most abundant
elements are, respectively, oxygen, silicon, aluminum, and iron
(see Table 4.1).   **8.** Sb (antimony); Cu (copper); Au (gold); Pb
(lead); Hg (mercury); K (potassium); Ag (silver); Na (sodium); Sn
(tin); W (tungsten)   **10.** (a) H   (b) Fe   (c) Mg   (d) Ca   (e) Au
(f) He   **12.** (a) N   (b) O   (c) Mn   (d) Hg   (e) Ne   (f) Ni
14. (a) cobalt   (b) silver   (c) chlorine   (d) aluminum   (e) zinc
(f) platinum   (g) chromium   (h) sodium   **16.** (a) False; According to Dalton, the atoms of a given element are always *different*
than the atoms of any other element.   (b) False; Atoms are *indivisible* during chemical reactions.   (c) False; Dalton's theory was
not accepted generally for many years.   **18.** According to Dalton,
all atoms of the same element were *identical;* in particular, every
atom of a given element has the same mass as every other atom of
that element. If a given compound always contains the *same relative numbers* of atoms of each kind, and those atoms always have
the *same masses,* then it follows that the compound made from
those elements would always contain the same relative masses of
its elements.   **20.** (a) SO_2 (b) Fe_2S_3 (c) FeO (d) $C_2H_4Cl_2$
(e) Ca_3N_2 (f) C_6H_6 **22.** (a) False; Rutherford's bombardment
experiments with metal foil suggested that the alpha particles were
being deflected by coming near a *dense, positively charged* atomic
nucleus.   (b) False; The proton and the electron have opposite
charges, but the mass of the electron is *much smaller* than the
mass of the proton.   (c) True   **24.** neutrons   **26.** neutron;
electron   **28.** electrons   **30.** False   **32.** mass   **34.** Atoms of
the same element (i.e., atoms with the same number of protons in
the nucleus) may have different numbers of neutrons, and so will
have different masses.   **36.** (a) strontium, Sr   (b) potassium, K
(c) uranium, U   (d) lithium, Li   (e) cadmium, Cd   (f) aluminum,

Al (g) cobalt, Co (h) iodine, I **38.** a, b, and c are all $^{27}_{13}$Al; d, e, and f are all $^{32}_{16}$S

40.

	protons	neutrons	electrons
(a)	27	33	27
(b)	16	17	16
(c)	4	6	4
(d)	18	22	18
(e)	11	12	11
(f)	36	48	36

42.

element	neutrons	atomic number	mass number	symbol
Cl	18	17	35	$^{35}_{17}$Cl
Ne	10	10	20	$^{20}_{10}$Ne
He	2	2	4	$^{4}_{2}$He
Na	12	11	23	$^{23}_{11}$Na
Ca	21	20	41	$^{41}_{20}$Ca

44. False; The *vertical* columns in the periodic table are referred to as groups or families. **46.** Metallic elements are found toward the *left* and *bottom* of the periodic table; there are far more metallic elements than there are nonmetals. **48.** hydrogen, nitrogen, oxygen, fluorine, chlorine, plus all the group 8 elements (noble gases) **50.** metalloids or semimetals **52.** (a) 8; noble gases (b) 7; halogens (c) 1; alkali metals (d) 3 (e) 1; alkali metals (f) 2; alkaline earth elements (g) 8; noble gases (h) 7; halogens **54.** (a) Cs, $Z = 55$; metal; alkali metals (b) I, $Z = 53$; nonmetal; halogens (c) Ra, $Z = 88$; metal; alkaline earth metals (d) Xe, $Z = 54$; nonmetal; noble gases **56.** (a) 6 (b) 1; alkali metals (c) transition metals (d) 2; alkaline earth elements (e) transition metals (f) transition metals (g) 8; noble gases (h) 2; alkaline earth elements

58.

	element	symbol	atomic number
Group 3	boron	B	5
	aluminum	Al	13
	gallium	Ga	31
	indium	In	49
Group 5	nitrogen	N	7
	phosphorus	P	15
	arsenic	As	33
	antimony	Sb	51
Group 8	helium	He	2
	neon	Ne	10
	argon	Ar	18
	krypton	Kr	36

60. Most of the mass of an atom is concentrated in the nucleus: the *protons* and *neutrons* which constitute the nucleus have similar masses, and these particles are nearly two thousand times heavier than electrons. The chemical properties of an atom depend on the number and location of the *electrons* it possesses. Electrons are found in the outer regions of the atom, and are the particles most likely to be involved in interactions between atoms. **62.** $C_6H_{12}O_6$ **64.** (a) 29 protons; 34 neutrons; 29 electrons (b) 35 protons; 45 neutrons; 35 electrons (c) 12 protons; 12

neutrons; 12 electrons **66.** The chief use of gold in ancient times was an *ornamentation,* whether in statuary or in jewelry. Gold possesses an especially beautiful lustre, and since it is relatively soft and malleable, it could be worked finely by artisans; among the metals, gold is particularly inert to attack by most substances in the environment. **68.** (a) Ba (b) Br (c) Bi (d) B (e) K (f) P **70.** (a) Zn (b) U (c) W (d) Sr (e) Ne (f) Ni **72.** (a) boron (b) radium (c) arsenic (d) calcium (e) mercury (f) bromine (g) fluorine (h) phosphorus **74.** (a) CO_2 (b) $AlCl_3$ (c) $HClO_4$ (d) SCl_6 (e) Al_2O_3 (f) NaN_3 **76.** a. $^{13}_{6}$C b. $^{13}_{6}$C c. $^{13}_{6}$C d. $^{44}_{19}$K e. $^{41}_{20}$Ca f. $^{35}_{19}$K **78.** a. $^{40}_{20}$Ca 20 21 41 b. $^{55}_{25}$Mn 25 30 55 c. $^{109}_{47}$Ag 47 62 109 d. $^{45}_{21}$Sc 21 24 45

CHAPTER **5**

2. rare **4.** noble gases **6.** single atoms **8.** chlorine **10.** diamond **12.** electrons **14.** 2+ **16.** -ide **18.** nonmetallic **20.** (a) 11 protons, 10 electrons (b) 33 protons, 36 electrons (c) 26 protons, 24 electrons (d) 20 protons, 18 electrons (e) 7 protons, 10 electrons (f) 21 protons, 20 electrons (g) 38 protons, 36 electrons (h) 19 protons, 18 electrons **22.** (a) Al: 13 p, 13 e, Al^{3+}: 13 p, 10 e (b) S: 16 p, 16 e, S^{2-}: 16 p, 18 e (c) Fe: 26 p, 26 e, Fe^{3+}: 26 p, 23 e (d) Cl: 17 p, 17 e, Cl^-: 17 p, 18 e (e) Na: 11 p, 11 e, Na^+: 11 p, 10 e (f) N: 7 p, 7 e, N^{3-}: 7 p, 10 e **24.** (a) Cs^+ (b) Ca^{2+} (c) S^{2-} (d) Br^- (e) Rb^+ (f) I^- **26.** Sodium chloride is an *ionic* compound, consisting of Na^+ and Cl^- ions. When NaCl is dissolved in water, these ions are *set free,* and can move independently to conduct the electrical current. Sugar crystals, although they may *appear* similar visually contain *no* ions. When sugar is dissolved in water, it dissolves as uncharged *molecules.* There are no electrically charged species present in a sugar solution to carry the electrical current. **28.** The total number of positive charges must equal the total number of negative charges so that there will be *no net charge* on the crystals of an ionic compound. A macroscopic sample of compound must ordinarily not have any net charge. **30.** (a) FeS (b) Rb_3N (c) BaO (d) Al_2S_3 (e) Ca_3N_2 (f) FeI_3 (g) PbO_2 (h) $AlBr_3$ **32.** cation (positive ion) **34.** sodium ions, Na^+, and chloride ions, Cl^- **36.** Roman numeral **38.** (a) aluminum chloride (b) sodium sulfide (c) magnesium oxide (d) barium chloride (e) lithium iodide (f) silver(I) oxide (g) radium fluoride (h) strontium sulfide **40.** a, b, and d are incorrect **42.** (a) lead(II) chloride (b) iron(III) oxide (c) tin(II) iodide (d) mercury(I) oxide (e) mercury(II) sulfide (f) copper(I) iodide **44.** (a) plumbic oxide (b) stannous bromide (c) cuprous sulfide (d) cuprous iodide (e) mercurous iodide (f) chromic chloride **46.** (a) carbon monoxide (b) sulfur trioxide (c) dinitrogen tetrachloride (d) carbon tetraiodide (e) phosphorus pentafluoride (f) diphosphorus pentoxide **48.** (a) diboron hexahydride—nonionic (b) calcium nitride—ionic (c) carbon tetrabromide—nonionic (d) silver(I) sul-

fide—ionic   (e) copper(II) chloride, cupric chloride—ionic
(f) chlorine monofluoride—nonionic   **50.** a. aluminum oxide—
ionic   b. diboron trioxide—nonionic   c. dinitrogen tetroxide—
nonionic   d. cobalt(III) sulfide, cobaltic sulfide—ionic
e. dinitrogen pentoxide—nonionic   f. aluminum sulfide—ionic
52. oxygen   **54.** perchlorate   **56.** hypobromite; IO_3^-; perio-
date; OI^- **58.** (a) NO_3^- (b) NO_2^- (c) NH_4^+ (d) CN^-
60. (a) PO_4^{3-} (b) P^{3-} (c) $H_2PO_4^-$ (d) HPO_4^{2-}
62. (a) hydrogen carbonate, bicarbonate   (b) nitrite
(c) hydrogen sulfate, bisulfate   (d) dihydrogen phosphate
(e) acetate   (f) chlorite   **64.** (a) ammonium nitrate
(b) potassium chlorate   (c) lead(II) sulfate, plumbous sulfate
(d) calcium phosphate   (e) sodium perchlorate   (f) copper(II)
hydroxide, cupric hydroxide   **66.** oxygen   **68.** (a) perchloric
acid   (b) iodic acid   (c) bromous acid   (d) hypochlorous acid
(e) sulfurous acid   (f) hydrocyanic acid   (g) hydrosulfuric acid
(h) phosphoric acid   **70.** (a) PbO_2 (b) $SnBr_2$ (c) CuS (d) CuI
(e) Hg_2Cl_2 (f) CrF_3 **72.** (a) CO_2 (b) SO_3 (c) N_2Cl_4 (d) CI_4
(e) PF_5 (f) P_2O_5 **74.** (a) $Ca_3(PO_4)_2$ (b) NH_4NO_3
(c) $Al(HSO_4)_3$ (d) $BaSO_4$ (e) $Fe(NO_3)_3$ (f) $CuOH$
76. (a) HCN (b) HNO_3 (c) H_2SO_4 (d) H_3PO_4 (e) $HClO$ or
$HOCl$ (f) HF (g) $HBrO_2$ (h) HBr **78.** Group 1: $M \rightarrow M^+ +$
e^-; Group 2: $M \rightarrow M^{2+} + 2e^-$; Group 6: $Y + 2e^- \rightarrow Y^{2-}$;
Group 7: $X + e^- \rightarrow X^-$ **80.** A moist paste of NaCl would contain Na^+ and Cl^- ions in solution and would serve as a *conductor*
of electric impulses.   **82.** $H \rightarrow H^+$ (hydrogen ion) $+ e^-$; $H +$
$e^- \rightarrow H^-$ (hydride ion)   **84.** (a) oxyanions: IO_3^-; ClO_2^-. oxyacids: $HClO_4$; $HClO$; $HBrO_2$ **86.** (a) gold(III) bromide, auric bromide   (b) cobalt(III) cyanide, cobaltic cyanide   (c) magnesium
hydrogen phosphate   (d) diboron hexahydride (diborane is its
common name)   (e) ammonia   (f) silver(I) sulfate (usually called
silver sulfate)   (g) beryllium hydroxide   **88.** (a) ammonium carbonate   (b) ammonium hydrogen carbonate, ammonium bicarbonate   (c) calcium phosphate   (d) sulfurous acid
(e) manganese(IV) oxide   (f) iodic acid   (g) potassium hydride
90. (a) MCl_4 (b) $M(NO_3)_4$ (c) MO_2 (d) $M_3(PO_4)_4$
(e) $M(CN)_4$ (f) $M(SO_4)_2$ (g) $M(Cr_2O_7)_2$ **92.** M^+ compounds:
MD, M_2E, M_3F; M^{2+} compounds: MD_2, ME, M_3F_2; M^{3+} compounds: MD_3, M_2E_3, MF   **94.** $CaBr_2$, $Ca(HCO_3)_2$, CaH_2,
$Ca(C_2H_3O_2)_2$, $Ca(HSO_4)_2$, $Ca_3(PO_4)_2$; $SrBr_2$, $Sr(HCO_3)_2$, SrH_2,
$Sr(C_2H_3O_2)_2$, $Sr(HSO_4)_2$, $Sr_3(PO_4)_2$; NH_4Br, NH_4HCO_3, NH_4H,
$NH_4C_2H_3O_2$, NH_4HSO_4, $(NH_4)_3PO_4$; $AlBr_3$, $Al(HCO_3)_3$, AlH_3,
$Al(C_2H_3O_2)_3$, $Al(HSO_4)_3$, $AlPO_4$; $FeBr_3$, $Fe(HCO_3)_3$, FeH_3,
$Fe(C_2H_3O_2)_3$, $Fe(HSO_4)_3$, $FePO_4$; $NiBr_2$, $Ni(HCO_3)_2$, NiH_2,
$Ni(HSO_4)_2$, $Ni(C_2H_3O_2)_2$, $Ni_3(PO_4)_2$; $AgBr$, $AgHCO_3$, AgH,
$AgC_2H_3O_2$, $AgHSO_4$, Ag_3PO_4; $AuBr_3$, $Au(HCO_3)_3$, AuH_3,
$Au(C_2H_3O_2)_3$, $Au(HSO_4)_3$, $AuPO_4$; KBr, $KHCO_3$, KH, $KC_2H_3O_2$,
$KHSO_4$, K_3PO_4; $HgBr_2$, $Hg(HCO_3)_2$, HgH_2, $Hg(C_2H_3O_2)_2$,
$Hg(HSO_4)_2$, $Hg_3(PO_4)_2$; $BaBr_2$, $Ba(HCO_3)_2$, BaH_2, $Ba(C_2H_3O_2)_2$,
$Ba(HSO_4)_2$, $Ba_3(PO_4)_2$ **96.** helium   **98.** iodine   **100.** $1-$
102. $1-$ **104.** (a) Mn (manganese), atomic number 25; Mn^{2+}
(25 protons, 23 electrons); Mn (25 protons, 25 electrons)   (b) Ni

(nickel), atomic number 28; Ni^{2+} (28 protons, 26 electrons); Ni
(28 protons, 28 electrons)   (c) N (nitrogen), atomic number 7;
N^{3-} (7 protons, 10 electrons); N (7 protons, 7 electrons)   (d) Co
(cobalt), atomic number 27; Co^{3+} (27 protons, 24 electrons); Co
(27 protons, 27 electrons)   **106.** (a) Na_2S (b) KCl   (c) BaO
(d) MgSe   (e) $CuBr_2$ (f) AlI_3 (g) Al_2O_3 (h) Ca_3N_2 **108.** b,
d, and e are incorrect   **110.** (a) cobaltic bromide   (b) plumbic
iodide   (c) ferric oxide   (d) ferrous sulfide   (e) stannic chloride
(f) stannous oxide   **112.** (a) iron(III) sulfide, ferric sulfide—ionic
(b) gold(III) chloride, auric chloride—ionic   (c) arsenic trihydride
(arsine)—nonionic   (d) chlorine monofluoride—nonionic
(e) potassium oxide—ionic   (f) carbon dioxide—nonionic
114. (a) CO_3^{2-} (b) HCO_3^- (c) $C_2H_3O_2^-$ (d) CN^-
116. (a) carbonate   (b) chlorate   (c) sulfate   (d) phosphate
(e) perchlorate   (f) permanganate   **118.** (a) $CaCl_2$ (b) Ag_2O
(c) Al_2S_3 (d) $BeBr_2$ (e) H_2S (f) KH   (g) MgI_2 (h) CsF
120. (a) Mg_3P_2 (b) CaF_2 (c) $CoBr_3$ (d) FeI_2 (e) BaO
(f) K_2S

CHAPTER 6

2. *Heat* is evolved as drain cleaners work (often boiling any water
in the clogged drain). Some drain cleaners containing small shav-
ings of magnesium also *bubble* (hydrogen gas is formed) as the reaction takes place.   **4.** The shiny metallic appearance of the
bumper is changed to a dull red color, and patches of the red
color will begin to flake off.   **6.** A gas is evolved during fermentation. The odor of the product (alcohol or vinegar) will be notice-
able.   **8.** atoms   **10.** the same   **12.** water   **14.** $H_2O_2(aq) \rightarrow$
$H_2O(l) + O_2(g)$ **16.** $(NH_4)_2CO_3(s) \rightarrow NH_3(g) + CO_2(g) +$
$H_2O(g)$ **18.** $H_2SO_4(aq) + Na_2SO(s) \rightarrow SO_2(g) + H_2O(l) +$
$Na_2SO_4(aq)$ **20.** $Ca(s) + H_2O(l) \rightarrow Ca(OH)_2(s) + H_2(g)$
22. $Pb(NO_3)_2(aq) + KI(aq) \rightarrow PbI_2(s) + KNO_3(aq)$
24. $NH_3(g) + HCl(g) \rightarrow NH_4Cl(s)$ **26.** $CaCO_3(s) +$
$H_2SO_4(aq) \rightarrow CaSO_4(s) + H_2O(l) + CO_2(g)$ **28.** $O_2(g) \rightarrow$
$O_3(g)$ **30.** $SiO_2(s) + C(s) \rightarrow Si(s) + CO(g)$ **32.** $Xe(g) +$
$F_2(g) \rightarrow XeF_4(s)$ **34.** $C_{12}H_{22}O_{11}(s) + O_2(g) \rightarrow CO_2(g) +$
$H_2O(g)$ **36.** whole numbers   **38.** (a) $Br_2(l) + 2KI(aq) \rightarrow$
$2KBr(aq) + I_2(s)$ (b) $4Co(s) + 3O_2(g) \rightarrow 2Co_2O_3(s)$
(c) $P_4(s) + 5O_2(g) \rightarrow 2P_2O_5(s)$ (d) $2Al(s) + 6HNO_3(aq) \rightarrow$
$2Al_2(NO_3)_3(aq) + H_2(g)$ (e) $PBr_3(l) + 3H_2O(l) \rightarrow$
$H_3PO_3(aq) + 3HBr(aq)$ (f) $2NO(g) + O_2(g) \rightarrow 2NO_2(g)$
(g) $2C_2H_6(g) + 7O_2(g) \rightarrow 4CO_2(g) + 6H_2O(g)$ (h) $CuO(s) +$
$H_2SO_4(aq) \rightarrow CuSO_4(aq) + H_2O(l)$ **40.** (a) $Ba(NO_3)_2(aq) +$
$2KOH(aq) \rightarrow Ba(OH)_2(s) + 2KNO_3(aq)$ (b) $Cu(s) +$
$2HNO_3(aq) \rightarrow Cu(NO_3)_2(aq) + H_2(g)$ (c) $2Cr(s) + 3S(s) \rightarrow$
$Cr_2S_3(s)$ (d) $2AgNO_3(aq) + Zn(s) \rightarrow Zn(NO_3)_2(aq) + 2Ag(s)$
(e) $H_2O(l) + Br_2(l) \rightarrow HBr(aq) + HOBr(aq)$ (f) $SnS(s) +$
$2O_2(g) \rightarrow SnO_2(s) + SO_2(g)$ (g) $Pb(NO_3)_2(aq) + 2KCl(aq) \rightarrow$
$PbCl_2(s) + 2KNO_3(aq)$ (h) $FeO(s) + C(s) \rightarrow Fe(s) + CO(g)$

42. (a) $SiI_4(s) + 2Mg(s) \rightarrow Si(s) + 2MgI_2(s)$ (b) $MnO_2(s) + 2Mg(s) \rightarrow Mn(s) + 2MgO(s)$ (c) $8Ba(s) + S_8(s) \rightarrow 8BaS(s)$ (d) $4NH_3(g) + 3Cl_2(g) \rightarrow 3NH_4Cl(s) + NCl_3(g)$ (e) $8Cu_2S(s) + S_8(s) \rightarrow 16CuS(s)$ (f) $2Al(s) + 3H_2SO_4(aq) \rightarrow Al_2(SO_4)_3(aq) + 3H_2(g)$ (g) $2NaCl + H_2SO_4(l) \rightarrow 2HCl(g) + Na_2SO_4(s)$ (h) $2CO(g) + O_2(g) \rightarrow 2CO_2(g)$ **44.** (a) $Ba(NO_3)_2(aq) + Na_2CrO_4(aq) \rightarrow BaCrO_4(s) + 2NaNO_3(aq)$ (b) $PbCl_2(aq) + K_2SO_4(aq) \rightarrow PbSO_4(s) + 2KCl(aq)$ (c) $C_2H_5OH(l) + 3O_2(g) \rightarrow 2CO_2(g) + 3H_2O(l)$ (d) $CaC_2(s) + 2H_2O(l) \rightarrow Ca(OH)_2(s) + H_2C_2(g)$ (e) $Sr(s) + 2HNO_3(aq) \rightarrow Sr(NO_3)_2(aq) + H_2(g)$ (f) $BaO_2(s) + H_2SO_4(aq) \rightarrow BaSO_4(s) + H_2O_2(aq)$ (g) $2AsI_3(s) \rightarrow 2As(s) + 3I_2(s)$ (h) $2CuSO_4(aq) + 4KI(s) \rightarrow 2CuI(s) + I_2(s) + 2K_2SO_4(aq)$ **46.** $Li(s) + H_2O(l) \rightarrow LiOH(aq) + H_2(g)$; $Na(s) + H_2O(l) \rightarrow NaOH(aq) + H_2(g)$; $K(s) + H_2O(l) \rightarrow KOH(aq) + H_2(g)$; $Rb(s) + H_2O(l) \rightarrow RbOH(aq) + H_2(g)$; $Cs(s) + H_2O(l) \rightarrow CsOH(aq) + H_2(g)$; $Fr(s) + H_2O(l) \rightarrow FrOH(aq) + H_2(g)$ **48.** $C_{12}H_{22}O_{11}(aq) + H_2O(l) \rightarrow 4C_2H_6O(aq) + 4CO_2(g)$ **50.** $2Al_2O_3(s) + 3C(s) \rightarrow 4Al(s) + 3CO_2(g)$ **52.** $Be(s) + F_2(g) \rightarrow BeF_2(s)$; $Be(s) + Cl_2(g) \rightarrow BeCl_2(s)$; $Mg(s) + F_2(g) \rightarrow MgF_2(s)$; $Mg(s) + Cl_2(g) \rightarrow MgCl_2(s)$; $Ca(s) + F_2(g) \rightarrow CaF_2(s)$; $Ca(s) + Cl_2(g) \rightarrow CaCl_2(s)$; $Sr(s) + F_2(g) \rightarrow SrF_2(s)$; $Sr(s) + Cl_2(g) \rightarrow SrCl_2(s)$; $Ba(s) + F_2(g) \rightarrow BaF_2(s)$; $Ba(s) + Cl_2(g) \rightarrow BaCl_2(s)$; $Ra(s) + F_2(g) \rightarrow RaF_2(s)$; $Ra(s) + Cl_2(g) \rightarrow RaCl_2(s)$ **54.** $(NH_4)_2Cr_2O_7(s) \rightarrow Cr_2O_3(s) + N_2(g) + 4H_2O(g)$ **56.** $2KClO_3(s) \rightarrow 2KCl(s) + 3O_2(g)$ **58.** $NH_3(g) + HCl(g) \rightarrow NH_4Cl(s)$ **60.** Most oils *decompose* at high temperatures, producing a foul-smelling chemical called acrolein that we associate with the smell and taste of burned food. Oil that has been heated to too high a temperature for too long a period "turns rancid" and is unfit for further use. Also, at high enough temperatures, the oil might ignite. **62.** $Fe(s) + S(s) \rightarrow FeS(s)$ **64.** $C_3H_8(g) + O_2(g) \rightarrow CO_2(g) + H_2O(g)$ **66.** $KI(aq) + H_2O(l) \rightarrow KOH(aq) + H_2(g) + I_2(s)$ **68.** $Mg(s) + H_2O(g) \rightarrow Mg(OH)_2(s) + H_2(g)$ **70.** $CuO(s) + H_2SO_4(aq) \rightarrow CuSO_4(aq) + H_2O(l)$ **72.** $Na_2SO_3(aq) + S(s) \rightarrow Na_2S_2O_3(aq)$ **74.** (a) $Ba(NO_3)_2(aq) + 2KF(aq) \rightarrow BaF_2(s) + 2KNO_3(aq)$ (b) $Zn(s) + 2HCl(aq) \rightarrow ZnCl_2(aq) + H_2(g)$ (c) $2Fe(s) + 3S(s) \rightarrow Fe_2S_3(s)$ (d) $C_6H_{12}O_6(s) + 6O_2(g) \rightarrow 6CO_2(g) + 6H_2O(g)$ (e) $H_2O + Cl_2 \rightarrow HCl + HOCl$ (f) $2ZnS(s) + 3O_2(g) \rightarrow 2ZnO(s) + 2SO_2(g)$ (g) $PbSO_4(s) + 4NaCl(aq) \rightarrow Na_2SO_4(aq) + Na_2PbCl_4(aq)$ (h) $3Fe_2O_3(s) + C(s) \rightarrow 2Fe_3O_4(s) + CO(g)$ **76.** (a) $Pb(NO_3)_2(aq) + K_2CrO_4(aq) \rightarrow PbCrO_4(s) + 2KNO_3(aq)$ (b) $BaCl_2(aq) + Na_2SO_4(aq) \rightarrow BaSO_4(s) + 2NaCl(aq)$ (c) $2CH_3OH(l) + 3O_2(g) \rightarrow 2CO_2(g) + 4H_2O(g)$ (d) $Na_2CO_3(aq) + S(s) + SO_2(g) \rightarrow CO_2(g) + Na_2S_2O_3(aq)$ (e) $Cu(s) + 2H_2SO_4(aq) \rightarrow CuSO_4(aq) + SO_2(g) + 2H_2O(l)$ (f) $MnO_2(s) + 4HCl(aq) \rightarrow MnCl_2(aq) + Cl_2(g) + 2H_2O(l)$ (g) $As_2O_3(s) + 6KI(aq) + 6HCl(aq) \rightarrow AsI_3(s) + 6KCl(aq) + 3H_2O(l)$ (h) $2Na_2S_2O_3(aq) + I_2(aq) \rightarrow Na_2S_4O_6(aq) + 2NaI(aq)$

CHAPTER 7

2. Driving forces are the types of *changes* in a system which pull a reaction in the *direction of product formation;* driving forces discussed in Chapter 7 include: formation of a *solid,* formation of *water,* formation of a *gas,* transfer of electrons. **4.** The net charge of a precipitate must be zero. The total number of positive charges equals the total number of negative charges. **6.** ions **8.** Consider the two new possible combinations of ions (when two ionic compounds are mixed the ions may "switch partners"); if one of these new possible combinations is insoluble in water it will form a precipitate having zero net charge (indicating in what proportions the ions must be combined). There is no foolproof method for predicting by inspection what precipitate may form: we rely on intuition, experience from similar reactions, and "rules of thumb" based on the results of experiments (such as the general solubility rules from Table 7.1). **10.** For most practical purposes, "insoluble" and "slightly" soluble mean the same thing. However, if a substance were highly toxic and were found in a water supply, for example, the difference between "insoluble" and "slightly soluble" could be crucial. **12.** (a) soluble (most nitrate salts are soluble) (b) soluble (most potassium salts are soluble) (c) soluble (most sodium salts are soluble) (d) insoluble (most hydroxide compounds are insoluble) (e) insoluble (exception for chloride salts) (f) soluble (most ammonium salts are soluble) (g) insoluble (most sulfide salts are insoluble) (h) insoluble (exception for sulfate salts) **14.** (a) Rule 6: most sulfide salts are insoluble (b) Rule 5: most hydroxide compounds are insoluble (c) Rule 4: exception (d) Rule 6: most carbonate salts are insoluble **16.** (a) HgS (b) $CaSO_4$ (c) $AgCl$ (d) $Ba(OH)_2$ (e) $Ni_3(PO_4)_2$ (f) $CuCO_3$ **18.** (a) No precipitate: $Ba(NO_3)_2$ and HCl are each soluble (b) $(NH_4)_2S(aq) + CoCl_2(aq) \rightarrow CoS(s) + 2NH_4Cl(aq)$ (c) $H_2SO_4(aq) + Pb(NO_3)_2(aq) \rightarrow PbSO_4(s) + 2HNO_3(aq)$ (d) $CaCl_2(aq) + H_2CO_3(aq) \rightarrow CaCO_3(s) + 2HCl(aq)$ (e) No precipitate: $NH_2C_2H_3O_2$ and $NaNO_3$ are each soluble (f) $Na_3PO_4(aq) + CrCl_3(aq) \rightarrow 3NaCl(aq) + CrPO_4(s)$ **20.** (a) $2AgNo_3(aq) + H_2SO_4(aq) \rightarrow Ag_2SO_4(s) + 2HNO_3(aq)$ (b) $Ca(NO_3)_2(aq) + H_2SO_4(aq) \rightarrow CaSO_4(s) + 2HNO_3(aq)$ (c) $Pb(NO_3)_2(aq) + H_2SO_4(aq) \rightarrow PbSO_4(s) + 2HNO_3(aq)$ **22.** (a) $NiCl_2(aq) + H_2S(aq) \rightarrow NiS(s) + 2HCl(aq)$ (b) $CuSO_4(aq) + 2NaOH(aq) \rightarrow Na_2SO_4(aq) + Cu(OH)_2(s)$ (c) $Ba(NO_3)_2(aq) + Na_2CO_3(aq) \rightarrow 2NaNO_3(aq) + BaCO_3(s)$ **24.** spectator **26.** (a) $Fe^{3+}(aq) + 3OH^-(aq) \rightarrow Fe(OH)_3(s)$ (b) $Ni^{2+}(aq) + S^{2-}(aq) \rightarrow NiS(s)$ (c) $Ag^+(aq) + Cl^-(aq) \rightarrow AgCl(s)$ (d) $Ba^{2+}(aq) + SO_4^{2-}(aq) \rightarrow BaSO_4(s)$ (e) $Hg_2^{2+}(aq) + 2Br^-(aq) \rightarrow Hg_2Br_2(s)$ (f) $Ba^{2+}(aq) + SO_4^{2-}(aq) \rightarrow BaSO_4(s)$ **28.** $Ba^{2+}(aq) + SO_4^{2-}(aq) \rightarrow BaSO_4(s)$ **30.** $Fe^{2+}(aq) + CO_3^{2-}(aq) \rightarrow FeCO_3(s)$; $Pb^{2+}(aq) + CO_3^{2-}(aq) \rightarrow PbCO_3(s)$; $2Al^{3+}(aq) + 3CO_3^{2-}(aq) \rightarrow Al_2(CO_3)_3(s)$ $Zn^{2+}(aq) + CO_3^{2-}(aq) \rightarrow ZnCO_3(s)$ **32.** acid **34.** HCl **36.** salt **38.** $RbOH(s) \rightarrow Rb^+(aq) + OH^-(aq)$; $CsOH(s) \rightarrow Cs^+(aq) + OH^-(aq)$ **40.** (a) $HCl(aq) +$

$KOH(aq) \rightarrow KCl(aq) + H_2O(l)$ (b) $H_2SO_4(aq) +$
$2NaOH(aq) \rightarrow Na_2SO_4(aq) + 2H_2O(l)$ (c) $HClO_4(aq) +$
$CsOH(aq) \rightarrow CsClO_4(aq) + H_2O(l)$ (d) $HNO_3(aq) +$
$KOH(aq) \rightarrow KNO_3(aq) + H_2O(l)$ **42**. A *molecular equation*
uses the normal, uncharged formulas for the compounds involved.
The *complete ionic equation* shows the compounds involved broken up into their respective ions (*all* ions present are shown). The
net ionic equation shows only those ions that combine to form a
precipitate, a gas, or a nonionic product such as water. The net
ionic equation shows most clearly the species that are combining
with each other. **44**. $Pb^{2+}(aq) + 2Cl^-(aq) \rightarrow PbCl_2(s)$;
$Pb^{2+}(aq) + CrO_4^{2-}(aq) \rightarrow PbCrO_4(s)$ **46**. (a) Rule 5:
$Co^{3+}(aq) + 3OH^-(aq) \rightarrow Co(OH)_3(s)$ (b) Rule 6: $2Ag^+(aq) +$
$CO_3^{2-}(aq) \rightarrow Ag_2CO_3(s)$ (c) no reaction (all combinations of
ions are soluble) (d) Rule 4: $Ba^{2+}(aq) + SO_4^{2-}(aq) \rightarrow BaSO_4(s)$
(e) no reaction (all combinations of ions are soluble) (f) Rule 6:
$3Ca^{3+}(aq) + 2PO_4^{3-}(aq) \rightarrow Ca_3(PO_4)_2(s)$ (g) Rule 5:
$Al^{3+}(aq) + 3OH^-(aq) \rightarrow Al(OH)_3(s)$ **48**. $Ca(OH)_2(s) \rightarrow$
$Ca^{2+}(aq) + 2OH^-(aq)$; $Mg(OH)_2(s) \rightarrow Mg^{2+}(aq) + 2OH^-(aq)$;
$Sr(OH)_2(s) \rightarrow Sr^{2+}(aq) + 2OH^-(aq)$; $Ba(OH)_2(s) \rightarrow Ba^{2+}(aq) +$
$2OH^-(aq)$ **50**. When *any* strong acid is reacted with *any* strong
base, the net ionic reaction is the *same*: $H^+(aq) + OH^-(aq) \rightarrow$
$H_2O(l)$. Since the net ionic reaction is the same for any strong
acid, the amount of heat liberated is also the same.
52. (a) soluble (Rule 2: most potassium salts are soluble)
(b) soluble (Rule 2: most ammonium salts are soluble)
(c) insoluble (Rule 6: most carbonate salts are only slightly soluble) (d) insoluble (Rule 6: most phosphate salts are only slightly
soluble) (e) soluble (Rule 2: most sodium salts are soluble)
(f) insoluble (Rule 5: most hydroxide salts are only slightly soluble) (g) soluble (Rule 3: most chloride salts are soluble)
54. (a) iron(III) hydroxide, $Fe(OH)_3$. Rule 5: most hydroxide salts
are only slightly soluble. (b) nickel(II) sulfide, NiS. Rule 6: most
sulfide salts are only slightly soluble. (c) silver chloride, AgCl.
Rule 3: Although most chloride salts are soluble, AgCl is a listed
exception. (d) barium carbonate, $BaCO_3$. Rule 6: most carbonate
salts are only slightly soluble. (e) mercury(I) chloride or mercurous chloride, Hg_2Cl_2. Rule 3: Although most chloride salts are soluble, Hg_2Cl_2 is a listed exception. (f) barium sulfate, $BaSO_4$. Rule
4: Although most sulfate salts are soluble, $BaSO_4$ is a listed exception. **56**. (a) $Co(NO_3)_2(aq) + (NH_4)_2S(aq) \rightarrow CoS(s) +$
$2NH_4NO_3(aq)$ (b) $CaCl_2(aq) + H_2SO_4(aq) \rightarrow CaSO_4(s) +$
$2HCl(aq)$ (c) $FeCl_3(aq) + Na_3PO_4(aq) \rightarrow FePO_4(s) +$
$3NaCl(aq)$ **58**. (a) $Ag^+(aq) + Cl^-(aq) \rightarrow AgCl(s)$
(b) $3Ca^{2+}(aq) + 2PO_4^{3-}(aq) \rightarrow Ca_3(PO_4)_2(s)$ (c) $Pb^{2+}(aq) +$
$2Cl^-(aq) \rightarrow PbCl_2(s)$ (d) $Fe^{3+}(aq) + 3OH^-(aq) \rightarrow Fe(OH)_3(s)$
60. $Fe^{2+}(aq) + S^{2-}(aq) \rightarrow FeS(s)$; $2Cr^{3+}(aq) + 3S^{2-}(aq) \rightarrow$
$Cr_2S_3(s)$; $Ni^{2+}(aq) + S^{2-}(aq) \rightarrow NiS(s)$ **62**. hydroxide
64. (a) $HClO_4(aq) + RbOH(aq) \rightarrow H_2O(l) + RbClO_4(aq)$
(b) $HNO_3(aq) + KOH(aq) \rightarrow H_2O(l) + KNO_3(aq)$
(c) $H_2SO_4(aq) + 2NaOH(aq) \rightarrow 2H_2O(l) + Na_2SO_4(aq)$
(d) $HBr(aq) + CsOH(aq) \rightarrow H_2O(l) + CsBr(aq)$ **66**. These

anions tend to form insoluble precipitates witn many metal ions.
The following are illustrative: (a) $CoCl_2(aq) + H_2S(aq) \rightarrow$
$CoS(s) + 2HCl(aq)$; $SnCl_2(aq) + H_2S(aq) \rightarrow SnS(s) +$
$2HCl(aq)$; $Ba(NO_3)_2(aq) + H_2S(aq) \rightarrow BaS(s) + 2HNO_3(aq)$
(b) $CoCl_2(aq) + Na_2CO_3(aq) \rightarrow CoCO_3(s) + 2NaCl(aq)$;
$SnCl_2(aq) + Na_2CO_3(aq) \rightarrow SnCO_3(s) + 2NaCl(aq)$;
$Ba(NO_3)_2(aq) + Na_2CO_3(aq) \rightarrow BaCO_3(s) + 2NaNO_3(aq)$
(c) $CoCl_2(aq) + 2NaOH(aq) \rightarrow Co(OH)_2(s) + 2NaCl(aq)$;
$SnCl_2(aq) + 2NaOH(aq) \rightarrow Sn(OH)_2(s) + 2NaCl(aq)$;
$Ba(NO_3)_2(aq) + 2NaOH(aq) \rightarrow Ba(OH)_2(s) + 2NaNO_3(aq)$
(d) $3CoCl_2(aq) + 2H_3PO_4(aq) \rightarrow Co_3(PO_4)_2(s) + 6HCl(aq)$;
$3SnCl_2(aq) + 2H_3PO_4(aq) \rightarrow Sn_3(PO_4)_2(s) + 6HCl(aq)$;
$3Ba(NO_3)_2(aq) + 2H_3PO_4(aq) \rightarrow Ba_3(PO_4)_2(s) + 6HNO_3(aq)$

CHAPTER **8**

2. transfer **4**. Metal atoms *lose* electrons and form *cations; non-*
metal atoms *gain* electrons and become *anions*. **6**. 1−; one; two
8. $AlBr_3$ is made up of Al^{3+} ions and Br^- ions. Aluminum atoms
each lose three electrons to become Al^{3+} ions. Bromine atoms
each gain one electron to become Br^- ions (so each Br_2 molecule
gains two electrons to become two Br^- ions). **10**. (a) $2Na(s) +$
$S(s) \rightarrow Na_2S(s)$ (b) $Cu(s) + 2AgNO_3(aq) \rightarrow Cu(NO_3)_2(aq) +$
$2Ag(s)$ (c) $Mg(s) + 2HCl(aq) \rightarrow MgCl_2(aq) + H_2(g)$
(d) $Pb(s) + 2S(s) \rightarrow PbS_2(s)$ (e) $2Li(s) + 2H_2O(l) \rightarrow$
$2LiOH(aq) + H_2(g)$ **12**. examples of formation of water:
$HCl(aq) + NaOH(aq) \rightarrow H_2O(l) + NaCl(aq)$, $H_2SO_4(aq) +$
$2KOH(aq) \rightarrow 2H_2O(l) + K_2SO_4(aq)$; examples of formation of a
gaseous product: $Mg(s) + 2HCl(aq) \rightarrow MgCl_2(aq) + H_2(g)$
$2KClO_3(s) \rightarrow 2KCl(s) + 3O_2(g)$ **14**. (a) oxidation−reduction
(b) oxidation−reduction (c) acid−base (d) acid−base, precipitation (e) precipitation (f) precipitation (g) oxidation−reduction
(h) oxidation−reduction (i) acid−base **16**. oxidation−reduction
18. decomposition **20**. (a) $C_3H_8(g) + 5O_2(g) \rightarrow 3CO_2(g) +$
$4H_2O(g)$ (b) $2C_2H_{10}(g) + 13O_2(g) \rightarrow 8CO_2(g) + 10H_2O(g)$
(c) $C_5H_{12}(g) + 8O_2(g) \rightarrow 5CO_2(g) + 6H_2O(g)$
22. (a) $C_{19}H_{40}(s) + 29O_2(g) \rightarrow 19CO_2(g) + 20H_2O(g)$
(b) $C_6H_{12}O_6(s) + 6O_2(g) \rightarrow 6CO_2(g) + 6H_2O(g)$
(c) $C_{12}H_{22}O_{11}(s) + 12O_2(g) \rightarrow 12CO_2(g) + 11H_2O(g)$
24. (a) $2Co(s) + 3S(s) \rightarrow Co_2S_3(s)$ (b) $2NO(g) + O_2(g) \rightarrow$
$2NO_2(g)$ (c) $FeO(s) + CO_2(g) \rightarrow FeCO_3(s)$ (d) $2Al(s) +$
$3F_2(s) \rightarrow 2AlF_3(s)$ (e) $2NH_3(g) + H_2CO_3(aq) \rightarrow (NH_4)_2CO_3(s)$
26. (a) $2NI_3(s) \rightarrow N_2(g) + 3I_2(s)$ (b) $BaCO_3(s) \rightarrow BaO(s) +$
$CO_2(g)$ (c) $C_6H_{12}O_6(s) \rightarrow 6C(s) + 6H_2O(g)$
(d) $Cu(NH_3)_4SO_4(s) \rightarrow CuSO_4(s) + 4NH_3(g)$ (e) $3NaN_3(s) \rightarrow$
$Na_3N(s) + 4N_2(g)$ **28**. (a) two; $O + 2e^- \rightarrow O^{2-}$ (b) one;
$F + e^- \rightarrow F^-$ (c) three; $N + 3e^- \rightarrow N^{3-}$ (d) one; $Cl +$
$e^- \rightarrow Cl^-$ (e) two; $S + 2e^- \rightarrow S^{2-}$ (f) one; $Br + e^- \rightarrow Br^-$
30. carbon dioxide gas and water vapor **32**. (a) $2C_3H_8O(l) +$
$9O_2(g) \rightarrow 6CO_2(g) + 8H_2O(g)$; oxidation−reduction, combustion
(b) $HCl(aq) + AgC_2H_3O_2(aq) \rightarrow AgCl(s) + HC_2H_3O_2(aq)$; pre-

cipitation (c) $3HCl(aq) + Al(OH)_3(s) \rightarrow AlCl_3(aq) + 3H_2O(l)$; acid–base (d) $2H_2O_2(aq) \rightarrow 2H_2O(l) + O_2(g)$; oxidation–reduction, decomposition (e) $N_2H_4(l) + O_2(g) \rightarrow N_2(g) + 2H_2O(g)$; oxidation–reduction, combustion **34.** $2Zn(s) + O_2(g) \rightarrow 2ZnO(s)$; $4Al(s) + 3O_2(g) \rightarrow 2Al_2O_3(s)$; $2Fe(s) + O_2(g) \rightarrow 2FeO(s)$, $4Fe(s) + 3O_2(g) \rightarrow 2Fe_2O_3(s)$; $2Cr(s) + O_2(g) \rightarrow 2CrO(s)$, $4Cr(s) + 3O_2(g) \rightarrow 2Cr_2O_3(s)$; $2Ni(s) + O_2(g) \rightarrow 2NiO(s)$ **36.** Al, 3+; Ba, 2+; Br, 1−; Ca, 2+; Cl, 1−; Cs, 1+; I, 1−; K, 1+; Li, 1+; Mg, 2+; Na, 1+; O, 2−; Rb, 1+; S, 2−; Sr, 2+ **38.** K^+, Ca^{2+} **40.** (a) $2Na(s) + O_2(g) \rightarrow Na_2O_2(s)$ (b) $Fe(s) + H_2SO_4(aq) \rightarrow FeSO_4(aq) + H_2(g)$ (c) $2Al_2O_3(s) \rightarrow 4Al(s) + 3O_2(g)$ (d) $2Fe(s) + 3Br_2(l) \rightarrow 2FeBr_3(s)$ (e) $Zn(s) + 2HNO_3(aq) \rightarrow Zn(NO_3)_2(aq) + H_2(g)$ **42.** (a) $C_5H_{12}(l) + 8O_2(g) \rightarrow 5CO_2(g) + 6H_2O(g)$ (b) $C_2H_6O(l) + 3O_2(g) \rightarrow 2CO_2(g) + 3H_2O(g)$ (c) $2C_6H_6(l) + 15O_2(g) \rightarrow 12CO_2(g) + 6H_2O(g)$ **44.** (a) $4FeO(s) + O_2(g) \rightarrow 2Fe_2O_3(s)$ (b) $2CO(g) + O_2(g) \rightarrow 2CO_2(g)$ (c) $H_2(g) + Cl_2(g) \rightarrow 2HCl(g)$ (d) $16K(s) + S_8(s) \rightarrow 8K_2S(s)$ (e) $6Na(s) + N_2(g) \rightarrow 2Na_3N(s)$ **46.** For simplicity, the physical states of the substances are omitted. $2Ba + O_2 \rightarrow 2BaO$; $Ba + S \rightarrow BaS$; $Ba + Cl_2 \rightarrow BaCl_2$; $3Ba + N_2 \rightarrow Ba_3N_2$; $Ba + Br_2 \rightarrow BaBr_2$; $4K + O_2 \rightarrow 2K_2O$; $2K + S \rightarrow K_2S$; $2K + Cl_2 \rightarrow 2KCl$; $6K + N_2 \rightarrow 2K_3N$; $2K + Br_2 \rightarrow 2KBr$; $2Mg + O_2 \rightarrow 2MgO$; $Mg + S \rightarrow MgS$; $Mg + Cl_2 \rightarrow MgCl_2$; $3Mg + N_2 \rightarrow Mg_3N_2$; $Mg + Br_2 \rightarrow MgBr_2$; $4Rb + O_2 \rightarrow 2Rb_2O$; $2Rb + S \rightarrow Rb_2S$; $2Rb + Cl_2 \rightarrow 2RbCl$; $6Rb + N_2 \rightarrow 2RB_3N$; $2Rb + Br_2 \rightarrow 2RbBr$; $2Ca + O_2 \rightarrow 2CaO$; $Ca + S \rightarrow CaS$; $Ca + Cl_2 \rightarrow CaCl_2$; $3Ca + N_2 \rightarrow Ca_3N_2$; $Ca + Br_2 \rightarrow CaBr_2$; $4Li + O_2 \rightarrow 2Li_2O$; $2Li + S \rightarrow Li_2S$; $2Li + Cl_2 \rightarrow 2LiCl$; $6Li + N_2 \rightarrow 2Li_3N$; $2Li + Br_2 \rightarrow 2 LiBr$ **47.** $F_2(g) + Cl_2(g) \rightarrow 2ClF(g)$; $F_2(g) + Br_2(l) \rightarrow 2BrF(l)$; $F_2(g) + I_2(s) \rightarrow 2IF(s)$ **48.** $Fe(s) + H_2SO_4(aq) \rightarrow FeSO_4(aq) + H_2(g)$; $Zn(s) + H_2SO_4(aq) \rightarrow ZnSO_4(aq) + H_2(g)$; $Cu(s) + H_2SO_4(aq) \rightarrow CuSO_4(aq) + H_2(g)$; $Co(s) + H_2SO_4(aq) \rightarrow CoSO_4(aq) + H_2(q)$; $Ni(s) + H_2SO_4(aq) \rightarrow NiSO_4(aq) + H_2(g)$ **49.** For simplicity, the physical states of the substances are omitted. $Mg + Cl_2 \rightarrow MgCl_2$; $Ca + Cl_2 \rightarrow CaCl_2$; $Sr + Cl_2 \rightarrow SrCl_2$; $Ba + Cl_2 \rightarrow BaCl_2$; $Mg + Br_2 \rightarrow MgBr_2$; $Ca + Br_2 \rightarrow CaBr_2$; $Sr + Br_2 \rightarrow SrBr_2$; $Ba + Br_2 \rightarrow BaBr_2$; $2Mg + O_2 \rightarrow 2MgO$; $2Ca + O_2 \rightarrow 2CaO$; $2Sr + O_2 \rightarrow 2SrO$; $2Ba + O_2 \rightarrow 2BaO$ **50.** For simplicity, the physical states of the substances are omitted. $2Na + S \rightarrow Na_2S$; $Ba + S \rightarrow BaS$; $2Al + 3S \rightarrow Al_2S_3$; $Fe + S \rightarrow FeS$; $2K + S \rightarrow K_2S$; $Ca + S \rightarrow CaS$; $Mg + S \rightarrow MgS$

CHAPTER 9

2. 307 corks; 116 stoppers; 2650 g **4.** the same **6.** (a) one (b) ten (c) 100 (d) 150 (e) 133 **8.** 1.97×10^6 amu; 1500. Au atoms **10.** Avogadro's number (6.022×10^{23}) **12.** 12.01 g carbon **14.** 42 g nitrogen **16.** 2.66×10^{-23} g

18. 0.50 mol O **20.** (a) 0.133 mol Au (b) 1.04 mol Ca (c) 2.59×10^3 mol Ba (d) 1.33×10^{-5} mol Pd (e) 5.20×10^{-13} mol Ni (f) 8.12 mol Fe (g) 1.00 mol C **22.** (a) 122 g Fe (b) 3.06 g Ni (c) 0.240 g Pt (d) 1.50×10^4 g Pb (e) 0.0200 g Mg (f) 1.31×10^5 g Al (g) 1468 g Li (h) 3.95×10^{-5} g Na **24.** (a) 1.08×10^{-24} Fe atoms (b) 1.79×10^{-2} mol Fe (c) 2.69 mol Fe (d) 150. g Fe (e) 2.07×10^{23} Fe atoms (f) 19.2 g Fe (g) 55.85 g Fe (h) 1.000 mol Fe **26.** adding (summing) **28.** (a) 58.12 g (b) 122.4 g (c) 407.7 g (d) 46.07 g (e) 78.11 g (f) 394.7 g **30.** (a) 310.2 g (b) 90.39 g (c) 146.1 g (d) 227.8 g (e) 84.01 g (f) 37.95 g **32.** (a) 0.0615 mol ClO_2 (b) 0.966 mol KF (c) 3.64×10^{-6} mol NH_4NO_3 (d) 30.8 mol FeO (e) 3.17×10^{-5} mol $CoCl_2$ **34.** (a) 2.36×10^{-6} mol $AlCl_3$ (b) 11.3 mol NaOH (c) 1.62×10^{-8} mol AsI_3 (d) 526 mol $NaHCO_3$ (e) 0.0357 mol K_2HPO_4 **36.** (a) 612 g AlI_3 (b) 0.149 g C_6H_6 (c) 721 g $C_6H_{12}O_6$ (d) 2.10×10^7 g C_2H_5OH (e) 463 g $Ca(NO_3)_2$ **38.** (a) 0.0559 g CO_2 (b) 4.96×10^5 g NCl_3 (c) 0.361 g NH_4NO_3 (d) 324 g H_2O (e) 1.00×10^4 g $CuSO_4$ **40.** (a) 2.58×10^{24} molecules NO_2 (b) 5.61×10^{22} molecules NO_2 (c) 1.17×10^{14} molecules HF (d) 5.87×10^{12} molecules HF (e) 1.63×10^{23} molecules NH_3 **42.** (a) 0.0142 mol S (b) 0.0159 mol S (c) 0.0258 mol S (d) 0.0254 mol S **44.** less **46.** (a) 32.38% Na; 22.58% S; 45.07% O (b) 36.49% Na; 25.44% S; 38.09% O (c) 58.92% Na; 41.08% S (d) 29.08% Na; 40.56% S; 30.36% O (e) 55.26% K; 14.59% P; 30.15% O (f) 44.89% K; 0.5786% H; 17.78% P; 36.74% O (g) 28.73% K; 1.481% H; 22.76% P; 47.02% O (h) 79.10% K; 20.88% P **48.** (a) 54.75% Na (b) 25.25% Ti (c) 27.93% Fe (d) 24.05% Co (e) 5.926% H (f) 46.56% Al (g) 69.59% Ba (h) 6.523% Li **50.** (a) 34.43% Fe (b) 29.63% O (c) 92.25% C (d) 11.92% N (e) 93.10% Ag (f) 45.40% Co (g) 30.45% N (h) 43.66% Mn **52.** (a) 61.2% PO_4^{3-} (b) 46.1% SO_4^{2-} (c) 72.1% SO_4^{2-} (d) 56.3% Cl^- **54.** The empirical formula represents the smallest whole number ratio of the elements present in a compound. The molecular formula indicates the actual number of atoms of each element found in a molecule of the substance. **56.** a, c **58.** BaO **60.** $BaSO_4$ **62.** $N_2H_8CO_3$ [actually $(NH_4)_2CO_3$] **64.** Co_2S_3 **66.** CuO **68.** AlF_3 **70.** Li_3N **72.** $Al_2S_3O_{12}$ [actually $Al_2(SO_4)_3$] **74.** $NaClO_3$ **76.** molar mass **78.** C_6H_6 **80.** N_2O_4 **82.** empirical formula, C_3H_3O; molecular formula, $C_6H_6O_2$

84.

5.00 g Al	0.185 mol	1.11×10^{23} atoms
0.140 g Fe	0.00250 mol	1.51×10^{21} atoms
2.7×10^2 g Cu	4.3 mol	2.6×10^{24} atoms
0.00250 g Mg	1.03×10^{-4} mol	6.19×10^{19} atoms
0.062 g Na	2.7×10^{-3} mol	1.6×10^{21} atoms
3.95×10^{-18} g U	1.66×10^{-20} mol	1.00×10^4 atoms

86. 24.8% X, 17.4% Y, 57.8% Z. If the molecular formula were actually $X_4Y_2Z_6$, the percentage composition would be the same: the *relative* mass of each element present would not change. The molecular formula is always a whole number multiple of the empirical formula. **88.** Cu_2O, CuO **90.** (a) 4.31×10^{21} Pb; 8.62×10^{21} N; 2.58×10^{28} O atoms (b) 8.43×10^{23} O atoms (c) 1.914×10^{21} S atoms (d) 7.47×10^{16} U; 4.48×10^{17} F atoms **92.** (a) 4.141 g C; 52.96% C; 2.076×10^{23} atoms C (b) 0.0305 g C; 42.9% C; 1.53×10^{21} atoms C (c) 14.4 g C; 76.6% C; 7.22×10^{23} atoms C **94.** 10.3 g Pb **96.** 7.86 g Hg **98.** 2.555×10^{-22} g **100.** (a) 0.9330 g N (b) 1.388 g N (c) 0.8537 g N (d) 1.522 g N **102.** MgN_2O_6 $[Mg(NO_3)_2]$ **104.** average **106.** 8.61×10^{11} sodium atoms; 6.92×10^{24} amu **108.** (a) 2.0×10^2 g K (b) 0.0612 g Hg (c) 1.27×10^{-3} g Mn (d) 325 g P (e) 2.7×10^6 g Fe (f) 868 g Li (g) 0.2290 g F **110.** (a) 44.09 g (b) 399.9 g (c) 108.0 g (d) 342.3 g (e) 96.09 g (f) 208.2 g **112.** (a) 0.311 mol (b) 0.270 mol (c) 0.0501 mol (d) 2.8 mol (e) 6.2 mol **114.** (a) 4.1 g (b) 3.05×10^5 g (c) 0.533 g (d) 1.99×10^3 g (e) 4.18×10^3 g **116.** (a) 1.15×10^{23} molecules (b) 2.08×10^{23} molecules (c) 4.95×10^{22} molecules (d) 2.18×10^{22} molecules (e) 6.32×10^{20} formula units (substance is ionic) **118.** (a) 38.7% Ca; 20.0% P; 41.3% O (b) 53.9% Cd; 15.4% S; 30.7% O (c) 27.9% Fe; 24.1% S; 48.0% O (d) 43.7% Mn; 56.3% Cl (e) 29.2% N; 8.39% H; 12.5% C; 50.0% O (f) 27.4% Na; 1.2% H; 14.3% C; 57.1% O (g) 27.3% C; 72.7% O (h) 63.5% Ag; 8.25% N; 28.3% O **120.** (a) 36.8% Fe (b) 93.1% Ag (c) 55.3% Sr (d) 55.8% C (e) 37.5% C (f) 52.9% Al (g) 36.7% K (h) 52.5% K **122.** $C_3H_7N_2O$ **124.** HgO **126.** $BaCl_2$

CHAPTER 10

2. The coefficients of the balanced chemical equation for a reaction indicate the *relative numbers of moles* of each reactant that combine during the process, as well as the number of moles of each product formed. **4.** Balanced chemical equations tell us in what proportions *on a mole basis* substances combine; since the molar masses of C(s) and $O_2(g)$ are different, 1 g of O_2 could not represent the same number of moles as 1 g of C.
6. (a) $2HC_2H_3O_2(aq) + Ca(OH)_2(aq) \rightarrow Ca(C_2H_3O_2)_2(aq) + 2H_2O(l)$; Two molecules of aqueous acetic acid react with one formula unit of aqueous calcium hydroxide, producing one formula unit of aqueous calcium acetate and two molecules of water. Two moles of aqueous acetic acid react with one mole of aqueous calcium hydroxide, to produce one mole of calcium acetate and two moles of water. (b) $Ba(OH)_2(s) \rightarrow BaO(s) + H_2O(g)$; One formula unit of barium hydroxide decomposes to produce one formula unit of barium oxide and one molecule of water vapor. One mole of barium hydroxide reacts to produce one mole of barium oxide and one mole of water vapor. (c) $P_4(s) + 6H_2(g) \rightarrow 4PH_3(l)$; One molecule of phosphorus solid (which consists of four atoms) combines with 6 molecules of hydrogen gas, producing 4 moles of liquid phosphorus trihydride (phosphine). One mole of phosphorus (which consists of molecules containing four phosphorus atoms) reacts with 6 moles of hydrogen gas, producing four

moles of phosphine. (d) $2Al(s) + 3H_2SO_4(l) \rightarrow Al_2(SO_4)_3(s) + 3H_2(g)$; Two aluminum atoms react with three sulfuric acid molecules, making one formula unit of aluminum sulfate, and releasing three molecules of hydrogen gas. Two moles of aluminum metal react with three moles of pure sulfuric acid, producing one mole of solid aluminum sulfate and three moles of gaseous elemental hydrogen. **8.** False; reactions take place on a *mole basis* **10.** $CH_3CH_2OH(l) + 3O_2(g) \rightarrow 2CO_2(g) + 3H_2O(g)$; 2 mol CO_2/1 mol CH_3CH_2OH; 3 mol H_2O/1 mol CH_3CH_2OH **12.** (a) 0.20 mol Ag_2SO_4; 0.20 mol $Ni(NO_3)_2$ (b) 0.10 mol $Al_2(SO_4)_3$; 0.30 mol H_2 (c) 0.10 mol N_2; 0.30 mol I_2 (d) 0.067 mol Na_3PO_4; 0.20 mol H_2O **14.** (a) 1.3×10^{-3} mol Ag_2SO_4; 1.3×10^{-3} mol $Ni(NO_3)_2$ (b) 3.7×10^{-3} mol $Al_2(SO_4)_3$; 5.6×10^{-3} mol H_2 (c) 2.5×10^{-4} mol N_2; 7.6×10^{-4} mol I_2 (d) 1.7×10^{-3} mol Na_3PO_4; 5.0×10^{-3} mol H_2O **16.** (a) 9.375 mol H_2 (b) 2.344 mol O_2 (c) 1.563 mol CO_3^{2-} (d) 25.00 mol O_2 **18.** Stoichiometry is the process of using a chemical equation to calculate the relative masses of reactants and products involved in a reaction. **20.** (a) 1.178 mol (b) 5.76×10^{-5} mol (c) 9.6 mol (d) 2.69×10^{-5} mol (e) 0.118 mol (f) 1.7×10^{-8} mol (g) 8.40×10^{-4} mol **22.** (a) 3.68 g (b) 789 g (c) 25.3 g (d) 5.50×10^3 g (e) 0.458 g (f) 4.75×10^4 g (g) 0.107 g **24.** (a) 0.0312 mol (b) 0.0141 mol (c) 8.96×10^{-3} mol (d) 2.17×10^{-2} mol **26.** (a) 7.63 mg $FeCO_3$; 11.5 mg K_2SO_4 (b) 30.4 mg $CrCl_3$; 17.1 mg Sn (c) 21.6 mg Fe_2S_3 (d) 20.2 mg $AgNO_3$; 1.43 mg H_2O; 1.19 mg NO **28.** 7.94 g O_2 **30.** 0.972 g **32.** 32.4 g Zn; 8.92 g Al; 12.1 g Mg **34.** 9.33×10^{-3} g **36.** 1.6×10^{-2} g **38.** 2.75 g **40.** 2.07 g **42.** To determine the limiting reactant, first calculate the number of moles of each reactant present. Then determine how these numbers of moles correspond to the stoichiometric ratio indicated by the balanced chemical equation for the reaction. **44.** A reactant is present *in excess* if there is more of that reactant present than is needed to combine with the limiting reactant for the process. An excess of any reactant does not affect the theoretical yield for a process: the theoretical yield is determined by the limiting reactant. **46.** (a) HCl is limiting reactant; 18.3 g $AlCl_3$; 0.423 g H_2 (b) NaOH is limiting reactant; 19.9 g Na_2CO_3; 3.38 g H_2O (c) $Pb(NO_3)_2$ is limiting reactant; 12.6 g $PbCl_2$; 5.71 g HNO_3 (d) I_2 is limiting reactant; 19.6 g KI **48.** (a) Na is limiting reactant; 84.7 g $NaNH_2$ (b) $BaCl_2$ is limiting reactant; 56.0 g $BaSO_4$ (c) NaOH is limiting reactant; 78.8 g Na_2SO_3 (d) H_2SO_4 is limiting reactant; 58.2 g $Al_2(SO_4)_3$ **50.** (a) CO is the limiting reactant; 11.4 mg CH_3OH (b) I_2 is the limiting reactant; 10.7 mg AlI_3 (c) HBr is the limiting reactant; 12.3 mg $CaBr_2$; 2.22 mg H_2O (d) H_3PO_4 is the limiting reactant; 15.0 mg $CrPO_4$; 0.309 mg H_2 **52.** 136 g urea **54.** 14.5 g $FeCl_3$ produced; 0.5 g Cl_2 remains **56.** 0.627 g CuI; 0.418 g I_2; 0.573 g K_2SO_4 **58.** 0.67 kg SiC **60.** If the reaction is performed in a solvent, the product may have a substantial solubility in the solvent; the reaction may come to equilibrium before the full yield of product is achieved (see Chapter 16); loss of product may occur through operator error. **62.** 94.60% yield

64. 46.7 kg Si; 36.8% yield **66.** 82.6% **68.** 28.6 g NaHCO$_3$ **70.** 1.47 g CO$_2$ **72.** 14 mg K$_2$CrO$_4$ **74.** a. UO$_2$(s) + 4HF(aq) → UF$_4$(aq) + 2H$_2$O(l); One molecule of uranium(IV) oxide will combine with four molecules of hydrofluoric acid, producing one uranium(IV) fluoride molecule and two water molecules. One mole of uranium(IV) oxide will combine with four moles of hydrofluoric acid to produce one mole of uranium(IV) fluoride and two moles of water. b. 2NaC$_2$H$_3$O$_2$(aq) + H$_2$SO$_4$(aq) → Na$_2$SO$_4$(aq) + 2HC$_2$H$_3$O$_2$(aq); Two molecules (formula units) of sodium acetate react exactly with one molecule of sulfuric acid, producing one molecule (formula unit) of sodium acetate and two molecules of acetic acid. Two moles of sodium acetate will combine with one mole of sulfuric acid, producing one mole of sodium sulfate and two moles of acetic acid. c. Mg(s) + 2HCl(aq) → MgCl$_2$(aq) + H$_2$(g); One magnesium atom will react with two hydrochloric acid molecules (formula units) to produce one molecule (formula unit) of magnesium chloride and one molecule of hydrogen gas. One mole of magnesium will combine with two moles of hydrochloric acid, producing one mole of magnesium chloride and one mole of gaseous hydrogen. d. B$_2$O$_3$(s) + 3H$_2$O(l) → 2B(OH)$_3$(s); One molecule (formula unit) of diboron trioxide will react exactly with three molecules of water, producing two molecules of boron trihydroxide (boric acid). One mole of diboron trioxide will combine with three moles of water to produce two moles of boron trihydroxide (boric acid). **76.** for O$_2$, 5 mol O$_2$/1 mol C$_3$H$_8$; for CO$_2$, 3 mol CO$_2$/1 mol C$_3$H$_8$; for H$_2$O, 4 mol H$_2$O/1 mol C$_3$H$_8$ **78.** (a) 0.0588 mol NH$_4$Cl (b) 0.0178 mol CaCO$_3$ (c) 0.0218 mol Na$_2$O; (d) 0.0322 mol PCl$_3$ **80.** (a) 3.2 × 10^2 g HNO$_3$ (b) 0.0612 g Hg (c) 4.49 × 10^{-3} g K$_2$CrO$_4$ (d) 1.400 × 10^3 g AlCl$_3$ (e) 7.2 × 10^6 g SF$_6$ (f) 2.13 × 10^3 g NH$_3$ (g) 0.9396 g Na$_2$O$_2$ **82.** 1.9 × 10^2 kg SO$_2$ **84.** 56.3 g Br$_2$ **86.** 0.0771 g H$_2$ **88.** (a) Br$_2$ is limiting reactant; 6.4 g NaBr (b) CuSO$_4$ is limiting reactant; 5.1 g ZnSO$_4$; 2.0 g Cu (c) NH$_4$Cl is the limiting reactant; 1.6 g NH$_3$; 1.7 g H$_2$O; 5.5 g NaCl (d) Fe$_2$O$_3$ is the limiting reactant; 3.5 g Fe; 4.1 g CO$_2$ **90.** 0.624 mol N$_2$, 17.5 g N$_2$; 1.25 mol H$_2$O, 22.5 g H$_2$O **92.** 5 g

CHAPTER **11**

2. The wavelength (λ) represents the distance between two corresponding points (peaks, troughs, etc.) on successive cycles of a wave. **4.** The molecule moves or rotates in space at a higher speed, and atoms in the molecule vibrate more vigorously. **6.** 10^{-7} **8.** photon **10.** lower (the frequency of red light is lower than that of blue light) **12.** Only *certain* energy levels are allowed to the electron in the hydrogen atom. These levels correspond to definite, distinct *energies*. When an electron moves from one allowed level to another, a characteristic photon of radiation is emitted. **14.** lower energy state [often the ground (lowest energy) state] **16.** The ground state of an atom is its lowest possible energy state. **18.** According to Bohr, electrons move in dis-

crete, fixed circular *orbits* around the nucleus. If the wavelength of the applied energy corresponds to the *difference in energy* between the two orbits, the atom absorbs a photon and the electron moves to a larger orbit. **20.** Bohr's theory *explained* the experimentally *observed* line spectrum of hydrogen *exactly*. Bohr's theory was ultimately discarded because when attempts were made to extend the theory to atoms other than hydrogen, the calculated properties did *not* correspond closely to experimental measurements. **22.** An orbit represents a definite, exact circular pathway around the nucleus in which an electron can be found. An orbital represents a region of space in which there is a high probability of finding the electron. **24.** Any experiment which sought to measure the exact location of an electron (such as shooting a beam of light at it) would cause the electron to move. Any measurement made would necessitate the application or removal of energy, which would disturb the electron from where it had been before the measurement. **26.** The principal energy levels represent sets of orbitals at a particular average distance from the nucleus and a particular average energy in which electrons may reside. With the Bohr theory, the nucleus was surrounded by a series of circular orbits of fixed radius in which electrons moved. **28.** The 2p orbitals have two lobes and are sometimes described as having a "dumbbell" shape. The individual 2p orbitals (2p$_x$, 2p$_y$, and 2p$_z$) are alike in shape and energy; they differ only in the direction in which the lobes of the orbital are oriented. **30.** increases **32.** four: 4s, 4p, 4d, and 4f **34.** The Pauli exclusion principle states that no two electrons in an atom can have the same four quantum numbers. As a result of this principle, only two electrons can occupy a given orbital, since there are only two possible values for the m$_s$ quantum number (electron spin). **36.** increases **38.** probability **40.** a, b, and e are incorrect **42.** Each of the three 2p orbitals of N will be occupied by a single, unpaired electron (according to Hund's rule);
(↿⇂) (↿⇂) (↑) (↑) (↑)
  1s   2s   2p
44. The elements in a given vertical column of the periodic table have the same valence electron configuration. **46.** (a) 1s^2 2s^2 2p^6 3s^2 3p^6 4s^2 3d^{10} 4p^6 5s^2 (b) 1s^2 2s^2 2p^6 3s^2 3p^6 4s^2 3d^{10} (c) 1s^2 (d) 1s^2 2s^2 2p^6 3s^2 3p^6 4s^2 3d^{10} 4p^5 **48.** (a) 1s^2 2s^2 2p^2 (b) 1s^2 2s^2 2p^6 3s^2 3p^3 (c)1s^2 2s^2 2p^6 3s^2 3p^4 (d)1s^2 2s^2 2p^1
50. (a) (↿⇂) (↿⇂) (↿⇂)(↿⇂)(↿⇂) (↿⇂) (↑)( )( )
       1s   2s   2p   3s   3p

(b) (↿⇂) (↿⇂) (↿⇂)(↿⇂)(↿⇂) (↿⇂) (↑)(↑)(↑)
     1s   2s   2p   3s   3p

(c) (↿⇂) (↿⇂) (↿⇂)(↿⇂)(↿⇂) (↿⇂) (↿⇂)(↿⇂)(↿⇂) (↿⇂)
     1s   2s   2p   3s   3p   4s

(↿⇂)(↿⇂)(↿⇂)(↿⇂)(↿⇂) (↿⇂)(↿⇂)(↑)
        3d   4p
(d) (↿⇂) (↿⇂) (↿⇂)(↿⇂)(↿⇂) (↿⇂) (↿⇂)(↿⇂)(↿⇂)
     1s   2s   2p   3s   3p

52. (a) two (b) four (c) seven (d) one **54.** The core electrons of an atom are those of the inner shells. The core electrons are those which are not valence electrons. **56.** (a) [Ar] $4s^2$ (b) [Rn] $7s^1$ (c) [Kr] $5s^24d^1$ (d) [Xe] $6s^24f^15d^1$ **58.** (a) [Ne] $3s^23p^2$ (b) [Ar] $4s^23d^{10}4p^1$ (c) [Ar] $4s^23d^3$ (d) [Ar] $4s^23d^1$ **60.** (a) one (b) two (c) zero (d) 10 **62.** (a) $5d$ (b) $4f$ (c) $5d$ (d) $3d$ **64.** (a) [Rn] $7s^25f^36d^1$ (b) [Ar] $4s^23d^5$ (c) [Xe] $6s^25d^{10}$ (d) [Rn] $7s^1$ **66.** The metallic elements *lose* electrons and form *positive* ions (cations); the nonmetallic elements *gain* electrons and form *negative* ions (anions). **68.** All exist as *diatomic* molecules (F_2, Cl_2, Br_2, I_2); all are *non-metals*; all have relatively high electronegativities; all form $1-$ ions in reacting with metallic elements. **70.** Elements at the *left* of a period (horizontal row) lose electrons more readily; at the left of a period (given principal energy level) the nuclear charge is the smallest and the electrons are least tightly held. **72.** The elements of a given period (horizontal row) have valence electrons in the same subshells. However, nuclear charge increases across a period going from left to right. Atoms at the left side have smaller nuclear charges, and hold onto their valence electrons less tightly. **74.** The *nuclear charge* increases from left to right within a period, pulling progressively more tightly on the valence electrons. **76.** (a) Cs (b) Li (c) I (d) Rb **78.** (a) Na (b) S (c) N (d) F **80.** speed of light **82.** photons **84.** quantized **86.** orbital **88.** transition metal **90.** spins (different m_s quantum numbers)

92. (a) $1s^22s^22p^63s^23p^64s^1$ [Ar] $4s^1$
(↑↓) (↑↓) (↑↓)(↑↓)(↑↓) (↑↓) (↑↓)(↑↓)(↑↓) (↑)
 $1s$ $2s$ $2p$ $3s$ $3p$ $4s$

(b) $1s^22s^22p^63s^23p^64s^23d^2$ [Ar] $4s^23d^2$
(↑↓) (↑↓) (↑↓)(↑↓)(↑↓) (↑↓) (↑↓)(↑↓)(↑↓) (↑↓) (↑)(↑)( )( )( )
 $1s$ $2s$ $2p$ $3s$ $3p$ $4s$ $3d$

(c) $1s^22s^22p^63s^23p^2$ [Ne] $3s^23p^2$
(↑↓) (↑↓) (↑↓)(↑↓)(↑↓) (↑↓) (↑)(↑)( )
 $1s$ $2s$ $2p$ $3s$ $3p$

(d) $1s^22s^22p^63s^23p^64s^23d^6$ [Ar] $4s^23d^6$
(↑↓) (↑↓) (↑↓)(↑↓)(↑↓) (↑↓) (↑↓)(↑↓)(↑↓) (↑↓) (↑↓)(↑)(↑)(↑)(↑)
 $1s$ $2s$ $2p$ $3s$ $3p$ $4s$ $3d$

(e) $1s^22s^22p^63s^23p^64s^23d^{10}$ [Ar] $4s^23d^{10}$
(↑↓) (↑↓) (↑↓)(↑↓)(↑↓) (↑↓) (↑↓)(↑↓)(↑↓) (↑↓) (↑↓)(↑↓)(↑↓)(↑↓)(↑↓)
 $1s$ $2s$ $2p$ $3s$ $3p$ $4s$ $3d$

94. (a) ns^2 (b) ns^2np^5 (c) ns^2np^4 (d) ns^1 (e) ns^2np^4 **96.** (a) 2.7×10^{-12} m (b) 4.4×10^{-34} m (c) 2×10^{-35} m.   The wavelengths for the ball and the person are infinitesimally small, whereas the wavelength for the electron is nearly the same order of magnitude as the diameter of an atom. **98.** Light is emitted from hydrogen atom only at certain fixed wavelengths. If the energy levels of hydrogen were *continuous*, a hydrogen atom would emit energy at all possible wavelengths. **100.** The third principal energy level of hydrogen is divided into

three sublevels ($3s$, $3p$, and $3d$); there is a *single* $3s$ orbital; there is a set of *three* $3p$ orbitals; there is a set of *five* $3d$ orbitals. **102.** a, c **104.** (a) $1s^22s^22p^63s^23p^5$ (b) $1s^22s^22p^4$ (c) $1s^22s^22p^63s^23p^1$ (d) $1s^22s^1$ **106.** (a) five ($2s$, $2p$) (b) seven ($2s$, $2p$) (c) one ($4s$) (d) three ($3s$, $3p$) **108.** (a) [Kr] $5s^24d^2$ (b) [Kr] $5s^24d^{10}5p^5$ (c) [Ar] $4s^23d^{10}4p^2$ (d) [Xe] $6s^1$ **110.** The *position* of the element (both in terms of the vertical column and the horizontal row) tells you which set of orbitals is being filled last. See Figure 11.27 for details. (a) $3d$ (b) $4d$ (c) $5f$ (d) $4p$ **112.** metals, low; nonmetals, high **114.** (a) Ba (b) Mg (c) Be

CHAPTER **12**

2. bond energy **4.** ionic **6.** polar **8.** electronegativity **10.** difference **12.** In each case, the element *higher up* within a group of the periodic table has the higher electronegativity. (a) K < Ca < Sc (b) At < Br < F (c) C < N < O **14.** (a) covalent (b) polar covalent (c) polar covalent (d) ionic **16.** c and d **18.** (a) H—O (1.4); H—N (0.9); the H—O bond is more polar. (b) H—N (0.9); H—F (1.9); the H—F bond is more polar. (c) H—O (1.4); H—F (1.9); the H—F bond is more polar. (d) H—O (1.4); H—Cl (0.9); the H—O bond is more polar. **20.** (a) Na—O (b) K—S (c) K—Cl (d) Na—Cl **22.** The presence of strong bond dipoles and a large overall dipole moment in water make it a very polar substance overall. Among those properties of water that are dependent on its dipole moment are its freezing point, melting point, vapor pressure, and ability to dissolve many substances. **24.** (a) H (b) Cl (c) I **26.** In the figures, the arrow points toward the more electronegative atom. (a) δ^+ Cl → F δ^- (b) δ^+ I → Br δ^- (c) δ^+ Br → Cl δ^- (d) δ^+ Br → F δ^- **28.** In the figures, the arrow points toward the more electronegative atom. (a) δ^+ P → S δ^- (b) δ^+ S → O δ^- (c) δ^+ S → N δ^- (d) δ^+ S → Cl δ^- **30.** previous **32.** Atoms in covalent molecules gain a configuration like that of a noble gas by sharing one or more pairs of electrons between atoms: such shared pairs of electrons "belong" to each of the atoms of the bond at the same time. In ionic bonding, one atom completely gives over one or more electrons to another atom, and then the resulting ions behave independently of one another (they are not "attached" to each other as in the case of a covalent bond, although they are attracted to each other). **34.** (a) Rb, $1s^22s^22p^63s^23p^64s^23d^{10}4p^65s^1$; Rb$^+$, $1s^22s^22p^63s^23p^64s^23d^{10}4p^6$; Kr has the same configuration as Rb$^+$ (b) O, $1s^22s^22p^4$; O^{2-}, $1s^22s^22p^6$; Ne has the same configuration as O^{2-} (c) Sr, $1s^22s^22p^63s^23p^64s^23d^{10}4p^65s^2$; Sr^{2+}, $1s^22s^22p^63s^23p^64s^23d^{10}4p^6$; Kr has the same configuration as Sr^{2+} (d) Cs, $1s^22s^22p^63s^23p^64s^23d^{10}4p^65s^24d^{10}5p^66s^2$; Cs$^+$, $1s^22s^22p^63s^23p^64s^23d^{10}4p^65s^24d^{10}5p^6$; Xe has the same configuration as Cs$^+$ (e) Ba, $1s^22s^22p^63s^23p^64s^23d^{10}4p^65s^24d^{10}5p^66s^2$; Ba^{2+}, $1s^22s^22p^63s^23p^64s^23d^{10}4p^65s^24d^{10}5p^6$; Xe has the same configuration as Ba^{2+} **36.** (a) Ca^{2+} (b) N^{3-} (c) Br$^-$ (d) Sc$^+$

38. (a) CaO (b) SrF_2 (c) Al_2S_3 (d) Rb_3N (e) Na_2Te
40. (a) Al^{3+}, [Ne]; S^{2-}, [Ar] (b) Mg^{2+}, [Ne]; N^{3-}, [Ne]
(c) Rb^+, [Kr]; O^{2-}, [Ne] (d) Cs^+, [Xe]; I^-, [Xe]   **42.** An ionic
solid such as NaCl consists of an array of alternating positively and
negatively charged ions: that is, each positive ion has as its nearest neighbors a group of negative ions, and each negative ion has
a group of positive ions surrounding it. In most ionic solids, the
ions are packed as tightly as possible.   **44.** A polyatomic ion
such as SO_4^{2-} can be thought of as a covalently bonded *molecule*
that carries a *net charge* (a polyatomic ion is an ion and can attract oppositely-charged ions).   **46.** (a) F^- (b) Cl^- (c) O^{2-}
(d) I^- **48.** (a) I^- (b) Na^+ (c) Cs^+ (d) O   **50.** When
atoms form covalent bonds, they try to attain a valence electronic
configuration similar to that of the nearest noble gas element.
When the elements in the first few horizontal rows of the periodic
table form covalent bonds, they will attempt to gain configurations
similar to the noble gases helium (2 valence electrons, duet rule),
and neon and argon (8 valence electrons, octet rule).   **52.** These
elements attain a total of eight valence electrons, making the valence electron configurations similar to those of the noble gases
Ne and Ar.   **54.** two

56. (a) Rb·   (d) Ba :

(b) : Cl ·   (e) : P ·

(c) :Kr: (f) : At ·

58. (a) 32 (b) 17 (c) 30 (d) 14
60. (a) NH_3 H—N—H
                          |
                          H

(b) CI_4 : I—C—I :

(c) NCl_3 : Cl—N—Cl :
                     |
                   : Cl :

(d) $SiBr_4$: Br—Si—Br :
                     |
                   : Br :

62. (a) H_2S H—S—H

(b) SiF_4 : F—Si—F :
                     |
                   : F :

(c) C_2H_4 H   H
                 \ /
                  C=C
                 / \
                H   H

(d) C_3H_8 H H H
                | | |
             H—C—C—C—H
                | | |
                H H H

64. (a) NO_2 O=N—O· ↔ ·O—N=O ↔ O=N—O: ↔ :O—N=

(b) H_2SO_4 H—O—S—O—H (with :O: above and :O: below S)

(c) N_2O_4 O=N—N=O ↔ :O—N—N— (with :O::O: above and :O::O: above)

66. (a) ClO_3^-

(:O—Cl—O:)⁻ with :O: below Cl

(b) O_2^{2-}

(:O—O:)²⁻

(c) $C_2H_3O_2^-$ (H—C—C—O:)⁻ ↔ (H—C=C=O)⁻ with H's and :O:

68. (a) CN^- (: C≡N :)⁻

(b) HSO_4^- (H—O—S—O:)⁻ with :O: above and :O: below S

(c) N_3^- (N=N=N)⁻

70. The geometric structure of NH_3 is that of a trigonal pyramid.
The nitrogen atom of NH_3 is surrounded by four electron pairs
(three are bonding, one is a lone pair). The H—N—H bond angle
is somewhat less than 109.5° (due to the presence of the lone
pair).   **72.** The geometric structure of CH_4 is that of a tetrahedron. The carbon atom of CH_4 is surrounded by four bonding electron pairs. The H—C—H bond angle is the characteristic angle of
the tetrahedron, 109.5°.   **74.** The general molecular structure of
a molecule is determined by how many electron pairs surround
the central atom in the molecule, and by which of those electron
pairs are used for bonding to the other atoms of the molecule.

76. You will remember from high school geometry, that two points in space are all that is needed to indicate a straight line. A diatomic molecule represents two points in space. **78.** In NF_3, the nitrogen atom has *four* pairs of valence electrons, whereas in BF_3, there are only *three* pairs of valence electrons around the boron atom. The nonbonding pair on nitrogen in NF_3 pushes the three F atoms out of the plane of the N atom. **80.** Each of the indicated atoms is surrounded by four pairs of electrons with a tetrahedral orientation. Although the electron pairs are arranged tetrahedrally, realize that the overall geometric shape of the molecules may *not* be tetrahedral. **82.** (a) tetrahedral (b) trigonal pyramid (c) nonlinear (bent) **84.** (a) trigonal pyramid (b) basically tetrahedral around P (distorted somewhat by the H atom) (c) trigonal pyramid **86.** (a) approximately 120° (for H_2CO_3) (b) approximately 109.5° (c) 109.5° d. approximately 120° **88.** In a covalent bond between two atoms of the same element, the electron pair is shared equally and the bond is nonpolar; with a bond between atoms of different elements, the electron pair is unequally shared and the bond is polar (assuming the elements have different electronegativities). **90.** double **92.** (a) S—F (b) P—O (c) C—H **94.** break **96.** (a) Be (b) N (c) F **98.** a, c **100.** (a) hydrogen (b) iodine (c) nitrogen **102.** (a) Al $1s^22s^22p^63s^23p^1$; $Al^{3+}1s^22s^22p^6$; Ne has the same configuration as Al^{3+}. (b) Br $1s^22s^22p^63s^23p^64s^23d^{10}4p^5$; Br^- $1s^22s^22p^63s^23p^64s^23d^{10}4p^6$; Kr has the same configuration as Br^-. (c) Ca $1s^22s^22p^63s^23p^64s^2$; Ca^{2+} $1s^22s^22p^63s^23p^6$; Ar has the same configuration as Ca^{2+}. (d) Li $1s^22s^1$; Li^+ $1s^2$; He has the same configuration as Li^+. (e) F $1s^22s^22p^5$; F^- $1s^22s^22p^6$; Ne has the same configuration as F^-. **104.** (a) Na_2Se (b) RbF (c) K_2Te (d) BaSe (e) KAt (f) FrCl **106.** (a) Na^+ (b) Al^{3+} (c) F^- (d) Na^+ **108.** (a) 8 (b) 48 (c) 34 (d) 26

110. (a) N_2H_4

H—N—N—H
(with H H below each N)

(b) C_2H_6

H—C—C—H
(with H H above and below)

(c) NCl_3 : Cl—N—Cl : (with : Cl : below)

(d) $SiCl_4$: Cl—Si—Cl : (with : Cl : above and : Cl : below)

112. (a) NO_3^-

$$\left[\ddot{O}=N-\ddot{O}: \right]^- \leftrightarrow \left[:\ddot{O}-N=\ddot{O} \right]^- \leftrightarrow \left[:\ddot{O}-N-\ddot{O}: \right]^-$$
(with : O : below each)

(b) CO_3^{2-}

$$\left[\ddot{O}=C-\ddot{O}: \right]^{2-} \leftrightarrow \left[:\ddot{O}-C=\ddot{O} \right]^{2-} \leftrightarrow \left[:\ddot{O}-C-\ddot{O}: \right]^{2-}$$
(with : O : below each)

(c) NH_4^+

$$\left[\begin{array}{c} H \\ H-N-H \\ H \end{array} \right]^+$$

114. (a) four pairs arranged tetrahedrally (b) four pairs arranged tetrahedrally (c) three pairs arranged trigonally (planar) **116.** (a) trigonal pyramid (b) nonlinear (V-shaped) (c) tetrahedral **118.** (a) nonlinear (V-shaped) (b) trigonal planar (c) basically trigonal planar around the S, distorted somewhat by the H (d) linear

CHAPTER 13

2. The "pressure of the atmosphere" represents the weight of the several-mile-thick layer of gases pressing down on every surface of the earth. Atmospheric pressure is most commonly measured with a mercury barometer (the pressure of the atmosphere is sufficient to maintain the height of a column of mercury to approximately 76 cm at sea level). The pressure of the atmosphere varies on the surface of the earth due to weather conditions, and also varies with altitude. **4.** 760 **6.** less **8.** (a) 1.088 atm (b) 0.976 atm (c) 0.580 atm (d) 1.21 atm **10.** (a) 775 mm Hg (b) 911 mm Hg (c) 792 mm Hg (d) 81.8 mm Hg **12.** (a) 2.07×10^3 kPa (b) 106.0 kPa (c) 1.10×10^3 kPa (d) 87.9 kPa **14.** increases **16.** $PV = k$; $P_1V_1 = P_2V_2$ **18.** (a) 588 mL (b) 2.53 L (c) 4.26×10^4 mm **20.** (a) 3.17×10^3 mL (b) 5.67×10^4 mL (c) 4.22×10^3 mL **22.** factor $(1.01/0.562) = 1.80$ **24.** 20.0 atm **26.** absolute zero, 0 K **28.** $V = bT$; $V_1/T_1 = V_2/T_2$ **30.** 569 mL **32.** (a) 546 K = 273 °C (b) 308 K = 35 °C (c) 298 mL **34.** (a) 35.4 K = -238 °C (b) 0 mL [absolute zero] (c) 40.5 mL **36.** 73.5 L **38.** 113 mL (238 K, -35 °C); 142 mL (299 K, 26 °C); 155 mL (326 K, 53 °C); 127 mL (267 K, -6 °C) **40.** $V = an$; $V_1/n_1 = V_2/n_2$ **42.** 44.8 L **44.** 10.8 L **46.** L atm/K mol (L atm K^{-1} mol^{-1}) **48.** Boyle's Law: constant amount of gas and constant temperature. $PV = nRT$; $PV = $ (constant)R(constant); $PV = $ constant. Charles's Law: constant pressure and constant amount of gas; $PV = nRT$; (constant)$V = $ (constant)RT; $V = $ (constant)T. Avogadro's Law: constant pressure and temperature. $PV = nRT$; (constant)$V = nR$(constant); $V = $ (constant)n **50.** (a) 5.02 L (b) 3.56 atm = 2.70×10^3 mm Hg (c) 334 K **52.** (a) 0.201 L (b) 0.338 atm (c) 2.48×10^{-2} mol **54.** 3.89 L **56.** 4.44×10^3 g He; 2.24×10^3 g H_2

58. 654 K = 381 °C **60.** 67.2 atm **62.** 85.2 mL **64.** 488 K = 215 °C **66.** The total pressure in a sample of gas which has been collected by bubbling through water is made up of two components: the pressure of the gas of interest and the pressure of water vapor. The partial pressure of the gas of interest is then the total pressure of the sample minus the vapor pressure of water. **68.** 7.94 L **70.** 2.47 atm **72.** 751.4 mm Hg **74.** $P_{hydrogen}$ = 0.990 atm; 9.56 × 10^{-3} mol H_2; 0.625 g Zn **76.** A theory is successful if it explains known experimental observations. Theories that have been successful in the past may not be successful in the future (for example, as technology evolves, more sophisticated experiments may be possible in the future). **78.** collisions **80.** no **82.** If the temperature of a sample of gas is increased, the average kinetic energy of the particles of gas increases. This means that the speeds of the particles increase. If the particles have a higher speed, they will hit the walls of the container more frequently and with greater force, thereby increasing the pressure. **84.** STP = 0 °C, 1 atm pressure. These conditions were chosen because they are easy to attain and reproduce *experimentally.* The barometric pressure within a laboratory is likely to be near 1 atm on most days, and 0 °C can be attained with a simple ice bath. **86.** 3.00 L **88.** 0.432 L **90.** 1.17 L **92.** 5.03 L (dry volume) **94.** 16.6 L **96.** 184 mL **98.** 40.5 L; P_{He} = 0.864 atm; P_{Ne} = 0.136 atm **100.** 24.5 L Cl_2 **102.** 0.708 L Cl_2 **104.** twice **106.** (a) $PV = k$; $P_1V_1 = P_2V_2$ (b) $V = bT$; $V_1/T_1 = V_2/T_2$ (c) $V = an$; $V_1/n_1 = V_2/n_2$ (d) $PV = nRT$ (e) $P_1V_1/T_1 = P_2V_2/T_2$ **108.** 125 balloons **110.** 12.5 mL **112.** 0.100 mol CO_2; 3.32 L wet; 2.68 L dry **114.** 18.1 L O_2 **116.** (a) 0.875 atm (b) 1.22 atm (c) 25.07 atm (d) 1.06 atm **118.** (a) 8.60 × 10^4 Pa (b) 2.21 × 10^5 Pa (c) 8.87 × 10^4 Pa (d) 4.3 × 10^3 Pa **120.** (a) 128 mL (b) 1.3 × 10^{-2} L (c) 9.8 L **122.** 8.34 × 10^5 mm Hg (1.10 × 10^3 atm) **124.** (a) 56.9 mL (b) 448 K = 175 °C (c) zero **126.** 15 L **128.** 2.59 g **130.** (a) 61.8 L (b) 0.993 atm (c) 1.67 × 10^4 L **132.** 487 mol gas needed; 7.79 kg CH_4; 13.6 kg N_2; 21.4 kg CO_2 **134.** 0.42 atm **136.** 2.51 × 10^3 K **138.** 17 L **140.** 3.43 L N_2; 10.3 L H_2 **142.** 5.8 L O_2; 3.9 L SO_2 **144.** 7.8 × 10^2 L **146.** 32 L; P_{He} = 0.86 atm; P_{Ar} = 0.017 atm; P_{Ne} = 0.12 atm **148.** 22.5 L O_2

CHAPTER 14

2. Water, as perspiration, helps cool the human body; on a hot day, you might take an extra shower just to cool off; large bodies of water (e.g., the oceans) have a cooling effect on nearby land masses. Water is used as a coolant in *many* commercial situations: for example, some nuclear power plants use water to cool the reactor core; many office buildings are air-conditioned by circulating cold water combined with fan systems. **4.** The fact that water expands when it freezes often results in broken water pipes during cold weather. The expansion of water when it freezes also makes ice float on liquid water. **6.** Sloped portions of a heating/cooling curve represent *changes in temperature* as heat is applied or removed; flat portions of such curves represent equilibrium transitions between states. **8.** As a sample of liquid water is heated, the motion of the molecules increases as the temperature rises. As the liquid reaches its boiling point, bubbles of vapor begin to form in the liquid, which rise to the surface of the liquid and burst. As the liquid remains at its boiling point, the additional heat energy being supplied to the liquid is converted to increased motion (kinetic energy) of the molecules. As heat energy continues to be applied, more and more molecules will be moving in the right direction and with sufficient energy to escape for the liquid's surface. **10.** The added energy ends up as the energy of motion (kinetic energy) of the particles, and serves to overcome the intermolecular forces in the liquid. **12.** In a vapor, the molecules are so far apart that intermolecular forces become negligible, whereas in a liquid, the intermolecular forces are still quite strong. It takes more heat to vaporize a liquid than to melt the same amount of solid because of the greater degree to which the intermolecular forces must be overcome. **14.** 0.48 kJ; 27 kJ **16.** 15 kJ **18.** 107 kJ/mol; 39.5 kJ; 1.02 × 10^3 kJ **20.** Dipole–dipole forces are relatively short-range forces. **22.** Water molecules are able to form strong hydrogen bonds with each other. These bonds are an especially strong form of dipole–dipole forces and are only possible when hydrogen atoms are bonded to the most electronegative elements (N, O, and F). The extra strong intermolecular forces in H_2O require much higher temperatures (high energies) to be overcome. **24.** The fact that such nonpolar, monatomic molecules can be liquefied and solidified indicates that there must be *some* sort of intermolecular forces possible between atoms in these substances. London dispersion forces arise when a temporary (instantaneous) dipolar arrangement of charge develops as the electrons of an atom move around its nucleus. The positive nucleus of one atom can then weakly attract the displaced electron cloud of another atom. **26.** (a) London dispersion forces (b) London dispersion forces (c) dipole–dipole forces (d) hydrogen bonding **28.** An increase in the heat of fusion is observed for an increase in the size of the halogen atom involved (the electron cloud of a larger atom is more easily polarized by an approaching dipole, thus giving larger London dispersion forces). **30.** For a homogeneous mixture to be able to form at all, the forces between molecules of the two substances being mixed must be at least *comparable in magnitude* to the intermolecular forces of each *separate* substance. Apparently in the case of a water–ethanol mixture, the forces that exist when water and ethanol are mixed are stronger than water–water or ethanol–ethanol forces in the separate substances. This allows ethanol and water molecules to approach each other more closely in the mixture than either substance's molecules could approach a like molecule in the separate substances. There is strong hydrogen bonding in both ethanol and water. **32.** Vapor pressure is the pressure of vapor present *at equilibrium* above a liquid in a sealed container at a particular temperature. When a liquid is placed in a closed container, molecules of the liquid evaporate freely into the empty space above the liquid. As the number of molecules present in the vapor state increases with time, vapor molecules begin to rejoin the liquid state (condense). Eventually a dynamic equilibrium is reached between evaporation and condensation in which the net number of molecules present in the vapor phase becomes *constant* with time.   **34**. At a given temperature, heavier molecules move more slowly than lighter molecules, and have less tendency to escape from a liquid. All other considerations being equal, low molar mass substances tend to be more volatile.   **36**. (a) HF. Although both substances are capable of hydrogen bonding, water has two O—H bonds which can be involved in hydrogen bonding versus only one F—H bond in HF. (b) CH_3OCH_3. Since there is no H attached to the O atom, no hydrogen bonding can exist. Since there is no hydrogen bonding possible, the molecule should be relatively more volatile than CH_3CH_2OH even though it contains the same number of atoms of each element.   (c) CH_3SH. Hydrogen bonding is not as important for a S—H bond (because S has a lower electronegativity than O). Since there is little hydrogen bonding, the molecule is relatively more volatile than CH_3OH.   **38**. Both substances have the same molar mass. However ethyl alcohol contains an hydrogen atom directly bonded to an oxygen atom. Therefore, hydrogen bonding can exist in ethyl alcohol, whereas only weak dipole–dipole forces can exist in dimethyl ether. Ethyl alcohol therefore is less volatile and has a higher boiling point.   **40**. *Ionic* solids have as their fundamental particles positive and negative *ions;* a simple example is sodium chloride, in which Na^+ and Cl^- ions are held together by strong electrostatic forces. *Molecular* solids have molecules as their fundamental particles, with the molecules being held together in the crystal by dipole–dipole forces, hydrogen bonding forces, or London dispersion forces (depending on the identity of the substance); simple examples of molecular solids include ice (H_2O) and ordinary table sugar (sucrose). *Atomic* solids have simple atoms as their fundamental particles, with the atoms being held together in the crystal either by covalent bonding (as in graphite or diamond) or by metallic bonding (as in copper or other metals).   **42**. The fundamental particles in molecular solids are individual *molecules;* ice (individual water molecules), table sugar (individual sucrose molecules)   **44**. Ionic solids consist of a crystal lattice of basically alternating positively and negatively charged ions. A given ion is surrounded by several ions of the opposite charge, all of which electrostatically attract it strongly. This pattern repeats itself throughout the crystal. Sometimes a crystal of an ionic substance is described as being like one giant molecule, with all the ions bonded to each other.   **46**. Ordinary ice contains nonlinear, highly polar water molecules. In addition, there is extensive, strong hydrogen bonding possible between water molecules in ordinary ice. Dry ice, on the other hand, consists of linear, nonpolar molecules, and only very weak intermolecular forces are possible.   **48**. In a metal, the valence electrons are mobile,

and can move throughout the entire metal crystal's lattice. In ionic solids, although there are positive and negative ions present, each ion is held rigidly in place by several ions of the opposite charge.   **50**. Alloys may be of two types: *substitutional* (in which one metal is substituted for another in the regular positions of the crystal lattice) and *interstitial* (in which a second metal's atoms fit into the empty space in a given metal's crystal lattice). The presence of atoms of a second metal in a given metal's crystal lattice changes the properties of the metal: frequently the alloy is stronger than either of the original metals because the irregularities introduced into the crystal lattice by the presence of a second metal's atoms prevent the crystal from being deformed as easily. The properties of iron may be modified by alloying with many different substances, particularly with carbon, nickel, and cobalt. Steels with relatively high carbon content are exceptionally strong, whereas steels with low carbon content are softer, more malleable, and more ductile. Steels produced by alloying iron with nickel and cobalt are more resistant to corrosion than iron itself.   **52**. j   **54**. f   **56**. d   **58**. a   **60**. l   **62**. Dimethyl ether has the larger vapor pressure. No hydrogen bonding is possible since the O atom does not have a hydrogen atom attached. Hydrogen bonding can occur *only* when a hydrogen atom is *directly* attached to a strongly electronegative atom (such as N, O, or F). Hydrogen bonding *is* possible in ethanol (ethanol contains an —OH group).   **64**. (a) H_2. London dispersion forces are the only intermolecular forces present in these nonpolar molecules; typically London forces become larger with increasing atomic size (as the atoms become bigger, the edge of the electron cloud lies farther from the nucleus and becomes more easily distorted).   (b) Xe. Only the relatively weak London forces could exist in a crystal of Xe atoms, whereas in NaCl strong ionic forces exist, and in diamond strong covalent bonding exists between carbon atoms.   (c) Cl_2. Only London forces exist among such nonpolar molecules. London forces become larger with increasing atomic size.   **66**. Steel is a general term applied to alloys consisting primarily of iron, but with small amounts of other substances added. Whereas pure iron itself is relatively soft, malleable, and ductile, steels are typically much stronger and harder, and much less subject to damage.   **68**. Water is the solvent in which cellular processes take place in living creatures. Water in the oceans moderates the earth's temperature. Water is used in industry as a cooling agent. Water serves as a means of transportation on the earth's oceans. The liquid range is 0°C to 100°C.   **70**. At higher altitudes, the boiling points of liquids, such as water, are lower because there is a lower atmospheric pressure above the liquid. The temperature at which food cooks is determined by the temperature to which the water in the food can be heated before it escapes as steam. Thus, food cooks at a lower temperature at high elevations where the boiling point of water is lowered.   **72**. Heat of fusion (melt); Heat of vaporization (boil). The heat of vaporization is always larger, because virtually all of the intermolecular forces must be overcome to form a gas. In a liquid, considerable intermolecular forces remain. Thus going from a solid to liquid requires less energy than going from the liquid to the gas.   **74.** Dipole–dipole interactions are typically about 1% as strong as a covalent bond. Dipole–dipole interactions represent electrostatic attractions between portions of molecules which carry only a *partial* positive or negative charge, and such forces require the molecules that are interacting to come *near* enough to each other.   **76.** London dispersion forces are relatively weak forces that arise among noble gas atoms and in nonpolar molecules. London forces are due to *instantaneous dipoles* that develop when one atom (or molecule) momentarily distorts the electron cloud of another atom (or molecule). London forces are typically weaker than either permanent dipole–dipole forces or covalent bonds.   **78.** For every mole of liquid water that evaporates, several kiloJoules of heat must be absorbed to provide kinetic energy for the molecules. This heat is absorbed by the water from its surroundings.   **80.** In NH_3, strong hydrogen bonding can exist. In CH_4, because the molecule is nonpolar, only the relatively weak London dispersion forces exist.   **82.** In a crystal of ice, strong *hydrogen bonding* forces are present, while in the crystal of a nonpolar substance like oxygen, only the much weaker *London* forces exist.

CHAPTER **15**

2. When an ionic solute dissolves in water, a given ion is pulled into solution by the attractive ion–dipole force exerted by several water molecules. For example, in dissolving a positive ion, the ion is approached by the negatively charged end of several water molecules: if the attraction of the water molecules for the positive ion is stronger than the attraction of the negative ions near it in the crystal, the ion leaves the crystal and enters solution. After entering solution, the dissolved ion is surrounded completely by water molecules, which tends to prevent the ion from reentering the crystal.   **4.** solid   **6.** In order for a molecular solid to dissolve in water, the forces between water molecules and solute molecules must be comparable to the forces in the crystal among the solute molecules themselves. Sugar molecules contain several —OH groups (making them similar in structure to water), which permit extensive hydrogen bonding between water and sugar molecules.   **8.** independently   **10.** unsaturated   **12.** large   **14.** nine   **16.** (a) 9.1% NH_4NO_3 (b) 3.8% NH_4NO_3 (c) 1.3% NH_4NO_3 (d) 0.99% NH_4NO_3 **18.** (a) 22.5 g NaCL   (b) 233 g solution   (c) 150. g NaCl   (d) 150. g NaCl   **20.** 3.44% Cu; 2.61% Zn; 93.9% Fe   **22.** 3.54% NaCl   **24.** 7.81 g KBr   **26.** 15.6 g $CuCl_2$ **28.** 4.7 g heptane; 2.7 g pentane; 86 g hexane   **30.** 0.50; 1.0   **32.** To say that a solution has a concentration of 5 *M* means that in 1 L of solution (*not* solvent) there would be 5 mol of solute: to prepare such a solution one would place 5 mol of NaCl in a 1 L flask, and then add whatever amount of water is necessary so that the *total* volume would be 1 L after mixing.   **34.** (a) 2.0 *M*   (b) 1.0 *M*   (c) 0.67 *M*   (d) 0.50 *M*

36. (a) 0.253 *M*   (b) 0.106 *M*   (c) 0.434 *M*   (d) 1.67 × 10^{-3} *M*   **38.** 0.405 *M*   **40.** 0.0902 *M*   **42.** 0.619 *M*   **44.** (a) 0.0550 mol HCl   (b) 0.281 mol KOH   (c) 4.4 × 10^{-3} mol sugar   (d) 659 mol NaCl   **46.** (a) 0.943 g EDTA   (b) 48.5 g $CuSO_4$ (c) 0.0452 g H_3PO_4 (d) 0.87 g HCl   **48.** 0.36 L   **50.** (a) 4.60 × 10^{-3} mol Al^{3+}; 1.38 × 10^{-3} mol Cl^- (b) 1.70 mol Na^+; 0.567 mol PO_4^{3-} (c) 2.19 × 10^{-3} mol Cu^{2+}; 4.38 × 10^{-3} mol Cl^- (d) 3.96 × 10^{-5} mol Ca^{2+}; 7.91 × 10^{-5} mol OH^- **52.** 1.68 g   **54.** half   **56.** (a) 0.0837 *M*   (b) 0.320 *M*   (c) 0.0964 *M*   (d) 0.622 *M*   **58.** 0.691 mL   **60.** 0.127 *M*   **62.** 1.0 *M*   **64.** 7.5 mL   **66.** 2.31 g   **68.** 0.300 g   **70.** 64.2 mL   **72.** 1.8 × 10^{-4} *M*   **74.** (a) 63.0 mL   (b) 2.42 mL   (c) 50.1 mL   (d) 1.22 L   **76.** 1 normal   **78.** 1.53 equivalents OH^- ion. By definition, one equivalent of OH^- ion exactly neutralizes one equivalent of H^+ ion.   **80.** (a) 0.277 *N*   (b) 3.37 × 10^{-3} *N*   (c) 1.63 *N*   **82.** (a) 1.20 × 10^{-2} *N*   (b) 2.02 × 10^{-4} *N*   (c) 0.101 *N*   **84.** 7.03 × 10^{-5} *M*; 1.41 × 10^{-5} *N*   **86.** 2.15 × 10^3 mL (2.15 L)   **88.** 0.05582 *M*; 0.1117 *N*   **90.** Molarity is defined as the number of moles of solute contained in 1 liter of *total* solution volume (solute plus solvent after mixing). In the first case, where 50. g of NaCl is dissolved in 1.0 L of water, the total volume after mixing is *not* known and the molarity cannot be calculated. In the second example, the final volume after mixing is known and the molarity can be calculated simply.   **92.** 3.8 × 10^2 g 5% NaCl solution   **94.** 0.47 g $BaSO_4$ forms; concentration of remaining Ba^{2+} ion = 0.15 *M*   **96.** 56 mol   **98.** 1.12 L HCl at STP   **100.** 26.3 mL   **102.** 1.71 *M* NaCl   **104.** (a) 6.3% KNO_3 (b) 0.25% KNO_3 (c) 11% KNO_3 (d) 18% KNO_3 **106.** (a) 4.7% C; 1.4% Ni; 93.9% Fe   **108.** 28 g Na_2CO_3 **110.** 9.4 g NaCl; 3.1 g KBr   **112.** (a) 4.0 *M*   (b) 1.0 *M*   (c) 0.73 *M*   (d) 3.6 *M*   **114.** 0.812 *M*   **116.** 0.026 *M*   **118.** (a) 4.2 × 10^2 g KCl   (b) 4.7 g NaCl   (c) 8.8 × 10^3 g HCl   (d) 0.25 g $HClO_4$ **120.** (a) 0.938 mol Na^+; 0.313 mol PO_4^{3-} (b) 0.042 mol H^+; 0.021 mol SO_4^{2-} (c) 0.0038 mol Al^{3+}; 0.011 mol Cl^- (d) 1.88 mol Ba^{2+}; 3.75 mol Cl^- **122.** (a) 0.0909 *M*   (b) 0.127 *M*   (c) 0.192 *M*   (d) 1.6 *M*   **124.** 0.90 *M*   **126.** 50. mL   **128.** 35.0 mL   **130.** (a) 0.822 *N* HCl   (b) 4.00 *N* H_2SO_4 (c) 3.06 *N* H_3PO_4 **132.** 0.083 *M* NaH_2PO_4; 0.17 *N* NaH_2PO_4 **134.** 9.6 × 10^{-2} *N* HNO_3

CHAPTER **16**

2. A nitrogen–nitrogen triple bond and three hydrogen–hydrogen bonds must be broken. Six nitrogen–hydrogen bonds must form.   **4.** The activation energy is the minimum energy two colliding molecules must possess in order for the collision to result in reaction. If molecules do not possess energies equal to or greater than E_a, a collision between these molecules will not result in a reaction.   **6.** $Cl + O_3 \rightarrow ClO + O_2$; $O + ClO \rightarrow Cl + O_2$. Most chlorine atoms in the upper atmosphere come from manmade sources (such as the decomposition of freons such as CF_2Cl_2). The ozone layer of the atmosphere is important because ozone is able to absorb harmful ultraviolet radiation from the sun. Although it may already be too late, it is likely that the ozone layer of the atmosphere will be replenished if the use of chlorinated aerosol compounds is stopped. **8.** A state of equilibrium is attained when two opposing processes are exactly balanced. The development of a vapor pressure above a liquid in a closed container is an example of a physical equilibrium. Any chemical reaction which appears to "stop" before completion serves as an example of a chemical equilibrium. **10.** Chemical equilibrium occurs when two opposing chemical reactions reach the same speed in a closed system. When a state of chemical equilibrium has been reached, the concentrations of reactants and products present in the system remain constant with time. A chemical reaction that reaches a state of equilibrium is indicated by using a *double arrow* ($\rightleftarrows$). The points of the double arrow point in opposite directions, indicating that two opposite processes are going on. **12.** Although we recognize a state of chemical equilibrium by the fact that the concentrations of reactants and products no longer change with time, the lack of change results from the fact that two opposing processes are going on at the same time (not because the reaction has "stopped"). Further reaction in the forward direction is cancelled out by an equal extent of reaction in the reverse direction. The reaction is still proceeding, but the opposite reaction is also proceeding at the same rate. **14.** The equilibrium constant is a *ratio* of the concentration of products to the concentration of reactants, all at equilibrium. Depending on how much reactant a particular experiment was begun with, there may be different amounts of reactants and products present at equilibrium, but the *ratio* will always be the same for a given reaction at a given temperature. For example, the ratios 4/2 and 6/3 are different absolutely in terms of the numbers involved, but each of these ratios has the value 2. **16.** (a) $K = [NCl_3(g)]^2/[N_2(g)][Cl_2(g)]^3$ (b) $K = [HI(g)]^2/[H_2(g)][I_2(g)]$ (c) $K = [N_2H_4(g)]/[N_2(g)][H_2(g)]^2$ **18.** (a) $K = [PCl_3(g)]^4/[P_4(g)][Cl_2(g)]^6$ (b) $K = [NO(g)]^2/[N_2(g)][O_2(g)]$ (c) $K = [CO_2(g)]^3[H_2O(g)]^4/[C_3H_8(g)][O_2(g)]^5$ **20.** $K = 2.5 \times 10^{10}$ **22.** $K = 9.0 \times 10^8$ **24.** Equilibrium constants represent ratios of the *concentrations* of products and reactants present at the point of equilibrium. The *concentration* of a pure solid or of a pure liquid is constant and is determined by the density of the solid or liquid. **26.** (a) $K = [SO_2(g)]/[O_2(g)]$ (b) $K = [CO_2(g)]^5/[O_2(g)]^8$ (c) $K = [NO_2(g)]^2$ **28.** (a) $K = 1/[N_2(g)]$ (b) $K = [SO_2(g)][HCl(g)]^2$ (c) $K = [H_2(g)]/[HI(g)]^2$ **30.** When an additional amount of one of the reactants is added to an equilibrium system, the system shifts to the right and adjusts so as to use up some of the added reactant. This results in a net *increase* in the amount of product, compared to the equilibrium system before the additional reactant was added. The numerical *value* of the equilibrium constant does *not* change when a reactant is added: the concentrations of all reactants and products adjust until the correct value of K is once again achieved. **32.** If heat is applied to an endother-

mic reaction (i.e., the temperature is raised), the equilibrium is shifted to the right. More product will be present at equilibrium than if the temperature had not been increased. The value of K increases. **34.** (a) shifts right (b) shifts right (c) no change (catalysts do not affect the position of equilibrium) **36.** (a) no effect (UO_2 is a solid) (b) no effect (Xe is not involved in the reaction) (c) shifts to left (if HF attacks glass, it is removed from system, causing reaction to replace the lost HF.) (d) shifts to right (e) shifts to left (4 mol gas versus 3 mol gas) **38.** increase: heat is a product of the reaction; removing heat (decreasing the temperature) favors the forward reaction **40.** For an *endo*thermic reaction, an increase in temperature will shift the position of equilibrium to the right (toward products). **42.** The reaction is *exo*thermic and should be performed at as low a temperature as possible (consistent with the molecule's still having sufficient energy to react). **44.** A small equilibrium constant means that the concentration of products is small, compared to the concentration of reactants. The position of equilibrium lies far to the left. Reactions with very small equilibrium constants are generally not very useful as a source of the products, unless Le Châtelier's principle can be applied to shift the position of equilibrium to the point where a sufficient amount of product can be isolated. **46.** $K = 6.3 \times 10^4$ **48.** $K = 2.1 \times 10^{-3}$ **50.** $[O_2(g)] = 8.0 \times 10^{-2}\,M$ **52.** $[NO_2(g)] = 6.1 \times 10^{-3}\,M$ **54.** solubility product, K_{sp} **56.** Stirring or grinding the solute increases the speed with which the solute dissolves, but the ultimate *amount* of solute that dissolves is fixed by the equilibrium constant for the dissolving process, K_{sp}. **58.** (a) $PbCl_2(s) \rightleftarrows Pb^{2+}(aq) + 2Cl^-(aq)$; $K_{sp} = [Pb^{2+}(aq)][Cl^-(aq)]^2$ (b) $NiS(s) \rightleftarrows Ni^{2+}(aq) + S^{2-}(aq)$; $K_{sp} = [Ni^{2+}(aq)][S^{2-}(aq)]$ (c) $Ba(OH)_2(s) \rightleftarrows Ba^{2+}(aq) + 2OH^-(aq)$; $K_{sp} = [Ba^{2+}(aq)][OH^-]^2$ (d) $Fe_2S_3(s) \rightleftarrows 2Fe^{3+}(aq) + 3S^{2-}(aq)$; $K_{sp} = [Fe^{3+}(aq)]^2[S^{2-}(aq)]$ **60.** $K_{sp} = 3.1 \times 10^{-10}$ **62.** $K_{sp} = 3.8 \times 10^{-15}$ **64.** $1.0 \times 10^{-5}\,M$ **66.** 1.7×10^{-4} g/L **68.** $K_{sp} = 1.9 \times 10^{-4}$ **70.** $2 \times 10^{-10}\,M$; 2×10^{-8} g/L **72.** temperature **74.** catalyst **76.** constant **78.** dynamic **80.** heterogeneous **82.** position **84.** exothermic **86.** An equilibrium reaction may come to many *positions* of equilibrium, but at each possible position of equilibrium, the numerical value of the equilibrium constant is fulfilled. If different amounts of reactants are taken in different experiments, the *absolute amounts* of reactants and products present at the point of equilibrium will differ from one experiment to another, but the *ratio* that defines the equilibrium constant will be the same. **88.** $9.0 \times 10^{-3}\,M$ **90.** 0.043 g/L **92.** Although a small solubility product generally implies a small solubility, comparisons of solubility based directly on K_{sp} values are only valid if the salts produce the same numbers of positive and negative ions per formula when they dissolve. For example, one can compare the solubilities of $AgCl(s)$ and $NiS(s)$ directly using K_{sp}, since each salt produces one positive and one negative ion per formula when dissolved. One could not directly compare $AgCl(s)$ with a salt such as $Ca_3(PO_4)_2$, however. **94.** At higher temperatures, the average kinetic energy of the reactant molecules is larger. At higher temperatures, the probability that a collision between molecules will be energetic enough for reaction to take place is larger. On a molecular basis, a higher temperature means a given molecule will be moving faster.   **96.** (a) $K = [HBr(g)]^2/[H_2(g)][Br_2(g)]$ (b) $K = [H_2S(g)]^2/[H_2(g)]^2[S_2(g)]$ (c) $K = [HCN(g)]^2/[H_2(g)][C_2N_2(g)]$ **98.** $K = 1.2 \times 10^{-3}$ **100.** (a) $K = 1/[O_2(g)]^3$ (b) $K = 1/[NH_3(g)][HCl(g)]$ (c) $K = 1/[O_2(g)]$ **102.** An *exo*thermic reaction is one which liberates heat energy. Increasing the temperature (adding heat) for such a reaction is fighting against the reaction's own tendency to liberate heat. The net effect of raising the temperature will be a shift to the left to decrease the amount of product. If it is desired to increase the amount of products in an exothermic reaction, heat must be *removed* from the system. Changing the temperature *does* change the numerical value of the equilibrium constant for a reaction.   **104.** The reaction is *exo*thermic as written. An increase in temperature (addition of heat) will shift the reaction to the left (toward reactants).   **106.** $[NH_3(g)] = 1.1 \times 10^{-3}\ M$ **108.** (a) $MgCO_3(s) \rightleftharpoons Mg^{2+}(aq) + CO_3^{2-}(aq)$; $K_{sp} = [Mg^{2+}(aq)][CO_3^{2-}(aq)]$ (b) $Al_2(SO_4)_3(s) \rightleftharpoons 2Al^{3+}(aq) + 3SO_4^{2-}(aq)$; $K_{sp} = [Al^{3+}(aq)]^2[SO_4^{2-}(aq)]^3$ (c) $Ca(OH)_2(s) \rightleftharpoons Ca^{2+}(aq) + 2OH^-(aq)$; $K_{sp} = [Ca^{2+}(aq)][OH^-(aq)]^2$ (d) $Ag_2SO_4(s) \rightleftharpoons 2Ag^+(aq) + SO_4^{2-}(aq)$; $K_{sp} = [Ag^+(aq)]^2[SO_4^{2-}(aq)]$ **110.** $K_{sp} = 3.9 \times 10^{-11}$ **112.** $K_{sp} = 1.4 \times 10^{-8}$

CHAPTER **17**

2. In the Arrhenius definition, an acid is a substance which produces ions (H^+) when dissolved in water, whereas a base is a substance that produces hydroxide ions (OH^-) in aqueous solution. These definitions proved to be too restrictive since the only base permitted was hydroxide ion, and the only solvent permitted was water.   **4.** The substances in a conjugate acid–base pair differ by the presence or absence of a proton (H^+).   **6.** When an acid is dissolved in water, the hydronium ion (H_3O^+) is formed. The hydronium ion is the conjugate acid of water (H_2O).   **8.** (a) a conjugate pair (HSO_4^- is the acid, SO_4^{2-} is the base)   (b) not a conjugate pair (Br^- is the conjugate base of HBr; BrO^- is the conjugate base of HBrO)   (c) not a conjugate pair ($H_2PO_4^-$ is the conjugate acid of HPO_4^- and also the conjugate base of H_3PO_4; HPO_4^- is the conjugate acid of PO_4^{3-})   (d) not a conjugate pair (NO_3^- is the conjugate base of HNO_3; NO_2^- is the conjugate base of HNO_2)   **10.** (a) HSO_4^-(acid), SO_4^{2-}(base); H_2O(base), H_3O^+(acid)   (b) CH_3NH_2(base), $CH_3NH_3^+$(acid); HCl(acid), Cl^- (base)   (c) CH_3COO^-(base), CH_3COOH(acid); H_2O(acid), OH^-(base)   **12.** (a) HSO_4^- (b) HPO_4^{2-} (c) $CH_3NH_3^+$ (d) HF   **14.** (a) NO_3^- (b) PO_4^{3-} (c) H_2O (d) HCO_3^- **16.** (a) $HBrO_4 + H_2O \rightleftharpoons H_3O^+ + BrO_4^-$ (b) $HF + H_2O \rightleftharpoons H_3O^+ + F^-$ (c) $CH_3NH_3^+ + H_2O \rightleftharpoons H_3O^+ + CH_3NH_2$ (d) $H_2PO_4^- + H_2O \rightleftharpoons H_3O^+ + HPO_4^-$ **18.** To say that an acid

is weak in aqueous solution means that the acid does not easily transfer protons to water (and does not fully ionize). If an acid does not lose protons easily, then the acid's anion must be a strong attractor of protons.   **20.** A strong acid is one which loses its protons easily and fully ionizes in water; this means that the acid's conjugate base must be poor at attracting and holding on to protons and is a relatively weak base. A weak acid is one which resists loss of its protons and does not ionize well in water; this means that the acid's conjugate base attracts and holds onto protons tightly and is a relatively strong base.   **22.** H_2SO_4 (sulfuric): $H_2SO_4 + H_2O \rightarrow HSO_4^- + H_3O^+$; HCl (hydrochloric): $HCl + H_2O \rightarrow Cl^- + H_3O^+$; HNO_3 (nitric): $HNO_3 + H_2O \rightarrow NO_3^- + H_3O^+$; $HClO_4$ (perchloric): $HClO_4 + H_2O \rightarrow ClO_4^- + H_3O^+$ **24.** oxyacids: $HClO_4$, HNO_3, H_2SO_4, CH_3COOH, etc.; non-oxyacids: HCl, HBr, HF, HI, HCN, etc.   **26.** (a) HSO_4^- is a *moderately* strong acid   (b) HBr is a strong acid   (c) HCN is a weak acid   (d) CH_3COOH is a weak acid   **28.** $H_2O \rightleftharpoons H^+(aq) + OH^-(aq)$; $K_w = [H^+(aq)][OH^-(aq)] = 1.00 \times 10^{-14}$ at 25 °C. The very small numerical value of K_w means pure water is not very much ionized at 25 °C.   **30.** If $[H^+(aq)] > [OH^-(aq)]$ the solution is acidic. If $[OH^-(aq)] > [H^+(aq)]$ the solution is basic. If $[H^+(aq)] = [OH^-(aq)]$ the solution is neutral. At 25 °C, if $[H^+(aq)] > 1.00 \times 10^{-7}\ M$ the solution is acidic. At 25 °C, if $[OH^-(aq)] > 1.00 \times 10^{-7}\ M$ the solution is basic.   **32.** (a) $[H^+(aq)] = 2.51 \times 10^{-10}\ M$; solution is basic   (b) $[H^+(aq)] = 3.43 \times 10^{-6}\ M$; solution is acidic   (c) $[H^+(aq)] = 1.38 \times 10^{-13}\ M$; solution is basic   (d) $[H^+(aq)] = 1.10 \times 10^{-8}\ M$; solution is basic   **34.** $[OH^-(aq)] = 7.87 \times 10^{-12}\ M$; solution is acidic; $[OH^-(aq)] = 2.22 \times 10^{-7}\ M$; solution is basic; $[OH^-(aq)] = 2.58 \times 10^{-10}\ M$; solution is acidic; $[OH^-(aq)] = 1.23 \times 10^{-4}\ M$; solution is basic   **36.** (a) $[OH^-(aq)] = 0.10\ M$ (b) $[H^+(aq)] = 0.00100\ M$ (c) $[H^+(aq)] = 1.15 \times 10^{-7}\ M$ **38.** household ammonia (pH 12); blood (pH 7–8); milk (pH 6–7); vinegar (pH 3); lemon juice (pH 2–3); stomach acid (pH 2).   **40.** The pH of a solution is defined as the *negative* of the logarithm of the hydrogen ion concentration, $pH = -\log[H^+]$. Mathematically, the *negative sign* in the definition causes the pH to *decrease* as the hydrogen ion concentration *increases*.   **42.** (a) $pH = 3.00$ (acidic)   (b) $pH = 3.660$ (acidic)   (c) $pH = 10.037$ (basic)   (d) $pH = 6.327$ (acidic)   **44.** (a) $pH = 11.625$ (basic)   (b) $pH = 7.299$ (basic)   (c) $pH = 10.000$ (basic)   (d) $pH = 3.393$ (acidic)   **46.** (a) $pOH = 7.51$ (acidic)   (b) $pOH = 12.07$ (acidic)   (c) $pOH = 2.79$ (basic)   (d) $pOH = 7.00$ (neutral)   **48.** (a) $[OH^-(aq)] = 1.00 \times 10^{-7}\ M$; $pH = 7.000$; $pOH = 7.000$ (b) $[H^+(aq)] = 2.28 \times 10^{-10}\ M$; $pH = 9.642$; $pOH = 4.358$ (c) $[OH^-(aq)] = 2.33 \times 10^{-4}\ M$; $pH = 10.368$; $pOH = 3.632$ (d) $[H^+(aq)] = 1.36 \times 10^{-13}\ M$; $pH = 12.867$; $pOH = 1.133$ **50.** (a) $[H^+(aq)] = 7.8 \times 10^{-5}\ M$ (b) $[H^+(aq)] = 2.9 \times 10^{-10}\ M$ (c) $[H^+(aq)] = 1.0 \times 10^{-7}\ M$ (d) $[H^+(aq)] = 6.2 \times 10^{-14}\ M$ **52.** (a) $[H^+(aq)] =$

$8.1 \times 10^{-11} M$ (b) $[H^+(aq)] = 3.6 \times 10^{-2} M$
(c) $[H^+(aq)] = 1.4 \times 10^{-13} M$ (d) $[H^+(aq)] = 5.9 \times 10^{-6} M$ **54.** (a) $[H^+(aq)] = 4.2 \times 10^{-5} M$ (b) $[H^+(aq)] = 2.4 \times 10^{-10} M$ (c) $[H^+(aq)] = 6.2 \times 10^{-8} M$
(d) $[H^+(aq)] = 2.5 \times 10^{-9} M$ **56.** The solution contains water molecules, H_3O^+ ions (protons), and NO_3^- ions. Because HNO_3 is a strong acid, which is completely ionized in water, there are no HNO_3 molecules present. **58.** (a) $[H^+(aq)] = 0.00010 M$; pH = 4.00 (b) $[H^+(aq)] = 0.0050 M$; pH = 2.30 (c) $[H^+(aq)] = 4.21 \times 10^{-5} M$; pH = 4.376 (d) $[H^+(aq)] = 6.33 \times 10^{-3} M$; pH = 2.199 **60.** A buffered solution consists of a mixture of a weak acid and its conjugate base; one example of a buffered solution is a mixture of acetic acid (CH_3COOH) and sodium acetate ($NaCH_3COO$). **62.** The weak acid component of a buffered solution is capable of reacting with added strong base. For example, using the buffered solution given as an example in Question 60, acetic acid would consume added sodium hydroxide as follows: $CH_3COOH(aq) + NaOH(aq) \rightarrow NaCH_3COO(aq) + H_2O(l)$. Acetic acid *neutralizes* the added NaOH and prevents it from having much effect on the overall pH of the solution. **64.** $CH_3COOH + NaOH \rightarrow NaCH_3COO + H_2O$; $CH_3COO^- + HCl \rightarrow CH_3COOH + Cl^-$; $HNO_2 + NaOH \rightarrow NaNO_2 + H_2O$; $NO_2^- + HCl \rightarrow HNO_2 + Cl^-$ **66.** (a) $[OH^-(aq)] = 0.10 M$; pOH = 1.00; pH = 13.00 (b) $[OH^-(aq)] = 2.0 \times 10^{-4} M$; pOH = 3.70; pH = 10.30 (c) $[OH^-(aq)] = 6.2 \times 10^{-3} M$; pOH = 2.21; pH = 11.79 (d) $[OH^-(aq)] = 0.001 M$; pOH = 4.0; pH = 10.0 **68.** a, b, d **70.** a, c, e **72.** Ordinarily in calculating the pH of strong acid solutions, the major contribution to the concentration of hydrogen ion present is from the dissolved strong acid; we ordinarily neglect the small amount of hydrogen ion present in such solutions due to the ionization of water. With $1.0 \times 10^{-7} M$ HCl solution, however, the amount of hydrogen ion present due to the ionization of *water* is *comparable* to that present due to the addition of *acid* (HCl) and must be considered in the calculation of pH. **74.** accepts **76.** base **78.** carboxyl (—COOH) **80.** 1.0×10^{-14} **82.** larger **84.** pH **86.** weak acid **88.** (a) H_2O and OH^- represent a conjugate acid–base pair (H_2O is the acid, having one more proton than the base, OH^-).
(b) H_2SO_4 and SO_4^{2-} are *not* a conjugate acid–base pair (they differ by two protons). The conjugate base of H_2SO_4 is HSO_4^-; the conjugate acid of SO_4^{2-} is also HSO_4^-. (c) H_3PO_4 and $H_2PO_4^-$ represent a conjugate acid–base pair (H_3PO_4 is the acid, having one more proton than the base $H_2PO_4^-$). (d) $HC_2H_3O_2$ and $C_2H_3O_2^-$ represent a conjugate acid–base pair ($HC_2H_3O_2$ is the acid, having one more proton than the base $C_2H_3O_2^-$).
90. (a) NH_4^+ (b) CH_3COOH (c) H_3PO_4 (d) $H_2C_2O_4$ **92.** (a) $CH_3CH_2COOH + H_2O \rightleftharpoons CH_3CH_2COO^- + H_3O^+$ (b) $NH_4^+ + H_2O \rightleftharpoons NH_3 + H_3O^+$ (c) $H_2SO_4 + H_2O \rightleftharpoons HSO_4^- + H_3O^+$ (d) $H_3PO_4 + H_2O \rightleftharpoons H_2PO_4^- + H_3O^+$ **94.** (a) $[H^+(aq)] = 2.37 \times 10^{-12} M$; solution is basic (b) $[H^+(aq)] = 9.90 \times 10^{-2} M$; solution is acidic

(c) $[H^+(aq)] = 3.28 \times 10^{-8} M$; solution is basic (d) $[H^+(aq)] = 1.66 \times 10^{-9} M$; solution is basic **96.** (a) $[OH^-(aq)] = 0.0000032 M$ (b) $[OH^-(aq)] = 1.54 \times 10^{-8} M$ (c) $[OH^-(aq)] = 4.02 \times 10^{-7} M$ **98.** (a) pH = 8.15; solution is basic (b) pH = 5.971; solution is acidic (c) pH = 13.344; solution is basic (d) pH = 2.902; solution is acidic **100.** (a) $[OH^-(aq)] = 1.75 \times 10^{-11} M$; pH = 3.243; pOH = 10.757 (b) $[H^+(aq)] = 1.12 \times 10^{-10} M$; pH = 9.950; pOH = 4.050 (c) $[OH^-(aq)] = 3.48 \times 10^{-3} M$; pH = 11.542; pOH = 2.458 (d) $[H^+(aq)] = 1.39 \times 10^{-7} M$; pH = 6.859; pOH = 11.141 **102.** (a) $[H^+(aq)] = 8.9 \times 10^{-7} M$ (b) $[H^+(aq)] = 1.0 M$ (c) $[H^+(aq)] = 1.0 \times 10^{-9}$ (d) $[H^+(aq)] = 8.9 \times 10^{-11} M$ **104.** (a) $[H^+(aq)] = 1.4 \times 10^{-3} M$; pH = 2.85 (b) $[H^+(aq)] = 3.0 \times 10^{-5} M$; pH = 4.52 (c) $[H^+(aq)] = 5.0 \times 10^{-2} M$; pH = 1.30 (d) $[H^+(aq)] = 0.0010 M$; pH = 3.00

CHAPTER **18**

2. Oxidation is defined as a *loss of electrons* by an atom, molecule, or ion. Reduction is defined as a *gain of electrons* by such a species. $Na \rightarrow Na^+ + e^-$ is an example of an oxidation process. $Cl + e^- \rightarrow Cl^-$ is an example of a reduction process. **4.** (a) sodium is oxidized, nitrogen is reduced (b) magnesium is oxidized, chlorine is reduced (c) aluminum is oxidized, bromine is reduced (d) iron is oxidized, oxygen is reduced **6.** (a) calcium is oxidized, sulfur is reduced (b) copper is oxidized, fluorine is reduced (c) calcium is oxidized, carbon is reduced (d) zinc is oxidized, oxygen is reduced **14.** (a) N, +3; Cl, −1 (b) S, +6; F, −1 (c) P, +5; Cl, −1 (d) Si, −4; H, +1 **16.** (a) Cl, 0 (b) Cl, +1; O, −2 (c) H, +1; Cl, −1 (d) Cl, +1; F, −1; **18.** (a) H, +1; O, −2; N, +5 (b) H, +1; O, −2; P, +5 (c) H, +1; O, −2; S, +6 (d) O, −1 **20.** (a) Cu, +2; Cl, −1 (b) Cr, +3; Cl, −1 (c) H, +1; O, −2; Cr, +6 (d) Cr, +3; O, −2 **22.** (a) Cl, +7; O, −2 (b) Cl, +5; O, −2 (c) Cl, +3; O, −2 (d) Cl, +1; O, −2 **24.** Electrons are negative; when an atom gains electrons, it gains one negative charge for each electron gained. For example, in the reduction reaction $Cl + e^- \rightarrow Cl^-$, the oxidation state of chlorine decreases from 0 to −1 as the electron is gained. **26.** An oxidizing agent *causes* another species to be oxidized (to lose electrons). In order to make another species lose electrons, the oxidizing agent must be capable of gaining the electrons; an oxidizing agent is itself reduced. On the contrary, a reducing agent is itself oxidized. **28.** An oxidizing agent oxidizes another species by gaining the electrons lost by the other species; therefore, an oxidizing agent itself decreases in oxidation state. A reducing agent increases its oxidation state when acting on another atom or molecule. **30.** (a) carbon is oxidized (0 → +2); oxygen is reduced (0 → −2) (b) carbon is oxidized (0 → +1); bromine is reduced $(0 \rightarrow -1)$ (c) carbon is oxidized $(0 \rightarrow +1)$; fluorine is reduced $(0 \rightarrow -1)$ (d) phosphorus is oxidized $(0 \rightarrow +3)$; chlorine is reduced $(0 \rightarrow -1)$ **32.** (a) copper is oxidized $(0 \rightarrow +2)$; silver is reduced $(+1 \rightarrow 0)$ (b) nitrogen is oxidized $(0 \rightarrow +3)$; fluorine is reduced $(0 \rightarrow -1)$ (c) sulfur is oxidized $(0 \rightarrow +4)$; iron is reduced $(+3 \rightarrow 0)$ (d) oxygen is oxidized $(-1 \rightarrow 0)$; oxygen is reduced $(-1 \rightarrow 2)$ **34.** Iron is reduced [$+3$ in $Fe_2O_3(s)$, 0 in $Fe(l)$]; carbon is oxidized [$+2$ in $CO(g)$, $+4$ in $CO_2(g)$]. $Fe_2O_3(s)$ is the oxidizing agent; $CO(g)$ is the reducing agent. **36.** Chlorine is reduced [0 in $Cl_2(g)$, -1 in $NaCl(s)$]; bromine is oxidized, [-1 in $NaBr(aq)$, 0 in $Br_2(l)$]. $Cl_2(g)$ is the oxidizing agent; $NaBr(aq)$ is the reducing agent. **38.** Oxidation–reduction reactions are often more complicated than "regular" reactions; frequently the coefficients necessary to balance the number of electrons transferred come out to be large numbers. **40.** Under ordinary conditions it is impossible to have "free" electrons that are not part of some atom, ion, or molecule. For this reason, the total number of electrons lost by the species being oxidized must equal the total number of electrons gained by the species being reduced. **42.** (a) $6e^- + N_2 \rightarrow 2N^{3-}$ (b) $O_2^{2-} \rightarrow O_2 + 2e^-$ (c) $Zn \rightarrow Zn^{2+} + 2e^-$ (d) $2e^- + F_2 \rightarrow 2F^-$ **44.** (a) $4e^- + 4H^+ + O_2 \rightarrow 2H_2O$ (b) $12H^+ + 2IO_3^- + 10e^- \rightarrow I_2 + 6H_2O$ (c) $2H^+ + VO^{2+} + e^- \rightarrow V^{3+} + H_2O$ (d) $2H^+ + BiO^+ + 3e^- \rightarrow Bi + H_2O$ **46.** (a) $12H^+ + 2BrO_3^- + 10Br^- \rightarrow 6Br_2 + H_2O$ (b) $4I^- + 4H^+ + O_2 \rightarrow 2I_2 + 2H_2O$ (c) $14H^+ + Cr_2O_7^{2-} + 6I^- \rightarrow 2Cr^{3+} + 3I_2 + 7H_2O$ (d) $2H_2O + 2Mn^{3+} \rightarrow MnO_2 + Mn^{2+} + 4H^+$ **48.** (a) $H_2O + H_3AsO_3 + 2Ce^{4+} \rightarrow H_3AsO_2 + 2H^+ + 2Ce^{3+}$ (b) $Ce^{4+} + Fe^{2+} \rightarrow Ce^{3+} + Fe^{3+}$ (c) $2Ce^{4+} + 2I^- \rightarrow 2Ce^{3+} + I_2$ **50.** A salt bridge typically consists of a U-shaped tube filled with an inert electrolyte (one involving ions that are not part of the oxidation–reduction reaction). A salt bridge is used to complete the electrical circuit in a cell. Any method that allows transfer of charge without allowing bulk mixing of the solutions may be used (another common method is to set up one half-cell in a porous cup, which is then placed in the beaker containing the second half-cell). **52.** In a galvanic cell, the anode is the electrode where oxidation occurs; the cathode is the electrode where reduction occurs.

54.

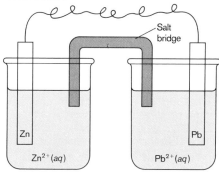

(a)

Pb^{2+} ion is reduced; $Zn(s)$ is oxidized. The anode reaction is $Zn(s) \rightarrow Zn^{2+}(aq) + 2e^-$ The cathode reaction is $Pb^{2+}(aq) + 2e^- \rightarrow Pb(s)$ **56.** Both normal and alkaline cells contain zinc as one electrode; zinc corrodes more slowly under alkaline conditions than in the highly acidic environment of a normal dry cell. anode: $Zn(s) + 2OH^-(aq) \rightarrow ZnO(s) + H_2O(l) + 2e^-$; cathode: $2MnO_2(s) + H_2O(l) + 2e^- \rightarrow Mn_2O_3(s) + 2OH^-(aq)$ **58.** Aluminum is a very reactive metal when freshly isolated in the pure state. However, on standing for even a relatively short period of time, aluminum metal forms a thin coating of Al_2O_3 on its surface from reaction with atmospheric oxygen. This coating of Al_2O_3 is much less reactive than the metal and serves to protect the surface of the metal from further attack. **60.** In cathodic protection of steel tanks and pipes, a more reactive metal than iron is connected to the item to be protected. The active metal is then preferentially oxidized rather than the iron of the tank or pipe. **62.** The main recharging reaction for the lead storage battery is $2PbSO_4(s) + 2H_2O(l) \rightarrow Pb(s) + PbO_2(s) + 2H^+(aq) + 2HSO_4^-(aq)$. A major side reaction is the electrolysis of water, $2H_2O(l) \rightarrow 2H_2(g) + O_2(g)$, resulting in the production of an explosive mixture of hydrogen and oxygen, which accounts for many accidents during the recharging of such batteries.
64. Electrolysis is applied in electroplating by making the item to be plated the cathode in a cell containing a solution of ions of the desired plating metal. **66.** loss; oxidation state
68. electronegative **70.** zero **72.** lose **74.** separate from **76.** anode **78.** electrolysis **80.** hydrogen; oxygen **82.** oxidation **84.** (a) $4Fe(s) + 3O_2(g) \rightarrow 2Fe_2O_3(s)$; iron is oxidized, oxygen is reduced (b) $2Al(s) + 3Cl_2(g) \rightarrow 2AlCl_3(s)$; aluminum is oxidized; chlorine is reduced (c) $6Mg(s) + P_4(s) \rightarrow 2Mg_3P_2(s)$; magnesium is oxidized, phosphorus is reduced **86.** (a) magnesium is oxidized, aluminum (as Al^{3+}) is reduced (b) magnesium is oxidized, aluminum (as Al^{3+} in $AlCl_3$) is reduced (c) oxygen is oxidized, chlorine is reduced **88.** (a) $C_3H_8(g) + 5O_2(g) \rightarrow 3CO_2(g) + 4H_2O(g)$ (b) $CO(g) + 2H_2(g) \rightarrow CH_3OH(l)$ (c) $SnO_2(s) + 2C(s) \rightarrow Sn(s) + 2CO(g)$ (d) $C_2H_5OH(l) + 3O_2(g) \rightarrow 2CO_2(g) + 3H_2O(g)$ **90.** (a) sodium is oxidized, oxygen is reduced (b) iron is oxidized, hydrogen is reduced (c) oxygen (O^{2-}) is oxidized, aluminum (Al^{3+}) is reduced (d) magnesium is oxidized, nitrogen is reduced **92.** (a) H $+1$; N -3 (b) C $+2$; O -2 (c) C $+4$; O -2 (d) N $+3$; F -1 **94.** (a) S $+4$; O -2 (b) S $+6$; O -2 (c) H $+1$; S -2 (d) H $+1$; S $+6$; O -2 **96.** (a) Bi $+3$; O -2 (b) P $+5$; O -2 (c) N $+3$; O -2 (d) Hg $+1$
98. (a) oxygen is oxidized; chlorine is reduced (b) germanium is oxidized; oxygen is reduced (c) carbon is oxidized; chlorine is reduced (d) oxygen is oxidized; fluorine is reduced
100. (a) $SiO_2(s) + 4H^+(aq) + 4e^- \rightarrow Si(s) + 2H_2O(l)$ (b) $S(s) + 2H^+(aq) + 2e^- \rightarrow H_2S(g)$ (c) $NO_3^-(aq) + 3H^+(aq) + 2e^- \rightarrow HNO_2(aq) + H_2O(l)$ (d) $NO_3^-(aq) + 4H^+(aq) + 3e^- \rightarrow NO(g) + 2H_2O(l)$ **102.** (a) $16H^+(aq) + 2MnO_4^-(aq) + 5C_2O_4^{2-}(aq) \rightarrow 2Mn^{2+}(aq) + 8H_2O(l) +$

$10CO_2(g)$ (b) $8H^+(aq) + MnO_4^-(aq) + 5Fe^{2+}(aq) \rightarrow$ $Mn^{2+}(aq) + 4H_2O(l) + 5Fe^{3+}(aq)$ (c) $16H^+(aq) +$ $2MnO_4^-(aq) + 10Cl^-(aq) \rightarrow 2Mn^{2+}(aq) + 8H_2O(l) + 5Cl_2(g)$

CHAPTER 19

2. The nucleus consists of protons (mass number 1, charge 1+) and neutrons (mass number 1, charge 0). 4. Isotopes represent atoms of the same element with different mass numbers. On a nuclear basis, isotopes differ in the number of neutrons present. For samples with large numbers of atoms, isotopes have the same chemical properties, but may have slightly different physical properties. 6. A nucleus is termed radioactive if it spontaneously decomposes, forming another nucleus and producing one or more subatomic particles. 8. Alpha particle: charge 2+, mass number 4, symbol ^4_2He 10. When a nucleus produces a beta particle, the atomic number of the parent nucleus is *increased* by *one* unit.
12. Gamma rays are high energy photons of electromagnetic radiation. Gamma rays are not normally considered to be particles. When a nucleus produces only gamma radiation, the atomic number and mass number of the nucleus do not change. 14. Electron capture occurs when one of the inner orbital electrons is pulled into and becomes part of the nucleus. 16. $^{235}_{92}U$ $^{238}_{92}U$ 18. $^{30}_{15}P$ (15 protons, 15 neutrons); $^{31}_{15}P$ (15 protons, 16 neutrons); $^{32}_{15}P$ (15 protons, 17 neutrons) 20. a. $_{-1}^0e$ or $_{-1}^0\beta$ (b) $_{-1}^0e$ or $_{+1}^0\beta$ (c) $_0^0\gamma$ 22. (a) $_{+1}^0e$ (positron) (b) $^{226}_{88}Ra$ (c) $^{234}_{90}Th$
24. (a) $_{-1}^0e$ (b) $_{-1}^0e$ (c) $^{210}_{83}Bi$ 26. (a) $^{59}_{26}Fe \rightarrow ^{59}_{27}Co + _{-1}^0e$ (b) $^{24}_{11}Na \rightarrow ^{24}_{12}Mg + _{-1}^0e$ (c) $^{198}_{79}Au \rightarrow ^{198}_{80}Hg + _{-1}^0e$ (d) $^{47}_{20}Ca \rightarrow$ $^{47}_{21}Sc + _{-1}^0e$ 28. (a) $^{226}_{88}Ra \rightarrow ^{222}_{86}Rn + ^4_2He$ (b) $^{222}_{86}Rn \rightarrow$ $^{218}_{84}Po + ^4_2He$ (c) $^{239}_{94}Pu \rightarrow ^{235}_{92}U + ^4_2He$ (d) $^8_4Be \rightarrow ^4_2He + ^4_2He$
30. There is often considerable repulsion between the target nucleus and the particles being used for bombardment (especially if the bombarding particle is positively charged like the target nucleus). Using accelerators to greatly speed up the bombarding particles can overcome this repulsion. 32. $^{27}_{13}Al + ^4_2He \rightarrow ^{30}_{15}P + ^1_0n$
34. The half-life of a nucleus is the time required for one-half of the original sample of nuclei to decay. A given isotope of an element always has the same half-life, although different isotopes of the same element may have greatly different half-lives. Nuclei of different elements have different half-lives. 36. $^{226}_{88}Ra$ is the most stable (longest half-life); $^{224}_{88}Ra$ is the "hottest" (shortest half-life).
38. highest activity
$^{87}Sr > ^{99}Tc > ^{24}Na > ^{99}Mo > ^{133}Xe > ^{131}I > ^{32}P > ^{51}Cr > ^{59}Fe$ lowest activity
40. For ^{223}Ra, the half-life is 12 days. After two half-lives (24 days), 250 mg remains; after three half-lives (36 days), 125 mg remains. For ^{224}Ra, the half-life is 3.6 days. One month would be approximately 8 half-life periods (29 days), and approximately 4 mg remains. For ^{225}Ra, the half-life is 15 days. One month would be two half-life periods, and 250 mg remains. 42. For an administered dose of 100 μg, 0.39 μg remains after 2 days. The

fraction remaining is $0.39/100 = 0.0039$; on a percentage basis, less than 0.4% of the original radioisotope remains.
44. Carbon-14 is produced in the upper atmosphere by the bombardment of ordinary nitrogen with neutrons from space: $^{14}_7N + ^1_0n \rightarrow ^{14}_6C + ^1_1H$. 46. We assume that the concentration of C-14 in the atmosphere is effectively constant. A living organism is constantly replenishing C-14 either through the processes of metabolism (sugars ingested in foods contain C-14), or photosynthesis (carbon dioxide contains C-14). When a plant dies, it no longer replenishes itself with C-14 from the atmosphere, and as the C-14 undergoes radioactive decay, its amount decreases with time. 48. These isotopes and their uses are listed in Table 19.4.
50. Combining two light nuclei to form a heavier, more stable nucleus is called nuclear *fusion*. Splitting a heavy nucleus into nuclei with smaller mass numbers is called nuclear *fission*.
52. $^1_0n + ^{235}_{92}U \rightarrow ^{142}_{56}Ba + ^{91}_{36}Kr + 3^1_0n$ 54. A critical mass of a fissionable material is the amount needed to provide a high enough internal neutron flux to sustain the chain reaction (enough neutrons are produced to cause the continuous fission of further material). A sample with less than a critical mass is still radioactive, but cannot sustain a chain reaction. 56. An actual nuclear explosion, of the type produced by a nuclear weapon, cannot occur in a nuclear reactor because the concentration of the fissionable materials is not sufficient to form a supercritical mass. However, since many reactors are cooled by water, a chemical explosion is possible which could scatter the radioactive material used in the reactor. 58. $^{235}_{92}U + ^1_0n \rightarrow ^{239}_{92}U; ^{239}_{92}U \rightarrow ^{239}_{93}Np + _{-1}^0e;$ $^{239}_{93}Np \rightarrow ^{239}_{94}Pu + _{-1}^0e$ 60. In one type of fusion reactor, two ^2_1H atoms are fused to produce ^4_2He. Because the hydrogen nuclei are positively charged, extremely high energies (temperatures of 40 million K) are needed to overcome the repulsion between the nuclei as they are shot into each other. 62. "Cold" fusion implies the fusing of hydrogen atoms into helium at or near ordinary laboratory temperatures. This method seems preferable since it should be cheaper and more easily controllable. Certain metals (palladium and platinum, for example) have been known for a long time to be capable of dissolving large quantities of hydrogen or deuterium (there is a great deal of space between the metal atoms in the metal's crystal structure). It is thought that if deuterium can be concentrated sufficiently inside the metal's lattice, then deuterium atoms may come close enough together to fuse. Experiments a few years ago indicated that this might be possible through electrolysis of heavy water (deuterium oxide) with palladium electrodes, though there is still great controversy over the results of these experiments. 64. Somatic damage is damage directly to the organism itself, causing nearly immediate sickness or death to the organism. Genetic damage is damage to the genetic machinery of the organism, which will be manifested in future generations of offspring. 66. Gamma rays penetrate long distances, but seldom cause ionization of biological molecules. Alpha particles, because they are much heavier although less penetrating, are very effective at ionizing biological molecules and leave a dense trail of damage in the organism. Alpha particles can be ingested or breathed into

the body where the damage will be more acute. **68**. The exposure limits given in Table 19.5 as causing no detectable clinical effect are 0–25 rem. The total yearly exposures from natural and human-induced radioactive sources are estimated in Table 19.6 as less than 200 *milli*rem (0.2 rem), which is well within the acceptable limits. **70**. radioactive **72**. mass **74**. neutron; proton **76**. radioactive decay **78**. mass number **80**. transuranium **82**. half-life **84**. radiotracers **86**. chain **88**. breeder **90**. 4.5×10^9 dollars (\$4.5 billion) **92**. 3.5×10^{-11} J/atom; 8.9×10^{10} J/g **94**. Despite the fact that nuclear waste has been generated for over 40 years, no permanent disposal plan has been implemented as yet. One proposal to dispose of such waste calls for the waste to be sealed in blocks of glass, which in turn are sealed in corrosion-proof metal drums, which would then be buried in deep, stable rock formations away from earthquake and other geologically active zones. In these deep storage areas, it is hoped that the waste could decay safely undisturbed until the radioactivity drops to "safe" levels. **96**. $^{27}_{13}$Al: 13 protons, 14 neutrons; $^{28}_{13}$Al: 13 protons, 15 neutrons; $^{29}_{13}$Al: 13 protons, 16 neutrons **98**. (a) $^1_0$n (b) $^{201}_{79}$Au (c) $^{\ 0}_{-1}$e **100**. For a decay of 10 μg to 1/1000 of this amount, we want to know when the amount of remaining ^{131}I is on the order of 0.01 μg.

time, days	0	8	16	24	32	40	48	56	64
mass, μg	10	5	2.5	1.25	0.625	0.313	0.156	0.078	0.039

time, days	72	80
mass, μg	0.020	0.01

Approximately 80 days are required.
102. Breeder reactors are set up to convert non-fissionable ^{238}U into fissionable ^{239}Pu. The material used for fission in a breeder reactor is a combination of U-235 (which undergoes fission in a chain reaction) and the more common U-238 isotope. Excess neutrons from the U-235 fission are absorbed by the U-238 converting it to the fissionable plutonium isotope Pu-239. Although Pu-239 is fissionable, its chemical and physical properties make it very difficult and expensive to handle and process.

CHAPTER **20**

2. A carbon atom can be bonded to at most four other atoms.
4. A triple bond represents the sharing of six electrons (three pairs of electrons). The simplest example of an organic molecule containing a triple bond is acetylene, H : C : : : C : H (H—C≡C—H). **6**. Each carbon is bonded to four other atoms. Each carbon atom has its electron pairs arranged tetrahedrally.
8. A saturated hydrocarbon is one in which all the bonds are *single* bonds, with each carbon atom forming bonds to four other atoms. The saturated chain hydrocarbons are called *alkanes*.
10. A "straight-chain" alkane is not really straight because the electron pairs on the carbon atoms have a tetrahedral orientation, separated by an angle of 109.5°. In order to give a truly straight chain, the angle between electron pairs would have to be 90° or multiples of 90°.

12. (a)

(b)

(c)

14. (a)

(b)

(c)

(d)

16. branch **18**. With eight carbon atoms, there are a lot of isomers possible. Here are carbon skeletons for some.

20. The root name is derived from the number of carbon atoms in the *longest continuous chain* of carbon atoms.   **22.** The position of a substituent is indicated by a number that corresponds to the carbon atom in the longest chain to which the substituent is attached.   **24.** Multiple substituents are listed in alphabetical order, disregarding any prefix.   **26.** (a) 3-ethylpentane (b) 2,2-dimethylbutane   (c) 2,2-dimethylpropane (d) 2,3,4-trimethylpentane

28. (a) $CH_3-CH_2-\underset{\underset{CH_3}{|}}{CH}-CH_2-CH_3$

(b) $CH_3-CH_2-CH_2-\underset{\underset{\underset{CH_3}{|}}{\overset{}{C}}}{\overset{\overset{CH_3}{|}}{C}}H-CH_2-CH_2-CH_3$

(c) $CH_3-CH_2-CH_2-CH_2-\underset{\underset{CH_2-CH_2-CH_3}{|}}{CH}-CH_2-CH_2-CH_2-CH_3$

(d) $CH_3-\underset{\underset{CH_3}{|}}{CH}-CH_2-\underset{\underset{CH_3}{|}}{CH}-CH_2-\underset{\underset{CH_3}{|}}{CH}-CH_2-CH_2-\underset{\underset{CH_3}{|}}{CH}-CH_3$

30.

Number of C atoms	Use
C_5-C_{12}	gasoline
$C_{10}-C_{18}$	kerosene, jet fuel
$C_{15}-C_{25}$	diesel fuel, heating oil, lubrication
$C_{25}-$	asphalt

32. Tetraethyl lead was added to gasolines to prevent "knocking" of high efficiency automobile engines. The use of tetraethyl lead is being phased out because of the danger to the environment of the lead in the substance.   **34.** Combustion represents the vigorous reaction of a hydrocarbon (or other substance) with oxygen. The combustion of alkanes has been made use of as a source of heat and light.   **36.** Dehydrogenation reactions involve the removal of hydrogen atoms from adjacent carbon atoms in an alkane (or other substance). When two hydrogen atoms are removed, a double bond is created.   **38.** (a) $2C_2H_6(g) + 7O_2(g) \rightarrow 4CO_2(g) + 6H_2O(g)$ (b) $CH_2Cl_2(l) + Cl_2(g) \rightarrow HCl(g) + CHCl_3(l)$ (c) $CH_2=CH-CH_2-CH_3(l) + Br_2(l) \rightarrow \underset{\underset{Br}{|}}{CH_2}-\underset{\underset{Br}{|}}{CH}-CH_2-CH_3(l)$

40. An alkyne is a hydrocarbon containing a carbon–carbon triple bond. The general formula is C_nH_{2n-2}.   **42.** The location of a double or triple bond in the longest chain of an alkene or alkyne is indicated by giving the *number* of the lowest number carbon atom involved in the bond.   **44.** hydrogenation   **46.** (a) propyne   (b) 3-hexyne   (c) 3-hexene

48. $C\equiv C-C-C-C-C \qquad C-C\equiv C-C-C-C-$

$C-C-C\equiv C-C-C \qquad C\equiv C-\underset{\underset{C}{|}}{C}-C-C$

$C-C\equiv C-\underset{\underset{C}{|}}{C}-C \qquad C\equiv C-\underset{\underset{C}{|}}{C}-C-C \qquad C\equiv C-\underset{\underset{C}{|}}{\overset{\overset{C}{|}}{C}}-C$

50. For benzene, a *set* of equivalent Lewis structures can be drawn, differing only in the location of the three double bonds in the ring. Experimentally, however, benzene does not demonstrate the chemical properties expected for molecules having *any* double bonds.   **52.** When named as a substituent, the benzene ring is called the phenyl group. Two examples are

$CH_2=CH-CH-CH_3$

$CH_3-CH-CH_2-CH_2-CH_2-CH_3$

3-phenyl-1-butene   2-phenylhexane

54. ortho- *refers to adjacent substituents (1,2-); meta- refers to two substituents with one unsubstituted carbon atom between them (1,3-); para- refers to two substituents with two unsubstituted carbon atoms between them (1,4-).*
56. (a) 1,2-dimethylbenzene (*o*-xylene)   (b) 1,2,3,4,5,6-hexachlorobenzene   (c) anthracene   (d) 3,5-dichloro-1-methylbenzene (3,5-dichlorotoluene)   **58.** (a) ether   (b) alcohol   (c) alcohol (d) organic (carboxylic) acid   **60.** Primary alcohols have one hydrocarbon fragment (alkyl group) bonded to the carbon atom where the —OH group is attached. Secondary alcohols have two such alkyl groups attached, and tertiary alcohols contain three such alkyl groups. Examples are

ethanol (primary) CH_3-CH_2-OH

2-propanol (secondary) $CH_3-\underset{\underset{OH}{|}}{CH}-CH_3$

2-methyl-2-propanol (tertiary) $CH_3-\underset{\underset{OH}{|}}{\overset{\overset{CH_3}{|}}{C}}-CH_3$

62. (a) $CH_3-\underset{\underset{OH}{|}}{CH}-CH_2-CH_3$ secondary

(b) $CH_3-CH_2-CH_2-CH_2-CH_2-OH$ primary

(c) $CH_3-CH_2-\underset{\underset{CH_3}{|}}{CH}-CH_2-OH$ primary

(d) $CH_3-CH_2-\underset{\underset{OH}{|}}{\overset{\overset{CH_3}{|}}{C}}-CH_3$ tertiary

64. $C_6H_{12}O_6 \xrightarrow{\text{yeast}} 2CH_3{-}CH_2{-}OH + CO_2$ The yeast necessary for the fermentation process are killed if the concentration of ethanol is over 13%. More concentrated ethanol solutions are most commonly made by distillation.   **66.** methanol (CH_3OH)—starting material for synthesis of acetic acid and many plastics; ethylene glycol ($CH_2OH{-}CH_2OH$)—automobile antifreeze; isopropyl alcohol (2-propanol, $CH_3{-}CH(OH){-}CH_3$)—rubbing alcohol   **68.** Aldehydes and ketones both contain the carbonyl

group ($\overset{\diagdown}{\underset{\diagup}{C}}{=}O$). Aldehydes and ketones differ in the *location* of

the carbonyl function: aldehydes contain the carbonyl group at the end of a hydrocarbon chain (the carbon atom of the carbonyl group is bonded only to at most one other carbon atom); the carbonyl group of ketones represents one of the interior carbon atoms of a chain (the carbon atom of the carbonyl group is bonded to two other carbon atoms).   **70.** Aldehydes and ketones are produced by the oxidation of primary and secondary alcohols, respectively.

$$CH_3{-}CH_2{-}OH \xrightarrow{\text{oxidation}} CH_3{-}\overset{}{\underset{H}{C}}{=}O$$

$$CH_3{-}\underset{\underset{CH_3}{|}}{C}H{-}OH \xrightarrow{\text{oxidation}} CH_3{-}\underset{\underset{CH_3}{|}}{C}{=}O$$

72. In addition to their systematic names (based on the hydrocarbon root, with the ending *-one*), ketones can also be named by naming the groups attached to either side of the carbonyl carbon as alkyl groups, followed by the word "ketone."

74. (a) $CH_3{-}\underset{\underset{O}{\|}}{C}{-}CH_2{-}CH_2{-}CH_3$

(b) $CH_3{-}\underset{\underset{O}{\|}}{C}{-}CH_2{-}CH_2{-}CH_3$

(c) $CH_3{-}\underset{\underset{H_3C}{|}}{\overset{\overset{CH_3}{|}}{C}}{-}\underset{\underset{O}{\|}}{C}{-}\underset{\underset{CH_3}{|}}{\overset{\overset{CH_3}{|}}{C}}{-}CH_3$

(d) $CH_3{-}CH_2{-}\underset{\underset{Cl}{|}}{C}H{-}CH_2{-}CH_2{-}\underset{\underset{O}{\|}}{C}H$

(e) $CH_3{-}CH_2{-}\underset{\underset{O}{\|}}{C}{-}CH_2{-}\underset{\underset{CH_3}{|}}{C}H{-}CH_2{-}CH_2{-}CH_3$

76. Carboxylic acids are typically *weak* acids.
$$CH_3{-}CH_2{-}COOH(aq) \rightleftharpoons H^+(aq) + CH_3{-}CH_2{-}COO^-(aq)$$

78. (a) $CH_3{-}COOH$

(b) $CH_3{-}\underset{\underset{O}{\|}}{C}{-}CH_3$

(c) $CH_3{-}\overset{\overset{}{}}{C}{=}O$
$\quad\quad\underset{O{-}CH_2CH_2CH_2CH_2CH_3}{|}$

80. Acetylsalicylic acid is synthesized from salicylic acid (behaving as an alcohol through its —OH group) and acetic acid.

82. (a) $CH_3{-}\overset{\overset{}{}}{C}{=}O$
$\quad\quad\underset{O{-}CH_3}{|}$

(b)

(c) $CH_3{-}CH_2{-}CH_2{-}CH_2{-}COOH$

(d) $CH_3{-}CH_2{-}CH_2{-}CH_2{-}CH_2{-}CH_2{-}CH_2{-}\underset{\underset{O}{\|}}{C}{=}O$
$\quad\quad\quad\quad\quad\quad\quad\quad\quad\quad\quad\quad\quad\quad\quad\underset{CH_2{-}CH{-}CH_3}{}$

(e) $CH_2{-}CH{-}CH_2{-}CH_2{-}CH{-}COOH$
$\quad\underset{Cl}{|}\quad\quad\quad\quad\quad\quad\quad\quad\underset{Cl}{|}$

84. In addition polymerization, the monomer units simply add together to form the polymer, with no other products. Polyethylene and polytetrafluoroethylene (Teflon) are common examples.

86. A polyester is formed from the reaction of a dialcohol (two —OH groups) with a diacid (two —COOH groups). One —OH group of the alcohol forms an *ester linkage* with one of the —COOH groups of the acid. Since the resulting dimer still possesses an —OH and a —COOH group, the dimer can undergo further esterification reactions. Dacron is a common polyester.

88. nylon

Dacron

90. saturated   **92.** straight-chain or normal
94. -ane   **96.** number   **98.** anti-knocking   **100.** substitution
102. hydrogenation   **104.** functional group   **106.** carbon monoxide   **108.** carbonyl   **110.** carboxyl   **112.** addition
114. The carbon skeletons are

$$C{-}C{-}C{-}C{-}C \quad\quad C{-}\underset{\underset{C}{|}}{C}{-}C{-}C \quad\quad C{-}\overset{\overset{C}{|}}{\underset{\underset{C}{|}}{C}}{-}C$$

116. (a) 2-chlorobutane   (b) 1,2-dibromoethane
(c) triiodomethane (common name: iodoform)
(d) 2,3,4-trichloropentane   (e) 2,2-dichloro-4-isopropylheptane

118. (a) $CH_3-CH_2-\underset{\underset{Cl}{|}}{C}=CH-CH_2-CH_3$

(b) $\underset{\underset{Br}{|}}{CH_2}-\underset{\underset{Br}{|}}{CH_2}$

(c) $\underset{\underset{Cl}{|}}{\overset{\overset{Cl}{|}}{CH}}-\underset{\underset{Cl}{|}}{\overset{\overset{Cl}{|}}{CH}}$

(d) $CH\equiv C-\underset{\underset{Br}{|}}{CH}-CH_2-CH_2-CH_3$

(e) $\underset{\underset{Cl}{|}}{CH_2}-\underset{\underset{Cl}{|}}{\overset{\overset{CH_3}{|}}{C}}-\underset{\underset{CH_3}{|}}{\overset{\overset{CH_3}{|}}{C}}-CH-CH_2-CH_2-CH_2-CH_3$

120. primary $CH_3-CH_2-CH_2-CH_2-CH_2-CH_2-OH$

secondary $CH_3-CH_2-CH_2-CH_2-\underset{\underset{OH}{|}}{CH}-CH_3$

tertiary $CH_3-CH_2-\underset{\underset{OH}{|}}{\overset{\overset{CH_3}{|}}{C}}-CH_2-CH_3$

(there are other possibilities for the secondary and tertiary alcohols)

122. $\underset{\underset{OH}{|}}{CH_2}-\underset{\underset{OH}{|}}{CH}-\underset{\underset{OH}{|}}{CH}-\underset{\underset{OH}{|}}{CH}-\underset{\underset{OH}{|}}{CH}-\underset{\underset{H}{|}}{C}=O$

124. (a) benzophenone structure

(b) $CH_3-\underset{\underset{OH}{|}}{\overset{\overset{CH_3}{|}}{C}}-CH_3$

(c) $\underset{\underset{CH_3}{|}}{\overset{\overset{CH_3}{|}}{CH}}-\overset{\overset{O}{\|}}{C}-CH_3$

(d) $\underset{\underset{OH}{|}}{CH_2}-\underset{\underset{OH}{|}}{CH_2}$

(e) $CH_3-CH_2-\overset{\overset{O}{\|}}{\underset{\underset{O-CH_3}{|}}{C}}$

126. $HO-C=O$

$CH_3-\underset{\underset{H}{|}}{CH}-\underset{\underset{CH_2-NH_2}{|}}{N}-C=O + H_2O$

One end has $-NH_2$, which can react with the $-COOH$ end of another of these dipeptides.

128. (a) octane structure   octane $CH_3CH_2CH_2CH_2CH_2CH_2CH_2CH_3$

(b) hexane structure   hexane $CH_3CH_2CH_2CH_2CH_2CH_3$

(c) butane structure   butane $CH_3CH_2CH_2CH_3$

(d) pentane structure   pentane $CH_3CH_2CH_2CH_2CH_3$

130. a. 2,3-dimethylbutane   b. 3,3-diethylpentane
c. 2,3,3-trimethyhexane   d. 2,3,4,5,6-pentamethylheptane

132. a. $CH_3Cl(g)$ b. $H_2(g)$ c. $HCl(g)$

134. $CH\equiv C-CH_2-CH_2-CH_2-CH_2-CH_2-CH_3$ 1-octyne
$CH_3-C\equiv C-CH_2-CH_2-CH_2-CH_2-CH_3$ 2-octyne
$CH_3-CH_2-C\equiv C-CH_2-CH_2-CH_2-CH_3$ 3-octyne
$CH_3-CH_2-CH_2-C\equiv C-CH_2-CH_2-CH_3$ 4-octyne

136. a. carboxylic acid   b. ketone   c. ester   d. alcohol (phenol)

138. (a) 3-methylpentanal $CH_3-CH_2-\underset{\underset{CH_3}{|}}{CH}-CH_2-\underset{\underset{H}{|}}{C}=O$

(b) 3-methyl-2-pentanone $CH_3-CH_2-\underset{\underset{CH_3}{|}}{\overset{}{CH}}-\overset{\overset{}{}}{\underset{\underset{O}{\|}}{C}}-CH_3$

(c) methyl phenyl ketone $CH_3-\underset{\underset{O}{\|}}{C}-$ (phenyl ring)

(d) 2-hydroxybutanal $CH_3-CH_2-\underset{\underset{OH}{|}}{CH}-\underset{\underset{H}{|}}{C}=O$

(e) propanal $CH_3-CH_2-\underset{\underset{H}{|}}{C}=O$

140. (a) CH_3—CH—CH_2—COOH
         |
         CH_3

(b)

(c) CH_3—CH_2—CH_2—CH_2—CH_2—COOH

(d) CH_3—COOH

CHAPTER **21**

2. Molar masses of proteins range from a few thousand amu to over 1 million amu. Such molar masses are consistent with proteins being large polymeric molecules. 4. Globular proteins have an approximately spherical shape. Globular proteins perform functions including the transport and storage of oxygen, the catalysis of many cellular processes, the fighting of infection due to foreign bodies, and the transport of electrons in metabolic processes.
6. The structures of the amino acids are given in detail in Figure 21.2. Generally a side chain is nonpolar if it is mostly hydrocarbon in nature (e.g., alanine, in which the side chain is a methyl group). Side chains are polar if they contain the hydroxyl group (—OH), the sulfhydryl group (—SH), or a second amino (—NH_2) or carboxyl (—COOH) group.

8. C=O
    |
    N—H
    |

10. There is a very large number of polypeptides possible. A *few* are indicated here. From this exercise, you can certainly appreciate the complexities of protein structures.

gly-ile-leu-pro-phe	gly-leu-pro-phe-ile	gly-pro-phe-ile-leu
gly-phe-ile-leu-pro	ile-leu-pro-phe-gly	ile-pro-phe-gly-leu
ile-phe-gly-leu-pro	ile-gly-leu-pro-phe	leu-gly-ile-pro-phe
leu-ile-pro-phe-gly	leu-pro-phe-gly-ile	leu-phe-gly-ile-pro

12. phe-ala-gly

phe-gly-ala

ala-phe-gly

ala-gly-phe

gly-phe-ala

gly-ala-phe

14. The secondary structure, in general, describes the arrangement of the long polypeptide chain of the protein. In the alpha-helical secondary structure, the chain forms a coil or spiral, which gives proteins consisting of such structures an elasticity or resilience. Such proteins are found in wool, hair, and tendons.
16. Long, thin, resilient proteins, such as hair, typically contain elongated, elastic alpha-helical protein molecules. Other proteins, such as silk, which in bulk form sheets or plates, typically contain protein molecules having the beta pleated sheet secondary structure. Proteins that do not have a structural function in the body, such as hemoglobin, typically have a globular structure.
18. Cysteine, an amino acid containing the sulfhydryl (—SH) group in its side chain, is capable of forming disulfide linkages (—S—S—) with other cysteine molecules in the same polypeptide chain. If such a disulfide linkage is formed, this effectively ties together two portions of the polypeptide, producing a kink or knot in the chain, which leads in part to the protein's overall 3-dimen-

sional shape (tertiary structure). Cysteine, and the disulfide linkages it forms, is responsible for the curling of hair (whether naturally or by a permanent wave). **20.** Collagen has an alpha-helical secondary structure. Collagen's function in the body is as the raw material of which tendons and muscles are constructed. The long, springy structure of the alpha-helix is responsible for collagen's strength and elasticity. **22.** Antibodies are special proteins which are synthesized in the body in response to foreign substances such as bacteria or viruses. Although there are usually specific antibodies for specific invaders, the antibody interferon offers general protection against invasion by viruses. **24.** In a permanent wave, cross-linkages between adjacent polypeptide chains of the protein are broken chemically, and then reformed chemically in a new location. The primary cross-linkage involved is a disulfide linkage between cysteine units in the polypeptide chains. It is primarily the tertiary structure of the hair protein which is affected by a permanent wave, although if the waving lotion is left on the hair too long, the secondary structure can also be affected (making the hair very "frizzy"). **26.** substrate **28.** Figure 21.11 illustrates the lock and key model clearly. The most important point of the model is that a given enzyme and its substrate have structures which are complementary, which allow the enzyme and substrate to approach each other and to fit together closely enough for a reaction to take place. **30.** The type names of sugars contain a root name which implies the number of carbon atoms present: aldo*hexose*–6 carbon atoms; keto*pentose*–5 carbon atoms; aldo*triose*–3 carbon atoms; keto*tetrose*–4 carbon atoms

32. (a)

```
        CHO
         |
   H —— C —— OH
         |
  HO —— C —— H
         |
   H —— C —— OH
         |
   H —— C —— OH
         |
        CH2OH
```

(b)

```
        CHO
         |
   H —— C —— OH
         |
   H —— C —— OH
         |
   H —— C —— OH
         |
        CH2OH
```

(c)

```
        CH2OH
         |
         C == O
         |
   H —— C —— OH
         |
   H —— C —— OH
         |
        CH2OH
```

(d)

```
        CHO
         |
   H —— C —— OH
         |
  HO —— C —— H
         |
  HO —— C —— H
         |
   H —— C —— OH
         |
        CH2OH
```

34. A polysaccharide is a large polymeric substance, containing as its building blocks repeating simple sugar (monosaccharide) monomer units. Both starch and cellulose consist of long chains of bonded *glucose* molecules: the differences in properties between starch and cellulose are derived from the different manner in which the glucose units are attached to one another to form the polysaccharide chains. **36.** ribose (aldopentose); arabinose (aldopentose); ribulose (ketopentose); glucose (aldohexose); mannose (aldohexose); galactose (aldohexose); fructose (ketohexose) **38.** A nucleotide consists of three components: a five-carbon sugar, a nitrogen-containing organic base, and a phosphate group. The phosphate and nitrogen base are each bonded to respective sites on the sugar, but are not bonded to each other. **40.** uracil (RNA only); cytosine (DNA, RNA); thymine (DNA only); adenine (DNA, RNA); guanine (DNA, RNA) **42.** When the two strands of a DNA molecule are compared, it is found that a given base in one strand is always found paired with a particular base in the other strand. Because of the shapes and side atoms along the rings of the nitrogen bases, only certain pairs are able to approach and hydrogen-bond with each other in the double helix. Adenine is always found paired with thymine; cytosine is always found paired with guanine. When a DNA helix unwinds for replication during cell division, only the appropriate complementary bases are able to approach and bond to the nitrogen bases of each strand. For example, for a guanine–cytosine pair in the original DNA, when the two strands separate, only a new cytosine molecule can approach and bond to the original cytosine. **44.** Messenger RNA molecules are synthesized to be complementary to a portion (gene) of the DNA molecule in the cell, and serve as the template or pattern upon which a protein will be constructed (a particular group of nitrogen bases on *m*-RNA is able to accommodate and specify a particular amino acid in a particular location in the protein). Transfer RNA molecules are much smaller than *m*-RNA, and their structure accommodates only a single specific amino acid molecule: transfer RNA molecules "find" their specific amino acid in the cellular fluids, and bring it to *m*-RNA where it is added to the protein molecule being synthesized. **46.** A triglyceride typically consists of a glycerine backbone, to which three separate fatty acid molecules are attached by ester linkages.

```
              O
              ||
  CH2 —— O —— C —— R
              O
              ||
  CH  —— O —— C —— R'
              O
              ||
  CH2 —— O —— C —— R''
```

48. Saponification is the production of a *soap* by treatment of a triglyceride with a strong base such as NaOH.

triglyceride + 3NaOH → glycerol + $3Na^+soap^-$

```
              O
              ||
  CH2 —— O —— C —— R                CH2 —— OH    RCOONa
              O
              ||
  CH  —— O —— C —— R'  + 3NaOH →    CH  —— OH  + R'COONa
              O
              ||
  CH2 —— O —— C —— R''               CH2 —— OH    R''COONa
```

50. Soaps have both an organic nature (due to the long chain of the fatty acid) and an ionic nature (due to the charge on the carboxyl group). In water, soap anions form aggregates called micelles, in which the water-repelling hydrocarbon chains are oriented toward the interior of the aggregate, with the ionic, water-attracting carboxyl groups oriented toward the outside. Most dirt has a greasy nature. A soap micelle is able to intereact with a grease molecule, pulling the grease molecule into the hydrocarbon interior of the micelle. When the clothing is rinsed, the micelle containing the grease is washed away. See Figure 21.22. **52**. Cholesterol is the naturally occurring steroid from which the body synthesizes other needed steroids. Since cholesterol is insoluble in water, it is thought that having too large a concentration of this substance in the bloodstream may lead to its deposition and buildup on the walls of blood vessels, causing their eventual blockage. **54**. Phospholipids have a similar structure to triglycerides, except that one of the fatty acids attached to the glycerol backbone is replaced by a phosphate group (phospholipids are sometimes referred to as phosphodiglycerides). An important phospholipid is lecithin, which behaves very strongly as an emulsifying agent (allowing polar and nonpolar substances to mix). Lecithin is found in high concentration in egg yolks, and is being used in baking in place of whole eggs: the emulsifying properties of the egg yolk are due to the lecithin, and using lecithin itself avoids the cholesterol problem associated with whole eggs. **56**. i **58**. m

60. u **62**. f **64**. g **66**. r **68**. p **70**. o **72**. b **74**. d **76**. a **78**. nucleotides **80**. ester **82**. adenine, cytosine **84**. transfer, messenger **86**. insoluble, soluble **88**. unsaturated, saturated **90**. ionic, nonpolar **92**. waxes **94**. progesterone **96**. The primary structure of a protein refers to the specific identity and ordering of amino acids in a protein's polypeptide chain. The primary structure is sometimes referred to as the protein's amino acid *sequence*. **98**. tendons, bone (with mineral constituents), skin, cartilage, hair, fingernails **100**. Proteins contain both acidic (—COOH) and basic (—NH$_2$) groups in their side chains, which can neutralize both acids and bases. **102**. pentoses (5 carbons); hexoses (6 carbons); trioses (3 carbons) **104**. In a strand of DNA, the phosphate group and the sugar molecule of adjacent nucleotides become bound to each other. The chain-portion of the DNA molecule, therefore, consists of alternating phosphate groups and sugar molecules. The nitrogen bases are found sticking out from the side of this phosphate-sugar chain, being bonded to the sugar molecules. **106**. The primary human bile acid is cholic acid, which helps to emulsify fats in the intestinal tract. In order to be digested by enzymes for absorption into the bloodstream, large clumps of fat must be dispersed into fine droplets in the liquid of the small intestine. Cholic acid basically acts as a detergent.

INDEX

Table 5.2 Common Simple Cations and Anions

Cation	Name	Anion*	Name
H^+	hydrogen	H^-	hydride
Li^+	lithium	F^-	fluoride
Na^+	sodium	Cl^-	chloride
K^+	potassium	Br^-	bromide
Cs^+	cesium	I^-	iodide
Be^{2+}	beryllium	O^{2-}	oxide
Mg^{2+}	magnesium	S^{2-}	sulfide
Ca^{2+}	calcium		
Ba^{2+}	barium		
Al^{3+}	aluminum		
Ag^+	silver		

*The root is given in color.

Table 5.3 Common Type II Cations

Ion	Systematic Name	Older Name
Fe^{3+}	iron(III)	ferric
Fe^{2+}	iron(II)	ferrous
Cu^{2+}	copper(II)	cupric
Cu^+	copper(I)	cuprous
Co^{3+}	cobalt(III)	cobaltic
Co^{2+}	cobalt(II)	cobaltous
Sn^{4+}	tin(IV)	stannic
Sn^{2+}	tin(II)	stannous
Pb^{4+}	lead(IV)	plumbic
Pb^{2+}	lead(II)	plumbous
Hg^{2+}	mercury(II)	mercuric
Hg_2^{2+}*	mercury(I)	mercurous

*Note that mercury(I) ions always occur bound together to form Hg_2^{2+}.

Table 5.5 Names of Common Polyatomic Ions

Ion	Name	Ion	Name
NH_4^+	ammonium	CO_3^{2-}	carbonate
NO_2^-	nitrite	HCO_3^-	hydrogen carbonate
NO_3^-	nitrate		(bicarbonate is a widely
SO_3^{2-}	sulfite		used common name)
SO_4^{2-}	sulfate	ClO^-	hypochlorite
HSO_4^-	hydrogen sulfate	ClO_2^-	chlorite
	(bisulfate is a widely	ClO_3^-	chlorate
	used common name)	ClO_4^-	perchlorate
OH^-	hydroxide	$C_2H_3O_2^-$	acetate
CN^-	cyanide	MnO_4^-	permanganate
PO_4^{3-}	phosphate	$Cr_2O_7^{2-}$	dichromate
HPO_4^{2-}	hydrogen phosphate	CrO_4^{2-}	chromate
$H_2PO_4^-$	dihydrogen phosphate	O_2^{2-}	peroxide